Manual of Chronic Total Occlusion Percutaneous Coronary Interventions

A Step-by-Step Approach

Manual of Chronic Total Occlusion Percutaneous Coronary Interventions

A Step-by-Step Approach

Third Edition

Emmanouil Brilakis

Allina Health Minneapolis Heart Institute, Minneapolis, MN,
United States; Minneapolis Heart Institute Foundation, Minneapolis,
MN, United States

ACADEMIC PRESS

An imprint of Elsevier

ELSEVIER

For Information on all Academic Press publications
visit our website at https://www.elsevier.com/books-and-journals

Publisher: Stacy Masucci
Acquisitions Editor: Ana Claudia A. Garcia
Editorial Project Manager: Pat Gonzalez
Production Project Manager: Swapna Srinivasan
Cover Designer: Miles Hitchen

Typeset by MPS Limited, Chennai, India

Dedication

To Nicole, Stelios, and Thomas.

To my parents and my brother.

Contents

Part E
How to develop a CTO PCI program

31. How to build a successful chronic total occlusion program

Part F
Appendices

List of contributors

Nidal Abi Rafeh North Oaks Health System and Tulane University Heart and Vascular Institute, New Orleans, LA, United States

Pierfrancesco Agostoni Hartcentrum, Ziekenhuis Netwerk Antwerpen (ZNA) Middelheim, Antwerp, Belgium

Sukru Akyuz HEARTist CLINIC, Istanbul, Turkey

Khaldoon Alaswad Henry Ford Hospital, Detroit, MI, United States

Ziad A. Ali DeMatteis Cardiovascular Institute, St Francis Hospital & Heart Center, Roslyn, NY, United States; Angiographic Core Laboratory, Cardiovascular Research Foundation, New York, NY, United States

Salman S. Allana Allina Health Minneapolis Heart Institute and Minneapolis Heart Institute Foundation, Minneapolis, MN, United States

Chadi Alraies Detroit Medical Center, Detroit, MI, United States

Mario Araya Clinica Alemana, Hospital Militar de Santiago, Santiago, Chile

Alexandre Avran Pasteur Clinic, Essey-lès-Nancy, France

Lorenzo Azzalini University of Washington, Seattle, WA, United States

Avtandil Babunashvili Department of Interventional Cardioangiology, Center of Endosurgery, Moscow State Medical Academy, Moscow, Russia

Subhash Banerjee Baylor Scott & White Heart and Vascular Hospital, Dallas, TX, United States; Baylor University Medical Center, Dallas, TX, United States

Sripal Bangalore Department of Medicine, New York University Grossman School of Medicine, New York, NY, United States

Baktash Bayani Mehr Hospital, Mashhad, Iran

Michael Behnes First Department of Medicine, University Medical Centre Mannheim, Faculty of Medicine Mannheim, University of Heidelberg, Heidelberg, Germany

Ravinay Bhindi Royal North Shore Hospital, University of Sydney, Sydney, NSW, Australia

Nicolas Boudou Interventional Cardiology, Clinique Saint Augustin, Bordeaux, France

Nenad Ž. Božinović University Clinical Center Nis, Nis, Serbia

Leszek Bryniarski 2nd Department of Cardiology, Institute of Cardiology, Jagiellonian University Medical College, Kraków, Poland; Department of Cardiology and Cardiovascular Interventions, University Hospital, Kraków, Poland

Alexander Bufe Heart Center Krefeld, Krefeld, Germany; Witten/Herdecke University, Witten, Germany

Christopher E. Buller Teleflex Interventional, Maple Grove, MN, United States

M. Nicholas Burke Allina Health Minneapolis Heart Institute and Minneapolis Heart Institute Foundation, Minneapolis, MN, United States

Pedro Pinto Cardoso Structural and Coronary Heart Disease, Centro Cardiovascular da Universidade de Lisboa (CCUL@RISE), Faculdade de Medicina, Universidade de Lisboa, Av. Prof. Egas Moniz, Lisbon, Portugal; Cardiology Service, Heart and Vascular Department, Hospital Santa Maria (CHULN), Av. Prof. Egas Moniz, Lisbon, Portugal

Mauro Carlino Interventional Cardiology Unit, Cardio-Thoracic-Vascular Department, IRCCS San Raffaele Scientific Institute, Milan, Italy

Joao L. Cavalcante Allina Health Minneapolis Heart Institute, Minneapolis, MN, United States; Minneapolis Heart Institute Foundation, Minneapolis, MN, United States

Tarek Chami First Coast Cardiovascular Institute PA, Jacksonville, FL, United States

Raj H. Chandwaney Oklahoma Heart Institute, Tulsa, OK, United States

Konstantinos Charitakis University of Texas Health Science Center at Houston, Houston, TX, United States

Victor Y. Cheng Allina Health Minneapolis Heart Institute, Minneapolis, MN, United States

James W. Choi Texas Health Heart & Vascular Specialists, Presbyterian Hospital, Dallas, TX, United States; Department of Internal Medicine, Texas A&M College of Medicine, Dallas, TX, United States

Evald Høj Christiansen Department of Cardiology, Aarhus University Hospital, Aarhus, Denmark

Yashasvi Chugh Baylor Heart and Vascular Hospital at Baylor University Medical Center, Dallas, TX, United States

Antonio Colombo Cardiology, Humanitas University, Humanitas IRCCS, Milan, Italy

Claudia Cosgrove St George's Hospital, London, United Kingdom

Kevin Croce Cardiovascular Division, Brigham and Women's Hospital, Boston, MA, United States

Ramesh Daggubati West Virginia University School of Medicine, Morgantown, WV, United States

Félix Damas de los Santos Interventional Cardiology Department Instituto Nacional de Cardiologia Ignacio Chávez, Mexico City, Mexico; Deputy Head of Endovascular Therapy Centro Medico ABC, Mexico City, Mexico

Rustem Dautov The Prince Charles Hospital, Brisbane, Australia

Rhian E. Davies WellSpan Health, York, PA, United States

Tony de Martini Midwest Cardiovascular Institute, Naperville, IL, United States; SIU School of Medicine, Springfield, IL, United States

Ali E. Denktas Baylor College of Medicine; Cardiac Catheterization Laboratories, Houston, TX, United States; Cardiology Section, MEDVA Medical Center, Houston, TX, United States

Joseph Dens Department of Cardiology, Ziekenhuis Oost-Limburg, Genk, Belgium

Carlo di Mario Department of Clinical & Experimental Medicine, University Hospital Careggi, Florence, Italy

Roberto Diletti Interventional Cardiology Department, Thoraxcenter, Erasmus MC, Rotterdam, The Netherlands

Zisis Dimitriadis Department of Cardiology, Center of Internal Medicine, Goethe University Frankfurt, Frankfurt, Germany

Darshan Doshi Massachusetts General Hospital, Boston, MA, United States

Parag Doshi Chicago Cardiology Institute, Schaumburg, IL, United States

Kefei Dou Research Center for Coronary Heart Disease, State Key Laboratory of Cardiovascular Disease, Beijing, P.R. China; Fuwai Hospital, National Center for Cardiovascular Diseases, Chinese Academy of Medical Sciences, Beijing, P.R. China

Mohaned Egred Freeman Hospital, University of Sunderland & Newcastle University, Newcastle upon Tyne, United Kingdom

Basem Elbarouni St. Boniface Hospital & University of Manitoba, Winnipeg, MB, Canada

Ahmed M. ElGuindy Department of Cardiology, Aswan Heart Centre, Magdi Yacoub Foundation, Aswan, Egypt

Amr Elhadidy Cairo University, Cairo, Egypt

Stephen Ellis Heart, Thoracic, Vascular Institute, Cleveland Clinic, Cleveland, OH, United States

Javier Escaned Hospital Clinico San Carlos, IdISSC, Complutense University of Madrid, Madrid, Spain

Panayotis Fasseas Division of Cardiovascular Medicine, Medical College of Wisconsin, Milwaukee, WI, United States

Farshad Forouzandeh Harrington Heart and Vascular Institute, University Hospitals, Case Western Reserve University, Cleveland, OH, United States

Sergey Furkalo National Institute of Surgery and Transplantology, NAMS, Kiev, Ukraine

Andrea Gagnor Maria Vittoria Hospital, Turin, Italy

Alfredo R. Galassi University of Palermo, Palermo, Italy

Robert Gallino MedStar Cardiology Associates, Washington, DC, United States

Roberto Garbo Maria Pia Hospital, GVM Care & Research, Turin, Italy

Santiago Garcia The Carl and Edyth Lindner Center for Research and Education at The Christ Hospital, Cincinnati, OH, United States

Gabriele Gasparini Department of Invasive Cardiology, Humanitas Clinical and Research Center, IRCCS, Rozzano, Italy

Junbo Ge Department of Cardiology, Zhongshan Hospital, Fudan University, Shanghai, P.R. China; Shanghai Institute of Cardiovascular Diseases, National Clinical Research Center for Interventional Medicine, Shanghai, P.R. China

Lei Ge Department of Cardiology, Zhongshan Hospital, Fudan University, Shanghai, P.R. China; Shanghai Institute of Cardiovascular Diseases, National Clinical Research Center for Interventional Medicine, Shanghai, P.R. China

Pravin Kumar Goel Interventional Cardiology, Medanta, The Medicity, Lucknow, India

Omer Goktekin Memorial Hospital, Istanbul, Turkey

Nieves Gonzalo Interventional Cardiology, IdISSC, Hospital Clinico San Carlos, Universidad Complutense, Madrid, Spain

Sevket Gorgulu Biruni University School of Medicine, Cardiology Department, Istanbul, Turkey

Luca Grancini Centro Cardiologico Monzino, IRCCS, Milan, Italy

J. Aaron Grantham Saint Luke's Mid America, Heart Institute, Kansas City, MO, United States

Raviteja Guddeti Campbell University School of Osteopathic Medicine, Lillington, NC, United States; Division of Cardiology, Cape Fear Valley Medical Center, Fayetteville, NC, United States

Elias V. Haddad Ascension Saint Thomas Heart Institute, Nashville, TN, United States

Allison B. Hall Eastern Health/Memorial University of Newfoundland, St. John's, NL, Canada

Jack J. Hall Prairie Heart Institute of Illinois, Springfield, IL, United States

Sean Halligan Avera Heart Hospital, Sioux Falls, SD, United States

Franklin Leonardo Hanna Quesada "Clinica Comfamiliar", Pereira, Colombia

Colm Hanratty Mater Private Hospital, Dublin, Ireland

Stefan Harb Medical University of Graz, Cardiology Division, University Heart Center Graz, Austria

Scott A. Harding Wellington Hospital, Capital and Coast District Health Board, Wellington, New Zealand

Raja Hatem Hôpital du Sacré-Coeur de Montréal, Université de Montréal, Montreal, QC, Canada

David Hildick-Smith University Hospitals Sussex, Brighton, United Kingdom

Jonathan M. Hill King's College Hospital, London, United Kingdom

Taishi Hirai University of Missouri, Columbia, MO, United States

Mario Iannaccone San Giovanni Bosco Hospital, ASL Città di Torino, Turin, Italy

Wissam Jaber Emory University, Atlanta, GA, United States

Farouc A. Jaffer Massachusetts General Hospital, Harvard Medical School, Boston, MA, United States

Yangsoo Jang Severance Cardiovascular Hospital, Yonsei University College of Medicine, Seoul, South Korea

Brian K. Jefferson Hospital Corporation of America/Tristar Centennial Medical Center, Nashville, TN, United States

Allen Jeremias St Francis Hospital & Heart Center, Roslyn, NY, United States

Risto Jussila Helsinki Heart Hospital, Helsinki, Finland

Nikolaos Kakouros UMass Memorial Medical Center, UMass Chan Medical School, Worcester, MA, United States

Artis Kalnins Riga East University Clinical Hospital, Riga, Latvia

Sanjog Kalra Peter Munk Cardiac Centre, Toronto General Hospital, Toronto, ON, Canada

Arun Kalyanasundaram Promed Hospital, Chennai, India

David E. Kandzari Piedmont Heart Institute, Atlanta, GA, United States

Hsien-Li Kao Division of Cardiology, Department of Internal Medicine and Cardiovascular Center, National Taiwan University Hospital, Taipei, Taiwan

Judit Karacsonyi Minneapolis Heart Institute Foundation, Minneapolis, MN, United States

Dimitri Karmpaliotis Gagnon Cardiovascular Institute, Morristown Medical Center, Morristown, NJ, United States

Hussien Heshmat Kassem Cairo University, Cairo, Egypt

Kathleen Kearney University of Washington, Seattle, WA, United States

Jimmy Kerrigan Ascension Saint Thomas Heart Institute, Nashville, TN, United States

Jaikirshan Khatri Cleveland Clinic, Cleveland, OH, United States

Dmitrii Khelimskii Meshalkin National Medical Research Center, Novosibirsk, Russia

Ajay J. Kirtane Division of Cardiology, Columbia University Irving Medical Center, New York, NY, United States; New York—Presbyterian Hospital, New York, NY, United States; Cardiovascular Research Foundation, New York, NY, United States

Paul Knaapen Amsterdam University Medical Center, Amsterdam, The Netherlands

Spyridon Kostantinis Minneapolis Heart Institute Foundation, Minneapolis, MN, United States

Michalis Koutouzis Hygeia Hospital, Athens, Greece

Mihajlo Kovacic Interventional Cardiology Department, County Hospital Cakovec, Cakovec, Croatia

Oleg Krestyaninov Meshalkin Novosibirsk Research Institute, Novosibirsk, Russia

A.V. Ganesh Kumar Head, Department of Cardiology, DR LH Hiranandani Hospital, Mumbai, India

Prathap Kumar N. Meditrina Hospitals, Thiruvananthapuram, Kerala, India

Katherine J. Kunkel Piedmont Heart Institute, Atlanta, GA, United States

Pablo Manuel Lamelas Instituto Cardiovascular de Buenos Aires, Buenos Aires, Argentina; Health Research Methods, Evidence, and Impact, McMaster University, Hamilton, ON, Canada

Seung-Whan Lee Department of Cardiology, Asan Medical Center, Seoul, South Korea

Thierry Lefevre Institut Cardiovasculaire Paris Sud, Hôpital PriveJacques Cartier, Massy, France

Gregor Leibundgut University Hospital Basel, Basel, Switzerland

Nicholas J. Lembo Columbia Interventional Cardiovascular Care, Columbia University Irving Medical Center / New York - Presbyterian Hospital, New York, NY, United States

Martin Leon Division of Cardiology, Columbia University Irving Medical Center/ New York–Presbyterian Hospital, New York, NY, United States; Cardiovascular Research Foundation, New York, NY, United States

John R. Lesser Allina Health Minneapolis Heart Institute, Minneapolis, MN, United States

Raymond Leung Royal Alexandra Hospital, Edmonton, AB, Canada

Soo-Teik Lim National Heart Centre, Singapore, Singapore

Sidney Tsz Ho Lo Liverpool Hospital and University of New South Wales, Sydney, NSW, Australia

William Lombardi University of Washington, Seattle, WA, United States

Michael Luna University of Texas Southwestern Medical Center, Dallas, TX, United States

Ehtisham Mahmud UCSD Cardiovascular Institute-Medicine, University of California, San Diego, CA, United States

Madeline K. Mahowald Division of Cardiology, University of Florida College of Medicine, Jacksonville, FL, United States

Anbukarasi Maran Medical University of South Carolina, Ralph H. Johnson VA Medical Center, Charleston, SC, United States

Konstantinos Marmagkiolis Tampa Heart, Tampa, FL, United States; Department of Internal Medicine, University of South Florida, Tampa, FL, United States

Evandro Martins Filho Hospital Santa Casa de Misericórdia de Maceió, Alagoas, Brazil

Kambis Mashayekhi Internal Medicine and Cardiology, MediClin Heart Center Lahr, Lahr/Schwarzwald, Germany

Margaret B. McEntegart Columbia University Irving Medical Center, New York, NY, United States

Michael Megaly CHIP/CTO Fellow, Henry Ford Hospital, Detroit, MI, United States

Perwaiz Meraj Sandra Atlas Bass Heart Hospital, Northwell Health, Manhasset, NY, United States

Lampros Michalis University of Ioannina, Ioannina, Greece

Anastasios N. Milkas Acute Cardiac Care Unit, Athens Naval and Veterans Hospital, Athens, Greece

Owen Mogabgab Cardiovascular Institute of the South, Houma, LA, United States

Jeffrey Moses NewYork-Presbyterian/Columbia University Irving Medical Center, New York, NY, United States; St Francis Hospital & Heart Center, Roslyn, NY, United States

Muhammad Munawar Binawaluya Cardiac Center, Jakarta, Indonesia; Department of Cardiology and Vascular of Medicine, Faculty of Medicine, Universitas Indonesia, West Java, Indonesia; Department of Cardiology and Vascular Medicine, Faculty of Medicine, Universitas Gadjah Mada, Yogyakarta, Indonesia

Bilal Murad Allina Health Minneapolis Heart Institute, St. Paul, MN, United States

Alexander Nap Amsterdam University Medical Center, Amsterdam, The Netherlands

Andres Navarro QRA Medicina Especializada, Hospital de los Valles, Universidad San Francisco De Quito, Quito, Ecuador

William J. Nicholson Emory University, Atlanta, GA, United States

Anja Øksnes Haukeland University Hospital, Bergen, Norway

Göran Olivecrona Department of Cardiology, SUS-Lund/Lund University, Lund, Sweden

Mohamed A. Omer Mayo Clinic, Rochester, MN, United States

Jacopo Andrea Oreglia Interventional Cardiology, Cardio Center, Niguarda Hospital, Milan, Italy

Lucio Padilla Department of Interventional Cardiology and Endovascular Therapeutics, ICBA, Instituto Cardiovascular, Buenos Aires, Argentina

Mitul P. Patel UC San Diego and VA San Diego Health Systems, San Diego, CA, United States

Rajan A.G. Patel Ochsner Medical Center, New Orleans, LA, United States; University of Queensland, Australia

Taral Patel Tristar Centennial Medical Center, Nashville, TN, United States

Ashish Pershad Chandler Regional and Mercy Gilbert Medical Center, Chandler, AZ, United States

Duane Pinto Division of Cardiology, Beth Israel Deaconess Medical Center, Boston, MA, United States

Paul Poommipanit Case Western Reserve University, Cleveland, OH, United States; UH Parma Medical Center, Parma, OH, United States; University Hospitals Harrington Heart and Vascular Institute, Cleveland, OH, United States

Marin Postu IECVD "C.C. Iliescu", Bucharest, Romania

Srini Potluri Baylor Scott and White The Heart Hospital Plano, Plano, TX, United States

Stylianos Pyxaras I. Med. Klinik, Klinikum Fuerth, Fuerth, Germany

Alexandre Schaan de Quadros Instituto de Cardiologia do Rio Grande do Sul and Hospital Divina Providência, Porto, Alegre, Brazil

Michael Ragosta University of Virginia Health System, Charlottesville, VA, United States

Sunil V. Rao NYU Langone Health System, New York, NY, United States

Vithala Surya Prakasa Rao Care Hospitals, Hyderabad, India

Sudhir Rathore Frimley Health NHS Foundation Trust, Surrey, United Kingdom

Joerg Reifart Department of Cardiology, Kerckhoff-Klinik, Bad Nauheim, Germany

Athanasios Rempakos Minneapolis Heart Institute Foundation, Minneapolis, MN, United States

Jeremy Rier Interventional Cardiology, WellSpan Health York Hospital, York, PA, United States

Robert Riley Overlake Medical Center, Bellevue, WA, United States

Stéphane Rinfret Emory University, Atlanta, GA, United States

Juan J. Russo University of Ottawa Heart Institute, Ottawa, ON, Canada

Meruzhan Saghatelyan Erebouni Medical Center, Yerevan, Armenia

Gurpreet S. Sandhu Mayo Clinic, Rochester, MN, United States

Yader Sandoval Allina Health Minneapolis Heart Institute, Minneapolis, MN, United States; Minneapolis Heart Institute, Minneapolis, MN, United States

Ricardo Santiago PCI Cardiology Group, Bayamon, Puerto Rico, United States

James Sapontis MonashHeart, Monash Medical Centre, Monash University, Melbourne, Australia

Alpesh Shah Houston Methodist Hospital, Houston, TX, United States

Evan Shlofmitz Intravascular Imaging, St Francis Hospital, Roslyn, NY, United States

Kendrick A. Shunk University of California, San Francisco, CA, United States; San Francisco VA Medical Center, San Francisco, CA, United States

George Sianos AHEPA University Hospital, Thessaloniki, Greece

Bahadir Simsek Minneapolis Heart Institute Foundation, Minneapolis, MN, United States

Elliot J. Smith BartsHeart Centre, St. Bartholomew's Hospital, London, United Kingdom

Anthony Spaedy University of Kansas Medical Center, Kansas City, KS, United States

James Spratt St. George's University Hospitals NHS Foundation Trust, London, United Kingdom

Julian W. Strange Bristol Royal Infirmary, University Hospitals Bristol NHS Trust, Bristol, United Kingdom

Bradley Strauss Sunnybrook Health Sciences Centre, Toronto, ON, Canada

Péter Tajti Gottsegen György National Cardiovascular Center, Budapest, Hungary

Hector Tamez Division of Cardiology, Beth Israel Deaconess Medical Center, Boston, MA, United States

Khalid O. Tammam International Medical Center, Jeddah, Saudi Arabia

Craig A. Thompson Interventional Cardiology, Hartford HealthCare Heart & Vascular Institute, Hartford, CT, United States

Aurel Toma University Heart Center Freiburg, Bad Krozingen, Germany

Catalin Toma University of Pittsburgh Medical Center, Heart and Vascular Institute, Pittsburgh, PA, United States

Ioannis Tsiafoutis Red Cross Hospital, Athens, Greece

Etsuo Tsuchikane Toyohashi Heart Center, Aichi, Japan

Imre Ungi Department of Invasive Cardiology, University of Szeged, Szeged, Hungary

Barry F. Uretsky University of Arkansas for Medical Sciences/Central Arkansas Veterans Health System, Little Rock, AR, United States

Georgios J. Vlachojannis Interventional Cardiology, University Medical Center Utrecht, Utrecht, The Netherlands

Minh Nhat Vo Royal Columbian Hospital, New Westminster, BC, Canada

Hoang Vu Vu University Medical Center, University of Medicine and Pharmacy at Ho Chi Minh City, Ho Chi Minh City, Vietnam

Simon Walsh Belfast Health and Social Care Trust, Belfast, United Kingdom

Daniel Weilenmann Department of Cardiology, Kantonsspital St. Gallen, St. Gallen, Switzerland

Gerald Werner Medizinische Klinik I Klinikum Darmstadt, Darmstadt, Germany

Jarosław Wójcik Invasive Cardiology Hospital IKARDIA, Nałęczów, Poland

Jason Wollmuth Providence Heart and Vascular Institute, Portland, OR, United States

Eugene B. Wu Prince of Wales Hospital, Chinese University of Hong Kong, Hong Kong

R. Michael Wyman Torrance Memorial Medical Center, Torrance, CA, United States

Iosif Xenogiannis Second Department of Cardiology, Attikon University Hospital, National and Kapodistrian University of Athens Medical School, Athens, Greece

Bo Xu Catheterization Laboratories, National Clinical Research Center for Cardiovascular Diseases, Fuwai Hospital, National Center for Cardiovascular Diseases, Chinese Academy of Medical Sciences and Peking Union Medical College, Beijing, P.R. China; Fuwai Hospital Chinese Academy of Medical Sciences, Shenzhen, Shenzhen, Guangdong Province, P.R. China

Masahisa Yamane Cardiology Division, Saitama-Sekishinkai Hospital, Saitama, Japan

Luiz F. Ybarra London Health Sciences Centre, Schulich School of Medicine & Dentistry, Western University, London, ON, Canada

Robert W. Yeh Richard A. and Susan F. Smith Center for Outcomes Research in Cardiology, Division of Cardiovascular Medicine, Beth Israel Deaconess Medical Center, Harvard Medical School, Boston, MA, United States

Manual of CTO PCI cases

The videos listed here are provided by the author. Elsevier does not own nor is responsible for the content or maintenance of the videos.

Case number	Case title
1	Retrograde for balloon uncrossable lesion
2	Scratch and go for proximal cap ambiguity
3	Large vessel perforation
4	CTO PCI for recurrent SVG failure
5	IVUS guidance for retrograde CTO PCI
6	OCT for thrombus and dissection
7	In-stent CTO with bifurcation at the distal cap
8	Investment for ISR CTO
9	Subintimal plaque modification (Investment)
10	Retrograde via SVG and aortocoronary dissection
11	PCI of the last remaining vessel
12	Impenetrable proximal cap + distal cap at bifurcation
13	Epicardial collateral perforation
14	Circumflex CTO in a patient with left main disease
15	External crush for balloon uncrossable CTO
16	CTO PCI for SVG aneurysm
17	Perforation of heavily calcified CTO
18	Unable to engage CTO vessel
19	CrossBoss for in-stent CTO
20	Retrograde with hemodynamic support
21	Ensuring true lumen wire position
22	Donor vessel injury
23	Retrograde microcatheter will not cross
24	Guidewire entrapment
25	ADR for circumflex CTO
26	Distal vessel perforation treated with covered stent
27	Carlino for balloon uncrossable CTO
28	Guide issues
29	Retrograde with VA-ECMO support
30	Circumflex CTO with ambiguous proximal cap
31	PCI of two CTOs with hemodynamic support
32	Scratch and go and stent crush for in-stent CTO
33	Retrograde for proximal cap ambiguity
34	IVUS for proximal cap ambiguity
35	IVUS facilitated ADR
36	Dual retrograde for proximal cap ambiguity
37	Flush RCA CTO
38	Retrograde for acute vessel closure

Manual of PCI cases

The videos listed here are provided by the author. Elsevier does not own nor is responsible for the content or maintenance of the videos.

Case number	Title
1	Guide extension for visualization and equipment delivery
2	What to do when you cannot engage the PCI target vessel
3	Challenges associated with severe calcification
4	STEMI in anomalous RCA
5	Aortic Tortuosity and DK crush
6	Severe calcification and distal vessel perforation
7	Wiggle wire to the rescue
8	Angiosculpt balloon fracture
9	MINOCA?
10	Stent loss
11	"The culprit"
12	"Neoluminal" revascularization
13	"Around the world"
14	Large coronary thrombus
15	"Balloon undilatable"
16	Septal perforation
17	Bifurcation and calcification
18	Radial PCI - when to switch?
19	Provisional stenting complication
20	Heavily calcified right coronary artery
21	Acute stent thrombosis
22	Acute closure
23	Atherectomy regret
24	LAD calcified bifurcation lesion
25	Reverse Crush
26	Provisional bifurcation stenting
27	STEMI and shock
28	Ad hod left main PCI
29	Microcatheter tip fracture
30	iFR guided PCI
31	PCI and extreme obesity
32	Guidewire entrapment
33	Guidewire fracture and unraveling
34	Delivery in tortuosity
35	Large RCA thrombus
36	Challenges with Impella-assisted PCI
37	Challenges with engagement and support

(Continued)

(*Continued*)

(Continued)

Prologue

It has been 10 years since the publication of the first edition of the *Manual of CTO Interventions*. Many things have changed since, in CTO PCI and in the world...

CTO PCI (and many other things in life) can often appear to be very difficult or even "impossible":

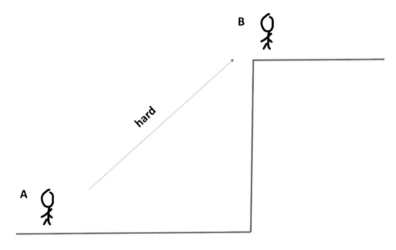

Breaking the "impossible" tasks down to smaller doable "chunks," "bites," or "steps" can "make the impossible possible":

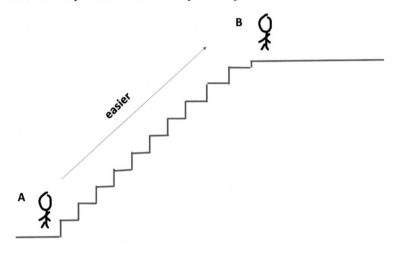

The third edition of the *Manual of CTO Interventions* aims to break down the approach to each challenge of CTO PCI and each CTO PCI technique into discrete, doable steps, based on our current understanding of the field. Often, several treatment strategies need to be implemented before finding a solution ("failing to success"): knowledge and sequential application of the possible treatment options is key for successful completion of CTO PCI.

Reading a book is not enough for creating an expert CTO PCI operator, but it could help you get better by understanding what is currently known in this area. **The eventual goal is to help the patients who need CTO PCI, by increasing the success and safety of the procedure**.

I am grateful to the many collaborators who helped "break down" the task of writing the book into smaller, doable chunks and who helped me (and hopefully you too) learn and improve. I am also grateful to the patients that have entrusted us with their care and whose procedures formed the basis of a large portion of the book.

Please reach out to me (esbrilakis@gmail.com) for any corrections and suggestions on how to make the fourth edition better.

Emmanouil Brilakis

Introduction

1 Chronic total occlusion definition

Coronary chronic total occlusions (CTOs) are defined as "coronary occlusions without antegrade flow through the lesion (TIMI [Thrombolysis In Myocardial Infarction] grade 0 flow) with a presumed or documented duration of ≥ 3 months."[1] Lesions with bridging collaterals that antegradely fill the target vessel can be classified as CTOs, as long as there is no antegrade flow through the lesion itself. Functional occlusions, defined as those with TIMI grade 1 antegrade flow through a severely stenosed but patent lumen, even if not visible on angiography, do not qualify as CTOs. In addition to TIMI grade 0 flow, the typical appearance of a CTO includes angiographically visible mature collaterals and absence of thrombus or staining at the proximal cap.

CTOs are classified as **definite** if they have typical appearance (TIMI grade 0 flow through the lesion with no thrombus, no staining at the proximal cap, and presence of mature collaterals) and definitive corroborating evidence of occlusion duration ≥ 3 months, otherwise they are classified as **probable**.[1]

The **duration** of occlusion may be difficult to determine if there has been no prior angiogram demonstrating presence of the CTO. In such cases, estimation of occlusion duration is based upon the first onset of symptoms and/or prior history of myocardial infarction in the target vessel territory.

Occluded arteries discovered within 30 days from a myocardial infarction, such as those included in the Open Artery Trial (OAT),[2] are not considered to be CTOs. Therefore the lack of benefit observed with PCI in these subacute total occlusions should not be extrapolated to CTO PCI.

2 Prevalence of chronic total occlusions

Coronary CTOs are common, being found in approximately **one in three patients** undergoing diagnostic coronary angiography (Table 1), but the prevalence of coronary CTOs varies widely depending on the population studied.[3-10]

TABLE 1 Prevalence of coronary chronic total occlusions (CTOs).

First author	Country	Year	Number of sites	Patient number	CTO prevalence (%)	CTO prevalence among prior CABG pts (%)
Kahn[3]	United States	1993	1	287	35	–
Christofferson[4]	United States	2005	1	8004	52	–
Werner[5]	Germany	2009	64	2002	35	–
Fefer[6]	Canada	2012	3	14,439	18	54
Jeroudi[7]	United States	2013	1	1669	31	89
Azzalini[8]	Canada	2015	1	2514	20	87
Tomasello[9]	Italy	2015	12	13,423	13	–
Ramunddal[10]	Sweden	2016	30	89,872	16	–

CABG, Coronary artery bypass graft surgery.

3 Outline of the *Manual of CTO PCI*

The third edition of the *Manual of CTO PCI* is restructured to reflect the format of the *Manual of Percutaneous Coronary Interventions*[11] and consists of six parts:

Part A: Presentation of each of the 14 steps of PCI as they apply to CTO PCI

Part B: Challenging lesion subgroups

Part C: Complications

Part D: Equipment

Part E: How to start and improve a CTO PCI program

Part F: Appendices

3.1 Part A: The steps

As outlined in the *Manual of Percutaneous Coronary Interventions*, PCI is performed using the following 14 steps (Fig. 1):

The following steps are performed in all PCI (and CTO PCI) cases:

● Planning (Chapter 1).
● Monitoring (Chapter 2).

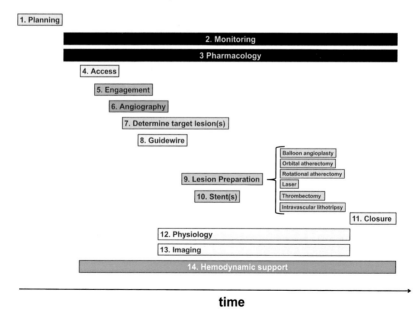

FIGURE 1 **The 14 steps of PCI**. *Reproduced with permission from Brilakis ES*. Manual of Percutaneous Coronary Interventions: a step-by-step approach. *Elsevier; 2021.*

- Medications (Chapter 3).
- Access (Chapter 4).
- Engagement (Chapter 5).
- Angiography (with the exception of the "zero contrast PCI," although the latter still requires a prior angiogram) (Chapter 6).
- Determining the target lesion(s) (Chapter 7).
- Wiring (Chapter 8). This is the most challenging part of CTO PCI and can be performed using antegrade wiring (Section 8.2), antegrade dissection and reentry (Section 8.3) or retrograde (Section 8.4). Several algorithms have been developed to guide selection of wiring strategy, culminating in the global CTO crossing algorithm (Section 8.5).
- Vascular closure (Chapter 11).

The following steps are not always performed:

- Lesion preparation (sometimes direct stenting is performed without predilation, although this is generally discouraged) (Chapter 9)
- Stenting (rarely balloon angioplasty, including drug-coated balloons, or thrombectomy only are performed) (Chapter 10)
- Coronary physiology (Chapter 12)
- Intravascular imaging (Chapter 13)
- Hemodynamic support (Chapter 14)

The first 14 chapters of this Manual discuss each of the 14 aforementioned steps, as they apply to CTO PCI.

3.2   Part B: Challenging lesion subgroups

The following 10 complex patient and lesion subgroups are discussed:

15. Ostial CTOs
16. Bifurcations in CTO PCI
17. Left main CTOs
18. CTO PCI in prior coronary artery bypass graft patients
19. CTO PCI in severely calcified lesions
20. CTO PCI in acute coronary syndromes patients
21. Multiple CTOs—prior CTO PCI failure
22. In-stent CTOs
23. Balloon uncrossable and undilatable lesions
24. CTO PCI in prior transcatheter aortic valve replacement patients and heart failure patients

3.3   Part C: Complications

Complications of CTO PCI can be classified according to timing (as acute and long-term) and according to location (cardiac coronary, cardiac non-coronary

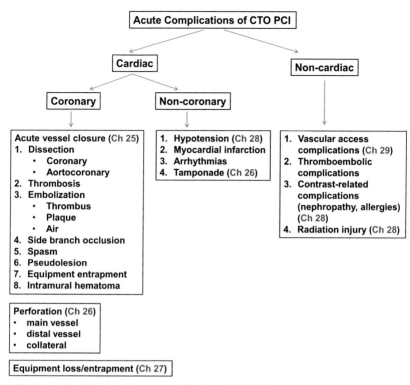

FIGURE 2   **Classification of acute complications of CTO PCI.**

and non-cardiac). The acute complications of CTO PCI are summarized in Fig. 2.[12]

The following complications are discussed in the *Manual of CTO PCI*:

Chapter 25: Acute vessel closure

Chapter 26: Perforation

Chapter 27: Equipment loss or entrapment

Chapter 28: Other complications (hypotension, contrast-induced acute kidney injury, radiation skin injury).

Chapter 29: Access site complications

3.4   Part D: Equipment

Availability of dedicated equipment and familiarity with its use are critical for successfully and safely performing CTO PCI. Overall, PCI equipment can be grouped into 22 categories (the ones relevant to CTO PCI are discussed in Chapter 30):

1. Sheaths
2. Catheters

3. Guide catheter extensions
4. Support catheters
5. Y-connectors
6. Microcatheters
7. Guidewires
8. Embolic protection devices (discussed in the Manual *of Percutaneous Coronary Interventions*)
9. Balloons
10. Atherectomy devices
11. Laser
12. Thrombectomy devices (discussed in the Manual *of Percutaneous Coronary Interventions*)
13. Ostial lesion treatment equipment
14. Stents
15. Arterial closure devices (discussed in the *Manual of Percutaneous Coronary Interventions*)
16. CTO PCI dissection/reentry equipment
17. Intravascular imaging (intravascular ultrasound and optical coherence tomography)
18. Complication management equipment: covered stents, coils, pericardio-centesis trays, and snares
19. Radiation protection equipment
20. Hemodynamic support
21. Contrast management equipment
22. Intracoronary brachytherapy (discussed in the *Manual of Percutaneous Coronary Interventions*)

3.5   Part E: How to start and improve a CTO PCI program

Part E provides advice about how to start and grow a successful CTO PCI program.

3.6   Part F: Appendices

There are two appendices: Appendix 1 provides the names of equipment commonly used in CTO PCI and Appendix 2 explains some of the common acronyms in CTO PCI.

References

1. Ybarra LF, Rinfret S, Brilakis ES, et al. Definitions and clinical trial design principles for coronary artery chronic total occlusion therapies: CTO-ARC consensus recommendations. *Circulation* 2021;**143**:479−500.
2. Hochman JS, Lamas GA, Buller CE, et al. Coronary intervention for persistent occlusion after myocardial infarction. *N Engl J Med* 2006;**355**:2395−407.

3. Kahn JK. Angiographic suitability for catheter revascularization of total coronary occlusions in patients from a community hospital setting. *Am Heart J* 1993;**126**:561−4.

4. Christofferson RD, Lehmann KG, Martin GV, Every N, Caldwell JH, Kapadia SR. Effect of chronic total coronary occlusion on treatment strategy. *Am J Cardiol* 2005;**95**:1088−91.

5. Werner GS, Gitt AK, Zeymer U, et al. Chronic total coronary occlusions in patients with stable angina pectoris: impact on therapy and outcome in present day clinical practice. *Clin Res Cardiol* 2009;**98**:435−41.

6. Fefer P, Knudtson ML, Cheema AN, et al. Current perspectives on coronary chronic total occlusions: the Canadian Multicenter Chronic Total Occlusions Registry. *J Am Coll Cardiol* 2012;**59**:991−7.

7. Jeroudi OM, Alomar ME, Michael TT, et al. Prevalence and management of coronary chronic total occlusions in a tertiary veterans affairs hospital. *Catheter Cardiovasc Interv* 2014;**84**:637−43.

8. Azzalini L, Jolicoeur EM, Pighi M, et al. Epidemiology, management strategies, and outcomes of patients with chronic total coronary occlusion. *Am J Cardiol* 2016;**118**:1128−35.

9. Tomasello SD, Boukhris M, Giubilato S, et al. Management strategies in patients affected by chronic total occlusions: results from the Italian Registry of Chronic Total Occlusions. *Eur Heart J* 2015;**36**:3189−98.

10. Ramunddal T, Hoebers LP, Henriques JP, et al. Prognostic impact of chronic total occlusions: a report from SCAAR (Swedish Coronary Angiography and Angioplasty Registry). *JACC Cardiovasc Interv* 2016;**9**:1535−44.

11. Brilakis ES. *Manual of Percutaneous Coronary Interventions: a step-by-step approach.* Elsevier; 2021.

12. Brilakis ES, Karmpaliotis D, Patel V, Banerjee S. Complications of chronic total occlusion angioplasty. *Interv Cardiol Clin* 2012;**1**:373−89.

Part A

The steps

Chapter 1

Planning

Planning is essential for every procedure, including percutaneous coronary intervention (PCI), especially complex PCI, such as chronic total occlusion (CTO) PCI. Thoughtful planning and appropriate preparation before performing PCI improves the success, safety, and efficiency, and optimizes the cost of the procedure.

The following items, that correspond to each of the 14 steps of PCI (Table 1.1), should be reviewed, with items of special importance for CTO PCI highlighted in **bold** font. While planning is in itself the first of the 14 steps, it primarily serves as a preview of each of the subsequent steps.

1.1 Planning

Obtain consent

- Informed consent for CTO PCI needs to be obtained and documented prior to the procedure.
- Consent for CTO PCI differs from the "standard" PCI consent because CTO PCI has lower success and higher complication rates, such as perforation, radiation skin injury, and contrast-induced acute kidney injury. Various scores (Section 6.3) can be used to provide an individualized numerical estimate of the likelihood of success (such as the J-CTO[1], Progress-CTO[2], CASTLE[3]) and the risk of complications (such as the PROGRESS-CTO COMPLICATIONS scores[4,4a]).
- CTO PCI can sometimes require >1 procedure if the initial attempt fails or is partially successful (see "Investment procedure," Section 21.2).
- Ad hoc CTO PCI should not be attempted in most patients in order to:
  - Allow for a detailed discussion with the patient and family
  - Allow time for thorough procedural planning and preparation
  - Obtain additional imaging
  - Minimize contrast volume (Section 28.3)
  - Minimize radiation dose (Section 28.2 — the Joint Commission identifies peak skin doses >15 Gray as a sentinel event)
  - Minimize patient and operator fatigue
  - Allow time to investigate the myocardial viability (in patients with left ventricular dysfunction, Chapter 14 and Section 24.2) and/or the

Manual of Chronic Total Occlusion Percutaneous Coronary Interventions.
DOI: https://doi.org/10.1016/B978-0-323-91787-2.00005-8

TABLE 1.1 Pre-procedure checklist for cardiac catheterization and PCI.[16]

Patient Name: _____ MRN: _____ Procedure Date: _____

Planned Procedure: Diagnostic Cardiac Catheterization

Diagnostic Cardiac Catheterization with possible PCI

Percutaneous Coronary Intervention

History:

Elective Outpatient Procedures: history and physical (H&P) documented within 30 days?   Yes ☐ No ☐

Inpatient Procedures: H&P documented within 24 hours of admission?   Yes ☐ No ☐

NPO per institutional protocol prior to procedure?   Yes ☐ No ☐

NPO GUIDELINE RECOMMENDATIONS (by the American Society of Anesthesiologists [ASA][16]:

2 hours prior to scheduled procedure time	Clear liquids, including clear/hard candies and drinks without pulp or dairy
6 hours prior to scheduled procedure time	Light solids, including toast/oatmeal/granola bar, liquids with dairy, hard candies, pulp, and infant formula
8 hours prior to scheduled procedure time	Regular diet
Chewing Tobacco	No chewing tobacco 6 hours prior to procedure

History of prior PCI or CABG: Yes ☐ No ☐ If yes, were reports and films obtained and reviewed? Yes ☐ No ☐

Prior radial artery harvesting for CABG?   Yes ☐ No ☐

Arteriovenous fistula for dialysis?   Yes ☐ No ☐

Clinical presentation with STEMI?   Yes ☐ No ☐

Severe peripheral arterial disease   Yes ☐ No ☐

Prior abdominal aortic aneurysm endograft?   Yes ☐ No ☐

Prior iliofemoral surgery?   Yes ☐ No ☐

Other comorbidities: Yes ☐ No ☐

Candidacy for stenting:

1. Is there significant anemia (i.e., hematocrit <30)?   Yes ☐ No ☐

(if yes, has RBC [red blood cell] type and cross been performed?)   Yes ☐ No ☐

2. Any major surgery in the past month or planned for next year?   Yes ☐ No ☐

3. Is there any clinically overt bleeding?   Yes ☐ No ☐

4. Is patient on chronic anticoagulation (e.g., warfarin, direct oral anticoagulant [DOAC])?   Yes ☐ No ☐

5. Is there history of medication non-adherence?   Yes ☐ No ☐

(Continued)

TABLE 1.1 (Continued)

<u>Allergies</u>:

1. Contrast:	Yes ☐ No ☐	If yes, was the patient pre-treated?	Yes ☐ No ☐
2. Aspirin:	Yes ☐ No ☐	If yes, does the patient need desensitization?	Yes ☐ No ☐
3. Heparin	Yes ☐ No ☐	If yes, consider alternative anti-thrombotic agents	
(Heparin-induced thrombocytopenia)			
4. Latex	Yes ☐ No ☐	If yes, remove all latex products from procedural area	
5. Multiple allergies	Yes ☐ No ☐	If yes, consider prednisone pretreatment	

<u>Medications</u>:

1. Did the patient take aspirin within the past 24 hours?	Yes ☐ No ☐
2. Did the patient take a P2Y12 inhibitor within the past 24 hours?	Yes ☐ No ☐
3. Did the patient take metformin within the past 24 hours?	Yes ☐ No ☐
4. Did the patient receive LMWH within the past 24 hours?	Yes ☐ No ☐
☐ If yes for LMWH, dose given and time of last dose _____	
5. Is the patient on immunosuppressants?	Yes ☐ No ☐
☐ If yes, avoid use of arterial closure devices	
6. Is the patient on chronic opioids?	Yes ☐ No ☐
☐ If yes, sedation can be challenging	

<u>Informed Consent</u>:

Was informed consent obtained within 30 days?	Yes ☐ No ☐
Was informed consent documented in the medical record?	Yes ☐ No ☐

<u>Is there a healthcare proxy?</u>	Yes ☐ No ☐

Is the patient DNR or DNI?   Yes ☐   No ☐   Yes, but revoked for procedure
(DNR= do not resuscitate, DNI=do not intubate)

<u>Sedation, Anesthesia and Analgesia</u>:

Are ASA and Mallampati Class documented?	Yes ☐ No ☐
Is there any contraindication to sedation?	Yes ☐ No ☐
Prior adverse reaction to sedation?	Yes ☐ No ☐
Adequate intravenous access?	Yes ☐ No ☐

(Continued)

TABLE 1.1 (Continued)

<u>**Physical Examination:**</u>

Abnormal cardiac examination?	Yes ☐ No ☐
Signs of congestive heart failure?	Yes ☐ No ☐
Signs of cardiogenic shock?	Yes ☐ No ☐
Distal pulses decreased?	Yes ☐ No ☐
Hostile groin (massive obesity, scar from prior procedure, infection, ulcer)?	Yes ☐ No ☐
Arteriovenous fistula for dialysis?	Yes ☐ No ☐
Severe arm tremor or involuntary movements?	Yes ☐ No ☐

<u>**Laboratories and Studies:**</u>

Complete blood count and basic electrolytes within 14 days (outpatient) or 24 hours (inpatient)?	Yes ☐ No ☐
Was ECG performed within 24 hours?	Yes ☐ No ☐
INR performed within 24 hours (for patients on vitamin K antagonists (VKA, such as warfarin))?	Yes ☐ No ☐
Serum beta hCG for women of childbearing potential?	Yes ☐ No ☐
Does the patient require pre-procedure hydration?	Yes ☐ No ☐
If available, prior angiograms reviewed?	Yes ☐ No ☐

(location of femoral bifurcation, peripheral arterial disease, extent of peripheral vessel and coronary calcification, prior stents, old severe lesions which were untreated, access issues-crossover, groin scar, kissing iliac stents or stenosis, radial loops, subclavian stenosis or tortuosity, need for a long sheath, anomalous origin of left or right coronary artery, issues with LIMA engagement, diagnostic and guide catheters used [and whether these provided optimal support based on report and angiographic images], guidewires used and issues with stent delivery, whether atherectomy was needed, etc.)

If available, echocardiogram reviewed?	Yes ☐ No ☐
If available, coronary computed tomography angiography (CCTA) reviewed?	Yes ☐ No ☐

extent of ischemia of the territory supplied by the occluded vessel (Section 7.3.1).[5,6]

In some cases, however, ad hoc CTO PCI may be necessary, such as in patients who present with an acute coronary syndrome due to failure of a highly diseased saphenous vein graft that cannot be recanalized and who have a CTO of the corresponding native coronary artery.[7] Sometimes, CTO PCI techniques can help treat acute vessel closure during non-CTO PCI.[8]

History

- **Clinical presentation (angina, dyspnea, fatigue, and other symptoms potentially suggestive of ischemia).**
- If stable coronary artery disease, is the indication for the procedure appropriate? (review Appropriate Use Criteria[9]).
- **Prior cardiac catheterization? If yes, are the prior images and report available?**
- **Prior coronary artery bypass graft surgery (CABG)? If yes, is the surgical report available?**
- Current medications (see Pharmacology Section 1.3 below).
- Comorbidities
  - Valvular heart disease
  - Congestive heart failure
  - Arrhythmias
  - Peripheral arterial disease (PAD)
  - Renal failure
  - Significant lung disease
  - Obstructive sleep apnea
  - Bleeding disorders
  - Back pain or other musculoskeletal disorders that could affect lying flat on the cardiac catheterization table
  - Diabetes mellitus
  - Advanced age
  - Untreated hyperthyroidism
  - Neurologic or psychologic disorder impeding the patient's ability to stay immobile during a long procedure.
- Is the patient likely to be noncompliant with medications or require noncardiac surgery in the upcoming 6–12 months? If yes, PCI may best be avoided to minimize the risk of stent thrombosis (due to increased coagulation induced by surgery and the early discontinuation of dual antiplatelet therapy). Medical therapy only or CABG may be preferred. If deferral is not feasible due to severe symptoms and CABG is not possible, use of stents with indications for shorter dual antiplatelet therapy (DAPT) regimens may be preferred.
- Chronic renal failure or chronic anticoagulant use? In these patients, nonemergent PCI should be staged, if possible.
- Contrast reaction or latex allergy? If yes, how will these risks be mitigated?

Physical examination

- **Radiation skin injury on the back?** If yes, may need to postpone CTO PCI to allow healing of the affected area.
- Cardiovascular examination that includes all pulses in the upper and lower extremities.

- Signs of congestive heart failure (pulmonary rales, high jugular venous pressure, lower extremity edema).

Labs

- Hemoglobin
- White blood cell count
- Platelet count
- International Normalized Ratio (INR)
- Potassium level
- Creatinine + estimated glomerular filtration rate (GFR) (limit contrast to a maximum of 3x GFR for patients at increased risk for contrast-induced acute kidney injury, such as patients with chronic kidney disease, Section 28.3)[10]
- Pregnancy test (for women of childbearing potential).

Prior imaging

- **Review prior coronary angiograms and PCIs. This is critical for CTO PCI and is discussed in detail in** Section 6.2.
- **Review coronary computed tomography angiography (CCTA), if available.** Any prior chest computed tomography (for example performed for evaluation for pulmonary embolism) should be reviewed, as with current (temporal and spatial) scanner resolution they can still provide information on coronary calcification and also gross/proximal graft patency.
- Review noninvasive testing results (echocardiography, MRI, stress testing). In most cases, CTO PCI should not be performed if the CTO-supplied territory is not viable.

1.2 Monitoring

- **Assess baseline ECG, heart rate, and arterial waveform. This is critical for any PCI and especially for CTO PCI, which can be prolonged, challenging, and/or result in complications.**
- Assess the patient's baseline vital signs and pulse oximetry.

1.3 Pharmacology

- Allergies?
- Has the patient received aspirin?
- For patients with a well-documented aspirin allergy: have they been desensitized?
- For patients with prior contrast reaction: have they been premedicated (Section 3.3)?
- For planned CTO PCI: have they received a P2Y12 inhibitor before the procedure?

- For patients on metformin: in patients with chronic kidney disease, hold metformin on the day of the procedure and do not restart until at least 48 h after the procedure. In patients without chronic kidney disease, metformin does not necessarily need to be discontinued; instead, renal function can be monitored after the procedure and metformin withheld if the renal function deteriorates.
- For patients on insulin: reduce insulin to adjust for fasting status before the procedure.

CTO PCI carries an increased risk of perforation; hence, anticoagulation should be held prior to the procedure, as follows:

- *On vitamin K antagonist (VKA, e.g., warfarin):* discontinue 2−5 days prior to CTO PCI depending on clinical assessment and baseline INR, and check the INR on the day of the procedure (goal INR <1.5). Bridging with low molecular weight heparin may be needed in some patients at very high thrombotic risk, such as patients with a mechanical mitral valve, prior stroke, or recent pulmonary embolism/deep venous thrombosis.[11]
- *On direct oral anticoagulants (DOAC):* discontinue prior to elective procedures, as follows (Table 1.2).

TABLE 1.2 How long to stop a DOAC before a cardiac catheterization procedure.

Direct factor Xa inhibitors	Days to hold
Apixaban (Eliquis)	2
Edoxaban (Savaysa - Lixiana)	
Creatinine clearance 50−95	2
Creatinine clearance 15−49	3
Rivaroxaban (Xarelto)	
Creatinine clearance >50	2
Creatinine clearance 15−49	3
Direct thrombin factor IIa inhibitor	**Days to hold**
Dabigatran (Pradaxa)	
Creatinine clearance >80	2
Creatinine clearance 50−79	3
Creatinine clearance 30−49	4
Creatinine clearance 15−29	5

Creatinine clearance calculator: https://www.mdcalc.com/calc/43/creatinine-clearance-cockcroft-gault-equation.

1.4 Access

History

- Prior radial artery harvesting for CABG?
- Arteriovenous (AV) fistula for dialysis? If so, avoid using the arm with the fistula for cardiac catheterization.
- Prior chest/abdomen/pelvis computed tomography (CT) to inform vascular access including large-bore access for potential mechanical circulatory support?
- Access site(s) used for any prior procedures? Has a closure device been used? Consider using contralateral femoral or radial access if a collagen plug based closure device, such as the Angio-Seal, was used within 90 days.
- Prior access site complications? If yes, what was the complication and how was it managed? If yes, avoid using the same access site.
- History of PAD? Access through severely diseased or occluded iliofemoral or subclavian arteries should be avoided.
- Clinical presentation: radial access currently has a class Ia recommendation for all PCI[12], but is especially favored in ACS patients.
- **On VKA or DOAC: radial access is preferred, but anticoagulation should be discontinued prior to CTO PCI, as outlined in section 1.3.**
- High risk of bleeding: radial access is preferred.
- Patient preference (patients who work extensively with their hands/arms or use them for support may prefer femoral approach).

Physical examination

- Good distal pulses?
- Morbid obesity? (favors radial access)

Labs: High INR and/or low platelets favor radial access.

Prior imaging

- Review prior cardiac and/or peripheral catheterization films: is there disease or tortuosity in aortoiliac and/or upper extremity vessels?
- Coronary CT angiography: anomalous coronary arteries? Proximal cap, occlusion length, quality of the distal vessel.[13]
- CT chest:
  - Anomalous aortic arch?
  - Size of iliac/subclavian vessels and presence of disease.
  - Arteria lusoria? (anomalous origin of right subclavian from the aortic arch). Arteria lusoria favors use of left radial or femoral access.
  - Ulcerated plaque and thrombus in the thoracic aorta.
- CT of abdomen/pelvis: location of the common femoral artery bifurcation and disease/tortuosity in the iliofemoral vessels.
- Ultrasound of peripheral arteries

Desired outcome: Decide on access site and size/length of the sheaths. When using femoral access, a 45-cm long sheath can facilitate equipment manipulations, especially in patients with significant aorto-iliac tortuosity.

1.5   Engagement

- Prior CABG: what is the bypass graft anatomy (surgical report, prior coronary angiograms)?
- Catheters used in prior coronary angiograms/PCIs? If there was significant difficulty or inability to engage the coronary arteries well from one access site, one should consider switching to a different access site (such as contralateral radial or femoral).
- Aortic CT angiography: aortic dilation? Anomalous coronary arteries?
- Aortic stenosis or regurgitation (associated with dilated ascending aorta that may require larger catheter for coronary engagement)?

1.6   Angiography

- Chronic kidney disease? If yes:
  - *Limit contrast volume* by using biplane cineangiography if available; performing limited angiographic projections; using IVUS; using 2:1 contrast dilution in low body mass index (BMI) patients; and potentially using contrast savings systems, such as the Dyevert Plus. Ad hoc CTO PCI should be avoided as described in Section 7.3.1.
  - Consider using iso-osmolar contrast agents (Section 29.3).
  - Administer pre- and post-procedural hydration depending on left ventricular function, the presence of severe valvular disease, and hemodynamic status.
  - The European Society of Cardiology/European Association for Cardio-Thoracic Surgery (ESC/EACTS) revascularization guidelines recommend 1 mL/kg/h starting 12 h before and continued for 24 h after the procedure, except for patients with left ventricular ejection fraction ≤ 35% or NYHA >2 for whom 0.5 mL/kg/h is recommended.[14]
  - The POSEIDON trial used 5 mL/kg/h if left ventricular end-diastolic pressure (LVEDP) was <13 mmHg, 3 mL/kg/h if LVEDP was 13−18 mmHg, and 1.5 mL/kg/h if LVEDP was >18 mmHg.[15]

1.7   Determine target lesion(s)

History: The presence and severity of symptoms is critical for deciding on the need for CTO PCI. CTO PCI is most often performed for symptom improvement, as described in Chapter 7.

Prior imaging: Prior coronary angiography and CCTA can help determine the culprit lesions. In patients who have both CTO and non-CTO lesions, the non-CTO lesions may need to be treated first to help improve the success and safety of CTO PCI. Alternatively, the CTO lesions may be attempted first, with non-CTO PCI following successful CTO recanalization vs. referral for CABG if CTO PCI fails. This latter point is especially important if the CTO involves a proximal left anterior descending artery (LAD) or right coronary artery (RCA) that perfuses a large viable myocardial mass, making complete revascularization important.

1.8   Lesion wiring

History: Prior challenges wiring the target lesion(s)?
  Prior imaging: Select optimal branch for an anchor wire.

1.9   Lesion preparation

History

- Prior challenges dilating the target lesion(s)?

Prior imaging

- Severe calcification: consider atherectomy, laser, intravascular lithotripsy, or the SIS/OPN balloon if available (Chapter 19).
- Assess proximal vessel tortuosity; proximal and distal CTO cap characteristics; collateral vessel supply and suitability of collaterals for retrograde wiring (Chapter 6).

1.10   Stenting

History

- Able to take DAPT? (Prior bleeding, high risk of bleeding, compliance with medications, allergies).
- Planned noncardiac surgery in the upcoming months.

1.11   Arterial closure

History

- Active infection or immunocompromised? May be best to avoid use of vascular closure devices (especially collagen-based) to minimize the risk of infection.

1.12   Physiology

Coronary physiology is infrequently used during CTO PCI, but may be useful for assessing intermediate lesions in non-CTO vessels or for assessing the CTO PCI result, as described in Chapter 12.

History: If symptoms are equivocal and there is no preprocedural noninvasive testing showing ischemia, physiologic coronary assessment can be useful.

- Prior adverse reaction or contraindication to coronary vasodilators such as adenosine?

1.13   Imaging

History

- Intravascular imaging should be performed in all CTO PCIs if feasible, but is especially important for prior stent failure cases to determine the mechanism of failure and choose optimal treatment.

1.14   Hemodynamic support

History

- Congestive heart failure symptoms.
- Low ejection fraction.
- Potential of causing myocardial ischemia, for example by occluding collaterals with retrograde equipment or by performing atherectomy.

Physical examination

- Elevated jugular venous pressure.
- Lower extremity edema.
- Lung crackles.
- Femoral and radial pulses.

Labs

- B-type natriuretic peptide (BNP) and N-terminal-pro-BNP (NT-pro-BNP).
- Lactate in patients with cardiogenic shock.

Prior imaging

- Echocardiography (left and right ventricular ejection fraction, left and right ventricular size, valvular abnormalities).
- Access site imaging to determine feasibility of hemodynamic support.

Hemodynamics

- Right heart catheterization measurements, if available (right atrial, pulmonary artery, and pulmonary artery capillary wedge pressure, cardiac output).
- Consider monitoring the pulmonary artery pressure during CTO PCI in high-risk patients with standby hemodynamic support after inserting a 4

French femoral arterial sheath. Insert a hemodynamic support device if the pulmonary artery pressure increases during the procedure.

Planned procedure

- PCI of the last remaining vessel.
- Use of the retrograde approach.
- Planned atherectomy.
- High risk for global ischemia.

Consider hemodynamic support in patients with reduced ejection fraction, poor or deteriorating hemodynamics (high RA, mean PA, and/or wedge pressure and/or low cardiac output), and/or complex/high-risk planned interventions (Chapter 14).

References

1. Morino Y, Abe M, Morimoto T, et al. Predicting successful guidewire crossing through chronic total occlusion of native coronary lesions within 30 minutes: the J-CTO (Multicenter CTO Registry in Japan) score as a difficulty grading and time assessment tool. *JACC Cardiovasc Interv* 2011;**4**:213−21.
2. Christopoulos G, Kandzari DE, Yeh RW, et al. Development and validation of a novel scoring system for predicting technical success of chronic total occlusion percutaneous coronary interventions: the PROGRESS CTO (prospective global registry for the study of chronic total occlusion intervention) score. *JACC Cardiovasc Interv* 2016;**9**:1−9.
3. Szijgyarto Z, Rampat R, Werner GS, et al. Derivation and validation of a chronic total coronary occlusion intervention procedural success score from the 20,000-patient EuroCTO Registry: the EuroCTO (CASTLE) score. *JACC Cardiovasc Interv* 2019;**12**:335−42.
4. Danek BA, Karatasakis A, Karmpaliotis D, et al. Development and validation of a scoring system for predicting periprocedural complications during percutaneous coronary interventions of chronic total occlusions: the prospective global registry for the study of chronic total occlusion intervention (PROGRESS CTO) complications score. *J Am Heart Assoc* 2016;5.
4a. Simsek B, Kostantinis S, Karacsonyi J, Alaswad K, Krestyaninov O, Khelimskii D, Davies R, Rier J, Goktekin O, Gorgulu S, ElGuindy A, Chandwaney RH, Patel M, Abi Rafeh N, Karmpaliotis D, Masoumi A, Khatri JJ, Jaffer FA, Doshi D, Poommipanit PB, Rangan BV, Sanvodal Y, Choi JW, Elbarouni B, Nicholson W, Jaber WA, Rinfret S, Koutouzis M, Tsiafoutis I, Yeh RW, Burke MN, Allana S, Mastrodemos OC, Brilakis ES. Predicting Periprocedural Complications in Chronic Total Occlusion Percutaneous Coronary Intervention: The PROGRESS-CTO Complication Scores. JACC Cardiovasc Interv. 2022 Jul 25;15(14):1413−1422.
5. Blankenship JC, Gigliotti OS, Feldman DN, et al. Ad hoc percutaneous coronary intervention: a consensus statement from the society for cardiovascular angiography and interventions. *Catheter Cardiovasc Interv* 2013;**81**:748−58.
6. Xenogiannis I, Nikolakopoulos I, Vemmou E, Brilakis ES. Chronic total occlusion recanalization for myocardial infarction. *Catheter Cardiovasc Interv* 2020;**95**:1133−5.
7. Brilakis ES, Banerjee S, Lombardi WL. Retrograde recanalization of native coronary artery chronic occlusions via acutely occluded vein grafts. *Catheter Cardiovasc Interv* 2010;**75**:109−13.

8. Shaukat A, Mooney M, Burke MN, Brilakis ES. Use of chronic total occlusion percutaneous coronary intervention techniques for treating acute vessel closure. *Catheter Cardiovasc Interv* 2018;**92**:1297−300.
9. Patel MR, Calhoon JH, Dehmer GJ, et al. *ACC/AATS/AHA/ASE/ASNC/SCAI/SCCT/STS 2017 appropriate use criteria for coronary revascularization in patients with stable ischemic heart disease: a report of the American College of Cardiology Appropriate Use Criteria Task Force, American Association for Thoracic Surgery, American Heart Association, American Society of Echocardiography, American Society of Nuclear Cardiology, Society for Cardiovascular Angiography and Interventions, Society of Cardiovascular Computed Tomography, and Society of Thoracic Surgeons. J Am Coll Cardiol*, 69. 2017. p. 2212−41.
10. Wu EB, Brilakis ES, Mashayekhi K, et al. Global chronic total occlusion crossing algorithm: JACC state-of-the-art review. *J Am Coll Cardiol* 2021;**78**:840−53.
11. Otto CM, Nishimura RA, Bonow RO, et al. 2020 ACC/AHA guideline for the management of patients with valvular heart disease: a report of the American College of Cardiology/American Heart Association Joint Committee on clinical practice guidelines. *Circulation* 2021;**143**:e72−e227.
12. Lawton JS, Tamis-Holland JE, Bangalore S, et al. 2021 ACC/AHA/SCAI guideline for coronary artery revascularization: a report of the American College of Cardiology/American Heart Association Joint Committee on clinical practice guidelines. *Circulation* 2022;**145**: e18−e114.
13. Hong SJ, Kim BK, Cho I, et al. Effect of coronary CTA on chronic total occlusion percutaneous coronary intervention: a randomized trial. *JACC Cardiovasc Imaging* 2021;**14**:1993−2004.
14. Neumann FJ, Sousa-Uva M, Ahlsson A, et al. 2018 ESC/EACTS guidelines on myocardial revascularization. *Eur Heart J* 2019;**40**:87−165.
15. Brar SS, Aharonian V, Mansukhani P, et al. Haemodynamic-guided fluid administration for the prevention of contrast-induced acute kidney injury: the POSEIDON randomised controlled trial. *Lancet* 2014;**383**:1814−23.
16. Practice guidelines for preoperative fasting and the use of pharmacologic agents to reduce the risk of pulmonary aspiration: application to healthy patients undergoing elective procedures: an updated report by the American Society of Anesthesiologists Task Force on preoperative fasting and the use of pharmacologic agents to reduce the risk of pulmonary aspiration. Anesthesiology 2017;126:376−93.

Chapter 2

Monitoring

Monitoring the patient should be continually performed from the beginning to the end of chronic total occlusion (CTO) percutaneous coronary intervention (PCI), so that potential complications are promptly identified and corrected. The following parameters are assessed (Fig. 2.1):

2.1 Patient

1. Patient comfort level: patient discomfort can lead to movement, potentially leading to complications. It can also lead to tachycardia and tachypnea, which may worsen ischemia.
2. Chest pain, abdominal pain, groin pain? Is the pain anticipated based on the procedure or is it unexpected? The pain could be due to ischemia, perforation, hematoma, etc.
3. Level of consciousness and breathing: Is breathing assistance needed (BiPAP or intubation)?
4. Ability to move all extremities (no stroke) or conversely excessive movements that may hinder performance of the procedure.
5. Signs of allergic reactions: skin rash; itching and hives; swelling of the lips, tongue, or throat; hypotension.

2.2 Electrocardiogram

1. Assess ECG morphology and heart rate at the beginning (to establish a baseline) and throughout the case (especially before and after each contrast injection) to identify ECG changes, such as:
   - New ST segment depression (Fig. 2.2).
   - New ST segment elevation (Fig. 2.2).
   - Pseudonormalization: change of the baseline ST segment pattern (Fig. 2.3).
   - Bradycardia (Figs. 2.4 and 2.5).
   - Tachycardia (Figs. 2.6 and 2.7).
   - QRS widening (Fig. 2.8).
   - Ventricular premature beats (may be due to inadvertent entry of the 0.035 inch wire in the ventricle during catheter advancement, wire entry into small branches, coronary guidewire exiting the vessel during CTO PCI, etc.) (Fig. 2.9).
   - Ventricular fibrillation (Fig. 2.10).

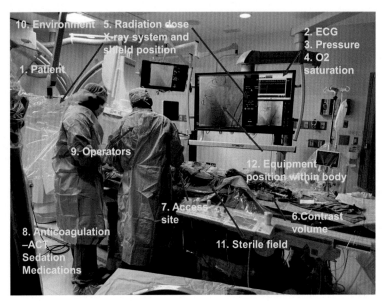

FIGURE 2.1   **What to monitor during CTO PCI**. *Reproduced with permission from pcimanual.org.*

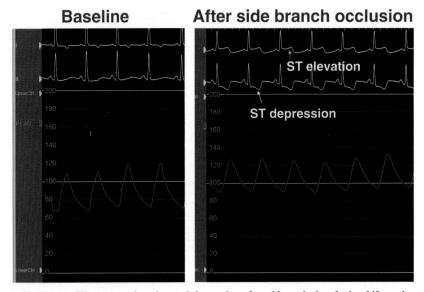

FIGURE 2.2   **ST segment elevation and depression after side occlusion during bifurcation PCI**. *Reproduced with permission from pcimanual.org.*

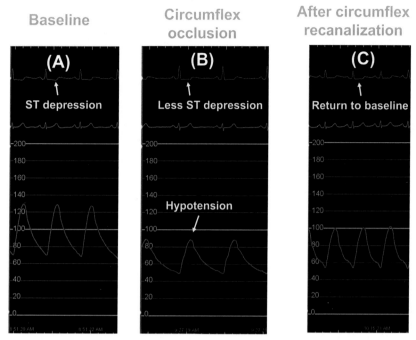

FIGURE 2.3   Baseline ST segment depression in a patient who underwent LAD PCI (**panel A**). During the procedure, ST segment depression decreased (**panel B**), prompting angiography that showed circumflex occlusion due to thrombus migration from the LAD. After restoration of flow into the circumflex the ST segments returned to baseline (**panel C**). This case highlights the importance of continuous ECG (and pressure) monitoring during cardiac catheterization. *Reproduced with permission from pcimanual.org.*

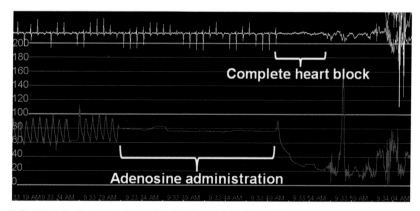

FIGURE 2.4   **Complete heart block after adenosine administration**. *Reproduced with permission from pcimanual.org.*

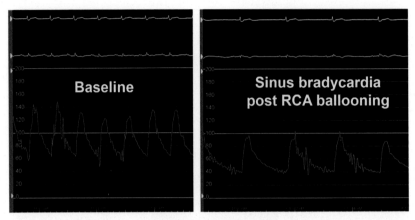

FIGURE 2.5 **Bradycardia after balloon angioplasty of the right coronary artery in a patient with inferior STEMI**. *Reproduced with permission from pcimanual.org.*

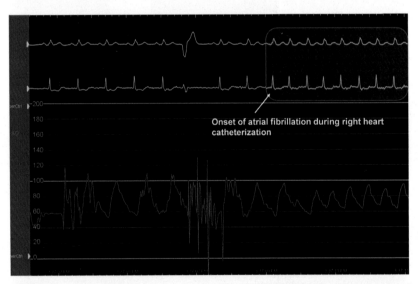

FIGURE 2.6 **Atrial fibrillation onset during right heart catheterization**. *Reproduced with permission from pcimanual.org.*

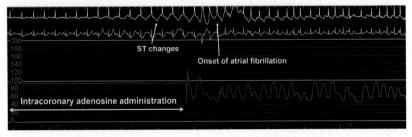

FIGURE 2.7 **Atrial fibrillation after adenosine administration**. *Reproduced with permission from pcimanual.org.*

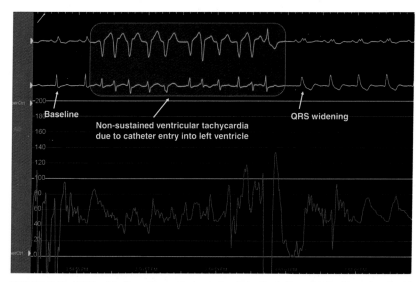

FIGURE 2.8   **QRS widening after catheter entry into the left ventricle.** *Reproduced with permission from pcimanual.org.*

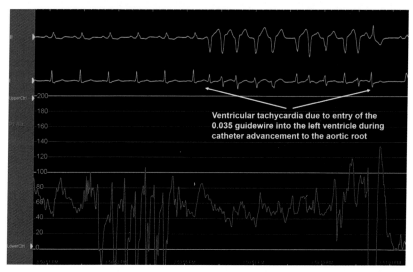

FIGURE 2.9   **Nonsustained ventricular tachycardia due to inadvertent entry of a 0.035 inch guidewire into the left ventricle during catheter advancement.** *Reproduced with permission from pcimanual.org.*

2.3   Pressure waveform

Monitor the systemic (and pulmonary artery, if available) pressure waveform during the case. Also monitor intravenous fluid administration, especially in patients with low ejection fraction or high filling pressures at baseline.

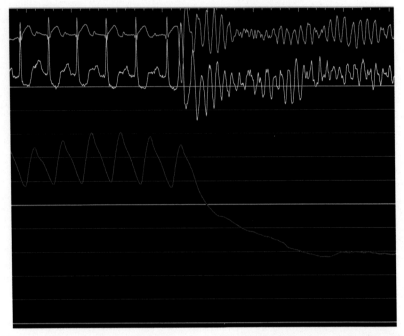

FIGURE 2.10 **Ventricular fibrillation in a patient with ST segment elevation MI**. *Reproduced with permission from pcimanual.org.*

Pressure waveform changes of concern include:

1. Hypotension (Figs. 2.11–2.15) (see Section 28.1)
2. Pulsus paradoxus
3. Hypertension
4. Pressure waveform dampening or disappearance (see Section 28.1.1.1)

Potential causes of pressure dampening:

- Deep guide catheter engagement (Figs. 2.16 and 2.17).
- Engagement of coronary arteries with aorto-ostial lesions: Injections should not be performed while the pressure is dampened, as it can lead to coronary or aortocoronary dissection and/or air embolism.
- Air entrainment within the guide catheter (for example, when using the trapping technique for equipment exchange).
- Thrombus formation within the catheter (Fig. 2.18): Injecting in such cases can lead to coronary or systemic thromboembolism.
- Guide catheter kinking (Fig. 2.19).
- Equipment insertion: for example, inserting an aspiration catheter, such as an Export or Penumbra Indigo CAT RX catheter into a 6 French guide catheter can cause pressure dampening.
- Disconnection of the pressure transducer.

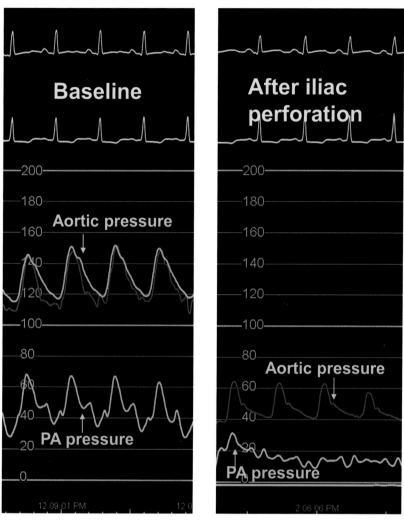

FIGURE 2.11   **Acute hypotension due to iliac artery perforation**. PA = pulmonary artery. *Reproduced with permission from pcimanual.org.*

2.4   Oxygen saturation

Oxygen desaturation may be caused by hypoventilation due to heavy sedation, but it can also be due to early pulmonary edema, artifact, or other causes. End tidal CO_2 monitoring can also be very useful in monitoring ventilation. Full arterial blood gas can provide more comprehensive information.

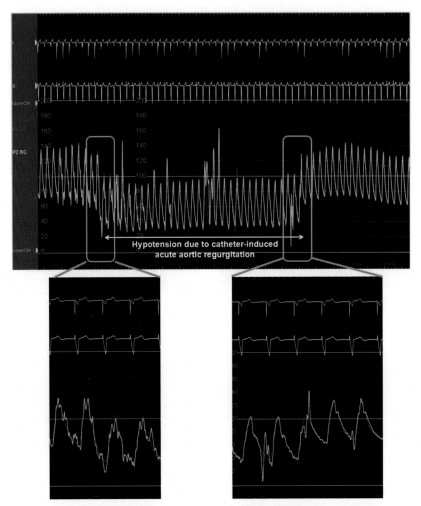

FIGURE 2.12 **Acute transient hypotension due to catheter-induced acute aortic regurgitation from an AL catheter.** *Reproduced with permission from pcimanual.org.*

2.5 Radiation dose—X-ray system and shield positioning

The cumulative air kerma (AK) and dose area product (DAP) radiation dose should be continuously monitored (Fig. 2.20). Usually, the procedure is stopped if the air kerma dose exceeds 5—7 Gray. More than 15 Gray air kerma dose is a sentinel event that requires reporting to the Joint Commission in the USA (Section 28.2).

The dose rate is another dynamic parameter that can be tracked.

There are also continuous operator dose monitoring devices (such as the DoseAware, Philips and the RaySafe i2 and i3 Real-time Radiation

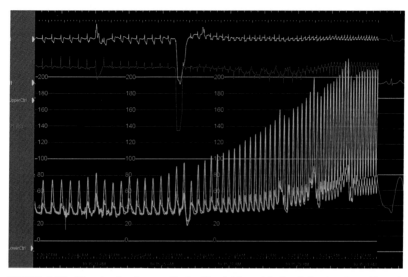

FIGURE 2.13 **Hypotension in a patient with right ventricular infarct and cardiogenic shock with rapid increase in systemic blood pressure after administration of epinephrine.** *Reproduced with permission from pcimanual.org.*

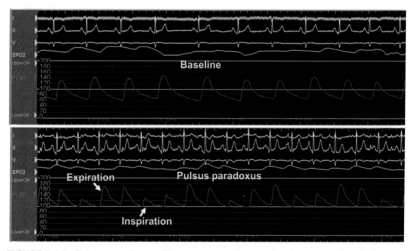

FIGURE 2.14 **Pulsus paradoxus after coronary artery perforation.** *Reproduced with permission from pcimanual.org.*

Dosimeter) that can alert to high operator doses in real time, enabling immediate action to reduce radiation dose.

The fluoroscopy intensity, frames per second, collimation, and the position of the various shields and the image receptor should also be continually monitored to minimize radiation dose.[1,2]

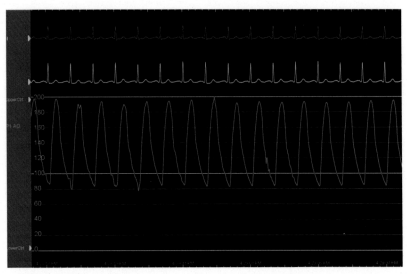

FIGURE 2.15   Systemic hypertension. *Reproduced with permission from pcimanual.org.*

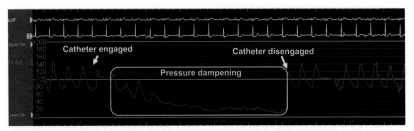

FIGURE 2.16 **Pressure waveform dampening after coronary artery engagement.** *Reproduced with permission from pcimanual.org.*

2.6   Contrast volume

The contrast volume administered can be tracked automatically by some systems (such as the ACIST injector and the DyeVert Plus system) or can be tracked and announced by the cath lab team (for example after reaching 100, 200, 300 mL, etc.). The procedure should generally be stopped before the amount of contrast exceeds 3x glomerular filtration rate (GFR),[3] although a lower threshold is preferable in patients with chronic kidney disease or single kidney (Section 28.3) and in patients with high left ventricular filling pressures due to risk of precipitating acute pulmonary edema. Recent contrast administration should be taken into consideration when determining the contrast threshold.

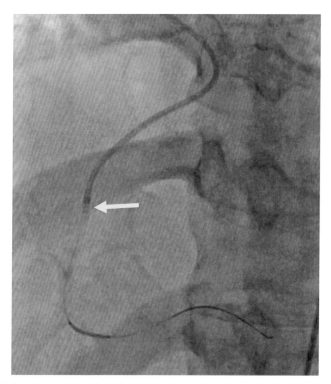

FIGURE 2.17   **Deep guide catheter intubation (arrow) to facilitate stent delivery engagement.** *Reproduced with permission from pcimanual.org.*

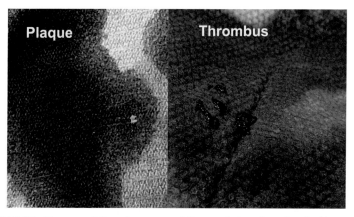

FIGURE 2.18   **Plaque and thrombus removed from catheter.** *Reproduced with permission from pcimanual.org.*

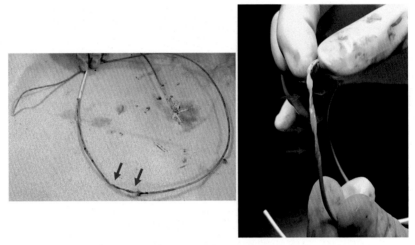

FIGURE 2.19 **Guide catheter kinking**. *Reproduced with permission from pcimanual.org.*

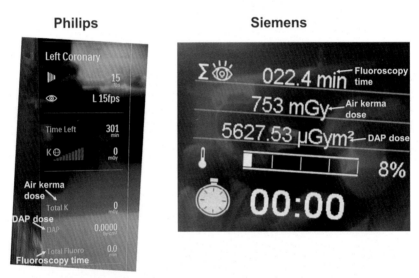

FIGURE 2.20 **Fluoroscopy time and radiation dose measurements in Philips and Siemens X-ray systems**. *Reproduced with permission from pcimanual.org.*

2.7 Access site

The pulse at the access site and distally should be assessed at the beginning and the end of the case.

Bleeding and hematoma formation can occur at the access site(s)—continuous inspection and palpation can help in early identification (Chapter 4, Access).

2.8 Medication administration (anticoagulation—ACT, sedation, other medications)

Sedation (Section 3.1) is given in every patient and is titrated to achieve acceptable patient comfort without compromising respiratory or hemodynamic status.

Anticoagulation (Section 3.4) is achieved with unfractionated heparin in most procedures and monitored using the ACT (activated clotting time). Goal ACT is >300 sec for antegrade CTO PCI and >350 sec for retrograde CTO PCI. Glycoprotein IIb/IIIa inhibitors should be avoided in CTO PCI.

Other medications may be required, such as vasopressors (Section 3.6), atropine, etc.

2.9 Operator and team performance

Paying attention to the operator's and team's operational state can help identify conditions that may lead to suboptimal outcomes, such as excessive fatigue, that could decrease the efficacy and safety of the procedure. CTO PCIs should be scheduled in the beginning of the day and should not be started by an exhausted personnel.

2.10 Cath lab environment

Avoiding excessive noise and distractions is important for better outcomes.

A rule analogous to the "sterile cockpit rule" for flying should be implemented during the critical parts of the procedure. The **Sterile Cockpit Rule** is an informal name for the Federal Aviation Administration (FAA) regulation that no flightcrew may engage in, nor may any pilot in command permit, any activity during a critical phase of flight which could distract any flightcrew from the performance of his or her duties or which could interfere in any way with the proper conduct of those duties. During CTO PCI the "pilot in command" would be the first operator and "crew members" would be the cath lab members.

2.11 Sterile field and equipment

Keeping the equipment and table organized facilitates equipment identification and use. This is especially true in CTO PCI, which often requires numerous pieces of equipment (Fig. 2.21). Reintroduction of the guidewires into their original plastic coil can help keep the table organized.

Dried blood and contrast can make the operator gloves and overall equipment (guidewires, catheters, balloons, stents, etc.) "sticky" and also could create risk of embolism if debris enters the manifold. Regularly wiping the gloves and equipment and flushing the catheters with heparinized saline will

FIGURE 2.21 **Keeping the guidewires organized.** *Reproduced with permission from pcimanual.org.*

facilitate equipment handling and reduce the risk of complications. Guidewires may be best wiped with clean and wet gloves instead of a gauze pad to reduce the risk of fiber embolism.

2.12 Equipment position within the body

The position of equipment inserted into the body (such as sheaths, guide catheters, guidewires, balloons, stents, etc.) should be continually monitored for both efficacy and safety.

A classic example is guide catheter disengagement while attempting to deliver balloons and stents, when the operator often focuses on the equipment that needs to be delivered (such as the stent) and does not pay attention to the guide catheter, which may become completely disengaged with both guide and guidewire position being lost. Conversely, deep guide engagement may result in dissection and acute vessel closure (Section 25.2.1), especially if contrast is injected.

Another example is not monitoring the location of the guidewire tip (especially when the shutters are used to minimize radiation dose), which may enter into small branches and cause distal vessel perforation (Section 26.4). Intermittently increasing the field of view and removing collimation can help prevent this complication.

Who is assessing the above parameters:

1. The primary and secondary operators.
2. The cath lab technician (traditionally a technician is constantly monitoring the ECG and pressure tracings).
3. The cath lab RN (who monitors the ECG, pressure, ACT, intravenous fluid administration and O_2 saturation). The cath lab RN is usually administering the various medications (sedation, anticoagulation, antiplatelet agents, etc.).

References

1. Karatasakis A, Brilakis ES. Shields and garb for decreasing radiation exposure in the cath lab. *Expert Rev Med Devices* 2018;**15**:683−8.
2. Christopoulos G, Makke L, Christakopoulos G, et al. Optimizing radiation safety in the cardiac catheterization laboratory: a practical approach. *Catheter Cardiovasc Interv* 2016;**87**:291−301.
3. Wu EB, Brilakis ES, Mashayekhi K, et al. Global chronic total occlusion crossing algorithm: JACC state-of-the-art review. *J Am Coll Cardiol* 2021;**78**:840−53.

Chapter 3

Medications

3.1 Sedatives and analgesics

Video 3.1. PCI Pharmacology: Sedation and Analgesia.

Goals:
- To improve patient comfort.

How?
- Midazolam: 0.5−1 mg IV—can be repeated. Duration of action: 15−80 min.
- Fentanyl: 25−100 mcg IV—can be repeated. Duration of action: 30−60 min. Other opioids, such as morphine can also be used, especially in STEMI patients.

What can go wrong?
1. **Respiratory failure—hypopnea**
   Causes:
   - Excessive sedation may suppress respiratory drive.
   Prevention:
   - Avoid excessive sedation.
   - Monitor oxygen saturation during the procedure.
   Treatment:
   - Do not administer more sedative medications.
   - Flumazenil (Romazicon) for reversing midazolam: 0.2 mg IV over 15 s. If there is no response after 45 s, administer 0.2 mg again over 1 min. Can repeat at 1 min intervals up to a total of 1 mg.
   - Naloxone (Narcan) for reversing opioids (fentanyl, morphine, etc.): 0.1−0.2 mg intravenously; can repeat at 2−3-min intervals until the desired degree of reversal is achieved.
   - Intubation may be required for severe respiratory depression.
2. **Delayed response to oral P2Y12 inhibitors,** potentially leading to thrombotic complications
   Causes:
   - Opioids can reduce gastric emptying and slow drug absorption.

Manual of Chronic Total Occlusion Percutaneous Coronary Interventions.
DOI: https://doi.org/10.1016/B978-0-323-91787-2.00017-4

Prevention:
- Avoid opioid use in STEMI within the first hour of being loaded with oral P2Y12 inhibitors.

Treatment:
- Consider use of intravenous antiplatelet agents (e.g., cangrelor, GP IIb/IIIa inhibitors).

3.2   Vasodilators

Video 3.2. PCI Pharmacology: Vasodilators.

Vasodilators can be subdivided into those that mainly cause large vessel vasodilation (nitroglycerin) and those that mainly cause microvascular vasodilation (nicardipine, nitroprusside, adenosine).

3.2.1   Nitroglycerin

Goals:
- To dilate coronary arteries (administer intracoronary nitroglycerin routinely before coronary angiography, both to prevent coronary spasm and to enable accurate interpretation of the coronary anatomy).
- To treat hypertension.
- To treat pulmonary edema.

How?
- Intracoronary/intragraft: 100−300 mcg.
- Administration through a single or dual lumen microcatheter can concentrate the effect of the medication where it is needed most, for example in cases of no reflow.
- Intravenous: start nitroglycerin drip at 10 mcg/min and increase by 10 mcg/min at 5-min intervals until the desired effect is achieved keeping systolic blood pressure above 100 mmHg. Maximum dose is 400 mcg/min.
- Sublingual: 0.4 mg.

What can go wrong?

1. Hypotension

Causes:
- Excessive peripheral venous dilation, reducing blood return to the heart (decreased preload) and excessive peripheral artery dilation (decreased afterload).
- Coadministration of nitroglycerin and phosphodiesterase type 5 (PDE-5) inhibitors (avanafil, sildenafil, vardenafil, and tadalafil).
- Hypertrophic obstructive cardiomyopathy (HOCM): nitroglycerin can worsen left ventricular outflow obstruction by decreasing both preload and afterload.

Prevention:

- Avoid high and multiple doses of nitroglycerin.
- Do not administer in patients with hypotension or in patients with a recent right ventricular infarction.
- Do not administer in patients who have recently received a phosphodiesterase type 5 inhibitor, such as avanafil (Stendra, within prior 12 hours), sildenafil (Viagra—within prior 24 h), vardenafil (Levitra, within 24 h), and tadalafil (Cialis, within 48 h).
- Do not administer to patients with hypertrophic obstructive cardiomyopathy.
- Use with caution in patients with severe aortic stenosis.

Treatment:

- Stop nitroglycerin administration. Half-life ranges from 1.5 to 7.5 min.
- Normal saline administration.
- Waiting (hypotension often resolves after a few minutes).
- Administer vasopressors (such as norepinephrine or phenylephrine) in cases of extreme or persistent hypotension.

2. **Headache, flushing, dizziness.**
3. **Tachycardia.**

3.2.2   Nicardipine

Goals:

- To prevent and treat no reflow. Nicardipine is a calcium channel blocker that is used intracoronary to achieve vasodilation of small arteries. Nicardipine is the preferred agent for treating or preventing no reflow (Section 25.2.3.2), for example during atherectomy and during saphenous vein graft PCI, as it causes less hypotension compared with nitroprusside and verapamil and has a shorter duration of action.

How?

- Intracoronary: 100−300 mcg.

What can go wrong?

1. **Hypotension.** This is treated as described in Section 3.2.1.

3.2.3   Nitroprusside

Goals:

- To prevent and treat no reflow.

How?

- Intracoronary: 100−300 mcg.

What can go wrong?

1. **Hypotension.** This is treated as described in Section 3.2.1.

3.2.4   Verapamil

Goals:
- To prevent radial spasm.
- To prevent and treat no reflow.

How?
- In radial artery: 2−3 mg (dilute with blood and administer slowly to minimize burning sensation of the patient's hand).
- Intracoronary: 1 mg intracoronary over 2 min.[1]

What can go wrong?
1. **Hypotension.** This is treated as described in Section 3.2.1.
2. **Bradycardia.**

3.2.5   Adenosine

Goals:
- To prevent and treat no reflow.
- To cause vasodilation during physiologic testing (Section 12.2.6).

How?
- Intracoronary in the RCA: 50−100 mcg
- Intracoronary in the left main: 100−200 mcg—several thousand mcg could be administered (slowly) in case of no reflow.
- Intragraft: 100−200 mcg
- Intravenous: 140 mcg/kg/min, administered through a central vein or a large peripheral vein for coronary physiologic assessment.
- Regadenoson or papaverine can also be administered for inducing vasodilation. Regadenoson is costly and papaverine (not available in the US) carries risk of ventricular fibrillation.

What can go wrong?
1. **Heart block** (Fig. 2.4).
   Causes:
   - Adenosine effect on atrioventricular node.
   - Heart block is more likely after injection in the right coronary artery.
   Prevention:
   - Avoid high doses of adenosine in the right coronary artery.
   - Slow adenosine administration.
   - Aminophylline administration (250−300 mg intravenously over 10 min) may be used to prevent bradycardia in patients undergoing coronary atherectomy. Aminophylline is an A1 adenosine receptor antagonist (Section 19.3)
   Treatment:
   - Watchful waiting (adenosine has short half-life).

2. **Atrial fibrillation** (Fig. 2.7).

   Atrial fibrillation is the most commonly documented adenosine-induced arrhythmia (2.7% after intravenous administration) and is usually well-tolerated except in patients with accessory pathways.

   Causes:
   - The exact mechanisms of adenosine-induced atrial fibrillation are unclear but likely include shortening of the atrial action potential through activation of A1 receptors and stimulatory effects of adenosine on pulmonary venous ectopic foci.[2]

   Prevention:
   - Same as for heart block above.

   Treatment:
   - DC cardioversion.

3. **Ventricular fibrillation**

   Causes:
   - Torsades des pointes or ventricular fibrillation can be triggered by adenosine administration. Torsades des pointes or ventricular fibrillation usually occur after a ventricular pause due to the R on T phenomenon but torsades des pointes may also occur without a pause.[3]

   Prevention:
   - Same as for heart block above.

   Treatment:
   - Ask patient to cough, as forceful coughing can generate sufficient blood flow to the brain to maintain consciousness until definitive treatment (defibrillation) can be initiated.
   - Defibrillation.

3.3   Contrast media

Video 3.3. PCI Pharmacology: Contrast Media.

Goals:
- To visualize coronary and peripheral arteries under X-ray.
- To clear blood from coronary arteries in order to perform optical coherence tomography (OCT) (Chapter 13)

How?
- Contrast media are injected through a manifold or through an automated injection system, such as the ACIST.
- The lowest possible volume of contrast should be administered, as described in Section 28.3.2.
- Iso-osmolar contrast media have been associated with lower risk of contrast-induced acute kidney injury.

What can go wrong?
1. **Contrast-induced acute kidney injury** (discussed in Section 28.3).
2. **Allergic reactions**

Allergic reactions to contrast agents may occur immediately after administration or after a delay (usually 6−12 h after contrast administration) and can range in severity from mild (skin rash) to life-threatening (angioedema, anaphylactic shock).

Causes:
- IgE mediated reactions or direct mast cell degranulation.

Prevention:
- Use iso-osmolar contrast media (but iso-osmolar contrast media carry higher risk of delayed skin reactions).
- Pretreatment (for patients known to have allergic reactions to contrast administration):
  - Steroids (administer prednisone 50 mg at 13, 7, and 1 h before the procedure) or methylprednisolone 32 mg 12, and 2 h prior to the procedure.
  - Diphenhydramine (Benadryl) 50 mg 1 h prior to the procedure
  - Cimetidine 300 mg orally or ranitidine 150 mg orally 1 h prior
- If the patient needs an emergent procedure, hydrocortisone sodium succinate (Solu-Cortef) 100 mg should be administered intravenously as soon as possible before the procedure. An intravenous H2 antihistamine, such as ranitidine 50 mg IV over 15 min, can also be used.

Treatment:
- Diphenhydramine (Benadryl) 25−50 mg intravenously.
- Cimetidine 300 mg IV or ranitidine 50 mg IV over 15 min.
- Steroids (hydrocortisone sodium succinate [Solu-Cortef] 100−400 mg administered intravenously over 1 min).
- Epinephrine (for anaphylactic shock) (0.3 mg of 1:10,000 solution intravenously - can repeat to a total dose of 1 mg).
- Intravenous normal saline.

3. **Thyroid dysfunction**

Causes:
- Contrast media contain high doses of iodine and may lead to hypersecretion of thyroid hormones in patients with hyperthyroidism (Jod−Basedow effect; it usually occurs 2−12 weeks after iodine administration). Iodine administration can also lead to hypothyroidism.

Prevention:
- Do not administer contrast in patients with significant hyperthyroidism or hypothyroidism.

Treatment:
- Refer to endocrinology.

4. **Contrast-induced sialadenitis**
5. **Contrast-induced transient cortical blindness**

3.4   Anticoagulants

Video 3.4. PCI Pharmacology: Anticoagulants.

Unfractionated heparin is the most commonly used anticoagulant for chronic total occlusion (CTO) percutaneous coronary intervention (PCI) and PCI in general. Bivalirudin has similar outcomes as unfractionated heparin and is used infrequently, but should not be used in CTO PCI, except in patients with heparin-induced thrombocytopenia.

Goals:
- To prevent thrombus formation within the coronary artery or within equipment inserted into the body (sheaths, guide catheters, wires, balloon, stents etc.).
- To reduce the risk of radial artery occlusion (when radial access is used).

How?
Unfractionated heparin

1. **Dose**:
   a. Without concomitant glycoprotein IIb/IIIa inhibitor or cangrelor administration: 70−100 units per kg.
   b. With concomitant glycoprotein IIb/IIIa inhibitor or cangrelor administration: 50−70 units per kg.
2. Renal failure: no adjustment needed
3. Half-life: 1−2 h—the half-life of unfractionated heparin increases with higher heparin doses (30 min after an intravenous bolus of 25 U/kg to 60 min with a bolus of 100 U/kg and to 150 min with a bolus of 400 U/kg).[4]
4. **Monitoring**: is performed using the activated clotting time (ACT) every 20−30 min. Goal ACT:
   a. without concomitant glycoprotein IIb/IIIa inhibitor or cangrelor administration:
      i. Hemochron device: 300−350 s (>350 s is often suggested for retrograde CTO interventions).[5]
      ii. HemoTec device: 250−300 s (>300 s is often suggested for retrograde CTO interventions).[5]
   b. with concomitant glycoprotein IIb/IIIa inhibitor or cangrelor administration: 200−250 s.[5]
5. **Reversal**: reverse heparin with protamine: 1−1.5 mg per 100 units of heparin, not to exceed 50 mg. Administer protamine slowly to minimize the risk of hypersensitivity reaction or anaphylaxis. Prior NPH insulin or protamine zinc insulin administration, fish hypersensitivity, vasectomy, severe left ventricular dysfunction, and abnormal preoperative pulmonary hemodynamics increase the risk of such reactions.

Protamine can cause severe systemic hypotension and relative pulmonary hypertension (due to vasoconstriction of the pulmonary vasculature), usually 5−10 min after administration. Methylene blue may help recovery from a protamine reaction.[6,7]

Bivalirudin

Bivalirudin is a direct thrombin inhibitor that carries no risk of heparin-induced thrombocytopenia.

1. **Dose**:
   a. Normal renal function: Bolus: 0.75 mg/kg IV, followed by infusion of 1.75 mg/kg/h IV for up to 4 h after the procedure.[8]
   b. Renal failure: Bolus: 0.75 mg/kg IV, followed by infusion. Infusion rate depends on creatinine clearance, as follows:
      i. Moderate (CrCl 30–59 mL/min): 1.75 mg/kg/h
      ii. Severe (CrCl <30 mL/min): 1 mg/kg/h
      iii. Hemodialysis: 0.25 mg/kg/h
2. **Half-life**: 25 min.
3. **Monitoring**: is performed using the activated clotting time (ACT). If ACT is at goal when checked ∼5 min after starting bivalirudin, no further ACTs are usually needed. If ACT remains too low, follow the algorithm as below. Goal ACT on a Hemochron device (Hemotec device goals are 50 s lower):
   a. without concomitant glycoprotein IIb/IIIa inhibitor or cangrelor administration: 300–350 s (>350 s is often suggested for retrograde CTO interventions).
   b. with concomitant glycoprotein IIb/IIIa inhibitor or cangrelor administration: 250-250 s.
4. **Reversal**: there is no reversal agent available. The infusion is discontinued.

Challenges
1. **ACT is low after anticoagulant administration**
   Causes:
   - Anticoagulant did not reach the circulation (for example because of malfunction of the intravenous line).
   - Low anticoagulant dose.
   - Decreased antithrombin III levels (familial or acquired).
   Prevention:
   - Dose anticoagulants appropriately.
   - Ensure intravenous lines are working properly.
   Treatment:
   - Administer additional bivalirudin (potentially through an arterial sheath if there are concerns about the IV line) followed by repeat ACT measurements before proceeding with PCI.
   - Avoid inserting devices such as guidewires, balloons or stents in the coronary arteries until after the ACT is in the therapeutic range to avoid thrombosis.

What can go wrong?
1. **Bleeding**
   Causes:
   - Excessive anticoagulation.

- Vascular injury.
- Hypertension.

Prevention:
- Avoid high doses of anticoagulants.
- Monitor anticoagulation level (perform ACT every 20−30 min).
- In patients receiving oral anticoagulation (for example for atrial fibrillation), oral anticoagulation should ideally be stopped prior to the procedure as described in Chapter 1, unless the procedure is emergent.

Treatment:
- Reverse anticoagulation with protamine if unfractionated heparin was administered. Do not reverse until after all intracoronary equipment has been removed to prevent vessel thrombosis.
- Tailor additional treatment to the site of bleeding.

2. **Thrombosis**

    Causes:
    - Low dose of heparin.
    - Administered heparin did not reach circulation (for example due to infiltration of the intravenous line).
    - Variable potency of unfractionated heparin batches.
    - Heparin-induced thrombocytopenia.
    - Resistance to heparin (for example in patients with antithrombin III deficiency).

    Prevention:
    - Assess platelet count prior to the procedure (to ensure that platelet count has not been decreasing, which could be due to HIT).
    - Administer correct dose of anticoagulants.
    - Ensure that the IV line is working properly.
    - Check ACT 5 min after heparin administration to ensure that adequate anticoagulation has been achieved before proceeding with PCI.

    Treatment:
    - Thrombus is managed as described in the *Manual of Percutaneous Coronary Interventions.*[9]
    - Administer additional doses of heparin to achieve a therapeutic ACT.

3. **Heparin-induced thrombocytopenia**

    Heparin administration can cause heparin-induced thrombocytopenia (HIT). There are two types of HIT:
    - Type 1 HIT: develops within the first 2 days after heparin administration, and the platelet count normalizes with continued heparin therapy.
    - Type 2 HIT: usually develops after 4−10 days from heparin administration and can lead to life-threatening venous and arterial thromboembolism.

        Causes:
        - Type 1: direct effect of heparin on platelet activation.
        - Type 2: immune-mediated disorder.

Prevention:
- Avoid administration of unfractionated heparin.

Treatment:
- Type 1: no specific treatment needed.
- Type 2: discontinuation of all heparin administration and anticoagulation with nonheparin agents (such as bivalirudin or argatroban).

3.5   Antiplatelet agents

Video 3.5. PCI Pharmacology: Antiplatelet Agents.

Both oral (aspirin and oral P2Y12 inhibitors) and intravenous (glycoprotein IIb/IIIa inhibitors and cangrelor [cangrelor is an intravenous P2Y12 inhibitor]) antiplatelet agents are currently available for use during PCI.

Dual antiplatelet therapy (DAPT) with aspirin and a P2Y12 inhibitor is the current standard of care for PCI, except in patients on concurrent oral anticoagulants (warfarin or direct oral anticoagulants) in whom aspirin is often discontinued within a few days or weeks from PCI. Such patients only receive clopidogrel in addition to the oral anticoagulant for one year, followed by discontinuation of clopidogrel and only administration of the oral anticoagulant.[10]

Patients planned to undergo CTO PCI should be preloaded with a P2Y12 inhibitor prior to the procedure.

3.5.1   Dual antiplatelet therapy

Goals:
- To prevent stent thrombosis.
- To prevent future acute coronary syndromes

How?
   Medication type: in general, all patients (except for those receiving vitamin K antagonists or direct oral anticoagulants or those who have allergic reactions) should receive aspirin (although some studies have advocated administration of a P2Y12 inhibitor alone without aspirin). Clopidogrel is the P2Y12 inhibitor of choice for stable coronary artery disease (CAD) patients and prasugrel or ticagrelor for acute coronary syndrome (ACS) patients, unless contraindicated.

Dose:
- **Aspirin:** 325 mg loading dose ideally prior to cardiac catheterization, then 81–100 mg daily. Aspirin is commonly discontinued immediately or 1–4 weeks post PCI in patients who receive oral anticoagulants and a P2Y12 inhibitor.

- **Clopidogrel**: 600 mg load, ideally given at least 6 h prior to PCI for stable patients, followed by 75 mg daily thereafter.
- **Ticagrelor**: 180 mg load, followed by 90 mg bid. Ticagrelor is contraindicated in patients with prior intracranial hemorrhage and should be avoided in patients receiving strong CYP3A inhibitors (e.g., ketoconazole, itraconazole, voriconazole, clarithromycin, nefazodone, ritonavir, saquinavir, nelfinavir, indinavir, atazanavir, and telithromycin), strong CYP3A inducers (e.g., rifampin, phenytoin, carbamazepine, and phenobarbital), and patients receiving >40 mg daily of simvastatin or lovastatin. Ticagrelor should only be used with low dose aspirin (81−100 mg daily).
- **Prasugrel**: 60 mg loading dose, followed by 10 mg daily maintenance dose (5 mg daily dose for patients with body weight <60 kg or patients aged >75 years). Prasugrel is contraindicated in patients with prior transient ischemic attack or stroke.

Pretreatment:
  Stable CAD patients: no before diagnostic angiography; yes if the patients have planned PCI.
  Non-ST-segment elevation ACS: no
  STEMI: yes, either before or at the time of PCI; ticagrelor 180 mg and prasugrel 60 mg are preferred over clopidogrel 600 mg, unless the patient is at high risk of bleeding or has contraindications, as described above.

Switching between oral P2Y12 inhibitors:
When to switch:
  When the risk of bleeding exceeds the potential benefits of more potent P2Y12 inhibitors, for example:

- Patients who experience major bleeding while receiving ticagrelor or prasugrel may benefit from changing to clopidogrel.
- Patients who experience stent thrombosis while on clopidogrel may benefit from changing to ticagrelor or prasugrel unless they have high bleeding risk.
- ACS patients on clopidogrel who are found to be carriers of CYP2C19*2 or CYP2C19*3 loss-of-function alleles on genetic testing may benefit from switching to ticagrelor or prasugrel.
- Patients who are intolerant of high potency P2Y12 inhibitors due to adverse reactions (i.e. dyspnea with ticagrelor).

How to switch: (Fig. 3.1)

Dual antiplatelet therapy (DAPT) duration (Fig. 3.2): Individualize DAPT duration to optimize the balance between benefits (fewer ischemic events) versus risks (bleeding).

SWITCHING BETWEEN ORAL P2Y12 INHIBITORS

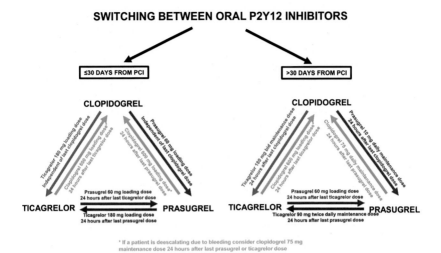

FIGURE 3.1 **How to switch between various P2Y12 inhibitors.** *Reproduced with permission from pcimanual.org.*

DAPT DURATION

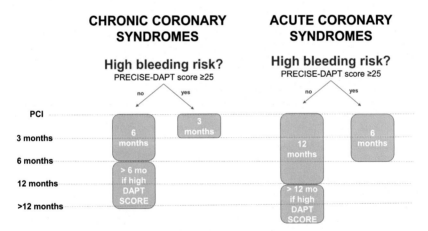

FIGURE 3.2 **How to determine the duration of DAPT post PCI.** *Reproduced with permission from pcimanual.org.*

- **Standard DAPT duration** after PCI is 6 months for patients with stable CAD and 12 months in patients who presented with acute coronary syndromes. In patients at high bleeding risk (PRECISE-DAPT score ≥ 25), DAPT duration can be shortened to 3 and 6 months, respectively, for stable angina and ACS patients.

Link to online calculator of the PRECISE-DAPT score: http://www.preci-sedaptscore.com/predapt/webcalculator.html

Based on the TWILIGHT trial, an alternative strategy of aspirin with-drawal after 3 months and continuing with ticagrelor monotherapy decreases the risk of bleeding without an increase in major adverse events.

- **Short DAPT duration** (1 month) can be used with high-bleeding risk patients,[11,12] but has not been specifically studied in CTO PCI.
- **Prolonged DAPT duration** (>12 months) can be considered for patients who do not have high bleeding risk and have high risk for recurrent ischemic events. This can be assessed using the DAPT score (DAPT score ≥ 2 favors >12 months DAPT duration up to 30 months).

Link to online calculator of the DAPT score: https://tools.acc.org/DAPTriskapp/#!/content/calculator/

What can go wrong?
1. Bleeding
Causes:
- Antiplatelet effect of P2Y12 inhibitors and aspirin.
- Concomitant use of anticoagulants, such vitamin K antagonists (VKAs) or direct oral anticoagulants (DOACs).

Prevention:
- Reduce aspirin treatment duration in patients receiving VKA or a DOAC together with a P2Y12 inhibitor. Adjust P2Y12 inhibitor duration based on the predicted bleeding risk.
- Avoid high potency P2Y12 inhibitors (prasugrel and ticagrelor) in patients at high risk of bleeding, such as patients receiving VKA or DOACs.
- Routinely administer a proton-pump inhibitor in patients receiving DAPT who are at increased risk of bleeding.

Treatment:
- Platelet transfusion in case of life-threatening bleeding and irre-versible P2Y12 inhibition (i.e., clopidogrel, prasugrel, aspirin).
- Platelet transfusion may not be successful with ticagrelor given its reversible effect on the P2Y12 receptor.
- De-escalate to clopidogrel if the patient was taking ticagrelor or prasugrel.
- Apply specific measures to control local bleeding.

2. Dyspnea
Causes:
- Ticagrelor is more likely to cause dyspnea (likely adenosine-medi-ated) than other P2Y12 inhibitors.
- Other cardiac (such as heart failure) and noncardiac (such as pri-mary pulmonary disorders) causes.

Prevention:
- Do not use ticagrelor.

Treatment:
- Evaluate potential causes of dyspnea.
- Potential treatments currently being studied are caffeine, aminophylline, and/or theophylline.
- Switch ticagrelor to prasugrel or clopidogrel if no other reversible cause is identified.

3. **Allergic reactions**
   Causes:
   - All medications, including aspirin and P2Y12 inhibitors, can cause allergic reactions.
   - Clopidogrel can very rarely cause thrombotic thrombocytopenic purpura.

   Prevention:
   - N/A

   Treatment:
   - Evaluate whether aspirin or P2Y12 inhibitors are causing the reaction—if yes, replace with another P2Y12 inhibitor or perform aspirin desensitization.

3.5.2 Intravenous antiplatelet agents

Goals:
- Prevent thrombus propagation in patients with large intracoronary thrombus or no reflow.
- Prevent stent thrombosis.

How?

Three intravenous antiplatelet agents are currently available in the US: two glycoprotein IIb/IIIa inhibitors (eptifibatide and tirofiban) and one intravenous, short-acting P2Y12-inhibitor (cangrelor).

Administration of these medication **should be avoided in CTO PCI**, as they can convert an otherwise insignificant wire perforation to a potentially life-threating pericardial effusion.

Cangrelor

Medication	Dose (normal renal function)	Dose in renal failure
Cangrelor	Bolus of 30 μg/kg IV, followed by 4 μg/kg/min infusion for at least 2 h or duration of procedure, whichever is longer.	No dose adjustment is needed in patients with renal or hepatic insufficiency.

The advantage of cangrelor over GP IIb/IIIa inhibitors is the short half-life, that allows faster return of platelet function in case of bleeding.

Administration of oral P2Y12 inhibitors in patients receiving cangrelor should be as follows (cangrelor inhibits binding of clopidogrel and prasugrel to the platelets):

Ticagrelor: can be administered at any time.

Clopidogrel and prasugrel: administer loading dose immediately after cangrelor discontinuation.

Glycoprotein IIb/IIIa inhibitors

Medication	Dose (normal renal function)	Dose in renal failure
Eptifibatide	Double bolus of 180 μg/kg i.v. (given at a 10 min interval) followed by an infusion of 2.0 μg/kg/min for up to 18 h.	In patients with creatinine clearance ≤ 60 mL/min, the post loading infusion is decreased by 50% to 1 mcg/kg/min.
Tirofiban	Bolus of 25 μg/kg over 3 min i.v., followed by an infusion of 0.15 μg/kg/min for up to 18 h.	In patients with creatinine clearance ≤ 60 mL/min the post loading infusion is decreased by 50% to 0.075 mcg/kg/min.

What can go wrong?
1. Bleeding

Causes:
- Antiplatelet effect of intravenous antiplatelet agents.
- Thrombocytopenia (caused by glycoprotein IIb/IIIa inhibitors).

Prevention:
- Selective use of GP IIb/IIIa inhibitors only in patients with large thrombus burden or other thrombotic complications and acceptable bleeding risk.
- Keep ACT low

Treatment:
- Tirofiban/eptifibatide: there is no reversal agent for eptifibatide or tirofiban. The infusion is stopped followed by a gradual return of platelet reactivity. Emergent dialysis can be effective in removing the drug from the circulation in cases of life-threatening bleeding.

3.6   Vasopressors and inotropes

Video 3.6. PCI Pharmacology: Vasopressors and Inotropes.

3.6.1   Vasopressors

Goals:
- To increase blood pressure in case of hypotension

How?

- Most IV adrenergic agents provide both vasopressor and inotropic effect, except for phenylephrine which is a selective alpha receptor agonist and has a purely vasopressor effect (as a result it is the vasopressor of choice in patient with hypertrophic cardiomyopathy, in whom an inotropic effect can worsen the intracavitary obstruction).
- Vasopressin is also a pure vasopressor.
- Norepinephrine is mainly a vasopressor with less inotropic effect.
- Dopamine at doses >10 mcg/kg/min also has a predominant vasopressor effect.
- Vasopressors can be used to maintain a mean arterial pressure is >60 mmHg.
- Administration through a central line is recommended, if possible.

Dosage:

Phenylephrine: 50−100 mcg IV bolus, followed by infusion at 0.5−1.4 mcg/kg/min.

Norepinephrine (Levophed): 0.1−0.5 mcg/kg/min IV infusion (norepinephrine is the preferred vasopressor for both cardiogenic and septic shock).

Vasopressin: initial dose: 0.03 units/min IV infusion; can titrate up by 0.005 units/min at 10−15 min intervals up to a maximum dose of 0.1 units/min.

Dopamine: 10−20 mcg/kg/min (at lower doses it mainly has an inotropic and chronotropic effect).

What can go wrong?

1. **Hypertension**

   Causes:
   - Excessive dose of vasopressors.
   - Correction of the initial event that caused hypotension.

   Prevention:
   - Avoid high doses of vasopressors
   - Use mechanical circulatory support.

   Treatment:
   - The half-life of catecholamines is short; hence, blood pressure will usually decrease within a few min.

2. **Bradycardia**

   Causes:
   - Phenylephrine administration: phenylephrine can cause reflex bradycardia due to unopposed vagal action on the heart.

   Prevention:
   - Avoid phenylephrine in patients with baseline bradycardia.

   Treatment:
   - Continued monitoring; atropine and temporary pacing may be needed in case of prolonged severe bradycardia.

3.6.2   Inotropes

Goals:
● To increase myocardial contractility.

How?
● Epinephrine provides equal vasopressor and inotropic effect.
● Dopamine at low doses (5–10 mcg/kg/min) has a predominant inotropic effect.
● Dobutamine has a pure inotropic effect.
● Administration through a central line is recommended, if possible.

Dosage:
  Epinephrine:
    Cardiac arrest: 1 mg IV every 3–5 min
    Cardiogenic shock: 2–10 mcg/min IV infusion
    Anaphylactic shock or angioedema: 0.3–0.5 mL of 1 mg/mL solution subcutaneously or IM every 1–2 h. In case of anaphylaxis, epinephrine should be administered immediately without waiting for steroids or other medications to take effect.

  Dopamine: 5–10 mcg/kg/min (at higher doses it acts mainly as a vasopressor).
  Dobutamine: 10–40 mcg/kg/min

What can go wrong?
1. Arrhythmias
  Causes:
    ● Adrenergic stimulation
    ● Underlying cardiac disorder, such as ischemia.
  Prevention:
    ● Use the lowest possible doses of inotropes.
    ● Use mechanical circulatory support.
  Treatment:
    ● Antiarrhythmics, such as lidocaine and amiodarone.
    ● Mechanical circulatory support, including Veno-Arterial Extracorporeal Membrane Oxygenation (VA-ECMO), may be needed in case of refractory arrhythmias.

3.7   Antiarrhythmics

Video 3.7. PCI Pharmacology: Antiarrhythmics.

3.7.1   Amiodarone

Goals:
● To treat ventricular and atrial tachyarrhythmias.
● To end refractory ventricular fibrillation.

How?

Dosage:

Non life-threatening arrhythmias: 150 mg over 10 min (may repeat if tachyarrhythmia recurs) followed by maintenance infusion of 1 mg/min for 6 h and 0.5 mg/min for an additional 18 h.

Ventricular fibrillation: 300 mg IV push, followed by maintenance infusion of 1 mg/min for 6 h and 0.5 mg/min for an additional 18 h.

What can go wrong?

1. **Hypotension**

   Causes:
   - Allergic reaction (in this scenario the patient often also develops additional signs of allergic reaction, such as angioedema or urticaria).
   - Vasodilation and depression of myocardial contractility; this may be partly due to the solvent (polysorbate 80 or benzyl alcohol) used for dissolving the drug.

   Prevention:
   - Use caution when administering to patients with low baseline blood pressure.

   Treatment:
   - Stop amiodarone administration.
   - If hypotension is due to anaphylactic reaction, treat accordingly.
   - If hypotension is not due to anaphylactic reaction: administer vasopressors and inotropes.

2. **Bradycardia and/or atrioventricular block**

   Causes:
   - Direct action of the medication on conduction through the AV node.

   Prevention:
   - Do not administer amiodarone to patients with second- or third-degree heart block who do not have pacemakers.

   Treatment:
   - Discontinue amiodarone.
   - Temporary pacing.

3. **Torsades des pointes or ventricular fibrillation**

   Causes:
   - Prolonged QT interval.
   - Administration in patients with preexcitation (Wolff-Parkinson-White syndrome) and concurrent atrial fibrillation.

   Prevention:
   - Do not administer to patients with prolonged QT.
   - Do not administer to patients with preexcitation (Wolff-Parkinson-White syndrome) and concurrent atrial fibrillation.

Treatment:
- Defibrillation
- Stop amiodarone administration.

3.7.2   Atropine

Goals:
- To increase heart rate.

How?

Dosage: 0.5 mg IV push - may repeat up to a total dose of 3 mg.

What can go wrong?
1. Rebound tachycardia.
   Causes:
   - Resolution of the underlying cause of bradycardia.
   - High atropine dose.
   Prevention:
   - Use the lowest possible dose of atropine.
   Treatment:
   - Tachycardia will usually resolve without specific treatment.
2. **Dry mouth.**
3. **Blurry vision due to mydriasis.**
4. **Flushing.**

References

1. Werner GS, Lang K, Kuehnert H, Figulla HR. Intracoronary verapamil for reversal of no-reflow during coronary angioplasty for acute myocardial infarction. *Catheter Cardiovasc Interv* 2002;**57**:444−51.
2. Ip JE, Cheung JW, Chung JH, et al. Adenosine-induced atrial fibrillation: insights into mechanism. *Circ Arrhythm Electrophysiol* 2013;**6**:e34−7.
3. Smith JR, Goldberger JJ, Kadish AH. Adenosine induced polymorphic ventricular tachycardia in adults without structural heart disease. *Pacing Clin Electrophysiol* 1997;**20**:743−5.
4. Zeymer U, Rao SV, Montalescot G. Anticoagulation in coronary intervention. *Eur Heart J* 2016;**37**:3376−85.
5. O'Gara PT, Kushner FG, Ascheim DD, et al. ACCF/AHA guideline for the management of ST-elevation myocardial infarction: a report of the American College of Cardiology Foundation/American Heart Association Task Force on Practice Guidelines. *Circulation* 2013;**2013**(127):e362−425.
6. Lutjen DL, Arndt KL. Methylene blue to treat vasoplegia due to a severe protamine reaction: a case report. *AANA J* 2012;**80**:170−3.
7. Del Duca D, Sheth SS, Clarke AE, Lachapelle KJ, Ergina PL. Use of methylene blue for catecholamine-refractory vasoplegia from protamine and aprotinin. *Ann Thorac Surg* 2009;**87**:640−2.

8. Neumann FJ, Sousa-Uva M, Ahlsson A, et al. ESC/EACTS Guidelines on myocardial revascularization. *Eur Heart J* 2018;**2019**(40):87−165.

9. Brilakis ES. *Manual of percutaneous coronary interventions: a step-by-step approach.* Elsevier; 2021.

10. Angiolillo DJ, Bhatt DL, Cannon CP, et al. Antithrombotic therapy in patients with atrial fibrillation treated with oral anticoagulation undergoing percutaneous coronary intervention: a North American perspective: 2021 update. *Circulation* 2021;**143**:583−96.

11. Windecker S, Latib A, Kedhi E, et al. Polymer-based or polymer-free stents in patients at high bleeding risk. *N Engl J Med* 2020;**382**:1208−18.

12. Valgimigli M, Cao D, Angiolillo DJ, et al. Duration of dual antiplatelet therapy for patients at high bleeding risk undergoing PCI. *J Am Coll Cardiol* 2021;**78**:2060−72.

Chapter 4

Access

4.1 Dual (or triple) arterial access

Arterial access is essential for every percutaneous coronary intervention (PCI). Chronic total occlusion (CTO) PCI is unique in that dual access is required in most cases, except rarely in patients with no contralateral collaterals. Even in such patients:

- contralateral collateral "recruitment" can occur during the procedure if ipsilateral collaterals are compromised, and
- the dual-guide (ping-pong) technique is often required for retrograde wire externalization if crossing is achieved using the retrograde approach via ipsilateral collaterals.[1]

Rarely, triple arterial access is needed for CTO PCI.[2] An example is a right coronary artery CTO with distal filling via the left anterior descending artery (LAD), which is supplied by a left internal mammary artery (LIMA) graft (Fig. 4.1), when a retrograde approach is planned via a septal collateral that originates proximal to the LAD occlusion. Similarly, triple access is required when trying to cross a RCA CTO whose distal bed is visualized via the left coronary artery, but the retrograde approach is performed via an occluded SVG.

Dual coronary angiography and in-depth and structured review of the angiogram (and, if available, coronary computed tomography angiography) are key for planning and safely performing CTO PCI and is one of the seven global guiding principles of CTO PCI.[4]

4.2 Femoral versus radial access

Both femoral and radial access have been successfully used for CTO PCI with several studies showing similar success rates.[5-9] Radial (proximal or distal[10,11]) or ulnar access[12] is associated with fewer bleeding and access site complications (which is especially important in CTO PCI due to often long procedure duration and use of high doses of heparin with high ACTs, especially for retrograde CTO PCI) and greater patient satisfaction.[13] However, engaging the coronary arteries can be more challenging and guide catheter support may be suboptimal (due to greater respiratory motion, subclavian tortuosity, or inability to insert large guide catheters in some patients). In the ever-growing complexity of the

Manual of Chronic Total Occlusion Percutaneous Coronary Interventions.
DOI: https://doi.org/10.1016/B978-0-323-91787-2.00042-3

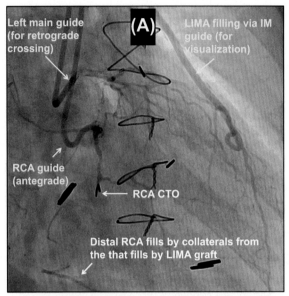

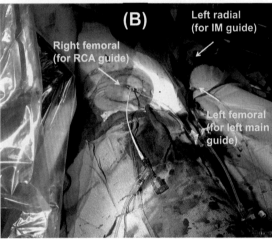

FIGURE 4.1 Bifemoral and left radial access were obtained to enable retrograde crossing of a right coronary artery CTO (**panel B**). The right femoral artery was used for engaging the right coronary artery, the left femoral artery for engaging the left main (through which retrograde crossing was performed) and the left radial artery was used for engaging the left internal mammary artery (through which visualization of the right posterior descending artery was achieved) (**panel A**).[3] *Reproduced from Michael et al. J. Invasive Cardiol. 2012;24:359−62 with permission from HMP Communications.*

CTO patient population, radial/ulnar arteries might become occluded due to multiple catheterizations, and femoral/iliac arteries may suffer aggressive peripheral arterial disease, hence various combinations of access points might be required in different patients. Developing expertise in both radial/ulnar and femoral access is, therefore, essential for the CTO PCI operator. Using meticulous access technique with fluoroscopic and ultrasound guidance can minimize the risk of arterial access complications.

Bifemoral access is used less often at present, but can be especially useful in highly complex CTOs and early in the learning curve (as it may allow easier vessel engagement and enhanced guide catheter support with use of large 8 Fr guide catheters). If there is pressure dampening with 8 French guide catheters, they can be exchanged for 7 French guide catheters. In a Japanese study of complex CTO PCI cases (J-CTO score ≥ 3) the transradial group had a significantly lower success rate than the transfemoral group (35.7% vs 58.2%; p = 0.04)[14] although other studies showed similar success rates with wrist versus femoral access.[5,6,8,9,12,15] Some operators use ipsilateral dual femoral sheaths (parallel sheath technique).[16]

Biradial access can be successfully used at centers with radial access expertise (Section 2.1).[15,17] The FORT-CTO randomized trial showed similar success and lower access site complication rate with only radial access (61% biradial) compared with femoral access in CTO PCI.[12]

Femoral—radial is a common configuration for CTO PCI, especially when retrograde crossing is not planned: usually femoral access (7 or 8 French) is used for the antegrade guide catheter and radial access (6 or 7 French) for the retrograde guide catheter.

4.3   Femoral access

Video 4.2. Femoral arterial access step-by-step.

To optimize success and safety, femoral access should be obtained using the following 14 steps as described in detail in the *Manual of Percutaneous Coronary Interventions*.[18]

Step 1. Palpation of the femoral pulse
Step 2. Sterile preparation and draping of the groin
Step 3. Fluoroscopy of the femoral head
Step 4. Ultrasound guidance
Step 5. Local anesthetic administration
Step 6. Femoral artery puncture
Step 7. Insertion of a 0.018 inch guidewire
Step 8. Fluoroscopy of the 0.018 inch guidewire
Step 9. Skin nick (optional)
Step 10. Insertion of the micropuncture-dilator assembly (some operators perform femoral angiography through the micropuncture sheath prior to advancing the sheath that will be used for PCI).
Step 11. Advancement of a 0.035 inch guidewire
Step 12. Fluoroscopy of the 0.035 inch guidewire

Step 13. Sheath insertion
Step 14. Sheath aspiration and flushing
Step 15. Femoral angiography

Long sheaths (usually 45 cm) enhance support and facilitate guide catheter manipulations, especially in patients with significant aortoiliac tortuosity.

4.4 Radial access

Video 4.3. Radial access step-by-step.

Radial access is obtained using the following 10 steps as described in detail in the *Manual of Percutaneous Coronary Interventions*.[18]

Step 1. Palpation of the radial pulse
Step 2. Sterile preparation and draping of the wrist
Step 3. Ultrasound guidance
Step 4. Local anesthetic administration
Step 5. Radial artery puncture
Step 6. Insertion of a 0.018 inch guidewire
Step 7. Skin nick (optional)
Step 8. Sheath insertion
Step 9. Sheath aspiration and flushing
Step 10. Securing the sheath

Proximal radial access, distal radial access,[10,11,19] and ulnar access, can be used for CTO PCI. Distal radial has been associated with lower risk of radial artery occlusion[19] and can be more comfortable for both the operator (no need to bend over patient if left radial access is needed, while maintaining total visual control of the entry point and the sheath) and the patient (they can place their hands over their stomach minimizing discomfort, Fig. 4.2).

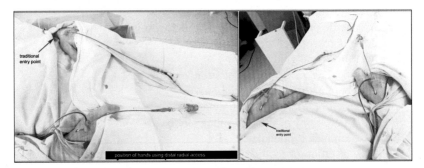

FIGURE 4.2 **Bilateral distal radial access for CTO PCI**. Bilateral distal radial access allows better visual control over the sheath entry point compared with the traditional proximal radial entry point and better comfort for both the patient and the interventionalist. *Courtesy of Dr. Avtandil Babunashvili.*

Seven French thin walled (such as the Terumo slender and the Merit Predule Ideal) sheaths are most commonly used for radial access in CTO PCI, either in the proximal or the distal radial artery. In patients with small radial arteries, sheathless guide catheters can sometimes be used. Balloon-assisted tracking or use of the Railway system can facilitate sheathless insertion of large diameter guide catheters without injuring the radial artery.

When using radial access, it is easier to engage the contralateral coronary artery (for example, it is easier to engage the left main via right radial access and the right coronary artery via left radial access).

References

1. Brilakis ES, Grantham JA, Banerjee S. "Ping-pong" guide catheter technique for retrograde intervention of a chronic total occlusion through an ipsilateral collateral. *Catheter Cardiovasc Interv* 2011;**78**:395−9.
2. Tsiafoutis I, Liontou C, Antonakopoulos A, Katsanou K, Koutouzis M, Katsivas A. Triple-access retrograde chronic total occlusion intervention through vein graft and epicardial collaterals. *J Invasive Cardiol* 2020;**32**:E172−3.
3. Michael TT, Banerjee S, Brilakis ES. Role of internal mammary artery bypass grafts in retrograde chronic total occlusion interventions. *J Invasive Cardiol* 2012;**24**:359−62.
4. Brilakis ES, Mashayekhi K, Tsuchikane E, et al. Guiding principles for chronic total occlusion percutaneous coronary intervention. *Circulation* 2019;**140**:420−33.
5. Meijers TA, Aminian A, van Wely M, et al. Randomized comparison between radial and femoral large-bore access for complex percutaneous coronary intervention. *JACC Cardiovasc Interv* 2021;**14**:1293−303.
6. Tajti P, Alaswad K, Karmpaliotis D, et al. Procedural outcomes of percutaneous coronary interventions for chronic total occlusions via the radial approach: insights from an international chronic total occlusion registry. *JACC Cardiovasc Interv* 2019;**12**:346−58.
7. Megaly M, Karatasakis A, Abraham B, et al. Radial vs femoral access in chronic total occlusion percutaneous coronary intervention. *Circ Cardiovasc Interv* 2019;**12**:e007778.
8. Gorgulu S, Kalay N, Norgaz T, Kocas C, Goktekin O, Brilakis ES. Femoral or radial approach in treatment of coronary chronic total occlusion: a randomized clinical trial. *JACC Cardiovasc Interv* 2022;**15**:823−30.
9. Simsek B, Gorgulu S, Kostantinis S, et al. PROGRESS-CTO investigators. Radial access for chronic total occlusion percutaneous coronary intervention: Insights from the PROGRESS-CTO registry. *Catheter Cardiovasc Interv* 2022. Available from: https://doi.org/10.1002/ccd.30347. Epub ahead of print.
10. Nikolakopoulos I, Patel T, Jefferson BK, et al. Distal radial access in chronic total occlusion percutaneous coronary intervention: insights from the PROGRESS-CTO registry. *J Invasive Cardiol* 2021;**33**:E717−e22.
11. Gasparini GL, Garbo R, Gagnor A, Oreglia J, Mazzarotto P. First prospective multicentre experience with left distal transradial approach for coronary chronic total occlusion interventions using a 7 Fr Glidesheath Slender. *EuroIntervention* 2019;**15**:126−8.
12. Poletti E, Azzalini L, Ayoub M, et al. Conventional vascular access site approach vs fully trans-wrist approach for chronic total occlusion percutaneous coronary intervention: a multicenter registry. *Catheter Cardiovasc Interv* 2020;**96**:E45−52.

13. Ferrante G, Rao SV, Juni P, et al. Radial vs femoral access for coronary interventions across the entire spectrum of patients with coronary artery disease: a meta-analysis of randomized trials. *JACC Cardiovasc Interv* 2016;**9**:1419–34.

14. Tanaka Y, Moriyama N, Ochiai T, et al. Transradial coronary interventions for complex chronic total occlusions. *JACC: Cardiovasc Interven* 2017;**10**:235–43.

15. Alaswad K, Menon RV, Christopoulos G, et al. Transradial approach for coronary chronic total occlusion interventions: Insights from a contemporary multicenter registry. *Catheter Cardiovasc Interv* 2015;**85**:1123–9.

16. Reifart N, Sotoudeh N. The parallel sheath technique in severe iliac tortuosity: a simple and novel technique to improve catheter manoeuvrability. *EuroIntervention* 2014;**10**:231–5.

17. Burzotta F, De Vita M, Lefevre T, Tommasino A, Louvard Y, Trani C. Radial approach for percutaneous coronary interventions on chronic total occlusions: technical issues and data review. *Catheter Cardiovasc Interv* 2014;**83**:47–57.

18. Brilakis ES. *Manual of percutaneous coronary interventions: a step-by-step approach.* Elsevier; 2021.

19. Tsigkas G, Papageorgiou A, Moulias A, et al. Distal or traditional transradial access site for coronary procedures: a single-center, randomized study. *JACC Cardiovasc Interv* 2022;**15**:22–32.

Chapter 5

Coronary and graft engagement

After obtaining arterial access (Chapter 4) catheters are advanced over a 0.035 or 0.038 inch guidewire from the access site to the coronary ostia to engage the coronary arteries or bypass grafts. Chronic total occlusion (CTO) percutaneous coronary intervention (PCI) cannot be performed without engaging the coronary artery and/or bypass graft ostia. Flush aorto-ostial occlusions and prior transcatheter aortic valve replacement (TAVR) may preclude coronary engagement, but such lesions can sometimes be approached by retrograde crossing, followed by snaring of the retrograde guidewire with an antegrade guide catheter (Chapter 15).

Coronary and bypass graft engagement can be achieved using the following nine steps as described in detail in the *Manual of Percutaneous Coronary Interventions*[1]:

Step 1. Catheter selection
Step 2. Advance guidewire to the aortic root
Step 3. Advance catheter to the aortic root and remove the 0.035/0.038 inch guidewire
Step 4. Aspirate the guide catheter
Step 5. Connect with manifold
Step 6. Ensure there is good pressure waveform
Step 7. Manipulate the catheter to engage coronary ostia
Step 8. Ensure there is a good pressure waveform
Step 9. Proceed with contrast injection as described in Chapter 6.

Step 1: Catheter selection is of critical importance for CTO PCI, which often requires strong guide catheter support. Catheter selection is based on arterial access site (radial vs. femoral), the target coronary vessel and the size of the aorta, as described in the *Manual of Percutaneous Coronary Interventions*.[1] The right coronary artery is most often engaged with Amplatz Left (AL) guide catheters and the left main with XB/EBU guide catheters.

Step 7: Usually, the right coronary artery guide catheter is inserted first (before inserting the left coronary artery guide catheter) to allow for unimpeded torqueing necessary to engage the right coronary artery ostium and to prevent guide-guide interaction with the left main guide catheter during the procedure. In case of difficult cannulation of the left main, inserting a

Manual of Chronic Total Occlusion Percutaneous Coronary Interventions.
DOI: https://doi.org/10.1016/B978-0-323-91787-2.00007-1

guidewire into the right coronary artery can prevent disengagement of the right coronary artery guide catheter.

Cannulation with the second guide catheter can be difficult due to interference with the first guide catheter. Pullback, rotation, and repositioning of the second guide catheter in the aortic root can often solve this problem. Inadvertent removal of the initially used guide catheter can be prevented by introducing a stabilizing 0.014 inch guidewire before manipulating the second guide catheter.

In case of difficult cannulation a guidewire can sometimes be introduced from a guide catheter floating in the aortic root ("fishing" or "air mail" method) followed by use of a distal anchor balloon for selective guide catheter engagement.

Poor guide catheter support is a much more frequent problem in the right coronary artery (RCA) than in the left main. If the anatomy of the RCA ostium is not favorable for an AL catheter, the support of the Judkins Right guide can be increased by using the side branch achor technique, a guide catheter extension or by deep cannulation. To achieve deep cannulation a simultaneous clockwise rotation and gentle push usually are usually sufficient, although sometimes a distal anchor balloon may be needed. When using a deeply seated guide catheter, pay special attention to preventing a wedged catheter position that increases the risk of vessel injury or coronary embolism.

Step 8: Equalize the pressure in both guide catheters to allow early detection of pressure dampening in case of inadvertent deep guide catheter position.

Reference

1. Brilakis ES. *Manual of percutaneous coronary interventions: a step-by-step approach.* Elsevier; 2021.

Chapter 6

Coronary angiography and coronary computed tomography angiography

6.1 Dual injection

Coronary angiography for planning and performing chronic total occlusion (CTO) percutaneous coronary intervention (PCI) differs from coronary angiography of non-CTO lesions as *dual injection* should be performed in CTO PCI unless there are no contralateral collaterals.[1]

Dual coronary angiography and a structured, in-depth review of the angiogram (and, if available, coronary computed tomography angiography) are key for planning and safely performing CTO PCI and are part of the seven global guiding principles of CTO PCI.[2]

Prior to dual angiography, even in antegrade-only cases, placing a workhorse (Section 30.6.1) or a support (Section 30.6.4) guidewire in the donor vessel (non-CTO vessel that supplies flow to the distal true lumen) may stabilize and prevent disengagement of the guide catheter and allow prompt treatment in case of a complication.

6.1.1 Why is dual injection important for chronic total occlusion percutaneous coronary intervention?

Dual injection provides the following benefits for CTO PCI[3]:
Before percutaneous coronary intervention

1. Nonsimultaneous, single catheter injection often provides suboptimal visualization of the CTO segment and a limited ability to assess both the proximal and distal cap, as well as the distal vessel beyond the CTO, due to competitive flow from collaterals (Fig. 6.1). Occasionally, dual injection will reveal that the "CTO" is not a total occlusion, but rather a "functional" occlusion with a central patent channel. In other cases, there may be more than one CTOs in tandem (CTO Manual online case 6). Dual injection also provides a more accurate assessment of the true length of the CTO.

Manual of Chronic Total Occlusion Percutaneous Coronary Interventions.
DOI: https://doi.org/10.1016/B978-0-323-91787-2.00025-3
© 2023 Elsevier Inc. All rights reserved.

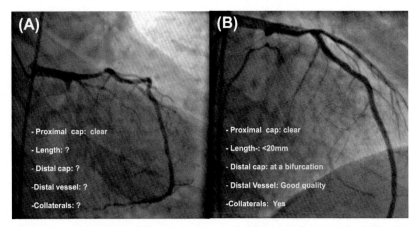

FIGURE 6.1   **Example of how dual injection can significantly improve the understanding of the CTO anatomy and CTO crossing options**. Injection of the left main coronary artery demonstrates a proximal circumflex CTO (**panel A**), but the key characteristics of the lesion remain unknown. Using dual injection (**panel B**) the characteristics of the CTO (proximal cap ambiguity, lesion length, bifurcation at the distal cap, quality of distal vessel, and presence of collaterals) are clarified. *Courtesy of Dr. Santiago Garcia.*

During percutaneous coronary intervention

1. Contralateral injection during CTO PCI allows *visualization of the guide-wire position* during antegrade crossing attempts. If the guidewire is outside the vessel or in a side branch, it can be repositioned before advancing equipment, thereby reducing the risk of perforation or other complications.

2. In patients whose distal vessel is filling via ipsilateral collaterals at baseline, the collateral flow direction and strength of flow can shift from one source to another due to ipsilateral collateral damage during CTO PCI, which may in turn hinder assessment of the guidewire position during antegrade or retrograde crossing attempts.

3. When using dissection/reentry techniques (Section 8.2) antegrade contrast injections may result in hydraulic enlargement of the extraplaque space and, as a result, reduce the likelihood of successful reentry. We recommend removing the injecting syringe from the antegrade guide manifold when performing antegrade dissection/reentry to prevent inadvertent contrast injection and expansion of the extraplaque space. Another way to protect the vessel from contrast-induced dissection propagation is selective contrast injection into ipsilateral collaterals through a microcatheter (e.g., selective injection into the conus branch communicating with right ventricular branches and supplying the distal right coronary artery [RCA] during RCA CTO PCI).

6.1.2   Dual injection technique

Dual injection should be performed as follows:

1. Engage both the CTO and the donor coronary artery, as described in Chapter 5.
2. Administer sublingual or intracoronary nitroglycerine before injections to maximally dilate the vessels and maximize collateral flow.
3. Use lower magnification (for example 13 instead of 8 inch) to enable visualization of the entire coronary circulation.
4. Do not pan the table, in order to facilitate recognition of collaterals.
5. Obtain long cine acquisition to allow for the contrast to travel through the collateral vessels and fill the distal vessel.
6. Inject the donor vessel (vessel that supplies the territory distal to the CTO) first, followed by injection of the occluded vessel after collaterals have filled the distal vessel. After the initial injection, the donor vessel can be injected before cine is recorded to reduce radiation dose.
7. Avoid using a side-hole guide catheter in the donor vessel to achieve better distal opacification and decrease contrast administration.
8. Use of 7 or 8 Fr guide catheters provides better vessel filling, thereby improving visualization, especially of small collateral vessels. Too little volume and low rate of contrast injection can mask important information about the lesion and result in image artifacts. Contrast volume, and flow rates may be easier to adjust when using an automated injector, such as the ACIST system.
9. In complex cases several contralateral injections might be necessary. The amount of contrast required may be decreased by inserting a microcatheter or guide extension selectively into the donor vessel or into the collateral and administering only 1−2 cc of contrast for each image acquisition. A thrombus aspiration catheter can also be used for this purpose in patients with well-developed collaterals as it allows higher flow and better image quality. This method cannot be applied in lesions with multiple collaterals due to the competitive blood flow from other branches.
10. Similar to standard coronary angiography, looking at the electrocardiogram and pressure waveform *before* and *after* each contrast injection is a must. Electrocardiographic changes can provide early warning of an impending complication, such as ischemia during collateral vessel crossing. Injection in the setting of severe pressure dampening can cause coronary artery dissection and/or aortocoronary dissection.
11. If collateral visualization is suboptimal, cine recording at 30 fps (frames per second) can enhance visualization. Also temporarily increasing magnification for these recordings can further improve collateral visualization. This technique uses more radiation than 15 fps and should be

used sparingly, remembering to reset the cine acquisition setting back to 15 fps and to low magnification.

12. When using bifemoral access, most interventionalists use the right femoral artery to cannulate the right coronary artery and the left femoral artery to cannulate the left main. This is done to avoid confusion when inserting and withdrawing equipment through the guide catheters. If the right coronary artery is not being used, the right femoral artery is often chosen for the antegrade guide catheter and the left femoral artery for the retrograde guide catheter. This choice is easy to remember if done the same way every time. Advancing or pulling on the wrong wire or catheter can easily negate any progress that has been made.

One technique for simultaneous injection of both coronary arteries *during the diagnostic angiogram without requiring a second point of access* is to upsize the femoral arterial access point to an 8 French sheath. Two 4 French catheters can be passed through a single 8 Fr sheath for simultaneous injection of both coronary arteries (Figs. 6.2 and 6.3).[4] However,

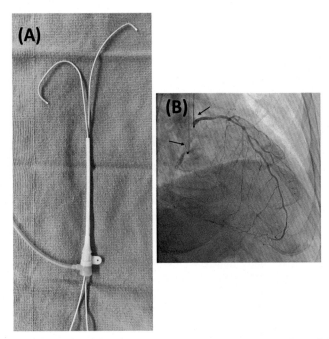

FIGURE 6.2 Illustration of using two 4 Fr diagnostic catheters through an 8 Fr sheath (**panel A**) and a dual angiogram obtained using this technique (**panel B**). *Courtesy of Dr. William Nicholson.*

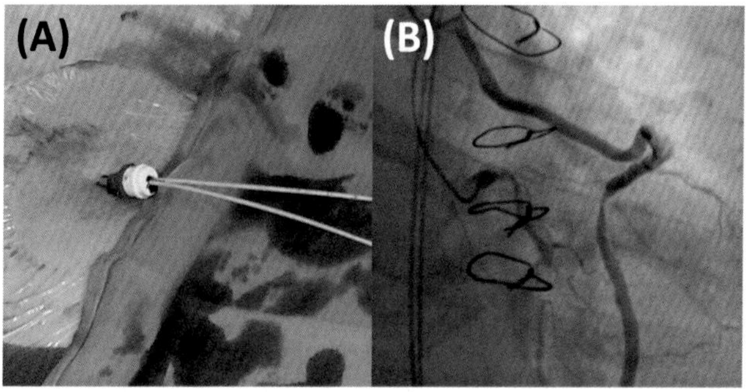

FIGURE 6.3 Illustration of dual angiography (**panel B**) using two 4 Fr diagnostic catheters through an 8 Fr sheath (**panel A**). *Courtesy of Dr. Gabriele Gasparini.*

coronary visualization may be poor with 4 Fr catheters even with use of automatic injectors, and bleeding may occur through the sheath's hemostatic valve.

6.2 Studying the lesion

6.2.1 How to evaluate the lesion

Spending enough time studying the CTO angiographic parameters will make the procedure easier, increasing the likelihood of success and decreasing the risk of complications. Previous angiograms should also be located and reviewed carefully.

When? Before the case begins.

By whom? Ideally the films should be reviewed by the entire CTO team, including the physicians, technicians, nurses, and fellows.

How long? Usually 15−30 min per patient. This much time is necessary to fully understand the CTO anatomy and determine the best action plan. With each repeat viewing of the images, new anatomic information becomes evident. For example the course of collaterals or the location of the proximal cap often become more evident with repeated views.

Which parameters should be assessed?

Four key parameters should be evaluated during angiographic review (Fig. 6.4)[3]:

1. Proximal cap and proximal vessel
2. Occlusion length and quality
3. Quality of the distal target vessel
4. Collateral circulation

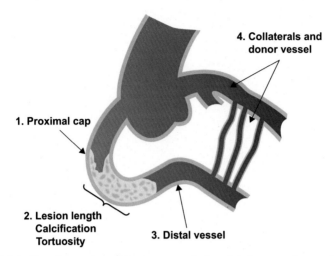

4. Collaterals and
donor vessel

1. Proximal cap

2. Lesion length
   Calcification
   Tortuosity

3. Distal vessel

FIGURE 6.4   **Four key angiographic parameters that need to be assessed as part of CTO PCI planning**.

These parameters help the operator understand:

1. where the CTO starts and the characteristics of the vessel proximal to the CTO;
2. the course and characteristics of the occluded segment;
3. the end point of the CTO and the quality of the distal vessel; and
4. potential retrograde pathways to access the distal cap.

Angiographic assessment tips and tricks:

- Slow replays and magnified views may help clarify the CTO vessel course and collateral connections. In case of multiple collaterals, a frame-by-frame replay can help determine the direction of flow in the branches of the distal vessel and identify the dominant collateral.
- Occasionally, tracing the collaterals backwards helps identify their origin and course.
- Some image postprocessing techniques, such as color inversion and increasing the image contrast, may help discover additional usable collaterals.
- In some patients, preprocedural coronary computed tomography angiography may help elucidate the course of the occluded vessel and evaluate the presence and extent of calcification and tortuosity (Section 6.4). This is especially helpful in patients with long CTO segments.

Why?

To understand the CTO anatomy and collateral circulation, which enables the operator to map out all possible options for crossing the occlusion in order to create a strategic plan (Fig. 6.5). Such procedural plans are often provided by proctors before planned cases.

EXAMPLE OF PROCEDURAL PLAN FOR RIGHT CORONARY ARTERY CTO

Proximal cap: ambiguous with side branch

Occlusion length: approximately 30-40mm
Distal vessel: diffusely diseased

Collaterals: multiple septal collaterals – would need to assess proximal and mid LAD to make sure there are no significant lesions before going retrograde.

There is also an occluded SVG-PDA that could be used for retrograde approach if needed.

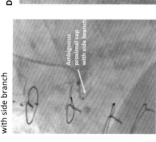

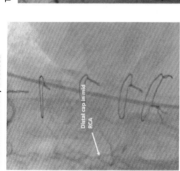

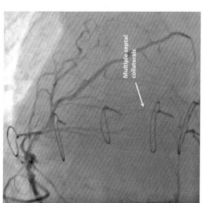

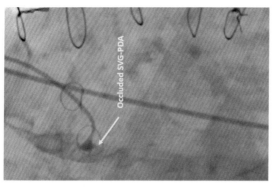

Plan

1. Would do lateral view to see if a stump is present in proximal RCA (unlikely)

2. Primary retrograde (AFTER CHECKING THAT PROXIMAL/MID LAD ARE OK) via septal collaterals

3. BASE (balloon-assisted subintimal entry) (balloon to dissect the proximal RCA then go extraplaque with knuckle wire)

4. Retrograde via occluded SVG

FIGURE 6.5 **Example of a procedural plan.** Case #1, RCA CTO. **Panel A:** proximal cap: ambiguous with side branch. **Panel B:** occlusion length: approximately 30–40 mm, Distal vessel: diffusely diseased. **Panel C:** collaterals: multiple septal collaterals - would need to assess proximal and mid LAD to make sure there are no significant lesions before going retrograde. **Panel D:** There is also an occluded SVG-PDA that could be used for the retrograde approach if needed. **Plan:** (1) Would do lateral view to see if a stump is present in the proximal RCA (unlikely). (2) Primary retrograde (after checking that proximal/mid LAD are fine) via septal collaterals. (3) BASE (balloon-assisted subintimal entry) (balloon to dissect the proximal RCA then go extraplaque with knuckle wire). (4) Retrograde via occluded SVG.

6.2.1.1 Proximal cap and vessel (Fig. 6.6)

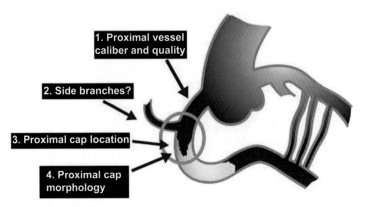

FIGURE 6.6 **Assessment of the proximal cap and vessel.**

6.2.1.1.1 Proximal vessel

Diffuse disease in the proximal vessel can result in pressure dampening upon guide catheter engagement. Pressure dampening may not cause ischemia in proximally occluded vessels supplying a small distal territory. Conversely, dampening in large vessels that have multiple side branches proximal to the occlusion, could cause ischemia and/or hypotension, and may as a result require intermittent guide catheter disengagement, use of a smaller guide catheter, or use of a guide catheter with side holes. Injection through a guide catheter with dampened pressure waveform could cause coronary and/or aortocoronary dissection (Chapter 25) and should be avoided. Precautionary removal of the injection syringe helps prevent inadvertent contrast injections. Prolonged pressure dampening also increases the risk of air embolization and thrombus formation within the guide catheter.

Careful wiring of diffusely diseased proximal vessels using workhorse guidewires with standard tip bends is critical to minimize the risk of proximal vessel dissection and of inadvertent modification of the CTO wire bend. A microcatheter is then advanced to the vicinity of the proximal cap and the workhorse wire is exchanged for specialty CTO wires with much smaller CTO-specific bends (advancing wires with CTO-specific bends through the proximal vessel can be challenging, especially when the proximal vessel is large).

6.2.1.1.2 Side branches

Side branches near or at the proximal cap may hinder antegrade wiring, as guidewires (especially polymer-jacketed guidewires) may preferentially enter those branches instead of engaging the proximal cap of the CTO. In some cases, side branches can be used to insert and inflate a balloon that increases guide catheter support (side branch anchoring, Fig. 6.7).

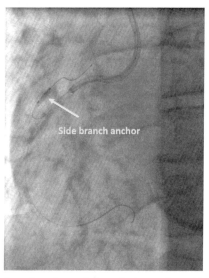

FIGURE 6.7   **The side-branch anchor technique for increasing guide catheter support.**

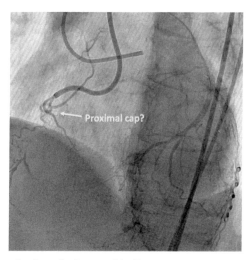

FIGURE 6.8   **Example of proximal cap ambiguity.**

6.2.1.1.3   Proximal cap location

Understanding the start of the CTO (proximal cap) is critical for both proce-dural success and safety. If the proximal cap is clearly defined and unambig-uous, upfront use of an antegrade approach is favored. CTOs with poorly defined, ambiguous proximal caps (Fig. 6.8) may be best approached with:

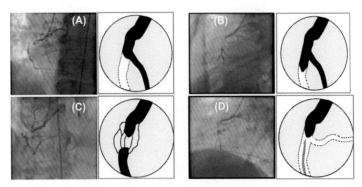

FIGURE 6.9   What appeared to be an ambiguous proximal cap (**panels A, B, and C**) with bridging collaterals (**panel C**), became a clear proximal cap after careful frame-by-frame evaluation. *Courtesy of Dr. Pedro Cardoso.*

(a) a primary retrograde approach; (b) intravascular ultrasound (IVUS)-guided puncture; or (c) extraplaque techniques ("move the cap" techniques)[5] (Section 16.1.4). Flush aorto-ostial occlusions (Chapter 15) also require a primary retrograde approach.

A frame-by-frame review can occasionally convert what appears to be an ambiguous cap into an unambiguous one (Fig. 6.9).

Multiple angiographic projections, including unconventional angles, may be needed to clarify the location of the proximal cap. For example, the straight lateral projection is often useful to define the proximal cap in right coronary artery CTOs. In some cases when angiography alone is inconclusive, intravascular ultrasound can help elucidate the proximal part of the CTO. A full discussion of how to approach a CTO with an ambiguous proximal cap is presented in Section 16.1.4.

6.2.1.1.4   Proximal cap morphology

Tapered proximal caps are more favorable than blunt caps, as they facilitate guidewire entry into the occlusion.[6,7] The proximal cap is usually more resistant than the distal cap, likely due to exposure to arterial pressure. Severe calcification can make wire penetration challenging, requiring highly penetrating guidewires, such as the Confianza Pro 12, Infiltrac and Infiltrac Plus, and Hornet 14.

6.2.1.2 Occlusion length and quality (Fig. 6.10)

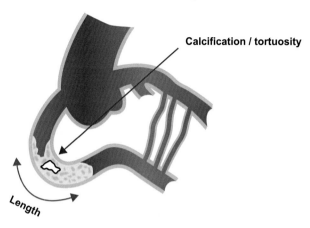

Lesion Length and Quality

FIGURE 6.10 **How to evaluate the occlusion length and quality of the occluded segment.**

6.2.1.2.1 Length

Lesion length is almost always overestimated with single injections, hence dual injections (Section 6.1) are important for accurate estimation of the lesion length. Lesion length and quality assessment can also be performed using coronary computed tomography. Longer lesion length is associated with more difficult CTO crossing and lower procedural success.

6.2.1.2.2 Quality

Intra-occlusion calcification and tortuosity also increase the difficulty of CTO crossing. Calcification (Chapter 19) increases the likelihood of extra-plaque guidewire entry and may hinder equipment advancement. Tortuosity increases the likelihood of guidewire exit and perforation; hence, highly tortuous occlusions may be best crossed through the extraplaque space with a knuckled guidewire (Section 8.2). The guidewire knuckle is less traumatic than the tip of a guidewire and more likely to remain "within the vessel architecture," which is reflected in the motto "trust the knuckle." Occasionally, small contrast-filled "islands" can be discovered in the body of long occlusions. These "islands" are supplied by side branches communicating with collaterals and can be helpful for wire-based strategies to follow the course of the vessel and minimize the risk of wire exit from the vessel architecture.

6.2.1.3   Quality of the distal vessel (Fig. 6.11)

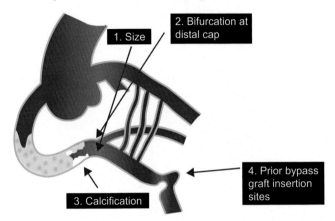

FIGURE 6.11   **How to evaluate the quality of the vessel distal to the CTO.**

Evaluating the *size* and quality of the vessel distal to the occlusion is important for deciding on a procedural plan and estimating the likelihood of success. Large vessels distal to the CTO are associated with easier crossing and higher likelihood of success. Conversely, small, diffusely diseased distal vessels can be much harder to recanalize and are associated with lower procedural success rates, in part due to difficulty reentering into the distal true lumen if the antegrade guidewire crosses the CTO through the extraplaque space. The size of the distal vessel may be small due to chronic hypoperfusion and can increase significantly both acutely and within a few months after successful CTO recanalization. In some patients, the size of the distal vessel may be underestimated due to partial filling caused by competitive flow via ipsilateral and/or contralateral collaterals. Having access to prior films can be useful for determining the "true" size of the distal vessel. The size of the proximal coronary artery segment might also provide a rough estimate about the true size of a collapsed distal segment.

Distal vessel *calcification* may hinder reentry attempts when the guidewire crosses the CTO through the extraplaque space and increases the risk of perforation during balloon inflation or stent deployment. Stent postdilatation with oversized balloons or at high pressures should be avoided in heavily calcified vessels.

The morphology of the distal cap often cannot be visualized as clearly as the proximal cap because of lower perfusion pressure through collaterals. In such cases selective collateral injection (Fig. 6.12) or a gentle tip injection through a retrograde microcatheter after ensuring intraluminal position of the catheter tip can clarify the anatomy of the distal cap. The presence of a *bifurcation* at the distal cap makes CTO crossing harder, especially in cases of antegrade extraplaque guidewire crossing where reentry into the distal true lumen distal to the

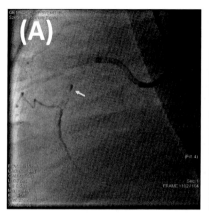 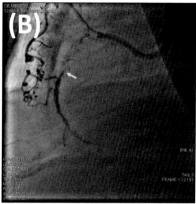

FIGURE 6.12   Visualization of the CTO distal cap (**arrow, panel A**) by contralateral (**panel A**) versus ipsilateral (**panel B**) injection. The distal cap is best visualized via ipsilateral injection of a bridging collateral. *Courtesy of Dr. Imre Ungi.*

bifurcation may result in loss of one of the branches.[8] Such lesions may be best approached with a primary retrograde approach, or by performing "double reentry" into both branches of the bifurcation (Section 16.5, CTO Manual online case 80). A retrograde dual lumen microcatheter might also help puncture a hard and blunt distal cap. Another option is to combine an antegrade approach for one of the branches with a retrograde approach for the second branch.

In patients with prior coronary artery bypass graft surgery (CABG) the distal vessel may be distorted at the *distal anastomotic site* ("tenting" of the native vessel from the bypass graft).

6.2.1.4   Collaterals/bypass grafts and donor vessel

6.2.1.4.1   Source of the collaterals

Collaterals may arise from the CTO artery itself, proximal to the occlusion or from another branch of the occluded artery, such as collaterals to the left anterior descending artery (LAD) from the circumflex, (ipsilateral collaterals), or from another coronary artery (contralateral collaterals) (Fig. 6.13). These would include right to left collaterals or left to right collaterals. Aortocoronary bypass grafts (patent or occluded) are not "collaterals" per se, but often serve as retrograde conduits. For example, in patients with severely degenerated saphenous vein grafts, recanalization of the native coronary artery provides better short- and long-term outcomes as compared with PCI of the saphenous vein graft (SVG).[9]

6.2.1.4.2   Course of the collaterals

As their name implies, septal collaterals course through the septum, whereas epicardial collaterals course at the heart's surface. Septal collaterals and bypass

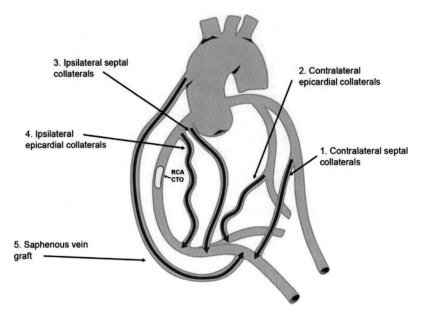

3. Ipsilateral septal
collaterals

2. Contralateral
epicardial collaterals

4. Ipsilateral
epicardial collaterals

1. Contralateral septal
collaterals

RCA
CTO

5. Saphenous vein
graft

FIGURE 6.13   **Examples of various sources and courses of collaterals for a right coronary artery CTO.**

grafts are preferred over epicardial collaterals, because they are safer to cross. Septal collateral perforations are often not clinically significant and can be managed conservatively, whereas epicardial collateral perforations often result in cardiac tamponade. In patients with prior coronary artery bypass graft surgery, perforation of epicardial collaterals can cause loculated effusions that can compress cardiac chambers and cause "dry tamponade." Such loculated effusions can be difficult or impossible to access through pericardiocentesis. Treatment of epicardial collateral perforations requires occlusion from both sides of the perforation for successful sealing. Not all epicardial collaterals are equal. For example, Mashayekhi et al. demonstrated that ipsilateral epicardial collaterals from acute marginal to acute marginal branch (Type B, Fig. 6.14) have a high risk of perforation (Fig. 6.15) and should not be used for retrograde crossing.[10]

Both patent and occluded SVGs can be used for retrograde crossing (Section 18.3).[11,12] Retrograde crossing through internal mammary artery (IMA) grafts (Section 18.4) is feasible, but may result in significant ischemia and hemodynamic compromise.

6.2.1.4.3   Collateral evaluation (Figs. 6.16 and 6.17)

Collateral vessels can be assessed for:

1. Quality of the vessel proximal to the collateral
2. Entry angle

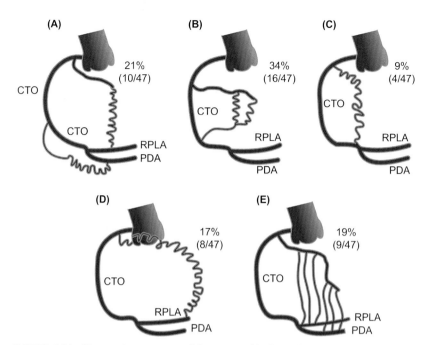

FIGURE 6.14 **The varying courses and frequency of ipsilateral collateral channels (CCs) of the RCA. Panel A**: type A: CCs originating from a high marginal branch (RM) inserting to the right posterolateral artery (RPLA) or from a lower RM to the posterior descending artery (PDA). **Panel B**: type B: CCs linking distal ends of higher and lower RMs, thereby bridging the CTO of the RCA. **Panel C**: type C: CCs originating directly from the proximal RCA and inserting close to the crux cordis. **Panel D**: type D: CCs with a longer epimyocardial course inserting at the distal part of the RPLA. **Panel E**: type E: septal intramyocardial, ipsilateral CCs, also called right superior descending artery. (Curved lines correspond to an epimyocardial course, straight lines correspond to an intramyocardial course of CCs).[10] *Reproduced with permission from Mashayekhi K, Behnes M, Akin I, Kaiser T, Neuser H. Novel retrograde approach for percutaneous treatment of chronic total occlusions of the right coronary artery using ipsilateral collateral connections: a European centre experience. EuroIntervention 2016;11:e1231−e1236.*

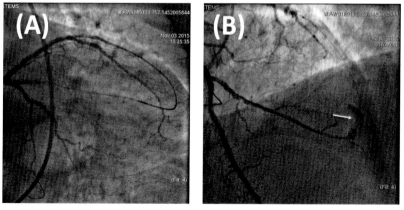

FIGURE 6.15 Example of an epicardial collateral that was successfully crossed with a Corsair microcatheter (**panel A**), but developed a perforation upon microcatheter withdrawal (**arrow, panel B**). *Courtesy of Dr. Imre Ungi.*

Collateral assessment

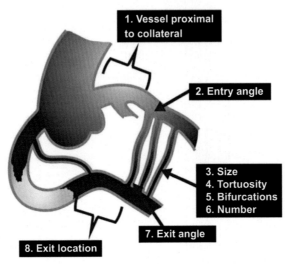

FIGURE 6.16 **How to evaluate the suitability of a collateral vessel for retrograde crossing**.

3. Size (CC classification)
4. Tortuosity
5. Bifurcations
6. Number of collaterals
7. Exit angle into distal vessel
8. Entry location in the distal vessel

Scores have been developed to determine the likelihood of collateral crossing and technical success, such as the score by McEntegart et al.[13], by Nagamatsu et al. (J-channel score)[14], and by Huang et al.[15] (the last score only includes collateral size and tortuosity).

6.2.1.4.3.1   Quality of the donor vessel proximal to the collateral In some cases, extensive disease in the donor vessel may lead to selection of an alternative revascularization strategy, such as coronary artery bypass graft surgery (CTO Manual online case 75).

When the collaterals originate distal to a severe lesion in the donor vessel, advancement of a microcatheter through the lesion can result in ischemia proportionate to the size of the affected myocardium. For example, in patients with severe left main disease, advancing microcatheters through the stenosis could cause a large area of ischemia and possibly hemodynamic collapse. In such cases, it is preferable to first treat the vessel proximal to the collateral takeoff before attempting retrograde CTO PCI. Also allow enough time for the proximal vessel stent to "heal" or "endothelialize" prior to bringing the patient back for

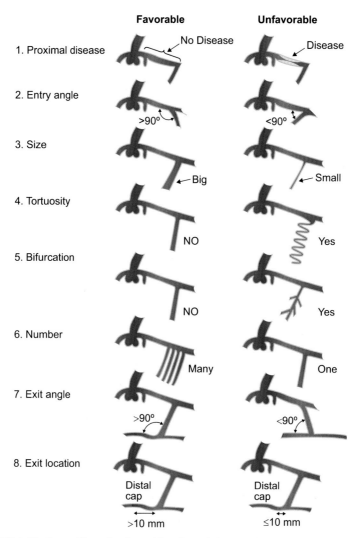

FIGURE 6.17   **Favorable and unfavorable collateral characteristics.**

CTO PCI. Balloon angioplasty of the ostium of a jailed collateral can facilitate subsequent equipment advancement into the collateral. Avoid occluding or jeopardizing the integrity of collaterals during PCI to the proximal diseased segment.

Accessing septal collaterals through previously implanted stents can be difficult.[16] Even when septal branch wiring is successful, advancing a microcatheter through the stent cells may not be possible and/or may damage the tip or the hydrophilic coating of the microcatheter. This can usually be avoided by predilating the stent struts at the collateral origin with a low-profile balloon. In such cases high-pressure postdilatation of the main vessel stent at the end of the CTO PCI may correct any deformation of the main vessel stent that could increase the risk of stent thrombosis.

6.2.1.4.3.2   *Entry angle* Obtuse angle (> 90 degrees) facilitates entry into the collateral — acute angle (< 90 degrees) makes entry more difficult and may require use of angulated microcatheters (such as the Venture and Supercross) or the reversed guidewire (hairpin wire) technique (see section 8.2.2.6, Challenge 3).

6.2.1.4.3.3   *Size (CC classification)* Larger collaterals are easier to cross and less likely to have compromised flow during wire and microcatheter advancement. The most commonly used collateral size classification is the Werner classification (Table 6.1, Fig. 6.18).[17]

6.2.1.4.3.4   *Tortuosity* Less tortuous collaterals are preferred, especially when they are small. In a single center analysis of 157 retrograde CTO PCIs, collateral tortuosity was one of the strongest predictors of failure.[18]

Extremely tortuous collaterals pose challenges in crossing with both the guidewire and microcatheter and have higher risk of perforation. "Z curve"

TABLE 6.1 Werner classification of coronary collateral circulation.

Werner collateral connection grade[17]	
CC0	No continuous connection
CC1	Thread like continuous connection
CC2	Side branch–like connection (≥0.4 mm)

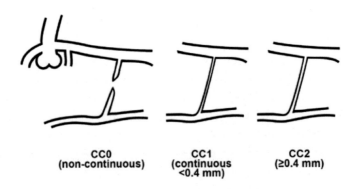

CC0
(non-continuous)

CC1
(continuous
<0.4 mm)

CC2
(≥0.4 mm)

Werner Classification

FIGURE 6.18   **Werner classification of collateral size.**

tortuosity should generally be avoided. Evaluation of the degree of tortuosity of a collateral during the cardiac cycle can provide useful information as to whether the collateral will be navigable with wires and microcatheters.

6.2.1.4.3.5 Bifurcations The presence of a bifurcation within a collateral makes distal crossing of the collateral more challenging. Bifurcations pose a particular challenge during septal crossing if they are located in the first 3 mm of the septal branch. Bifurcations in epicardial collaterals are of particular concern as inadvertent entry into a branch may result in perforation and pericardial tamponade.

6.2.1.4.3.6 Number Presence of multiple collaterals not only offers more retrograde crossing options, but also minimizes the risk of ischemia from compromised collateral flow during retrograde crossing attempts. Compromising the flow of a solitary collateral vessel supplying a large area of myocardium could result in profound ischemia, chest pain, and even hemodynamic collapse. However, at times, other nonvisible collaterals are recruited to supply the distal vessel when a dominant collateral gets compromised. That is another reason why dual injection is important, even for CTOs without apparent contralateral collaterals.

Conversely, in patients with multiple septal collaterals it can be difficult to select the best path for crossing with the wire. This difficulty can be resolved by repeated tip injections. In cases of multiple epicardial collaterals, perforation of one branch carries high risk, because even if the antegrade route is successfully sealed, there can be continuous bleeding from retrograde flow through the other collaterals, potentially resulting in tamponade.

6.2.1.4.3.7 Exit angle Similar to entry angle, an obtuse (>90 degrees) exit angle facilitates guidewire advancement into the distal vessel in the direction of the distal cap. Acute angles can be difficult to navigate as the guidewire may preferentially advance towards the distal vessel instead of towards the distal cap. This is commonly seen when using saphenous vein grafts for retrograde access. Another challenge can be in distal septal collaterals that have a big curve towards the apex. This anatomy sometimes hinders following the guidewire with the microcatheter and a forceful push might cause injury of the collateral and the septum.

6.2.1.4.3.8 Location of entry into the distal vessel Ideal collaterals for the retrograde approach connect to the distal vessel >10 mm distal to the distal cap in a straight coronary segment. Collateral insertion close to the distal cap (Fig. 6.19) makes CTO crossing challenging or impossible as the guidewire may preferentially enter the distal vessel or the proximal extraplaque space instead of engaging the CTO distal cap.

Examples of favorable and unfavorable collateral branches are shown in Figs. 6.20 and 6.21.

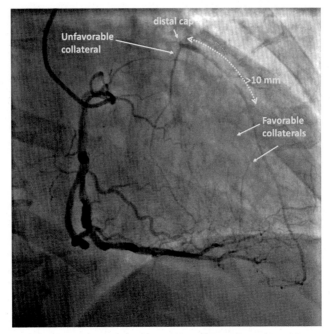

FIGURE 6.19   Example of the impact of collateral entry location into the distal vessel.

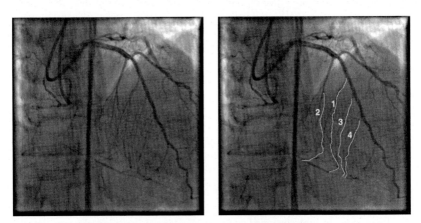

FIGURE 6.20   Example of the multiple favorable septal collaterals in a patient with a coronary artery CTO. In such patients, collaterals can be "graded" according to suitability, with crossing of more favorable collaterals attempted first $(1 > 2 > 3 > 4)$.

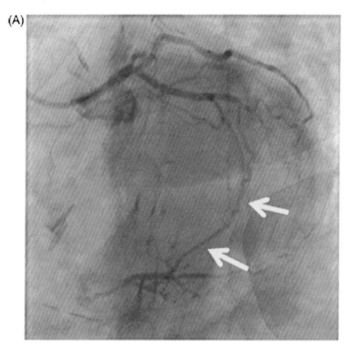

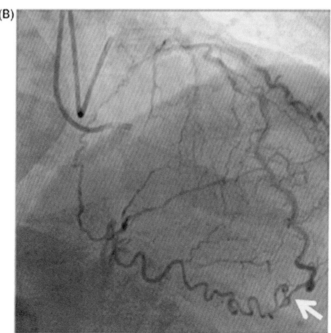

FIGURE 6.21 **Examples of favorable and unfavorable epicardial collateral vessels.** Favorable epicardial collateral from the circumflex to the right posterolateral branch with minimal tortuosity and adequate size (**panel A**). Unfavorable epicardial collateral from the left anterior descending artery to the right coronary artery because of extreme tortuosity and because it enters the right coronary artery very close to the distal cap (**panel B**).[19] *Modified with permission from Joyal D, Thompson CA, Grantham JA, Buller CEH, Rinfret S. The retrograde technique for recanalization of chronic total occlusions: a step-by-step approach. JACC Cardiovasc Interv 2012;5:1−11.*

Note: the presence of collaterals does not necessarily signify viability of the collateralized myocardial territory, as collaterals can also develop to non-viable territories.[20] Conversely, lack of good collateral circulation does not necessarily mean that the myocardium is nonviable.[21]

6.2.1.4.4 Optimal views for visualizing collaterals

1. Septal collaterals
   a. RAO (right anterior oblique) cranial is best for determining the part of the collateral that is closer to the LAD.
   b. Straight RAO or RAO caudal is best for visualizing the collateral segment closer to the PDA, which is usually more tortuous than the segment closer to the LAD.
   c. LAO (left anterior oblique) view may be helpful during wiring attempts.
2. Epicardial collaterals in the lateral wall (diagonal—obtuse marginal vessels)
   a. LAO cranial
   b. RAO cranial
   c. Lateral
3. Epicardial collaterals between the proximal circumflex and RCA
   a. AP caudal
   b. RAO

6.2.1.4.5 Rentrop classification

The Rentrop classification is commonly used to describe the filling of the distal vessel (Table 6.2).[22]

TABLE 6.2 The Rentrop classification of distal vessel filling.[22]

Rentrop classification (developed for occluded and nonoccluded arteries)	
0	No filling of collateral vessels
1	Filling of collateral vessels without any epicardial filling of the target artery
2	Partial epicardial filling of the target artery by collateral vessels
3	Complete epicardial filling of the target artery by collateral vessels (in CTOs, Rentrop 3 filling is observed 85% of lesions)

6.3 Chronic total occlusion percutaneous coronary intervention angiographic scores

The success and safety of CTO PCI depends both on operator experience and lesion complexity. Many angiographic and clinical parameters have been included in various scoring systems that were developed in diverse CTO PCI cohorts and can be used to estimate the likelihood of success (9 scores, Table 6.3)[6,23−30] and the risk of complications (one score)[31] of CTO PCI.

An online calculator for several CTO PCI scores is available at: http://www.progressscto.org/cto-scores.

6.3.1 Success and efficiency scores

Table 6.3 presents various scores developed for estimating the likelihood of success and efficiency of CTO PCI.[6,23−30]

The most commonly used scores are the J-CTO, PROGRESS-CTO, and CASTLE scores, but they all have moderate predictive capacity[32−34] and are best used in the populations in which they were developed.

The J-CTO score (Multicenter Chronic Total Occlusion [CTO] Registry in Japan) was the first score developed for CTO PCI. It uses five variables (occlusion length ≥ 20 mm, blunt stump, CTO calcification, CTO tortuosity and prior failed attempt) to create a 5-point score that is associated with successful guidewire crossing within the first 30 min (Fig. 6.22).[6] The J-CTO score is also associated with the likelihood of technical success[35−37] and the risk of major cardiovascular and cerebrovascular events after 1 year.[38]

The PROGRESS-CTO score (Prospective Global Registry for the Study of Chronic Total Occlusion Intervention) uses four variables (proximal cap ambiguity, moderate/severe tortuosity, circumflex artery CTO, and absence of "interventional" collaterals) to create a 4-point score that predicts technical success (Fig. 6.23).[24]

The CASTLE score was derived by cases performed by EuroCTO Club experienced operators and uses six variables: coronary artery bypass graft history, age (≥70 years), stump anatomy [blunt or invisible], tortuosity degree [severe or unseen], length of occlusion [≥ 20 mm], and extent of calcification [severe].[29]

6.3.2 Complications scores

CTO PCI carries increased risk of complications as compared with standard PCI. In the NCDR (National Cardiovascular Data Registry) Cath PCI registry the risk for major adverse cardiac events (MACE) was 1.6% for CTO PCI versus 0.8% for non-CTO PCI ($P < .001$).[39] In the Prospective Global Registry for the Study of Chronic Total Occlusion Intervention (PROGRESS CTO) the risk for MACE was 2.8%. Age >65 years, occlusion length ≥ 23 mm, and use of the retrograde approach was associated with a higher risk of complications.[31] These parameters

TABLE 6.3 Comparison of various scores for estimating the success and efficiency of CTO PCI.

	J-CTO[6]	CL[23]	PROGRESS-CTO[24]	ORA[25]	Ellis[26]	RECHARGE[27]	W-CTO[28]	CASTLE[29]	E-CTO[30]
Year of publication	2011	2015	2016	2016	2017	2018	2018	2019	2021
Number of variables	5	6	4	3	7	6	5	6	4
Number of cases	494	1,657	781	1,073	436 patients (456 lesions)	880	404 patients (408 lesions)	>20,000	457 patients (540 lesions)
Setup	12 Japanese centers	2 French centers	7 US Centers	Single expert operator	9 US operators	European centers	Single Indian center – antegrade only cases	Expert European operators	Single European center
Dates	2006–07	2004–13	2012–15	2005–14	2014–15	2014–15	2009–15	2008–16	2007–21
Overall success	88.6% (guidewire crossing)	72.5% (procedural success)	92.9% (technical success)	91.9% (technical success)	78.6% (technical success)	84% (technical success)	83.6% (procedural success)	87.8% (technical success)	80.1% (procedural success)
Clinical									
Age				≥75 years				≥70 years	
Prior CABG		+				+		+	
Prior MI		+							
Prior CTO PCI failure	+								
Angiographic									
Proximal cap ambiguity			+		+				

Blunt stump	+			+	+	+		+
Ostial location						+	+	+
Calcification	+		+	+	+	+	+	+
Proximal tortuosity			+		+	+	+	
Within occlusion tortuosity	+				+	+		+
CTO length	≥ 20 mm		≥ 10 mm	≥ 20 mm	>20 mm	≥ 20 mm	≥ 20 mm	≥ 20 mm
Diseased distal landing zone					+			
CTO target vessel	+ (non-LAD)	+ (circumflex)						
Collaterals	+ (interventional)		+ (Rentrop <2)	+	+ (Rentrop <2)	+ (Rentrop <2)		
Other								
Operator			+			+		

CABG, coronary artery bypass graft surgery; *CASTLE*, Coronary artery bypass graft history, Age (≥70 years), Stump anatomy [blunt or invisible], Tortuosity degree [severe or unseen], Length of occlusion [≥ 20 mm], and Extent of calcification [severe]; *CL*, Clinical and Lesion-related score; *CTO*, chronic total occlusion; *E-CTO*, operator experience–chronic total occlusion; *J-CTO*, Multicenter CTO Registry in Japan score; *LAD*, left anterior descending artery; *MI*, myocardial infarction; *ORA score*, Ostial location, Rentrop <2, Age ≥75 years score; *PCI*, percutaneous coronary intervention; *PROGRESS-CTO*, Prospective Global Registry for the Study of Chronic Total Occlusion Intervention score; *RECHARGE*, REgistry of CrossBoss and Hybrid procedures in FrAnce, the NetheRlands, BelGium and UnitEd Kingdom; *W-CTO*, Weighted angiographic scoring model.

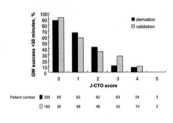

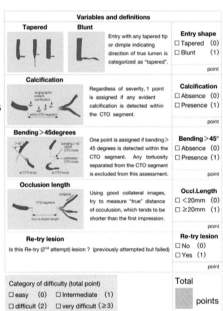

FIGURE 6.22   **The J-CTO score.** Description of the components of the J-CTO score that was developed to predict the likelihood of successful guidewire crossing of the occlusion within 30 min. *Reproduced with permission from Morino Y, Abe M, Morimoto T, et al. Predicting successful guidewire crossing through chronic total occlusion of native coronary lesions within 30 minutes: the J-CTO (Multicenter CTO Registry in Japan) score as a difficulty grading and time assessment tool. JACC Cardiovasc Interv 2011;4:213–221.*

were used to develop a score predicting the risk for MACE during CTO PCI (PROGRESSCTO-CTO Complications score) (Fig. 6.24).[31] Four updated PROGRESS-CTO complications scores were developed in 2022 for estimating the risk of major adverse cardiac events (MACE), mortality, pericardiocentesis, and acute myocardial infarction.[40]

There are two scores for predicting the risk of coronary perforation during CTO PCI, the OPEN-CLEAN (Outcomes, Patient health status, and Efficiency IN CTO hybrid procedures-CABG, Length of occlusion, EF <50%, Age, severe calcificatioN) perforation score[41] and the PROGRESS-CTO perforation score.[42] The OPEN-CLEAN score is computed from five parameters: prior coronary artery bypass graft (CABG), occlusion length, ejection fraction (EF), age, and severe calcification. The score range is 0 to 7 and a higher score is associated with a higher risk of perforation. The PROGRESS-CTO perforation score includes 5 parameters: patient age ≥65 years, moderate/severe calcification, blunt/no stump, use of antegrade dissection and reentry, and use of the retrograde approach. A higher score is associated with higher risk of clinical perforation.

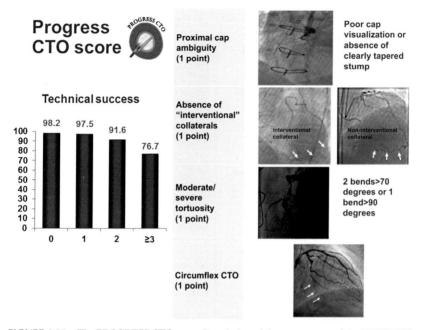

FIGURE 6.23   **The PROGRESS-CTO score.** Description of the components of the PROGRESS CTO score that was developed to predict technical success of CTO PCI. *Reproduced with permission from Christopoulos G, Kandzari DE, Yeh RW, et al. Development and validation of a novel scoring system for predicting technical success of chronic total occlusion percutaneous coronary interventions: the PROGRESS CTO (Prospective Global Registry for the Study of Chronic Total Occlusion Intervention) Score. JACC Cardiovasc Interv 2016;9:1−9.*

6.4   Use of coronary computed tomography angiography

Preprocedural coronary computed tomography angiography (CCTA) can provide accurate assessment of the proximal cap location, calcification, tortuosity, vessel course and length of the occluded segment. Preprocedural CCTA was associated with higher CTO PCI success rates in a randomized-controlled trial.[43] CCTA is more accurate for assessing lesion length, as coronary angiography often overestimates the occlusion length (Fig. 6.25).

CCTA is the gold standard for the diagnosis of anomalous coronary arteries (Figs. 6.26 and 6.27). In aorto-ostial right coronary occlusions with no visible calcification, CT angiography may help identify the location of the ostium or an aberrant origin of the vessel. CT angiography can be very helpful in cases with poor visualization of the distal target vessel. CCTA can help clarify proximal cap ambiguity (Fig. 6.28). CCTA is, however, limited in identifying collateral circulation.

Several CCTA-specific CTO PCI scores have been developed, such as the CT-RECTOR (Computed Tomography Registry of Chronic Total Occlusion Revascularization) score[44] (Figs. 6.29−6.31) and the KCCT (Korean Multicenter

Progress CTO complications score

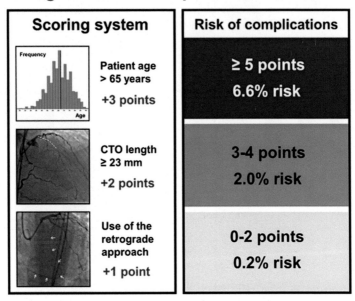

Scoring system	Risk of complications
Patient age > 65 years +3 points	≥ 5 points 6.6% risk
CTO length ≥ 23 mm +2 points	3-4 points 2.0% risk
Use of the retrograde approach +1 point	0-2 points 0.2% risk

FIGURE 6.24 **The PROGRESS-CTO complications score.** Description of the components of the Progress-CTO complications score that was developed to predict periprocedural complications during CTO PCI. Periprocedural complications included any of the following adverse events prior to hospital discharge: death, myocardial infarction, recurrent symptoms requiring urgent repeat target vessel revascularization with PCI or CABG, tamponade requiring either pericardiocentesis or surgery, and stroke. *Reproduced with permission from Danek BA, Karatasakis A, Karmpaliotis D et al Development and validation of a scoring system for predicting periprocedural complications during percutaneous coronary interventions of chronic total occlusions: the prospective global registry for the study of chronic total occlusion intervention (PROGRESS CTO) complications score. J Am Heart Assoc 2016;5.*

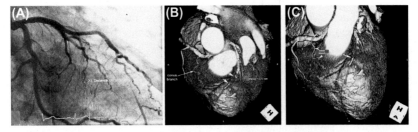

FIGURE 6.25 **CTO length mismatch between angiography and CCTA** (32 and 10.7 mm respectively) due to multisource collateral competitive filling of the distal LAD post-occlusion segment. Due to very proximal origin of the conus branch and deep engagement of the diagnostic catheter tip into the right coronary artery, collateral filling from the conus branch was not visualized during diagnostic angiography. *Courtesy of Dr. Avtandil Babunashvili.*

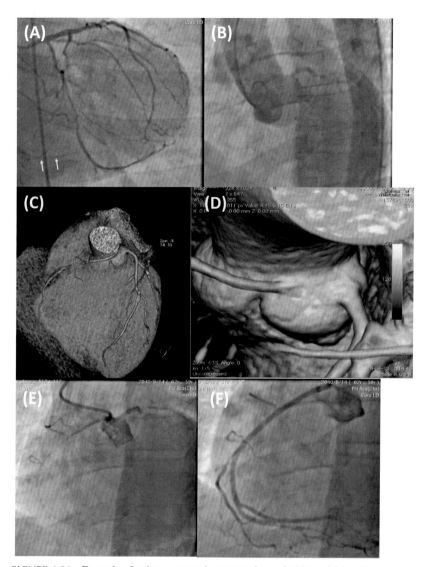

FIGURE 6.26   **Example of using computed tomography to facilitate CTO PCI**. A patient presented with angina and inferior ischemia. Although collaterals to the right posterior descending artery were visualized (*arrows*, **panel A**), it was impossible to find the ostium of the right coronary artery (**panel B**). Computed tomography demonstrated an anomalous right coronary artery originating from the left sinus of Valsalva (**panels C and D**). The right coronary artery had a short CTO (**panel C**) but did not have any calcification. The right coronary artery was engaged with a 3D right coronary artery guide (**panel E**) and successfully recanalized (panel F) using antegrade wiring. *Courtesy of Dr. Leszek Bryniarski.*

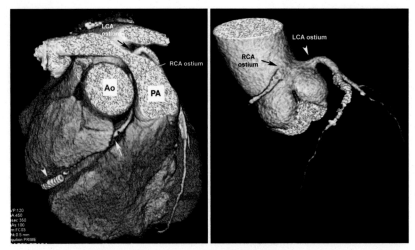

FIGURE 6.27   Complex in-stent CTO of an anomalous right coronary artery arising from the left cusp with an intra-arterial course of its proximal segment (possible compression of the arterial lumen between the aorta and the pulmonary artery). These CCTA findings affect treatment strategy (PCI vs CABG). *Courtesy of Dr. Avtandil Babunashvili.*

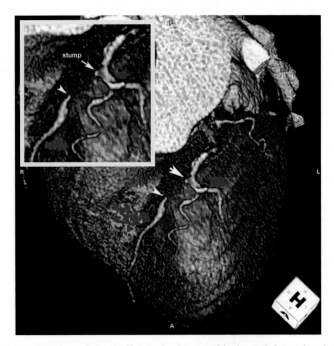

FIGURE 6.28   CCTA can help clarify proximal cap ambiguity and determine the optimal angulation for visualization of the proximal cap. *Courtesy of Dr. Avtandil Babunashvili.*

CT-RECTOR Score Calculator

Predictors Definitions

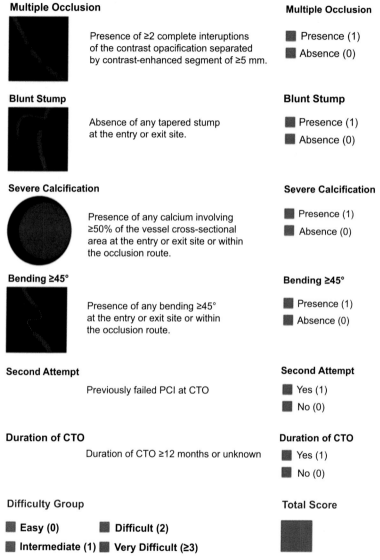

Multiple Occlusion

Presence of ≥2 complete interruptions of the contrast opacification separated by contrast-enhanced segment of ≥5 mm.

Multiple Occlusion
■ Presence (1)
■ Absence (0)

Blunt Stump

Absence of any tapered stump at the entry or exit site.

Blunt Stump
■ Presence (1)
■ Absence (0)

Severe Calcification

Presence of any calcium involving ≥50% of the vessel cross-sectional area at the entry or exit site or within the occlusion route.

Severe Calcification
■ Presence (1)
■ Absence (0)

Bending ≥45°

Presence of any bending ≥45° at the entry or exit site or within the occlusion route.

Bending ≥45°
■ Presence (1)
■ Absence (0)

Second Attempt

Previously failed PCI at CTO

Second Attempt
■ Yes (1)
■ No (0)

Duration of CTO

Duration of CTO ≥12 months or unknown

Duration of CTO
■ Yes (1)
■ No (0)

Difficulty Group
■ Easy (0)   ■ Difficult (2)
■ Intermediate (1) ■ Very Difficult (≥3)

Total Score

FIGURE 6.29 **The CT-RECTOR Score Calculator**. Calculation sheet for the CT-RECTOR (Computed Tomography Registry of Chronic Total Occlusion Revascularization) score with illustrated definitions of each variable and listing of the difficulty groups. *CTO*, chronic total occlusion; *PCI*, percutaneous coronary intervention. *Reproduced with permission from Opolski MP, Achenbach S, Schuhback A, et al. Coronary computed tomographic prediction rule for time-efficient guidewire crossing through chronic total occlusion: insights from the CT-RECTOR multicenter registry (Computed Tomography Registry of Chronic Total Occlusion Revascularization). JACC Cardiovasc Interv 2015;8:257–267.*

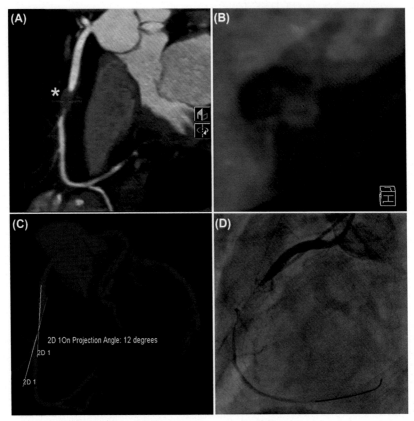

FIGURE 6.30   CTO with favorable angiographic and CCTA characteristics that was easily crossed with a guidewire. **Panel A**: curved multiplanar reconstruction of a short noncalcified occlusion (*yellow asterisk*) with tapered entry and exit sites in the mid-segment of the right coronary artery. **Panel B**: cross-sectional view of the noncalcified occlusion site. **Panel C**: three-dimensional reconstruction displaying nonrelevant bending within the occlusion route. **Panel D**: coronary angiography showing successful recanalization after 2 min of guidewire manipulation. *Reproduced with permission from Opolski MP, Achenbach S, Schuhback A, et al. Coronary computed tomographic prediction rule for time-efficient guidewire crossing through chronic total occlusion: insights from the CT-RECTOR multicenter registry (Computed Tomography Registry of Chronic Total Occlusion Revascularization). JACC Cardiovasc Interv 2015;8:257—267.*

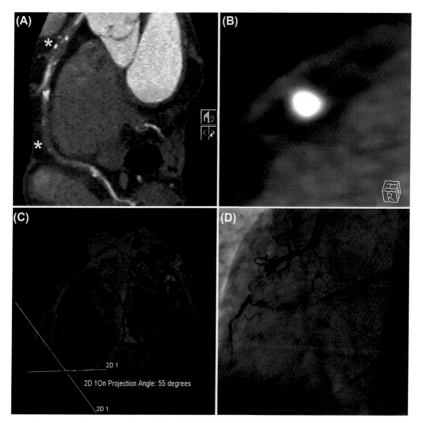

FIGURE 6.31   Highly complex CTO that could not be crossed. Panel A: curved multiplanar reconstruction displaying 2 occlusion sites (*yellow asterisks*) and severe calcification in the midsegment of the right coronary artery. **Panel B**: cross-sectional view of the calcification occupying ≥50% vessel area. **Panel C**: three-dimensional reconstruction displaying bending of 55 degrees within the occlusion route. **Panel D**: coronary angiography after failed attempt of guidewire crossing within 55 min. *Reproduced with permission from Opolski MP, Achenbach S, Schuhback A, et al. Coronary computed tomographic prediction rule for time-efficient guidewire crossing through chronic total occlusion: insights from the CT-RECTOR multicenter registry (Computed Tomography Registry of Chronic Total Occlusion Revascularization). JACC Cardiovasc Interv 2015;8:257–267.*

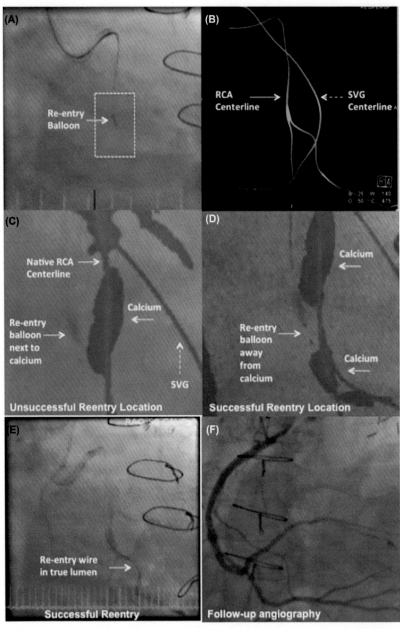

(A) Re-entry Balloon

(B) RCA Centerline   SVG Centerline

(C) Native RCA Centerline — Calcium — Re-entry balloon next to calcium — SVG — Unsuccessful Reentry Location

(D) Calcium — Re-entry balloon away from calcium — Calcium — Successful Reentry Location

(E) Re-entry wire in true lumen — Successful Reentry

(F) Follow-up angiography

(*Continued*)

◀ FIGURE 6.32 **Analysis of unsuccessful then successful antegrade dissection reentry during CTO PCI.** In this example, coronary CTA fusion data display was chosen to show both the centerline and the arterial calcium. **Panel A**: fluoroscopic image showing a reentry balloon placed distal to an RCA CTO (as confirmed by contralateral contrast injection; image not shown). The initial attempt at reentry from this location was unsuccessful. **Panel B**: centerlines from CTA segmentation showing the RCA and SVG. **Panel C**: magnified CTA/fluoroscopy fusion of the initial unsuccessful reentry site showing that the reentry balloon resided next to a large zone of calcium (*yellow arrow*). The reentry balloon was then advanced more distally approximately 1 cm, and successful reentry was performed using a stick-and-swap technique. **Panel D**: magnified CTA/fluoroscopy fusion revealed that the successful, reentry zone was between two areas of calcification. **Panel E**: fluoroscopic image during contralateral contrast injection shows the reentry wire (*white arrow*) in the distal RCA true lumen. **Panel F**: follow-up angiogram showing successful CTO PCI revascularization of the RCA. *RCA*, right coronary artery; *SVG*, saphenous vein graft; *CTA*, computed tomography angiography; *PCI*, percutaneous coronary intervention. *Reproduced with permission from Ghoshhajra BB, Takx RAP, Stone LL, et al. Real-time fusion of coronary CT angiography with x-ray fluoroscopy during chronic total occlusion PCI. Eur Radiol 2017;27:2464−2473.*

CTO CT Registry) score[45]. In one study the CCTA-derived J-CTO score was a stronger predictor of both procedural success and wire crossing within 30 min than the J-CTO score derived from conventional angiography.[36]

CCTA can also help identify the best angiographic projection for CTO crossing.[46]

Pre-procedural CCTA can help with prediction of a successful antegrade technique[44,47] and with selection of patients that may be best served by referral to more experienced CTO operators. In addition, improvements in coregistration of coronary CT angiography with real-time fluoroscopy (Fig. 6.32)[48,49] might help the operator better identify wire position in real time.

References

1. Singh M, Bell MR, Berger PB, Holmes Jr. DR. Utility of bilateral coronary injections during complex coronary angioplasty. *J Invasive Cardiol* 1999;**11**:70−4.
2. Brilakis ES, Mashayekhi K, Tsuchikane E, et al. Guiding principles for chronic total occlusion percutaneous coronary intervention. *Circulation* 2019;**140**:420−33.
3. Brilakis ES, Grantham JA, Rinfret S, et al. A percutaneous treatment algorithm for crossing coronary chronic total occlusions. *JACC Cardiovasc Interv* 2012;**5**:367−79.
4. Nicholson WJ, Rab T. Simultaneous diagnostic coronary angiography utilizing a single arterial access technique. *Catheter Cardiovasc Interv* 2006;**68**:718.
5. Vo MN, Karmpaliotis D, Brilakis ES. "Move the cap" technique for ambiguous or impenetrable proximal cap of coronary total occlusion. *Catheter Cardiovasc Interv* 2016;**87**:742−8.
6. Morino Y, Abe M, Morimoto T, et al. Predicting successful guidewire crossing through chronic total occlusion of native coronary lesions within 30 minutes: the J-CTO (Multicenter CTO Registry in Japan) score as a difficulty grading and time assessment tool. *JACC Cardiovasc Interv* 2011;**4**:213−21.
7. Nombela-Franco L, Urena M, Jerez-Valero M, et al. Validation of the J-chronic total occlusion score for chronic total occlusion percutaneous coronary intervention in an independent contemporary cohort. *Circ Cardiovasc Interv* 2013;**6**:635−43.

8. Kotsia A, Christopoulos G, Brilakis ES. Use of the retrograde approach for preserving the distal bifurcation after antegrade crossing of a right coronary artery chronic total occlusion. *J Invasive Cardiol* 2014;**26**:E48−9.

9. Brilakis ES, O'Donnell CI, Penny W, et al. Percutaneous coronary intervention in native coronary arteries vs bypass grafts in patients with prior coronary artery bypass graft surgery: insights from the veterans affairs clinical assessment, reporting, and tracking program. *JACC Cardiovasc Interv* 2016;**9**:884−93.

10. Mashayekhi K, Behnes M, Akin I, Kaiser T, Neuser H. Novel retrograde approach for percutaneous treatment of chronic total occlusions of the right coronary artery using ipsilateral collateral connections: a European centre experience. *EuroIntervention* 2016;**11**:e1231−6.

11. Kahn JK, Hartzler GO. Retrograde coronary angioplasty of isolated arterial segments through saphenous vein bypass grafts. *Cathet Cardiovasc Diagn* 1990;**20**:88−93.

12. Brilakis ES, Banerjee S, Lombardi WL. Retrograde recanalization of native coronary artery chronic occlusions via acutely occluded vein grafts. *Catheter Cardiovasc Interv* 2010;**75**:109−13.

13. McEntegart MB, Badar AA, Ahmad FA, et al. The collateral circulation of coronary chronic total occlusions. *EuroIntervention* 2016;**11**:e1596−603.

14. Nagamatsu W, Tsuchikane E, Oikawa Y, et al. Successful guidewire crossing via collateral channel at retrograde percutaneous coronary intervention for chronic total occlusion: the J-Channel score. *EuroIntervention* 2020;**15**:e1624−32.

15. Huang CC, Lee CK, Meng SW, et al. Collateral channel size and tortuosity predict retrograde percutaneous coronary intervention success for chronic total occlusion. *Circ Cardiovasc Interv* 2018;**11**:e005124.

16. Dash D. Complications encountered in coronary chronic total occlusion intervention: Prevention and bailout. *Indian Heart J* 2016;**68**:737−46.

17. Werner GS, Ferrari M, Heinke S, et al. Angiographic assessment of collateral connections in comparison with invasively determined collateral function in chronic coronary occlusions. *Circulation* 2003;**107**:1972−7.

18. Rathore S, Katoh O, Matsuo H, et al. Retrograde percutaneous recanalization of chronic total occlusion of the coronary arteries: procedural outcomes and predictors of success in contemporary practice. *Circ Cardiovasc Interv* 2009;**2**:124−32.

19. Joyal D, Thompson CA, Grantham JA, Buller CEH, Rinfret S. The retrograde technique for recanalization of chronic total occlusions: a step-by-step approach. *JACC Cardiovasc Interv* 2012;**5**:1−11.

20. Heil M, Schaper W. Influence of mechanical, cellular, and molecular factors on collateral artery growth (arteriogenesis). *Circ Res* 2004;**95**:449−58.

21. Schumacher SP, Everaars H, Stuijfzand WJ, et al. Coronary collaterals and myocardial viability in patients with chronic total occlusions. *EuroIntervention* 2020;**16**:e453−61.

22. Rentrop KP, Cohen M, Blanke H, Phillips RA. Changes in collateral channel filling immediately after controlled coronary artery occlusion by an angioplasty balloon in human subjects. *J Am Coll Cardiol* 1985;**5**:587−92.

23. Alessandrino G, Chevalier B, Lefevre T, et al. A clinical and angiographic scoring system to predict the probability of successful first-attempt percutaneous coronary intervention in patients with total chronic coronary occlusion. *JACC Cardiovasc Interv* 2015;**8**:1540−8.

24. Christopoulos G, Kandzari DE, Yeh RW, et al. Development and validation of a novel scoring system for predicting technical success of chronic total occlusion percutaneous

coronary interventions: the PROGRESS CTO (prospective global registry for the study of chronic total occlusion intervention) score. *JACC Cardiovasc Interv* 2016;**9**:1−9.

25. Galassi AR, Boukhris M, Azzarelli S, Castaing M, Marza F, Tomasello SD. Percutaneous coronary revascularization for chronic total occlusions: a novel predictive score of technical failure using advanced technologies. *JACC Cardiovasc Interv* 2016;**9**:911−22.

26. Ellis SG, Burke MN, Murad MB, et al. Predictors of successful hybrid-approach chronic total coronary artery occlusion stenting: an improved model with novel correlates. *JACC Cardiovasc Interv* 2017;**10**:1089−98.

27. Maeremans J, Spratt JC, Knaapen P, et al. Towards a contemporary, comprehensive scoring system for determining technical outcomes of hybrid percutaneous chronic total occlusion treatment: The RECHARGE score. *Cathete Cardiovasc Interv* 2018;**91**:192−202.

28. Khanna R, Pandey CM, Bedi S, Ashfaq F, Goel PK. A weighted angiographic scoring model (W-CTO score) to predict success of antegrade wire crossing in chronic total occlusion: analysis from a single centre. *AsiaIntervention* 2018;**4**:18−25.

29. Szijgyarto Z, Rampat R, Werner GS, et al. Derivation and validation of a chronic total coronary occlusion intervention procedural success score from the 20,000-patient eurocto registry: the EuroCTO (CASTLE) score. *JACC Cardiovasc Interv* 2019;**12**:335−42.

30. Mohandes M, Moreno C, Fuertes M, et al. New scoring system for predicting percutaneous coronary intervention of chronic total occlusion success: impact of operator's experience. *Cardiol J* 2021.

31. Danek BA, Karatasakis A, Karmpaliotis D, et al. Development and validation of a scoring system for predicting periprocedural complications during percutaneous coronary interventions of chronic total occlusions: the prospective global registry for the study of chronic total occlusion intervention (PROGRESS CTO) complications score. *J Am Heart Assoc* 2016;**5**.

32. Karacsonyi J, Stanberry L, Alaswad K, et al. Predicting technical success of chronic total occlusion percutaneous coronary intervention: comparison of 3 scores. *Circ Cardiovasc Interv* 2021;**14**:e009860.

33. Karatasakis A, Danek BA, Karmpaliotis D, et al. Comparison of various scores for predicting success of chronic total occlusion percutaneous coronary intervention. *Int J Cardiol* 2016;**224**:50−6.

34. Kalogeropoulos AS, Alsanjari O, Keeble TR, et al. CASTLE score versus J-CTO score for the prediction of technical success in chronic total occlusion percutaneous revascularisation. *EuroIntervention* 2020;**15**:e1615−23.

35. Christopoulos G, Wyman RM, Alaswad K, et al. Clinical utility of the Japan-chronic total occlusion score in coronary chronic total occlusion interventions: results from a multicenter registry. *Circ Cardiovasc Interv* 2015;**8**:e002171.

36. Fujino A, Otsuji S, Hasegawa K, et al. Accuracy of J-CTO score derived from computed tomography vs angiography to predict successful percutaneous coronary intervention. *JACC Cardiovasc Imaging* 2018;**11**:209−17.

37. Forouzandeh F, Suh J, Stahl E, et al. Performance of J-CTO and PROGRESS CTO scores in predicting angiographic success and long-term outcomes of percutaneous coronary interventions for chronic total occlusions. *Am J Cardiol* 2018;**121**:14−20.

38. Ebisawa S, Kohsaka S, Muramatsu T, et al. Derivation and validation of the J-CTO extension score for pre-procedural prediction of major adverse cardiac and cerebrovascular events in patients with chronic total occlusions. *PLoS One* 2020;**15**:e0238640.

39. Brilakis ES, Banerjee S, Karmpaliotis D, et al. Procedural outcomes of chronic total occlusion percutaneous coronary intervention: a report from the NCDR (National Cardiovascular Data Registry). *JACC Cardiovasc Interv* 2015;**8**:245−53.

40.   Simsek B, Kostantinis S, Karacsonyi J, et al. Predicting periprocedural complications in chronic total occlusion percutaneous coronary intervention: the PROGRESS-CTO complication scores. *JACC Cardiovasc Interv* 2022;**15**:1413−22.

41.   Hirai T, Grantham JA, Sapontis J, et al. Development and validation of a prediction model for angiographic perforation during chronic total occlusion percutaneous coronary intervention: OPEN-CLEAN perforation score. *Catheter Cardiovasc Interv* 2022;**99**:280−5.

42.   Kostantinis S, Simsek B, Karacsonyi J, et al. Development and validation of a scoring system for predicting clinical coronary artery perforation during percutaneous coronary intervention of chronic total occlusions: the PROGRESS-CTO perforation score. *EuroIntervention* 2022 Oct 24:EIJ-D-22-00593.

43.   Hong SJ, Kim BK, Cho I, et al. Effect of coronary CTA on chronic total occlusion percutaneous coronary intervention: a randomized trial. *JACC Cardiovasc Imaging* 2021;**14**:1993−2004.

44.   Opolski MP, Achenbach S, Schuhback A, et al. Coronary computed tomographic prediction rule for time-efficient guidewire crossing through chronic total occlusion: insights from the CT-RECTOR multicenter registry (computed tomography registry of chronic total occlusion revascularization). *JACC Cardiovasc Interv* 2015;**8**:257−67.

45.   Yu CW, Lee HJ, Suh J, et al. Coronary computed tomography angiography predicts guidewire crossing and success of percutaneous intervention for chronic total occlusion: korean multicenter CTO CT registry score as a tool for assessing difficulty in chronic total occlusion percutaneous coronary intervention. *Circ Cardiovasc Imaging* 2017;**10**.

46.   Magro M, Schultz C, Simsek C, et al. Computed tomography as a tool for percutaneous coronary intervention of chronic total occlusions. *EuroIntervention* 2010;**6**(Suppl. G): G123−31.

47.   Luo C, Huang M, Li J, et al. Predictors of interventional success of antegrade PCI for CTO. *JACC Cardiovasc Imaging* 2015;**8**:804−13.

48.   Ghoshhajra BB, Takx RAP, Stone LL, et al. Real-time fusion of coronary CT angiography with x-ray fluoroscopy during chronic total occlusion PCI. *Eur Radiol* 2017;**27**:2464−73.

49.   Xenogiannis I, Jaffer FA, Shah AR, et al. Computed tomography angiography co-registration with real-time fluoroscopy in percutaneous coronary intervention for chronic total occlusions. *EuroIntervention* 2021;**17**:e433−5.

Chapter 7

Selecting target lesion(s)

Chronic total occlusion (CTO) percutaneous coronary intervention (PCI) is a revascularization tool to be used in selected patients. The following decisions need to be made for every patient who is diagnosed with coronary artery disease (CAD) on coronary angiography.

1. *Is coronary revascularization needed?*

   Coronary revascularization should be done when the anticipated benefits exceed the potential risks.[1,2] Potential benefits are decreasing symptoms and improving prognosis. This is discussed separately for patients with stable angina (Section 7.1) and for patients with acute coronary syndromes (ACS) (Section 7.2).

2. *If yes, should it be done with PCI or coronary artery bypass graft surgery (CABG)?*

   PCI and CABG have advantages and disadvantages: PCI is generally easier to perform and carries lower upfront risk, but is associated with higher need for repeat revascularization compared with CABG. The choice of revascularization modality is discussed in Sections 7.1 and 7.2.

3. *If PCI is selected, which lesions should be treated, in which sequence, and with what techniques?*

   Target lesion selection depends on clinical presentation (e.g., culprit lesions should be treated first in ACS patients), lesion location and lesion complexity. This is discussed in Section 7.3.

   Algorithms for determining the need, modality, and sequence (in case of CTO PCI) of coronary revascularization are discussed in the following sections.

7.1 Stable angina

Most patients who undergo CTO PCI present with stable angina (chronic coronary syndromes). Fig. 7.1 illustrates an algorithm for selecting revascularization strategy in such patients.

7.1.1 Symptoms?

The goal of coronary revascularization is to improve symptoms (help patients feel better) or improve prognosis (help patients live longer or reduce their risk of subsequent unwanted events, such as myocardial infarction).

Manual of Chronic Total Occlusion Percutaneous Coronary Interventions.
DOI: https://doi.org/10.1016/B978-0-323-91787-2.00040-X

99

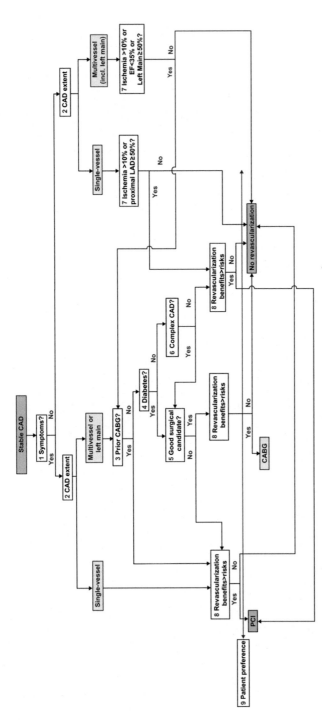

FIGURE 7.1 **Algorithm for deciding about coronary revascularization in stable angina patients.**[3] *Reproduced with permission from the Manual of PCI.*

Symptoms caused by coronary artery disease (CAD) are as follows:

1. Chest pain that is provoked by exertion and relieved by rest. Chest pain at rest is the hallmark of ACS.
2. Dyspnea is a frequent "angina equivalent," especially in patients with CTOs.[4]
3. Easy fatigue with exertion and lack of energy can also be "angina equivalents" in some patients.

Symptoms can only be improved if they are present at baseline! In other words, asymptomatic patients cannot feel better after coronary revascularization. Some patients, however, may deny symptoms because they have gradually limited their activities. Obtaining information from their family and performing an exercise stress test can help determine if a patient is truly asymptomatic or not.

In patients with stable ischemic heart disease[5] (also called chronic coronary syndromes in the European revascularization guidelines[6]), coronary revascularization is more effective in relieving angina than medical therapy,[7] reducing or eliminating the need for antianginal medications and improving quality of life, but it remains controversial whether it can also improve prognosis (i.e., lower the risk for myocardial infarction or death),[7,8] except possibly in patients with high-risk coronary anatomy[6] as described in Sections 7.1.2 and 7.1.7.

Symptom improvement is the key indication for CTO PCI.[2] For patients with medically refractory angina caused by a CTO, successful CTO recanalization can reduce or eliminate angina and the need for antianginal medications, as well as improve exercise capacity.[9–12] Three[13–15] of four randomized-controlled trials of CTO PCI vs. no CTO PCI demonstrated symptom improvement in the CTO PCI arm (the DECISION CTO[16] trial did not find difference in symptoms between the CTO PCI and the no CTO PCI arm, but had multiple limitations). Patients who undergo successful CTO PCI usually require fewer or no antianginal medications, obviating the medication-related costs and side effects. Eliminating nitrate intake can also allow patients to take phosphodiesterase inhibitors (e.g., avanafil, sildenafil, vardenafil, tadalafil) for erectile dysfunction, which is common in CAD patients. Several patients with CTOs also suffer from undiagnosed major depression, and such patients derived the most benefit from successful CTO PCI in one study.[17]

7.1.2 CAD extent

The extent of CAD is a key determinant of whether coronary revascularization is needed and of the optimal type of coronary revascularization (PCI or CABG).

With the exception of highly stenotic coronary lesions (diameter stenosis ≥ 90%) that are almost always hemodynamically significant, coronary physiology (Chapter 12) can more accurately help determine the severity of CAD.[18]

Patients with single vessel disease are in most cases treated with PCI. This is especially true in patients with RCA lesions, including RCA CTOs.

Patients with multivessel (or left main) coronary artery disease can be treated with either PCI or CABG:

- PCI is preferred for prior CABG patients, poor surgical candidates, and patients with less complex CAD.
- CABG is preferred for patients with diabetes, reduced ejection fraction, and patients with complex multivessel CAD, as discussed in more detail below.

7.1.3 Prior CABG

Due to increased risk of death and complications, redo CABG is performed infrequently in patients who have already had CABG.[19] However, redo CABG could be considered in patients with severely diseased or occluded bypass grafts, native vessels that are not amenable to PCI, depressed left ventricular ejection fraction, absence of patent arterial grafts, or need for valve surgery. Internal mammary artery grafts should be used, if feasible, in patients undergoing reoperation.

Nearly all prior CABG patients have coronary CTOs.[20] Even though the success of CTO PCI in prior CABG patients is lower compared with non-prior CABG patients,[21] CTO PCI is preferred to redo CABG in most such patients who require coronary revascularization.

7.1.4 Diabetes mellitus

Patients with diabetes mellitus and multivessel CAD had better outcomes with CABG (including lower mortality) in several trials[22–26] (although most, but not all,[26] such studies included first generation drug-eluting stents [DES]). This is likely due to higher risk of restenosis and disease progression in nonstented coronary segments or vessels in diabetic patients and has been linked to the use of left internal mammary (LIMA) to left anterior descending artery (LAD) grafts. Hence, CABG is generally preferred for diabetic patients with multivessel CAD, who are good surgical candidates.[6]

7.1.5 Good surgical candidate?

The risk of CABG depends on the patients' cardiac status and noncardiac comorbidities.

This is often assessed using the Society of Thoracic Surgeons (STS) score (http://riskcalc.sts.org/stswebriskcalc/) that predicts in-hospital or 30-day mortality and in-hospital morbidity, or the Euroscore II that predicts in-hospital mortality (http://www.euroscore.org/calc.html).

The following factors are assessed:

Cardiac
1. Recent ACS
2. Urgency of the operation
3. Angina severity
4. Left ventricular function
5. Congestive heart failure
6. Pulmonary hypertension
7. Valvular heart disease
8. Active endocarditis
9. Prior cardiac surgery
10. Porcelain aorta
11. Atrial fibrillation
12. Heart block
13. Cardiogenic shock

Noncardiac
1. Age
2. Gender
3. Height and weight
4. Mobility
5. Renal disease
6. Liver disease
7. Chronic lung disease
8. Peripheral arterial disease
9. Cerebrovascular disease
10. Diabetes mellitus
11. Immunocompromised status
12. Prior mediastinal radiation
13. Severe chest deformation or scoliosis
14. Illicit drug use
15. Cancer
16. Hematocrit
17. Platelet count
18. Current medications (P2Y12 inhibitors, glycoprotein IIb/IIIa inhibitors, etc.)
19. Frailty

TABLE 7.1 Revascularization decision making based on Syntax score.

# of diseased coronary vessels	Syntax score	Revascularization modality[6,28]
2		PCI or CABG
3	0–22	PCI or CABG
3	>22	CABG
Left main	0–32	PCI or CABG
Left main	>32	CABG

7.1.6   Complex CAD?

The complexity of CAD is most commonly assessed using the SYNTAX score (http://www.syntaxscore.org),[27] which incorporates the extent of disease and angiographic characteristics, such as lesion location, diameter stenosis, presence of bifurcations or trifurcations, aorto-ostial location, severe tortuosity, lesion length, calcification, thrombus, and vessel size (Table 7.1).[6,28,29]

7.1.7   Factors associated with improved prognosis after coronary revascularization

Coronary revascularization in stable CAD patients, including patients with coronary CTOs, is mainly done to improve symptoms as discussed in Section 7.1.1, but might also improve prognosis in the following patient subgroups[6]:

- Left main disease with stenosis >50%
- Proximal LAD stenosis >50%
- Two- or three-vessel disease with stenosis >50% with impaired LV function (LVEF ≤ 35%)
- Large area of ischemia detected by functional testing (>10% LV) or abnormal invasive FFR
- Single remaining patent coronary artery with stenosis >50%
- Sudden cardiac death due to ventricular fibrillation or ventricular tachycardia

Several observational studies[30–32] and meta-analyses[10,33,34] have reported better long-term survival after successful vs. failed CTO PCI, even among patients with well-developed collateral circulation.[35] Suboptimal CTO recanalization has also been associated with poor long-term outcomes.[36] Patients with a CTO had higher mortality than patients without a CTO in a large registry.[37] Two randomized-controlled trials, the EuroCTO[13] and DECISION-CTO[16], did not demonstrate any difference in the incidence of major adverse cardiac events between the CTO PCI and the no CTO PCI groups, but both studies had multiple limitations (they were both underpowered and 20% of patients randomized to no CTO PCI in DECISION CTO crossed over to CTO PCI).

A potential beneficial effect of CTO recanalization on long-term survival may be related to:

1. *Protection from future ACSs* in vessels supplying collateral perfusion to the ischemic CTO territory. Patients who have a CTO and present with myocardial infarction have higher risk of out-of-hospital cardiac arrest[38] and arrhythmias,[39] and higher early and late mortality[40–46] (Fig. 7.2) likely due to more severe ischemia ("double jeopardy" Fig. 7.3). It remains unproven, however, whether prophylactic CTO PCI could improve outcomes in patients who subsequently develop ACSs.[47]

2. *Improved myocardial contractility.* Several retrospective studies have demonstrated that successful CTO revascularization can improve left ventricular systolic function,[48–57] provided that the CTO-supplied myocardium is viable[53,54] and the vessel remains patent during followup.[51,52] Two randomized-controlled trials (EXPLORE[58] and REVASC[59]) did not demonstrate improvement in ejection fraction post CTO PCI, but the baseline ejection fraction was not significantly impaired in either study. In a meta-analysis of 12 studies, successful CTO PCI was associated with a 2.6% mean increase in left ventricular ejection fraction, but the mean increase was 5.5% in patients with baseline left ventricular ejection fraction <45%.[60]

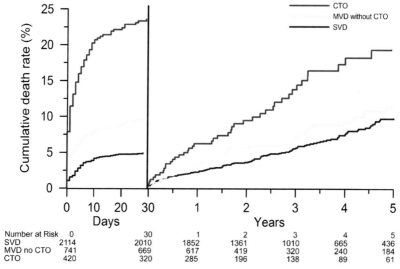

FIGURE 7.2 **Impact of the presence of CTO on outcomes of patients presenting with ST-segment elevation acute myocardial infarction.** *Reproduced with permission from Claessen BE, van der Schaaf RJ, Verouden NJ, et al. Evaluation of the effect of a concurrent chronic total occlusion on long-term mortality and left ventricular function in patients after primary percutaneous coronary intervention. JACC Cardiovasc Interv 2009;2:1128–1134.*

Why CTO impacts MI outcomes

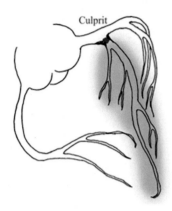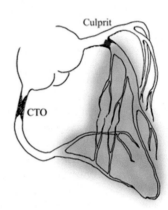

FIGURE 7.3 **How having CTO increases myocardial ischemia and infarct size in MI.** Having an MI culprit lesion in a collateral donor vessel to a CTO increases myocardial ischemia (shaded area) and infarct size. *CTO*, chronic total occlusion; *MI*, myocardial infarction. *Reproduced with permission from Kosugi S, Shinouchi K, Ueda Y, et al. Clinical and angiographic features of patients with out-of-hospital cardiac arrest and acute myocardial infarction.* J Am Coll Cardiol 2020;76:1934−1943.

Myocardial viability can be assessed using several techniques, such as cardiac magnetic resonance imaging (MRI) and positron emission tomography - computed tomography (PET-CT). If the affected myocardial segment is hypokinetic but not akinetic and if there are no Q-waves in the corresponding region of the electrocardiogram,[61] then viability is highly likely. Lack of good collateral circulation does not necessarily mean that there is no viability in the affected myocardium.[62]

3. *Reduction in the risk of ischemic arrhythmias.* Ischemia may predispose to ventricular arrhythmias. Among 162 patients with ischemic cardiomyopathy who received an implantable cardioverter defibrillator in the VACTO study, 44% had at least one CTO.[63] During a median followup of 26 months, the presence of CTO was associated with higher rates of ventricular arrhythmias and death ($P < .01$),[63] although a subsequent study failed to replicate these findings.[64] Infarct-related artery CTO was independently associated with ventricular tachycardia recurrence after successful catheter ablation.[65] Patients with ischemia-induced arrhythmias could benefit from CTO recanalization.[66,67]

4. *Reduction in ischemia.* Reducing the extent and severity of ischemia might improve subsequent clinical outcomes.

As discussed in Chapter 12, nearly all CTO lesions have fractional flow reserve (FFR) <0.80 when assessed after guidewire crossing.[68,69] In a study of 301 patients who underwent myocardial perfusion imaging before and after CTO PCI, a baseline ischemic burden of $>12.5\%$ was optimal in identifying patients most likely to have a significant decrease in ischemic burden post-CTO PCI.[70] Moreover, the presence of a CTO was the strongest independent predictor of incomplete revascularization in the PCI arm of patients treated for multivessel CAD in the SYNTAX trial. Irrespective of surgical or percutaneous revascularization strategy, incomplete revascularization and consequent ischemic burden was associated with significantly higher 4-year clinical event rates including mortality.[71] The success of CTO PCI was relatively low in this cohort compared with current standards.

7.1.8 Benefits versus risks (Figs. 7.4 and 7.5)

Before proceeding with PCI the risks and benefits need to be assessed to ensure that benefits outweigh the risks. Estimation of the risks and benefits can be based on the following four areas:

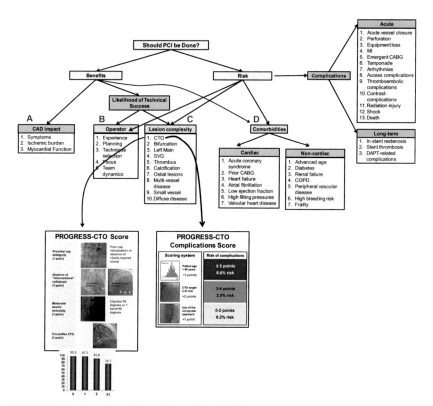

FIGURE 7.4 **Assessing risks and benefits of PCI.** *Reproduced with permission from the Manual of Percutaneous Coronary Interventions.*

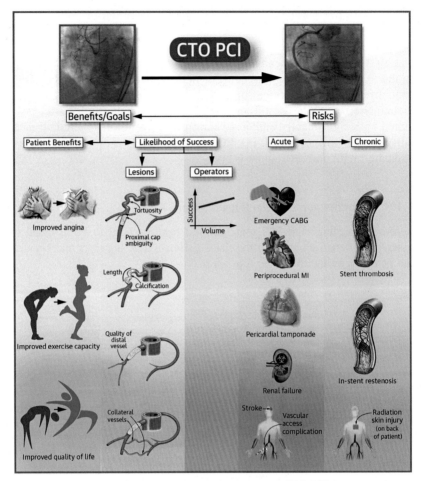

FIGURE 7.5 **Overview of the potential risks and benefits of CTO PCI**. Parameters that can help determine the risks and benefits of chronic total occlusion percutaneous coronary intervention. *CABG*, coronary artery bypass grafting; *CTO*, chronic total occlusion; *MI*, myocardial infarction; *PCI*, percutaneous coronary intervention. *Reproduced with permission from Tajti P, Burke MN, Karmpaliotis D, et al. Update in the percutaneous management of coronary chronic total occlusions.* JACC Cardiovasc Interv 2018;11:615−625.

1. Impact of CAD

   The more severe the impact of CAD on the patient's clinical condition (e.g., severe angina even with minimal exertion), the greater the potential benefit of PCI.

2. Operator competence

   Greater operator experience is associated with higher success and lower complication rates.[2,44] This is especially important in CTO PCI, given large discrepancy between success rates achieved at experienced centers (85%−90%)[72−75] vs. less experienced centers (approximately 60%)[44,76] (Table 7.2).

Experienced operators may also be more adept in managing complications should they occur.[2] Moreover, the operator and staff condition (e.g., well rested and not sleep deprived, etc.) can affect the outcome of the procedure.[81−83]

3. Lesion complexity

More complex lesions (such as CTOs, heavily calcified lesions, bifurcations, etc.) can be more challenging to recanalize and carry increased risk.

Several angiographic and coronary computed tomography angiography scores have been developed to assess CTO lesion complexity. The most commonly used scores are the J-CTO,[84] PROGRESS-CTO,[85] and CASTLE[86] scores for predicting success and ease of crossing and the PROGRESS-Complications scores[87,88] for assessing the risk of complications. These scores are discussed in Section 6.3.

4. Comorbidities

More comorbidities may increase the risk of the procedure and decrease the benefits (e.g., coronary revascularization should not be performed with the goal to improve prognosis in patients with nocardiac terminal disease, such as cancer).

7.1.9 Patient preference

The final decision should always be the patient's. The interventionalist's and the heart team's role is to educate the patients and help them in the decision making process.

Fig. 7.6 presents an algorithm on the role of CTO PCI for coronary revascularization.

The 2018 European Society of Cardiology/European Association for Cardio-Thoracic Surgery myocardial revascularization guidelines give a class IIa, level of evidence B recommendation to CTO PCI: "Percutaneous revascularization of CTOs should be considered in patients with angina resistant to medical therapy or with a large area of documented ischemia in the territory of the occluded vessel."[6]

The 2021 American College of Cardiology/American Heart Association/ Society of Cardiovascular Angiography and Interventions guideline for coronary artery revascularization give a class IIb, level of evidence B recommendation to CTO PCI: "In patients with suitable anatomy who have refractory angina on medical therapy, after treatment of non-CTO lesions, the benefit of PCI of a CTO to improve symptoms is uncertain,"[5] citing the conflicting results of the EuroCTO[13] and DECISION-CTO[16] trials and the lack of ventricular function improvement in the EXPLORE[58] and REVASC[59] trials.

TABLE 7.2 Procedural outcomes of CTO PCI in all-comer registries and at experienced centers.

Study	n	Procedural success	Technique use overall	Technique used as final successful strategy	MACE	Death
All comer registries						
NCDR[44]	22,365	59%			1.6%	0.4%
British Cardiovascular Society[77]	28,050	67%			0.73%	0.2%
BMC2[76]	7,389	1st tertile*: 44.9% 2nd tertile: 45.5% 3rd tertile: 64.5%			1st tertile: 4% 2nd tertile: 2.9% 3rd tertile: 3.2%	1st tertile: 1.8% 2nd tertile: 0.8% 3rd tertile: 1.7%
Experienced centers						
OPEN CTO[78]	1,000	90%		AWE 40.8% ADR 24.3% Retrograde dissection and re-entry 24.6% RWE 10.3%	7%	0.9%
PROGRESS-CTO[73]	10,019	86%	AWE 86% ADR 21% Retrograde 32%	AWE 55% ADR 12% Retrograde 19%	2.1%	0.5%
EURO CTO[74]	4,314	88%		AWE 76.9% ADR 3.6% Retrograde 19.5%	0.5%	0.1%

RECHARGE[79]	1,253	89%	ADR 23% retrograde 34%	AWE 58% ADR 18% Retrograde 24%	2.6%	0.2%
Japanese[75]	3,229	88%	Primary antegrade approach: HC: 78.4% LC: 76.8% Primary retrograde approach: HC: 21.6% LC: 23.2%	Primary antegrade approach: HC: 65% LC: 60% Primary retrograde approach: HC: 16%, LC: 16%	0.53%	0.2%
LATAM[80]	1,040	81%		AWE 81% ADR 8% Retrograde 11%	3%	1%

ADR, antegrade dissection and re-entry; AWE, antegrade wire escalation; HC, high volume center; LC, low volume center; RWE, retrograde wire escalation. BMC2, Blue Cross Blue Shield of Michigan Cardiovascular Consortium; EURO CTO, Evaluate the Utilization of Revascularization or Optimal Medical Therapy for the Treatment of Chronic Total Coronary Occlusions; NCDR, National Cardiovascular Data Registry; OPEN-CTO, Outcomes, Patient Health Status, and Efficiency in Chronic Total Occlusion progress CTO; RECHARGE, REgistry of Crossboss and Hybrid procedures in FrAnce, the NetheRlands, BelGium, and UnitEd Kingdom *Michigan registry divided procedural success based on operator experience into tertiles, with 3rd tertile being the most experienced operators.

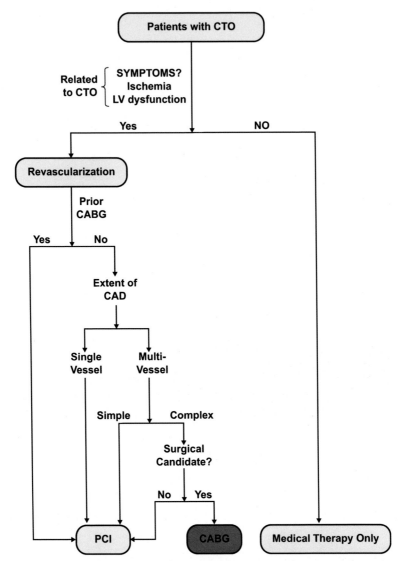

FIGURE 7.6 **Revascularization options for patients with coronary chronic total occlusions.**
Algorithm for determining the need for coronary revascularization in patients with coronary
chronic total occlusions. Revascularization is indicated in patients with symptoms, significant
ischemia and left ventricular dysfunction attributable to the CTO(s). Patients with prior coronary
bypass graft surgery (CABG) are almost always treated with PCI given the increased risk of redo
CABG. In patients without prior CABG, CTO percutaneous coronary intervention (PCI) and coronary artery bypass graft surgery are both treatment options, with coronary bypass graft surgery
(CABG) preferred for patients with complex multivessel disease and PCI (including CTO PCI)
preferred for patients with simple multivessel or single vessel disease or patients who are poor candidates for CABG.[89] *Modified with permission from Azzalini L, Torregrossa G, Puskas JD, et al.
Percutaneous revascularization of chronic total occlusions: rationale, indications, techniques,
and the cardiac surgeon's point of view.* Int J Cardiol 2017;231:90−96.

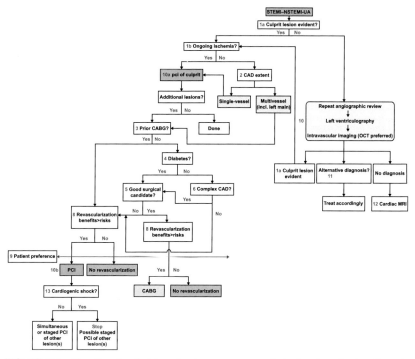

FIGURE 7.7 Algorithm for deciding about coronary revascularization in patients with acute coronary syndromes. *Reproduced with permission from the Manual of PCI.*

7.2 Acute coronary syndromes (Fig. 7.7)

7.2.1 Culprit lesion evident?

Treatment of the culprit lesion takes priority in ACS patients. A culprit lesion is the stenosis that is causing the patient's symptoms and clinical syndrome.

Sometimes the culprit lesion is easy to identify (vessel occlusion with evident thrombus), but sometimes there may be no evident culprit lesion (see Section 7.2.10) and at other times there may be multiple possible culprit lesions. The following criteria can help determine the culprit lesion(s):

1. Artery supplies myocardium that corresponds to area of ST-elevation
2. Artery supplies myocardium that is hypokinetic by left ventriculography or echocardiography
3. Thrombus
4. Contrast staining
5. Poor or no collaterals
6. Ease of wiring

Differentiating acute occlusions from CTOs can sometimes be challenging. Significant difficulty crossing the occlusion may sometimes (but not always) suggest that the occlusion is chronic.

7.2.2 Active ischemia?

The following signs/symptoms may suggest active ischemia:

1. Ongoing symptoms (such as ongoing chest pain).
2. Persistent ST-segment elevation (unless it is considered to be due to an aneurysm, pericarditis, or other cause)
3. Refractory arrhythmias
4. Refractory heart failure
5. Cardiogenic shock

The classic example of active ischemia is ST-segment elevation myocardial infarction, when "time is muscle." Another example is patients with non-ST segment elevation ACS and ongoing chest pain. Immediate PCI of the culprit lesion(s) is indicated in patients with active ischemia.

7.2.3 CAD extent

Same concepts apply as discussed in Section 7.1.2.

7.2.4 Prior CABG

Same concepts apply as discussed in Section 7.1.3.

7.2.5 Diabetes mellitus

Same concepts apply as discussed in Section 7.1.4.

7.2.6 Good surgical candidate?

Same concepts apply as discussed in Section 7.1.5. Patients with ACS, especially ST-segment elevation acute myocardial infarction (STEMI) are often preloaded with oral P2Y12 inhibitors that increase the risk of bleeding if emergency CABG is performed.

7.2.7 Complex CAD?

Same concepts apply as discussed in Section 7.1.6.

7.2.8 CAD extent

Same concepts apply as discussed in Section 7.1.7.

7.2.9   Revascularization benefits and risks

Same concepts apply as discussed in Section 7.1.8. However, the urgency for performing PCI is often higher in ACS patients.

7.2.10   Patient preference

Same concepts apply as discussed in Section 7.1.9. However, the urgency for making a decision is higher in ACS patients, given the potential adverse consequences of delayed revascularization.

7.2.11   Assessment for culprit lesion(s)

When there is no immediate culprit lesion (or lesions) on initial angiographic assessment, the following steps are recommended:

1. Repeat careful angiographic review: sometimes the angiographic findings, such as "staining" or thrombus can be very subtle.
2. Left ventriculography can be diagnostic of related conditions (such as takotsubo cardiomyopathy) or mechanical complications of myocardial infarction (such as papillary rupture or free wall rupture). Left ventriculography can also demonstrate areas of myocardial hypokinesis that can help determine the culprit lesion(s). Alternatively, a transthoracic echocardiogram can be performed in the cardiac catheterization laboratory.
3. Intravascular imaging. Due to its high spatial resolution, optical coherence tomography (OCT) is the intravascular imaging modality of choice for determining the presence (or confirming the absence) of culprit lesion(s). Hallmark OCT findings of culprit lesions are thrombus and plaque rupture.

7.2.12   Alternative diagnoses

Alternative diagnoses of myocardial infarction include pericarditis, myocarditis, and stress cardiomyopathy (takotsubo).

7.2.13   Cardiac MRI

Cardiac MRI can clarify the area of myocardial infarction or suggest alternative diagnoses, such as myocarditis or stress cardiomyopathy (takotsubo).

7.2.14   Cardiogenic shock

Based on the results of the CULPRIT-SHOCK trial,[90,91] PCI should only be performed in the culprit lesion(s) in ACS patients presenting with cardiogenic shock. *CTO PCI should not be performed in such patients during the index procedure.*[92]

7.3   Percutaneous coronary intervention timing

7.3.1   Ad hoc versus delayed percutaneous coronary intervention

PCI can be performed "ad hoc" (i.e., immediately following diagnostic coronary angiography) or at a later time. The advantage of ad hoc PCI is avoidance of a second procedure with any concomitant risks. Disadvantages of ad hoc PCI include higher contrast and radiation dose, less planning, and less time devoted to a detailed discussion about the risks and benefits of the procedure.

In most cases, CTO PCI should *not* be performed ad hoc in order to[93]:

1. Allow time for thorough procedural planning and preparation for both the operator and the cardiac catheterization laboratory staff, which greatly enhances the likelihood of success and decreases the procedural risks.
2. Minimize contrast volume and radiation dose.
3. Minimize patient and operator fatigue.
4. Allow time to collect sufficient information on the viability and/or the extent of ischemia of the territory supplied by the occluded vessel.
5. Allow for a detailed discussion with the patient and his or her family about the indications, goals, risks, and alternatives (such as medical therapy and coronary artery bypass graft surgery) to the procedure. Risks that may be higher in CTO PCI compared with non-CTO PCI include radiation injury and perforation.

In some cases ad hoc PCI may be needed, such as in patients who present with an ACS due to failure of a highly diseased saphenous vein graft that cannot be recanalized.[94] In most such cases, however, PCI of the SVG first, followed by staged PCI of the native coronary artery CTO is preferred, if technically feasible.[95]

7.3.2   Immediate versus staged multivessel/multilesion revascularization

In patients with STEMI and multivessel coronary artery disease, complete revascularization (either immediate or staged) improves hard outcomes (death or myocardial infarction).[96,97] In patients with STEMI and multivessel CAD it remains controversial whether complete revascularization should be performed at the time of the index PCI or later. Complete revascularization during the index procedure or during the index hospitalization has shown better outcomes compared with culprit-only revascularization in STEMI patients.[98] Better outcomes with complete revascularization have also been demonstrated in observational studies of stable coronary artery disease patients.[99]

Factors favoring staged PCI in patients with multivessel CAD include:

1. High complexity of the remaining lesion(s). In general, coronary CTOs should be treated as a staged procedure, given high lesion complexity and risk of complications.
2. No active ischemia caused by the remaining lesion(s) in ACS patients.
3. Excessive contrast or radiation during the first part of PCI or anticipated during PCI.
4. Complication.
5. Operator fatigue.
6. Patient fatigue.

7.4 Percutaneous coronary intervention lesion sequence selection

In ACS patients with multivessel CAD, the culprit lesion(s) should be treated first, especially if they are causing active ischemia (e.g., the STEMI culprit lesion).

In stable patients with multiple lesions in the same coronary artery, the more distal lesions are generally treated first, unless there is severe pressure dampening or ischemia caused by more proximal or ostial lesion or inability to deliver devices through a tight proximal lesion. In general, all significant lesions within the same target vessel should be treated during the index procedure.

In stable patients with multivessel disease, parameters to consider for choosing the sequence of PCI include:

1. Lesion complexity: for example, treating the less complex lesions first may facilitate subsequent treatment of the more complex lesions.
2. Contrast and radiation dose: in patients who have received high radiation or contrast dose, treating the easier lesion(s) first and deferring the more complex lesions may be preferable.
3. Alternative revascularization options: for example, in a patient with LAD CTO and nonocclusive lesions in the RCA and circumflex, some operators may choose to attempt LAD CTO PCI first and refer the patient for CABG if the attempt is unsuccessful. However, other operators may choose to treat the nonocclusive RCA and circumflex lesions first to increase the safety of the LAD CTO PCI attempt, since the success of CTO PCI is 85%−90% at experienced centers.[1]

References

1. Tajti P, Burke MN, Karmpaliotis D, et al. Update in the percutaneous management of coronary chronic total occlusions. *JACC Cardiovasc Interv* 2018;**11**:615−25.
2. Brilakis ES, Mashayekhi K, Tsuchikane E, et al. Guiding principles for chronic total occlusion percutaneous coronary intervention. *Circulation* 2019;**140**:420−33.

3. Rathore S, Khanra D, Galassi AR, et al. Procedural characteristics and outcomes following chronic total occlusion coronary intervention: pooled analysis from 5 registries. *Expert Rev Cardiovasc Ther* 2021;1−10.
4. Safley DM, Grantham J, Jones PG, Spertus J. Heatlh status benefits of angioplasty for chronic total occlusions—an analysis from the OPS/PRISM studies. *J Am Coll Cardiol* 2012;**59** E101-E.
5. Lawton JS, Tamis-Holland JE, Bangalore S, et al. 2021 ACC/AHA/SCAI guideline for coronary artery revascularization: a report of the american college of cardiology/american heart association joint committee on clinical practice guidelines. *Circulation* 2022;**145**: e18−e114.
6. Neumann FJ, Sousa-Uva M, Ahlsson A, et al. 2018 ESC/EACTS guidelines on myocardial revascularization. *Eur Heart J* 2019;**40**:87−165.
7. Al-Lamee R, Thompson D, Dehbi HM, et al. Percutaneous coronary intervention in stable angina (ORBITA): a double-blind, randomised controlled trial. *Lancet* 2018;**391**:31−40.
8. Boden WE, O'Rourke RA, Teo KK, et al. Optimal medical therapy with or without PCI for stable coronary disease. *N Engl J Med* 2007;**356**:1503−16.
9. Olivari Z, Rubartelli P, Piscione F, et al. Immediate results and one-year clinical outcome after percutaneous coronary interventions in chronic total occlusions: data from a multicenter, prospective, observational study (TOAST-GISE). *J Am Coll Cardiol* 2003;**41**:1672−8.
10. Christakopoulos GE, Christopoulos G, Carlino M, et al. *Meta*-analysis of clinical outcomes of patients who underwent percutaneous coronary interventions for chronic total occlusions. *Am J Cardiol* 2015;**115**:1367−75.
11. Joyal D, Afilalo J, Rinfret S. Effectiveness of recanalization of chronic total occlusions: a systematic review and *meta*-analysis. *Am Heart J* 2010;**160**:179−87.
12. Rossello X, Pujadas S, Serra A, et al. Assessment of inducible myocardial ischemia, quality of life, and functional status after successful percutaneous revascularization in patients with chronic total coronary occlusion. *Am J Cardiol* 2016;**117**:720−6.
13. Werner GS, Martin-Yuste V, Hildick-Smith D, et al. A randomized multicentre trial to compare revascularization with optimal medical therapy for the treatment of chronic total coronary occlusions. *Eur Heart J* 2018;**39**:2484−93.
14. Obedinskiy AA, Kretov EI, Boukhris M, et al. The IMPACTOR-CTO trial. *JACC Cardiovasc Interv* 2018;**11**:1309−11.
15. Juricic SA, Tesic MB, Galassi AR, et al. Randomized controlled comparison of optimal medical therapy with percutaneous recanalization of chronic total occlusion (COMET-CTO). *Int Heart J* 2021;**62**:16−22.
16. Lee SW, Lee PH, Ahn JM, et al. Randomized trial evaluating percutaneous coronary intervention for the treatment of chronic total occlusion. *Circulation* 2019;**139**:1674−83.
17. Bruckel JT, Jaffer FA, O'Brien C, Stone L, Pomerantsev E, Yeh RW. Angina severity, depression, and response to percutaneous revascularization in patients with chronic total occlusion of coronary arteries. *J Invasive Cardiol* 2016;**28**:44−51.
18. Tonino PAL, Fearon WF, De Bruyne B, et al. Angiographic vs functional severity of coronary artery stenoses in the FAME study: fractional flow reserve vs angiography in multivessel evaluation. *J Am Coll Cardiol* 2010;**55**:2816−21.
19. Morrison DA, Sethi G, Sacks J, et al. Percutaneous coronary intervention vs repeat bypass surgery for patients with medically refractory myocardial ischemia: AWESOME randomized trial and registry experience with post-CABG patients. *J Am Coll Cardiol* 2002;**40**:1951−4.

20. Jeroudi OM, Alomar ME, Michael TT, et al. Prevalence and management of coronary chronic total occlusions in a tertiary veterans affairs hospital. *Catheter Cardiovasc Interv* 2014;**84**:637−43.

21. Tajti P, Karmpaliotis D, Alaswad K, et al. In-hospital outcomes of chronic total occlusion percutaneous coronary interventions in patients with prior coronary artery bypass graft surgery. *Circ Cardiovasc Interv* 2019;**12**:e007338.

22. Farkouh ME, Domanski M, Sleeper LA, et al. Strategies for multivessel revascularization in patients with diabetes. *N Engl J Med* 2012;**367**:2375−84.

23. Farkouh ME, Domanski M, Dangas GD, et al. Long-term survival following multivessel revascularization in patients with diabetes: the FREEDOM follow-on study. *J Am Coll Cardiol* 2019;**73**:629−38.

24. BARI investigators. The final 10-year follow-up results from the BARI randomized trial. *J Am Coll Cardiol* 2007;**49**:1600−6.

25. Kamalesh M, Sharp TG, Tang XC, et al. Percutaneous coronary intervention vs coronary bypass surgery in United States veterans with diabetes. *J Am Coll Cardiol* 2013;**61**:808−16.

26. Park SJ, Ahn JM, Kim YH, et al. Trial of everolimus-eluting stents or bypass surgery for coronary disease. *N Engl J Med* 2015;**372**:1204−12.

27. Sianos G, Morel MA, Kappetein AP, et al. The SYNTAX score: an angiographic tool grading the complexity of coronary artery disease. *EuroIntervention* 2005;**1**:219−27.

28. Mohr FW, Morice MC, Kappetein AP, et al. Coronary artery bypass graft surgery vs percutaneous coronary intervention in patients with three-vessel disease and left main coronary disease: 5-year follow-up of the randomised, clinical SYNTAX trial. *Lancet* 2013;**381**:629−38.

29. Lin S, Guan C, Wu F, et al. Coronary artery bypass grafting and percutaneous coronary intervention in patients with chronic total occlusion and multivessel disease. *Circ Cardiovasc Interv* 2022;**15**:e011312.

30. Mehran R, Claessen BE, Godino C, et al. Long-term outcome of percutaneous coronary intervention for chronic total occlusions. *JACC Cardiovasc Interv* 2011;**4**:952−61.

31. Jones DA, Weerackody R, Rathod K, et al. Successful recanalization of chronic total occlusions is associated with improved long-term survival. *JACC Cardiovasc Interv* 2012;**5**:380−8.

32. George S, Cockburn J, Clayton TC, et al. Long-term follow-up of elective chronic total coronary occlusion angioplasty: analysis from the U.K. Central Cardiac Audit Database. *J Am Coll Cardiol* 2014;**64**:235−43.

33. Khan MF, Wendel CS, Thai HM, Movahed MR. Effects of percutaneous revascularization of chronic total occlusions on clinical outcomes: a *meta*-analysis comparing successful vs failed percutaneous intervention for chronic total occlusion. *Catheter Cardiovasc Interv* 2013;**82**:95−107.

34. Hoebers LP, Claessen BE, Elias J, Dangas GD, Mehran R, Henriques JP. *Meta*-analysis on the impact of percutaneous coronary intervention of chronic total occlusions on left ventricular function and clinical outcome. *Int J Cardiol* 2015;**187**:90−6.

35. Jang WJ, Yang JH, Choi SH, et al. Long-term survival benefit of revascularization compared with medical therapy in patients with coronary chronic total occlusion and well-developed collateral circulation. *JACC Cardiovasc Interv* 2015;**8**:271−9.

36. Guan C, Yang W, Song L, et al. Association of acute procedural results with long-term outcomes after CTO PCI. *JACC Cardiovasc Interv* 2021;**14**:278−88.

37. Ramunddal T, Hoebers LP, Henriques JP, et al. Prognostic impact of chronic total occlusions: a report from SCAAR (Swedish Coronary Angiography and Angioplasty Registry). *JACC Cardiovasc Interv* 2016;**9**:1535−44.

38. Kosugi S, Shinouchi K, Ueda Y, et al. Clinical and angiographic features of patients with out-of-hospital cardiac arrest and acute myocardial infarction. *J Am Coll Cardiol* 2020;**76**:1934−43.

39. Saad M, Fuernau G, Desch S, et al. Prognostic impact of non-culprit chronic total occlusions in infarct-related cardiogenic shock: results of the randomised IABP-SHOCK II trial. *EuroIntervention* 2018;**14**:e306−13.

40. Claessen BE, Dangas GD, Weisz G, et al. Prognostic impact of a chronic total occlusion in a non-infarct-related artery in patients with ST-segment elevation myocardial infarction: 3-year results from the HORIZONS-AMI trial. *Eur Heart J* 2012;**33**:768−75.

41. Claessen BE, van der Schaaf RJ, Verouden NJ, et al. Evaluation of the effect of a concurrent chronic total occlusion on long-term mortality and left ventricular function in patients after primary percutaneous coronary intervention. *JACC Cardiovasc Interv* 2009;**2**:1128−34.

42. Hoebers LP, Vis MM, Claessen BE, et al. The impact of multivessel disease with and without a co-existing chronic total occlusion on short- and long-term mortality in ST-elevation myocardial infarction patients with and without cardiogenic shock. *Eur J Heart Fail* 2013;**15**:425−32.

43. Lexis CP, van der Horst IC, Rahel BM, et al. Impact of chronic total occlusions on markers of reperfusion, infarct size, and long-term mortality: a substudy from the TAPAS-trial. *Catheter Cardiovasc Interv* 2011;**77**:484−91.

44. Brilakis ES, Banerjee S, Karmpaliotis D, et al. Procedural outcomes of chronic total occlusion percutaneous coronary intervention: a report from the NCDR (National Cardiovascular Data Registry). *JACC Cardiovasc Interv* 2015;**8**:245−53.

45. O'Connor SA, Garot P, Sanguineti F, et al. *Meta*-analysis of the impact on mortality of noninfarct-related artery coronary chronic total occlusion in patients presenting with ST-segment elevation myocardial infarction. *Am J Cardiol* 2015;**116**:8−14.

46. Shinouchi K, Ueda Y, Kato T, et al. Relation of chronic total occlusion to in-hospital mortality in the patients with sudden cardiac arrest due to acute coronary syndrome. *Am J Cardiol* 2019;**123**:1915−20.

47. Yang ZK, Zhang RY, Hu J, Zhang Q, Ding FH, Shen WF. Impact of successful staged revascularization of a chronic total occlusion in the non-infarct-related artery on long-term outcome in patients with acute ST-segment elevation myocardial infarction. *Int J Cardiol* 2013;**165**:76−9.

48. Melchior JP, Doriot PA, Chatelain P, et al. Improvement of left ventricular contraction and relaxation synchronism after recanalization of chronic total coronary occlusion by angioplasty. *J Am Coll Cardiol* 1987;**9**:763−8.

49. Danchin N, Angioi M, Cador R, et al. Effect of late percutaneous angioplastic recanalization of total coronary artery occlusion on left ventricular remodeling, ejection fraction, and regional wall motion. *Am J Cardiol* 1996;**78**:729−35.

50. Van Belle E, Blouard P, McFadden EP, Lablanche JM, Bauters C, Bertrand ME. Effects of stenting of recent or chronic coronary occlusions on late vessel patency and left ventricular function. *Am J Cardiol* 1997;**80**:1150−4.

51. Sirnes PA, Myreng Y, Molstad P, Bonarjee V, Golf S. Improvement in left ventricular ejection fraction and wall motion after successful recanalization of chronic coronary occlusions. *Eur Heart J* 1998;**19**:273−81.

52. Piscione F, Galasso G, De Luca G, et al. Late reopening of an occluded infarct related artery improves left ventricular function and long term clinical outcome. *Heart* 2005;**91**:646–51.

53. Baks T, van Geuns RJ, Duncker DJ, et al. Prediction of left ventricular function after drug-eluting stent implantation for chronic total coronary occlusions. *J Am Coll Cardiol* 2006;**47**:721–5.

54. Kirschbaum SW, Baks T, van den Ent M, et al. Evaluation of left ventricular function three years after percutaneous recanalization of chronic total coronary occlusions. *Am J Cardiol* 2008;**101**:179–85.

55. Cheng AS, Selvanayagam JB, Jerosch-Herold M, et al. Percutaneous treatment of chronic total coronary occlusions improves regional hyperemic myocardial blood flow and contractility: insights from quantitative cardiovascular magnetic resonance imaging. *JACC Cardiovasc Interv* 2008;**1**:44–53.

56. Werner GS, Surber R, Kuethe F, et al. Collaterals and the recovery of left ventricular function after recanalization of a chronic total coronary occlusion. *Am Heart J* 2005;**149**:129–37.

57. Cardona M, Martin V, Prat-Gonzalez S, et al. Benefits of chronic total coronary occlusion percutaneous intervention in patients with heart failure and reduced ejection fraction: insights from a cardiovascular magnetic resonance study. *J Cardiovasc Magn Reson* 2016;**18**:78.

58. Henriques JP, Hoebers LP, Ramunddal T, et al. Percutaneous intervention for concurrent chronic total occlusions in patients with STEMI: the EXPLORE trial. *J Am Coll Cardiol* 2016;**68**:1622–32.

59. Mashayekhi K, Nuhrenberg TG, Toma A, et al. A randomized trial to assess regional left ventricular function after stent implantation in chronic total occlusion: the REVASC trial. *JACC Cardiovasc Interv* 2018;**11**:1982–91.

60. Megaly M, Brilakis ES, Abdelsalam M, et al. Impact of chronic total occlusion revascularization on left ventricular function assessed by cardiac magnetic resonance. *JACC Cardiovasc Imaging* 2021;**14**:1076–8.

61. Surber R, Schwarz G, Figulla HR, Werner GS. Resting 12-lead electrocardiogram as a reliable predictor of functional recovery after recanalization of chronic total coronary occlusions. *Clin Cardiol* 2005;**28**:293–7.

62. Schumacher SP, Everaars H, Stuijfzand WJ, et al. Coronary collaterals and myocardial viability in patients with chronic total occlusions. *EuroIntervention* 2020;**16**:e453–61.

63. Nombela-Franco L, Mitroi CD, Fernandez-Lozano I, et al. Ventricular arrhythmias among implantable cardioverter-defibrillator recipients for primary prevention: impact of chronic total coronary occlusion (VACTO primary study). *Circ Arrhythm Electrophysiol* 2012;**5**:147–54.

64. Raja V, Wiegn P, Obel O, et al. Impact of chronic total occlusions and coronary revascularization on all-cause mortality and the incidence of ventricular arrhythmias in patients with ischemic cardiomyopathy. *Am J Cardiol* 2015;**116**:1358–62.

65. Di Marco A, Paglino G, Oloriz T, et al. Impact of a chronic total occlusion in an infarct-related artery on the long-term outcome of ventricular tachycardia ablation. *J Cardiovasc Electrophysiol* 2015;**26**:532–9.

66. Mixon TA. Ventricular tachycardic storm with a chronic total coronary artery occlusion treated with percutaneous coronary intervention. *Proc (Bayl Univ Med Cent)* 2015;**28**:196–9.

67. Myat A, Patel M, Silberbauer J, Hildick-Smith D. Impact of chronic total coronary occlusion revascularisation on infarct-related myocardial scars responsible for recurrent ventricular tachycardia. *EuroIntervention* 2021;**16**:1204−6.

68. Werner GS, Surber R, Ferrari M, Fritzenwanger M, Figulla HR. The functional reserve of collaterals supplying long-term chronic total coronary occlusions in patients without prior myocardial infarction. *Eur Heart J* 2006;**27**:2406−12.

69. Sachdeva R, Agrawal M, Flynn SE, Werner GS, Uretsky BF. The myocardium supplied by a chronic total occlusion is a persistently ischemic zone. *Catheter Cardiovasc Interv* 2014;**83**:9−16.

70. Safley DM, Koshy S, Grantham JA, et al. Changes in myocardial ischemic burden following percutaneous coronary intervention of chronic total occlusions. *Catheter Cardiovasc Interv* 2011;**78**:337−43.

71. Farooq V, Serruys PW, Garcia-Garcia HM, et al. The negative impact of incomplete angiographic revascularization on clinical outcomes and its association with total occlusions: the SYNTAX (Synergy Between Percutaneous Coronary Intervention with Taxus and Cardiac Surgery) trial. *J Am Coll Cardiol* 2013;**61**:282−94.

72. Tajti P, Karmpaliotis D, Alaswad K, et al. The hybrid approach to chronic total occlusion percutaneous coronary intervention: update from the PROGRESS CTO registry. *JACC Cardiovasc Interv* 2018;**11**:1325−35.

73. Kostantinis S, Simsek B, Karacsonyi J et al. In-hospital outcomes and temporal trends of percutaneous coronary interventions for chronic total occlusion. EuroIntervention 2022.

74. Konstantinidis NV, Werner GS, Deftereos S, et al. Temporal trends in chronic total occlusion interventions in Europe. *Circ Cardiovasc Interv* 2018;**11**:e006229.

75. Habara M, Tsuchikane E, Muramatsu T, et al. Comparison of percutaneous coronary intervention for chronic total occlusion outcome according to operator experience from the Japanese retrograde summit registry. *Catheter Cardiovasc Interv* 2016;**87**:1027−35.

76. Othman H, Seth M, Zein R, et al. Percutaneous coronary intervention for chronic total occlusion-the Michigan experience: insights from the BMC2 registry. *JACC Cardiovasc Interv* 2020;**13**:1357−68.

77. Kinnaird T, Gallagher S, Cockburn J, et al. Procedural success and outcomes with increasing use of enabling strategies for chronic total occlusion intervention. *Circ Cardiovasc Interv* 2018;**11**:e006436.

78. Sapontis J, Salisbury AC, Yeh RW, et al. Early procedural and health status outcomes after chronic total occlusion angioplasty: a report from the OPEN-CTO registry (outcomes, patient health status, and efficiency in chronic total occlusion hybrid procedures). *JACC Cardiovasc Interv* 2017;**10**:1523−34.

79. Maeremans J, Walsh S, Knaapen P, et al. The hybrid algorithm for treating chronic total occlusions in europe: the RECHARGE registry. *J Am Coll Cardiol* 2016;**68**:1958−70.

80. Quadros A, Belli KC, de Paula JET, et al. Chronic total occlusion percutaneous coronary intervention in Latin America. *Catheter Cardiovasc Interv* 2020;**96**:1046−55.

81. Lobo AS, Sandoval Y, Burke MN, et al. Sleep deprivation in cardiology: a multidisciplinary survey. *J Invasive Cardiol* 2019;**31**:195−8.

82. Sandoval Y, Lobo AS, Somers VK, et al. Sleep deprivation in interventional cardiology: implications for patient care and physician-health. *Catheter Cardiovasc Interv* 2018;**91**:905−10.

83. Iverson A, Stanberry L, Garberich R, et al. Impact of sleep deprivation on the outcomes of percutaneous coronary intervention. *Catheter Cardiovasc Interv* 2018;**92**:1118−25.

84. Morino Y, Abe M, Morimoto T, et al. Predicting successful guidewire crossing through chronic total occlusion of native coronary lesions within 30 minutes: the J-CTO (Multicenter CTO Registry in Japan) score as a difficulty grading and time assessment tool. *JACC Cardiovasc Interv* 2011;**4**:213−21.

85. Christopoulos G, Kandzari DE, Yeh RW, et al. Development and validation of a novel scoring system for predicting technical success of chronic total occlusion percutaneous coronary interventions: the PROGRESS CTO (prospective global registry for the study of chronic total occlusion intervention) score. *JACC Cardiovasc Interv* 2016;**9**:1−9.

86. Szijgyarto Z, Rampat R, Werner GS, et al. Derivation and validation of a chronic total coronary occlusion intervention procedural success score from the 20,000-patient EuroCTO registry: the EuroCTO (CASTLE) score. *JACC Cardiovasc Interv* 2019;**12**:335−42.

87. Danek BA, Karatasakis A, Karmpaliotis D, et al. Development and validation of a scoring system for predicting periprocedural complications during percutaneous coronary interventions of chronic total occlusions: the prospective global registry for the study of chronic total occlusion intervention (PROGRESS CTO) complications score. *J Am Heart Assoc* 2016;5.

88. Simsek B, Kostantinis S, Karacsonyi J, et al. Predicting Periprocedural Complications in Chronic Total Occlusion Percutaneous Coronary Intervention: The PROGRESS-CTO Complication Scores. *JACC Cardiovasc Interv* 2022;**15**:1413−22.

89. Azzalini L, Torregrossa G, Puskas JD, et al. Percutaneous revascularization of chronic total occlusions: rationale, indications, techniques, and the cardiac surgeon's point of view. *Int J Cardiol* 2017;**231**:90−6.

90. Thiele H, Akin I, Sandri M, et al. One-year outcomes after PCI strategies in cardiogenic shock. *N Engl J Med* 2018;**379**:1699−710.

91. Thiele H, Akin I, Sandri M, et al. PCI strategies in patients with acute myocardial infarction and cardiogenic shock. *N Engl J Med* 2017;**377**:2419−32.

92. Braik N, Guedeney P, Behnes M, et al. Impact of chronic total occlusion and revascularization strategy in patients with infarct-related cardiogenic shock: a subanalysis of the culprit-shock trial. *Am Heart J* 2021;**232**:185−93.

93. Blankenship JC, Gigliotti OS, Feldman DN, et al. Ad hoc percutaneous coronary intervention: a consensus statement from the society for cardiovascular angiography and Interventions. *Catheter Cardiovasc Interv* 2013;**81**:748−58.

94. Brilakis ES, Banerjee S, Lombardi WL. Retrograde recanalization of native coronary artery chronic occlusions via acutely occluded vein grafts. *Catheter Cardiovasc Interv* 2010;**75**:109−13.

95. Xenogiannis I, Tajti P, Burke MN, Brilakis ES. Staged revascularization in patients with acute coronary syndromes due to saphenous vein graft failure and chronic total occlusion of the native vessel: a novel concept. *Catheter Cardiovasc Interv* 2019;**93**:440−4.

96. Mehta SR, Wood DA, Storey RF, et al. Complete revascularization with multivessel PCI for myocardial infarction. *N Engl J Med* 2019;**381**:1411−21.

97. Bainey KR, Engstrom T, Smits PC, et al. Complete vs culprit-lesion-only revascularization for ST-segment elevation myocardial infarction: a systematic review and *meta*-analysis. *JAMA Cardiol* 2020;**5**:881−8.

98. Elgendy IY, Mahmoud AN, Kumbhani DJ, Bhatt DL, Bavry AA. Complete or culprit-only revascularization for patients with multivessel coronary artery disease undergoing

percutaneous coronary intervention: a pairwise and network *meta*-analysis of randomized trials. *JACC Cardiovasc Interv* 2017;**10**:315−24.

99. Farooq V, Serruys PW, Bourantas CV, et al. Quantification of incomplete revascularization and its association with five-year mortality in the synergy between percutaneous coronary intervention with taxus and cardiac surgery (SYNTAX) trial validation of the residual SYNTAX score. *Circulation* 2013;**128**:141−51.

Chapter 8

Wiring

Section 8.1

CTO Wiring classification/definitions

8.1.1 CTO crossing strategies definitions

Crossing the occlusion with a guidewire is usually the most difficult part of chronic total occlusion (CTO) percutaneous coronary intervention (PCI). There are four CTO crossing techniques that are classified according to crossing direction and device course within the occluded segment[1,2] (Fig. 8.1.1 and Table 8.1.1):

Direction of crossing

Antegrade is approaching the occlusion segment in the original direction of blood flow, that is, from the proximal to the distal CTO cap into the distal true lumen.

Retrograde is approaching the occlusion segment against the original direction of blood flow, that is, from the distal to the proximal CTO cap into the proximal true lumen.

FIGURE 8.1.1 Illustration of CTO crossing techniques *ADR*, antegrade dissection and reentry; *AW*, antegrade wiring; *RDR*, retrograde dissection and reentry; *RW*, retrograde wiring.

Manual of Chronic Total Occlusion Percutaneous Coronary Interventions.
DOI: https://doi.org/10.1016/B978-0-323-91787-2.00009-5

TABLE 8.1.1 Classification of CTO crossing strategies.[2]

Approach and crossing technique	Definition	Retrograde approach contribution
1. Antegrade wiring (AW)	Wire-based technique with the intention of traversing from the proximal vessel true lumen through the CTO to the distal vessel true lumen	No: AW-0 Yes: AW-R
2. Antegrade dissection and reentry (ADR)	Dissection technique (wire-based or device-based with a dedicated dissection device or equivalent) with the intention of passing from the proximal vessel lumen through a dissection plane followed by reentry into the distal vessel lumen at or beyond the distal cap of the occlusion	No: ADR-0 Yes: ADR-R
3. Retrograde wiring (RW)	Wire-based technique with the intention of traversing from the distal vessel true lumen to the proximal vessel true lumen	
4. Retrograde dissection and reentry (RDR)	Dissection technique (usually with knuckled wires) with the intention of connecting an antegrade dissection plane and a retrograde dissection plane, with wires advanced antegrade and/or retrograde	

Source: Reproduced with permission from Ybarra LF, Rinfret S, Brilakis ES, et al. Definitions and clinical trial design principles for coronary artery chronic total occlusion therapies: CTO-ARC consensus recommendations. *Circulation* 2021;143:479–500.

Device course within the occluded segment

Vessel "architecture" is the vascular space contained by the naturally resistant adventitia, including the occlusive plaque and vessel wall (Fig. 8.1.2).[2]

Normal coronary arteries consist of intimal, medial, and adventitial layers. However, identifying the 3-layer structure of the vessel wall may be challenging in a CTO vessel, even by intravascular imaging because of extensive architectural disruption. As a result, the CTO-Academic Research Consortium (CTO-ARC) recommends dividing the occlusion segment into occlusive plaque (composed of the former true lumen now occupied by atherosclerotic plaque and organized thrombus) and what lies outside it (media and adventitia).[2] Plaque or atherosclerosis is a disease of the intima that is bound by the internal elastic lamina. Device course within the occluded segment can be described as:

Intraplaque: wire tracking within the occlusive intima-based plaque.

Extraplaque: device tracking outside the plaque but still contained within the adventitial layer (often called "subintimal" or through a "false

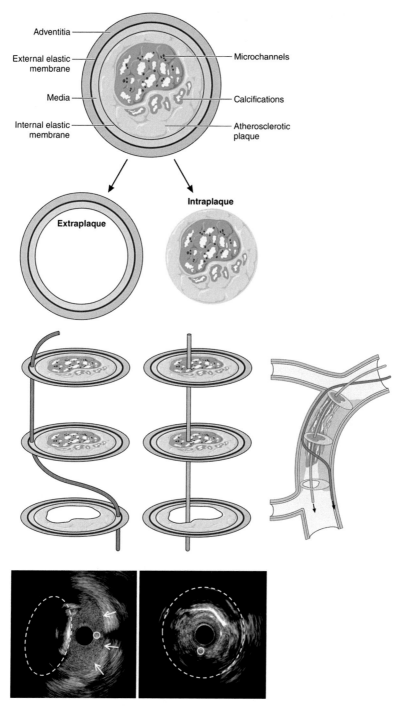

FIGURE 8.1.2 Vessel architecture: intraplaque and extraplaque structures. Intravascular ultrasound images: white dotted circles indicate plaque; yellow circles indicate the guidewire; and white arrows show a monolayer appearance of the vessel, indicating absence of the intimal layer. *Reproduced with permission from Ybarra LF, Rinfret S, Brilakis ES, et al. Definitions and clinical trial design principles for coronary artery chronic total occlusion therapies: CTO-ARC consensus recommendations.* Circulation *2021;143:479−500.*

lumen" in the past). Crossing the CTO through the extraplaque space is called **dissection and reentry**.[3]

The terms "true lumen" and "false lumen" can be used for the vessel segments proximal and distal to the CTO caps; wherein, a wire or a dedicated dissection device may track within the artery wall rather than within the true lumen (such as in the "scratch and go" technique or when a device is used to reenter into the true lumen).

The intended location of crossing (intraplaque vs extraplaque) may be different from the actual location as determined by intravascular imaging, but CTO-ARC recommends reporting the intended technique to traverse the CTO segment regardless of the actual position of the wire.[2]

8.1.2 CTO crossing strategies classification and terminology

The terminology utilized in dissection/reentry CTO strategies can be confusing (Fig. 8.1.3).[3]

In the **antegrade** approach, dissection can be achieved by a:

1. Wire-based strategy, that is, inadvertent wiring or knuckle wire
2. Catheter-based strategy, using the CrossBoss catheter[4]

In the **antegrade** approach, reentry can be achieved by:

1. **Wire-based strategies**, such as the Subintimal Tracking And reentry (STAR)[5] techniques and its modifications, that is, "contrast-guided STAR," "**mini-STAR**,"[6] and Limited Antegrade Subintimal Tracking (**LAST**)[7] (Section 8.3). These techniques, however, tend to have lower success rates because of difficulty in reliably reentering into the true lumen, often due to extensive uncontrolled dissection with extraplaque hematoma formation and true lumen compression.
2. **Dedicated reentry systems (preferred)**, such as the Stingray balloon (Boston Scientific, Section 30.16.2) and the ReCross dual lumen microcatheter. (IMDS, Section 30.6.4).

In the **retrograde approach**, dissection is usually performed using a knuckle wire, and reentry is achieved using the techniques described in Section 8.4.

Sections 8.2−8.4 describe the three key wiring strategies (retrograde wiring, retrograde dissection, and reentry are discussed together in Section 8.4). Section 8.5 discusses CTO crossing strategy selection.

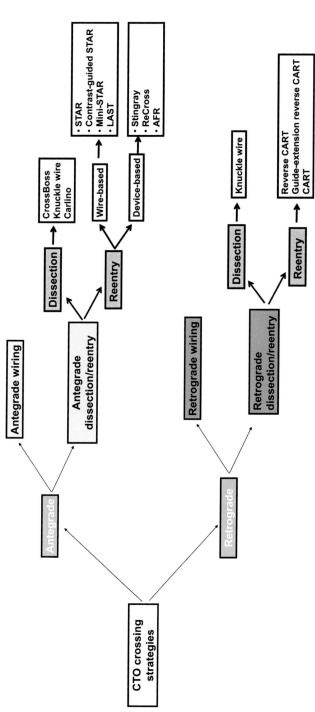

FIGURE 8.1.3 **CTO crossing strategy classification and nomenclature.** *CART*, controlled antegrade and retrograde tracking; *LAST*, limited antegrade subintimal tracking; *STAR*, subintimal tracking and reentry.

References

1. Brilakis ES, Mashayekhi K, Tsuchikane E, et al. Guiding principles for chronic total occlusion percutaneous coronary intervention. *Circulation* 2019;**140**:420−33.
2. Ybarra LF, Rinfret S, Brilakis ES, et al. Definitions and clinical trial design principles for coronary artery chronic total occlusion therapies: CTO-ARC consensus recommendations. *Circulation* 2021;**143**:479−500.
3. Michael TT, Papayannis AC, Banerjee S, Brilakis ES. Subintimal dissection/reentry strategies in coronary chronic total occlusion interventions. *Circ Cardiovasc Interv* 2012;**5**:729−38.
4. Whitlow PL, Burke MN, Lombardi WL, et al. Use of a novel crossing and reentry system in coronary chronic total occlusions that have failed standard crossing techniques: results of the FAST-CTOs (Facilitated Antegrade Steering Technique in Chronic Total Occlusions) trial. *JACC Cardiovasc Interv* 2012;**5**:393−401.
5. Colombo A, Mikhail GW, Michev I, et al. Treating chronic total occlusions using subintimal tracking and reentry: the STAR technique. *Catheter Cardiovasc Interv* 2005;**64**:407−11 discussion 12.
6. Galassi AR, Tomasello SD, Costanzo L, et al. Mini-STAR as bail-out strategy for percutaneous coronary intervention of chronic total occlusion. *Catheter Cardiovasc Interv* 2012;**79**:30−40.
7. Lombardi WL. Retrograde PCI: what will they think of next? *J Invasive Cardiol* 2009;**21**:543.

Section 8.2

Antegrade wiring

Antegrade wiring is the simplest and most widely used CTO crossing technique.[8–10] At least 50% of CTOs are currently successfully recanalized using antegrade wiring.[11–13] Familiarity and confidence with this technique provides the foundation upon which all other CTO PCI techniques (antegrade dissection/reentry and retrograde) are built.

The term "antegrade wire escalation" was used in the past but is no longer recommended, because antegrade wiring can involve both escalation (use of stiffer tip guidewires) and de-escalation (use of softer tip guidewires after penetration of a highly resistant proximal cap with a stiff tip guidewire).

8.2.1 Antegrade wiring: when?

A detailed description of an initial and subsequent crossing strategy selection is provided in Section 8.5. Antegrade wiring is most appropriate for:

- Occlusions with a clear proximal cap
- Short occlusions, that is, <20 mm length
- Longer occlusions of straight segments and/or where a through-and-through microchannel is suspected
- Select cases of occlusive in-stent restenosis

8.2.2 Antegrade wiring: how?

As discussed in the *Manual of Percutaneous Coronary Interventions*, wiring is performed in 10 steps[14]:

1. Determining whether a microcatheter is needed
2. Selecting a guidewire
3. Shaping the wire tip
4. Inserting the guidewire into the guide catheter
5. Advancing the guidewire to the tip of the guide catheter
6. Advancing the guidewire from the tip of the guide catheter to the target lesion
7. Crossing the lesion with the guidewire
8. Advancing the guidewire distal to the target lesion
9. Removing the microcatheter (if one was used)
10. Monitoring guidewire position

Next, each step is discussed separately, as it applies to antegrade wiring of CTOs.

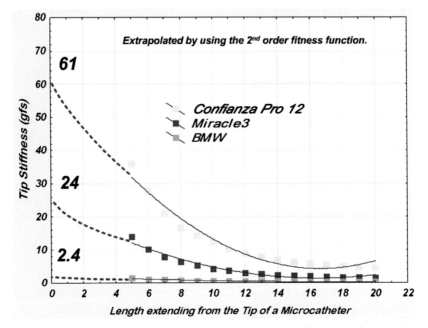

FIGURE 8.2.1 **Change in guidewire tip stiffness with various guidewire lengths extending past a microcatheter tip.** *Reproduced with permission from Waksman R, Saito S.* Chronic total occlusions: a guide to revascularization. *Wiley-Blackwell; 2013.*

8.2.2.1 Step 1. Selecting a microcatheter

Goal: To optimize the success and safety of guidewire manipulations.

How?

A microcatheter (or over-the-wire balloon if a microcatheter is not available) should be **used in all CTO PCIs** for both antegrade and retrograde crossing.

CTO crossing should not be attempted with unsupported guidewires[15] because a microcatheter:

1. Improves the precision of both rotational and longitudinal guidewire movements both in fluid (blood-filled vessels) and in tissue (the occlusion itself).
2. Allows the penetration force of the wire to be dynamically altered by changing the distance between the tip of the guidewire and the microcatheter, with guidewires becoming stiffer when the microcatheter tip is positioned closer to the guidewire tip (Fig. 8.2.1).
3. Allows wire tip reshaping without losing wire position.
4. Facilitates wire exchanges.
5. Prevents twisting of wires when using the parallel wire technique.

6. Protects the proximal part of the vessel from guidewire-induced injuries during both antegrade and retrograde CTO PCI.
7. Allows delivery of contrast, either for visualization or to accomplish the Carlino technique[16] (intralesional injection of <0.5−1 mL of contrast to elucidate microcatheter position and facilitate crossing, especially in wire-resistant lesions, by modifying plaque compliance).
8. Allows delivery of coils, fat, or thrombin in case of perforation.

A microcatheter is preferred rather than an over-the-wire balloon[17] because it:

1. Allows accurate assessment of the microcatheter tip location (because the marker is located at the tip, whereas in 1.0−1.5 mm balloons the marker is located in mid shaft and the tip is not angiographically visible, Fig. 30.27);
2. Has better wire-to-lumen internal diameter ratio;
3. Provides better support due to metal braiding and coil construction (Section 30.5);
4. Is more resistant to kinking due to metal braiding and coil construction. Over-the-wire balloons are prone to kinking upon wire removal, thus hindering reliable wire exchanges; and
5. Is less likely to cause proximal vessel injury.[17]

These advantages are particularly important in cases of **tortuosity or poor guide catheter support**.

8.2.2.2 Step 2. Selecting a guidewire

Goal: To select a guidewire capable of fulfilling the desired task while minimizing the risk of complications.

How?

Guidewire selection (Section 30.6) depends on the CTO lesion morphology and the task that needs to be achieved:

1. **Delivering a microcatheter to the proximal cap**: workhorse guidewire
2. **Proximal cap penetration** (Fig. 8.2.2) (if the location of the proximal cap is clear—ambiguous proximal caps are approached as discussed in Section 16.1):
   a. **Tapered proximal cap**
      i. **Soft tip, tapered, polymer-jacketed** guidewires (such as Fielder XT, Fielder XT-A, Fielder XT-R, Fighter, Bandit, Section 30.6.2.2).
      ii. If this fails, escalate to a stiff tip, polymer-jacketed guidewire (such as Gladius, Gladius Mongo, Pilot 200, Raider; Section 30.6.2.3) or to an intermediate tip stiffness, tapered-tip guidewire (such as Gaia 1, 2, or 3; Gaia Next 1, 2, or 3; or Judo 1, 3, or 6; Section 30.6.3.2).

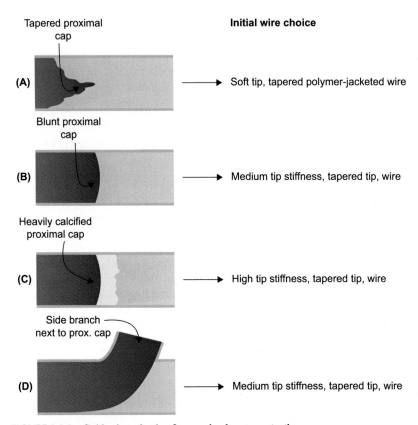

FIGURE 8.2.2   **Guidewire selection for proximal cap penetration.**

b. **Blunt proximal cap**
   i. **Intermediate tip stiffness, tapered-tip** guidewires (such as Gaia 1, 2 or 3, Gaia Next 1, 2, or 3, or Judo 1, 3, or 6; Section 30.6.3.2).
   ii. In case of failure, escalate to a high tip-stiffness guidewire (such as Confianza Pro 12, Hornet 14, Infiltrac and Infiltrac Plus, Warrior, or Astato 20; Section 30.6.3.3).
c. **Heavily calcified proximal cap**
   i. **Intermediate tip stiffness, tapered-tip** guidewires (such as Gaia 1st and 2nd, Gaia Next 1 or 2, or Judo 1 and 3), or
   ii. **High-tip stiffness** guidewires (such as Confianza Pro 12, Hornet 14, Infiltrac, Infiltrac Plus, Warrior, Astato 20).
   iii. Once the proximal cap is crossed, de-escalate to a softer tip guidewire, either polymer-jacketed or medium tip stiffness, non-polymer-jacketed guidewire.

Advancement through occlusion   Initial guidewire choice

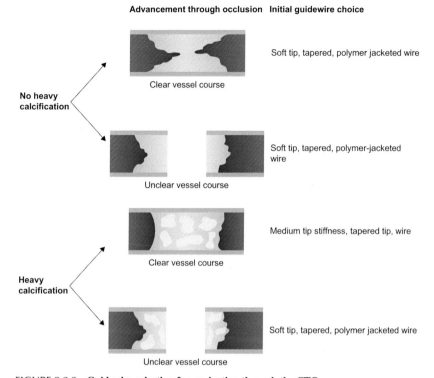

Clear vessel course

Soft tip, tapered, polymer jacketed wire

No heavy
calcification

Unclear vessel course

Soft tip, tapered, polymer-jacketed
wire

Heavy
calcification

Clear vessel course

Medium tip stiffness, tapered tip, wire

Unclear vessel course

Soft tip, tapered, polymer jacketed wire

FIGURE 8.2.3   Guidewire selection for navigating through the CTO.

    d. Side branch at the proximal cap

       i. Do NOT use polymer-jacketed guidewires, as they are likely to prolapse into the side branch.

      ii. **Intermediate tip stiffness, tapered-tip guidewires** (such as Gaia 1st and 2nd, Gaia Next 1 or 2, or Judo 1 and 3) are preferred, especially when used through a dual lumen microcatheter.

3. Advancement through the occlusion (Fig. 8.2.3)

  a. No heavy calcification

       i. **Well-understood CTO course**: **soft tip, tapered, polymer-jacketed** guidewires; if this fails, escalate to an intermediate tip stiffness tapered tip guidewire.

      ii. **Not well understood CTO course**: **soft tip, tapered, polymer-jacketed** guidewires; if this fails, escalate to stiff tip, polymer-jacketed guidewires (such as Gladius, Gladius Mongo, Pilot 200, Raider).

  b. Heavy calcification or tortuosity

       i. **Well-understood CTO course**: **intermediate tip stiffness, tapered-tip** guidewires (such as Gaia 1, 2, or 3, Gaia Next 1, 2, or 3, or Judo

1, 3, or 6). If this fails, escalate to stiff tip, polymer-jacketed guide-wires (such as Gladius, Gladius Mongo, Pilot 200, Raider).

   ii. **Not well understood CTO course: soft tip, tapered, polymer-jacketed** guidewire; if this fails, escalate to stiff tip, polymer-jacketed guidewires (such as Gladius, Gladius Mongo, Pilot 200, Raider)

4. **Entry into the distal true lumen**
   a. **Wire redirection**: medium or high tip stiffness, non-polymer-jacketed, tapered-tip guidewires.
   b. **Parallel wire**: medium or high tip stiffness, non-polymer-jacketed, tapered-tip guidewires.
   c. **LAST** (limited antegrade subintimal tracking, Section 8.3.5.2): usually intermediate or high tip stiffness, non-polymer-jacketed, tapered-tip guidewires. Polymer-jacketed guidewires can also be used but may enlarge the dissection and tend to remain in the extraplaque (subintimal) space.
   d. **Stingray balloon** (Section 8.3.5.3) **or ReCross** (Section 8.3.5.4) **reentry**: reentry is done using the "stick and drive" or the "stick and swap" technique (initial puncture with a high tip stiffness guidewire, such as Confianza Pro 12, Infiltrac and Infiltrac Plus, Hornet 14, Warrior, Astato 20, followed by swap to a stiff tip, polymer-jacketed guidewire, such as Pilot 200, Gladius, Raider).
   e. **STAR (Subintimal tracking and reentry)** (Section 8.3.5.1): Gladius Mongo or soft tip, polymer-jacketed guidewire, such as Fielder XT.

5. **Balloon and stent delivery**
   After CTO crossing, the microcatheter should be advanced through the occlusion and the guidewire that crossed should be exchanged for a work-horse guidewire, or a highly supportive guidewire (such as Grand Slam, Iron Man, Wiggle wire, Sion blue extra support; Section 30.6.4), or an atherectomy guidewire (Rotawire Drive Floppy or Rotawire Drive extra support for rotational atherectomy [Section 30.9.1] or ViperWire Advance or ViperWire Advance with flex tip for orbital atherectomy [Section 30.9.2] if atherectomy is planned). Guidewire exchanges using a microcatheter are best performed using the trapping technique (Section 8.2.2.9.1).

8.2.2.3 Step 3. Shaping the wire tip

Goal. To shape the wire tip in the optimal way for performing the desired function.

How?

The optimal shape of the wire tip depends on the desired wire function, as follows (Fig. 8.2.4):

1. **Advancing to the CTO proximal cap:** a standard bend is used on a work-horse guidewire, depending on the size and tortuosity of the proximal vessel.

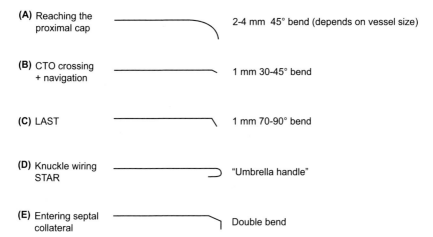

(A) Reaching the proximal cap ——————— 2-4 mm 45° bend (depends on vessel size)

(B) CTO crossing + navigation ——————— 1 mm 30-45° bend

(C) LAST ——————— 1 mm 70-90° bend

(D) Knuckle wiring STAR ——————— "Umbrella handle"

(E) Entering septal collateral ——————— Double bend

FIGURE 8.2.4   **Various guidewires shapes to achieve various tasks during CTO PCI.**

2. Crossing the CTO proximal cap + navigation through CTO:

A small (1 mm long, 30−45 degrees) distal bend (Fig. 8.2.4) is preferred for crossing the proximal cap and navigating through the occlusion because it:

a. Enhances the penetrating capacity of the guidewire.
b. Facilitates entry into microchannels.
c. Reduces the likelihood of deflection outside the vessel architecture or into branches arising within the occlusion.
d. Improves steerability within tight spaces, such as the CTO segment, which straighten larger bends.

Some guidewires, such as the Gaia and Gaia Next (Section 30.6.3.2), come preshaped.

Creating a small bend at the tip of a guidewire can only be accomplished by inserting the guidewire through an introducer, rather than using the side of the introducer, as is commonly done for workhorse guidewires (Fig. 8.2.5).

1. The guidewire is inserted through an introducer with approximately 1 mm protruding through the tip.
2. The guidewire tip is bent by 30−45 degrees (sometimes a syringe is used to bend the tip of very stiff guidewires, such as the Confianza Pro 12, Hornet 14, Infiltrac and Infiltrac Plus, or Astato 20 guidewires, as they can puncture the operator's glove).
3. The guidewire tip is inspected to verify optimal shaping.
4. The guidewire is withdrawn into the introducer and advanced into the microcatheter or over-the-wire balloon (it is best to insert the shaped guidewires into a microcatheter using an introducer to prevent potential tip damage or deformation).

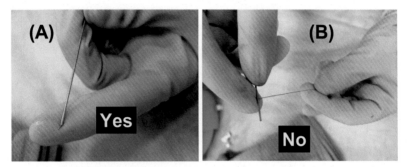

FIGURE 8.2.5 **Guidewire tip shaping**. **Panel A**: how to shape the tip of a guidewire for CTO crossing (by inserting it through an introducer). **Panel B**: what to avoid when shaping a guidewire for CTO crossing (i.e., using the side of the introducer).[18] *Reproduced with permission from the Brilakis ES.* Manual of coronary chronic total occlusion interventions. A step-by-step approach. *2nd ed. Elsevier; 2017.*

3. **Entry into the distal true lumen**:
   a. **Wire redirection**: medium or high tip stiffness, non-polymer-jacketed, tapered-tip guidewire.
   b. **Parallel wire**: medium or high tip stiffness, non-polymer-jacketed, tapered-tip guidewire.
   c. **LAST** (limited antegrade subintimal tracking): usually medium or high tip stiffness, non-polymer-jacketed, tapered-tip guidewire. Polymer-jacketed guidewires can also be used but may enlarge the dissection and tend to remain in the extraplaque (subintimal) space.
   d. **Stingray balloon** (Section 8.3.5.3) **or ReCross** (Section 8.3.5.4) **reentry**: usually reentry is achieved using the "stick and swap" technique, with an initial puncture with a high tip stiffness guidewire (such as Confianza Pro 12, Infiltrac and Infiltrac Plus, Hornet 14, Warrior, Astato 20) followed by swap to a stiff, polymer-jacketed guidewire (such as Pilot 200, Gladius Mongo, or Raider).
   e. **STAR (Subintimal Tracking And Reentry)** (Section 8.3.5.1): Gladius Mongo or soft tip, polymer-jacketed guidewire, such as Fielder XT.
4. **Balloon and stent delivery:** a standard bend is used on a workhorse guidewire. Sometimes a support guidewire (such as Grand Slam or Wiggle) may be needed when equipment delivery is challenging.

What can go wrong?
1. **Excessive guidewire bending or tip fracture.**

Causes
- Forceful or excessive wire manipulation.
- Removing the guidewire from its hoop by pulling it from the tip.

Prevention

- Insert the wire through the introducer for shaping the wire tip (Fig. 8.2.5).
- Do not hold the tip of the guidewire while removing it from the hoop.
- Manipulate the wire tip gently, especially with medium and high tip stiffness non-polymer jacketed guidewires. Polymer jacketed guidewires require more force as they are resistant to shaping and are prone to straightening during manipulations. Excessive force may lead to fracture of the guidewire tip.
- Initially create a small bend and change it if it fails to advance. It is always possible to create additional bend(s) on the guidewire, but it can be hard to remove them.

Treatment

- If the wire cannot be reshaped, it may need to be discarded and another guidewire used.

8.2.2.4   Step 4. Inserting the guidewire into the guide catheter

Goal: to insert the guidewire into the guide catheter through the hemostatic valve of the Y-connector.

How: the guidewire tip is withdrawn into the introducer needle. The introducer needle is inserted into the hemostatic valve, followed by guidewire advancement.

Alternatively, the guidewire can be preloaded (using the same introducer that was used for shaping the tip) into a microcatheter or over-the-wire balloon, the tip of which is then inserted through the hemostatic valve.

What can go wrong?
1. Guidewire tip deformation

Causes
1. Guidewire tip not fully withdrawn into the introducer needle.
2. Guidewire enters side the arm of the Y-connector.
3. Guidewire comes in contact with devices in the guide catheter, such as guide catheter extensions or balloons.

Prevention
1. Ensure that the guidewire tip is not protruding from the tip of the introducer needle or the microcatheter.
2. Do not force the guidewire against resistance.
3. Ensure that the introducer is advanced all the way through the Y-connector.
4. Remove other devices from the guide catheter whenever possible.
5. Use fluoroscopy when advancing guidewires through the proximal collar of guide catheter extensions.

Treatment
1. Exchange the guidewire for a new one (if the damaged guidewire cannot be reshaped).

8.2.2.5 Step 5. Advancing the guidewire to the tip of the guide catheter

Goal: To advance the guidewire to the tip of the guide catheter.
How: the guidewire is advanced to the tip of the guide catheter. Fluoroscopy is used to check the guidewire position.

Some guidewires, such as the BMW, have length markers on their shaft at 90 and 100 cm from the tip that can be used to minimize use of fluoroscopy during wire advancement.

Knowing the length of the guide catheter is important, as guidewires will exit sooner when used in 90 cm as compared with 100 cm long guide catheters.

Caution should be used when wiring through side hole guide catheters, as the guidewire can exit through the side hole, instead of the catheter tip.

What can go wrong?

1. Inadvertent advancement into the coronary artery.

The guidewire may be advanced into the coronary artery inadvertently or during contrast or saline injections without fluoroscopy guidance, which may lead to dissection, perforation, or loss of guide catheter position.

Causes
1. Too distal guidewire advancement.
2. Unsecured wire during contrast injections.

Prevention
1. Careful monitoring of the position of the guidewire tip.
2. Some guidewires (such as the BMW), have a proximal marker that can help prevent too distal advancement.
3. The torquer can be tightened on the guidewire at approximately 90 cm from the guidewire tip to prevent excessive guidewire advancement.
4. Securing the wire before contrast injection, especially when using automated contrast injectors, such as the ACIST device.

Treatment
1. Do NOT immediately remove the guidewire (sometimes the guidewire enters the intended branch).
2. Inject contrast to check guidewire position, followed by guidewire redirection, if needed.

2. Guidewire tip deformation

Causes
1. Advancement through a guide catheter extension.
2. Advancement past balloons/stent previously inserted in the guide catheter.
3. Advancement through side holes of the guide catheter.

Prevention

1. Remove guide extensions and/or balloons or stents before inserting another guidewire.
2. If this is not feasible, guidewire advancement through the collar of the guide extension or past the balloons/stents should be done either under fluoroscopy without forcing the wire or through a microcatheter.
3. Balloons can be advanced into the coronary artery while advancing additional wires through the guide catheter and then retracted into the guide catheter once wire advancement is completed.
4. Use a dual lumen microcatheter: the monorail lumen of the dual lumen microcatheter is advanced over the initially placed guidewire, followed by insertion of the new guidewire through the over-the-wire lumen. The trapping technique (Section 8.2.2.9.1) is then used to remove the dual lumen microcatheter.

Treatment

1. Attempt to reshape the guidewire tip.
2. If reshaping fails, the guidewire is discarded and a new guidewire is used.

8.2.2.6 Step 6. Advancing the guidewire and microcatheter from the tip of the guide catheter to the CTO proximal cap

Goal: To advance the guidewire and microcatheter from the tip of the guide catheter to the proximal cap of the target CTO.

How: The guidewire is advanced from the tip of the guide catheter to the lesion under fluoroscopic guidance with intermittent contrast injections. A still frame image of the coronary anatomy can be used as reference to assist with guidewire advancement. Another way to guide wiring is a dynamic roadmap that is available in some X-ray systems.

The microcatheter can be advanced into the target vessel either after the tip of the guidewire has reached the CTO proximal cap, or earlier to facilitate wire advancement, especially in highly tortuous vessels.

Guidewire selection

As described in Step 1, a workhorse guidewire should be used to deliver a microcatheter to the CTO proximal cap, unless the CTO proximal cap is ostial or very proximal. CTO wires with high penetrating power and tapered tips should not be used to traverse the proximal vessel to get to the CTO segment because:

1. They can cause vessel injury, especially in diffusely diseased vessels (Fig. 8.2.6).
2. The wire bend required to reach the CTO is usually different (much larger) than the wire bend used when entering and crossing the CTO (much smaller) (Fig. 8.2.7).

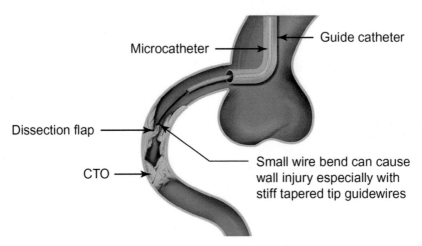

FIGURE 8.2.6 Illustration of proximal vessel injury during attempts to reach the proximal cap of the chronic total occlusion. *Reproduced with permission from the Brilakis ES.* Manual of coronary chronic total occlusion interventions. A step-by-step approach. *2nd ed. Elsevier; 2017.*

Challenges

1. Difficulty advancing the guidewire through the tip of the guide catheter

Causes
- Suboptimal guide engagement.

Prevention
- Optimal guide engagement (coaxial, not too deep, no pressure waveform dampening)

Solutions
1. Reposition the guide catheter tip to obtain optimal vessel engagement.
2. Change the shape of the guidewire tip.
3. Leave the original guidewire in place and wire with a second guidewire.

2. Guidewire inadvertently enters a side branch

Causes
- Challenging coronary anatomy (tortuosity).
- Suboptimal guidewire tip shaping.

Prevention
- Optimally shape the guidewire tip.
- Use guidewires that provide 1:1 torque response.
- Use a torquer for guidewire manipulation.
- Use a microcatheter to facilitate torque transmission and allow reshaping of the guidewire tip.

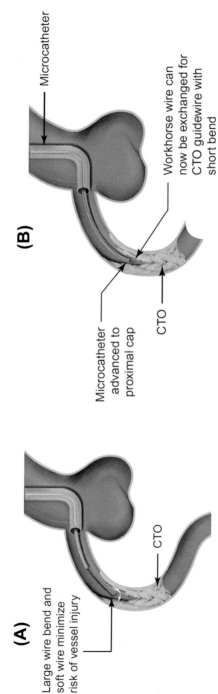

(A)

Large wire bend and
soft wire minimize
risk of vessel injury

CTO

(B)

Microcatheter

Microcatheter
advanced to
proximal cap

Workhorse wire can
now be exchanged for
CTO guidewire with
short bend

CTO

FIGURE 8.2.7 Panel A: reaching the CTO proximal cap with a workhorse guidewire and a large bend without causing trauma to the diseased target vessel proximal to the occlusion. **Panel B:** successful delivery of the microcatheter to the CTO proximal cap. *Reproduced with permission from the Brilakis ES. Manual of coronary chronic total occlusion interventions. A step-by-step approach. 2nd ed. Elsevier; 2017.*

Treatment

- Reposition the guide catheter tip. For example, if the guidewire keeps entering into a conus branch in the right coronary artery, the guide may be withdrawn, followed by attempts to obtain coaxial engagement.
- Change the shape of the guidewire tip.
- Use a torquer, if not used during initial wiring attempts.
- Change guide catheter: for example, if a guidewire keeps entering the circumflex instead of the left anterior descending artery (LAD), using a smaller curve guide catheter will facilitate advancement into the LAD. Alternatively, a guidewire can be left in the initially wired vessel, stabilizing the guide catheter, followed by insertion of a second guidewire into the target vessel.

3. Guidewire cannot be advanced through tortuosity

Causes

- Challenging coronary anatomy (tortuosity).
- Suboptimal guidewire tip shaping.

Prevention

- Optimally shape the guidewire tip.
- Use guidewires that provide 1:1 torque response.
- Use hydrophilic or polymer-jacketed guidewires (which, however, may increase the risk of dissection and perforation).

Treatment (Fig. 8.2.8).

- **Solution 1: Use a different guidewire** (with hydrophilic-coating or polymer-jacket) and **change the shape of the guidewire tip**. Some guidewires (such as the Pilot family) are less likely to prolapse compared with other wires.
- **Solution 2: Use a microcatheter** (with straight or angulated tip, Section 30.5) to facilitate advancement. The microcatheter also allows guidewire exchanges without the need to rewire from the ostium of the vessel. A dual lumen microcatheter can also be used, wiring the angulated vessel through the over-the-wire lumen.
- **Solution 3: Reversed (also called "hairpin") guidewire technique (CTO Manual online case 71, 81, 178; PCI Manual online cases 119, 123).**[19-21] In this technique a polymer-jacketed wire is bent approximately 3 cm from the wire tip and the bend is inserted through the hemostatic valve of the Y-connector (Fig. 8.2.9).

The hairpin wire is then advanced into the main vessel (Fig. 8.2.10, panels B and C), and pulled back (Fig. 8.2.10, panels D and E), entering the main branch (Fig. 8.2.10, panel F).

Alternatively, the hairpin wire can be advanced through a dual lumen microcatheter (Section 30.6.4, Fig. 30.55).

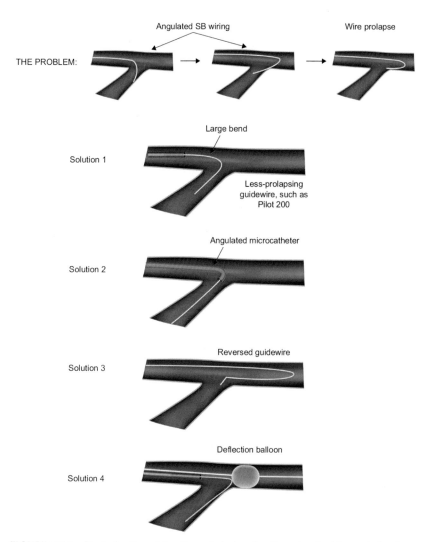

FIGURE 8.2.8 **Strategies for wiring through tortuosity**. *Reproduced with permission from the Brilakis ES.* Manual of percutaneous coronary interventions: a step-by-step approach. *Elsevier; 2021.*

In a variation of this technique called "streamlined reverse wire technique"[22] (Fig. 8.2.11) the polymer-jacketed guidewire with a bend 3 cm from the tip is advanced through the over-the-wire lumen of a dual lumen microcatheter that is already placed in the coronary vessel until it enters into another side branch. The dual-lumen microcatheter and guidewire are then advanced, creating the hairpin at the tip of the wire, which is then pulled back in order to enter the angulated target branch.

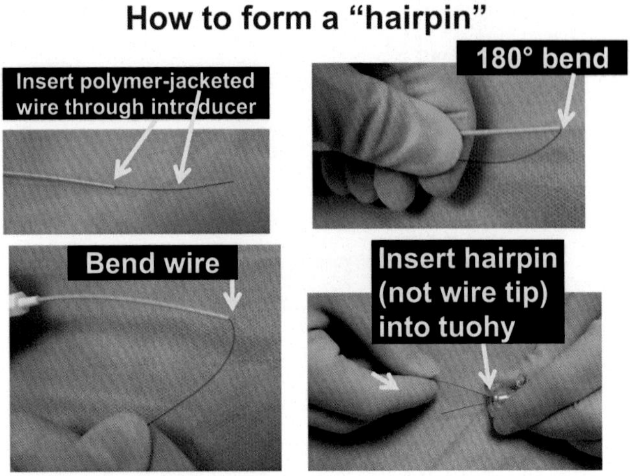

How to form a "hairpin"

FIGURE 8.2.9   How to form a "hairpin" wire. *Reproduced with permission from the Brilakis ES.* Manual of percutaneous coronary interventions: a step-by-step approach. *Elsevier; 2021.*

What Can Go Wrong?

Use of the reversed guidewire technique may cause vessel dissection. Also, after the "hairpin" enters the main vessel, further advancement may be challenging due to the bend in the wire. A single lumen microcatheter is advanced over the hairpin wire into the SB, followed by exchange for another guidewire.

- **Solution 4**: Deflection balloon technique. A balloon is inflated in the main vessel immediately distal to the takeoff of the angulated branch, allowing wires and microcatheters to be "deflected" into the side branch. The disadvantages of this technique include ischemia (since the main vessel is occluded for the duration of the maneuver) and the risk of vessel injury, such as dissection.

What can go wrong?

1. Coronary dissection

Causes
- Severe or ulcerated lesions.
- Use of stiff tip or polymer-jacketed guidewires.
- Aggressive guidewire manipulation.

Prevention
- Avoid using aggressive guidewires in severe lesions.
- Manipulate guidewires gently.

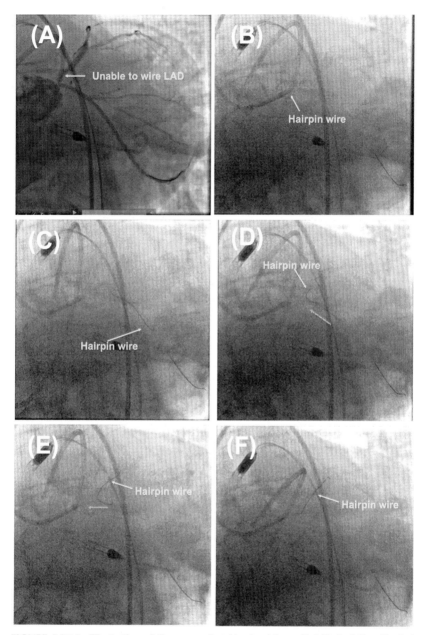

FIGURE 8.2.10 **Illustration of the reversed guidewire (also called "hairpin" guidewire) technique. Panel A**: difficulty wiring LAD due to left main and ostial LAD lesions. **Panel B and C**: a hairpin Sion black guidewire is advanced into the circumflex. **Panel D–F**: the hairpin wire is withdrawn, successfully entering the LAD.

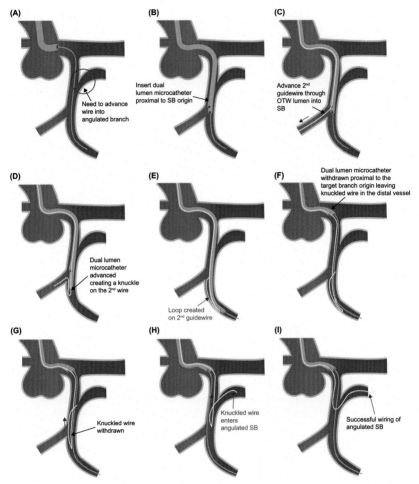

FIGURE 8.2.11 **The "streamlined reverse wire technique."** **Panel A**: a workhorse guidewire is inserted into the main vessel. **Panel B**: a dual lumen microcatheter is advanced with its Rx port over the main vessel wire. **Panel C**: a second guidewire is advanced through the OTW port of the dual lumen microcatheter and engages a nontarget distal side branch (e.g., septal branch, atrial branch, etc). **Panels D, E**: the duel lumen microcatheter is pushed distally to allow the wire to buckle and form the hairpin shape inside the main vessel. **Panel F**: the dual lumen microcatheter is pulled back proximal to the target difficult angulated SB. **Panels G, H, I**: the hairpin wire is pulled back very gently with minimal rotation into the side branch.

Treatment

- Advance another wire into the distal true lumen, followed by stenting (see Section 25.2.1).
- Use of CTO techniques, such as the retrograde approach or the Stingray balloon can facilitate reentry into the distal true lumen, but significant expertise is needed for successful implementation of these techniques.

8.2.2.7   Step 7. Crossing the lesion with the guidewire

Goal: Cross CTO into the distal true lumen.

How?
Three steps of CTO crossing
   Crossing of the CTO can be divided into three stages: (1) penetrating the proximal cap; (2) navigating through the occlusion; and (3) entering into the distal true lumen.

Stage 1: Penetrating the proximal cap
   A clear understanding of the location of the proximal cap is necessary before attempting to penetrate it. If the location of the proximal cap is unclear, additional steps should be undertaken as described in Section 16.1.
   Guidewire selection for proximal cap penetration depends on proximal cap morphology, as described in Step 2 (Guidewire selection, Fig. 8.2.2).

a. **Tapered proximal cap**
   Soft tip, tapered-tip, polymer-jacketed guidewires (such as Fielder XT, Fielder XT-A, Fielder XT-R, Fighter, Bandit) are used first, but they should not be used for an extended period of time. Within seconds/minutes it will become apparent whether these wires will advance through the proximal cap or not. If no progress is achieved (usually within a minute) escalation to a stiffer tip guidewire is recommended, either a stiff tip, polymer-jacketed guidewire (such as Gladius, Gladius Mongo, Pilot 200, Raider) or to intermediate tip stiffness, tapered-tip guidewires (such as Gaia 1st and 2nd, Gaia Next 1 or 2, or Judo 1 and 3).

b. **Blunt proximal cap**
   Intermediate tip stiffness, tapered-tip guidewires (such as Gaia 2nd, or 3rd, Gaia Next 2 or 3, or Judo 3 or 6) are usually used first. If they fail to advance escalation can be done to high tip-stiffness guidewires (such as Confianza Pro 12, Hornet 14, Infiltrac, Infiltrac Plus, Warrior, Astato 20).

c. **Heavily calcified proximal cap**
   Intermediate tip stiffness, tapered-tip guidewires (such as Gaia 1st and 2nd, Gaia Next 1 or 2, or Judo 1 and 3) are usually used first. If they fail to penetrate the cap, **high tip stiffness** guidewires (such as Confianza Pro 12, Hornet 14, Infiltrac, Infiltrac Plus, Warrior, Astato 20) can be used. This is often done with simultaneous use of techniques to increase guidewire support (Section 19.1).
   High tip stiffness guidewires should be de-escalated to a softer guidewire immediately after penetration of the proximal cap.

d. **Side branch at the proximal cap**
   A side branch next to the proximal cap hinders proximal cap puncture by providing an easy route for the guidewire to prolapse into (Fig. 8.2.12). As a result, polymer-jacketed guidewires are generally

avoided in such lesions and **intermediate tip stiffness, tapered-tip guidewires** (such as Gaia 1st and 2nd, Gaia Next 1 or 2, or Judo 1, 2, and 3) are used first. A workhorse guidewire should be inserted into the side branch (if the branch is significant) to protect it in case of dissection during attempts to puncture the proximal cap. A dual lumen microcatheter inserted over the side branch guidewire may facilitate proximal cap puncture.

Stage 2: Navigating through the occlusion

Guidewire selection

Guidewire selection for navigating through the occlusion depends on the occlusion characteristics, such as calcification, clarity of the vessel course, tortuosity, and length.

After crossing the proximal cap with the microcatheter, the initially used guidewire is often exchanged for a softer guidewire to track the occluded segment ("escalation—de-escalation"), especially when a stiff, tapered guidewire was used.

CTOs without heavy calcification and well understood vessel course: A **soft tip, tapered-tip, polymer-jacketed** guidewire is usually used first for a brief period of time. If it fails to advance escalate to a medium tip stiffness, tapered tip guidewire. Alternatively, a stiff tip, polymer-jacketed guidewire can be used.

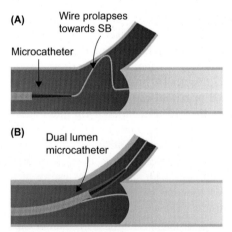

FIGURE 8.2.12 **Antegrade crossing in CTOs with side branch at the proximal cap.** **Panel A**: wire prolapse into the side branch during antegrade wiring attempts. **Panel B**: use of a dual lumen microcatheter with the Monorail lumen advanced over the side branch guidewire facilitates wire penetration of the proximal cap with a wire advanced through the over-the-wire lumen.

CTOs without heavy calcification and poorly understood vessel course: A **soft tip, tapered polymer-jacketed** guidewire is used first, followed by escalation to a **stiff tip, polymer-jacketed** guidewire (such as Gladius, Gladius Mongo, Pilot 200, Raider). Stiff tip, non-polymer-jacketed guidewires should be avoided through tortuous vessels or when the vessel course is poorly understood, as they carry an increased risk of perforation (Fig. 8.2.13).

CTOs with heavy calcification or tortuosity and well-understood vessel course: A **medium tip stiffness, tapered-tip** guidewire (such as Gaia 1st and 2nd, Gaia Next 1 and 2, or Judo 1 and 3) is usually used first, followed by escalation to a stiff tip, polymer-jacketed guidewire (such as Gladius Mongo, Pilot 200, Raider).

CTOs with heavy calcification or tortuosity and poorly understood vessel course: A **soft tip, tapered-tip, polymer-jacketed** guidewire is usually used first, followed by escalation to a stiff tip, polymer-jacketed guidewire (such as Gladius Mongo, Pilot 200, Raider). Stiff tip, non-polymer-jacketed guidewires should be avoided due to increased risk of perforation (Fig. 8.2.13).

Wire advancement techniques

Traditionally, three guidewire handling techniques have been described, two (sliding and drilling) used for both non-CTO and CTO lesions and one (penetration) used specifically for CTO lesions. Combinations of these techniques are often used.

1. **Sliding**, is usually the first step in CTO crossing and consists of forward movement of a tapered-tip polymer-jacketed guidewire (such as the Fielder XT, Bandit, or Fighter wires), aiming to track microchannels within the CTO. The wire is advanced with gentle tip rotation and probing, that is, a modest controlled drilling movement. These wires provide limited tactile feedback, hence visual assessment of the wire course is important. An apparent deflection of the tip must be avoided because it

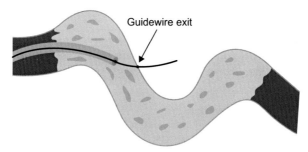

FIGURE 8.2.13 **Wire perforation caused by the advancement of a stiff tip guidewire though a tortuous CTO**.

may lead to wire entry into the extraplaque position. If the wire fails to progress within a few minutes, the guidewire and advancement technique should be changed.

2. **Drilling** consists of controlled rotation of the guidewire in both directions. Usually guidewires with moderate tip stiffness (3−6 grams) are used (such as the Gaia, Gaia Next, and Ultimate 3 guidewires), followed by escalation to stiffer tip wires (stiffer tip wires provide less tactile feedback). A small tip bend is crucial for this technique to avoid the creation of a large extraplaque space. Non-polymer-jacketed wires are recommended as the first choice (such as the Ultimate 3 g or Miracle 6 g) for this method because of their better tactile feedback.

3. **Penetration** consists of forward guidewire advancement intentionally steering (directing) the wire, not blindly rotating it, usually using a medium (such as Gaia, Gaia Next and Judo) or high (such as Miracle 12, Hornet 14, Infiltrac and Infiltrac Plus, and Confianza Pro 12) tip stiffness guidewire. The guidewire is used as a "needle" to penetrate the occlusion. This technique is important for lesions with a calcified, hard to penetrate, proximal cap and for steering through shorter occlusions when the vessel course is well understood. After crossing the area of resistance, wire de-escalation is performed and transition to the other two wire advancement techniques.

Wire advancement tips and tricks

1. The microcatheter should be as close as possible to the tip of the wire, but not too close to avoid biasing the direction of the wire and also to prevent a perforation in case the guidewire exits the vessel architecture. Advancing and retracting the microcatheter can change the penetration force of the wire.

2. Flexibility is important: if no progress is achieved within a few minutes, the wire and wire advancement strategy should be modified.

3. Direction of advancement: **alternating between approximately orthogonal views** can prevent inadvertent departure from the intended guidewire course, especially when directed penetration is employed. Vessel wall calcifications or previously implanted stents can also be very helpful as an aid for determining the vessel course and guidewire location. Contralateral injections are also very important in determining the guidewire location, as discussed in Section 6.1.

4. In the case of noncalcified, long lesions and/or tortuosity of the occluded vessel,[23] the vessel course can be determined if coronary CT images are postprocessed and shown at the same angle as that of the gantry during wire advancement (Section 6.4).[24,25]

How to assess the wire position

It is **critical** to understand the guidewire course before advancing the microcatheter or other equipment. Wire exit in most cases does not cause significant complications, whereas microcatheter or other equipment exit usually causes significant perforations.

How can the guidewire location be ascertained?

Best method

1. Contralateral contrast injection is the best method for understanding the guidewire position when collaterals arise primarily from the contralateral coronary artery. Two orthogonal views are needed. In patients with only ipsilateral collaterals, ipsilateral contrast injection can also be helpful, but it could enlarge an extraplaque dissection caused by wiring attempts.

   Less reliable methods (require that the operator is certain [based on dual injection], that the guidewire is NOT outside the vessel architecture):

1. Distal wiring with a workhorse guidewire: if the operator is certain that the CTO crossing guidewire is within the vessel architecture, a microcatheter is advanced over it, followed by exchange for a workhorse guidewire. Easy distal advancement of the workhorse wire, especially if branches can be intentionally selected, suggests a true lumen position. A workhorse wire will not traverse easily in the extraplaque space and will likely coil up or prolapse on itself indicating an extraplaque position.
2. Aspirating through the microcatheter: blood return is suggestive of distal true lumen crossing, however extraplaque hematoma formation can also result in blood return. Aspiration can help reduce the size of extraplaque hematoma.
3. Transducing the microcatheter pressure can be helpful (but requires microcatheter advancement that should be avoided before ascertaining that the guidewire has not exited from the vessel architecture): an arterial waveform suggests distal true lumen position (**CTO Manual online case 131**) (Fig. 8.2.14).

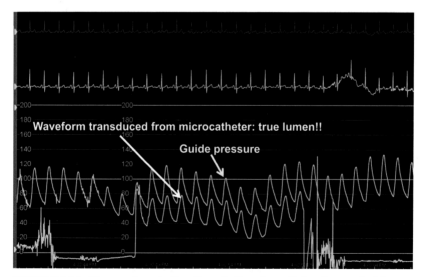

FIGURE 8.2.14   **Pressure transduction through the microcatheter shows arterial waveform, suggesting distal true lumen position**.

4. Intravascular ultrasound, but advancing the intravascular ultrasonography (IVUS) catheter into the extraplaque space can extend the dissection and hinder wire reentry attempts.

Not recommended methods

1. Contrast injection through the microcatheter. This maneuver can cause expansion of the extraplaque space if the microcatheter has not crossed into the distal true lumen, hence it is NOT recommended.
2. Sudden, spontaneous freedom of the wire tip as one passes the distal cap should NOT be used as a clue to guidewire position: it may suggest that distal true lumen wire position has been achieved, but can also signify guidewire exit into the pericardium or entry into the extraplaque space.

Assessing wire position: tips and tricks

1. **During CTO crossing, equipment should NEVER be advanced over a guidewire if the operator is unsure about the position of the wire, as the guidewire may be outside the vessel architecture. Contrast injection from the donor vessel in orthogonal projections can often clarify the wire position. Wire perforations are usually well tolerated, but catheter perforation can be catastrophic.**
2. The position of the guidewire tip should be checked when it reaches 2—3 mm proximal to the distal cap. The guidewire should not be advanced further until after confirmation that it is pointing towards the distal true lumen in two orthogonal views (to reduce the likelihood of extraplaque guidewire entry).
3. In the case of exclusively or dominantly ipsilateral collateral supply, introducing an OTW balloon or microcatheter into the donor branch for selective contrast injection is a useful technique with a very low risk of antegrade hydraulic dissection. A thrombus aspiration catheter can also be used and is more effective for this purpose if the branch is large enough.

There are six possible outcomes of antegrade wiring (Fig. 8.2.15):

1. Crossing into a side branch
2. Exit from the vessel architecture (wire perforation)
3. Extraplaque position within the occlusion
4. Distal false lumen position
5. Intraplaque position
6. Crossing into the distal true lumen

Outcome 1: Crossing into a side branch

- Occasionally the guidewire may enter a side branch or a bypass graft and appear as if it has exited the vessel architecture. Therefore, careful assessment of the diagnostic angiogram is critical for understanding the coronary anatomy and guidewire position.
- In case of side branch entry the guidewire can (Fig. 8.2.16):

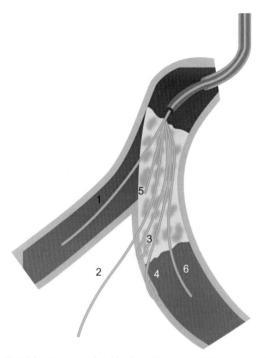

FIGURE 8.2.15 **Possible outcomes of guidewire advancement through the CTO**.

a. be withdrawn and redirected, or
b. be left in place followed by crossing attempts with another guidewire (parallel wire technique described below, which is often facilitated by use of a dual lumen microcatheter). An intravascular ultrasound catheter can sometimes be used in the side branch to help clarify the course of the main vessel and assist with subsequent wire crossing attempts.

Outcome 2: Wire exits the vessel architecture

- If the guidewire exits the vessel architecture (vessel's adventitia), it should be withdrawn, followed by repeat crossing attempts using the same or a different guidewire or another guidewire could be used parallel to the initial one (Fig. 8.2.17).
- Wire exit without advancing a microcatheter or other equipment is very unlikely to cause a clinically important perforation or tamponade due to the small caliber of the guidewire.

Outcome 3: Extraplaque position proximal to the distal cap

The CTO-ARC document recommends the use of the terms intraplaque (for wire tracking within the occlusive intima-based plaque) and extraplaque (for wire tracking outside the plaque but still contained within the adventitial layer) when describing the device course within the occluded CTO segment.[26]

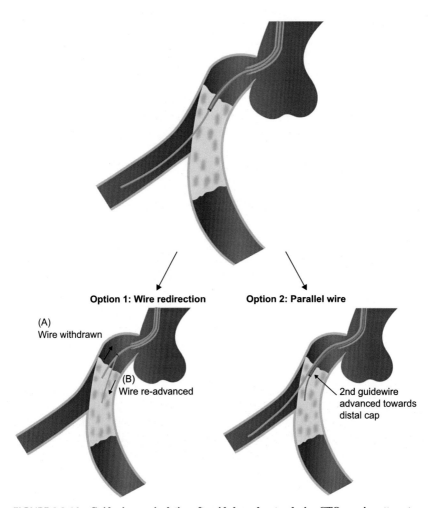

FIGURE 8.2.16   Guidewire manipulation after side branch entry during CTO crossing attempts.

The following options exist after a guidewire has been advanced to the extraplaque position within the occluded CTO segment (Fig. 8.2.18):

Option 1. Wire redirection: Retract and redirect the guidewire. Usually moderate tip stiffness, tapered tip guidewires are used, such as the Gaia, Gaia Next, and Judo series.

Option 2. Parallel wire and see-saw (CTO Manual online case 68, 73): Leave the original guidewire in place and advance a second guidewire "parallel" to the first one. The parallel wire technique is one of the oldest and most popular techniques for CTO PCI.[27,28] In the parallel wire technique (Fig. 8.2.19) when the

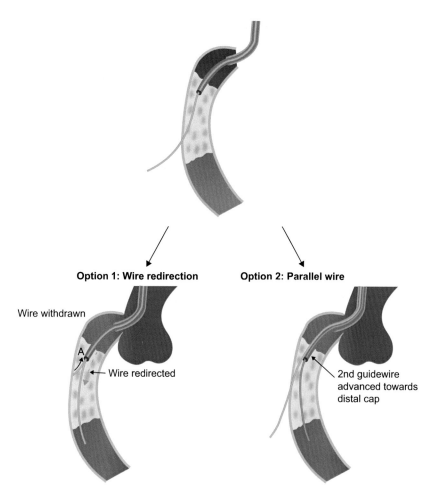

Option 1: Wire redirection

Wire withdrawn

A

← Wire redirected

Option 2: Parallel wire

2nd guidewire
advanced towards
distal cap

FIGURE 8.2.17   **Managing guidewire exit from the vessel architecture during CTO crossing attempts**.

guidewire enters the extraplaque space (or occasionally a side branch), it is left in place, the microcatheter is removed and a second guidewire (usually stiffer than the initial wire) is advanced "parallel" to the first wire through the microcatheter until it enters into the distal true lumen. In the parallel wire technique, a single support catheter is used, whereas in a variation of the parallel-wire technique called the "see-saw" technique (Fig. 8.2.20) two microcatheters (or over-the-wire balloons) are used to support both guidewires. Alternatively, a dual lumen microcatheter, (Section 30.3) can be used to direct the second guidewire (Fig. 8.2.21).[29]

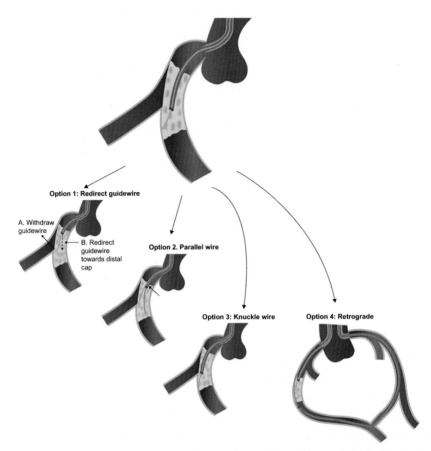

FIGURE 8.2.18   **How to manage extraplaque wire position within the CTO during CTO crossing attempts**.

Parallel wire technique tips and tricks

a. **Avoid prolonged parallel wire attempts as these can cause enlargement of the extraplaque space and hinder reentry attempts**.

b. Use of a dual lumen microcatheter (Fig. 8.2.21) is useful because it: (a) keeps the position of the first guidewire stable, (b) enhances the second guidewire penetration capacity, (c) straightens the vessel, and (d) allows easier reshaping of the second guidewire.

c. Similarly, the use of two microcatheters (see-saw technique; Fig. 8.2.20) is advantageous, as it allows better support and easier reshaping of the second guidewire.

d. Use of contrast and fluoroscopy may be decreased with parallel wire techniques, as the first guidewire acts as a "marker" that

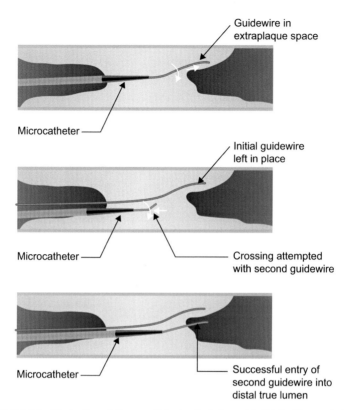

Guidewire in
extraplaque space

Microcatheter

Initial guidewire
left in place

Microcatheter

Crossing attempted
with second guidewire

Microcatheter

Successful entry of
second guidewire into
distal true lumen

FIGURE 8.2.19 **Illustration of the parallel wire technique.** *Modified with permission from the Brilakis ES.* Manual of coronary chronic total occlusion interventions. A step-by-step approach. *2nd ed. Elsevier; 2017.*

guides the advancement of the second guidewire. The first guidewire may also help "straighten" the vessel.

e. The wires most commonly used as second "parallel" guidewires are intermediate tip stiffness and highly torqueable wires, such as the Gaia, Gaia Next, and Judo series.[30]

f. The second guidewire should usually be stiffer than the first one.

g. The second guidewire should diverge from the course of the first wire at the presumed entry point into the extraplaque space (estimated based on vessel bending, calcification, side branch ostium, etc.).

h. Rotation of the second guidewire should be limited (i.e., the wire should not be "spinned") to minimize the likelihood of wrapping around the first guidewire. This complication can be prevented by using a dual lumen microcatheter.

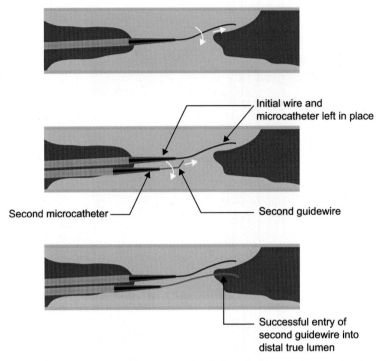

Initial wire and
microcatheter left in place

Second microcatheter

Second guidewire

Successful entry of
second guidewire into
distal true lumen

FIGURE 8.2.20 **Illustration of the "see-saw" technique**. *Modified with permission from the Brilakis ES*. Manual of coronary chronic total occlusion interventions. A step-by-step approach. 2nd ed. Elsevier; 2017.

i. Although leaving the first guidewire in place may block the entrance of the second guidewire into the same space, occasionally the second guidewire may follow the path of the first guidewire; hence, early redirection of the guidewire within the CTO is important to create a new pathway. The use of intravascular ultrasound in a side branch next to the proximal cap can facilitate determining the extraplaque or true lumen position of the first guidewire.

j. It may sometimes be difficult to adequately visualize both guidewires, but this can be facilitated by orthogonal angiographic views, which also allow understanding of the exact guidewire position during advancement.

k. Occasionally >2 guidewires can be used in parallel wire techniques, but wire visualization can be challenging due to overlap.

l. Sometimes the tip of the second guidewire may require a more acute bend to find the distal true lumen.

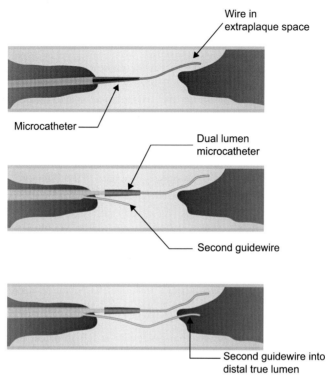

FIGURE 8.2.21   Illustration of the use of a dual lumen microcatheter for parallel wiring.
Modified with permission from the Brilakis ES. Manual of coronary chronic total occlusion interventions. A step-by-step approach. *2nd ed. Elsevier; 2017.*

       m. The parallel wire technique may have a special advantage in case of intra-occlusion tortuosity, because the first wire stretches the vessel, favorably modifying the curve and making tracking with the second wire easier.

Option 3. **Knuckle a polymer-jacketed guidewire** distal to the distal cap, followed by antegrade reentry, as described in outcome 4. This strategy is preferred for long occlusion length and heavily calcified occlusions.

Option 4. **Retrograde crossing**: the original guidewire is left in place, followed by retrograde guidewire advancement in the extraplaque space and CTO crossing using the reverse CART technique.

Outcome 4: Distal extraplaque position

Advancing a guidewire into the extraplaque space distal to the distal cap should be avoided, as it may lead to compression of the distal true lumen, hindering distal true lumen wire entry.

If the guidewire is found to be in the extraplaque space space when advanced past the distal cap, the following techniques can be used (Fig. 8.2.22):

1. **Wire redirection**: retract and redirect the guidewire: this is generally not recommended, as the original guidewire will usually follow the path of least resistance and remain in the extraplaque space. Moreover, wire manipulations may create or enlarge an extraplaque hematoma.
2. **Parallel wire and see-saw**: Leave the original guidewire in place and advance a second guidewire "parallel" to the first one (parallel wire technique, as described in Option 2 above).
3. **Device-based antegrade reentry**: advance a Stingray balloon (Section 30.16.2) or ReCross microcatheter (Section 30.6.4) to the extraplaque position followed by reentry.
4. **Wire-based antegrade reentry**: bring a microcatheter into the extraplaque space and position its tip adjacent to a well visualized segment of the distal true lumen. In a projection that provides optimal visualization, use directed penetration to reenter into the true lumen. Avoid extending the extraplaque track in an uncontrolled fashion. IVUS can be used to guide reentry, but it can be difficult to advance and may lead to extraplaque hematoma enlargement.
5. **Retrograde crossing**: the original guidewire is left in place, followed by retrograde guidewire advancement in the extraplaque space and CTO crossing, usually using the reverse CART technique.

Outcome 5. Intraplaque position
For intraplaque wire position there are several options (Fig. 8.2.23):
1. Advance the microcatheter close to the tip of the guidewire and redirect the guidewire towards the distal true lumen. The guidewire tip can be reshaped or different guidewires may be used. Intermediate tip stiffness, tapered-tip guidewires are usually preferred for this task.
2. Parallel wiring, as described above.

Outcome 6. Wire crosses into the distal true lumen
CTO Manual online case 67

After distal true lumen crossing has been achieved (Fig. 8.2.24):
1. Advance the microcatheter into the distal true lumen. Additional measures such as the use of a guide catheter extension or balloon anchoring in proximal side branches may be required to assist crossing of the CTO with a microcatheter, as described in Section 9.3.1.1.3.
2. Remove the CTO crossing guidewire and exchange it for a workhorse guidewire. CTO crossing wires are more likely to cause distal vessel perforation and dissection as compared with workhorse guidewires when used for equipment delivery. Sometimes, use of a supportive non-

The problem: extraplaque wire position distal to the distal cap

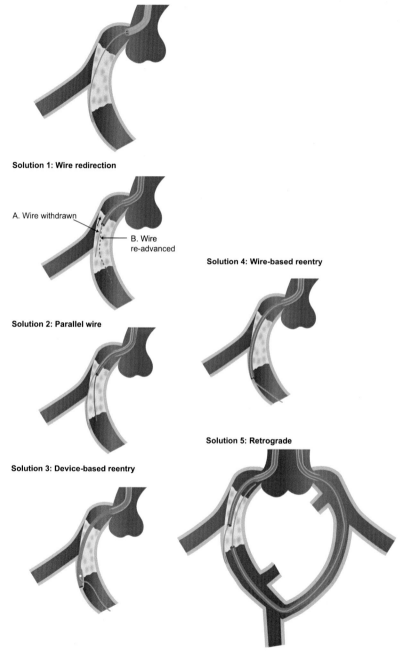

Solution 1: Wire redirection

A. Wire withdrawn

B. Wire re-advanced

Solution 2: Parallel wire

Solution 3: Device-based reentry

Solution 4: Wire-based reentry

Solution 5: Retrograde

FIGURE 8.2.22   **How to manage extraplaque guidewire position distal to the distal cap during CTO crossing attempts**.

**The problem: intraplaque guidewire
position during antegrade wiring**

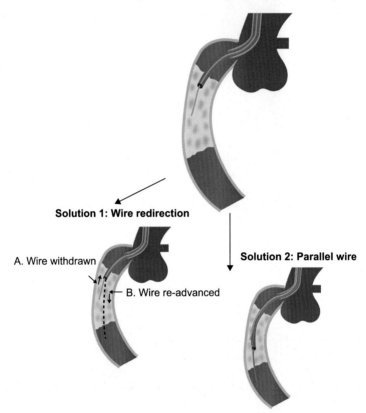

Solution 1: Wire redirection

A. Wire withdrawn

B. Wire re-advanced

Solution 2: Parallel wire

FIGURE 8.2.23   How to manage intraplaque guidewire position during CTO crossing attempts.

polymer-jacketed wire, such as the BHW, Ironman, or Grand Slam, can facilitate equipment delivery even with limited guide catheter support or in calcified or tortuous vessels.

3. Remove the microcatheter (ideally using the trapping technique to minimize wire motion and use of fluoroscopy, as described in Section 8.2.2.9.1).

4. Proceed with lesion preparation and stenting.

Stage 3. Puncture of the distal cap

Although in many patients the distal cap is softer than the proximal cap, in some patients (such as prior CABG patients in whom the distal cap is exposed to systemic pressure) the distal cap can be heavily calcified and difficult to penetrate, potentially deflecting the guidewire towards the extraplaque space.

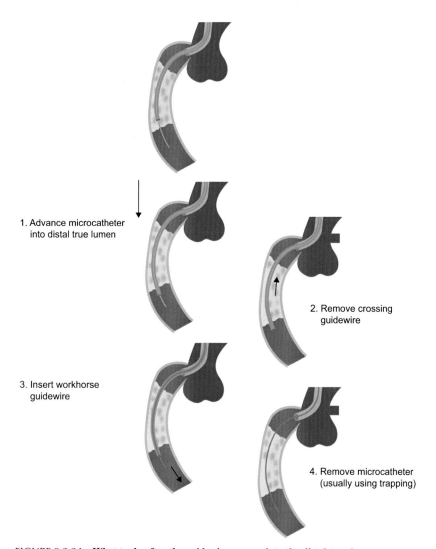

1. Advance microcatheter
into distal true lumen

2. Remove crossing
guidewire

3. Insert workhorse
guidewire

4. Remove microcatheter
(usually using trapping)

FIGURE 8.2.24   **What to do after the guidewire crosses into the distal true lumen**.

In such cases, the microcatheter is advanced close to the distal cap and a highly steerable guidewire (such as the Gaia and Gaia Next series) is used to direct the wire towards the distal true lumen (Fig. 8.2.25). Occasionally, stiffer tip guidewires (such as the Confianza Pro 12, Hornet 14, Infiltrac and Infiltrac Plus, and Warrior) may be needed if the distal cap is calcified and resistant to penetration. Contralateral injection should be used early to determine whether the guidewire is entering the true or a false lumen.

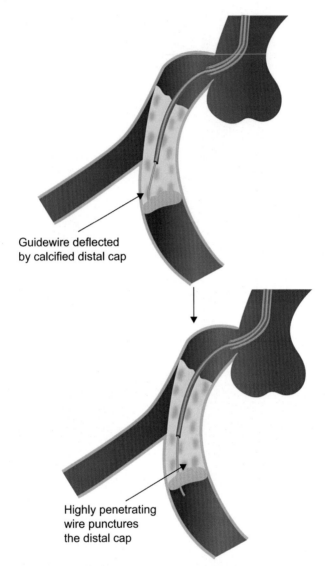

Guidewire deflected
by calcified distal cap

Highly penetrating
wire punctures
the distal cap

FIGURE 8.2.25 **Guidewire deflection by a heavily calcified distal cap. The distal cap was punctured using a highly penetrating guidewire.**

8.2.2.8 Step 8. Advancing the guidewire and microcatheter distal to the target lesion and exchanging the CTO crossing guidewire for a workhorse (or other) guidewire

Goal: To advance the guidewire that crossed the CTO as far distally as safely possible from the CTO distal cap in order to facilitate advancement of the microcatheter distally. The CTO crossing guidewire is subsequently exchanged for a workhorse guidewire over which, balloon and stent delivery will subsequently take place.

How (Fig. 8.2.26): The CTO crossing guidewire is advanced distally under fluoroscopic guidance. The microcatheter is then advanced over the guidewire, followed by exchange of the CTO crossing guidewire for another guidewire (usually workhorse, but sometimes it can be a support guidewire, such as the Grand Slam or Wiggle wires, or an atherectomy guidewire). Contralateral contrast injection can be used to confirm that the guidewire is in a good distal location.

What can go wrong?

1. **CTO crossing wire advanced too distally causing a perforation or dissection**
a. **Prevention:**
    i. Monitor distal guidewire position and prevent very distal advancement. This is especially important for CTO PCI, since stiff tip and polymer-jacketed guidewires are often used for crossing and such wires carry a high risk of perforation if advanced too distally through a small branch (**CTO Manual online case 172**).
    ii. If a stiff tip or polymer-jacketed guidewire is used to cross the lesion, exchange it for a workhorse guidewire immediately after crossing.
b. **Treatment**
    i. Start with balloon inflation to stop bleeding into the pericardium, followed by fat or coil embolization if extravasation continues. This is described in detail in Section 26.4.

2. **Unable to advance the microcatheter over the guidewire that crossed the CTO**

a. **Causes**
    i. Heavy tortuosity and calcification within the occlusion or the CTO distal cap.
    ii. Microcatheter deformation or fatigue.
    iii. Poor guide catheter support.
b. **Prevention**
    i. Obtain strong guide catheter support.
    ii. Exchange the microcatheter if excessive friction is experienced, suggesting microcatheter fatigue.
c. **Treatment**
    i. Treatment is described in the balloon uncrossable lesion section (Section 23.1).

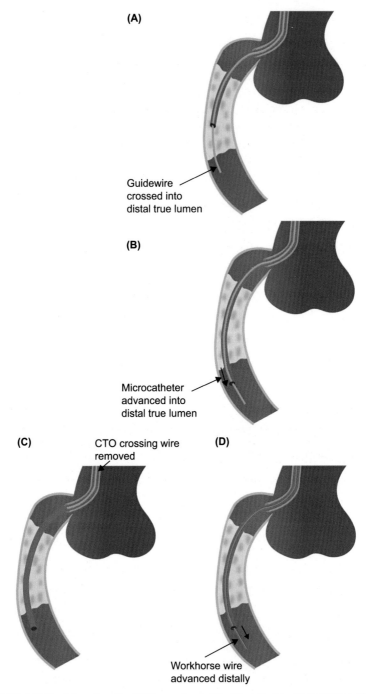

(A)

Guidewire
crossed into
distal true lumen

(B)

Microcatheter
advanced into
distal true lumen

(C) CTO crossing wire
removed

(D)

Workhorse wire
advanced distally

FIGURE 8.2.26 **Advancing the CTO crossing guidewire and the microcatheter distally and exchanging it for a workhorse (or other) guidewire**. **Panel A**: the antegrade guidewire crosses into the distal true lumen. **Panel B**: the microcatheter is advanced over the guidewire into the distal true lumen. **Panel C**: the crossing guidewire is removed. **Panel D**: a workhorse guidewire is advanced into the distal true lumen.

8.2.2.9 Step 9. Removing the microcatheter (after exchanging the CTO crossing guidewire for a workhorse guidewire)

Goal: To remove the microcatheter used for lesion wiring while maintaining distal wire position without movement.

How: There are four techniques for removing or exchanging microcatheters or any over-the-wire system when using a short (180–190 cm) guidewire[18]:

1. **Trapping (preferred technique)**
2. Hydraulic exchange
3. Use of a guidewire extension
4. "Circumcision" of the over-the-wire system

8.2.2.9.1 *Trapping*

Trapping is the best technique for removing or exchanging microcatheters or any over-the-wire system when using a short (180–190 cm) or even long (300 cm) guidewire, because it:

1. Minimizes guidewire movement, which can result in guidewire position loss, distal vessel injury, and/or perforation.
2. Minimizes radiation exposure (as fluoroscopy is only needed during the initial phase of microcatheter withdrawal).

The small inner diameter of a 6 Fr guide catheter prevents utilization of the trapping technique for over-the-wire balloons, the CrossBoss catheter, and the Stingray balloon. However, 6 Fr guides with 0.071 inch inner diameter (such as Medtronic Launcher guides) can allow trapping of low profile catheters such as Turnpike, SuperCross, Corsair, Caravel, Sasuke, and Finecross, especially if a specialized trapping balloon is used (Section 30.9.4).

The TrapLiner catheter (Section 30.2) combines guide extension functionality with a built-in trapping balloon to pin short guidewires. This catheter is available in 6, 7, and 8 Fr sizes. The advantages of the TrapLiner are less need of fluoroscopy to position a trapping balloon and reduced equipment use, since standard balloons used for trapping can be difficult to reinsert through the Y-connector after multiple inflations during prolonged procedures. The TrapLiner catheter works best with short (135 cm) microcatheters, as 150 cm long microcatheters are often too long to expose the guidewire for trapping with the TrapLiner balloon.

Trapping technique (Fig. 8.2.27):
Step 1: Withdraw the over-the-wire balloon or microcatheter into the guide catheter just proximal to the position where the trapping balloon will be inflated (Fig. 8.2.27, panel B).

Step 2: Insert the trapping balloon through the Y-connector, next to (but not over) the guidewire. Either standard semicompliant balloons or dedicated

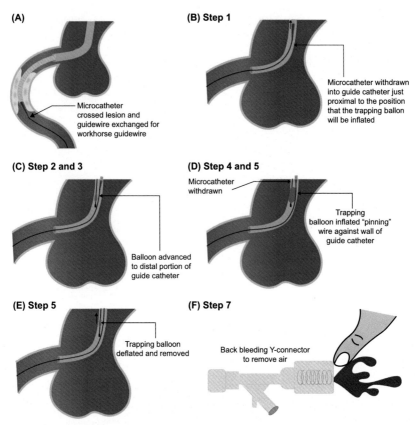

FIGURE 8.2.27 Illustration of the trapping technique. *Modified with permission from the Brilakis ES.* Manual of coronary chronic total occlusion interventions. A step-by-step approach. *2nd ed. Elsevier; 2017.*

"trapping" balloons can be used. Keeping the stylet in the end of the balloon catheter can assist in advancement through the hemostatic valve of the Y-connector; the stylet should be removed before advancing the balloon further.

Type of balloon: Over-the-wire balloons have larger profiles compared with rapid exchange balloons and may not fit into the guide catheter along with other equipment, such as the Stingray balloon and the Venture catheter. Dedicated trapping balloons (Section 30.9.4) have smaller profiles and can trap equipment in smaller guide cattheters.

Size of the trapping balloon: 2.5 mm for 6 or 7 French guide catheters, 3.0 mm for 7 and 8 French guide catheters.

Length of the trapping balloon: Ideally ≥ 20 mm (longer balloon length provides more area of contact and "pins" the wire better, which is especially important for trapping polymer-jacketed wires that are more slippery).

Caveats: A balloon previously used for trapping may get deformed and be difficult to reinsert through the Y-connector on subsequent attempts. In such cases, the trapping balloon should be exchanged for a new one. Attempting to trap multiple devices within a single guide catheter is more likely to fail, resulting in loss of wire position.

Step 3: Advance the trapping balloon to the distal portion of the guide catheter beyond the marker of the microcatheter or over-the-wire balloon, usually at or near the primary curve (Fig. 8.2.27, panel C).
Caveat: when shortened guide catheters are used, balloon markers cannot be relied upon to prevent inadvertent advancement of the trapping balloon into the target vessel.

Step 4: Inflate the trapping balloon (Fig. 8.2.27, panel D).
Pressure: 15−20 atm to provide adequate "trapping." The guide catheter pressure should be dampened during trapping.
Caveat: Balloon may rupture if inflated next to "handmade" side holes.

Step 5: Withdraw the microcatheter (or over-the-wire balloon) from the guide catheter (Fig. 8.2.27, panel D). Once the trapping balloon is in place and inflated, fluoroscopy is generally not required to maintain the distal wire position. Extra care should, however, be utilized when working with polymer-jacketed guidewires, as they have a tendency to slip past the inflated trapping balloon, during aggressive withdrawal of the microcatheter (or the over-the-wire balloon).

Step 6: Deflate and remove the trapping balloon (Fig. 8.2.27, panel E).

Step 7: Back bleed the Y-connector (Fig. 8.2.27, panel F).
This is a very important step because often air is entrained into the guide catheter during trapping. If contrast is injected without back bleeding, coronary air embolization is likely to occur. An alternative technique to prevent air embolization is to use continuous flushing when removing the catheters while the trapping balloon is inflated.

Step 8: The microcatheter has been successfully removed while maintaining the distal wire position.
Tip: if an anchor balloon is kept in a side branch and its size is appropriate, that is, ≥ 2.5 mm in a 6 or 7 F guide catheter, this balloon can be deflated and temporarily withdrawn into the guide catheter for trapping. This maneuver can save time and cost.

What can go wrong during trapping?

1. **Air embolization**
    Prevention
    a. Back bleed the guide after using the trapping technique.
    Treatment.
    a. Treatment of air embolism is described in Section 25.2.3.3.
2. **Loss of wire position**
    Prevention
    a. Apply trapping technique with meticulous attention to detail.
    Treatment
    a. Rewire the vessel.

8.2.2.9.2   Hydraulic exchange (Fig. 8.2.28)

Hydraulic exchange (also called "jet exchange" or "Nanto technique" is easier to perform than trapping, but less reliable and may not maintain the distal position of the guidewire.[31,32]
Caveats
Use of the hydraulic exchange should be avoided in CTO PCI in general and especially when using stiff tip, tapered-tip guidewires or stiff polymer-jacketed guidewires in order to minimize the risk of inadvertent guidewire advancement and distal guidewire perforation or dissection. Hydraulic exchange should also be avoided for removing over-the-wire balloon catheters and the Stingray balloon because increased friction between the inner surface of the catheter and the guidewire may result in loss of guidewire position.

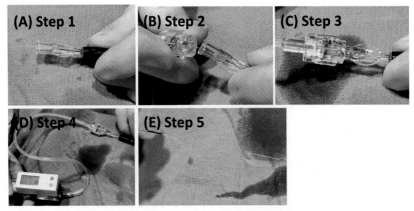

FIGURE 8.2.28 **Illustration of the hydraulic exchange technique.** *Courtesy Dr. William Nicholson. Reproduced with permission from Brilakis ES.* Manual of coronary chronic total occlusion interventions. A step-by-step approach. *Cambridge, MA: Elsevier, 2nd ed. 2017.*

Hydraulic exchange can be performed through any size guide catheter as follows:

Step 1: Fill an inflation device with normal saline (alternatively a standard mixture of contrast and saline can be used). The inflating device should be filled with the maximum volume possible. Alternatively, a 3 cc luer lock syringe can be used, however, an inflation device is preferred.

Step 2: Withdraw the microcatheter or over-the-wire balloon as far as possible until the back end of the short guidewire is at the hub of the microcatheter/over-the-wire balloon (Fig. 8.2.28, panel A).

Step 3: Connect the saline-filled inflation device to the hub of the microcatheter/over-the-wire balloon (wet to wet connection) (Fig. 8.2.28, panels B and C).

Step 4: Pressurize the inflation device to 14−20 atm, while performing fluoroscopy (Fig. 8.2.28, panel D). Open the valve of the Y-connector to minimize friction on the microcatheter.

Step 5: Under fluoroscopy, when the inflation device pressure reaches 14−20 atm withdraw the microcatheter/over-the-wire balloon (Fig. 8.2.28, panel E), while maintaining the inflation pressure at 14−20 atm as the microcatheter is withdrawn. Otherwise, if the pressure is lost, the wire will be withdrawn. If partial guidewire withdrawal occurs, slowing down the speed of microcatheter withdrawal while maintaining the inflation device pressure and opening the hemostatic valve of the Y-connector can often correct/salvage the situation.

8.2.2.9.3   Use of a guidewire extension

How?

By inserting the back end of the short guidewire into the guidewire extension (Fig. 8.2.29). Using a guidewire extension can be performed with any size guide catheter.[18]

Caveats

1. Each guidewire type has a specific guidewire extension that should be available in the catheterization laboratory. For example:
   a. Abbott guidewires: "DOC" guidewire extension (145 cm long).
   b. Asahi guidewires: "Asahi" guidewire extension (150/165 cm long).
   c. Cordis guidewire: "Cinch" guidewire extension (145 cm long).
2. The connection should be tightened as much as possible to minimize the risk of the guidewire and guidewire extension coming apart during withdrawal of the microcatheter or over-the-wire balloon.
3. Be careful to avoid kinking or bending of the wire when tightening the connection.

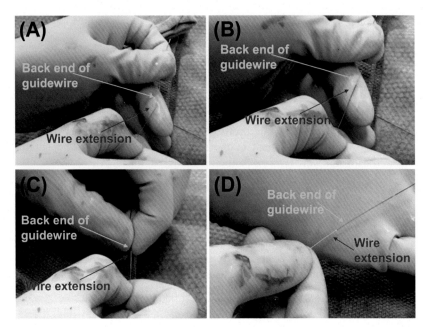

FIGURE 8.2.29   **Using a guidewire extension to remove a microcatheter over a short guide-wire**. The wire extension is aligned with the back end of the guidewire to the extended (**panel A**) and advanced over it until a tight connection is achieved (**panels B—D**).

8.2.2.9.4   Circumcision technique (Fig. 8.2.30)

This is an advanced and cumbersome technique but can be very useful, especially in cases where the guidewire becomes entrapped within the microcatheter/over-the-wire balloon, or when appropriate extension guidewires are not available and trapping is not possible. The circumcision technique can be performed with any size guide catheter.

Step 1: Withdraw the microcatheter or over-the-wire balloon as far as possible until the back end of the short guidewire is at the hub of the microcatheter/over-the-wire balloon.

Step 2: Using a scalpel and a hard surface, a circumferential cut is made as close to the Y-introducer as possible.
(Fig. 8.2.30, panels A and B).

Step 3: Once circumferential cutting is complete, the microcatheter/over-the-wire balloon fragment is removed.
(Fig. 8.2.30, panel C).

Step 4: Steps 1—3 are repeated until the entire microcatheter/over-the-wire balloon is removed. Cutting 2—3 segments is usually necessary to remove the entire microcatheter or over-the-wire balloon.

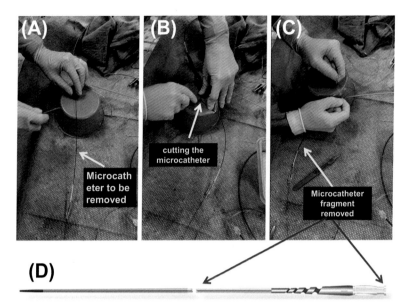

FIGURE 8.2.30   **Illustration of the circumcision technique for removing a microcatheter.**

Caveats

1. Cutting should be done very carefully to minimize the risk of injuring the patient, or the operator, and to avoid damaging the guidewire making it difficult, if not impossible, to pass other equipment over it.
2. Consider using a hard surface, such as an upside-down saline bowl upon which to perform the cutting.
3. The microcatheter/over-the-wire balloon will be destroyed during this maneuver and hence cannot be reused.

8.2.2.10   Step 10. Monitoring the guidewire position

Goal

To avoid inadvertent guidewire withdrawal (that can be very frustrating in cases where CTO crossing was laborious) or distal guidewire advancement (that can lead to perforation).

How?

The tip of the guidewire is monitored. If the tip position changes the guidewire position is adjusted.

What can go wrong?

8.2.2.10.1   Perforation

Perforation can occur if the guidewire tip is advanced too distally. The risk is increased if polymer-jacketed or stiff tip and/or tapered-tip guidewires are used.

Prevention
- Continuous (or at least intermittent) monitoring of the guidewire position. This is especially important if collimation is used, not allowing continuous monitoring of the guidewire tip position and when there is significant difficulty with equipment delivery or withdrawal that often results in excessive guidewire movement.

Treatment
- Distal guidewire perforation is discussed in Section 26.4. The first step is to inflate a balloon proximal to the perforation site to prevent further bleeding into the pericardium, while preparations are made for definitive treatment, which usually involves fat or coil embolization.

8.2.2.10.2   Guidewire withdrawal proximal to the target lesion
This may lead to failure of the procedure despite successful guidewire crossing.

Prevention
- Continuous monitoring of the guidewire position.
- Good lesion preparation.
- Avoid aggressive attempts to advance equipment through a lesion. For example, if a microcatheter, balloon, or stent cannot cross a lesion, repeat balloon angioplasty (or sometimes atherectomy) can facilitate crossing.

References

8. Grantham JA, Marso SP, Spertus J, House J, Holmes Jr. DR, Rutherford BD. Chronic total occlusion angioplasty in the United States. *JACC Cardiovasc Interv* 2009;**2**:479−86.
9. Morino Y, Kimura T, Hayashi Y, et al. In-hospital outcomes of contemporary percutaneous coronary intervention in patients with chronic total occlusion insights from the J-CTO Registry (Multicenter CTO Registry in Japan). *JACC Cardiovasc Interv* 2010;**3**:143−51.
10. Sianos G, Werner GS, Galassi AR, et al. Recanalisation of chronic total coronary occlusions: 2012 consensus document from the EuroCTO club. *EuroIntervention* 2012;**8**:139−45.
11. Christopoulos G, Karmpaliotis D, Alaswad K, et al. Application and outcomes of a hybrid approach to chronic total occlusion percutaneous coronary intervention in a contemporary multicenter US registry. *Int J Cardiol* 2015;**198**:222−8.
12. Wilson WM, Walsh SJ, Yan AT, et al. Hybrid approach improves success of chronic total occlusion angioplasty. *Heart* 2016;**102**:1486−93.
13. Maeremans J, Walsh S, Knaapen P, et al. The hybrid algorithm for treating chronic total occlusions in Europe: the RECHARGE registry. *J Am Coll Cardiol* 2016;**68**:1958−70.
14. Brilakis ES. *Manual of percutaneous coronary interventions: a step-by-step approach.* Elsevier; 2021.
15. Brilakis ES, Mashayekhi K, Tsuchikane E, et al. Guiding principles for chronic total occlusion percutaneous coronary intervention. *Circulation* 2019;**140**:420−33.
16. Azzalini L, Uretsky B, Brilakis ES, Colombo A, Carlino M. Contrast modulation in chronic total occlusion percutaneous coronary intervention. *Catheter Cardiovasc Interv* 2019;**93**: E24−9.

17. Brilakis ES, Grantham JA, Rinfret S, et al. A percutaneous treatment algorithm for crossing coronary chronic total occlusions. *JACC Cardiovasc Interv* 2012;**5**:367−79.

18. Brilakis ES. *Manual of coronary chronic total occlusion interventions. A step-by-step approach.* 2nd ed Elsevier; 2017.

19. Ide S, Sumitsuji S, Kaneda H, Kassaian SE, Ostovan MA, Nanto S. A case of successful percutaneous coronary intervention for chronic total occlusion using the reversed guidewire technique. *Cardiovasc Interv Ther* 2013;**28**:282−6.

20. Kawasaki T, Koga H, Serikawa T. New bifurcation guidewire technique: a reversed guide-wire technique for extremely angulated bifurcation−a case report. *Catheter Cardiovasc Interv* 2008;**71**:73−6.

21. Suzuki G, Nozaki Y, Sakurai M. A novel guidewire approach for handling acute-angle bifurcations: reversed guidewire technique with adjunctive use of a double-lumen micro-catheter. *J Invasive Cardiol* 2013;**25**:48−54.

22. Hasegawa K, Yamamoto W, Nakabayashi S, Otsuji S. Streamlined reverse wire technique for the treatment of complex bifurcated lesions. *Catheter Cardiovasc Interv* 2020;**96**: E287−91.

23. Luo C, Huang M, Li J, et al. Predictors of interventional success of antegrade PCI for CTO. *JACC Cardiovascular Imaging* 2015;**8**:804−13.

24. Ghoshhajra BB, Takx RAP, Stone LL, et al. Real-time fusion of coronary CT angiogra-phy with x-ray fluoroscopy during chronic total occlusion PCI. *Eur Radiol.* 2017;**27**:2464−73.

25. Xenogiannis I, Jaffer FA, Shah AR, et al. Computed tomography angiography co-registration with real-time fluoroscopy in percutaneous coronary intervention for chronic total occlusions. *EuroIntervention* 2021;**17**:e433−5.

26. Ybarra LF, Rinfret S, Brilakis ES, et al. Definitions and clinical trial design principles for coronary artery chronic total occlusion therapies: CTO-ARC consensus recommendations. *Circulation* 2021;**143**:479−500.

27. Rathore S, Matsuo H, Terashima M, et al. Procedural and in-hospital outcomes after percutaneous coronary intervention for chronic total occlusions of coronary arteries 2002 to 2008: impact of novel guidewire techniques. *JACC Cardiovasc Interv* 2009;**2**:489−97.

28. Mitsudo K, Yamashita T, Asakura Y, et al. Recanalization strategy for chronic total occlu-sions with tapered and stiff-tip guidewire. The results of CTO new techniQUE for STandard procedure (CONQUEST) trial. *J Invasive Cardiol* 2008;**20**:571−7.

29. Chiu CA. Recanalization of difficult bifurcation lesions using adjunctive double-lumen microcatheter support: two case reports. *J Invasive Cardiol* 2010;**22**:E99−103.

30. Khalili H, Vo MN, Brilakis ES. Initial experience with the gaia composite core guidewires in coronary chronic total occlusion crossing. *J Invasive Cardiol* 2016;**28**:E22−5.

31. Nanto S, Ohara T, Shimonagata T, Hori M, Kubori S. A technique for changing a PTCA balloon catheter over a regular-length guidewire. *Cathet Cardiovasc Diagn* 1994;**32**:274−7.

32. Feiring AJ, Olson LE. Coronary stent and over-the-wire catheter exchange using standard length guidewires: jet exchange (JEX) practice and theory. *Cathet Cardiovasc Diagn* 1997;**42**:457−66.

Section 8.3

Antegrade dissection and reentry (ADR)

Antegrade dissection/reentry can be an efficient strategy for crossing long CTOs. Antegrade dissection leverages the distensibility of the extraplaque space for traversing the CTO, followed by reentry into the distal true lumen.

8.3.1 Antegrade dissection/reentry (ADR) nomenclature

As described in Section 8.1, the terminology utilized in dissection/reentry CTO crossing strategies can be confusing (Fig. 8.3.1).[33] CTO crossing can occur either in the antegrade or the retrograde direction. In either direction, crossing can be achieved either from true lumen to true lumen or by first entering the extraplaque space, followed by reentry into the true lumen (dissection/reentry strategies) (Fig. 8.3.1).[33]

In the **antegrade** approach, dissection can be achieved by:

1. **Wire-based strategies (inadvertent or intentional wire entry into the extraplaque space or knuckle wiring).** Sometimes during antegrade wiring attempts the guidewire will enter the extraplaque (formerly called "subintimal") space. At other times, a knuckle (prolapsed) guidewire is intentionally inserted into the extraplaque space. Polymer-jacketed guidewires, such as the Fielder XT, Fighter or Bandit (Section 30.6.2.2) or the Gladius Mongo, Pilot 200 and Raider (Section 30.6.2.3) are usually used for knuckling. A knuckle wire can be formed by creating an "umbrella handle" shape before inserting it into the coronary artery or by pushing (without rotating) the wire with a conventional bend at its tip until it forms a "tight loop"[34] (Fig. 8.3.2). The Gladius Mongo wire is engineered to facilitate the creation of a "tight" knuckle at its tip. Tight knuckles are also likely to form with soft tip, polymer-jacketed guidewires (such as the Fielder XT, Fighter, and Bandit) whereas stiffer tip polymer-jacketed guidewires (such as the Pilot 200, the Gladius, and the Raider) are more likely to form larger knuckles. The knuckle is then advanced into the extraplaque space. Compared with trying to advance the tip of a wire, advancing a knuckle can be faster, safer (the loop minimizes the risk of vessel perforation), and the larger loops allow you to stay out of side branches, enhancing the efficiency of the procedure. The knuckle size and shape can be controlled by periodically advancing the microcatheter (chasing the knuckle), which also increases the support for advancing the knuckled wire more distally.

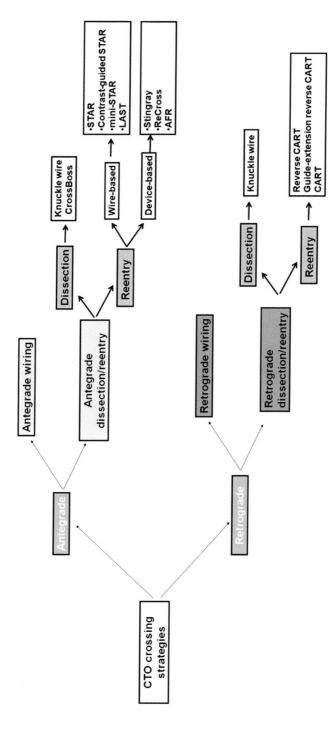

FIGURE 8.3.1 **CTO crossing strategy classification and nomenclature.** *AFR,* antegrade fenestration and reentry; *CART,* controlled antegrade and retrograde tracking; *STAR,* subintimal tracking and reentry.

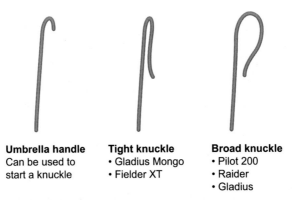

Umbrella handle
Can be used to
start a knuckle

Tight knuckle
• Gladius Mongo
• Fielder XT

Broad knuckle
• Pilot 200
• Raider
• Gladius

FIGURE 8.3.2 **Illustration of knuckle wires.** Smaller diameter knuckles are preferred, as large knuckles may enlarge the extraplaque (subintimal) space and hinder reentry. Large diameter knuckles can be useful for advancing around side branches.

2. **Catheter-based   strategy,   using   the   CrossBoss   catheter** (Section 30.16.1).[35]

In the **antegrade** approach, reentry can be achieved by (Fig. 8.3.3):

1. **Wire-based strategies**
   a. **Uncontrolled distal reentry (STAR technique) leading to extensive dissection** (Section 8.3.5.1). A knuckled guidewire is advanced distally until it spontaneously reenters into the distal true lumen (usually at a distal bifurcation). This technique was first described by Antonio Colombo in 2005 and is called **S**ubintimal **T**racking **A**nd **R**eEntry (STAR).[37] A modification of the STAR technique called "contrast-guided STAR" developed by Mauro Carlino uses extraplaque/intraplaque contrast injection through a microcatheter inserted into the proximal cap to create/visualize a dissection plane that can sometimes fenestrate into the distal true lumen.[38] The STAR technique (and its subsequent iterations): (a) often results in side branch loss which is suboptimal, especially in the left anterior descending artery, (b) is less predictably successful, and (c) has high reocclusion rates (likely due to long stent length and limited vessel outflow) and is rarely used as an upfront crossing technique.[39]

   STAR without stent placement or STAR with deferred stenting is increasingly being used after failure of other crossing attempts, followed by a repeat CTO PCI attempt after 2−3 months. This is called an "investment procedure," Section 8.3.5.1). Another type of investment procedure is "subintimal plaque modification" [SPAM]).

   b. **Distal true lumen reentry close to the distal cap leading to limited dissection**. Distal true lumen reentry as close as possible to the

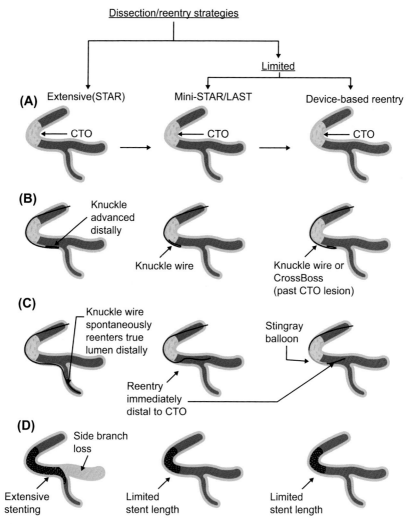

FIGURE 8.3.3 **Illustration of antegrade dissection/reentry techniques**. Limited dissection/reentry is preferred to minimize side branch loss and the risk of in-stent restenosis. **Panel A**: starting point: CTO of the mid right coronary artery. **Panel B**: dissection is performed with an antegrade knuckled guidewire. With the STAR technique the knuckled wire is advanced until it spontaneously reenters into the distal true lumen, which usually occurs at a bifurcation. With the limited dissection/reentry strategies the knuckle is advanced just distal to the distal cap. **Panel C**: with the STAR technique reentry occurs in a distal vessel, whereas with limited dissection/reentry techniques reentry is performed immediately distal to the distal cap. **Panel D**: after stenting side branches are occluded with the STAR technique, whereas they remain patent with limited dissection/reentry.[36] *Modified with permission from Brilakis ES. Manual of coronary chronic total occlusion interventions. A step-by-step approach. 2nd ed.; Elsevier; 2017.*

distal cap using a guidewire can be achieved by the "**mini-STAR**"[34] or the Limited Antegrade Subintimal Tracking (**LAST**)[40,41] technique (Section 8.3.5.2). These techniques are unpredictable and have variable success rates in reentering the distal true lumen, due to extensive uncontrolled dissection planes causing extraplaque hematoma formation and true lumen compression. Consequently, they are used infrequently, as rescue technique when other approaches have failed.

2. **Dedicated reentry systems (preferred)**
   a. Stingray (Boston Scientific) balloon and guidewire.[42,43]
   b. ReCross (IMDS) dual lumen microcatheter.
   c. Antegrade fenestration and reentry (AFR).[44–46]

Extensive dissection/reentry techniques lead to stenting of long coronary segments, often sacrificing side branches, and have been associated with poor long-term outcomes with high rates of in-stent restenosis.[37,39,47] Limited dissection/reentry (using wire-based strategies or with dedicated reentry systems, such as the Stingray balloon, the ReCross, and AFR) minimizes side branch loss and requires shorter stent lengths. Due to poor long-term outcomes associated with wire-based reentry techniques,[37,39,47] dedicated reentry systems, such as the Stingray balloon, the ReCross microcatheter, and AFR have become the preferred way to perform antegrade dissection/reentry.

8.3.2   Antegrade dissection/reentry: when?

Antegrade dissection/reentry can be used:

1. After failure of antegrade wiring or failure of the retrograde approach.
2. After extraplaque wire entry during antegrade wiring (other options are wire redirection, parallel wiring, and the retrograde approach).
3. As the initial crossing strategy (**primary dissection/reentry**). Good candidate lesions for primary dissection/reentry are those with:
   a. Well-defined proximal cap.
   b. Large caliber distal vessel.
   c. No large branches within the occluded segment or, more importantly, at the distal cap.
   d. Long occlusion length.
   e. Ambiguous vessel course or severe tortuosity within the occlusion.
   f. Severe calcification within the occlusion.
   g. Lack of interventional collaterals.

Moreover, primary dissection/reentry can be used in lesions with proximal cap ambiguity when other crossing strategies (such as IVUS-guidance or use of the retrograde approach) are not feasible or have failed (Section 16.1).

Antegrade dissection/reentry versus antegrade wiring

The optimal role and timing of antegrade dissection/reentry in CTO PCI has been a subject of debate. Some operators perform antegrade dissection/ reentry early or in complex lesions to improve the efficiency of the procedure. Other operators argue that dissection/reentry should only be used as a last resort after other crossing strategies fail.

Several studies have shown that more complex lesions are more likely to require the use of advanced crossing techniques, that is, antegrade dissection/ reentry and the retrograde approach (Fig. 8.5.10).[48−50] However, antegrade dissection/reentry carries lower risk than the retrograde approach;[51] hence, if feasible, ADR is for many operators the preferred initial advanced crossing technique in challenging CTOs.

Several studies have shown similar restenosis rates with antegrade dissection/reentry and antegrade wiring. Antegrade dissection/reentry can also provide unique solutions to anatomic challenges, such as proximal cap ambiguity ("move the cap" techniques, such as balloon-assisted subintimal entry [BASE][52] and side-BASE,[53] "scratch and go," and power knuckle technique, Section 16.1.4.3), wire uncrossable lesions (Carlino technique, Section 8.3.4.3), excessive tortuosity in the occluded segment, balloon uncrossable lesions (subintimal lesion modification or subintimal distal anchor, Section 23.1) and crossing of in-stent CTOs (using the CrossBoss catheter, Section 22.1, **CTO Manual online cases 8, 19, 180**). However, the cost of antegrade dissection/reentry equipment can be high, which along with a perception of ADR being challenging to master, limit adoption in some parts of the world.

If technically and economically feasible and if performed by operators with experience in the technique, **limited** antegrade dissection/reentry (such as the use of the Stingray system for reentry close to the distal cap) is a favorable advanced crossing technique for many complex lesions. In contrast, **extensive** dissection/reentry techniques (such as STAR, Section 8.3.5.1) have been associated with high restenosis and reocclusion rates[39,52,66] and should only be used as a final "bailout" maneuver after the failure of other crossing techniques (investment procedure).

8.3.3 Antegrade dissection/reentry: preparation

Achieving success with ADR requires: (1) strong guide catheter support and (2) techniques to prevent subintimal hematoma formation.

8.3.3.1 Strong guide catheter support

Strong guide catheter support is especially important for ADR because ADR equipment (such as the CrossBoss catheter and the Stingray balloon) can be

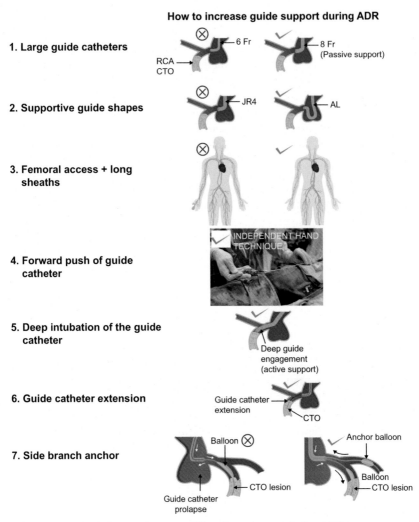

How to increase guide support during ADR

1. Large guide catheters

2. Supportive guide shapes

3. Femoral access + long sheaths

4. Forward push of guide catheter

5. Deep intubation of the guide catheter

6. Guide catheter extension

7. Side branch anchor

FIGURE 8.3.4   **Techniques to increase guide catheter support during ADR.**

bulky and difficult to deliver. Strong guide catheter support can be achieved in several ways (Fig. 8.3.4):

8.3.3.1.1   *Large guide catheters*

Large guide catheters (such as 7 or 8 French) should be routinely used for ADR and CTO PCI in general, as they provide stronger support than smaller ones (5 or 6 French) and allow easier trapping of equipment, application of different CTO crossing techniques, and treatment of complications.

8.3.3.1.2   Supportive guide shapes

EBU/XB guides provide stronger support than JL4 guides for left coronary interventions, and AL1 or 3D Right guide catheters provide stronger support than JR4 guide catheters for right coronary artery (RCA) interventions.

8.3.3.1.3   Femoral access

Femoral access often provides stronger support than radial access. Long sheaths increase guide catheter support. Left radial access provides stronger support for PCI of the right coronary artery guide while right radial access provides stronger support for PCI of the left coronary artery.

8.3.3.1.4   Forward push of the guide catheter

Forward push of the guide catheter can increase support, but may also in some cases lead to guide catheter prolapse and disengagement. The "**independent hand**" **technique** (the left hand pushes the guide, while the right hand both fixes the guidewire and advances the balloon, Fig. 9.1) allows a single operator to simultaneously manipulate the guide and advance the equipment.

8.3.3.1.5   Deep intubation of the guide catheter

Deep intubation of the guide catheter increases support, but can potentially cause proximal vessel dissection. Deep intubation is usually done in the RCA or in saphenous vein grafts (SVGs) and rarely in the left main.

8.3.3.1.6   Guide catheter extensions

Guide catheter extensions (Section 30.2) are the most commonly used tools for increasing guide catheter support.[54] They can also block blood entry into the dissection plane, minimizing the risk of extraplaque hematoma as described below. The TrapLiner guide extension can both increase support and allow trapping of equipment and is preferred for ADR. However, wedged position of the guide catheter extension may predispose to thrombus formation.

8.3.3.1.7   Anchor techniques

The most commonly used anchor technique is the **side-branch anchor** technique, in which a balloon is inflated in a proximal side branch of the target vessel.[55-64] Using a long balloon increases support by increasing the "grip" on the vessel, especially if the hydrophilic coating is wiped off the surface of the balloon with a wet gauze before insertion.

8.3.3.1.8   Use of a support catheter, such as the NovaCross

Various support catheters (Section 30.3) can provide strong support by using a nitinol wire against the vessel wall to anchor the guide catheter.

8.3.3.2   Prevention of extraplaque hematoma

The creation of a large false lumen distal to the distal cap can compress the distal true lumen and hinder reentry. Sometimes, dilatation of the proximal cap with small balloons is required to allow advancement of the Stingray balloon (which has a high crossing profile), predisposing to formation of extraplaque hematoma.

Prevention of extraplaque hematoma is key and can be achieved by:

1. Use of a guide catheter extension. The guide extension provides additional guide support as discussed above, but can also block the entry of blood into the extraplaque (subintimal) space.[54] In contemporary ADR and reverse CART, guide catheter extensions are strongly recommended to avoid expansion of the extraplaque space by hematoma formation.
2. Avoid antegrade injections after starting a dissection (Fig. 8.3.5). To prevent inadvertent injections, disconnect the contrast-containing syringe from the antegrade guide catheter manifold (Fig. 8.3.6) or cover it with a towel. Automated injection devices, such as the ACIST, should be avoided.
3. If the collaterals (mostly epicardial) enter the dissected extraplaque space, retrograde injections should be minimized to avoid hematoma formation.

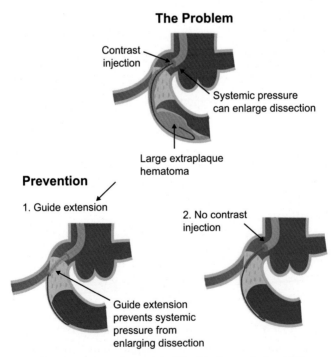

FIGURE 8.3.5   **How to prevent enlargement of the false lumen during ADR.**

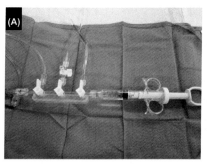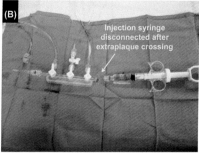

FIGURE 8.3.6 Example of disconnecting the injection syringe from the manifold (**Panel B**) after antegrade extraplaque crossing to prevent inadvertent contrast injection that could enlarge the extraplaque space. *Reproduced with permission from Brilakis ES.* Manual of coronary chronic total occlusion interventions. A step-by-step approach. *2nd ed.; Elsevier; 2017.*

8.3.4 Antegrade dissection

8.3.4.1 Antegrade dissection: CrossBoss

CrossBoss catheter: when?

The CrossBoss catheter (Section 30.16.1) is a stiff, metallic, over-the-wire catheter with a 1 mm blunt, rounded, hydrophilic-coated distal tip that can advance through the occlusion when the catheter is rotated rapidly using a proximal torque device ("fast spin" technique).

The CrossBoss first trial compared upfront use of the CrossBoss versus antegrade wiring for CTO crossing and found similar overall success, safety, and efficiency with both approaches, but faster crossing of in-stent CTOs.[65]

The CrossBoss catheter is currently used infrequently, mainly for:

1. Crossing in-stent CTOs
2. During the final stage of knuckle wiring (Section 8.3.4.2.3) to minimize the size of the extraplaque space. The creation of large extraplaque and distal false lumen hematomas can compress the distal true lumen and hinder reentry attempts.

CrossBoss catheter: how?

The CrossBoss can be used in four steps: (Fig. 8.3.7)

8.3.4.1.1 Step 1. CrossBoss delivery to the proximal cap (Fig. 8.3.8)

Unless the CTO proximal cap is ostial or very proximal, the CrossBoss catheter should be advanced over a workhorse guidewire or over the CTO crossing guidewire after an antegrade dissection has been created.

After the CrossBoss catheter is advanced to the proximal cap or the antegrade dissection plane the guidewire is retracted within the CrossBoss

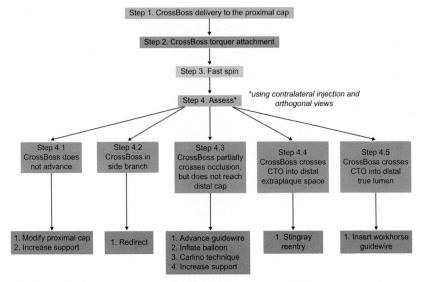

FIGURE 8.3.7 **Use of the CrossBoss catheter: step-by-step**. *Modified with permission from Brilakis ES. Manual of coronary chronic total occlusion interventions. A step-by-step approach. 2nd ed.; Elsevier; 2017.*

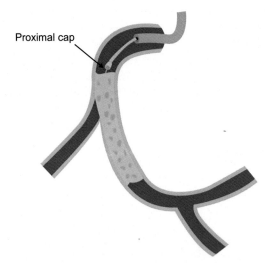

FIGURE 8.3.8 **CrossBoss step 1: CrossBoss delivery to the proximal cap**. *Modified with permission from Brilakis ES. Manual of coronary chronic total occlusion interventions. A step-by-step approach. 2nd ed.; Elsevier; 2017.*

catheter (but not removed, as it may help prevent blood entry and thrombus formation within the CrossBoss catheter lumen). The guidewire may also be useful for advancing the CrossBoss catheter through areas of resistance or side branches.

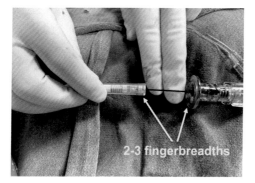

FIGURE 8.3.9   **Where to attach the CrossBoss torquer.**

8.3.4.1.2   Step 2. CrossBoss torquer attachment (Fig. 8.3.9)

The CrossBoss catheter torquer is positioned 2−3 cm (2−3 finger widths) proximal to the Y-connector and tightened.

The torquer should not be attached too proximal on the CrossBoss catheter because:

1. Torque transmission is better when the torquer is closer to the hemostatic valve.
2. To prevent excessive forward movement of the CrossBoss catheter (so called "CrossBoss jump") during catheter spinning. As the CrossBoss catheter engages and penetrates tissue, it may store torsional energy and jump during advancement, potentially causing a perforation.

8.3.4.1.3   Step 3. Fast spin (Fig. 8.3.10)

The CrossBoss catheter is rotated using the *fast spin* technique and gentle forward pressure.

The Y-connector is held between the little finger and the palm of the left hand and the torque is rotated using the index finger and thumb of both hands.

The catheter can be spun by hand in either direction, rotating **as fast as possible (until crossing or significant operator discomfort!)**. Faster spinning decreases friction and increases the likelihood of advancement and crossing. During catheter advancement keep the CrossBoss torque device close (2-3 finger breadths) to the hemostatic valve, to prevent excessive forward movement of the CrossBoss. A guidewire should be kept within the CrossBoss catheter lumen during spinning to increase support and prevent blood entry into the catheter. A clicking sound suggests that the tip of the CrossBoss is stuck.

FIGURE 8.3.10   **The fast spin technique for advancing the CrossBoss catheter.**

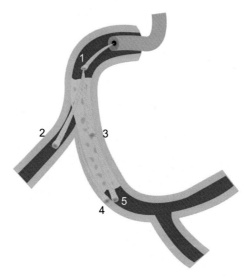

FIGURE 8.3.11   **Assessment of the CrossBoss catheter position after using the fast spin technique**. *Modified with permission from Brilakis ES.* Manual of coronary chronic total occlusion interventions. A step-by-step approach. *2nd ed.; Elsevier; 2017.*

8.3.4.1.4   Step 4. Assess the position of the CrossBoss catheter (Fig. 8.3.11)

The outcome of the fast spin technique is assessed using contralateral injection (antegrade injections should not be performed after antegrade dissection is started to minimize the risk of extending a proximal dissection) and orthogonal views (to detect possible entry of the CrossBoss catheter into a side branch). Advancement of the CrossBoss catheter can lead to five possible outcomes.

8.3.4.1.4.1   Outcome 4.1. CrossBoss fails to advance (Fig. 8.3.12)

Failure of the CrossBoss to advance can be due to poor guide catheter support or a hard, calcified proximal cap. Potential solutions include:

1. Increase guide catheter support (for example by using a more supportive guide catheter, a side-branch anchor technique, or a guide catheter extension, Fig. 8.3.4).
2. Modify the proximal cap. In patients with a hard, calcified proximal cap a stiff guidewire can be used to puncture the proximal cap. This wire should be immediately withdrawn and a softer wire advanced, followed by modification of the proximal cap with a small balloon. Large balloons should be avoided as they can lead to large extraplaque hematomas, potentially extending distal to the distal cap and hindering reentry. The CrossBoss catheter is then reinserted. Other techniques, called "move the cap" techniques (Section 16.1.4) can be used for entering the extraplaque space, followed by advancement of the CrossBoss catheter.
3. Change to a guidewire crossing strategy, either intraplaque or extraplaque.

8.3.4.1.4.2   Outcome 4.2. CrossBoss enters a side branch (Fig. 8.3.13)

Although the CrossBoss catheter does not exit the vessel adventitia due to its blunt tip, it can enter side branches. If undetected, entry of the CrossBoss into a side branch can be a catastrophic complication, as continued advancement can make it exit the side branch, causing a perforation.

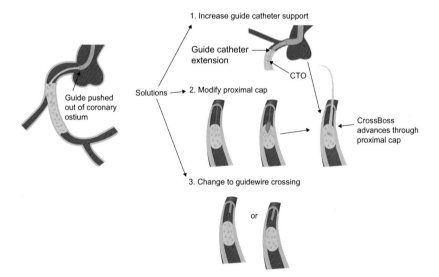

FIGURE 8.3.12   **Solutions to failure to advance the CrossBoss catheter through the proximal cap.** *Modified with permission from Brilakis ES.* Manual of coronary chronic total occlusion interventions. A step-by-step approach. *2nd ed.; Elsevier; 2017.*

Solutions

1. Knucke wire redirection

2. Blocking balloon in side branch

Blocking balloon

3. Stiff guidewire redirection

FIGURE 8.3.13 Solutions to CrossBoss catheter entry into a side branch.

Detection: Side branch course of the CrossBoss catheter is detected by using imaging in various projections and contralateral injections. Side branch course should be suspected when the CrossBoss catheter is not "dancing" in sync with the CTO target vessel.

Management: The CrossBoss catheter is retracted and redirected, usually using a knuckled polymer-jacketed guidewire (stiff tip, polymer-jacketed guidewires form larger knuckles and are less likely to enter side branches) or less commonly a using medium tip stiffness, non-polymer-jacketed penetrating guidewire. Highly penetrating guidewires, such as Confianza Pro 12, Infiltrac Plus, and Hornet 14, should not be used with the CrossBoss catheter due to very high penetrating power that increases the risk of perforation along with lack of tactile feedback. Inflation of a blocking balloon into the side branch can sometimes deflect the guidewire enabling distal advancement.

The CrossBoss catheter is then delivered distal to the origin to the side branch, the guidewire is withdrawn within the CrossBoss catheter, and the CrossBoss catheter is readvanced using the fast spin technique.

8.3.4.1.4.3   Outcome 4.3. CrossBoss partially crosses the occlusion (Fig. 8.3.14)

Partial CrossBoss crossing can be due to intra-occlusion calcification and/or tortuosity or due to interaction with a previously deployed stent.

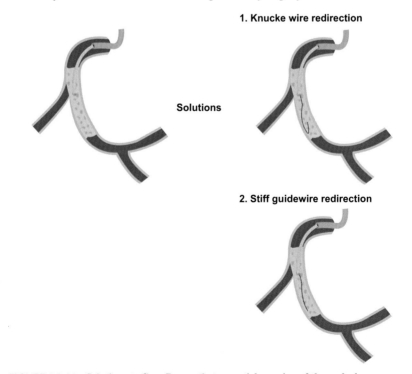

1. Knucke wire redirection

Solutions

2. Stiff guidewire redirection

FIGURE 8.3.14   **Solutions to CrossBoss catheter partial crossing of the occlusion**.

Management: The CrossBoss catheter is retracted and redirected, usually using a knuckled polymer-jacketed guidewire or, less commonly, a stiff tip guidewire. Withdrawing the CrossBoss catheter to allow balloon inflations or use of the Carlino technique at the site of resistance may allow subsequent advancement of the CrossBoss. Increasing guide catheter support may also help advance the CrossBoss.

8.3.4.1.4.4   Outcome 4.4. CrossBoss crosses into false lumen distal to the distal cap (Fig. 8.3.15)

- Entry of the CrossBoss catheter in the false lumen distal to the distal cap creates favorable conditions for reentry, as the CrossBoss catheter has a low profile and does not enlarge the false lumen like a knuckled guide wire would ("dry dissection").
- Reentry is optimally performed using the Stingray balloon (Section 30.16.2) without attempting reentry with guidewires alone, as the latter may enlarge the area of dissection and cause large false lumen hematomas that can compress the distal true lumen, and hinder reentry.
- The CrossBoss is ideally removed over a supportive, nonlubricious guidewire, such as a Miracle 12, using the trapping technique to prevent wire movement and maintain distal position without enlarging the dissection.
- As described in Section 8.3.3, after extraplaque crossing with the CrossBoss, it is best to disconnect the contrast-containing syringe from the antegrade guide catheter manifold (or cover the manifold) to minimize the risk of hydraulic dissection. Inadvertent contrast injection could enlarge the false lumen and hinder reentry.

8.3.4.1.4.5   Outcome 4.5. CrossBoss crosses into the distal true lumen (Fig. 8.3.16)

In $25\%^{65}-37\%^{35}$ of the cases, the CrossBoss catheter crosses into the distal true lumen. A workhorse guidewire is then inserted into the distal true lumen and the CrossBoss catheter is removed (ideally using the trapping technique to minimize the risk of losing guidewire position), followed by lesion preparation and stenting.

8.3.4.2   Antegrade dissection: knuckle wire

Antegrade dissection with knuckle wire: when?

Advancing a knuckled guidewire (Fig. 8.3.2) is the most common technique for achieving dissection through a CTO, in both the antegrade and the retrograde direction.

Common uses of antegrade dissection with a knuckle wire include:

1. Inadvertent guidewire entry in the extraplaque space.
2. In long and heavily calcified CTOs.

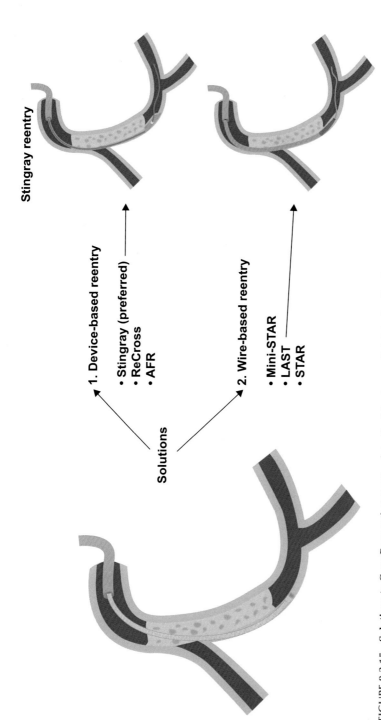

Stingray reentry

Solutions

1. Device-based reentry
- Stingray (preferred)
- ReCross
- AFR

2. Wire-based reentry
- Mini-STAR
- LAST
- STAR

FIGURE 8.3.15 Solutions to CrossBoss catheter entry into the false lumen distal to the distal cap.

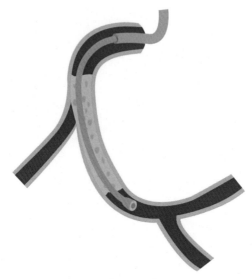

FIGURE 8.3.16   CrossBoss crosses into distal true lumen.

Antegrade dissection with knuckle wire: how?

Similar to the use of the CrossBoss catheter, antegrade dissection with a knuckle wire is performed in four steps (Fig. 8.3.17)

8.3.4.2.1   Step 1. Decision to proceed with antegrade dissection using a knuckle wire (Fig. 8.3.18)

A knuckle wire can be used for antegrade dissection:

1. If the antegrade guidewire enters the extraplaque (formerly called subintimal) space. Alternative strategies to knuckled guidewire are: (1) redirecting the initial guidewire into the true lumen, (2) using a parallel wire technique, or (3) retrograde guidewire crossing. The use of a knuckle wire is preferred to antegrade wiring in long and heavily calcified CTOs with an ambiguous vessel course, since it is safer and more efficient.
2. To allow redirection of the CrossBoss catheter, as discussed in step 4.2 of Section 8.3.4.1.4.2.
3. In cases with proximal cap ambiguity (move the cap techniques) (Section 16.1). A dissection is created intentionally proximal to the proximal cap, followed by insertion and advancement of a knuckled guidewire.

Extraplaque crossing: knuckle wire versus CrossBoss

The CrossBoss catheter has advantages over a knuckle wire:

1. It creates a smaller and more controlled extraplaque dissection space (Fig. 8.3.19), enabling a more predictable and controlled reentry into the distal true lumen, and

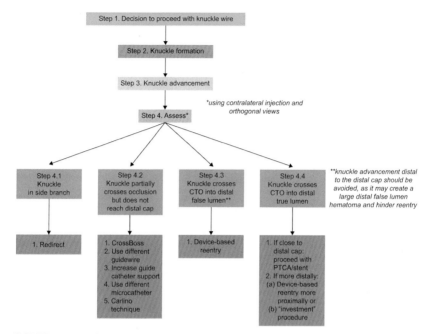

FIGURE 8.3.17 **Step-by-step knuckle wire crossing.** *Reproduced with permission from Brilakis ES.* Manual of coronary chronic total occlusion interventions. A step-by-step approach. *2nd ed.; Elsevier; 2017.*

2. The relatively stiff CrossBoss catheter tends to advance along a longitudinal path parallel to the artery axis, whereas knuckled guidewires sometimes wrap around the artery circumference (in a "barber pole" fashion, Fig. 8.3.20) potentially hindering subsequent advancement of other devices and reentry into the distal true lumen.

The CrossBoss catheter also has disadvantages compared with the knuckle wire:

1. It is more expensive.
2. May be more likely to enter side branches, especially in tortuous segments, because its stiff shaft tends to follow the outer curvature of the vessel.
3. Can be harder to advance around areas of severe calcification and/or tortuosity.
4. CrossBoss is not a good "beginner" but is a good "finisher" during extra-plaque crossing.

In some cases, the CrossBoss catheter and a knuckle wire may be used simultaneously ("knuckle-Boss" technique) to navigate beyond side branches or advance beyond calcific or tortuous anatomy.

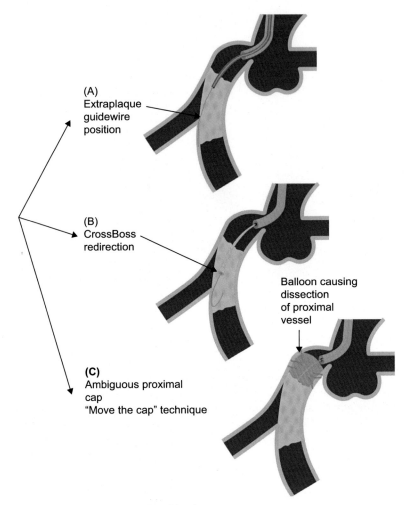

(A)
Extraplaque
guidewire
position

(B)
CrossBoss
redirection

Balloon causing
dissection
of proximal
vessel

(C)
Ambiguous proximal
cap
"Move the cap" technique

FIGURE 8.3.18   **When to use a knuckle wire.**

8.3.4.2.2   *Step 2. Knuckle formation (Fig. 8.3.21)*
How to knuckle a wire

1. The usual starting point of knuckling is when a guidewire and a micro-catheter have been advanced into the extraplaque space. Ensure that the microcatheter is located within the extraplaque (or intraplaque) space by observing that the guidewire is moving ("dancing") in sync with the coronary artery.
2. Larger microcatheters (such as the Corsair, TurnPike, Teleport, and Mamba) may provide stronger support for forming a knuckle as compared with lower profile microcatheters (such as the FineCross or Caravel).

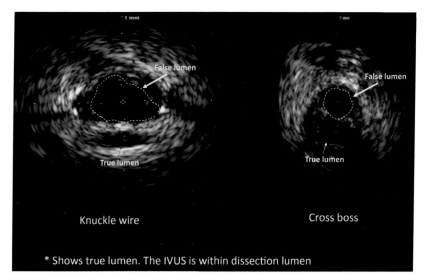

FIGURE 8.3.19 **Comparison of the dissection plane created using a knuckled guidewire (larger size) and created with a CrossBoss catheter (smaller size).** *Courtesy of Dr. Craig Thompson. Reproduced with permission the Brilakis ES.* Manual of coronary chronic total occlusion interventions. A step-by-step approach. *2nd ed.; Elsevier; 2017.*

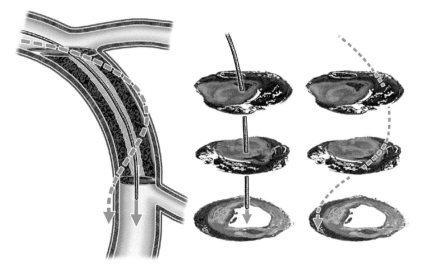

FIGURE 8.3.20 **Antegrade intraplaque tracking vs extraplaque tracking**. Antegrade intraplaque tracking (solid line) and extraplaque tracking (dotted line). Once the wire migrates into the extraplaque space, the wire easily advances into the extraplaque space and wraps around the intraplaque space (in a barber-pole fashion). It is difficult for the extraplaque wire to cross into the distal true lumen because of lower resistance to advance into the extraplaque space coupled with increased resistance to traversing the plaque. *Reproduced with permission from Sumitsuji S, Inoue K, Ochiai M, Tsuchikane E, Ikeno F. Fundamental wire technique and current standard strategy of percutaneous intervention for chronic total occlusion with histopathological insights.* JACC Cardiovasc Interv *2011;4:941−51.*

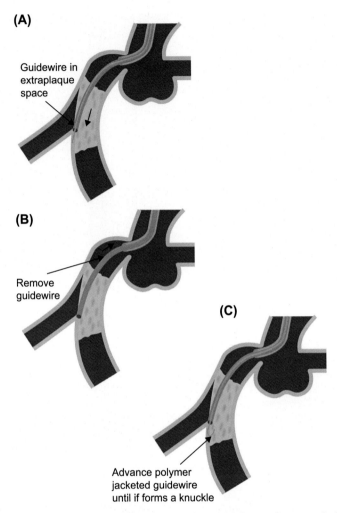

FIGURE 8.3.21   How to create a knuckle wire after extraplaque wire entry during ante-grade wiring. Panel A: the original guidewire and the microcatheter are advanced into the extraplaque space. Panel B: the original guidewire is removed. Panel C: a polymer-jacketed guidewire is advanced through the microcatheter and pushed into the extraplaque space until it forms a knuckle. Preshaping the guidewire tip as "umbrella handle" can help with knuckling.

3. Remove the original guidewire and insert a polymer-jacketed wire (such as Fielder XT, Fighter, Bandit, Gladius Mongo, Pilot 200 and Raider). Softer wires (such as Fielder XT, Fighter, and Bandit) usually form smaller, tighter loops. Stiffer polymer-jacketed wires (such as the Pilot 200 and Raider) form larger loops. The Gladius Mongo wire is engineered to create tight knuckles.
4. Shaping the wire tip as "umbrella bend" with the introducer facilitates knuckling (Fig. 8.3.22).

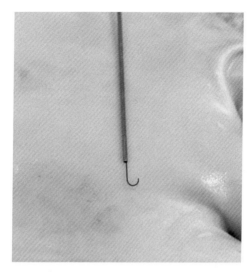

FIGURE 8.3.22   Example of "umbrella handle" shaping of the guidewire tip.

5. Push WITHOUT rotation (to minimize risk of wire fracture or entanglement).
6. Do not be afraid to **push hard**!
7. A loop often forms at the junction of the radiopaque and the stiffer radiolucent parts of the wire.
8. Try to keep the knuckle small—may need to withdraw the wire into the microcatheter and re-advance it if the knuckle size becomes too large.
9. If a loop does not form, withdraw and re-advance.

Troubleshooting: unable to form a knuckle
Solutions
1. Confirm microcatheter position in the extraplaque space within the vessel architecture, using orthogonal projections.
2. Use a new umbrella shaped wire.
3. Try a different guidewire. For example change between soft tip, tapered polymer-jacketed guidewires (such as Fielder XT, Fighter or Bandit) and stiff tip, polymer-jacketed guidewires (such as Pilot 200, Raider or Gladius Mongo).
4. Try a different microcatheter.
5. Reposition the microcatheter more proximal or more distal within the occlusion.
6. Use the Carlino technique (Section 8.3.4.3) to start a microdissection plane, that can subsequently be followed by a guidewire.
7. Use the "power knuckle" technique (Fig. 8.3.23).

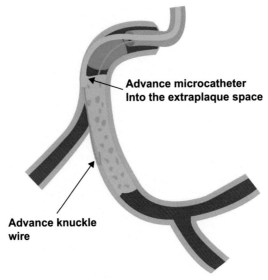

FIGURE 8.3.23   Illustration of the "power knuckle."

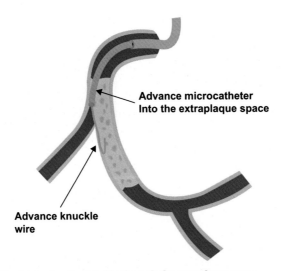

FIGURE 8.3.24   Advancing the knuckle through the extraplaque space.

8.3.4.2.3   Step 3. Advance the knuckle (Fig. 8.3.24)

1. Advancement is made in intermittent, forceful forward pushes.
2. After confirmation of wire position within the vessel architecture, the microcatheter is advanced closer to the tip of the knuckled guidewire.

Troubleshooting: unable to advance knuckle
Causes
1. Calcification and tortuosity
2. Poor guide catheter support

Solutions
1. Increase guide catheter support:
   a. Advance microcatheter to the tip of the knuckle.
   b. Use supportive guide catheter, such as Amplatz Left 1 for RCA CTOs.
   c. Guide catheter extension.
   d. Side branch anchor.
   e. NovaCross catheter.
   f. Co-axial anchor (the knuckle is advanced through an inflated over-the-wire balloon).
   g. Power knuckle (Fig. 8.3.23): a second guidewire is advanced adjacent to the microcatheter and a 1:1 sized balloon is inflated over the second guidewire, "anchoring" the microcatheter against the vessel wall (**CTO Manual online case 57**).
   h. Use a different guidewire (for example, Pilot 200 is stiffer than the Fielder XT and forms bigger knuckles, whereas the Fielder XT is softer and forms tighter knuckles).
   i. Microcatheter knuckle: a tapered-tip, kink-resistant microcatheter (such as Corsair or Turnpike) is advanced towards the knuckled tip of the guidewire and the guidewire is simultaneously withdrawn, creating a knuckle-shaped microcatheter tip that is subsequently advanced more distally.[66] Using a supportive guidewire (such as the Miracle 12), can help forward advancement of the knuckled shaped microcatheter tip.

8.3.4.2.4   Step 4. Assess the knuckle position (Fig. 8.3.25)

The result of knuckle advancement is assessed using contralateral injection and orthogonal views. There are four possible outcomes with knuckle advancement:

8.3.4.2.4.1   Outcome 4.1. Knuckle enters a side branch (Fig. 8.3.26)

The knuckle wire is extremely unlikely to exit the vessel adventitia, due to its blunt bend, but can enter side branches, such as acute marginal branches during right coronary artery CTO PCI. Perforation can occur if side branch entry is not recognized and the knuckle continues to be advanced.

Detection: side-branch course of the knuckled guidewire is detected by using imaging in various projections and contralateral injection. Side branch course is suspected when the knuckled guidewire is not "dancing" in sync with the CTO target vessel.

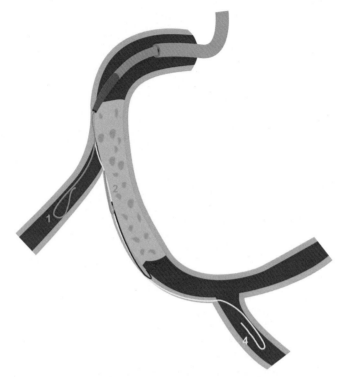

FIGURE 8.3.25 Possible outcomes of knuckle advancement.

Management: As a first step, the knuckled guidewire is retracted and redirected. Using a stiffer polymer-jacketed guidewire (such as the Pilot 200 or Raider) can create larger loops, which are less likely to enter side branches. Alternatively, a stiff tip guidewire (such as the Gaia or Judo series) can be used to advance past the origin of the side branch, or a balloon can be inflated into the side branch, deflecting the knuckled guidewire).

8.3.4.2.4.2 Outcome 4.2. Knuckle partially crosses the occlusion but does not reach the distal cap (Fig. 8.3.27)

Sometimes the knuckled guidewire cannot cross to the distal cap, due to calcification, poor guide catheter support, or other reasons.

Detection: Contralateral injection shows intra-occlusion location of the guidewire, with the loop "dancing" in sync with the vessel.

Management: Before proceeding with aggressive knuckling advancement maneuvers it is critical to ascertain (using contralateral injection) that the guidewire is within the vessel architecture and not within a side branch or in the pericardium. After confirmation that the knuckled guidewire is

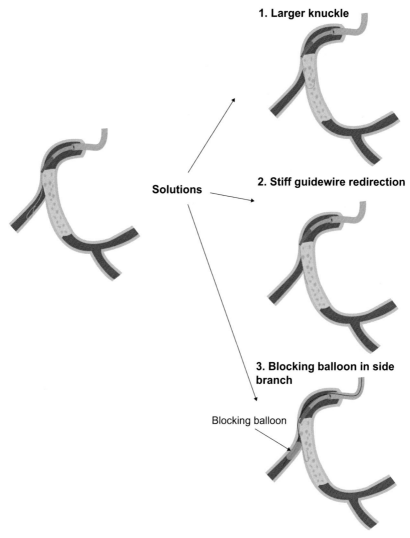

FIGURE 8.3.26 **Possible outcomes of knuckle advancement.**

within the occlusion and not within a branch, the following actions can be taken:

1. Use a different polymer-jacketed guidewire to form a knuckle, such as the Pilot 200, Raider or Gladius Mongo.
2. Exchange the knuckled guidewire for a CrossBoss catheter. Knuckles (especially large ones) can cause extensive dissection and extraplaque hematoma that can impair reentry attempts. This can be prevented by crossing the distal CTO segment with the CrossBoss catheter

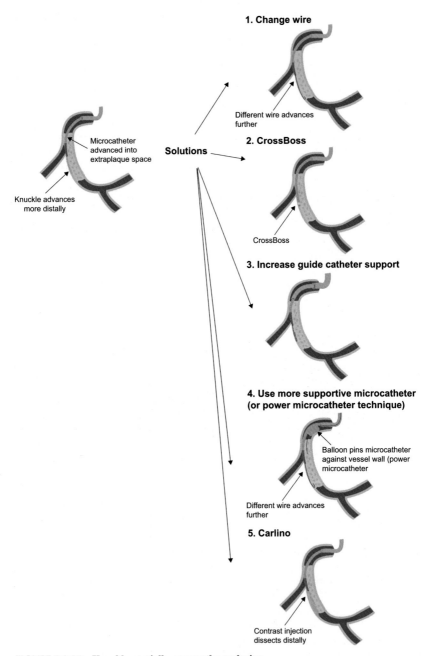

FIGURE 8.3.27   **Knuckle partially crosses the occlusion**.

("finish with the Boss"), or by starting a knuckle in a more proximal location in the CTO target vessel.

3. Increase guide catheter support, for example by using a side branch anchor or a guide catheter extension.
4. Use a more supportive microcatheter, such as the Corsair, Turnpike, or Turnpike spiral.
5. Use the Carlino technique (Section 8.3.4.3): advance the microcatheter as far as possible within the occlusion and inject <0.5−1 mL of contrast under cine-angiography. The contrast injection can often modify the compliance of the plaque by softening and recruiting loose tissue, facilitating subsequent guidewire crossing (**CTO Manual online case 48**).

8.3.4.2.4.3 Outcome 4.3. Knuckle crosses into the distal false lumen (Fig. 8.3.28)

This is the intended outcome of knuckle guidewire advancement. After confirmation that the guidewire is indeed in the distal false lumen (by using contralateral injection), the next step is to achieve reentry into the distal true lumen.

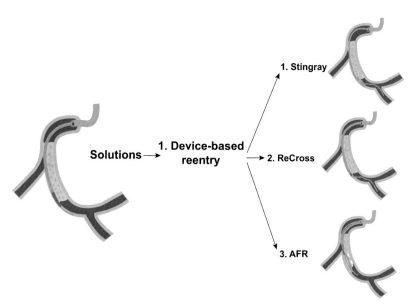

FIGURE 8.3.28   Knuckle crosses into the distal false lumen.

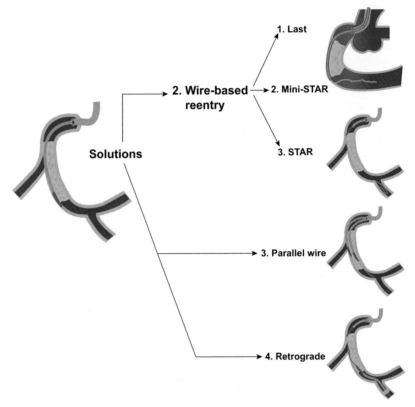

FIGURE 8.3.28 *(Continued)*

Caution: Too distal advancement of the knuckled guidewire should be avoided, as it can cause side branch occlusion and may require long stent length with high restenosis rates. The goal is to reenter the distal true lumen as close as possible to the distal cap, to minimize the length of dissection.

Management: There are four broad categories of techniques to allow crossing into the distal true lumen: (1) device based-reentry, such as using the Stingray balloon (Section 8.3.5.3), the ReCross Microcatheter (Section 8.3.5.4) or Antegrade Fenestration and reentry (AFR) (8.3.5.5); (2) wire-based reentry techniques, such as LAST and mini-STAR (Section 8.3.5.2) and STAR (Section 8.3.5.1); (3) parallel wire (Section 8.2.2.7); and (4) the retrograde approach (Section 8.4).

1. **Device-based reentry:** Many operators favor immediate use of the Stingray system for reentry because: (1) use of guidewires alone for reentry can cause a large extraplaque hematoma that can hinder Stingray-based reentry, and (2) it is faster and more efficient. The ReCross catheter can be used if Stingray is not available or if it fails to deliver to the reentry zone. AFR can also be used, but antegrade balloon inflation could enlarge the false lumen and lead to reentry failure, especially if no retrograde option is available.
2. **Wire-based reentry techniques** should be avoided in most cases, due to the unpredictability of reentry and the risks of causing or enlarging a hematoma.
3. **Parallel wire:** The use of the parallel wire technique (leaving the knuckle in place and advancing a second guidewire proximal in the occlusion, often through a dual lumen microcatheter) is an acceptable option.
4. **Retrograde approach:** If feasible, retrograde CTO crossing can be attempted with the antegrade knuckle wire facilitating the performance of the reverse CART technique.

8.3.4.2.4.4   Outcome 4.4. Knuckle enters the distal true lumen (Fig. 8.3.29)

Infrequently, the knuckled guidewire may enter the distal true lumen, as demonstrated by contralateral injection. Most commonly this occurs in a distal branch, which is suboptimal, as stenting will result in loss of side branches proximal to the reentry location, decreasing the benefit of revascularization and is associated with very high restenosis and reocclusion rates. This is the original STAR technique (Section 8.3.5.1) that should only be used as a last resort, when every other attempt has failed; stents should not be implanted when the STAR technique is used.

Management:
1. If knuckle reentry is achieved close to the distal cap, balloon angioplasty and stenting are performed.
2. If reentry is more distal beyond major side branches (which occurs much more frequently) stenting should be avoided, as it will occlude all more proximal side branches. Treatment options include:
   a. Reentry close to the distal cap (using device-based, wire-based, parallel wiring or the retrograde approach as shown in Fig. 8.3.29).
   b. Reentry into important side branches (that can be facilitated by a dual lumen microcatheter), followed by bifurcation stenting using 2-stent techniques.

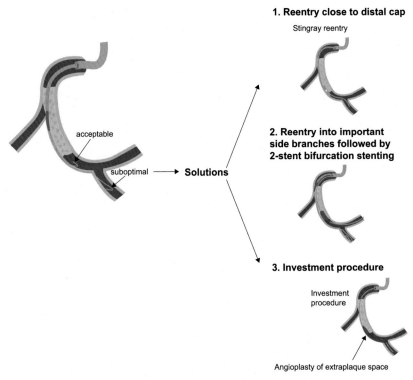

1. **Reentry close to distal cap**

Stingray reentry

2. **Reentry into important side branches followed by 2-stent bifurcation stenting**

3. **Investment procedure**

Investment procedure

Angioplasty of extraplaque space

acceptable

suboptimal ⟶ **Solutions**

FIGURE 8.3.29   **Knuckle crosses into the distal true lumen**.

   c. Investment procedure (balloon angioplasty of the dissected segment without stent implantation, Section 21.2) (**CTO Manual online cases 9, 181**).[49] Repeat angiography is then performed in 2−3 months, often showing restoration of antegrade flow, allowing stenting without compromising multiple side branches. Maintenance of patency is more likely if there is good antegrade flow after extraplaque ballooning.[67]

8.3.4.3   Antegrade dissection: Carlino technique

CTO Manual online cases 13, 27, 37, 45, 48, 49, 53, 60, 62, 84, 107, 123, 144.

Historical perspective

   Mauro Carlino is a highly creative interventional cardiologist from Milan, Italy who pioneered the technique of using contrast to cause subintimal dissection. In the original version of the technique a significant amount of contrast was used, essentially performing a variation of the STAR technique, that is, causing extensive dissection with the goal of distal true lumen reentry (contrast-guided STAR). As with the original STAR technique, contrast-guided STAR had high risk of complications, such as perforation ("storm cloud" dissection) and high restenosis rates.[38,68] The technique was subsequently modified to its current form (the so-called "Carlino technique"), in which only a small volume of contrast ($<$0.5-1.0 mL) is injected very gently with the goal of modifying plaque compliance to facilitate guidewire and microcatheter advancement trough a fibrocalcific plaque, limiting the extent of dissection and the associated risk.[69,70]

Carlino technique: when?

The original Carlino technique (contrast-guided STAR) is no longer used as an upfront technique.

The modified Carlino technique is currently used for specialized indications, such as:

1. To modify plaque compliance in case of failure to advance a guidewire or a microcatheter through the occlusion.
2. To resolve proximal of distal (for retrograde PCI) cap ambiguity.
3. To clarify the location of the microcatheter: a "storm-cloud dissection" indicates microcatheter position within a small branch, whereas a "tubular dissection" suggests microcatheter position in a large vessel.

Modified Carlino technique: how?

(Fig. 8.3.30)

8.3.4.3.1 Step 1

An antegrade microcatheter (or over-the-wire balloon) is advanced into the extraplaque space. A microcatheter is preferred because the microcatheter tip has a marker and can be visualized, whereas small over-the-balloons have a marker in the middle of the balloon and the balloon tip location is unclear.

8.3.4.3.2 Step 2

The coronary guidewire is removed and a small syringe is attached to the microcatheter. Gentle aspiration is performed through the syringe. The presence or absence of blood in the aspirate does not change strategy, as the purpose of aspiration is to "de-air" the system.

8.3.4.3.3 Step 3

A small amount of contrast ($<$0.5$-$1 mL) is injected gently into the occlusion, under cine-angiography.

8.3.4.3.4 Step 4

Assess contrast course. There are four possible outcomes:

1. Failure to opacify the vessel distally: suggests limited lesion penetration by the microcatheter or insufficient contrast injection. A guidewire is used to advance the microcatheter more distally, followed by repeat contrast injection.

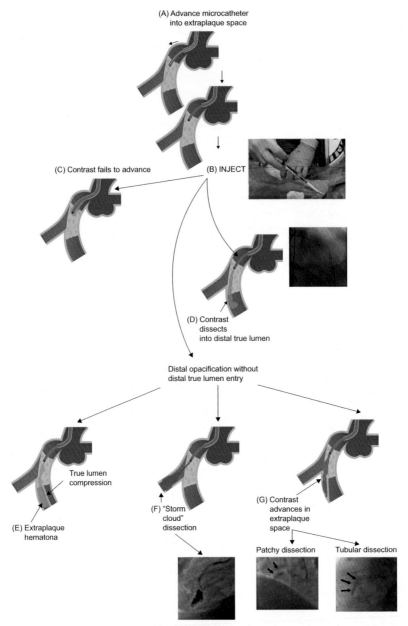

FIGURE 8.3.30 **Illustration of the modified Carlino technique. Panel A**: a microcatheter is advanced over the guidewire into the extraplaque space and the guidewire is removed. **Panel B**: a small amount of contrast (approximately 0.5-1.0 mL) is injected through the microcatheter. **Panel C**: the contrast fails to advance into the occlusion. **Panel D**: the contrast dissects into the distal true lumen. **Panel E**: the contrast creates an extraplaque hematoma distal to the distal cap. **Panel F**: the contrast enters into a side branch. **Panel G**: the contrast advances distally into the extraplaque space.

2. Distal opacification without entry into the distal true lumen. This can take several forms[70]:

    a. Large spiral dissection: this may lead to an extraplaque hematoma, creating a false lumen that may compress the distal true lumen, potentially hindering reentry.

    b. "Storm cloud" dissection: implies position of the microcatheter within a small branch: the microcatheter should be withdrawn and redirected.

    c. "Patchy" or "tubular" appearance: both suggest contrast location within a larger vessel (i.e., the CTO vessel architecture) and provide a road map to aid in further wire manipulation.[69] "Patchy" appearance suggests contrast tracking within patches of loose tissue adjacent to highly calcific occlusion. A knuckled guidewire can often be advanced through the dissection plane.

3. Dissection into the distal true lumen, creating a pathway that can be subsequently tracked by a guidewire (usually a polymer-jacketed guidewire).

8.3.5 Antegrade reentry

Antegrade reentry: when?

If the antegrade guidewire is advanced through the occlusion into the distal false lumen, reentry into the distal true lumen can be achieved using:

1. Device-based techniques (such as Stingray, ReCross and AFR) (preferred), or
2. Wire-based techniques, such as STAR, mini-STAR, and LAST)

Device-based reentry with the Stingray system is the most reliable and commonly used technique, but awareness of other techniques is important as the Stingray balloon may fail to achieve reentry and is not available in many cardiac catheterization laboratories.

Antegrade reentry: how?

8.3.5.1 Antegrade reentry: STAR

CTO Manual online cases 9, 41, 104, 147, 166.

In the STAR (Subintimal Tracking And reentry) technique, a knuckled polymer-jacketed guidewire is advanced distally until it spontaneously re-enters into the distal true lumen (usually at a distal bifurcation).[37] The advantage of STAR is its simplicity. The disadvantage is distal wire reentry, leading to side branch loss and the need for long stent length with high rates of restenosis and reocclusion. STAR is currently used as an "investment" procedure, that is, with balloon angioplasty only without stent implantation, followed by a repeat CTO crossing attempt after 2 months.[71-75]

How to perform the STAR technique (Fig. 8.3.31)

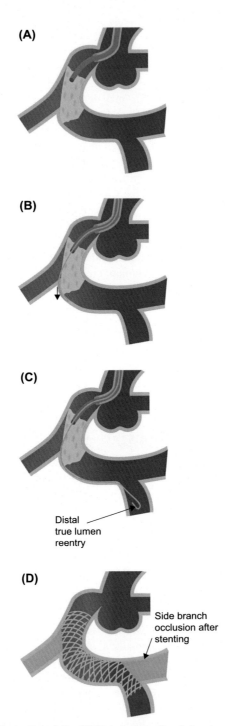

(A)

(B)

(C)

Distal
true lumen
reentry

(D)

Side branch
occlusion after
stenting

FIGURE 8.3.31 **Illustration of the STAR technique. Panel A**: microcatheter advanced past the proximal cap. **Panel B**: advancement of a knuckled guidewire. **Panel C**: knuckled guidewire enters distal true lumen (usually at a bifurcation). **Panel D**: stent implantation causes side branch occlusion.

8.3.5.1.1 Step 1

A polymer-jacketed guidewire, with either soft tip (such as Fielder XT, Fighter, Bandit) or stiff tip (such as Gladius Mongo, Raider and Pilot 200) is inserted into the extraplaque space and a loop is formed at its distal tip.

8.3.5.1.2 Step 2

The knuckled guidewire is advanced with swift forward movement (as described in Section 8.3.4.2) until it spontaneously reenters the true lumen (usually at a distal bifurcation).[39]

What can go wrong

1. Extensive shearing and occlusion of side branches, resulting in limited outflow. The STAR technique should be avoided in the left anterior descending artery, as it can lead to occlusion of multiple septal and diagonal branches (and may also preclude future coronary artery bypass graft surgery due to extensive stent implantation). It should also be avoided in very distal occlusions, as reentry might cause a perforation.
2. Inability to reenter into the distal true lumen. If this occurs, use of the Stingray system may facilitate reentry, although reentry will likely be challenging because of the extensive dissections caused by the STAR technique.
3. If stenting is done after the STAR procedure, the risk of restenosis and reocclusion is high, likely due to the long-stent length and limited outflow, although studies are limited by presenting data with both drug-eluting and bare-metal stents. When considering TIMI 2 flow post PCI as a successful procedure there was 52% need for target vessel revascularization in the original Colombo series,[37] 54% restenosis with contrast-guided STAR[47] and 57% reocclusion rate in an Italian registry.[39]

Contemporary use of STAR

At present, the STAR technique is usually performed when other techniques fail to cross the CTO, prior to stopping the procedure (investment procedure). Balloon dilatation of the extraplaque space is then done (sometimes with a drug-coated balloon[76]) aiming to establish antegrade flow. No stents are implanted to minimize the risk of side branch occlusion and restenosis/reocclusion. Many patients have symptomatic improvement after an investment procedure.[71] For patients whose symptoms do not improve or for a longer-lasting result, CTO crossing can be reattempted usually after 2 months to allow time for healing of the dissections.[71–74]

8.3.5.2 Antegrade reentry: LAST and mini-STAR

Mini-STAR[34,77] and LAST[40] are antegrade wire-based techniques that aim to reenter into the distal true lumen as close as possible to the distal cap (in contrast to STAR where the guidewire typically reenters into the distal true

lumen much more distally). Mini-STAR and LAST are performed infrequently as success is limited and guidewire manipulations may lead to
enlargement of the false lumen, hindering reentry. If available, the Stingray
system is preferred for reentry just beyond the distal cap.

The starting point for both mini-STAR and LAST is after a guidewire (or
the CrossBoss catheter) is advanced in the false lumen distal to the distal cap
(Fig. 8.3.32, panel A). Subsequent steps are as follows (Fig. 8.3.32):

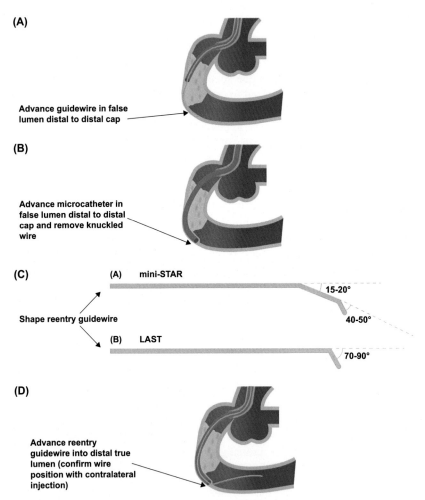

FIGURE 8.3.32   Illustration of the mini-STAR and LAST techniques. Panel A: a knuckled
guidewire is advanced past the distal cap. Panel B: a microcatheter is advanced into the false
lumen distal to the distal cap. Panel C: a reentry guidewire is shaped to facilitate reentry.
Panel D: the reentry guidewire is advanced into the distal true lumen. Wire position is confirmed using contralateral injection.

8.3.5.2.1 Step 1

The antegrade microcatheter is advanced to the reentry zone (Fig 8.3.32, panel B).

8.3.5.2.2 Step 2. A guidewire is selected to use for reentry

In "mini-STAR"[34] a soft polymer-jacketed guidewire (such as Fielder XT, Fighter, or Bandit) is used. In LAST a stiff-tip, highly penetrating guidewire (such as the Confianza Pro 12, Infiltrac and Infiltrac Plus, and Hornet 14) is used.[41]

8.3.5.2.3 Step 3. The reentry guidewire is shaped to facilitate reentry, as follows (Fig 8.3.32, panel C)

Mini STAR: A 40–50 degrees curve is created 1–2 mm proximal to the tip and a second 15–20 degrees curve is created 3–5 mm proximal to the tip. The guidewire bend can be modified by advancing and withdrawing the guidewire through the microcatheter.
LAST: A 70–90 degrees curve is created 2–3 mm proximal to the tip.

8.3.5.2.4 Step 4

The reentry wire is manipulated until reentry is achieved into the distal true lumen (Fig 8.3.32, panel D). This can be sometimes facilitated by the use of an angulated microcatheter[78,79] or a dual lumen microcatheter.

What can go wrong
1. Failure to reenter: either device-based antegrade reentry (such as the Stingray system, the ReCross dual lumen microcatheter, or AFR) or retrograde crossing can be used in such cases.
2. Distal true lumen compression by extraplaque hematoma. Use of the STRAW technique (Section 8.3.5.3.4.3) can be used to decompress the hematoma.
3. Wire perforation: wire perforation alone will only rarely cause tamponade, but it is important to detect it early to prevent the advancement of equipment, such as balloons and microcatheters, that will enlarge the exit point and may cause a significant perforation.
4. Distal reentry, that can lead to similar consequences as the STAR technique (loss of side branches and long stent length leading to high restenosis and reocclusion rates).

8.3.5.3 Antegrade reentry: Stingray

**CTO Manual online cases 9, 16, 17, 21, 25, 27, 32, 34, 35, 43, 45, 48, 72, 29, 80, 82, 83, 92, 93, 97, 98, 100, 105, 120, 129, 131, 135, 142, 144, 146, 148, 149, 152, 154, 155, 159, 161, 164, 165, 166, 168, 173, 181, 184, 187, 189, 191, 198.
PCI Manual online cases 40, 105, 120, 128, 135, 153**

Stingray-based distal true lumen reentry is more reproducible and reliable than wire-based techniques and is currently the preferred antegrade reentry technique.

8.3.5.3.1 Step 1. Preparation of the Stingray balloon (Fig. 8.3.33)

1. Stingray preparation steps (Fig. 8.3.33):
   a. Attach a new, completely dry, 3-way stopcock to the end of the Stingray balloon port (Fig. 8.3.33, panel A).
   b. Using a new, completely dry, 20 cc Luer-lock syringe, aspirate negative 2−3 times (Fig. 8.3.33, panel B) and turn the stopcock each time

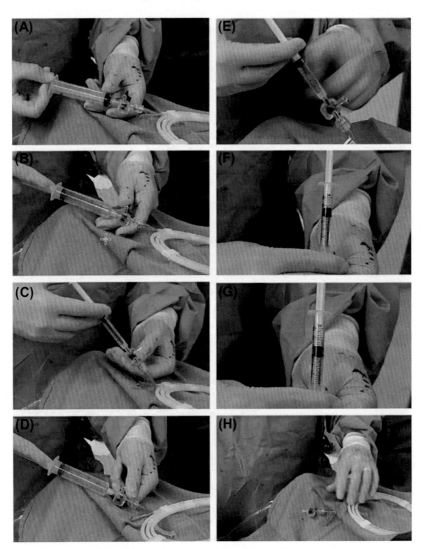

FIGURE 8.3.33 Illustration of the Stingray balloon preparation.

keeping the balloon port closed (Fig. 8.3.33, panel C), to retain vacuum in the Stingray balloon.

c. Remove the 20 cc syringe, replacing it with a 3 cc Luer-lock syringe that contains 100% contrast (Fig. 8.3.33, panel D).

d. Flush contrast through a 3-way stopcock, "priming the pump," ensuring that there are no air bubbles in the stopcock (Fig. 8.3.33, panel E).

e. Open stopcock to syringe—the plunger will advance by 2−3 mm (Fig. 8.3.33, panel F is before opening stopcock—Fig. 8.3.33, panel G is after opening stopcock). The contrast syringe can remain attached to the Stingray balloon until it is ready to connect with the indeflator. If the plunger does not move, repeat Steps 1−5. Do NOT apply negative pressure.

f. The Stingray balloon is now ready for use (Fig. 8.3.33, panel H).

8.3.5.3.2 Step 2. Delivery of the Stingray balloon to the reentry zone (Fig. 8.3.34)

The reentry zone should ideally have little or no calcification and a large distal true lumen. Coronary angiography and CT coregistration can help identify an optimal reentry zone.[80−82]

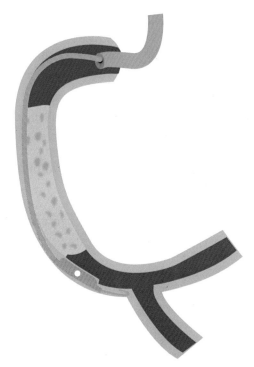

FIGURE 8.3.34 **Delivery of the Stingray balloon to the reentry zone.**

The Stingray balloon can be delivered to the reentry zone over the original knuckled guidewire or over a more supportive guidewire, such as the Miracle 12. Delivery of the Stingray balloon can be challenging, especially in heavily calcified and tortuous lesions. Occasionally the original guidewire may enter the distal true lumen during attempts to deliver the Stingray balloon.

The wire trapping technique should be used to prevent inadvertent guidewire movement during Stingray balloon delivery that may extend the extraplaque hematoma and prevent successful reentry. The Stingray LP balloon can be trapped in a 7 Fr guide catheter or even a 6 French guide catheter using the TrapLiner guide extension.

What to do if the Stingray balloon cannot be delivered to the reentry zone?

1. Advance a microcatheter or a CrossBoss catheter to the reentry zone, to create a track for the Stingray balloon.
2. Predilate the guidewire track with a small (1.0−1.5 mm) balloon, although this may lead to enlargement of the false lumen. The more proximal the balloon inflation is to the reentry site, the lower the risk of hematoma expansion.
3. Use a supportive guidewire, such as a 300 cm long Miracle 12 wire.
4. Increase guide catheter support, for example by using the side branch anchor technique or a guide catheter extension (preferably a TrapLiner that can also be used for trapping).
5. Use a new Stingray balloon, as the Stingray balloon can get damaged during repeated attempts to deliver it to the reentry zone or after it is used for reentry.
6. Use an alternative antegrade reentry strategy, such as the ReCross dual lumen microcatheter or AFR.
7. Use retrograde crossing.
8. If everything fails, attempt to create a new extraplaque space, which was be difficult to do as the guidewires will preferentially enter the existing dissection plane.

8.3.5.3.3 Step 3. Reentry into the distal true lumen

After the Stingray balloon is delivered to the reentry zone, it is inflated at 2−4 atm. Upon inflation, the balloon assumes a flat shape and self-orients, with one surface of the balloon facing the true lumen and the other facing the adventitia.

Reentry can be achieved using:

1. angiographic guidance using only a stiff guidewire ("stick and drive" technique);
2. angiographic guidance, first using a stiff guidewire to puncture the tissue separating the false from the true lumen, followed by exchange for a polymer-jacketed guidewire ("stick and swap" technique); or
3. without angiographic guidance ("double-blind stick and swap technique").[83]

8.3.5.3.3.1 Stingray reentry: "stick and drive" technique

8.3.5.3.3.1.1 Obtain optimal angiographic view (Fig. 8.3.35)

A contralateral injection is performed using orthogonal angiographic projections to select the optimal view for reentry (Fig. 8.3.36). The ideal view is the one in which the Stingray balloon is seen as one line located at the side of the vessel lumen (Fig. 8.3.36, panel B). The side view is necessary to determine in which direction the wire leaves the Stingray balloon, to direct it towards the vessel lumen. Finding this view may require extreme angulations of the image receptor, leading to difficult visualization; moving the Stingray balloon slightly more distally or more proximally can potentially favorably change this angle, as the dissection plane usually spirals around the vessel.

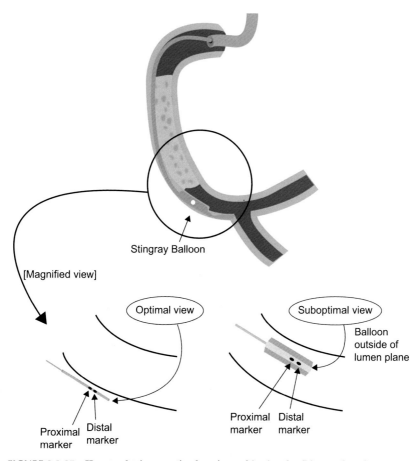

FIGURE 8.3.35 **How to obtain an optimal angiographic view for Stingray-based reentry.**

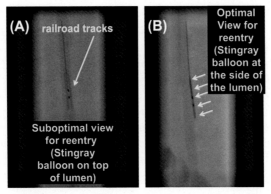

FIGURE 8.3.36   **Selection of the optimal view for Stingray-based reentry**. *Reproduced with permission from the Brilakis ES*. Manual of coronary chronic total occlusion interventions. A step-by-step approach. *2nd ed.; Elsevier; 2017.*

8.3.5.3.3.1.2   Advance a penetrating guidewire into the distal true lumen (Fig. 8.3.37)

A highly penetrating guidewire (such as Astato 20, Hornet 14, Infiltrac, Infiltrac Plus, Confianza Pro 12, Conquest Pro 12 Sharpened Tip [ST], Stingray guidewire, or Gaia Next 3) is shaped into a 1 mm, 30−45 degrees distal bend and is advanced through the side ports of the Stingray balloon under fluoroscopic guidance into the distal true lumen. The Stingray balloon has three exit ports: two of them are offset and opposed 180 degrees apart on the flat surface of the Stingray balloon for vessel reentry, and the third is the end hole. The proximal exit port is proximal to the two markers and the middle exit port is between the two markers. If the wire enters the exit port that faces away from the true lumen, it is withdrawn and redirected into the other exit port facing the true lumen. A "pop" or "release" sensation is often felt when the guidewire penetrates into the true lumen. Once the wire exits from the correct port, it is advanced until it punctures back into the true lumen.

8.3.5.3.3.1.3   Confirm distal true lumen wire position   Contralateral injection is used to determine whether true lumen reentry has been achieved. If the wire has successfully reentered into the distal true lumen and the distal vessel is not severely diseased with a large lumen, the reentry guidewire can be rotated 180 degrees and advanced further down the vessel ("stick and drive" method).

Another way to confirm distal true lumen wire position is with intravascular ultrasound (IVUS) (CTO Manual online case 35).

8.3.5.3.3.1.4   Remove Stingray balloon and exchange it for a workhorse guidewire   The Stingray balloon should be removed using the

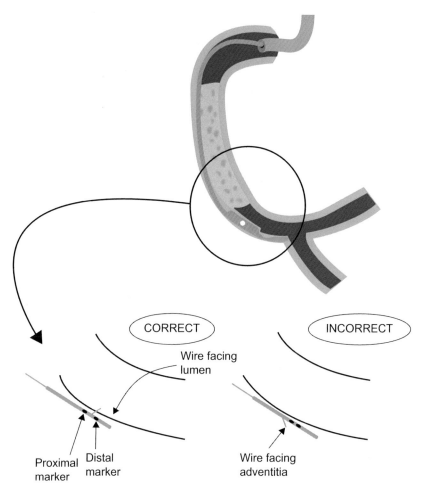

CORRECT

INCORRECT

Wire facing
lumen

Proximal Distal
marker marker

Wire facing
adventitia

FIGURE 8.3.37   **How to advance a penetrating guidewire through the side port of the Stingray balloon into the distal true lumen.**

trapping technique, to minimize the risk of losing distal wire position or causing distal vessel injury. The Stingray balloon can be trapped within a 7 Fr guide catheter. After removal of the Stingray balloon, a microcatheter is advanced over the guidewire that has reentered into the distal true lumen followed by exchange for a workhorse guidewire for subsequent equipment delivery.

8.3.5.3.3.2 "Stick and swap" technique (Fig. 8.3.38)

CTO Manual online cases 2, 16, 17, 25, 34, 35, 72, 83, 97, 100, 111, 117, 139, 144, 147, 149, 152, 154

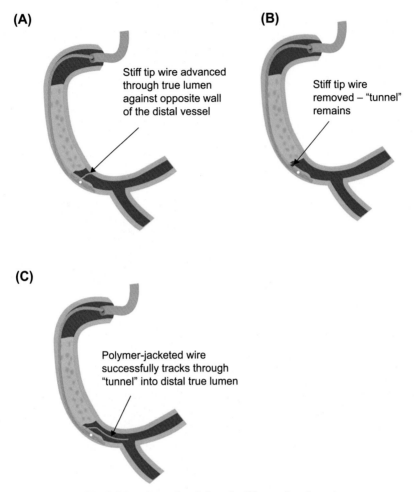

(A)

Stiff tip wire advanced through true lumen against opposite wall of the distal vessel

(B)

Stiff tip wire removed – "tunnel" remains

(C)

Polymer-jacketed wire successfully tracks through "tunnel" into distal true lumen

FIGURE 8.3.38   The "stick and swap" technique for Stingray-based reentry.

8.3.5.3.3.2.1 Obtain optimal angiographic view This is performed as described in Section 8.3.5.3.3.1.1.

8.3.5.3.3.2.2 Advance a penetrating guidewire This is performed as described in Section 8.3.5.3.3.1.2.

8.3.5.3.3.2.3 Swap for a polymer-jacketed guidewire After the penetrating guidewire is advanced through the side port of the balloon it is rotated 180 degrees and withdrawn to enlarge the wire exit connection and is then removed from the Stingray balloon. Another guidewire (usually stiff tip, polymer-jacketed wire, such as the Pilot 200, Gladius, or Raider) is advanced through the same Stingray balloon exit port and rotated until it tracks the course of the vessel. Avoid buckling (knuckling) of the wire when it exits the Stingray balloon. The polymer-jacketed guidewire should be shaped similar to the penetrating guidewire with a 1 mm 30−45 degrees distal bend. Placing more bend on the wire decreases the chance of entering the Stingray balloon side port.

8.3.5.3.3.2.4 Confirm distal true lumen position This is done as described in Section 8.3.5.3.3.1.3.

8.3.5.3.3.2.5 Remove Stingray balloon and exchange it for a workhorse guidewire This is done as described in Section 8.3.5.3.3.1.4.

8.3.5.3.3.3   Double blind stick and swap technique
CTO Manual online cases 13, 30, 32, 43, 45, 48, 79, 80, 82, 85, 92, 93, 94, 100, 120, 129, 131, 135, 142, 146, 148, 155, 159, 161, 165, 166, 168, 173

8.3.5.3.3.3.1 Advancing a penetrating guidewire A highly penetrating guidewire (such as the Astato 20, Hornet 14, Confianza Pro 12, Stingray, Infiltrac, Infiltrac Plus, or Gaia Next 3) is advanced through both side exit ports of the Stingray balloon, for approximately 3−5 mm. Occasionally, a firm "hard" push of the guidewire is needed to achieve advancement.

There remain concerns among some operators that advancing a guidewire towards the adventitia can cause perforation. However, a guidewire exit alone through the adventitia without advancing a balloon or microcatheter will almost never cause clinically significant coronary perforation, in part due to small caliber of the wire and in part because the perfusion pressure is very low since the track proximal to the perforation is usually extraplaque.

8.3.5.3.3.3.2 Swap The penetrating guidewire is removed and a polymer-jacketed guidewire (such as Pilot 200, Raider or Gladius) is advanced through each exit port in random order. The wire is advanced rotating, observing for smooth guidewire movement down the vessel. Difficulty with guidewire advancement and/or guidewire buckling usually suggests extraplaque or distal false lumen guidewire position. The wire should be withdrawn and either advanced again through the same port or advanced through the contralateral exit port until there is smooth and free wire movement indicating true lumen entry.

8.3.5.3.3.3.3 Confirm distal true lumen position This is done as described in Section 8.3.5.3.3.1.3.

8.3.5.3.3.3.4 Remove Stingray balloon and exchange for a workhorse guidewire This is done as described in Section 8.3.5.3.3.1.4.

8.3.5.3.4 Stingray troubleshooting

8.3.5.3.4.1 Poor visualization of the Stingray balloon and the distal true lumen

Potential solutions:

1. Double blind stick and swap technique.
2. Meticulous preparation of the Stingray balloon.
3. Orthogonal views to identify an optimal reentry projection: the goal is to achieve a "sideways" projection (Fig. 8.3.36 panel B) and avoid a "railroad/tram track projection" (Fig. 8.3.36 panel A).
4. Magnified views (magnification is discouraged during the other portions of the CTO PCI procedure as it can significantly increase radiation dose).
5. Placement of a retrograde guidewire (and a retrograde balloon if feasible) to mark the position of the distal true lumen (retrograde facilitated ADR, **CTO Manual online cases 83, 111**)

8.3.5.3.4.2 Unable to enter a diffusely diseased distal vessel

Reentry from the false lumen into the distal true lumen can be difficult in patients with small, diffusely diseased distal vessels because the distal true lumen is small and the Stingray wire may go "through and through" into the opposite vessel wall (Fig. 8.3.38, panel A).

Solutions

1. Change the reentry site by moving the Stingray balloon to a healthier, straighter, and larger vessel segment (the horizontal part of the distal right coronary artery is usually preferable for reentry). This technique is called "**bobsled**" (**CTO Manual online cases 16, 48, 79, 93, 129, 142, 155, 173**).
2. Use the "stick and swap" or the "double blind stick and swap" technique, as described in Fig. 8.3.38.
3. IVUS-guided reentry.

8.3.5.3.4.3 Compression of distal true lumen by hematoma

Prevention of distal false lumen hematomas is key for successful reentry. Antegrade contrast injections should not be performed after antegrade dissection has started. The use of small knuckles and guide catheter extensions reduces the risk of hematoma formation. Use of the CrossBoss catheter (instead of a knuckle wire) for dissection (especially in the distal segment of the occlusion) can reduce the risk of enlarging the distal false lumen ("dry dissection"). This strategy is called "finish with the Boss."

If a hematoma develops, aspiration of the hematoma can be attempted through:

1. the Stingray balloon itself[84] (before attempting reentry or while attempting reentry by using a Y-connector at the Stingray balloon side port and using a syringe for aspiration, "Stingray-STRAW technique" Fig. 8.3.39).
2. an over-the-wire balloon (classic STRAW: Subintimal TRAnscatheter Withdrawal technique, Fig. 8.3.40)[85] advanced proximal to the Stingray balloon.
3. a microcatheter (modified STRAW, Fig. 8.3.41)[86] advanced proximal to the Stingray balloon.

The classic STRAW technique requires the use of 8 French guide catheters, whereas the modified STRAW technique can be performed through 7 French guide catheters. However, the modified STRAW technique may be less effective if there is a large proximal hematoma, since the small microcatheters are unlikely to fully occlude blood inflow to allow meaningful false lumen hematoma decompression at the site of reentry.[86] The ReCross microcatheter can also be used, allowing aspiration of the hematoma through one of the two over-the-wire lumens while trying to reenter through the other over-the-wire lumen.

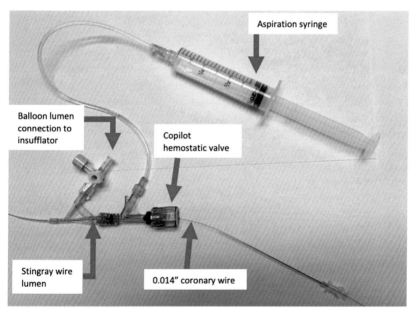

FIGURE 8.3.39 **Illustration of the Stingray-STRAW technique**: continuous aspiration is applied through the wire lumen of the Stingray balloon during reentry attempts. *Courtesy of Dr. Catalin Toma.*

A.   Classic STraW: Subintimal Transcatheter Withdrawal

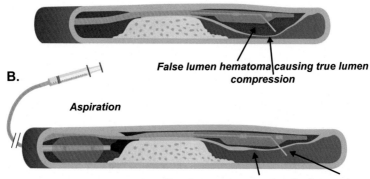

False lumen hematoma causing true lumen compression

B.

Aspiration

True lumen re-expansion & wire reentry

FIGURE 8.3.40   **Illustration of the subintimal transcatheter withdrawal (STRAW) technique. Panel A**: an extraplaque hematoma compresses the distal true lumen, hindering reentry attempts with the Stingray balloon. **Panel B**: insertion of an over-the-wire balloon into the extraplaque space, through which aspiration is performed, decompressing the hematoma and allowing re-expansion of the distal true lumen, facilitating Stingray-based reentry.

Another modification of the STRAW technique involves advancing a guide catheter extension (ideally TrapLiner that also allows trapping), into the vessel and aspirating at the manifold. Upfront placement of the guide extension into the lesion with pressure dampening, can also prevent blood flow into the extraplaque space and reduce the risk of hematoma formation. The STRAW technique is most effective if the proximal vessel is occluded (for example with a balloon) to prevent continuing expansion from proximal blood flow.[85]

Sometimes an extraplaque hematoma can cause distal true lumen compression even after successful reentry and stent implantation. In such cases, use of a cutting balloon may be useful to decompress the hematoma and restore antegrade flow.[87]

The STRAW technique has also been successfully used in patients with spontaneous coronary dissections.[88]

8.3.5.3.4.4   Distal vessel calcification

Reentry can be challenging in heavily calcified vessels where the reentry guidewire may "slide" inside the false lumen instead of puncturing into the true lumen. Potential solutions:

- Use a different stiff tip, tapered, highly penetrating guidewire (such as the Astato 20, Hornet 14, Infiltrac and Infiltrac Plus, Confianza Pro 12, Stingray, or Gaia 3rd) to puncture through the calcified vessel wall.
- Place a secondary bend proximally on the guidewire.

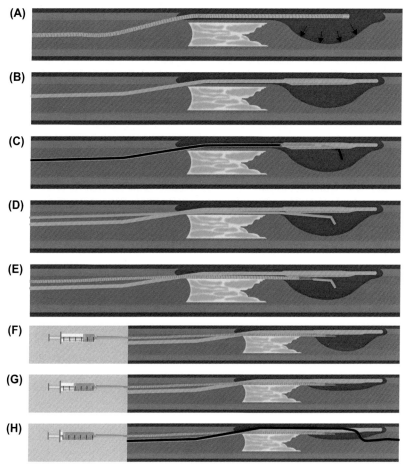

FIGURE 8.3.41 **Illustration of the modified subintimal transcatheter withdrawal (STRAW) technique.**[86] **Panel A**: CrossBoss advanced into the subintimal space complicated by compressive hematoma (arrows). **Panel B**: CrossBoss exchanged for a Stingray balloon to facilitate wire reentry. **Panel C**: failed wire reentry due to compressive hematoma. **Panel D**: another guidewire is advanced into the false lumen. **Panel E**: a microcatheter is advanced into the false lumen. **Panel F**: suction is applied to the microcatheter with a syringe. **Panel G**: decompression of the false lumen hematoma. **Panel H**: true lumen re-expansion allowing successful guidewire entry into the distal true lumen.

- Attempt reentry into a more distal location by advancing the Stingray balloon distally ("bobsled" technique).
- Retrograde crossing.

8.3.5.3.4.5 Occlusion of side branch at distal cap

Problem: In lesions with a bifurcation at the distal cap, reentry into one of the branches may result in occlusion of the other branch.

Solutions
1. Reenter proximal to the bifurcation.
2. After reentering into one branch, reenter into the other, often using the Stingray system again ("double reentry," **CTO Manual online cases 25 and 80**).[89]
3. Use the retrograde approach to restore the patency of the other branch.
4. Use a dual lumen microcatheter after the initial reentry has been completed.
5. STAR into side branch.

8.3.5.3.4.6   Inability to advance a penetrating guidewire through the Stingray balloon

Causes
- Severe vessel tortuosity
- Kinking of the Stingray balloon shaft.

Solutions
1. Slightly withdraw the Stingray balloon to straighten its shaft.
2. Advance a polymer-jacketed guidewire (such as the Pilot 200 or Mongo) through the Stingray balloon first.
3. Use a different penetrating guidewire.
4. Do NOT push aggressively, as the guidewire could perforate the Stingray balloon shaft.
5. If all else fails, remove the Stingray balloon and wire the lesion again.

8.3.5.4   Antegrade reentry: ReCross

The ReCross is a unique dual lumen microcatheter (Section 30.6.4), as it has two over-the-wire lumens, in contrast to all other dual lumen microcatheters that have a monorail and an over-the-wire lumen. The two lumens have 180 degrees offset exit ports distally, similar to the Stingray balloon. The distal section of the ReCross is oval, facilitating alignment of one exit port against the distal true lumen. Because of this specialized construction, the ReCross can be used for distal true lumen reentry after extraplaque CTO crossing, similar to the Stingray balloon.

Compared with the Stingray balloon, the ReCross microcatheter is more deliverable and allows the application of suction through one of the two over-the-wire lumens, decompressing the false lumen and preventing hematoma formation. Conversely, the Stingray balloon with its longer "wings" (2 mm in diameter) may allow a more favorable orientation of an exit port against the distal true lumen.

8.3.5.5   Antegrade fenestration and reentry: AFR

Antegrade fenestration and reentry (AFR) is an antegrade reentry technique that uses balloon inflation across the distal cap of the CTO to create transient fenestrations between the false lumen and the distal true lumen.[45,46] Upon balloon deflation a second wire is quickly advanced through the dissection tears into the distal true lumen before the dissection flap collapses. It is performed in five steps (Fig. 8.3.42):

8.3.5.5.1   Step 1

A guidewire has been advanced unintentionally into the extraplaque space or into the false lumen distal to the distal cap.

8.3.5.5.2   Step 2

The microcatheter is removed, leaving the first guidewire in place. The occlusion is wired again through the extraplaque space, keeping the second guidewire tip as close as possible to the first wire (checking in multiple projections), within the occlusion. This second wire is delivered to the level of the distal cap. This step can be facilitated by use of a dual lumen microcatheter.

8.3.5.5.3   Step 3

A balloon (sized 1:1 with the artery diameter) is advanced over the first guidewire and placed across the distal cap.

8.3.5.5.4   Step 4

The balloon is inflated, at least up to nominal pressure (if a noncompliant balloon is used, higher pressures can be used, but never exceeding a balloon-to-artery ratio of 1:1, to reduce the risk of perforation).

8.3.5.5.5   Step 5

Immediately after balloon deflation, the second guidewire (polymer-jacketed, low tip load guidewires are recommended, such as the Sion black, Fielder, Fighter and Bandit with a 2-mm, 45 degrees bend) is rapidly advanced through the fenestrations created by the balloon (before the collapse of the dissection flaps) between the false lumen and the true lumen, which is now accessible, since balloon dilatation took place across the distal cap. In case of fibrocalcification at the site of reentry, or excessive separation between the two wires, a higher tip load polymer-jacketed wire can be used (such as the Pilot 200, Gladius, or Raider).

A guidewire is advanced inadvertently antegradely in the extraplaque space. The distal tip of the guidewire is located beyond the distal cap.

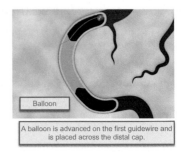

Balloon

A balloon is advanced on the first guidewire and is placed across the distal cap.

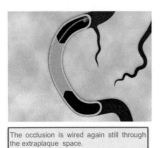

The occlusion is wired again still through the extraplaque space.

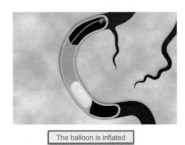

The balloon is inflated

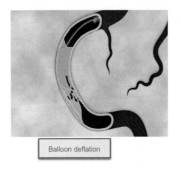

Balloon deflation

Wire into true lumen

Immediately after balloon deflation, the second wire is advanced into the true lumen through the fenestrations created by the balloon inflation.

FIGURE 8.3.42 **Illustration of the antegrade fenestration and reentry (AFR) technique**. *Courtesy of Dr. Mauro Carlino.*

An advantage of AFR compared with Stingray-based reentry is wide availability and lower cost, as it can be performed with wires, microcatheters, and balloons already available in the cardiac catheterization laboratory. A potential limitation of AFR is creation of unfenestrated extraplaque hematoma due to extraplaque balloon inflation that could compress the distal true lumen and hinder reentry. To solve this issue, repeated balloon inflations at multiple sites throughout the occlusion and at the distal cap are recommended. AFR may also be less predictable than Stingray reentry.

Similar to Stingray-based reentry, insertion of a retrograde guidewire can facilitate AFR (retrograde facilitated AFR).[44] A dual wire balloon is currently being developed to facilitate AFR.[90] If AFR fails, LAST, STAR, retrograde crossing, or subintimal plaque modification can be used.

References

33. Michael TT, Papayannis AC, Banerjee S, Brilakis ES. Subintimal dissection/reentry strategies in coronary chronic total occlusion interventions. *Circ Cardiovasc Interv* 2012;**5**:729−38.

34. Galassi AR, Tomasello SD, Costanzo L, et al. Mini-STAR as bail-out strategy for percutaneous coronary intervention of chronic total occlusion. *Catheter Cardiovasc Interv* 2012;**79**:30−40.

35. Whitlow PL, Burke MN, Lombardi WL, et al. Use of a novel crossing and reentry system in coronary chronic total occlusions that have failed standard crossing techniques: results of the FAST-CTOs (Facilitated Antegrade Steering Technique in Chronic Total Occlusions) trial. *JACC Cardiovasc Interv* 2012;**5**:393−401.

36. Brilakis ES. *Manual of coronary chronic total occlusion interventions. a step-by-step approach*. 2nd ed. Elsevier; 2017.

37. Colombo A, Mikhail GW, Michev I, et al. Treating chronic total occlusions using subintimal tracking and reentry: the STAR technique. *Catheter Cardiovasc Interv* 2005;**64**:407.

38. Carlino M, Godino C, Latib A, Moses JW, Colombo A. Subintimal tracking and reentry technique with contrast guidance: a safer approach. *Catheter Cardiovasc Interv* 2008;**72**:790−6.

39. Valenti R, Vergara R, Migliorini A, et al. Predictors of reocclusion after successful drug-eluting stent-supported percutaneous coronary intervention of chronic total occlusion. *J Am Coll Cardiol* 2013;**61**:545−50.

40. Azzalini L, Dautov R, Brilakis ES, et al. Procedural and longer-term outcomes of wire- vs device-based antegrade dissection and reentry techniques for the percutaneous revascularization of coronary chronic total occlusions. *Int J Cardiol* 2017;**231**:78−83.

41. Lombardi WL. Retrograde PCI: what will they think of next? *J Invasive Cardiol* 2009;**21**:543.

42. Werner GS. The BridgePoint devices to facilitate recanalization of chronic total coronary occlusions through controlled subintimal reentry. *Expert Rev Med Devices* 2011;**8**:23−9.

43. Brilakis ES, Badhey N, Banerjee S. "Bilateral knuckle" technique and Stingray reentry system for retrograde chronic total occlusion intervention. *J Invasive Cardiol* 2011;**23**:E37−9.

44. Azzalini L, Carlino M. A new combined antegrade and retrograde approach for chronic total occlusion recanalization: Facilitated antegrade fenestration and reentry. *Catheter Cardiovasc Interv* 2021;**98**:E85−90.

45. Azzalini L, Alaswad K, Uretsky BF, et al. Multicenter experience with the antegrade fenestration and reentry technique for chronic total occlusion recanalization. *Catheter Cardiovasc Interv* 2021;**97**:E40−50.

46. Carlino M, Azzalini L, Mitomo S, Colombo A. Antegrade fenestration and reentry: a new controlled subintimal technique for chronic total occlusion recanalization. *Catheter Cardiovasc Interv* 2018;**92**:497−504.

47. Godino C, Latib A, Economou FI, et al. Coronary chronic total occlusions: mid-term comparison of clinical outcome following the use of the guided-STAR technique and conventional anterograde approaches. *Catheter Cardiovasc Interv* 2012;**79**:20−7.
48. Christopoulos G, Wyman RM, Alaswad K, et al. Clinical utility of the Japan-chronic total occlusion score in coronary chronic total occlusion interventions: results from a Multicenter Registry. *Circ Cardiovasc Interv* 2015;**8**:e002171.
49. Wilson WM, Walsh SJ, Yan AT, et al. Hybrid approach improves success of chronic total occlusion angioplasty. *Heart* 2016;**102**:1486−93.
50. Tajti P, Karmpaliotis D, Alaswad K, et al. The hybrid approach to chronic total occlusion percutaneous coronary intervention: update from the PROGRESS CTO registry. *JACC Cardiovasc Interv* 2018;**11**:1325−35.
51. Stetler J, Karatasakis A, Christakopoulos GE, et al. Impact of crossing technique on the incidence of periprocedural myocardial infarction during chronic total occlusion percutaneous coronary intervention. *Catheter Cardiovasc Interv* 2016;**88**:1−6.
52. Vo MN, Karmpaliotis D, Brilakis ES. "Move the cap" technique for ambiguous or impenetrable proximal cap of coronary total occlusion. *Catheter Cardiovasc Interv* 2016;**87**:742−8.
53. Roy J, Hill J, Spratt JC. The "side-BASE technique": combined side branch anchor balloon and balloon assisted sub-intimal entry to resolve ambiguous proximal cap chronic total occlusions. *Catheter Cardiovasc Interv* 2018;**92**:E15−19.
54. Kaier TE, Kalogeropoulos A, Pavlidis AN. Guide-extension facilitated antegrade dissection reentry: a case series. *J Invasive Cardiol* 2020;**32**:E209−12.
55. Fujita S, Tamai H, Kyo E, et al. New technique for superior guiding catheter support during advancement of a balloon in coronary angioplasty: the anchor technique. *Catheter Cardiovasc Interv* 2003;**59**:482−8.
56. Hirokami M, Saito S, Muto H. Anchoring technique to improve guiding catheter support in coronary angioplasty of chronic total occlusions. *Catheter Cardiovasc Interv* 2006;**67**:366−71.
57. Kirtane AJ, Stone GW. The Anchor-Tornus technique: a novel approach to "uncrossable" chronic total occlusions. *Catheter Cardiovasc Interv* 2007;**70**:554−7.
58. Matsumi J, Saito S. Progress in the retrograde approach for chronic total coronary artery occlusion: a case with successful angioplasty using CART and reverse-anchoring techniques 3 years after failed PCI via a retrograde approach. *Catheter Cardiovasc Interv* 2008;**71**:810−14.
59. Fang HY, Wu CC, Wu CJ. Successful transradial antegrade coronary intervention of a rare right coronary artery high anterior downward takeoff anomalous chronic total occlusion by double-anchoring technique and retrograde guidance. *Int Heart J* 2009;**50**:531−8.
60. Lee NH, Suh J, Seo HS. Double anchoring balloon technique for recanalization of coronary chronic total occlusion by retrograde approach. *Catheter Cardiovasc Interv* 2009;**73**:791−4.
61. Saito S. Different strategies of retrograde approach in coronary angioplasty for chronic total occlusion. *Catheter Cardiovasc Interv* 2008;**71**:8−19.
62. Surmely JF, Katoh O, Tsuchikane E, Nasu K, Suzuki T. Coronary septal collaterals as an access for the retrograde approach in the percutaneous treatment of coronary chronic total occlusions. *Catheter Cardiovasc Interv* 2007;**69**:826−32.
63. Surmely JF, Tsuchikane E, Katoh O, et al. New concept for CTO recanalization using controlled antegrade and retrograde subintimal tracking: the CART technique. *J Invasive Cardiol* 2006;**18**:334−8.
64. Rathore S, Katoh O, Matsuo H, et al. Retrograde percutaneous recanalization of chronic total occlusion of the coronary arteries: procedural outcomes and predictors of success in contemporary practice. *Circ Cardiovasc Interv* 2009;**2**:124−32.

65. Karacsonyi J, Tajti P, Rangan BV, et al. Randomized comparison of a CrossBoss first vs standard wire escalation strategy for crossing coronary chronic total occlusions: the crossboss first trial. *JACC Cardiovasc Interv* 2018;**11**:225−33.

66. Carlino M, Demir OM, Colombo A, Azzalini L. Microcatheter knuckle technique: a novel technique for negotiating the subintimal space during chronic total occlusion recanalization. *Catheter Cardiovasc Interv* 2018;**92**:1256−60.

67. Visconti G, Focaccio A, Donahue M, Briguori C. Elective vs deferred stenting following subintimal recanalization of coronary chronic total occlusions. *Catheter Cardiovasc Interv* 2015;**85**:382−90.

68. Carlino M, Latib A, Godino C, Cosgrave J, Colombo A. CTO recanalization by intraocclusion injection of contrast: the microchannel technique. *Catheter Cardiovasc Interv* 2008;**71**:20−6.

69. Carlino M, Ruparelia N, Thomas G, et al. Modified contrast microinjection technique to facilitate chronic total occlusion recanalization. *Catheter Cardiovasc Interv* 2016;**87**:1036−41.

70. Azzalini L, Uretsky B, Brilakis ES, Colombo A, Carlino M. Contrast modulation in chronic total occlusion percutaneous coronary intervention. *Catheter Cardiovasc Interv* 2019;**93**: E24−9.

71. Hirai T, Grantham JA, Sapontis J, et al. Impact of subintimal plaque modification procedures on health status after unsuccessful chronic total occlusion angioplasty. *Catheter Cardiovasc Interv* 2018;**91**:1035−42.

72. Goleski PJ, Nakamura K, Liebeskind E, et al. Revascularization of coronary chronic total occlusions with subintimal tracking and reentry followed by deferred stenting: experience from a high-volume referral center. *Catheter Cardiovasc Interv* 2019;**93**:191−8.

73. Hall AB, Brilakis ES. Hybrid 2.0: subintimal plaque modification for facilitation of future success in chronic total occlusion percutaneous coronary intervention. *Catheter Cardiovasc Interv* 2019;**93**:199−201.

74. Xenogiannis I, Choi JW, Alaswad K, et al. Outcomes of subintimal plaque modification in chronic total occlusion percutaneous coronary intervention. *Catheter Cardiovasc Interv* 2020;**96**:1029−35.

75. Davies RE, Rier JD, McEntegart M, Riley RF, Kearney K, Lombardi W. Subintimal tracking and reentry as a tool in CTO-PCI: past, present, and future. *Catheter Cardiovasc Interv* 2021;**98**:1144−51.

76. Ybarra LF, Dandona S, Daneault B, Rinfret S. Drug-coated balloon after subintimal plaque modification in failed coronary chronic total occlusion percutaneous coronary intervention: a novel concept. *Catheter Cardiovasc Interv* 2020;**96**:609−13.

77. Galassi AR, Boukhris M, Tomasello SD, et al. Long-term clinical and angiographic outcomes of the mini-STAR technique as a bailout strategy for percutaneous coronary intervention of chronic total occlusion. *Can J Cardiol* 2014;**30**:1400−6.

78. Badhey N, Lombardi WL, Thompson CA, Brilakis ES, Banerjee S. Use of the venture wire control catheter for subintimal coronary dissection and reentry in chronic total occlusions. *J Invasive Cardiol* 2010;**22**:445−8.

79. Qureshi WT, Ogunsua AA, Kundu A, Sattar Y, Fisher DZ, Kakouros N. Angled microcatheter assisted antegrade dissection reentry technique for tortuous totally occluded coronary arteries. *Cardiovasc Revasc Med* 2021;**28S**:127−31.

80. Ghoshhajra BB, Takx RAP, Stone LL, et al. Real-time fusion of coronary CT angiography with x-ray fluoroscopy during chronic total occlusion PCI. *Eur Radiol* 2017;**27**: 2464−73.

81. Habara M, Tsuchikane E, Shimizu K, et al. Japanese multicenter registry evaluating the antegrade dissection reentry with cardiac computerized tomography for chronic coronary total occlusion. *Cardiovasc Interv Ther* 2022;**37**:116−27.

82. Opolski MP, Zysk A, Wolny R, Debski A, Witkowski A. Coronary CTA co-registration for guiding antegrade dissection reentry in chronic total occlusion percutaneous coronary intervention. *J Cardiovasc Comput Tomogr* 2022;**16**:e14−16.

83. Christopoulos G, Kotsia AP, Brilakis ES. The double-blind stick-and-swap technique for true lumen reentry after subintimal crossing of coronary chronic total occlusions. *J Invasive Cardiol* 2015;**27**:E199−202.

84. Wu EB, Brilakis ES, Lo S, et al. Advances in CrossBoss/Stingray use in antegrade dissection reentry from the Asia Pacific chronic total occlusion club. *Catheter Cardiovasc Interv* 2020;**96**:1423−33.

85. Smith EJ, Di Mario C, Spratt JC, et al. Subintimal TRAnscatheter Withdrawal (STRAW) of hematomas compressing the distal true lumen: a novel technique to facilitate distal reentry during recanalization of chronic total occlusion (CTO). *J Invasive Cardiol* 2015;**27**: E1−4.

86. Vo MN, Brilakis ES, Pershad A, Grantham JA. Modified subintimal transcatheter withdrawal: a novel technique for hematoma decompression to facilitate distal reentry during coronary chronic total occlusion recanalization. *Catheter Cardiovasc Interv* 2020;**96**:E98−101.

87. Vo MN, Brilakis ES, Grantham JA. Novel use of cutting balloon to treat subintimal hematomas during chronic total occlusion interventions. *Catheter Cardiovasc Interv* 2018;**91**:53−6.

88. Matsuura S, Otowa K, Maruyama M, Usuda K. Successful revascularization with percutaneous coronary intervention using a combination of the subintimal transcatheter withdrawal technique and coronary artery fenestration for spontaneous coronary artery dissection. *Clin Case Rep* 2021;**9**:e05045.

89. Tajti P, Doshi D, Karmpaliotis D, Brilakis ES. The "double stingray technique" for recanalizing chronic total occlusions with bifurcation at the distal cap. *Catheter Cardiovasc Interv* 2018;**91**:1079−83.

90. Galassi AR, Vadalà G, Testa G, Puglisi S, Sucato V, Diana D, Giunta R, Novo G. Dual guidewire balloon antegrade fenestration and reentry technique for coronary chronic total occlusions percutaneous coronary interventions. *Catheter Cardiovasc Interv.* 2022;100:492−501.

Section 8.4

Retrograde

8.4.1   Historical perspective

The retrograde technique differs from the standard antegrade approach in that the occlusion is approached from the distal vessel by advancing a wire against the original direction of blood flow, that is, retrograde.[91,92] The guidewire is advanced into the artery distal to the occlusion through either a bypass graft or through a collateral channel. This approach differs from the antegrade approach, in which all equipment is inserted only proximal to the occlusion and travels in the same direction as the original arterial flow, that is, antegrade.

The retrograde CTO PCI technique was first described by Kahn and Hartzler in 1990 who performed balloon angioplasty of a left anterior descending artery (LAD) CTO via a saphenous vein graft (SVG).[93] In 1996, Silvestri et al. reported retrograde stenting of the left main artery via a SVG.[94] In 2006, Surmely, Tsuchikane, Katoh et al. first reported retrograde crossing via septal collaterals,[95] starting the modern era of the retrograde techniques through septal[95–100] and epicardial[101] collaterals as well as through arterial bypass grafts.[100] The introduction of specialized equipment and further refinements of the technique started in Japan[102–104] with rapid adoption both in Europe[105–108] and in the US.[109,110]

8.4.2   Advantages of the retrograde approach

Crossing in the retrograde direction can sometimes be easier than antegrade crossing because the distal cap:

1. is easier to enter than the proximal cap, as it is more frequently tapered;[111]
2. is often softer than the proximal cap, likely because of exposure to lower pressure; and
3. is less frequently anatomically ambiguous.

Moreover, the antegrade approach may not be feasible or desirable in some CTOs, such as:

1. Ostial and stumpless CTOs.
2. CTOs with an ambiguous proximal cap or a bifurcation at the distal cap.
3. Long and tortuous CTOs.
4. Previously failed antegrade CTO PCI.
5. Extensive dissection or diffuse disease distal to the occlusion.

In cases where antegrade wiring is challenging because of ambiguous course, the retrograde wire can help direct the antegrade wire. Similarly, a retrograde guidewire can facilitate reentry into the distal true lumen when attempting antegrade dissection/reentry. Retrograde CTO PCI might also be advantageous in patients with severe renal insufficiency and clear retrograde channels, because most steps of the retrograde approach require limited contrast injections if retrograde access to the distal vessel can be achieved without difficulty.[112]

8.4.3 Special equipment

In addition to the standard equipment needed for the antegrade approach, the retrograde approach requires specialized equipment, that is, short (Section 30.2.2) or shortened (Section 30.2.3) guide catheters, 150−155 cm long microcatheters (Section 30.6) and externalization guidewires, such as the R350 and RG3 wires (Section 30.7.6).

1. **Short guide catheters (usually 90 cm long) and guide catheter extensions**

    The standard guide catheter length is 100 cm (shaft length, although the length from the hub to the guide tip is approximately 106 cm).[91] If standard guide catheters are used for the retrograde approach, equipment may not be long enough to reach the lesion retrogradely; the retrograde microcatheter might not reach the antegrade guide catheter, and wires advanced retrogradely may be too short to be externalized, especially with retrograde crossing through epicardial collaterals or bypass grafts. Utilizing a shorter guide catheter extends the reach of balloons, wires, and microcatheters advanced retrogradely, as the length of the catheter outside the body has been decreased by the shorter guide length.

    Shorter guide catheters (usually 90 cm long) are commercially available, but if they are not locally available, any guide can be shortened using an interposition segment from a sheath, as described in Section 30.2.3.[113]

    An alternative to using a shorter guide catheter is to use an antegrade guide catheter extension deeply intubated into the CTO target vessel that effectively decreases the lesion length that needs to be traversed by the retrograde equipment in order to reach the antegrade guide catheter. Another option is to use the "tip-in" technique (Section 8.4.4.8.2).

2. **Microcatheters**

    Several microcatheters are available for the retrograde approach, such as the Corsair and Corsair XS, Caravel, Turnpike, Turnpike LP, Teleport, Mamba Flex, Finecross, Nhancer ProX, and Telemark (Section 30.5). Long (150 cm, except for the Nhancer ProX that is 155 cm long) microcatheters should be used for nearly all retrograde crossing cases. Larger microcatheters, such as the Corsair and Turnpike provide collateral dilatation and have good penetration power into the distal cap, yet can be more challenging to deliver especially through small and tortuous collaterals. Lower profile microcatheters, such as the Caravel, Corsair XS, Turnpike LP, Telemark and Finecross may be easier to deliver, especially through small caliber epicardial collaterals, but provide less support for crossing complex (such as heavily calcified) lesions.

3. **Externalization guidewires**

Dedicated externalization guidewires are currently available (RG3 and R350) (Section 30.6.6) and should be used for externalization whenever possible. These wires are long (330 cm for the RG3 and 350 cm for the R350), thinner than standard guidewires, and have a hydrophilic coating over more than half of their shafts. The tip of the wire should not be bent, to facilitate antegrade equipment delivery after externalization.

If a dedicated externalization guidewire is not available or cannot be used (for example, when the guidewire crosses into the antegrade guide catheter but the microcatheter cannot be advanced to the antegrade guide catheter) any long (300 cm) guidewire or the Rotawire Drive Floppy and Viper Advance with flex tip guidewire can be used for externalization, especially if short guide catheters or guide catheter extensions are being used. However, externalization of standard guidewires can be more challenging and take longer time and more force, potentially leading to compression of the heart with hypotension, bradycardia, and occasionally asystole. Lubricating the microcatheter with Rotaglide (or Propofol 0.5−1 mL) may facilitate externalization of such wires. Another way to facilitate externalization of a conventional guidewire is to trap it with a balloon inside a guide catheter extension, followed by a gentle pull of this unit while simultaneously pushing the wire into the retrograde microcatheter. Guidewire extensions, such as the doc and cinch, should not be used to extend short wires for externalization.

4. **Collateral crossing guidewires**

Preferred wires for septal crossing are composite core, soft tip guidewires (such as the Sion, Suoh 03, and Samurai RC) (Section 30.6.1) or soft tip, polymer-jacketed wires (such as the Sion Black, Fielder XT-R, Fielder FC, and Pilot 50) (Section 30.6.2). The usual tip bend is very short (1 mm) and quite shallow (20−30 degrees, although some operators use 90-degree bends) to allow for tracking very small, tortuous collaterals.

8.4.4   Step-by-step description of the procedure

Retrograde CTO PCI is performed in 10 steps, as follows[114] (Fig. 8.4.1):

8.4.4.1   Step 1. Decide that retrograde is the next step

Goal: Decide whether the retrograde approach should be used; this is discussed in detail in Section 8.5. If retrograde crossing is selected, the

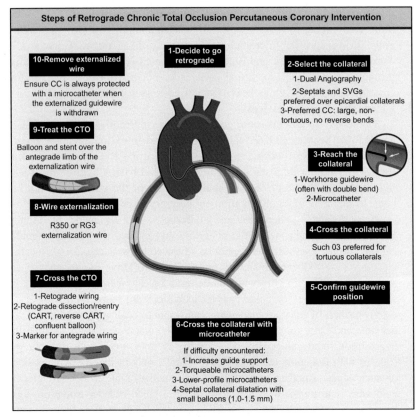

FIGURE 8.4.1   **The 10 steps of the retrograde approach**.

activated clotting time (ACT) should be kept >350 sec to minimize the risk of donor vessel and guide thrombosis.

How (Fig. 8.4.2)?

Three key criteria must be met before using the retrograde approach:

1. Feasible retrograde option
2. Local experience and expertise in CTO PCI
3. Appropriate anatomic substrate

8.4.4.1.1   Feasible retrograde option

The term "interventional" collateral has traditionally been used for retrograde collateral vessels considered appropriate for crossing by the operator.[115] Coronary bypass grafts, whether patent or occluded, are often used for the retrograde approach, even though bypass grafts are not technically "collateral" vessels. The CTO global crossing algorithm introduced the

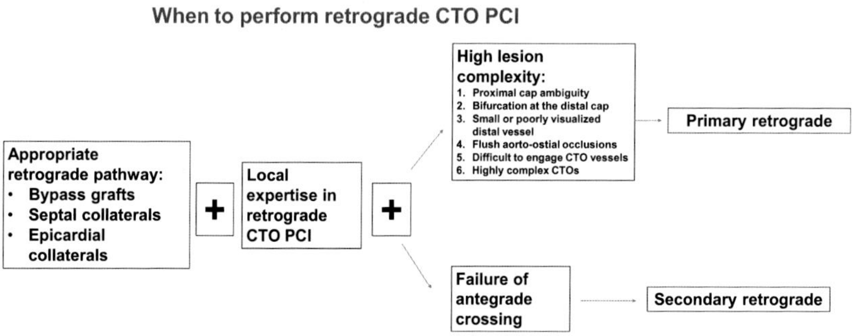

When to perform retrograde CTO PCI

FIGURE 8.4.2   **When to use the retrograde approach**.

more accurate term "feasible retrograde option" to describe the presence of any retrograde route considered appropriate and safe for retrograde CTO PCI.[116]

8.4.4.1.2   Local experience and expertise in retrograde CTO PCI

The retrograde approach requires several steps, can be challenging to perform, can lead to potentially catastrophic complications, such as donor vessel injury, and has higher complication rates compared with the antegrade approach.[117] As a result, it should be used by experienced operators and is best learned through proctorship or participation in dedicated CTO PCI training.

8.4.4.1.3   Appropriate anatomic substrate

The retrograde approach can be performed either (1) after failure of antegrade crossing or (2) as the initial crossing approach (primary retrograde) (Fig. 8.4.2).

Given increased complexity and risk, the retrograde approach is usually performed when antegrade crossing attempts fail or carry more risk. Some operators advocate always performing an antegrade crossing attempt before trying retrograde crossing because the antegrade attempt may be successful and even if it fails, antegrade preparation will likely be needed for completing the retrograde crossing.[114] However, the antegrade attempts should not be excessive, as use of too much contrast material, radiation, and time for antegrade crossing may necessitate termination of the procedure before attempting retrograde crossing.

The following anatomic substrates favor a primary retrograde approach:

1. Ambiguous proximal cap that cannot be resolved by other means, such as IVUS, antegrade dissection/reentry "move the cap" techniques, or coronary CT angiography.

2. CTOs with a bifurcation at the distal cap; the retrograde approach increases the likelihood of achieving patency of both branches.
3. CTOs with small or poorly visualized distal vessel.
4. Flush aorto-ostial CTOs.[118,119]
5. CTO vessels that are difficult to engage, such as anomalous coronary arteries,[118,120] flush aorto-ostial CTOs, in-stent aorto-ostial CTOs with the stent protruding into the aorta, or coronary arteries in patients with prior transcatheter aortic valve replacement (TAVR).
6. Highly complex CTOs, such as CTOs with long occlusion length, severe tortuosity, and severe calcification.

The risks and benefits of antegrade versus retrograde crossing should be continually and individually assessed for each target CTO throughout the procedure, as discussed in Section 8.5.

8.4.4.2   Step 2. Selecting the retrograde route (bypass graft or collateral channel)

Goal: Select the bypass graft(s) or collateral channel(s) that will be used for the retrograde approach.

How?
The usual preference order for selecting a retrograde pathway is: bypass graft, septal, then epicardial. However, due to availability, septal collaterals are most commonly used for retrograde CTO PCI (50%−65%), followed by epicardial collaterals (25%−38%) and bypass grafts (5%−13%).[114] It is easier to cross septal collaterals from the LAD to the RCA (for RCA CTOs) as compared from the RCA to the LAD (for LAD CTOs) because the inferior segment of the septal collaterals is often more tortuous than the superior segment.

The advantages and disadvantages of each retrograde pathway are shown in Fig. 8.4.3.[91] The classification and optimal angiographic views for evaluating collateral vessels are discussed in detail in Section 6.2.1.4.

Bypass grafts can be saphenous vein grafts (SVGs) and arterial grafts, either radial or internal mammary (free or in situ) grafts.

Saphenous vein grafts (patent or occluded) are large and easy to wire (Fig. 8.4.3), but are the least frequently available: they were used in 19% of retrograde CTO PCI in prior CABG patients in one series.[121] Although coronary artery bypass graft surgery causes scarring of the pericardium, it does not eliminate the likelihood of free pericardial effusion (or even worse, loculated pericardial effusion[122−125]) and tamponade in case of perforation during CTO PCI.[126] Even acutely[127] or chronically[121] occluded SVGs can serve as conduits to the distal arterial segment of chronically occluded native coronary arteries. In chronically occluded grafts, a tapered entry stump increases the likelihood of success navigating the occluded graft. Bypass grafts touchdowns may "tent" the vessel to which they are anastomosed, potentially

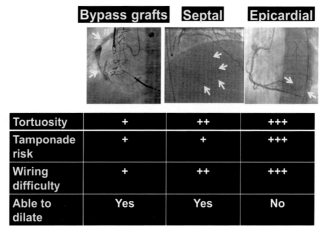

FIGURE 8.4.3   **Comparison of advantages and disadvantages of various collateral vessels that can be used for retrograde CTO interventions.**[91] *Reproduced with permission from Brilakis ES, Grantham JA, Thompson CA, DeMartini TJ, Prasad A, Sandhu GS, Banerjee S, Lombardi WL. The retrograde approach to coronary artery chronic total occlusions: a practical approach. Catheter Cardiovasc Interv. 2012;79:3–19.*

changing the expected course of the native coronary vessel. Advancing guidewires and other equipment through SVGs (especially recently occluded) carries risk of distal embolization.

There is currently controversy as to whether patent but degenerated saphenous vein grafts that were used as retrograde conduit for CTO PCI should be occluded (with coils or Amplatzer vascular plugs) after successful completion of PCI of the native vessel CTO.[121] SVG occlusion could stop competitive flow through the native stented segment and possibly decrease the risk of subsequent stent occlusion or thrombosis (**CTO Manual online cases 10, 127, 137, 162, 167, 169**).

Retrograde crossing via internal mammary artery grafts, such as the left internal mammary artery (LIMA), is feasible in many cases,[128] but has been associated with increased risk of dissection, ischemia, and hemodynamic compromise (**CTO Manual online case 46**). Ischemia can occur even without injury of the graft, likely due to straightening of the graft tortuosity by wires and microcatheters causing pseudolesions (**CTO Manual online cases 46, 37, 57**). Internal mammary artery grafts should, therefore, only be used as the last resort for retrograde CTO PCI (**CTO Manual online cases 29, 46**) with strong consideration of prophylactic hemodynamic support.

Septal collaterals[96] are preferred over epicardial collaterals, mainly because they have lower risk of perforation and tamponade compared with the more "fragile" epicardial collaterals.[129] Injury or perforation of a septal collateral is less likely to cause acute myocardial infarction, epicardial hematoma,[99] or tamponade[129] compared with perforation of an epicardial collateral. However, not all "septal" collaterals run their entire course in the septum. For example, very proximal LAD septal channels occasionally have a connection to the right posterolateral, and distal channels may run their distal course at the surface of the right ventricle; such collaterals should be recognized and treated with extreme caution, as they become epicardial collaterals at their distal segment. Treatment of collateral perforation is discussed in Section 26.5. Septal collaterals are usually less tortuous than epicardial collaterals, and are less likely to cause ischemia during crossing, as multiple septal collaterals usually exist.

Selecting the shorter collateral is preferred because (1) it provides better support and (2) it minimizes the risk of not being able to reach the target lesion. However, if a septal collateral enters the vessel close to the distal cap, there may not be enough space to allow for delivery of a wire and a microcatheter distal to the distal cap of the CTO; using a collateral that enters the vessel more distally is preferred in such cases. Collaterals with corkscrew morphology and <90 degrees angle with the recipient vessel may be challenging or impossible to wire;[102] whereas, nontortuous, large collaterals (CC1 or CC2 by the Werner classification[130] as described in Section 6.2.1.4.3.3) are the easiest to wire. Often invisible septal collaterals (CC 0 by the Werner classification) can be successfully crossed using the surfing technique.[131]

It is generally easier to advance a wire through a septal collateral from the LAD to the RCA compared from the RCA to the LAD, because the lower portion of the septal collaterals (closer to the PDA) has more acute turns and more tortuosity than the higher portion (closer to the LAD).[113] Furthermore, the PDA often has tortuosity and provides nonseptal branches that can hinder selection of the optimal septal branch.

Epicardial collaterals are the least preferred for retrograde CTO PCI,[97,132] because they are usually more tortuous than septal collaterals[111,112] and their perforation can rapidly lead to tamponade, especially in patients with an intact pericardium. For all practical purposes, any collateral that is not connecting the LAD and the right posterior descending artery, or ipsilateral collateral from the LAD to the left posterior descending artery and vice versa, is an epicardial collateral. A distinction is often made between AV groove epicardial collaterals and other epicardial collaterals because the former carry a higher risk of perforation and tamponade. In patients with prior coronary artery bypass graft surgery, epicardial collateral perforation can lead to hematoma and localized tamponade, not accessible with standard pericardiocentesis techniques (Section 26.6). Moreover, if epicardial collaterals are the only source of collateral blood flow and they become occluded during CTO PCI, acute ischemia and myocardial infarction may occur. Sometimes, due to severe tortuosity the control of the guidewire can be limited and advancement challenging or impossible (**CTO Manual online case 62**). Despite these limitations, with increasing experience and improvements in the retrograde equipment (wires and microcatheters), the use of epicardial collaterals (including ipsilateral epicardial collaterals[133,134]) has been increasing.

Ipsilateral collaterals: In patients with ipsilateral collaterals, such as septal collaterals from the proximal into the distal LAD (Fig. 8.4.4) (**CTO Manual online case 51**), or diagonal or obtuse marginal left-to-left collaterals (**CTO Manual online cases 56 and 66**) dual injection may not be required.[135,136]

- A special challenge with ipsilateral collaterals is that the retrograde wire often takes a fairly sharp turn to return into the proximal vessel, which can lead to kinking and difficulty advancing equipment,[135] or, more importantly, to collateral rupture (that may occur more frequently with ipsilateral than contralateral collaterals).
- If the CTO is successfully wired through the collateral, then a second ipsilateral guide catheter may be beneficial for trapping or externalizing the wire, because if the retrograde wire is inserted into the antegrade guide catheter, equipment delivery is often difficult. Stenting is also more difficult since it may jail and trap the retrograde equipment. Equipment delivery is easier using a "ping-pong" technique, in which engagement of the target vessel is alternated between the two guide catheters (**CTO Manual online case 51**).[137]
- Another option is to start the procedure with an 8 French guide catheter and use a 6 French guide catheter extension in it to trap the wire for advancement of the microcatheter into the guide catheter. This maneuver, however, might result in a high stress on the retrograde system, potentially injuring the collateral.
- The "tip in" (Section 8.4.4.8.2) or "rendezvous" technique decreases the risk of collateral injury. In this case, the retrograde wire is inserted into an antegrade microcatheter at the distal curve of the antegrade guide catheter, then the retrograde microcatheter is withdrawn and the antegrade one crosses the occlusion. This allows early removal of the retrograde system, decreasing the risk of retrograde complications (e.g., donor vessel injury or thrombosis).

"Invisible" collaterals: Some patients may appear to only have epicardial collaterals, but if those collaterals become occluded, septal collaterals may also appear (recruitable collaterals). Selective injection, the so called "tip injection," of the septal perforator branches (through an over-the-wire balloon, or preferably through a microcatheter) may also allow visualization of previously "invisible" collaterals. Alternatively, temporary occlusion of a dominant epicardial collateral for a few minutes and repeat angiography may reveal septal collaterals that were initially nonvisible. This technique also provides information regarding potential hemodynamic and ischemic consequences if crossing this dominant epicardial collateral is attempted. Crossing invisible septal collaterals

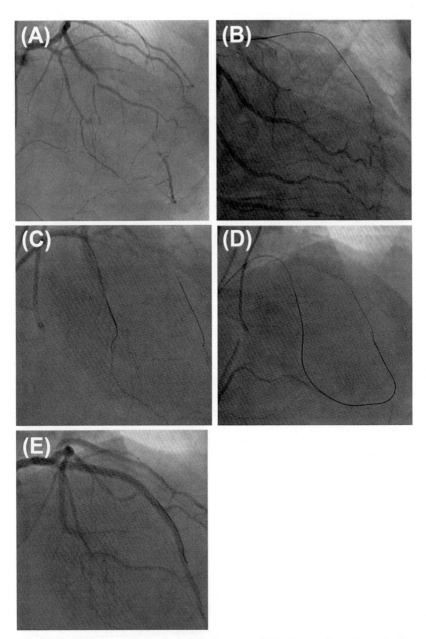

FIGURE 8.4.4   **Example of retrograde PCI of an LAD CTO via an ipsilateral septal collateral**. **Panel A**: mid LAD CTO; the distal LAD is filling via an ipsilateral septal collateral.
Panels B and C: antegrade wiring into distal false lumen. **Panel D**: retrograde wire advancement via the ipsilateral septal collateral. **Panel E**: successful recanalization of the LAD CTO
after stenting. *Courtesy of Dr. Marin Postu.*

is feasible in many cases using the "surfing" technique, in which septal collaterals are probed and crossed without contrast injection.[131] This technique increases the success rates of collateral crossing, but has the limitation that sometimes the collaterals are too small to be tracked by microcatheters.

Rarely, collateral vessels may not be apparent during diagnostic angiography. For example, an isolated conus branch can occasionally supply collaterals to an occluded LAD territory and has been used for retrograde PCI[138] or for facilitating antegrade crossing (**CTO Manual online case 2**) in such cases. Super selective angiography using a microcatheter can be useful in such vessels.

8.4.4.3   Step 3. Getting into the bypass graft or collateral channel

Goal: To advance a wire and microcatheter into the bypass graft or collateral vessel that will be used for retrograde crossing.

How?
1. Use a workhorse guidewire to minimize the risk of proximal vessel injury.
2. Larger, double bends on the workhorse guidewire are often needed to get into the collateral (Fig. 8.4.5). After microcatheter advancement, the wire is exchanged for a collateral crossing guidewire with a small distal bend.
3. Advancing a guidewire into collaterals with an acute takeoff from the parent vessel can be facilitated by (Fig. 8.4.6):
   a. Use of soft, polymer-jacketed guidewires, such as the Sion black.
   b. Use of an angulated microcatheter, such as the Venture catheter or the Supercross (**caution**: the over-the-wire Venture catheter has large profile and requires 8 Fr guide catheters for trapping).
   c. Use of a dual lumen microcatheter.

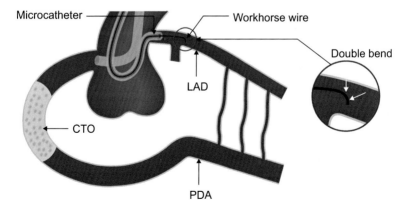

FIGURE 8.4.5   **Illustration of "double bent" shaping of the wire tip for entering a septal collateral.**[139] *Reproduced with permission from the Brilakis ES. Manual of coronary chronic total occlusion interventions. A step-by-step approach. 2nd ed. Elsevier; 2017.*

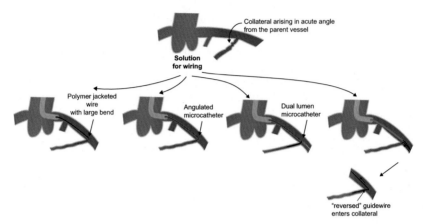

FIGURE 8.4.6   **Techniques for wiring a collateral that arises with acute takeoff from the parent vessel**.

    d. Use of the reversed guidewire (also called hairpin guidewire) technique (Fig. 8.4.7). Delivery of the reversed guidewire can be facilitated by use of a dual lumen microcatheter.[140]

    e. Balloon angioplasty or atherectomy of proximal calcified lesions or ostial calcified septal lesions that hinder device crossing.

    f. Use of the balloon deflection technique in which a balloon is inflated just distal to the collateral to deflect the guidewire into the collateral.

4. Similar techniques can also facilitate guidewire advancement into collaterals that have been jailed by a stent.

What can go wrong?

- **Injury (such as dissection) of the donor vessel**, while trying to enter the collateral (see Section 12.1.1) This can be a catastrophic complication, leading to rapid hemodynamic collapse, and requires immediate treatment (usually with stenting). Stenting of proximal vessel lesions should be considered prior to retrograde crossing to minimize the risk of ischemia and proximal vessel dissection (jailing of a septal collateral usually allows wiring of the collateral branch and subsequent equipment delivery through the stent struts). **Use of a safety wire in the donor vessel** can facilitate donor guide catheter engagement, straighten the artery potentially facilitating wiring of collaterals, and provide access to the vessel in case of a complication, such as dissection or thrombosis. Keeping an ACT >350 sec minimizes the risk of donor vessel thrombosis.

8.4.4.4   Step 4. Crossing the bypass graft or collateral channel with a guidewire

Goal: To cross the bypass graft or collateral channel with a guidewire.

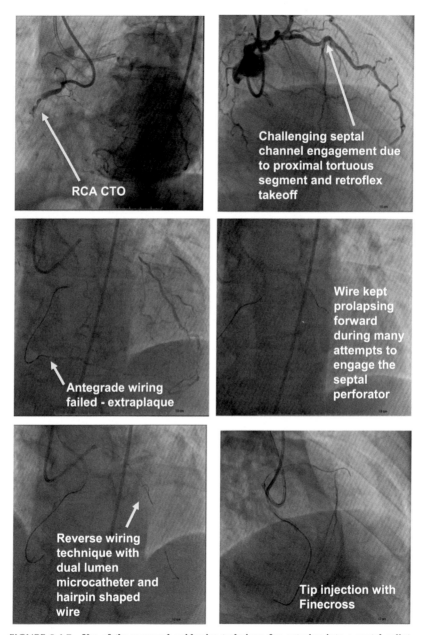

FIGURE 8.4.7   **Use of the reversed guidewire technique for entering into a septal collateral**. *Courtesy of Dr. Evandro Martins Filho.*

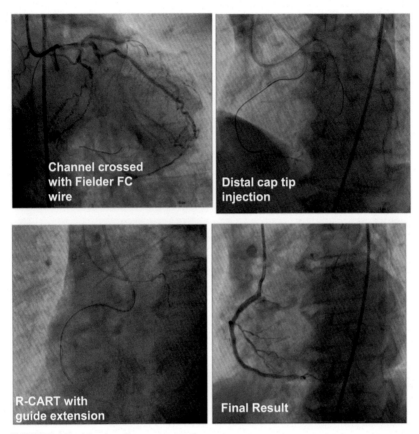

Channel crossed with Fielder FC wire

Distal cap tip injection

R-CART with guide extension

Final Result

FIGURE 8.4.7   (*Continued*)

How?

The technique depends on the type of collateral used (septal, epicardial, or saphenous vein graft).

8.4.4.4.1 Septal collaterals

CTO Manual online cases 5, 7, 9, 12, 17, 18, 20, 22, 23, 28, 31, 32, 33, 36, 40, 41, 42, 43, 44, 53, 57, 58, 59, 60, 64, 65, 69, 70, 71, 74, 77, 78, 79, 84, 90, 107, 109, 111, 115, 116, 119, 120, 123, 124, 129, 132, 133, 138, 139, 141, 151, 152, 154, 157, 160, 161, 164, 165, 168, 170, 177, 181.

Once the microcatheter is inserted into the septal collateral the workhorse wire is removed and exchanged for a guidewire that will be used for crossing the collateral into the distal true lumen (Fig. 8.4.8).

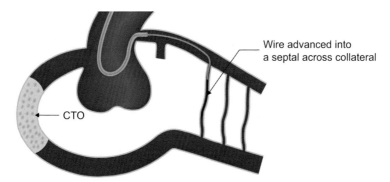

FIGURE 8.4.8 **Guidewire crossing of a septal collateral**. *Reproduced with permission from Brilakis ES. Manual of coronary chronic total occlusion interventions. A step-by-step approach. 2nd ed. Elsevier; 2017.*

8.4.4.4.1.1 Wire and microcatheter selection for septal collateral crossing

- **Sion** wire: The Sion wire has a soft tip and hydrophilic coating (unlike the Sion blue that does not have hydrophilic coating) and is also used often for septal collateral crossing.
- **Polymer-jacketed** guidewires (Section 30.6.2): The Sion black (soft tip, nontapered) and Fielder XT-R (soft tip, tapered) are commonly used guidewires for retrograde septal crossing, especially for small and tortuous septal collaterals, but carry higher risk of collateral injury as compared with the Suoh 03 and Sion wires.
- **Suoh 03**: The Suoh 03 (Section 30.6.1) is the softest tip guidewire currently available (0.3 gram tip load) and has hydrophilic coating. Suoh 03 is often used for crossing tortuous septal collateral crossing because it advances well through small and tortuous vessels and carries low risk of collateral injury. A drawback of the Suoh 03 wire in septal collaterals is that its torque control is inferior to polymer-jacketed composite core wires; hence, in case of several branching points, it can be difficult to find the optimal way to cross.
- Either large torqueable microcatheters (such as Corsair, Turnpike, Teleport, Mamba) or low profile torqueable (such as Corsair XS and Turnpike LP) or nontorqueable microcatheters can be used. Lower profile microcatheters can be easier to advance across the septal collateral but provide less support than the larger ones for subsequent CTO crossing.
- All microcatheters used for retrograde crossing should be 150−155 cm in length.

8.4.4.4.1.2 Wire tip shaping for septal collateral crossing

- A very small (1−2 mm), 30 degrees bend is usually created at the wire tip using the wire introducer. This shape facilitates navigation through tortuosity and allows steering away from small side branches.

8.4.4.4.1.3 Wire advancement technique

There are two techniques for septal collateral crossing: "surfing"[131] and "contrast-guided":

8.4.4.4.1.3.1 Technique 1: Septal "surfing"

- In septal "surfing" a guidewire is advanced through the septal collateral with rapid rotation without contrast visualization.
- Septal "surfing" was introduced by Dr. George Sianos and has gained widespread acceptance as an efficient and safe septal collateral crossing strategy.[131]
- Surfing should NOT be done in epicardial collaterals, because of high risk of perforation.
- In septal "surfing" the guidewire is advanced rapidly with simultaneous rotation until it either "buckles" or advances into the distal target vessel. If the wire "buckles", it is withdrawn and redirected. Sometimes, if the wire "buckles" it may release and cross in diastole. This is easier accomplished without using a torquer; the guidewire is manipulated using bare hands.
- If the wire repeatedly takes the same unsuccessful course, retract further back before re-advancing to select an alternate route.
- Do **NOT** push hard and stop immediately when you feel resistance! Excessive force will increase the risk of collateral injury without increasing crossing success.
- The odds of successful wiring are usually higher in proximal, straighter septals.
- Septal collaterals are usually straight in their upper half (LAD side), then bow toward the apex and turn again into the PDA (Fig. 8.4.9). Therefore, right anterior oblique (RAO) cranial is the best projection for initial

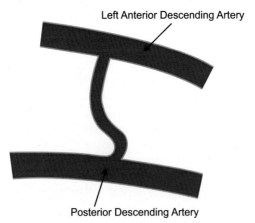

Left Anterior Descending Artery

Posterior Descending Artery

FIGURE 8.4.9 **Right anterior oblique caudal view of septal collaterals**. *Reproduced with permission from the Brilakis ES. Manual of coronary chronic total occlusion interventions. A step-by-step approach. 2nd ed. Elsevier; 2017.*

wiring from the LAD towards the RCA, and RAO caudal for wiring from the posterior descending artery to the LAD. Sometimes, a large septal curve can hinder microcatheter crossing. Forceful push of the microcatheter could injure the collateral and even cause a septal hematoma.

- Septal collaterals from the most proximal LAD tend to connect to the right posterolateral branch, whereas more distal septals usually connect to the posterior descending artery. Very distal septals may connect to a right ventricular branch and their course may be partly epicardial, hence they should be crossed with caution.

8.4.4.4.1.3.2   Technique 2: "Contrast-guided" septal crossing

- In contrast-guided septal crossing a guidewire is intentionally advanced along the course of the septal collateral after selective (through the microcatheter tip) or nonselective (through the guide catheter) contrast injection to delineate the course of the septal collateral.

How to perform "contrast-guided" septal crossing

1. Use a 3 mL Luer-lock syringe with 100% contrast. Medallion syringes (Merit Medical) are more resistant to breaking during forceful injection.
2. First aspirate until blood enters the syringe (to avoid air embolization and ensure that the microcatheter is not against the wall of the septal collateral). If no blood can be aspirated, pull back the microcatheter for a short distance until blood can be aspirated. This prevents air embolism and reduces the risk of hydraulic dissection.
3. Perform cine-angiography while gently injecting contrast with the 3 mL syringe.
4. Flush the microcatheter with saline before re-inserting the guidewire (to minimize subsequent "stickiness" of the microcatheter that may lead to guidewire entrapment). A second angiogram can be performed in an alternate angulation while flushing the contrast remaining in the microcatheter.
5. If a continuous connection to distal vessel is observed, re-attempt crossing through that connection.
6. Do NOT pan to avoid change in collateral road mapping.
7. Consider the right anterior oblique (RAO) caudal projection to evaluate the length and tortuosity of the distal part of the septal collateral. Also, left anterior oblique (LAO) projections can be useful if there is limited progress with the RAO views.
8. A description of various types of septal collaterals is shown in Fig. 8.4.10.

8.4.4.4.1.4   What can go wrong?

1. Collateral dissection (in most cases further attempts to cross into the distal true lumen can be performed via different collaterals).

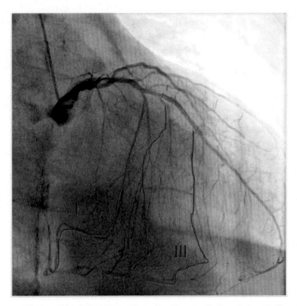

FIGURE 8.4.10 Types of septal collaterals according to their origin: (I) **Proximal Septal**: they often connect to the posterolateral system and have a partial epicardial course. (II) **Mid Septal**: they generally connect to the PDA and are often tortuous before entering into the posterior descending artery. (III) **Distal Septal**: attention is needed, because crossing through such collaterals can create high shear stress during externalization. *Courtesy of Dr. Kambis Mashayekhi. Reproduced with permission from the Brilakis ES.* Manual of coronary chronic total occlusion interventions. A step-by-step approach. *2nd ed. Elsevier; 2017.*

2. Collateral perforation (Section 26.5) is nearly always benign and only causes localized staining; however there are reported cases of septal hematoma formation and/or perforation into the pericardium causing hemodynamic compromise.[141,142] If a hemodynamically significant septal hematoma is causing left ventricular outflow tract (LVOT) obstruction, fenestration into the right ventricle (RV) and decompression can be considered. Septal collateral perforation has been treated in some patients with septal occlusion using covered stents[143] or with coils. Sometimes the wire gets into the ventricular cavity with a characteristic fluttering movement of the distal tip. This is a benign complication and does not require any other action than pullback of the wire.

3. Guidewire entrapment: to prevent this complication do not allow big (>1.5 mm) and acute (>75 degrees) bends to form at the tip of the guidewire during attempts for retrograde septal collateral crossing.[144] The wire should not be overtorqued by continuous spinning in one direction.

4. Microcatheter tip fracture and entrapment. If the tip of the microcatheter is stuck, avoid turning it multiple times; instead, withdraw the microcatheter and then rotate it again or exchange it for a new microcatheter.

8.4.4.4.2   Epicardial Collaterals

CTO Manual online cases 36, 54, 56, 63, 72, 88, 93, 97, 106, 107, 111, 112, 118, 119, 125, 128, 130, 152, 156, 161, 164, 168, 173, 181

8.4.4.4.2.1   Wire and microcatheter selection for epicardial collateral crossing

- The Suoh 03 guidewire is currently preferred for epicardial collateral crossing due to excellent crossing performance and low risk of collateral dissection or perforation.
- The Sion wire and soft, polymer-jacketed guidewires can also be used but should be advanced gently to minimize the risk of collateral injury.
- Low profile, 150-cm or 155-cm long microcatheters, such as the Caravel, Turnpike LP and Corsair XS are preferred in epicardial collaterals, especially if they are small and tortuous.

8.4.4.4.2.2   Wire tip shaping for septal collateral crossing

Similar to septal crossing, a very small (1−2 mm), 30 degrees bend is usually created at the wire tip using the wire introducer.

8.4.4.4.2.3   Wire advancement technique

- Epicardial collateral crossing should always be performed using contrast guidance (i.e., no "surfing").
- Perform contrast injection through the microcatheter to visualize the collateral vessel course. Prior to injecting, ensure that the microcatheter is not "wedged" (blood can be aspirated through the microcatheter) and inject gently to avoid collateral damage
- Orthogonal injections are important to determine the collateral vessel course.
- During wire advancement keep the working projection angiogram on the reference screen.
- Advance the wire first, then follow with the microcatheter—never let the microcatheter advance ahead of the guidewire.
- Microcatheter will "straighten" tortuosity and allow subsequent advancement.
- Rotate the wire (do not push) in tortuous segments. Crossing may be easier during diastole when the angle between collateral turns is wider (Fig. 8.4.11).
- If the wire becomes stuck, first pull it back then start rotation to prevent knuckle formation on the wire.
- Once the wire reaches the distal true lumen, it is advanced to the distal cap before following with the microcatheter.
- Sometimes, crossing epicardial collaterals can prove impossible due to severe tortuosity and small size.

Systole

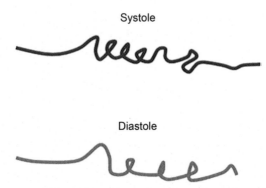

Diastole

FIGURE 8.4.11   Illustration of changes in epicardial collateral channel angulation during the cardiac cycle. In tortuous epicardial channels, wire crossing through the spiraling segments is the key to success. When the tip gets caught in the curve, quick torque of the wire tends to slide a little on a wider angle in diastole. Therefore, timely torqueing is necessary to go through the spiral segment of the channel. *Reproduced with permission from the Brilakis ES. Manual of coronary chronic total occlusion interventions. A step-by-step approach. 2nd ed. Elsevier; 2017.*

8.4.4.4.2.4   What can go wrong?

1. Ischemia of the myocardium supplied by the collateral, especially if there are no other collaterals supplying the same territory.
2. Collateral perforation can cause tamponade. In contrast to prior beliefs, perforation in patients with prior coronary artery bypass graft surgery can be MORE DANGEROUS than perforation in patients with an intact pericardium, as it can lead to loculated effusions compressing various cardiac structures[123] (such as the left atrium[122,124,145] or the right ventricle[125]) that cannot be drained with pericardiocentesis (Section 26.5).
3. Collateral dissection (in most cases further attempts to cross can be performed via a different collateral).
4. Guidewire entrapment: to prevent this, the operator should not allow a loop or "knuckle" to form at the tip of the guidewire during attempts for retrograde crossing of the collateral (**CTO Manual online case 13**), although occasionally tiny loops at the tip of a polymer-jacketed guidewire can help cross tortuous collaterals.

8.4.4.4.3   Bypass Grafts

8.4.4.4.3.1   Guidewire and microcatheter choice for crossing saphenous vein grafts

Patent SVGs: workhorse guidewires are typically used along with large microcatheters (such as Corsair and Turnpike) for more support.

Occluded SVGs: stiff tip, polymer-jacketed guidewires are usually used (such as the Pilot 200, Gladius, or Raider). Given the often large caliber of the SVGs, a large microcatheter is preferred. If there is difficulty with wire and microcatheter advancement, balloon dilatation of the proximal segment of the SVG can be performed to "free up" the microcatheter.

Bypass grafts for retrograde CTO PCI: tips and tricks

Both arterial grafts and SVGs (either patent or occluded[146]) can be used for retrograde CTO PCI:

1. There is a risk of distal embolization when advancing equipment through either patent or occluded SVGs. There is also a risk of perforation when crossing occluded SVGs.
2. Internal mammary artery (IMA) bypass grafts are the least preferred bypass grafts for retrograde wiring, because insertion of equipment in the graft could result in pseudolesion formation and even antegrade flow cessation[147] and because injury of the IMA graft could have catastrophic consequences. Moreover, crossing collaterals through IMA grafts can cause IMA dissection, especially in tortuous IMA grafts and when rotating the microcatheter through the collateral. If the LIMA to LAD is the only available collateral donor, consider using mechanical circulatory support (**CTO Manual online case 46**).
3. A challenge of retrograde wiring through bypass grafts is navigating severe angulation at the distal SVG anastomosis (**CTO Manual online case 47**). This can be overcome by several techniques (Fig. 8.4.12):
   a. Use of hydrophilic (Section 30.6.1) or polymer-jacketed (Section 30.6.2) guidewires
   b. Use of angulated microcatheters (such as the Venture deflectable tip catheter[148,149] or the Supercross microcatheter[150]) (Section 30.6.3) or dual lumen microcatheters (Section 30.6.4).
   c. Using the reversed guidewire (also called hairpin guidewire) technique (Section 8.2.2.6, challenge C)[151,152]
   d. Using a deflection balloon.
   e. Using a dual lumen microcatheter.
   f. Using a shapeable microcatheter (i.e., ev3 Echelon 10, Fig. 8.4.13).
4. After a native coronary CTO is recanalized, occlusion of the patent but diseased SVG can be considered (to minimize the risk of subsequent distal embolization and to decrease competitive flow and the risk of stent thrombosis), although this approach remains controversial.[153]

8.4.4.5   Step 5. Confirm guidewire position within the distal true lumen

Goal:

To confirm that the retrograde guidewire has crossed through the collateral into the distal true lumen

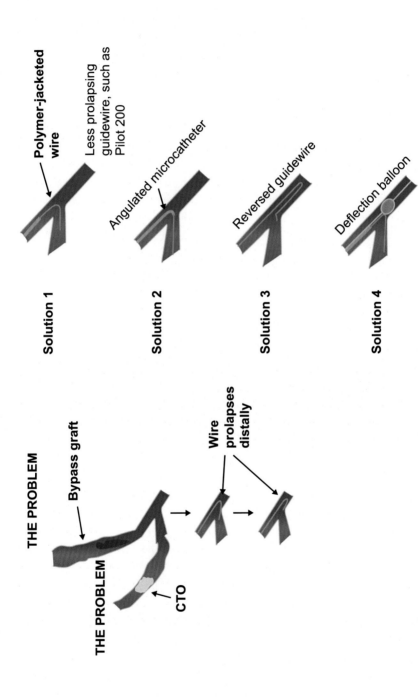

FIGURE 8.4.12 Retrograde guidewire advancement through the SVG distal anastomosis.

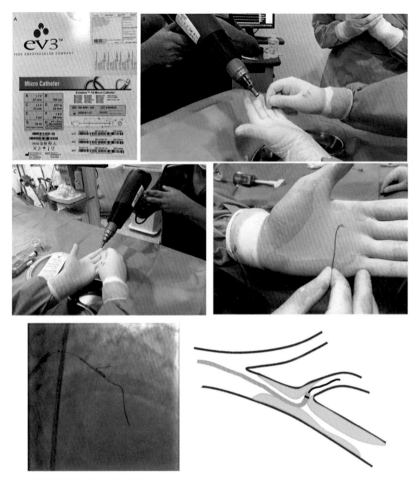

FIGURE 8.4.13 **Use of a heat gun to shape the tip of an Echelon 10 microcatheter.** *Courtesy of Dr. Evandro Martins Filho.*

How?

By injecting contrast through the retrograde guide catheter in two orthogonal projections (Figs. 8.4.14 and 8.4.15). Sometimes the retrograde guidewire may easily advance all the way to the distal cap close to an antegrade guidewire, confirming true lumen position.

Distal wire position confirmation

- Angiographic confirmation of distal guidewire position should always be done in orthogonal projections **before** advancing the microcatheter through the collateral (to prevent collateral rupture if the wire has exited the vessel) (Fig. 8.4.15).

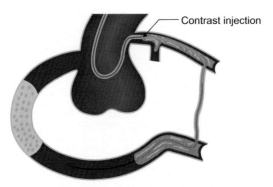

FIGURE 8.4.14 **Illustration of distal true lumen positioning of the retrograde guidewire,** which is an imperative step before advancing the microcatheter through the collateral.

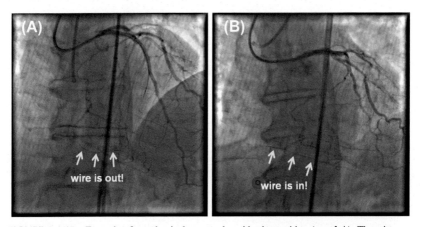

FIGURE 8.4.15   Example of extraluminal retrograde guidewire position (**panel A**). The wire was repositioned achieving intraluminal position (**Panel B**). In **panel A** the wire is located outside the distal true lumen—the microcatheter should not be advanced with the guidewire in this position. In **panel B** the wire is located inside the distal true lumen—the microcatheter can now be safely advanced over the wire.

- Possible wire positions
  - Distal true lumen.
  - Septum for septal collaterals (no crossing achieved).
  - Cavity (suspected if the wire starts making large back and forth movements). This is a common occurrence and is almost always benign.
  - Pericardium.
  - Non septal collateral (occasionally a collateral may appear to be septal in one projection but may in reality be epicardial; such collaterals have a higher risk of rupture; a classic example is an acute marginal collateral supplying the right coronary artery from the distal left

anterior descending artery). Obtaining an orthogonal view can help clarify the collateral type and location.

- The microcatheter should be advanced only if the guidewire is located in the distal true lumen. In all other cases, the wire should be retracted and redirected.
- Septal staining is almost always benign and does not cause tamponade (but can cause cardiac biomarker elevation) (Fig. 8.4.16).

8.4.4.6   Step 6. Crossing the collateral with the microcatheter

Goal: To advance the microcatheter into the distal true lumen

How?

After confirmation of guidewire position in the distal true lumen, the guidewire is advanced as far as possible close to the distal CTO cap (or deeply in another distal branch, such as the posterolateral branch) to provide sufficient backup for the advancement of the retrograde microcatheter (Fig. 8.4.17).

What to do if the microcatheter will not advance through the collateral (Fig. 8.4.18)?

(CTO Manual online cases 23, 59)

1. Rapid clockwise and counterclockwise rotation of the microcatheter using both hands. Rotate no more than 10 turns in one direction before releasing to prevent damage to the microcatheter. Counterclockwise rotation is preferred for the Corsair microcatheter and clockwise rotation is preferred for the Turnpike microcatheters. Continually monitor the microcatheter and donor guide catheter position, as aggressive microcatheter

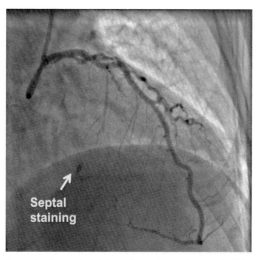

FIGURE 8.4.16   **Example of septal staining after retrograde septal wiring attempts.**

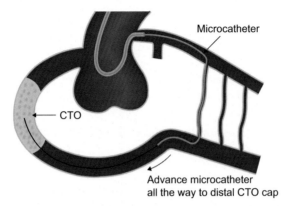

FIGURE 8.4.17 Microcatheter advancement through septal collateral after guidewire crossing.

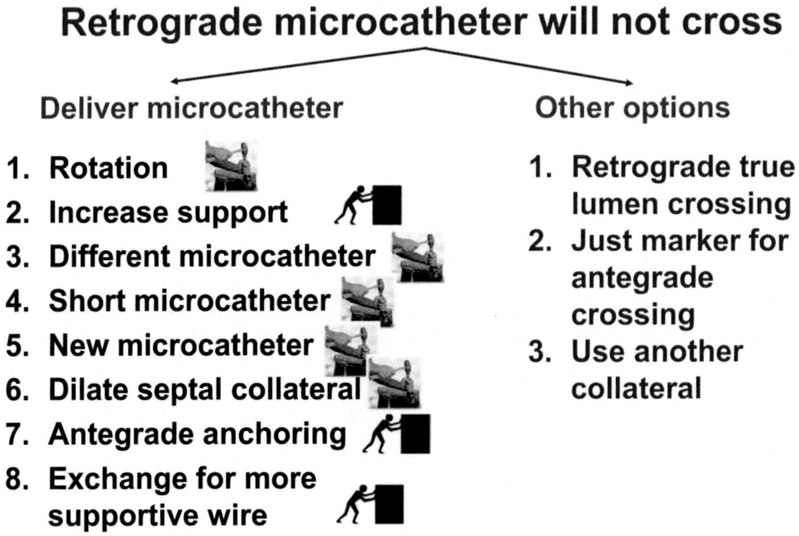

Retrograde microcatheter will not cross

Deliver microcatheter

1. **Rotation**
2. **Increase support**
3. **Different microcatheter**
4. **Short microcatheter**
5. **New microcatheter**
6. **Dilate septal collateral**
7. **Antegrade anchoring**
8. **Exchange for more supportive wire**

Other options

1. **Retrograde true lumen crossing**
2. **Just marker for antegrade crossing**
3. **Use another collateral**

FIGURE 8.4.18 What to do if the retrograde microcatheter will not cross the collateral.

advancement may result in microcatheter entanglement (Fig. 8.4.19) and donor guide catheter disengagement, potentially causing loss of microcatheter and wire position.

2. Increase retrograde guide catheter support, with active support (forward push of left-sided guides or clockwise rotation of right-sided guides), using additional extra support wires, a side branch anchor, or main vessel anchor (Fig. 8.4.20) technique, or a guide catheter extension. When performing anchoring with balloon inflation in the main vessel, keep balloon inflations brief and continually monitor the electrocardiogram and for development of symptoms, such as chest pain.

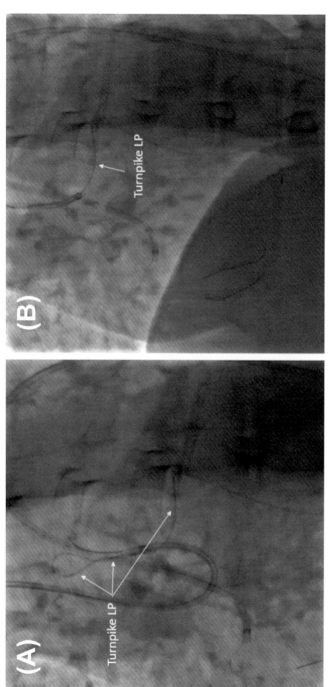

FIGURE 8.4.19 Complications of aggressive advancement of the retrograde microcatheter. Panel A: entanglement of the retrograde microcatheter after aggressive advancement through the donor vessel. **Panel B:** correction of the guide and guidewire position after gentle pulling of the microcatheter and counterclockwise rotation of the guide catheter. *Courtesy of Dr. Sevket Gorgulu.*

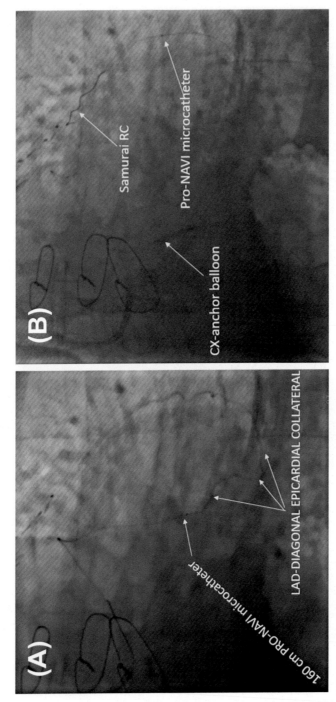

FIGURE 8.4.20 Main vessel anchoring for advancing the retrograde microcatheter. Panel A: inability to advance the retrograde microcatheter through an epicardial collateral. **Panel B:** successful advancement of the retrograde microcatheter after inflation of an anchoring balloon in the circumflex. *Courtesy of Dr. Sevket Gorgulu.*

3. Try a different microcatheter (lower profile microcatheters, such as the Caravel, Turnpike LP, Corsair XS, Finecross, 1.7 Fr Mizuki FX, and Telemark may be more likely to cross than the Corsair and Turnpike (Section 30.5).

4. Try a short (135 cm long) microcatheter, which allows for more transmission of torque.

5. Try a new microcatheter (sometimes the microcatheter can become "fatigued" with prolonged use or the tip may get deformed).

6. Dilate the septal collateral with a small (1.0−1.5 mm) balloon at low pressure (2−4 atm). Epicardial collaterals should NEVER be dilated.

7. If the retrograde guidewire is located adjacent to an antegrade extraplaque guidewire, the retrograde guidewire could be anchored by inflating a balloon over the antegrade guidewire (Fig. 8.4.21). Anchoring can sometimes provide enough support to advance the microcatheter through the collateral.

8. If the retrograde microcatheter is close to the distal vessel, the retrograde guidewire can be exchanged for a more supportive guidewire, such as Pilot 200 that could facilitate microcatheter advancement.

Alternatively, if the microcatheter cannot cross the septal collateral:

1. Attempt to cross the CTO with the retrograde guidewire into the antegrade guide catheter or guide catheter extension (more likely to be successful in short, noncalcified occlusions) (Fig. 8.4.22). Trapping this wire in the antegrade guide catheter or guide catheter extension will often facilitate the advancement of the retrograde microcatheter. Alternatively,

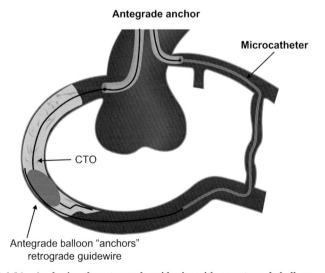

Antegrade anchor

Microcatheter

CTO

Antegrade balloon "anchors"
retrograde guidewire

FIGURE 8.4.21   **Anchoring the retrograde guidewire with an antegrade balloon.**

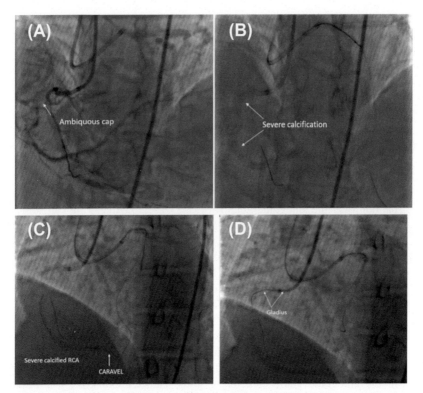

FIGURE 8.4.22   Retrograde wire crossing after failure to advance the retrograde micro-catheter into the distal vessel. Panel A: proximal right coronary artery CTO with ambiguous proximal cap. **Panel B**: the RCA CTO is heavily calcified. **Panel C**: a Turnpike LP could not be advanced to the distal RCA. A Caravel microcatheter was advanced to the distal RCA but could not advance any further. **Panel D**: successful crossing into the antegrade guide catheter with a Gladius Mongo ES. **Panel E**: tip in of a 135 cm long Mamba Flex microcatheter that was advanced to the distal RCA. **Panel F**: rotational atherectomy with a 1.5 mm burr. **Panel G**: successful final result after balloon angioplasty and stenting. *Courtesy of Dr. Sevket Gorgulu.*

a "tip-in" technique (Section 8.4.4.8.2) can be applied in the antegrade guide catheter or guide catheter extension.

2. Use the retrograde guidewire as "just marker" for antegrade crossing, either antegrade wiring or antegrade dissection/reentry (retrograde facilitated ADR).

3. Try to use another collateral. Leaving the first guidewire in place will increase support for the new guidewire.

What can go wrong?

1. Ischemia can occur, if most or all of the CTO target vessel perfusion comes from the wired collateral. This is most likely to occur with

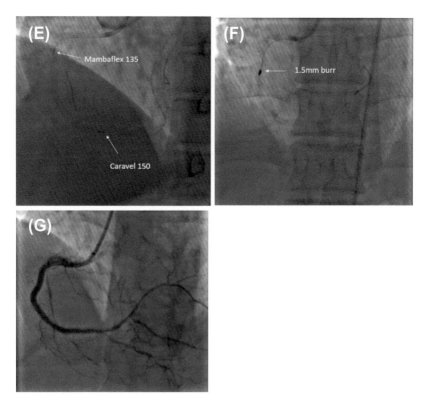

FIGURE 8.4.22   Continued.

epicardial collaterals, as there are usually multiple septal collaterals. Mild chest discomfort is, however, common during retrograde crossing.

2. Retrograde guide position loss, usually with excessive advancement of the retrograde microcatheter against resistance. To prevent this, pay careful attention to the retrograde guide position during attempts to deliver the retrograde microcatheter. Using a bigger field of view can allow simultaneous visualization of both the guide catheter and the microcatheter.

3. Injury of the donor vessel (especially if there is excessive back and forth movement of the retrograde guide catheter). In cases of retrograde intervention of right coronary artery CTOs, inserting a safety guidewire in both the left anterior descending artery and the circumflex will prevent the guide catheter from diving deeply into the LAD that could cause a dissection.

4. Donor vessel thrombosis (particularly in long retrograde cases and if the donor vessel is diseased). To avoid this, the ACT should be maintained at >350 seconds and checked every 30 minutes. Also regularly assess flow in the donor vessel.

8.4.4.7   Step 7. Crossing the CTO

Goal: To cross the CTO with a guidewire.

How?

Once the collateral branch has been successfully wired and the retrograde microcatheter advanced to the distal cap, the CTO is crossed using one of 4 techniques[91]:

1. Retrograde wiring into the proximal true lumen (also called retrograde "true lumen puncture" or retrograde "true to true").
2. Retrograde dissection/reentry (reverse CART is the most commonly used technique).
3. Antegrade wiring of the CTO (using the "kissing wire" or the "just marker" technique).
4. Antegrade dissection/reentry using the retrograde wire as marker of the distal true lumen (retrograde facilitated ADR)

All techniques require excellent guide support, that can be achieved as described in Section 9.3.1.8.[154]

8.4.4.7.1   Retrograde wiring

Retrograde true lumen puncture can be achieved in approximately 20%–40% of retrograde CTO PCIs (Fig. 8.4.23).[102]

The wire that crossed the collateral is advanced to the CTO distal cap, followed by advancement of the microcatheter or over-the-wire balloon for additional support. The CTO is then crossed from the distal into the proximal true lumen, either with the same guidewire (as the distal CTO cap may be softer and more tapered than the proximal cap) or with a stiffer tip

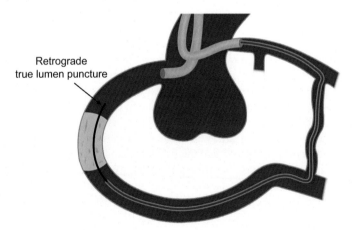

Retrograde
true lumen puncture

FIGURE 8.4.23   **Illustration of retrograde true lumen puncture**.

guidewire.[91] A useful technique to better define the distal cap is to perform distal tip contrast injection prior to wiring. In situations with blunt distal cap at the bifurcation of the right posterior descending artery and the right posterolateral a low profile dual lumen microcatheter can be advanced over the retrograde wire into the right posterolateral, followed by use of the over-the-wire lumen for distal tip injection and for inserting a second wire to penetrate the distal cap (Fig. 8.4.24). This technique has several advantages including increased support and protection of the right posterolateral.

Several maneuvers can be used to enhance the chance of crossing, such as inflating a retrograde balloon for more support (coaxial anchor) and using stiffer tip, tapered tip, and/or polymer-jacketed wires. Some operators recommend avoiding highly penetrating guidewires (such as the Hornet 14, Confianza Pro 12, Infiltrac and Infiltrac Plus, and Warrior) because retrograde perforations may be difficult to control. Antegrade IVUS can also facilitate directing the retrograde guidewire into the proximal true lumen and confirm successful true lumen crossing. [155,156]

What can go wrong?

1. While direct retrograde wire crossing can be an efficient method in short CTO segments, attempts to cross long, tortuous, heavily calcified or ambiguous CTO segments with stiff guidewires should be avoided due to risk of perforation.
2. Attempting to advance a retrograde guidewire into the left main coronary artery in an ostial or near ostial LAD CTO could result in extraplaque crossing, potentially compromising the origin of the circumflex and should, therefore, be avoided (Figs. 8.4.25 and 8.4.26). The same applies to retrograde crossing of ostial circumflex CTOs. If retrograde wiring is attempted in such lesions, a safety guidewire should be inserted into the patent vessel (circumflex in case of ostial LAD CTOs or LAD in case of ostial circumflex CTOs) to allow access to the vessel in case of compromise after CTO crossing (Section 16.2).

8.4.4.7.2 Retrograde dissection/reentry

If during manipulation of the antegrade, retrograde, or both guidewires they enter the CTO extraplaque space, reentry into the true lumen and CTO crossing can usually be achieved using the following two techniques (Figs. 8.4.27 and 8.4.28)[91,92]:

1. Inflating a balloon over the retrograde guidewire, followed by advancement of the antegrade guidewire into the distal true lumen (Controlled Antegrade and Retrograde subintimal Tracking—CART),[95] or
2. Inflating a balloon over the antegrade guidewire, followed by advancement of the retrograde guidewire into the proximal true lumen (reverse CART).

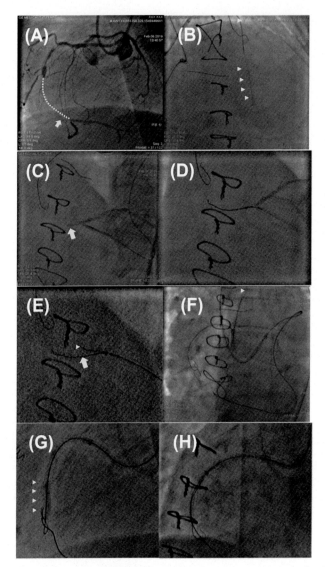

FIGURE 8.4.24 Retrograde and antegrade use of the Sasuke microcatheter for CTO PCI.
Panel A: dual angiography showing RCA CTO with bifurcation at the distal cap, and occluded
SVG at the crux (arrow). **Panel B**: retrograde microcatheter tip injection. **Panel C**: successful
wiring towards distal cap using Sion guidewire with the support of a Corsair Pro microcatheter.
Panel D: Confianza Pro 12 guidewire in the occluded SVG graft. **Panel E**: application of a dual
lumen microcatheter (Sasuke) to facilitate the distal cap puncture using the over-the-wire (OTW)
port. **Panel F**: antegrade and retrograde knuckle wires (Fielder XT, Pilot 200) in a coaxial position. **Panel G**: failed traditional reverse-CART in the mid-RCA. **Panel H**: GuideLiner assisted
reverse-CART. **Panel I**: kissing balloon inflation after rewiring in antegrade fashion using the
Sasuke MC. **Panel J**: GuideLiner facilitated angiography showing a dissection in the crux.
Panel K: rewiring the PDA using the Sasuke microcatheter. **Panel L**: crossover stenting towards
the right AV groove branch. **Panel M**: proximal stent deployment. **Panel N**: final angiographic
result. **Panel O**: schematic of the various approaches used for CTO recanalization in this case.
Courtesy of Dr. Peter Tajti and Dr. Imre Ungi.

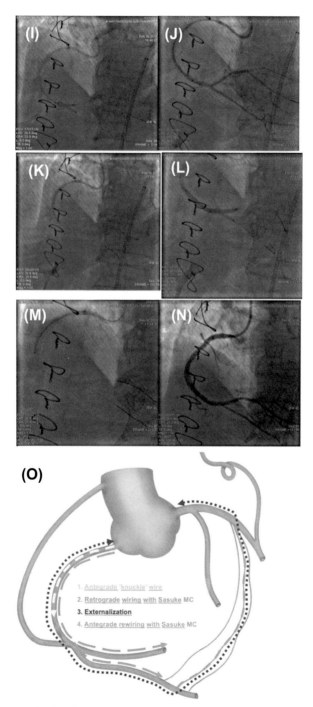

FIGURE 8.4.24   Continued.

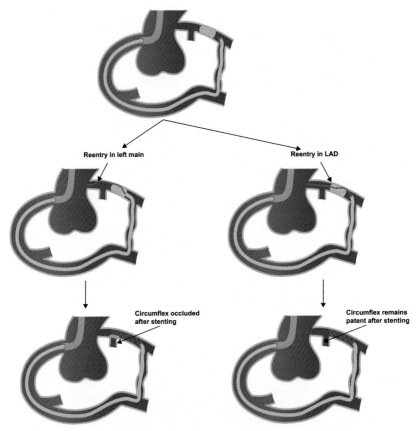

FIGURE 8.4.25 Potential complications of retrograde crossing of ostial LAD CTOs. If the guidewire partially crosses through the false lumen in the left main and subsequent reenters, a left main dissection flap could result in occlusion of the circumflex ostium.

Knuckling is especially important for crossing long occlusions, especially in heavily calcified and tortuous vessels.

Several variations of the CART techniques have been reported, such as the IVUS-guided CART,[103] the guide extension-assisted reverse CART,[157–159] the "stent reverse CART," the "confluent balloon"[160] technique, the "Deflate, Retract and Advance into the Fenestration Technique" (DRAFT) technique,[161] and the directed reverse CART (also called "contemporary reverse CART").[162]

8.4.4.7.2.1 CART technique
CTO Manual online cases 102, 103

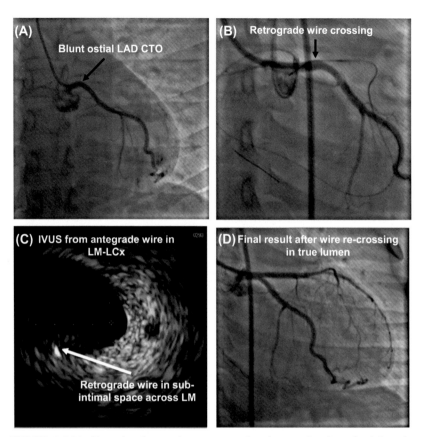

FIGURE 8.4.26 **Example of extraplaque retrograde wire crossing into the left main.** Retrograde wiring was repeated achieving true lumen crossing into the left main. LAD: left anterior descending artery; LCx: left circumflex; LM: left main. *Courtesy of Dr. Evandro Martins Filho.*

First described by Katoh in 2006,[95] the CART technique (Fig. 8.4.27) is based on the principle of creating an intraplaque or extraplaque space (ideally confined to the CTO segment) that is known to communicate with the true lumen. The space is enlarged by inflating a balloon inserted over the retrograde wire.[95] While the retrograde balloon is being deflated, the antegrade wire is directed towards the balloon crossing the lesion and entering the path taken by the balloon and wire. Usually 2.5−3.0 mm diameter, long, over-the-wire balloons are used for the CART technique. After CTO crossing with an antegrade guidewire, balloon angioplasty and stenting is performed in a standard manner.

This technique is often limited by the ability of the balloon to cross the collateral vessel and the associated risk of collateral channel injury. Early during the development of the CART technique, balloon dilatation of the

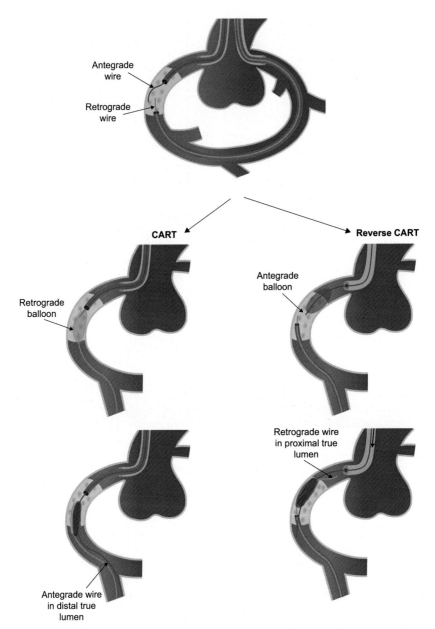

FIGURE 8.4.27   **Illustration of the CART and reverse CART techniques.**

septal collaterals was mandatory. In current practice, the use of novel micro-catheters obviates the need for retrograde balloon crossing and channel dila-tation. Since antegrade balloon advancement is usually easier and simpler,

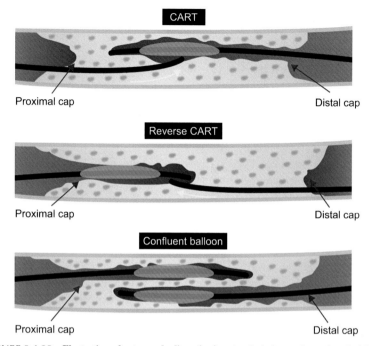

FIGURE 8.4.28   **Illustration of retrograde dissection/reentry techniques**. *Reproduced with permission from the* Manual of Chronic Total Occlusion Percutaneous Coronary Interventions. *2nd ed.*

the reverse CART technique has become the technique of choice. Use of the CART technique is currently reserved for cases in which the retrograde equipment is not long enough to reach the antegrade guide catheter (mainly in patients with long epicardial connections and very enlarged hearts), or in cases where the antegrade equipment (microcatheter / balloon) cannot be advanced to the site of wire overlap.

8.4.4.7.2.2   Reverse CART technique

CTO Manual online cases 1, 5, 36, 58, 59, 60, 61, 62, 64, 65, 66, 70, 78, 79, 84, 85, 86, 87, 88, 90, 105, 106, 115, 116, 118, 123, 130, 132, 134, 137, 147, 153, 157, 158, 160, 161, 162, 164, 167, 169, 171, 175, 177

The reverse CART (Fig. 8.4.27) is the most commonly used technique for retrograde CTO PCI.[110,163] The reverse CART technique is similar to the CART technique, with the difference that the balloon is inflated over the antegrade guidewire, creating a space into which the retrograde guidewire is advanced. It is usually easier to perform reverse CART and make the connection in a straight section of the vessel rather than in a bend.

8.4.4.7.2.2.1   *Reverse CART classification* Depending on the location of reentry and the size of the antegrade balloon, reverse CART is classified into 3 subtypes (Figs. 8.4.29 and 8.4.30)[162]:

1. **Conventional reverse CART**, that usually uses large balloons on the antegrade wire to achieve reentry within the CTO segment;
2. **"Directed" reverse CART (also often called "contemporary reverse CART")**, that uses small antegrade balloons and more active, intentional vessel tracking and penetration with a controllable retrograde wire, still within the CTO segment; and
3. **"Extended" reverse CART (previously known as "modified reverse CART")**, in which the intraplaque/extraplaque dissection is extended proximal or distal to the CTO segment, achieving reentry outside the CTO segment.

Some operators favor use of "directed" reverse CART to minimize the injury of the CTO vessel. However, "directed" reverse CART is not well suited for CTOs with severe tortuosity or calcification, CTOs with proximal cap and/or occlusion course ambiguity, CTOs with long occlusion length and when the torque control of the retrograde guidewire is poor, for example due to collateral tortuosity or cardiac motion.[162] "Extended" reverse CART is better suited for cases where antegrade preparation or retrograde cap penetration is not possible and where there are no significant side branches close to the reentry site.[162]

8.4.4.7.2.2.2   *Reverse CART troubleshooting* The following steps can help facilitate crossing if reverse CART is challenging to complete (Fig. 8.4.31):

1. Use a guide extension
CTO Manual online cases 12, 36, 60, 70, 86, 90, 105, 106, 115, 123, 130, 137, 153, 157, 158, 160, 161, 162, 164

In **guide extension reverse CART** (Fig. 8.4.32), a **guide catheter extension** is advanced over the antegrade guidewire to form a proximal target for the retrograde guidewire and facilitate its entry into the guide catheter.[157,159,164] Larger size guide catheter extensions create a bigger "target" for the retrograde guidewire
(CTO Manual online case 70).

This technique can be particularly useful when there is dissection or diffuse disease in the vessel proximal to the connection site or when the retrograde microcatheter is too short to reach the antegrade guide catheter.

Guide extension reverse CART is strongly recommended for ostial left anterior descending artery or ostial circumflex CTOs to minimize the risk of left main dissection during retrograde crossing attempts.

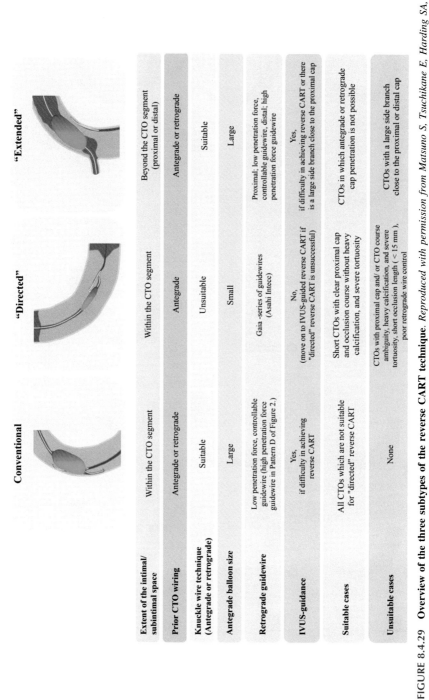

	Conventional	"Directed"	"Extended"
Extent of the intimal/ subintimal space	Within the CTO segment	Within the CTO segment	Beyond the CTO segment (proximal or distal)
Prior CTO wiring	Antegrade or retrograde	Antegrade	Antegrade or retrograde
Knuckle wire technique (Antegrade or retrograde)	Suitable	Unsuitable	Suitable
Antegrade balloon size	Large	Small	Large
Retrograde guidewire	Low penetration force, controllable guidewire (high penetration force guidewire in Pattern D of Figure 2.)	Gaia -series of guidewires (Asahi Intecc)	Proximal; low penetration force, controllable guidewire, distal; high penetration force guidewire
IVUS-guidance	Yes, if difficulty in achieving reverse CART	No, (move on to IVUS-guided reverse CART if "directed" reverse CART is unsuccessful)	Yes, if difficulty in achieving reverse CART or there is a large side branch close to the proximal cap
Suitable cases	All CTOs which are not suitable for "directed" reverse CART	Short CTOs with clear proximal cap and occlusion course without heavy calcification, and severe tortuosity	CTOs in which antegrade or retrograde cap penetration is not possible
Unsuitable cases	None	CTOs with proximal cap and/ or CTO course ambiguity, heavy calcification, and severe tortuosity, short occlusion length (< 15 mm), poor retrograde wire control	CTOs with a large side branch close to the proximal or distal cap

FIGURE 8.4.29 **Overview of the three subtypes of the reverse CART technique.** *Reproduced with permission from Matsuno S, Tsuchikane E, Harding SA, et al. Overview and proposed terminology for the reverse controlled antegrade and retrograde tracking (reverse CART) techniques. EuroIntervention 2018;14:94−101.*

Reverse CART subtypes

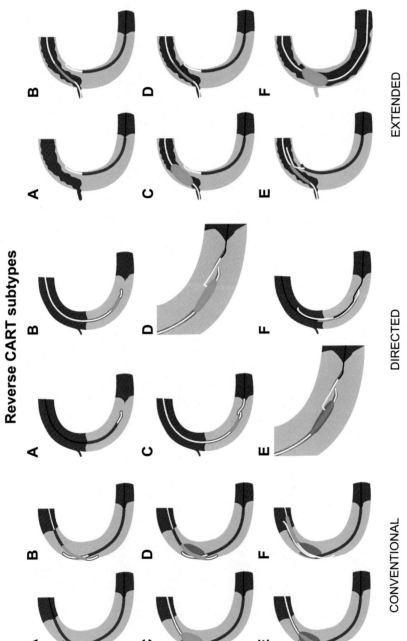

CONVENTIONAL   DIRECTED   EXTENDED

(Continued)

▼ FIGURE 8.4.30 **Procedural steps of the three reverse CART subtypes. Conventional reverse CART: Panel A:** a case of proximal RCA CTO. A longitudinal overlap of the antegrade and retrograde guidewires is made. **Panel B:** a balloon is delivered over the antegrade guidewire to the point of guidewire overlap. **Panel C:** the balloon is inflated. **Panel D:** after balloon dilatation, a connection is created between the spaces containing both guidewires. **Panel E:** A retrograde guidewire is advanced through the connection. **Panel F:** The retrograde guidewire is subsequently advanced into the proximal true lumen. **"Directed" reverse CART: Panel A:** a case of mid RCA CTO. Antegrade preparation: an antegrade guidewire is advanced through the occlusion until 5—10 mm proximal to the distal cap with the support of a microcatheter. **Panel B:** a small balloon is advanced to the tip of the antegrade guidewire. **Panel C:** after collateral channel crossing by a guidewire and a microcatheter, the retrograde guidewire is advanced towards the antegrade balloon. **Panel D:** the retrograde guidewire is directed towards the antegrade balloon and gently pushed. **Panel E:** after balloon deflation, the retrograde guidewire is advanced into the space created by the balloon. **Panel F:** the retrograde guidewire is advanced into the proximal true lumen with sequential antegrade balloon inflation within the CTO segment proximal to the connecting point. **"Extended" reverse CART: Panel A:** a case of mid RCA CTO with a small side branch around the proximal cap. A retrograde guidewire is advanced into the extraplaque space. **Panel B:** an antegrade balloon is delivered to the planned connecting point. **Panel C:** the antegrade balloon is inflated in an attempt to create a medial dissection. **Panel D:** after balloon deflation, a connection is created between the retrograde extraplaque space and the proximal true lumen. **Panel E:** the retrograde guidewire is advanced into the proximal true lumen through the created connection. **Panel F:** after guidewire externalization and balloon dilatation, the small branch is often compromised. *Reproduced with permission from Matsuno S, Tsuchikane E, Harding SA, et al. Overview and proposed terminology for the reverse controlled antegrade and retrograde tracking (reverse CART) techniques.* EuroIntervention 2018;14:94—101.

**Reverse CART
troubleshooting**

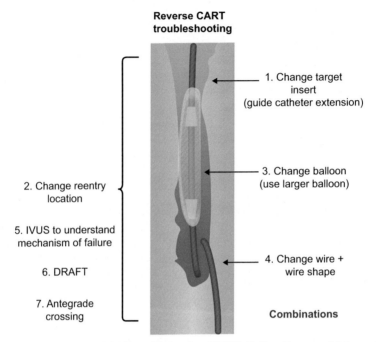

1. Change target
insert
(guide catheter extension)

2. Change reentry
location

3. Change balloon
(use larger balloon)

5. IVUS to understand
mechanism of failure

6. DRAFT

4. Change wire +
wire shape

7. Antegrade
crossing

Combinations

FIGURE 8.4.31 **Reverse CART troubleshooting**. *DRAFT:* Deflate, Retract and Advance into the Fenestration; *IVUS*, intravascular ultrasound.

2. Change reentry location Changing the site of reentry can facilitate reverse CART. Less tortuous and non-heavily calcified segments are preferred for reentry and the connection should be made at the segment of the occlusion where the antegrade and retrograde equipment are close to one another (optimal site for reverse CART). It is best to keep the reentry within the occluded segment, but "extended" reverse CART (reentry proximal to the proximal cap or distal to the distal cap) may be an option when the proximal/distal cap are hard to penetrate without adjacent major side branches (Section 8.4.4.7.2.2.1).

3. Change balloon size or type Use of undersized balloons is a common cause of reverse CART failure. Using larger balloons can significantly facilitate reentry. Balloon sizing can be facilitated by intravascular ultrasonography, that allows measurement of the media-to-media dimensions.

In heavily calcified lesions, intravascular lithotripsy balloons[165] can facilitate the creation of a connection between the antegrade and retrograde guidewire.

If the antegrade guidewire is located within the intraplaque space and the retrograde wire in the extraplaque space a cutting balloon can sometimes help (**CTO Manual online case 116**). The cutting balloon inflation should not be used in the extraplaque space.[166]

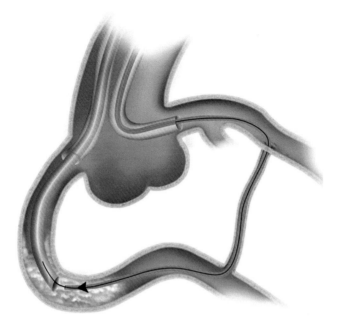

FIGURE 8.4.32   **Illustration of guide extension reverse CART**.

4. Change retrograde wire type and shape Changing guidewire and the tip bend can facilitate reentry.

Stiff tip, polymer-jacketed guidewires (such as Pilot 200, Raider, Gladius, and Gladius Mongo) and intermediate tip stiffness wires (such as Gaia or Judo) are most commonly used. In straight segments, stiff tip, non polymer-jacketed wires (Confianza 12, Hornet 14, Infiltrac and Infiltrac Plus, etc.) can be useful for puncturing into an antegrade guide catheter extension.

A larger bend (3–4 mm, ≥ 45 degrees) is preferred over smaller bends.

5. Intravascular ultrasound (IVUS) IVUS can be very useful when there is difficulty making the connection between the antegrade and retrograde spaces in reverse CART. An IVUS catheter is advanced over the antegrade wire. Although any IVUS catheter can be used, an IVUS catheter where the imaging transducer is located at the catheter tip, such as the Eagle Eye and Eagle Eye short tip (Philips) is preferred. IVUS examination allows:

1. Determination of **whether or not a connection has already been made**. If IVUS demonstrates that a connection between the retrograde and antegrade spaces has already been made then the problem is usually dissection, disease with recoil following ballooning, tortuosity or a combination of these factors in the vessel proximal to the connection, preventing passage of the retrograde wire. The solution in this case is guide extension reverse CART: advancing a guide extension catheter just proximal to where the connection has been made, eliminates challenges with wiring through the proximal vessel.

2. Precise sizing of the vessel and the balloon (to maximize the space created for reentry without risking vessel rupture).[103,167]

3. Definition of the location of the antegrade and retrograde wire in the vessel. There are four potential guidewire locations (Fig. 8.4.33). Connection of the antegrade and retrograde guidewire is more difficult when they are located **in different spaces** (one wire in the intraplaque and one in the extraplaque space) (Fig. 8.4.33)[168−170]:

   a. **Antegrade guidewire intraplaque + retrograde guidewire intraplaque**: this is the most favorable scenario with reentry usually achieved by inflating an antegrade balloon without needing wire repositioning.

   b. **Antegrade guidewire extraplaque + retrograde guidewire extraplaque**: this scenario is also favorable; connection can usually be achieved by antegrade balloon inflation and sometimes guidewire repositioning.

   c. **Antegrade guidewire intraplaque + retrograde guidewire extraplaque**: this is a more difficult scenario but increasing the size of the antegrade balloon often succeeds in making a connection. Sometimes creating a dissection with the antegrade guidewire is needed and can be achieved by using the "balloon-assisted subintimal entry" or the "scratch and go" techniques (**CTO Manual online case 59**). Another option is to use a cutting balloon.[166]

   d. **Antegrade guidewire extraplaque + retrograde guidewire intraplaque**: this is the most difficult situation, is seen in the most complex CTOs and is the hardest to resolve.[170] Balloon dilatation on the antegrade guidewire is usually not effective as it enlarges the extraplaque space without disrupting the plaque—plaque disruption is necessary to connect the antegrade and retrograde wires. Repositioning of the retrograde guidewire is needed in such cases, usually using a high penetration force guidewire, followed by antegrade balloon dilatation and retrograde wire crossing (Fig. 8.4.33).

6. Deflate, Retract and Advance into the Fenestration Technique—DRAFT
CTO Manual online cases 123, 132

In the DRAFT technique (Fig. 8.4.34) the antegrade balloon is withdrawn by one operator while the other operator advances *simultaneously* the retrograde *knuckled* guidewire into the guide catheter.[161]

Infrequently used reverse CART techniques
7. Stent reverse CART In the "**Stent reverse CART**" technique a stent is placed in the proximal true lumen into the extraplaque space to facilitate retrograde wiring into the stent (Fig. 8.4.35).

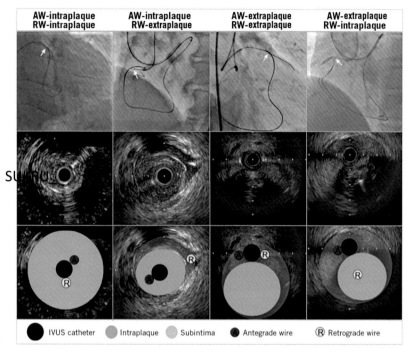

FIGURE 8.4.33 **Four patterns of antegrade and retrograde wire positions**. In the top panels, angiography shows overlapped AW and RW. White arrows indicate IVUS location. Middle panels show corresponding IVUS images; bottom panels show the same IVUS images with annotation. *AW*, antegrade wire; *IVUS*, intravascular ultrasound; *RW*, retrograde wire. *Reproduced with permission from Fan Y, Maehara A, Yamamoto MH, et al. Outcomes of retrograde approach for chronic total occlusions by guidewire location.* EuroIntervention *2021;17: e647−e655.*

The stent reverse CART can be used to overcome disease or dissection in the proximal vessel and to provide a large target for the retrograde wire, but has many limitations:

First, if the retrograde wire enters through the side of the stent, rather than the distal end of the stent, the retrograde microcatheter may not advance and if it does, it may deform the stent, hindering subsequent antegrade equipment delivery.

Second, if the stent is placed before a connection is established between the antegrade and retrograde spaces, the retrograde wire may not be able to puncture into the stent.

Third, the site of the reverse CART cannot be moved proximally because of the stent and if no connection is made, thrombosis of the proximal stent is likely to occur, which may compromise proximal branches covered by the stent. Therefore, it is recommended that IVUS be performed prior to stent placement to ensure there is a connection between the antegrade and retrograde space and to allow appropriate stent sizing. Once the stent has been

DRAFT (Deflate, Retract and Advance into the Fenestration Technique)

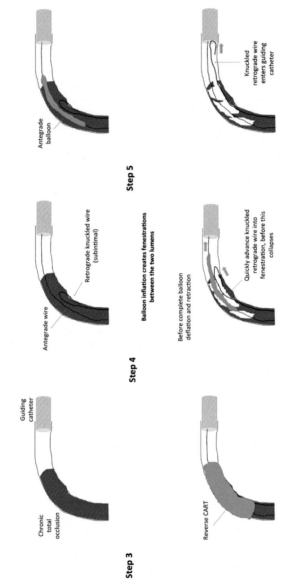

Courtesy of Mauro Carlino, MD

FIGURE 8.4.34 **Deflate, retract and advance into the fenestration (DRAFT) technique. Step 1:** antegrade and retrograde guidewires in the extraplaque space, during reverse CART. **Step 2:** a balloon is advanced over the antegrade guidewire. **Step 3:** inflation of the antegrade balloon (which is sized 1 to 1 with the vessel). **Steps 4 and 5:** balloon inflation over the antegrade guidewire creates multiple fenestrations and subsequent connections between the extraplaque space and the true lumen, so that, when quick advancement of the retrograde guidewire is performed before complete balloon deflation and retraction, the retrograde knuckled wire can easily penetrate into the true lumen. **Step 6:** the balloon and retrograde guidewire enter the antegrade guide catheter. *Courtesy of Dr. Mauro Carlino.*

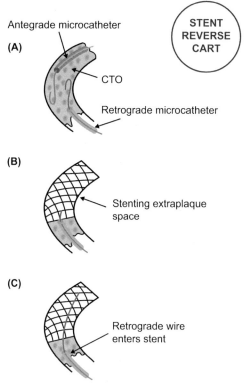

FIGURE 8.4.35   **Stent reverse CART technique. Panel A**: extraplaque location of both ante-grade and retrograde guidewires. **Panel B**. stenting over the antegrade guidewire into the subinti-mal space. **Panel C**. retrograde guidewire crossing into the stented segment.

placed and the retrograde wire has been advanced through the stent, further IVUS should be performed over the antegrade wire to ensure central passage of the retrograde wire through the stent.

Stent reverse CART is currently not recommended; it has largely been replaced by the guide extension reverse CART technique.

8. Confluent balloon technique
CTO Manual online case 96

In the "**confluent balloon technique**," an antegrade and a retrograde bal-loon are inflated simultaneously in a kissing fashion to cause the extraplaque (subintimal) space to become confluent, allowing wire passage through the CTO (Fig. 8.4.28).[160,171] The confluent technique is more suitable when attempting retrograde crossing via bypass grafts that often allow easy pas-sage of retrograde balloons.

9. Antegrade balloon puncture

Wu et al. proposed a modification of the reverse CART technique, in which the antegrade balloon remains inflated during retrograde crossing attempts and is "**punctured**" by the retrograde guidewire, which is then advanced while the punctured antegrade balloon is retracted under fluoroscopy.[156]

8.4.4.7.2.3 Retrograde dissection/reentry—what can go wrong?

1. As with all dissection strategies (antegrade and retrograde), side branches at the area of dissection may become occluded, with consequences dependent on the size of the supplied territory.
2. Antegrade injections should be avoided as they can cause extensive hydraulic dissection down to the distal vessel. To prevent this the injection syringe can be disconnected from the manifold.
3. Vessel perforation, if the extraplaque balloon is oversized (although it is usually undersized). IVUS can help select balloon size.

8.4.4.7.3 Antegrade wiring

"**Just marker**" (Fig. 8.4.36) is the simplest (but least reliable) form of the retrograde technique; the retrograde wire is advanced to the distal cap and acts as a marker of the distal true lumen, serving as a target for the antegrade wire.[98,154] This allows continuous visualization of the distal true lumen without contrast injections.

"Kissing wire" entails manipulation of both the antegrade and retrograde wires in the CTO until the ends of the wires meet; the antegrade wire then follows the channel made by the retrograde wire into the distal true lumen (Fig. 8.4.37).[91,154]

8.4.4.7.4 Antegrade dissection/reentry

CTO Manual online cases 83, 111

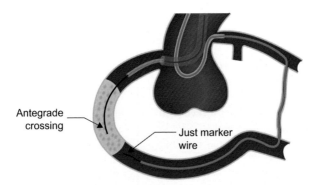

FIGURE 8.4.36 **Illustration of the "just marker" technique**. *Reproduced from the* Manual of Chronic Total Occlusion Percutaneous Coronary Interventions *2nd ed.*

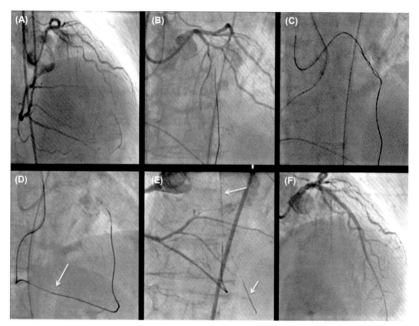

FIGURE 8.4.37 Long CTO of the left anterior descending artery (LAD) (**Panel A**) with contra-lateral collaterals from the RCA. A Sion black guidewire was advanced retrogradely within a Corsair microcatheter through septal collaterals from the posterior descending artery (PDA) towards the LAD. A Sion blue guidewire was advanced antegradely into the second septal branch of the LAD at the proximal cap of the LAD CTO (**Panel B**). The "kissing wire technique" was performed with an antegrade Gaia 2nd and a retrograde Ultimate 3 g guidewire (**Panel C**). The antegrade Gaia 2nd guidewire supported by an antegrade Finecross microcatheter intubated the tip of the retrograde Corsair (arrow, **Panel D**). Thereafter, a dual lumen Crusade microcatheter (Kaneka, Tokyo, Japan) was advanced antegradely to the mid LAD together with a second antegrade Sion black guidewire. The Sion black guidewire was then advanced into the distal LAD (**Panel E**, white arrows). An excellent final result was achieved after drug-eluting stent implantation (**Panel F**). *Courtesy of Dr. Kambis Mashayekhi. Reproduced from the* Manual of Chronic Total Occlusion Percutaneous Coronary Interventions *2nd ed.*

In cases of antegrade guidewire crossing into the distal extraplaque space, the retrograde wire can act as a marker of the distal true lumen, facilitating antegrade reentry (retrograde-facilitated ADR).[98,154] Sometimes (especially with retrograde crossing via bypass grafts) a retrograde balloon can be inflated into the distal true lumen, expanding it and facilitating antegrade reentry (Fig. 8.4.38).

8.4.4.8   Step 8. Wire externalization or antegrade wire crossing

If CTO crossing is achieved using antegrade wiring (Section 8.4.4.7.3) or antegrade dissection/reentry (Section 8.4.4.7.4), the next step is treatment of the CTO (lesion preparation and stenting) as described in Section 8.4.4.9.

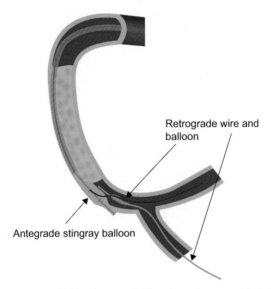

FIGURE 8.4.38   Retrograde facilitated antegrade dissection and reentry with the Stingray balloon.

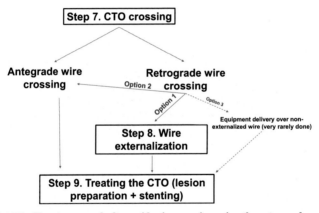

FIGURE 8.4.39   How to proceed after guidewire crossing using the retrograde approach.

If the CTO is crossed by the retrograde guidewire, there are 3 options for subsequent lesion treatment (Fig. 8.4.39):

1. Retrograde wire externalization (most common).
2. Antegrade guidewire crossing: an antegrade guidewire is advanced through the CTO, followed by CTO lesion treatment over the antegrade guidewire. This can be achieved using the "tip-in" technique if the retrograde micro-catheter cannot be advanced into the antegrade guide catheter.
3. Equipment delivery over the non-externalized retrograde guidewire (almost never performed currently).

Retrograde wire externalization is the preferred technique in most cases as it provides strong support for equipment delivery. However, in some cases wire externalization may not be possible or may not be the preferred approach, such as:

1. The wire may not easily advance into the antegrade guide catheter, usually due to tortuous and/or calcified retrograde course that hinders advancement of the retrograde microcatheter.
2. Retrograde equipment is causing ischemia and needs to be removed.

In such cases antegrade wire crossing is performed, usually using the "tip-in" technique (Section 8.4.4.8.2).

8.4.4.8.1 Retrograde wire externalization

After retrograde guidewire crossing, **wire externalization is performed in most cases**.

Goal: To externalize the retrograde guidewire, in order to use it as rail to advance balloons and stents in an antegrade direction, followed by safe removal of the externalized equipment.

This step is only applicable to cases in which the CTO is crossed in the retrograde direction. If the CTO is crossed in the antegrade direction this step is not needed.

How?

Two options are available for retrograde wire externalization depending on whether the retrograde guidewire enters the antegrade guide or not: (1) wiring the antegrade guide catheter and (2) snaring.

Wiring the antegrade guide catheter is simpler and preferable and can be facilitated by advancing a guide catheter extension into the antegrade vessel. Wiring the antegrade guide catheter may not always be possible, especially in the following circumstances:

1. Aorto-ostial CTOs.
2. Large caliber of the proximal CTO vessel.
3. Non coaxial guide positioning.

8.4.4.8.1.1 Option A: Retrograde guidewire enters into the antegrade guide catheter

Wiring the antegrade guide catheter with the retrograde guidewire (Figs. 8.4.40 and 8.4.41) is the simplest technique to externalize a guidewire and should be the first choice whenever possible.

Retrograde wire externalization is done using the following steps (Fig. 8.4.42):

1. After the retrograde wire (that crossed the CTO) enters the antegrade guide catheter, a trapping balloon is inflated within the antegrade guide next to the retrograde wire to facilitate the advancement of the retrograde microcatheter into the antegrade guide catheter (Fig. 8.4.42, panel A).

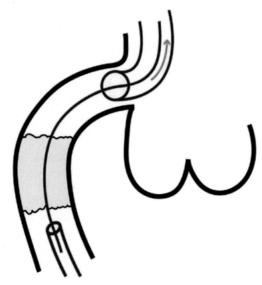

FIGURE 8.4.40   **Advancing the retrograde guidewire into the antegrade guide catheter.**

2. The retrograde guidewire is removed, after the trapping balloon is deflated, while the retrograde microcatheter remains within the antegrade guide catheter (Fig. 8.4.42, panel B).
3. The wire to be externalized (RG3 or R350) is advanced through the microcatheter retrogradely into the antegrade guide (Fig. 8.4.42, panel C).
4. The antegrade Y-connector is disconnected from the guide catheter and a finger is placed over the antegrade guide catheter hub, until the retrograde guidewire is felt "tapping" on the finger (Fig. 8.4.42, panel D).
5. After inflating a balloon in the antegrade guide catheter (or inflating the TrapLiner balloon if TrapLiner-facilitated reverse CART was performed) to minimize bleeding (Figs. 8.4.43 and 8.4.44), a wire introducer is inserted through the antegrade Y-connector and the retrograde guidewire tip is threaded through the introducer (Fig. 8.4.42, panel E).
6. The antegrade Y-connector is reconnected to the guide catheter hub, **without flushing** (to avoid an antegrade hydraulic dissection) (Fig. 8.4.42, panel F).
7. After deflating the balloon in the antegrade guide catheter, the retrograde guidewire is pushed until 20–30 cm has exited through the Y-connector.
8. If the tip of the externalized guidewire is damaged, it can be cut off to facilitate the loading of balloons/stents or other equipment. This is best achieved by using dedicated wire cutters if available.
9. To prevent "losing" the retrograde ending of the externalization wire during PCI, a torque device should be attached to the distal end of the externalized wire, so that it is not inadvertently pulled into the retrograde microcatheter (Fig. 8.4.45).

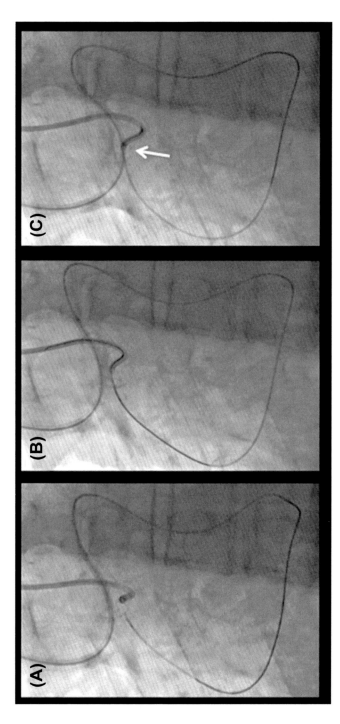

FIGURE 8.4.41 **Conventional wire externalization.** A retrograde microcatheter was advanced towards the proximal cap of the right coronary artery (RCA) CTO (**Panel A**). The retrograde guidewire entered the antegrade guide catheter (**Panel B**), followed by advancement of the retrograde microcatheter into the antegrade guide catheter (white arrow) (**Panel C**). After retracting the retrograde guidewire, an externalization wire was advanced retrogradely via the externalization route within the retrograde microcatheter towards the antegrade guide catheter (not shown). *Courtesy of Dr. Kambis Mashayekhi.*

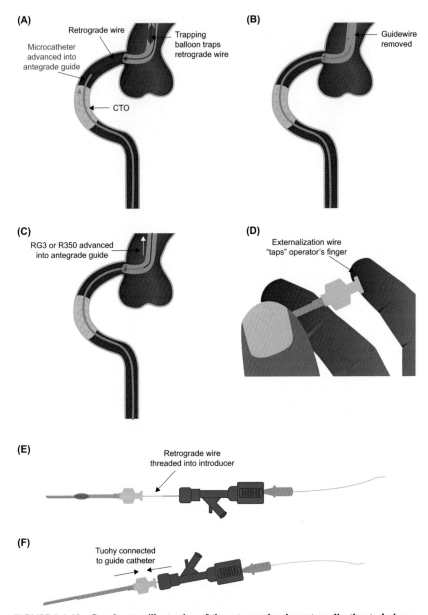

FIGURE 8.4.42 **Step-by-step illustration of the retrograde wire externalization technique.**

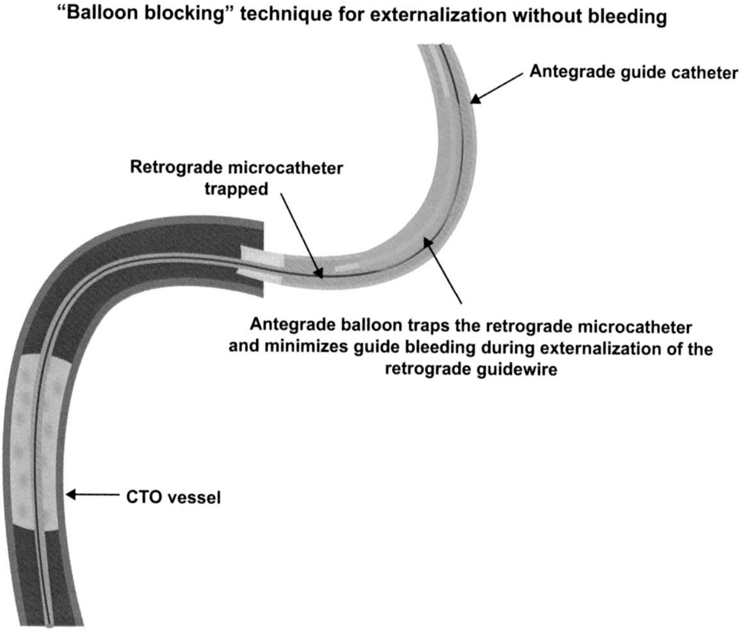

"Balloon blocking" technique for externalization without bleeding

FIGURE 8.4.43 **Balloon inflation inside the antegrade guide catheter to prevent bleeding during threading of the retrograde guidewire through the Y-connector.**

8.4.4.8.1.2 Option B. Snaring the retrograde guidewire

If wiring the antegrade guide catheter with the retrograde wire fails, then wire snaring can be performed. If the retrograde microcatheter successfully crossed the lesion into the aorta, then an externalization wire (RG3, R350) can be advanced through it and snared. If not, then the wire used for retrograde lesion crossing can be externalized, if it is 300 cm long.

Short guidewires (180−190 cm) should **never** be snared or docked with an extension wire and then externalized, because of the danger of the connection becoming detached, resulting in collateral channel or vessel injury or wire loss.

Snaring of the retrograde guidewire is performed as follows:

1. **Snare preparation** (illustrated in Section 30.18.5)
   a. Of the commercially available snares, the 27−45 or 18−30 mm EN Snare has three loops facilitating capture of the retrograde guidewire and is preferred over single loop snares, such as the Amplatz Goose Neck snares and microsnares and the Micro Elite snare.
   b. The snare is removed from the package and pulled back into the snare introducer.
   c. The snare delivery catheter is discarded (the guide catheter is used for snare delivery).

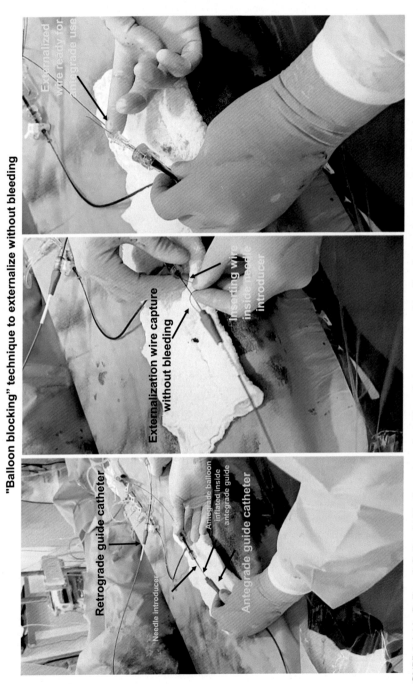

"Balloon blocking" technique to externalize without bleeding

Externalized wire ready for antegrade use

Externalization wire capture without bleeding

Inserting wire inside needle introducer

Retrograde guide catheter

Needle introducer

Antegrade balloon inflated inside antegrade guide

Antegrade guide catheter

FIGURE 8.4.44 Balloon inflation inside the antegrade guide catheter to prevent bleeding during threading of the retrograde guidewire through the Y-connector. *Courtesy of Dr. Evandro Martins Filho.*

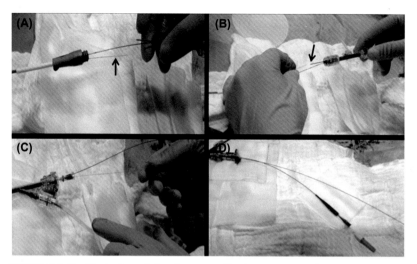

FIGURE 8.4.45 **Illustration of the wire externalization process outside of the body.** The RG3 externalization wire is externalized through the antegrade guide catheter (black arrow) (6 French) (**Panel A**). The externalization wire is advanced through the distal end of the introducer needle (black arrow), which in turn has been placed within the antegrade Y-connector (**Panel B**). After reconnecting the Y-connector with the antegrade guide catheter, guidewire externalization can be safely performed (**Panel C**). To avoid losing the retrograde end of the externalization wire, a torquer is attached to its tip, in front of the hub of the retrograde microcatheter (**Panel D**). *Courtesy of Dr. Kambis Mashayekhi.*

    d. The snare is inserted into the antegrade guide catheter by inserting the introducer through the Y-connector (a connector with an automatic hemostatic valve is preferred to minimize bleeding).

    e. If snaring is performed through a guide for the right coronary artery it is preferable to use a JR4 instead of an Amplatz guide catheter (which is used as the antegrade guide catheter in most antegrade RCA CTO PCIs), as the JR4 guide catheter poses less risk of ostial right coronary artery dissection.

2. The snare is advanced out of the antegrade guide and opened (Fig. 8.4.46, panel A).

3. The retrograde guidewire is advanced through the snare (Fig. 8.4.46, panel B).

It is preferable to snare the guidewire you plan to externalize (RG3 or R350), if possible. This is dependent on getting the microcatheter through the CTO segment. The radiopaque portion of these wires should be snared, followed by insertion into the antegrade guide catheter. More proximal snaring of the steel core portion of the wire may lead to wire kinking, potentially leading to failure to remove it from the body. Snaring is sometimes easier if performed in the brachiocephalic artery.

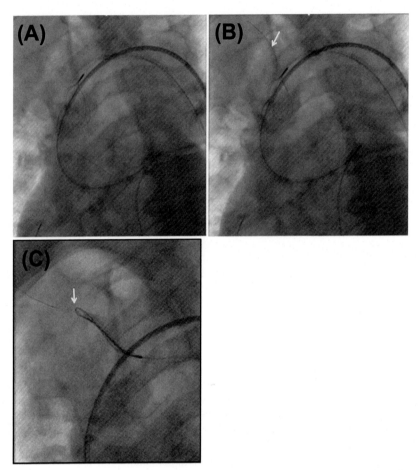

FIGURE 8.4.46 **Illustration of retrograde guidewire snaring. Panel A**: illustration of a three-loop snare advanced through the guide catheter and deployed in the ascending aorta. **Panel B**: the retrograde guidewire (arrow) is advanced through one or more of the snare loops. **Panel C**: the snare is withdrawn into the guide, capturing the retrograde guidewire.

If it is not possible to advance the microcatheter through the CTO and into the aorta, a standard long (300 cm) guidewire may need to be snared (hence, it is best to use long guidewires during retrograde crossing attempts). Short guidewires should never be snared. More care must be taken where these wires are snared to avoid fracture or unraveling of the distal part of the guidewire. The ideal snaring location is immediately proximal to the radiopaque portion of the wire.

4. **The snare is pulled back capturing the retrograde guidewire** (arrow, Fig. 8.4.46, panel C).

What can go wrong?

a. Snaring the distal flexible portion of the retrograde wire can result in wire fracture,[172,173] although this is highly unlikely with the externalization wires (RG3 and R350).

b. The snared wire may unravel (which is why snaring should be performed under continuous fluoroscopy). A polymer-jacketed guidewire is preferred as it has a lower risk of unraveling and more lubricity.

5. **The retrograde guidewire is pushed through the retrograde microcatheter** (while applying gentle traction on the snare) until it exits from the antegrade guide Y-connector (Fig. 8.4.47).

If the wire tip is deformed it can be cut to facilitate the loading of equipment over the externalized portion of the guidewire and a torque device is attached on the retrograde side of the wire to prevent the distal tip from inadvertently slipping into the microcatheter.

6. **Preparing for angioplasty and stenting**

a. The microcatheter is retracted distal to the CTO (but continues to cover the portion of the externalized guidewire that is coursing within the collateral vessel to prevent collateral injury). During withdrawal of the retrograde microcatheter, the donor guide catheter should be disengaged and watched closely to avoid deep engagement and dissection of the donor vessel.

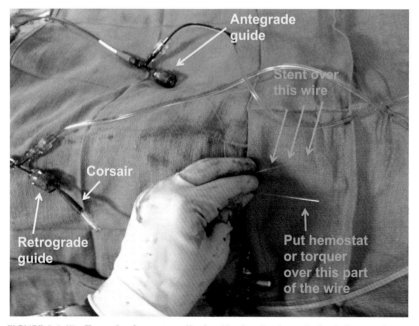

FIGURE 8.4.47   **Example of an externalized guidewire that is ready for balloon and stent delivery**.

298 PART | A The Steps

b. A torquer or hemostat is attached at the proximal end of the externalized guidewire (to reduce the risk of inadvertent withdrawal of the wire into the microcatheter) (Fig. 8.4.47).

8.4.4.8.2 Antegrade wire crossing

Externalization of the retrograde wire may fail or may not be desirable (for example when crossing small tortuous epicardial collaterals). In such cases antegrade wire crossing can be performed instead of retrograde wire externalization using the following techniques:

Retrograde guidewire inserted into antegrade guide catheter

1. The "**tip-in**" technique (Figs. 8.4.48 and 8.4.22),[174] in which an antegrade microcatheter is advanced over the tip of the retrograde guidewire that has entered the antegrade guide catheter. The antegrade microcatheter is then advanced over the retrograde guidewire through the CTO, followed by insertion of an antegrade guidewire and antegrade delivery of balloons and stents (**CTO Manual online cases 57, 156, 165, 170**). The "tip in" is the most commonly used technique for antegrade wiring after retrograde wire crossing into the antegrade guide catheter if the retrograde microcatheter cannot be advanced into the antegrade guide catheter. If advancement of the antegrade microcatheter is challenging, the back end of a 0.014 inch guidewire can be inserted into the antegrade microcatheter to stiffen it (facilitated tip-in technique).[175]

2. The "**antegrade probing of the retrograde microcatheter technique**" is the opposite of the tip-in technique: the retrograde microcatheter is advanced into the antegrade guide catheter, followed by removal of the retrograde guidewire and insertion of an antegrade wire into the retrograde microcatheter.[176]

3. The "**bridge**" or "**rendezvous**" method in which the retrograde microcatheter is inserted into the antegrade guide catheter and aligned with an antegrade microcatheter allowing insertion of an antegrade guidewire into the retrograde microcatheter, positioned within the antegrade guide catheter[143,154] or in the CTO segment (**CTO Manual online cases 31 and 66**).[155,177,178]

4. Externalization of the retrograde wire (as described in step 8, Section 8.4.4.8) followed by antegrade insertion of a microcatheter (single or dual lumen), over which an antegrade wire is inserted.[179]

Retrograde guidewire not inserted into antegrade guide catheter:

5. **Antegrade wire advancement along the course of the retrograde guidewire**. This could be facilitated by performing balloon angioplasty of the CTO segment using a balloon advanced over the retrograde guidewire to create a lumen through the CTO. Retrograde balloon angioplasty can be facilitated by advancing the retrograde wire far into the aorta, or,

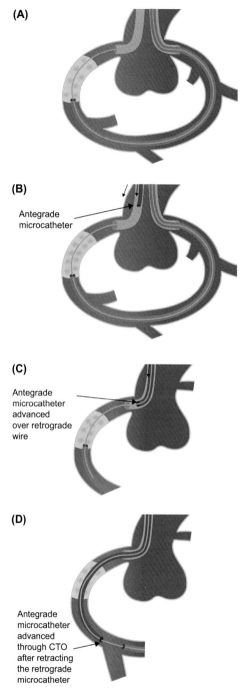

FIGURE 8.4.48 **Illustration of the "tip in" technique for antegrade wiring through CTO after successful retrograde wire crossing**. The retrograde guidewire enters into the antegrade guide catheter (**panel A**). An antegrade microcatheter is advanced over the tip of the retrograde guidewire (**panels B and C**) and through the occlusion (**panel D**). The retrograde wire is subsequently removed, followed by insertion of an antegrade guidewire.

if possible, into the antegrade guide where it can be "trapped" with an antegrade balloon inflated at 10−15 atm. Other support techniques can improve the retrograde deliverability of equipment, such as the double balloon anchoring technique, in which the retrograde wire is anchored into the antegrade guide catheter, and the retrograde guide catheter is anchored in the donor vessel ostium by inflating a balloon in a small vessel side branch.[180]

6. The "**reverse wire trapping**" technique that involves snaring of the retrograde guidewire followed by withdrawal of the retrograde guidewire pulling the antegrade snare through the CTO into the distal true lumen.[173]

The advantage of all these techniques is that they minimize retrograde guidewire and balloon manipulations after retrograde crossing, placing less strain on collaterals as compared with retrograde guidewire externalization. However, wiring a microcatheter can be challenging. Moreover, wire externalization can provide superior support for antegrade equipment delivery; equipment delivery attempts after antegrade wiring may result in loss of wire position. The reverse anchoring technique (anchoring the antegrade wire by inflation of a retrograde balloon) could provide strong backup support to facilitate antegrade equipment delivery.[176,181]

8.4.4.8.3 Retrograde equipment delivery over non-externalized guidewire

Retrograde stent delivery is often feasible through SVGs,[182] but is almost never performed currently through septal or epicardial collaterals due to high risk of collateral injury, stent loss, and stent entrapment.

Retrograde stent delivery has been reported through both septal[183] and epicardial[184] collaterals. It requires adequate predilatation of the collateral to minimize the risk of injury and stent entrapment or dislodgement.[185] After completion of the intervention the donor artery is imaged again to ascertain that no complication has occurred.

8.4.4.9 Step 9. Treatment of the CTO

After the retrograde guidewire has been successfully externalized, balloon angioplasty and stenting can be performed using rapid-exchange equipment, followed by removal of the externalized guidewire.

- The externalized guidewire provides outstanding support, facilitating delivery of virtually any device.
- The retrograde microcatheter is retracted to the distal vessel, after disengaging the retrograde guide catheter to avoid deep guide insertion and dissection of the donor vessel.
- The retrograde microcatheter should always be in the distal true lumen to protect the collateral from wire injury.

- The tip of the antegrade balloon/catheters should never be allowed to "meet" the tip of the retrograde microcatheter/balloon on the same guidewire, because they may "interlock", resulting in equipment entrapment that may require surgery for removal (Section 27.4).

8.4.4.10   Step 10. Externalized guidewire removal

1. Once stenting of the CTO is completed (Fig. 8.4.49, panel A), the retrograde wire should be removed in a safe manner.
2. The retrograde microcatheter is advanced back into the antegrade guide catheter (through the recently deployed stents, Fig. 8.4.49, panel B), unless significant resistance is encountered.
3. Both guide catheters are disengaged (Fig. 8.4.49, panel C) (to minimize the risk of the guides getting "sucked into" the coronary ostia, potentially causing dissection).
   A. The antegrade guide catheter is disengaged by pushing the externalized guidewire.
   B. The retrograde guide catheter is disengaged by fixing the microcatheter and using it as rail for retracting the guide catheter.
4. The retrograde guidewire is partially withdrawn (Fig. 8.4.49, panel D).
5. The retrograde microcatheter is withdrawn into the donor vessel leaving the retrograde guidewire through the collateral (Fig. 8.4.49, panel E).
6. Contrast is injected via the retrograde guide catheter to ensure that no injury (perforation or rupture) of the collateral vessel has occurred. If injury is detected, the microcatheter (or a new microcatheter) can be re-advanced over the retrograde guidewire to cover the collateral channel perforation and possibly deliver coils.
7. If no collateral vessel injury is detected, the guidewire is removed *after* re-advancing the microcatheter over the guidewire to minimize the risk of injury, especially in tortuous epicardial collaterals.

What can go wrong?
1. Collateral dissection: once the wire has crossed the collateral, it should always be covered with a microcatheter or an over-the-wire balloon to minimize the risk of collateral vessel injury.
2. Dissection of the target vessel ostium (beware of guide catheter movement during externalization; the externalized guidewire should be pushed, rather than pulled forward).
3. Dissection of the donor vessel ostium. Care must be taken to disengage the retrograde guide catheter when withdrawing the retrograde microcatheter and wire.
4. Collateral perforation or rupture: collateral injury can occur during attempts to cross the collateral, during snaring, or during equipment delivery.

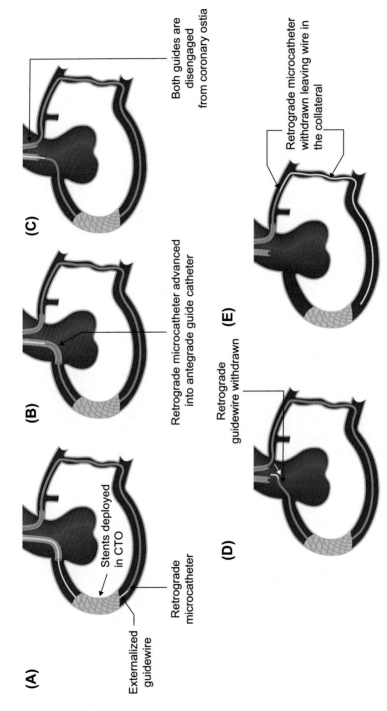

(A)

Stents deployed
in CTO

Externalized
guidewire

Retrograde
microcatheter

(B)

Retrograde microcatheter advanced
into antegrade guide catheter

(C)

Both guides are
disengaged
from coronary ostia

(D)

Retrograde
guidewire withdrawn

(E)

Retrograde microcatheter
withdrawn leaving wire in
the collateral

FIGURE 8.4.49 **How to remove the externalized guidewire.**

5. Equipment entrapment: the antegrade microcatheter/balloon/stents should not meet the retrograde microcatheter or balloon to minimize the risk of catheter "interlocking" and entrapment. If the microcatheter and the balloon become interlocked inadvertently, the hub of the microcatheter can be cut after advancing its tip into the antegrade guide, then the antegrade balloon should be pulled out, pulling the microcatheter through the collaterals. If the balloon and microcatheter become separated during the maneuver, the microcatheter should be trapped inside the antegrade guide catheter with a 3 mm semicompliant balloon, followed by removal of the guide catheter along with the microcatheter.

8.4.5   Contemporary role of the retrograde approach

8.4.5.1   Success

Use of the retrograde approach is critical for achieving high success rates,[104,108,110,131,133,170,186–190] although in some centers advanced antegrade techniques can result in fairly high success rates as well.

The retrograde approach is of particular importance for more complex CTOs, as shown in several contemporary CTO PCI registries.[191–193] Some CTOs (such as flush aorto-ostial CTOs) can only be approached using retrograde crossing.

8.4.5.2   Complications

As described in detail in part C of the book (Complications), the retrograde approach carries higher risk of complications,[110] such as periprocedural myocardial infarction,[194–197] perforation and tamponade, and donor vessel injury, a potentially lethal complication. Use of the retrograde approach is associated with a higher risk of periprocedural complications and is part of the PROGRESS-CTO Complications score.[198]

Therefore, the retrograde approach should be used cautiously and with meticulous attention to technique. Operators who perform retrograde CTO PCI should be equipped with the skills necessary to deal with the complications that might occur.

References

91. Brilakis ES, Grantham JA, Thompson CA, et al. The retrograde approach to coronary artery chronic total occlusions: a practical approach. *Catheter Cardiovasc Interv* 2012;**79**:3–19.
92. Joyal D, Thompson CA, Grantham JA, Buller CEH, Rinfret S. The retrograde technique for recanalization of chronic total occlusions: a step-by-step approach. *JACC Cardiovasc Interv* 2012;**5**:1–11.

93. Kahn JK, Hartzler GO. Retrograde coronary angioplasty of isolated arterial segments through saphenous vein bypass grafts. *Cathet Cardiovasc Diagn* 1990;**20**:88–93.

94. Silvestri M, Parikh P, Roquebert PO, Barragan P, Bouvier JL, Comet B. Retrograde left main stenting. *Cathet Cardiovasc Diagn* 1996;**39**:396–9.

95. Surmely JF, Tsuchikane E, Katoh O, et al. New concept for CTO recanalization using controlled antegrade and retrograde subintimal tracking: the CART technique. *J Invasive Cardiol* 2006;**18**:334–8.

96. Ozawa N. A new understanding of chronic total occlusion from a novel PCI technique that involves a retrograde approach to the right coronary artery via a septal branch and passing of the guidewire to a guiding catheter on the other side of the lesion. *Catheter Cardiovasc Interv* 2006;**68**:907–13.

97. Kumar SS, Kaplan B. Chronic total occlusion angioplasty through supplying collaterals. *Catheter Cardiovasc Interv* 2006;**68**:914–16.

98. Niccoli G, Ochiai M, Mazzari MA. A complex case of right coronary artery chronic total occlusion treated by a successful multi-step Japanese approach. *J Invasive Cardiol* 2006;**18**:E230–3.

99. Lin TH, Wu DK, Su HM, et al. Septum hematoma: a complication of retrograde wiring in chronic total occlusion. *Int J Cardiol* 2006;**113**:e64–6.

100. Rosenmann D, Meerkin D, Almagor Y. Retrograde dilatation of chronic total occlusions via collateral vessel in three patients. *Catheter Cardiovasc Interv* 2006;**67**:250–3.

101. Lane RE, Ilsley CD, Wallis W, Dalby MC. Percutaneous coronary intervention of a circumflex chronic total occlusion using an epicardial collateral retrograde approach. *Catheter Cardiovasc Interv* 2006;**69**:842–4.

102. Rathore S, Katoh O, Matsuo H, et al. Retrograde percutaneous recanalization of chronic total occlusion of the coronary arteries: procedural outcomes and predictors of success in contemporary practice. *Circ Cardiovasc Interv* 2009;**2**:124–32.

103. Rathore S, Katoh O, Tuschikane E, Oida A, Suzuki T, Takase S. A novel modification of the retrograde approach for the recanalization of chronic total occlusion of the coronary arteries intravascular ultrasound-guided reverse controlled antegrade and retrograde tracking. *JACC Cardiovasc Interv* 2010;**3**:155–64.

104. Okamura A, Yamane M, Muto M, et al. Complications during retrograde approach for chronic coronary total occlusion: sub-analysis of Japanese multicenter registry. *Catheter Cardiovasc Interv* 2016;**88**:7–14.

105. Sianos G, Barlis P, Di Mario C, et al. European experience with the retrograde approach for the recanalisation of coronary artery chronic total occlusions. A report on behalf of the euroCTO club. *EuroIntervention* 2008;**4**:84–92.

106. Biondi-Zoccai GG, Bollati M, Moretti C, et al. Retrograde percutaneous recanalization of coronary chronic total occlusions: outcomes from 17 patients. *Int J Cardiol* 2008;**130**:118–20.

107. Galassi AR, Tomasello SD, Reifart N, et al. In-hospital outcomes of percutaneous coronary intervention in patients with chronic total occlusion: insights from the ERCTO (European Registry of Chronic Total Occlusion) registry. *EuroIntervention* 2011;**7**:472–9.

108. Galassi AR, Sianos G, Werner GS, et al. Retrograde recanalization of chronic total occlusions in Europe: procedural, in-hospital, and long-term outcomes from the multicenter ERCTO Registry. *J Am Coll Cardiol* 2015;**65**:2388–400.

109. Thompson CA, Jayne JE, Robb JF, et al. Retrograde techniques and the impact of operator volume on percutaneous intervention for coronary chronic total occlusions an early U.S. experience. *JACC Cardiovasc Interv* 2009;**2**:834–42.

110. Karmpaliotis D, Karatasakis A, Alaswad K, et al. Outcomes with the use of the retrograde approach for coronary chronic total occlusion interventions in a contemporary multicenter US Registry. *Circ Cardiovasc Interv* 2016;**9**.

111. Sakakura K, Nakano M, Otsuka F, et al. Comparison of pathology of chronic total occlusion with and without coronary artery bypass graft. *Eur Heart J* 2014;**35**:1683−93.

112. Hatem R, Finn MT, Riley RF, et al. Zero contrast retrograde chronic total occlusions percutaneous coronary intervention: a case series. *Eur Heart J Case Rep* 2018;**2**:1−5.

113. Wu EB, Chan WW, Yu CM. Retrograde chronic total occlusion intervention: tips and tricks. *Catheter Cardiovasc Interv* 2008;**72**:806−14.

114. Megaly M, Xenogiannis I, Abi Rafeh N, Karmpaliotis D, Rinfret S, Yamane M, Burke MN, Brilakis ES. Retrograde approach to chronic total occlusion percutaneous coronary intervention. *Circ Cardiovasc Interv* 2020;**13**:e008900.

115. Brilakis ES, Grantham JA, Rinfret S, et al. A percutaneous treatment algorithm for crossing coronary chronic total occlusions. *JACC Cardiovasc Interv* 2012;**5**:367−79.

116. Wu EB, Brilakis ES, Mashayekhi K, et al. Global chronic total occlusion crossing algorithm: JACC state-of-the-art review. *J Am Coll Cardiol* 2021;**78**:840−53.

117. Wu EB, Tsuchikane E. The inherent catastrophic traps in retrograde CTO PCI. *Catheter Cardiovasc Interv* 2018;**91**:1101−9.

118. Fang HY, Wu CC, Wu CJ. Successful transradial antegrade coronary intervention of a rare right coronary artery high anterior downward takeoff anomalous chronic total occlusion by double-anchoring technique and retrograde guidance. *Int Heart J* 2009;**50**:531−8.

119. Nombela-Franco L, Werner GS. Retrograde recanalization of a chronic ostial occlusion of the left anterior descending artery: how to manage extreme takeoff angles. *J Invasive Cardiol* 2010;**22**:E7−12.

120. Kaneda H, Takahashi S, Saito S. Successful coronary intervention for chronic total occlusion in an anomalous right coronary artery using the retrograde approach via a collateral vessel. *J Invasive Cardiol* 2007;**19**:E1−4.

121. Dautov R, Manh Nguyen C, Altisent O, Gibrat C, Rinfret S. Recanalization of chronic total occlusions in patients with previous coronary bypass surgery and consideration of retrograde access via saphenous vein grafts. *Circ Cardiovasc Interv* 2016;**9**.

122. Aggarwal C, Varghese J, Uretsky BF. Left atrial inflow and outflow obstruction as a complication of retrograde approach for chronic total occlusion: report of a case and literature review of left atrial hematoma after percutaneous coronary intervention. *Catheter Cardiovasc Interv* 2013;**82**:770−5.

123. Karatasakis A, Akhtar YN, Brilakis ES. Distal coronary perforation in patients with prior coronary artery bypass graft surgery: the importance of early treatment. *Cardiovasc Revasc Med* 2016;**17**:412−17.

124. Wilson WM, Spratt JC, Lombardi WL. Cardiovascular collapse post chronic total occlusion percutaneous coronary intervention due to a compressive left atrial hematoma managed with percutaneous drainage. *Catheter Cardiovasc Interv* 2015;**86**:407−11.

125. Adusumalli S, Morris M, Pershad A. Pseudo-pericardial tamponade from right ventricular hematoma after chronic total occlusion percutaneous coronary intervention of the right coronary artery: successfully managed percutaneously with computerized tomographic guided drainage. *Catheter Cardiovasc Interv* 2016;**88**:86−8.

126. Marmagkiolis K, Brilakis ES, Hakeem A, Cilingiroglu M, Bilodeau L. Saphenous vein graft perforation during percutaneous coronary intervention: a case series. *J Invasive Cardiol* 2013;**25**:157−61.

127. Brilakis ES, Banerjee S, Lombardi WL. Retrograde recanalization of native coronary artery chronic occlusions via acutely occluded vein grafts. *Catheter Cardiovasc Interv* 2010;**75**:109−13.

128. Tajti P, Karatasakis A, Karmpaliotis D, et al. Retrograde CTO-PCI of native coronary arteries via left internal mammary artery grafts: insights from a multicenter U.S. registry. *J Invasive Cardiol* 2018;**30**:89−96.

129. Matsumi J, Adachi K, Saito S. A unique complication of the retrograde approach in angioplasty for chronic total occlusion of the coronary artery. *Catheter Cardiovasc Interv* 2008;**72**:371−8.

130. Werner GS, Ferrari M, Heinke S, et al. Angiographic assessment of collateral connections in comparison with invasively determined collateral function in chronic coronary occlusions. *Circulation* 2003;**107**:1972−7.

131. Dautov R, Urena M, Nguyen CM, Gibrat C, Rinfret S. Safety and effectiveness of the surfing technique to cross septal collateral channels during retrograde chronic total occlusion percutaneous coronary intervention. *EuroIntervention* 2017;**12**:e1859−67.

132. Brilakis ES, Badhey N, Banerjee S. "Bilateral knuckle" technique and Stingray reentry system for retrograde chronic total occlusion intervention. *J Invasive Cardiol* 2011;**23**:E37−9.

133. Mashayekhi K, Behnes M, Akin I, Kaiser T, Neuser H. Novel retrograde approach for percutaneous treatment of chronic total occlusions of the right coronary artery using ipsilateral collateral connections: a European centre experience. *EuroIntervention* 2016;**11**: e1231−6.

134. Mashayekhi K, Behnes M, Valuckiene Z, et al. Comparison of the ipsi-lateral vs contralateral retrograde approach of percutaneous coronary interventions in chronic total occlusions. *Catheter Cardiovasc Interv* 2017;**89**:649−55.

135. Otsuji S, Terasoma K, Takiuchi S. Retrograde recanalization of a left anterior descending chronic total occlusion via an ipsilateral intraseptal collateral. *J Invasive Cardiol* 2008;**20**:312−16.

136. Utsunomiya M, Mukohara N, Hirami R, Nakamura S. Percutaneous coronary intervention for chronic total occlusive lesion of a left anterior descending artery using the retrograde approach via a septal-septal channel. *Cardiovasc Revasc Med* 2010;**11**:34−40.

137. Brilakis ES, Grantham JA, Banerjee S. "Ping-pong" guide catheter technique for retrograde intervention of a chronic total occlusion through an ipsilateral collateral. *Catheter Cardiovasc Interv* 2011;**78**:395−9.

138. Kawamura A, Jinzaki M, Kuribayashi S. Percutaneous revascularization of chronic total occlusion of left anterior descending artery using contralateral injection via isolated conus artery. *J Invasive Cardiol* 2009;**21**:E84−6.

139. Brilakis ES. *Manual of coronary chronic total occlusion interventions. a step-by-step approach.* 2nd ed. Elsevier; 2017.

140. Wu K, Luo B, Huang Z, Zhang B. Simplified dual-lumen catheter-facilitated reverse wire technique for markedly angulated collateral channel entry in retrograde chronic total occlusion intervention. *Int Heart J* 2021;**62**:416−21.

141. Abdel-Karim AR, Vo M, Main ML, Grantham JA. Interventricular septal hematoma and coronary-ventricular fistula: a complication of retrograde chronic total occlusion intervention. *Case Rep Cardiol* 2016;**2016**:8750603.

142. Araki M, Murai T, Kanaji Y, et al. Interventricular septal hematoma after retrograde intervention for a chronic total occlusion of a right coronary artery: echocardiographic and magnetic resonance imaging-diagnosis and follow-up. *Case Rep Med* 2016;**2016**:8514068.

143. Abdelghani MS, Chapra A, Abed H, Al-Qahtani A, Alkindi F. Interventricular septal hematoma: a rare complication of retrograde chronic total occlusion intervention. *Heart Views* 2021;**22**:201−5.

144. Sianos G, Papafaklis MI. Septal wire entrapment during recanalisation of a chronic total occlusion with the retrograde approach. *Hellenic J Cardiol* 2011;**52**:79−83.

145. Franks RJ, de Souza A, Di Mario C. Left atrial intramural hematoma after percutaneous coronary intervention. *Catheter Cardiovasc Interv* 2015;**86**:E150−2.

146. Tsiafoutis I, Liontou C, Koutouzis M, Brilakis E. Use of a totally occluded graft as a conduit for retrograde native artery recanalization. *Hellenic J Cardiol* 2021;**62**:490−2.

147. Lichtenwalter C, Banerjee S, Brilakis ES. Dual guide catheter technique for treating native coronary artery lesions through tortuous internal mammary grafts: separating equipment delivery from target lesion visualization. *J Invasive Cardiol* 2010;**22**:E78−81.

148. Routledge H, Lefevre T, Ohanessian A, Louvard Y, Dumas P, Morice MC. Use of a deflectable tip catheter to facilitate complex interventions beyond insertion of coronary bypass grafts: three case reports. *Catheter Cardiovasc Interv* 2007;**70**:862−6.

149. Iturbe JM, Abdel-Karim AR, Raja VN, Rangan BV, Banerjee S, Brilakis ES. Use of the venture wire control catheter for the treatment of coronary artery chronic total occlusions. *Catheter Cardiovasc Interv* 2010;**76**:936−41.

150. Saeed B, Banerjee S, Brilakis ES. Percutaneous coronary intervention in tortuous coronary arteries: associated complications and strategies to improve success. *J Interv Cardiol* 2008;**21**:504−11.

151. Kawasaki T, Koga H, Serikawa T. New bifurcation guidewire technique: a reversed guidewire technique for extremely angulated bifurcation−a case report. *Catheter Cardiovasc Interv* 2008;**71**:73−6.

152. Shirai S, Doijiri T, Iwabuchi M. Treatment for LMCA ostial stenosis using a bifurcation technique with a retrograde approach. *Catheter Cardiovasc Interv* 2010;**75**:748−52.

153. Wilson SJ, Hanratty CG, Spence MS, et al. Saphenous vein graft sacrifice following native vessel PCI is safe and associated with favourable longer-term outcomes. *Cardiovasc Revasc Med* 2019;**20**:1048−52.

154. Saito S. Different strategies of retrograde approach in coronary angioplasty for chronic total occlusion. *Catheter Cardiovasc Interv* 2008;**71**:8−19.

155. Furuichi S, Satoh T. Intravascular ultrasound-guided retrograde wiring for chronic total occlusion. *Catheter Cardiovasc Interv* 2010;**75**:214−21.

156. Wu EB, Chan WW, Yu CM. Antegrade balloon transit of retrograde wire to bail out dissected left main during retrograde chronic total occlusion intervention−a variant of the reverse CART technique. *J Invasive Cardiol* 2009;**21**:e113−18.

157. Mody R, Dash D, Mody B, Saholi A. Guide extension catheter-facilitated reverse controlled antegrade and retrograde tracking for retrograde recanalization of chronic total occlusion. *Case Rep Cardiol* 2021;**2021**:6690452.

158. Kaier TE, Kalogeropoulos A, Pavlidis AN. Guide-extension facilitated antegrade dissection reentry: a case series. *J Invasive Cardiol* 2020;**32**:E209−12.

159. Xenogiannis I, Karmpaliotis D, Alaswad K, et al. Comparison between traditional and guide-catheter extension reverse controlled antegrade dissection and retrograde tracking: insights from the PROGRESS-CTO registry. *J Invasive Cardiol* 2019;**31**:27−34.

160. Wu EB, Chan WW, Yu CM. The confluent balloon technique−two cases illustrating a novel method to achieve rapid wire crossing of chronic total occlusion during retrograde approach percutaneous coronary intervention. *J Invasive Cardiol* 2009;**21**:539−42.

161. Carlino M, Azzalini L, Colombo A. A novel maneuver to facilitate retrograde wire externalization during retrograde chronic total occlusion percutaneous coronary intervention. *Catheter Cardiovasc Interv* 2017;**89**:E7−12.
162. Matsuno S, Tsuchikane E, Harding SA, et al. Overview and proposed terminology for the reverse controlled antegrade and retrograde tracking (reverse CART) techniques. *EuroIntervention* 2018;**14**:94−101.
163. Tsuchikane E, Katoh O, Kimura M, Nasu K, Kinoshita Y, Suzuki T. The first clinical experience with a novel catheter for collateral channel tracking in retrograde approach for chronic coronary total occlusions. *JACC Cardiovasc Interv* 2010;**3**:165−71.
164. Mozid AM, Davies JR, Spratt JC. The utility of a guideliner catheter in retrograde percutaneous coronary intervention of a chronic total occlusion with reverse cart-the "capture" technique. *Catheter Cardiovasc Interv* 2014;**83**:929−32.
165. Yeoh J, Hill J, Spratt JC. Intravascular lithotripsy assisted chronic total occlusion revascularization with reverse controlled antegrade retrograde tracking. *Catheter Cardiovasc Interv* 2019;**93**:1295−7.
166. Alshamsi A, Bouhzam N, Boudou N. Cutting balloon in reverse CART technique for recanalization of chronic coronary total occlusion. *J Invasive Cardiol* 2014;**26**:E115−16.
167. Dai J, Katoh O, Kyo E, Tsuji T, Watanabe S, Ohya H. Approach for chronic total occlusion with intravascular ultrasound-guided reverse controlled antegrade and retrograde tracking technique: single center experience. *J Interv Cardiol* 2013;**26**:434−43.
168. Galassi AR, Sumitsuji S, Boukhris M, et al. Utility of intravascular ultrasound in percutaneous revascularization of chronic total occlusions: an overview. *JACC Cardiovasc Interv* 2016;**9**:1979−91.
169. Xenogiannis I, Tajti P, Karmpaliotis D, et al. Intravascular imaging for chronic total occlusion intervention. *Curr Cardiovascular Imaging Rep* 2018;**11**:31.
170. Fan Y, Maehara A, Yamamoto MH, et al. Outcomes of retrograde approach for chronic total occlusions by guidewire location. *EuroIntervention* 2021;**17**:e647−55.
171. Michael TT, Papayannis AC, Banerjee S, Brilakis ES. Subintimal dissection/reentry strategies in coronary chronic total occlusion interventions. *Circ Cardiovasc Interv* 2012;**5**:729−38.
172. Ge JB, Zhang F, Ge L, Qian JY, Wang H. Wire trapping technique combined with retrograde approach for recanalization of chronic total occlusion. *Chin Med J (Engl)* 2008;**121**:1753−6.
173. Ge J, Zhang F. Retrograde recanalization of chronic total coronary artery occlusion using a novel "reverse wire trapping" technique. *Catheter Cardiovasc Interv* 2009;**74**:855−60.
174. Vo MN, Ravandi A, Brilakis ES. "Tip-in" technique for retrograde chronic total occlusion revascularization. *J Invasive Cardiol* 2015;**27**:E62−4.
175. Venuti G, D'Agosta G, Tamburino C, La Manna A. When antegrade microcatheter does not follow: the "facilitated tip-in technique.". *Catheter Cardiovasc Interv* 2020;**96**:E458−61.
176. Christ G, Glogar D. Successful recanalization of a chronic occluded left anterior descending coronary artery with a modification of the retrograde proximal true lumen puncture technique: the antegrade microcatheter probing technique. *Catheter Cardiovasc Interv* 2009;**73**:272−5.
177. Muramatsu T, Tsukahara RI. "Rendezvous in coronary" technique with the retrograde approach for chronic total occlusion. *J Invasive Cardiol* 2010;**22**:E179−82.
178. Kim MH, Yu LH, Mitsudo K. A new retrograde wiring technique for chronic total occlusion. *Catheter Cardiovasc Interv* 2010;**75**:117−19.
179. Ng R, Hui PY, Beyer A, Ren X, Ochiai M. Successful retrograde recanalization of a left anterior descending artery chronic total occlusion through a previously placed left anterior descending-to-diagonal artery stent. *J Invasive Cardiol* 2010;**22**:E16−18.

180. Lee NH, Suh J, Seo HS. Double anchoring balloon technique for recanalization of coronary chronic total occlusion by retrograde approach. *Catheter Cardiovasc Interv* 2009;**73**:791−4.

181. Matsumi J, Saito S. Progress in the retrograde approach for chronic total coronary artery occlusion: a case with successful angioplasty using CART and reverse-anchoring techniques 3 years after failed PCI via a retrograde approach. *Catheter Cardiovasc Interv* 2008;**71**:810−14.

182. Karacsonyi J, Koike H, Fukui M, Kostantinis S, Simsek B, Brilakis ES. Retrograde treatment of a right coronary artery perforation. *JACC Cardiovasc Interv* 2022;**15**:670−2.

183. Utunomiya M, Katoh O, Nakamura S. Percutaneous coronary intervention for a right coronary artery stent occlusion using retrograde delivery of a sirolimus-eluting stent via a septal perforator. *Catheter Cardiovasc Interv* 2009;**73**:475−80.

184. Bansal D, Uretsky BF. Treatment of chronic total occlusion by retrograde passage of stents through an epicardial collateral vessel. *Catheter Cardiovasc Interv* 2008;**72**:365−9.

185. Utsunomiya M, Kobayashi T, Nakamura S. Case of dislodged stent lost in septal channel during stent delivery in complex chronic total occlusion of right coronary artery. *J Invasive Cardiol* 2009;**21**:E229−33.

186. Wu EB, Tsuchikane E, Ge L, et al. Retrograde vs antegrade approach for coronary chronic total occlusion in an algorithm-driven contemporary Asia-Pacific multicentre registry: comparison of outcomes. *Heart Lung Circ* 2020;**29**:894−903.

187. Kalra S, Doshi D, Sapontis J, et al. Outcomes of retrograde chronic total occlusion percutaneous coronary intervention: a report from the OPEN-CTO registry. *Catheter Cardiovasc Interv* 2021;**97**:1162−73.

188. Tajti P, Xenogiannis I, Gargoulas F, et al. Technical and procedural outcomes of the retrograde approach to chronic total occlusion interventions. *EuroIntervention* 2020;**16**:e891−9.

189. Xu R, Shi Y, Chang S, et al. Outcomes of contemporary vs conventional reverse controlled and antegrade and retrograde subintimal tracking in chronic total occlusion revascularization. *Catheter Cardiovasc Interv* 2022;**99**:226−33.

190. Katoh H, Yamane M, Muramatsu T, et al. Safety of Percutaneous coronary intervention for chronic total occlusion in patients with multi-vessel disease: sub-analysis of the Japanese Retrograde summit registry. *Cardiovasc Revasc Med* 2021;**25**:36−42.

191. Christopoulos G, Wyman RM, Alaswad K, et al. Clinical utility of the Japan-chronic total occlusion score in coronary chronic total occlusion interventions: results from a multicenter registry. *Circ Cardiovasc Interv* 2015;**8**:e002171.

192. Wilson WM, Walsh SJ, Yan AT, et al. Hybrid approach improves success of chronic total occlusion angioplasty. *Heart* 2016;**102**:1486−93.

193. Tajti P, Karmpaliotis D, Alaswad K, et al. The hybrid approach to chronic total occlusion percutaneous coronary intervention: update from the PROGRESS CTO registry. *JACC Cardiovasc Interv* 2018;**11**:1325−35.

194. Kim SM, Gwon HC, Lee HJ, et al. Periprocedural myocardial infarction after retrograde approach for chronic total occlusion of coronary artery: demonstrated by cardiac magnetic resonance imaging. *Korean Circ J* 2011;**41**:747−9.

195. Lo N, Michael TT, Moin D, et al. Periprocedural myocardial injury in chronic total occlusion percutaneous interventions: a systematic cardiac biomarker evaluation study. *JACC Cardiovasc Interv* 2014;**7**:47−54.

196. Stetler J, Karatasakis A, Christakopoulos GE, et al. Impact of crossing technique on the incidence of periprocedural myocardial infarction during chronic total occlusion percutaneous coronary intervention. *Catheter Cardiovasc Interv* 2016;**88**:1−6.

197. Werner GS, Coenen A, Tischer KH. Periprocedural ischaemia during recanalisation of chronic total coronary occlusions: the influence of the transcollateral retrograde approach. *EuroIntervention* 2014;**10**:799–805.

198. Danek BA, Karatasakis A, Karmpaliotis D, et al. Development and validation of a scoring system for predicting periprocedural complications during percutaneous coronary interventions of chronic total occlusions: the prospective global registry for the study of chronic total occlusion intervention (PROGRESS CTO) complications score. *J Am Heart Assoc* 2016;**5**.

Section 8.5

Crossing algorithms

8.5.1   Historical perspectives

Several algorithms have been developed to provide guidance on when and how to use the various CTO crossing strategies (antegrade wiring; antegrade dissection/reentry; retrograde wiring; retrograde dissection/reentry) and on when to stop.

The first CTO crossing algorithm, the hybrid algorithm (Fig. 8.5.1), was published in 2012.[199] The hybrid algorithm emphasized the importance of dual angiography and careful angiographic review to guide the selection of initial and subsequent crossing strategies.[199] It also recommended prompt change of crossing strategy in case of failure to achieve progress as well as awareness of radiation dose, contrast volume, and procedure time for deciding when to stop. Application of the "hybrid" approach significantly improved success rates in CTO PCI in several studies in the US[200–207] and Europe.

Several CTO crossing algorithms were subsequently published, including the Asia Pacific (Fig. 8.5.2),[208] CTO Club China (Fig. 8.5.3),[209] EuroCTO Club (Fig. 8.5.4),[210] and Japan CTO Club (Fig. 8.5.5).[211] The similarities and differences of these algorithms are shown in Table 8.5.1.

Collaboration of several experienced CTO operators from around the world resulted in publication of a global CTO crossing algorithm in 2021 (Fig. 8.5.6).[212] The global CTO crossing algorithm has 10 steps that are discussed in detail in Section 8.5.2.

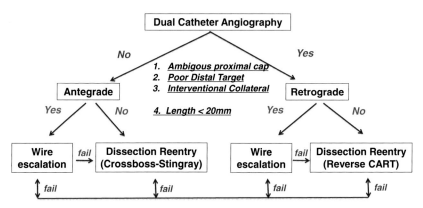

FIGURE 8.5.1   **The hybrid CTO crossing algorithm**. *Reproduced from the* Manual of Chronic Total Occlusion Percutaneous Coronary Interventions. *2nd ed.*

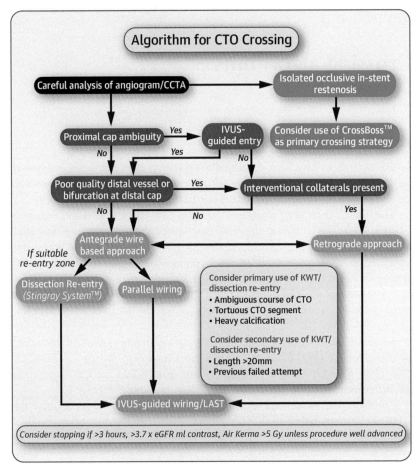

FIGURE 8.5.2 **The Asia Pacific CTO crossing algorithm.** *Reproduced with permission from Harding SA, Wu EB, Lo S, et al. A new algorithm for crossing chronic total occlusions from the asia pacific chronic total occlusion club.* JACC Cardiovasc Interv *2017;10:2135–2143.*

8.5.2 Steps of the global CTO crossing algorithm

8.5.2.1 Step 1. Dual Injection

The first and arguably most important step of CTO PCI is to perform dual coronary injection, in nearly all cases (unless the distal CTO vessel is filling exclusively from ipsilateral collaterals),[212] as described in detail in Chapter 6. Dual injection allows good visualization of the proximal and distal vessel and the collateral circulation, allowing selection of the most suitable initial crossing technique. It also clarifies the location of the guide-wire(s) during crossing attempts and can help detect CTO PCI complications.

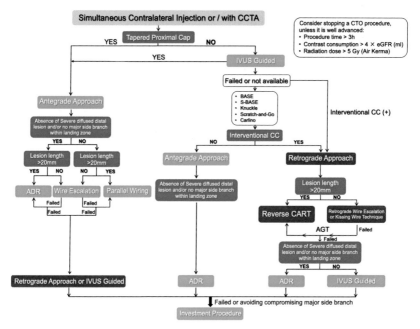

FIGURE 8.5.3 **The CTO Club China CTO crossing algorithm.** *Courtesy of Dr. Lei Ge.*

Routine performance of dual injection is the simplest and most important step to increase the success and safety of CTO PCI.

Coronary CT angiography is also increasingly used for planning CTO PCI and has been associated with increased success rates, especially for more complex occlusions.[213,214] CT coregistration with live fluoroscopy can provide live guidance for CTO crossing.[215,216]

8.5.2.2 Step 2. Assessment of CTO characteristics

In-depth review of diagnostic angiographic images prior to PCI is critical. Most experienced CTO operators recommend against performing ad-hoc CTO PCI (Section 7.3.1); instead, CTO PCI should be performed only after a well thought out procedural plan has been developed. Time spent studying the diagnostic film is an investment towards a successful CTO PCI procedure and reduces radiation and contrast utilization during PCI. Often dual injection is not performed until the time of CTO PCI.

Dual coronary angiography and an in-depth and structured review of the angiogram (and, if available, coronary computed tomography angiography) are key for planning and safely performing CTO-PCI and are one of the seven global guiding principles for CTO PCI.[217]

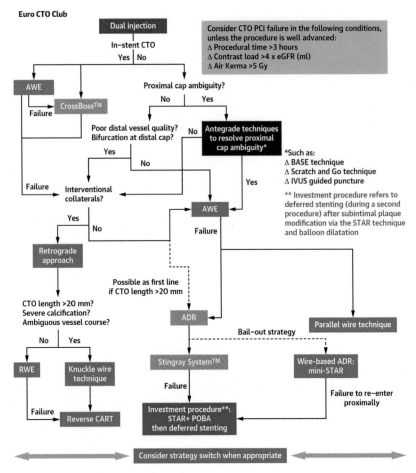

FIGURE 8.5.4   The EuroCTO Club CTO crossing algorithm. *Reproduced with permission from Galassi AR, Werner GS, Boukhris M, et al. Percutaneous recanalisation of chronic total occlusions: 2019 consensus document from the EuroCTO Club.* EuroIntervention *2019;15:198–208.*

Four angiographic parameters are assessed (Section 6.2): (1) the morphology of the proximal cap; (2) the length of the occlusion; (3) the size of the distal vessel and presence of bifurcations beyond the distal cap (i.e., "landing zone"); and (4) the location and suitability of collateral channels for retrograde access (Fig. 8.5.7).[199] Assessment and utilization of the above four angiographic characteristics is highly dependent on operator experience and skillset.

1. **Proximal cap location and morphology**. This characteristic refers to the ability to unambiguously localize the entry point to the CTO lesion by angiography or intravascular ultrasonography and to understand the course of the vessel in the CTO segment.

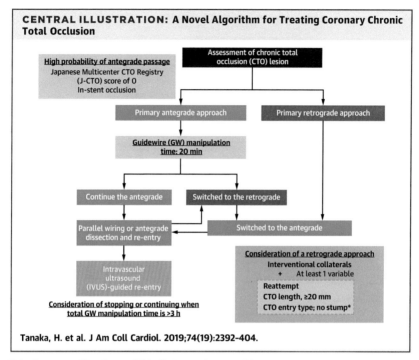

CENTRAL ILLUSTRATION: A Novel Algorithm for Treating Coronary Chronic Total Occlusion

Tanaka, H. et al. J Am Coll Cardiol. 2019;74(19):2392-404.

FIGURE 8.5.5 **The Japan CTO Club crossing algorithm**. *Reproduced with permission from Tanaka H, Tsuchikane E, Muramatsu T, et al. A novel algorithm for treating chronic total coronary artery occlusion.* J Am Coll Cardiol *2019;74:2392–2404.*

An ambiguous proximal cap (Section 16.1) increases the complexity of the procedure and decreases the likelihood of success.[218] A favorable proximal cap is one that is tapered, as opposed to blunt, and has no bridging collaterals or major side branches that would make engagement of the CTO segment difficult using antegrade wiring.

CTOs with ambiguous proximal caps may be approached using: (1) the retrograde approach (Section 8.4); (2) IVUS-guided proximal cap puncture (Section 13.2.1); or (3) the "move-the-cap" subintimal crossing techniques (Section 16.1.4.3).[219] A particularly challenging anatomic subset are flush aorto-ostial occlusions, which usually require a primary retrograde approach (Section 15.1).

2. **Lesion length and composition**. In most cases, lesion length can only be accurately assessed by using dual injection (the antegrade injection helps visualize the proximal cap and the retrograde injection the distal cap). Longer occlusions are usually more challenging to cross with more frequent extraplaque crossing.[220] Severe calcification and tortuosity favor the use of extraplaque crossing techniques during both antegrade and retrograde crossing attempts.

TABLE 8.5.1 Comparison of various CTO crossing algorithms.

	Hybrid	Asia Pacific	Euro CTO	CTO Club China	J-CTO	GLOBAL CTO
1. Dual angiography	+	+	+	+	+	+
1a. CCTA		+		+		+
2. Careful angiographic review	+	+	+	+	+	+
ISR	No specific recommendation	+ (CrossBoss)	+ (CrossBoss)	+ (CrossBoss)	Antegrade	No specific recommendation—assess lesion characteristics and treat accordingly
3. Proximal cap ambiguity	+	+	+	+	+	+
3a. Solutions to proximal ambiguity	Retrograde	Retrograde, IVUS	Retrograde, IVUS, move the cap	Retrograde, IVUS, move the cap	Retrograde	IVUS, retrograde, move the cap
4. Poor distal vessel quality or bifurcation at the distal cap	+ (Retrograde)	+ (Retrograde)	+ (Retrograde)	+ (Retrograde)		+ (Retrograde)

5. Retrograde option	+	+	+	+	+	+
6. Antegrade wiring strategies	ADR for length ≥20 mm	AW, Parallel wiring. Primary ADR for ambiguous CTO course, tortuous CTO segment, heavy calcification. Secondary ADR: length ≥20 mm, prior failed attempt	AW preferred—ADR possible as first line if length ≥20 mm	ADR preferred if no severe diffuse distal disease, no major side branch near landing zone and length >20 mm	ADR or parallel wiring after AW failure	AW preferred
7. Retrograde	Ambiguous proximal cap, poor distal vessel + INTERV COLLATERALS	Ambiguous proximal cap (if IVUS fails), poor distal vessel + INTERV COLLATERALS	Ambiguous proximal cap, poor distal vessel + INTERV COLLATERALS	Ambiguous proximal cap + no or failed IVUS-guided approach + INTERV COLLATERALS	Reattempt, CTO length of ≥20 mm, and no stump + INTERV COLLATERALS	Ambiguous proximal cap, poor distal vessel + FEASIBLE RETROGRADE OPTION
7a. RDR preferred over RWE	Length ≥20 mm	Length ≥15 mm	Length ≥20 mm Severe calcification Ambiguous vessel course	Length ≥20 mm		Length ≥20 mm Severe calcification Ambiguous vessel course

(Continued)

TABLE 8.5.1 (Continued)

	Hybrid	Asia Pacific	Euro CTO	CTO Club China	J-CTO	GLOBAL CTO
7b. RDR preferred technique	Reverse CART	Contemporary reverse CART	Reverse CART	Reverse CART		Reverse CART
8. Change	+	+	+	+	+ (after 20 min wire manipulation time)	+
9. Investment			+			+
10. When to stop	Air kerma radiation dose >10 Gray	• Procedure duration >3 h • Air kerma radiation dose >5 Gray • Contrast volume >3.7x eGFR	• Procedure duration >3 h • Air kerma radiation dose > 5 Gray • Contrast volume >4x eGFR		Procedure duration >3 h	• Procedure duration >3 h • Air kerma radiation dose >5 Gray • Contrast volume >3x eGFR • Complication

ADR, antegrade dissection and reentry; *AW*, antegrade wiring; *CART*, controlled antegrade and retrograde tracking; *CCTA*, coronary computed tomography angiography; *CTO*, chronic total occlusion; *eGFR*, estimated glomerular filtration rate; *ISR*, in-stent restenosis; *IVUS*, intravascular ultrasound; *RDR*, retrograde dissection and reentry; *RWE*, retrograde wire escalation.
Source: Reproduced with permission from Wu EB, Brilakis ES, Mashayekhi K, et al. Global chronic total occlusion crossing algorithm: JACC state-of-the-art review. *J Am Coll Cardiol* 2021;78:840−853.

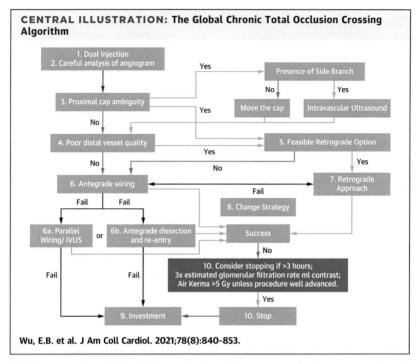

CENTRAL ILLUSTRATION: The Global Chronic Total Occlusion Crossing Algorithm

Wu, E.B. et al. J Am Coll Cardiol. 2021;78(8):840-853.

FIGURE 8.5.6 **The Global CTO crossing algorithm**. *Reproduced with permission from Wu EB, Brilakis ES, Mashayekhi K, et al. Global chronic total occlusion crossing algorithm: JACC state-of-the-art review.* J Am Coll Cardiol *2021;78:840−853.*

3. **Quality of distal vessel**. This refers to the size of the distal lumen, presence of significant side branches, vessel disease at the reconstitution point, and ability to adequately angiographically visualize this segment. A distal vessel of large caliber (≥ 2 mm), that fills well, does not have significant disease, and is free from major branches facilitates CTO recanalization.[212] Conversely, small and diffusely diseased distal vessels with significant bifurcations can be harder to recanalize, especially when using antegrade dissection/reentry.

4. **Retrograde crossing options**.
   The retrograde approach can be performed through bypass grafts (patent or recently occluded)[220,221] and septal or epicardial collaterals.
   Optimal collateral vessels for retrograde CTO PCI (Section 6.2.1.4):
   a. are sourced from a healthy (or repaired) donor vessel;
   b. can be easily accessed with wires and microcatheters;
   c. have minimal tortuosity and few or no bifurcations;
   d. are not the only source of flow to the CTO segment (which places the patient at risk of intraprocedural ischemia during crossing of the collateral);

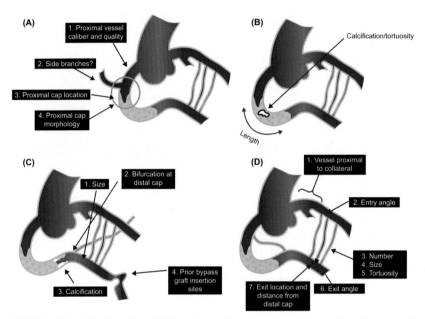

FIGURE 8.5.7 The four key CTO angiographic parameters that need to be assessed before CTO PCI. Panel A: Proximal cap and vessel; **Panel B**: lesion length and composition; **Panel C**: distal vessel; and **Panel D**: collateral circulation. *Reproduced with permission from the Manual of Chronic Total Occlusion Percutaneous Coronary Interventions 2nd ed.*

   e. enter the CTO vessel well beyond the distal cap; and
   f. have a favorable entry angle from the donor vessel into the collateral and exit angle from the collateral into the distal true lumen.

Favorable collateral circulation characteristics lower the barriers to utilizing retrograde techniques as an initial strategy or as an early crossover strategy. In-depth understanding of the collateral circulation is also useful during antegrade crossing attempts, because dissection/reentry techniques and the formation of extraplaque hematomas may compromise ipsilateral or bridging collaterals leading to poor visualization of the distal vessel and occasionally ischemia.

8.5.2.3 Step 3. Approaching proximal cap ambiguity

Proximal cap ambiguity is defined as inability to determine the location of the proximal cap.[212] Highly penetrating guidewires should not be used to puncture ambiguous proximal caps due to increased risk of perforation. Coronary CT angiography, coronary angiography in multiple projections, and contrast injection through a microcatheter advanced close to the proximal cap can sometimes help resolve proximal cap ambiguity. If the location of the proximal cap remains ambiguous, three strategies can be used, depending on the lesion characteristics:

1. IVUS-guided proximal cap puncture
2. Move-the-cap techniques
3. Retrograde crossing

How to approach CTOs with ambiguous proximal cap is discussed in detail in Section 16.1.

8.5.2.4 Step 4. Approaching poor distal vessel quality and bifurcation at the distal cap

In patients with poor distal vessel quality or a bifurcation at the distal cap, the retrograde approach may be safer and more efficient when performed by experienced operators.[212]

In patients with a bifurcation at the distal cap (Section 16.4), the use of a dual-lumen microcatheter is recommended for advancing a guidewire into the side branch to prevent side branch occlusion after balloon angioplasty and stenting. If the CTO crossing guidewire position is unclear, IVUS should be performed before stent placement to avoid side branch occlusion (especially if the side branch is large) due to extraplaque position of the main branch wire at the bifurcation.[212]

8.5.2.5 Step 5. Feasible retrograde option

The term "interventional" collateral has traditionally been used for retrograde collateral vessels considered appropriate for crossing by the operator. Coronary bypass grafts, whether patent or occluded, are often used for the retrograde approach, even though they are not truly "collateral" vessels. The term "feasible retrograde option" is used in the global CTO crossing algorithm to describe the presence of any retrograde route considered appropriate for retrograde CTO PCI based on the operator's experience. Bypass grafts and septal collateral vessels are preferred over epicardial collaterals, as perforation of the latter is more likely to cause tamponade or loculated hematomas.

8.5.2.6 Step 6. Antegrade wiring

Antegrade wiring (Section 8.2) is the most commonly used CTO crossing strategy and should be performed in most cases unless there is proximal cap ambiguity or poor quality distal vessel.[212] Antegrade wiring usually starts with a low tip load, tapered, polymer-jacketed guidewire followed by escalation to a stiffer tip, polymer-jacketed or a moderate tip stiffness tapered-tip guidewire if there is resistance to crossing. Escalation is usually followed by de-escalation to a softer tip, torqueable guidewire after advancing through the area of resistance, especially when the vessel course is ambiguous. If antegrade wiring fails to cross into the distal true lumen, either antegrade (parallel wiring, device-based reentry [such as use of the Stingray system], or IVUS-guided antegrade wiring), or the retrograde approach can be used.[212]

In antegrade dissection and reentry (ADR, Section 8.3) the extraplaque space is entered, followed by extraplaque crossing of the CTO with reentry into the distal true lumen. Antegrade dissection and reentry may be intentional or unintentional during antegrade wiring attempts. The hybrid algorithm uses occlusion length alone (≥ 20 mm) to determine whether to use wire escalation or ADR, whereas the Asia Pacific CTO Club (APCTO) and European CTO Club (EuroCTO) algorithms use a combination of length and other factors, such as tortuosity, calcification, and proximal cap ambiguity.

In the global CTO crossing algorithm, the following four parameters favor use of ADR[212]: ≥ 20 mm occlusion length, calcification, tortuosity, and presence of an appropriate reentry zone of large caliber and without major side branches. ADR may be less desirable in long left anterior descending artery CTOs with multiple branches (septal and diagonal) at risk of occlusion. Minimizing extraplaque hematoma formation improves the likelihood of successful reentry into the distal true lumen. Reentry should be performed as close to the distal cap as possible, which may be best achieved by using a dedicated reentry device, such as the Stingray balloon. If the subintimal tracking and reentry (STAR) technique is used, stenting should be avoided during the index procedure, as it is associated with high rates of restenosis and reocclusion (Section 8.3.5.1).

8.5.2.7 Step 7. Retrograde approach

The retrograde approach is a critical component of the contemporary CTO PCI armamentarium. It is critical for achieving high success rates, especially in more complex occlusions,[221–230] however it also carries increased risk of complications, such as myocardial infarction,[231–233] perforation and donor vessel injury[226,228] (Figs. 8.5.8, 8.5.9, and 8.5.10). Antegrade CTO crossing is, therefore, preferred, if feasible.

The retrograde approach (Section 8.4) can be used either upfront (primary retrograde when an antegrade approach is not feasible or carries high risk) or after a failed antegrade crossing attempt. Factors that favor a primary retrograde approach include an ambiguous proximal cap, poor distal vessel quality, or bifurcation at the distal cap (**CTO Manual online case 5**). Similar to antegrade wiring, retrograde crossing can be achieved via retrograde wiring versus retrograde dissection and reentry, with the latter favored in cases with long occlusion length (≥ 20 mm), severe tortuosity and calcification, and lack of large side branches that could be compromised with use of dissection

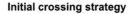

Initial crossing strategy

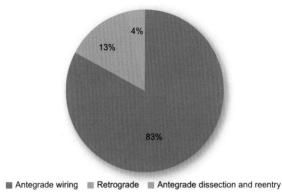

Final successful crossing strategy

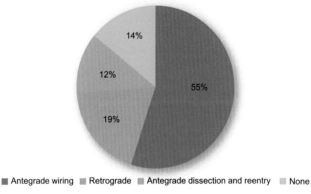

FIGURE 8.5.8 **Initial and final successful CTO crossing strategy in 10,000 CTO PCIs from the PROGRESS-CTO registry** (www.progresscto.org).

techniques. The reverse controlled antegrade and retrograde tracking (reverse CART) is the most commonly used retrograde crossing technique and can be facilitated by use of guide catheter extensions and IVUS.

8.5.2.8 Step 8. Change

Alternating between different CTO PCI techniques is recommended by all CTO crossing algorithms. When one approach fails, something different should be attempted ("don't get stuck in a failure mode"). For example, if antegrade wiring fails, then antegrade dissection/reentry should be tried, and if this fails too, retrograde crossing should be attempted (in patients with feasible retrograde option). Changing strategy is one of the seven global guiding principles of CTO PCI: "If the initially selected crossing strategy fails, efficient change to an alternative crossing technique increases the likelihood of eventual PCI success, shortens procedure time, and lowers radiation and contrast use."[217]

Every CTO is different and as a result may require different strategies for success (Fig. 8.5.9). More complex CTOs are more likely to require advanced crossing techniques, such as antegrade dissection/reentry and the retrograde approach (Fig. 8.5.9), but these techniques carry an increased risk of complications (Fig. 8.5.10).

Excessive persistence with one strategy in the face of minimal progress increases the chances of procedural failure due to utilization of limited resources (radiation, contrast, time). However, the operator should not change too early, but instead invest enough effort in the utilized strategy to maximize its chance for success. What constitutes an "adequate effort" varies from lesion to lesion and from operator to operator, and is best determined by experience in CTO PCI. The Japanese CTO algorithm recommends changing from antegrade to retrograde crossing after 20 minutes of guidewire manipulation;[211] however, the threshold for change varies depending on CTO anatomy and local expertise in various crossing techniques. Generally, no more than 5−10 minutes should be spent in a stagnant mode without a minor (such as reshaping the tip of the wire, or changing to a wire with significantly different properties), or major (such as switching from an antegrade to a retrograde approach) technique adjustment being made. Efficient change of strategy can result in a shorter procedure time and lower patient and staff radiation exposure, and contrast utilization. Changing approach requires a high level of familiarity and comfort with all crossing strategies so that there are no impediments to making a change.

8.5.2.9 Step 9. Investment procedure

If CTO crossing attempts fail, a variety of investment procedures (Section 21.2) should be considered if the anticipated benefit exceeds the potential harm.[212] For example, if an antegrade guidewire has been advanced intraplaque or extraplaque through an ambiguous proximal cap, balloon angioplasty of the proximal cap (external cap crush) is recommended. Other "investment" techniques are subintimal plaque modification (balloon inflation within the extraplaque space) and the subintimal tracking and reentry (STAR) technique. Repeat CTO crossing attempts are usually performed after ≥2 months to allow healing of any created dissections.

8.5.2.10 Step 10. When to stop

Deciding when to stop the CTO crossing attempts depends on the dynamic balance between the likelihood of success and patient safety. The global CTO algorithm recommends stopping the CTO PCI procedure if[212]:

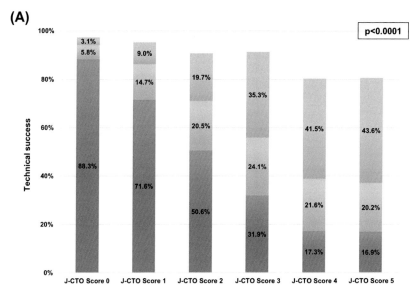

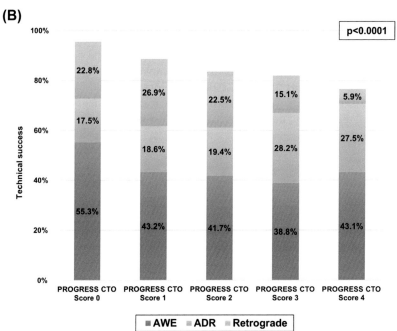

FIGURE 8.5.9 Technical success and crossing strategy use according to J-CTO and PROGRESS CTO score. Impact of chronic total occlusion (CTO) lesion complexity, as assessed by the J-CTO score (**panel A**) and PROGRESS CTO score (**panel B**) on technical success and use of various crossing strategies. *ADR*, antegrade dissection re-entry; *AWE*, antegrade wire escalation; *J-CTO*, Multicenter Chronic Total Occlusion Registry of Japan; *PROGRESS CTO*, Prospective Global Registry for the Study of Chronic Total Occlusion Intervention. *Reproduced with permission from Tajti P, Karmpaliotis D, Alaswad K, et al. The hybrid approach to chronic total occlusion percutaneous coronary intervention: update from the progress CTO registry. JACC Cardiovasc Interv 2018;11:1325−1335 (http://www.progressscto.org).*

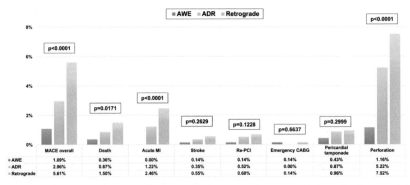

	AWE	ADR	Retrograde

	MACE overall	Death	Acute MI	Stroke	Re-PCI	Emergency CABG	Pericardial tamponade	Perforation
AWE	1.09%	0.36%	0.00%	0.14%	0.14%	0.14%	0.43%	1.16%
ADR	2.96%	0.87%	1.22%	0.35%	0.52%	0.00%	0.87%	5.22%
Retrograde	5.61%	1.50%	2.46%	0.55%	0.68%	0.14%	0.96%	7.52%

FIGURE 8.5.10 In-hospital major complications classified according to final successful crossing strategy. Use of the retrograde approach and antegrade dissection/reentry was associated with higher overall in-hospital major adverse cardiac events (MACE) and risk for perforation. *ADR*, antegrade dissection and reentry; *AWE*, antegrade wire escalation; *CABG*, coronary artery bypass graft; *MI*, myocardial infarction; *PCI*, percutaneous coronary intervention. *Reproduced with permission from Tajti P, Karmpaliotis D, Alaswad K, et al. The hybrid approach to chronic total occlusion percutaneous coronary intervention: update from the progress CTO registry.* JACC Cardiovasc Interv 2018;11:1325−1335.

- procedure time is >3 h (because of potential patient and operator fatigue that may increase the risk of errors and complications).
- contrast volume is >3x the estimated glomerular filtration rate. Even lower contrast volume thresholds may be used in patients with chronic kidney disease and comorbidities, such as diabetes.
- air kerma radiation dose is >5 Gray, unless the procedure is well advanced.[212]
- it becomes evident that advanced crossing strategies, such as retrograde crossing or antegrade dissection/reentry are needed for successful crossing but the operator does not have expertise in those techniques.
- there is significant operator or patient fatigue. In such cases referral to a CTO expert center or repeat attempt with a proctor could be considered.
- a serious complication occurs (unless recanalization of the CTO is essential for treating the complication).

8.5.2.11   How to apply the global CTO crossing algorithm

Full application of the global CTO crossing algorithm requires long-term commitment, ongoing training and experience with all types of CTO PCI techniques; thus, it may not be fully applicable to all operators at different stages of their learning curve. It does provide, however, a framework for effective communication between operators that can facilitate acquiring and building a comprehensive CTO PCI skillset. The use of the hybrid[207] and the other algorithms has resulted in procedural success close to 90% at experienced centers[207,234−238] with approximately 3% risk of major complications but success remains much lower in all comer registries[239−241] (Table 7.1).

References

199. Brilakis ES, Grantham JA, Rinfret S, et al. A percutaneous treatment algorithm for crossing coronary chronic total occlusions. *JACC Cardiovasc Interv* 2012;**5**:367−79.
200. Vo MN, McCabe JM, Lombardi WL, Ducas J, Ravandi A, Brilakis ES. Adoption of the hybrid CTO approach by a single non-CTO operator: procedural and clinical outcomes. *J Invasive Cardiol* 2015;**27**:139−44.
201. Christopoulos G, Karmpaliotis D, Alaswad K, et al. Application and outcomes of a hybrid approach to chronic total occlusion percutaneous coronary intervention in a contemporary multicenter US registry. *Int J Cardiol* 2015;**198**:222−8.
202. Michael TT, Mogabgab O, Fuh E, et al. Application of the "hybrid approach" to chronic total occlusion interventions: a detailed procedural analysis. *J Interv Cardiol* 2014;**27**:36−43.
203. Christopoulos G, Menon RV, Karmpaliotis D, et al. Application of the "hybrid approach" to chronic total occlusions in patients with previous coronary artery bypass graft surgery (from a Contemporary Multicenter US registry). *Am J Cardiol* 2014;**113**:1990−4.
204. Christopoulos G, Menon RV, Karmpaliotis D, et al. The efficacy and safety of the "hybrid" approach to coronary chronic total occlusions: insights from a contemporary multicenter US registry and comparison with prior studies. *J Invasive Cardiol* 2014;**26**:427−32.
205. Shammas NW, Shammas GA, Robken J, et al. The learning curve in treating coronary chronic total occlusion early in the experience of an operator at a tertiary medical center: the role of the hybrid approach. *Cardiovasc Revasc Med* 2016;**17**:15−18.
206. Pershad A, Eddin M, Girotra S, Cotugno R, Daniels D, Lombardi W. Validation and incremental value of the hybrid algorithm for CTO PCI. *Catheter Cardiovasc Interv* 2014;**84**:654−9.
207. Tajti P, Karmpaliotis D, Alaswad K, et al. The hybrid approach to chronic total occlusion percutaneous coronary intervention: update from the progress CTO registry. *JACC Cardiovasc Interv* 2018;**11**:1325−35.
208. Harding SA, Wu EB, Lo S, et al. A new algorithm for crossing chronic total occlusions from the asia pacific chronic total occlusion club. *JACC Cardiovasc Interv* 2017;**10**:2135−43.
209. Ge J-B, Ge L, Huo Y, Chen J-Y, Wang W-M. On behalf of chronic total occlusion club. Updated algorithm of chronic total occlusion percutaneous coronary intervention from chronic total occlusion Club China. *Cardiology Plus* 2021;**6**:81−7.
210. Galassi AR, Werner GS, Boukhris M, et al. Percutaneous recanalisation of chronic total occlusions: 2019 consensus document from the EuroCTO Club. *EuroIntervention* 2019;**15**:198−208.
211. Tanaka H, Tsuchikane E, Muramatsu T, et al. A novel algorithm for treating chronic total coronary artery occlusion. *J Am Coll Cardiol* 2019;**74**:2392−404.
212. Wu EB, Brilakis ES, Mashayekhi K, et al. Global chronic total occlusion crossing algorithm: JACC state-of-the-art review. *J Am Coll Cardiol* 2021;**78**:840−53.
213. Hong SJ, Kim BK, Cho I, et al. Effect of coronary CTA on chronic total occlusion percutaneous coronary intervention: a randomized trial. *JACC Cardiovascular Imaging* 2021;**14**:1993−2004.

214. Fujino A, Otsuji S, Hasegawa K, et al. Accuracy of J-CTO score derived from computed tomography vs angiography to predict successful percutaneous coronary intervention. *JACC Cardiovascular Imaging* 2018;**11**:209–17.

215. Ghoshhajra BB, Takx RAP, Stone LL, et al. Real-time fusion of coronary CT angiography with x-ray fluoroscopy during chronic total occlusion PCI. *Eur Radiology* 2017;**27**:2464–73.

216. Xenogiannis I, Jaffer FA, Shah AR, et al. Computed tomography angiography co-registration with real-time fluoroscopy in percutaneous coronary intervention for chronic total occlusions. *EuroIntervention* 2021;**17**:e433–5.

217. Brilakis ES, Mashayekhi K, Tsuchikane E, et al. Guiding principles for chronic total occlusion percutaneous coronary intervention. *Circulation* 2019;**140**:420–33.

218. Christopoulos G, Kandzari DE, Yeh RW, et al. Development and validation of a novel scoring system for predicting technical success of chronic total occlusion percutaneous coronary interventions: the PROGRESS CTO (Prospective Global Registry for the Study of Chronic Total Occlusion Intervention) Score. *JACC Cardiovasc Interv* 2016;**9**:1–9.

219. Vo MN, Karmpaliotis D, Brilakis ES. "Move the cap" technique for ambiguous or impenetrable proximal cap of coronary total occlusion. *Catheter Cardiovasc Interv* 2016;**87**:742–8.

220. Morino Y, Abe M, Morimoto T, et al. Predicting successful guidewire crossing through chronic total occlusion of native coronary lesions within 30 minutes: the J-CTO (Multicenter CTO Registry in Japan) score as a difficulty grading and time assessment tool. *JACC Cardiovasc Interv* 2011;**4**:213–21.

221. Christopoulos G, Wyman RM, Alaswad K, et al. Clinical utility of the japan-chronic total occlusion score in coronary chronic total occlusion interventions: results from a multicenter registry. *Circ Cardiovasc Interv* 2015;**8**:e002171.

222. Wilson WM, Walsh SJ, Yan AT, et al. Hybrid approach improves success of chronic total occlusion angioplasty. *Heart* 2016;**102**:1486–93.

223. Rathore S, Katoh O, Matsuo H, et al. Retrograde percutaneous recanalization of chronic total occlusion of the coronary arteries: procedural outcomes and predictors of success in contemporary practice. *Circ Cardiovasc Interv* 2009;**2**:124–32.

224. Karmpaliotis D, Michael TT, Brilakis ES, et al. Retrograde coronary chronic total occlusion revascularization: procedural and in-hospital outcomes from a multicenter registry in the United States. *JACC Cardiovasc Interv* 2012;**5**:1273–9.

225. Tsuchikane E, Yamane M, Mutoh M, et al. Japanese multicenter registry evaluating the retrograde approach for chronic coronary total occlusion. *Catheter Cardiovasc Interv* 2013;**82**:E654–61.

226. Karmpaliotis D, Karatasakis A, Alaswad K, et al. Outcomes with the use of the retrograde approach for coronary chronic total occlusion interventions in a contemporary Multicenter US Registry. *Circ Cardiovasc Interv* 2016;**9**.

227. Galassi AR, Sianos G, Werner GS, et al. Retrograde recanalization of chronic total occlusions in europe: procedural, in-hospital, and long-term outcomes from the Multicenter ERCTO Registry. *J Am Coll Cardiol* 2015;**65**:2388–400.

228. El Sabbagh A, Patel VG, Jeroudi OM, et al. Angiographic success and procedural complications in patients undergoing retrograde percutaneous coronary chronic total occlusion interventions: a weighted *meta*-analysis of 3,482 patients from 26 studies. *Int J Cardiol* 2014;**174**:243–8.

229. Yamane M, Muto M, Matsubara T, et al. Contemporary retrograde approach for the recanalisation of coronary chronic total occlusion: on behalf of the Japanese Retrograde Summit Group. *EuroIntervention* 2013;**9**:102−9.
230. Mashayekhi K, Behnes M, Akin I, Kaiser T, Neuser H. Novel retrograde approach for percutaneous treatment of chronic total occlusions of the right coronary artery using ipsilateral collateral connections: a European centre experience. *EuroIntervention* 2016;**11**: e1231−6.
231. Werner GS, Coenen A, Tischer KH. Periprocedural ischaemia during recanalisation of chronic total coronary occlusions: the influence of the transcollateral retrograde approach. *EuroIntervention* 2014;**10**:799−805.
232. Lo N, Michael TT, Moin D, et al. Periprocedural myocardial injury in chronic total occlusion percutaneous interventions: a systematic cardiac biomarker evaluation study. *JACC Cardiovasc Interv* 2014;**7**:47−54.
233. Stetler J, Karatasakis A, Christakopoulos GE, et al. Impact of crossing technique on the incidence of periprocedural myocardial infarction during chronic total occlusion percutaneous coronary intervention. *Catheter Cardiovasc Interv* 2016;**88**:1−6.
234. Sapontis J, Salisbury AC, Yeh RW, et al. Early procedural and health status outcomes after chronic total occlusion angioplasty: a report from the OPEN-CTO Registry (outcomes, patient health status, and efficiency in chronic total occlusion hybrid procedures). *JACC Cardiovasc Interv* 2017;**10**:1523−34.
235. Konstantinidis NV, Werner GS, Deftereos S, et al. Temporal trends in chronic total occlusion interventions in Europe. *Circ Cardiovasc Interv* 2018;**11**:e006229.
236. Maeremans J, Walsh S, Knaapen P, et al. The hybrid algorithm for treating chronic total occlusions in Europe: the RECHARGE registry. *J Am Coll Cardiol* 2016;**68**:1958−70.
237. Habara M, Tsuchikane E, Muramatsu T, et al. Comparison of percutaneous coronary intervention for chronic total occlusion outcome according to operator experience from the Japanese retrograde summit registry. *Catheter Cardiovasc Interv* 2016;**87**:1027−35.
238. Quadros A, Belli KC, de Paula JET, et al. Chronic total occlusion percutaneous coronary intervention in Latin America. *Catheter Cardiovasc Interv* 2020;**96**:1046−55.
239. Brilakis ES, Banerjee S, Karmpaliotis D, et al. Procedural outcomes of chronic total occlusion percutaneous coronary intervention: a report from the NCDR (National Cardiovascular Data Registry). *JACC Cardiovasc Interv* 2015;**8**:245−53.
240. Kinnaird T, Gallagher S, Cockburn J, et al. Procedural success and outcomes with increasing use of enabling strategies for chronic total occlusion intervention. *Circ Cardiovasc Interv* 2018;**11**:e006436.
241. Zein R, Seth M, Othman H, et al. Association of operator and hospital experience with procedural success rates and outcomes in patients undergoing percutaneous coronary interventions for chronic total occlusions: insights from the blue cross blue shield of michigan cardiovascular consortium. *Circ Cardiovasc Interv* 2020;**13**:e008863.

Chapter 9

Lesion preparation

9.1 Goal

To adequately prepare the target lesion after successful wire crossing in order to facilitate stent delivery and adequate stent expansion.

9.2 When is lesion preparation needed?

Lesion preparation (in most cases with balloon angioplasty) should be performed in nearly all lesions (especially in CTOs) because it:

1. facilitates stent delivery and decreases the risk of stent loss;
2. helps determine optimal stent diameter and length (especially when no intracoronary imaging is used and when there is poor flow of contrast distal to a severe lesion);
3. informs the operator as to the need of additional lesion modification (for example with atherectomy or intravascular lithotripsy in heavily calcified lesions) to prevent "stent-regret"; and
4. helps achieve optimal stent expansion (ideally full balloon expansion balloon should be confirmed in two orthogonal views before stent placement).

9.3 How to prepare a chronic total occlusion

Before performing lesion preparation, it is essential to confirm that the guidewire has successfully crossed into the distal true lumen. Advancement of balloons, microcatheters, or other devices over a guidewire that has exited the vessel architecture can cause a severe perforation.

Lesion preparation can be performed with:

1. Balloon angioplasty
2. Atherectomy (orbital and rotational) and intravascular lithotripsy for heavily calcified lesions. Use of intravascular imaging (Chapters 13 and 23) can help determine the need for atherectomy. The following four IVUS findings have been associated with stent underexpansion[1]: Superficial calcium angle >270 degrees longer than 5 mm; 360 degrees of superficial calcium; calcified nodule; vessel diameter <3.5 mm. The following 3 OCT findings have been associated with stent

Manual of Chronic Total Occlusion Percutaneous Coronary Interventions.
DOI: https://doi.org/10.1016/B978-0-323-91787-2.00027-7

underexpansion[2]: maximum calcium angle >180 degrees; maximum calcium thickness >0.5 mm; and calcium length >5.0 mm.
3. Thrombectomy (aspiration thrombectomy or with the Penumbra system). Thrombectomy is rarely needed during CTO PCI in case of luminal thrombus formation (for example in cases of donor vessel thrombosis) (CTO Manual online case 6).

9.3.1 Balloon angioplasty

Balloon angioplasty is the most commonly used lesion preparation technique for both CTO and non-CTO lesions and is performed in 13 steps, as described in detail in the *Manual of Percutaneous Coronary Interventions: a step-by-step approach*[3]:

1. Confirm that the guidewire is optimally positioned through the target lesion.
2. Confirm that the guide catheter is aspirated and flushed.
3. Select balloon type and size.
4. Prepare the balloon.
5. Load the balloon on the guidewire.
6. Advance the balloon monorail segment through the Y-connector.
7. Advance the balloon to the tip of the guide catheter.
8. Advance the balloon to the target lesion.
9. Inflate the balloon.
10. Deflate the balloon.
11. Withdraw the balloon into the guide catheter.
12. Remove balloon from the guide catheter.
13. Check the balloon angioplasty result.

9.3.1.1 Step 1. Confirm that the guidewire is optimally positioned through the target lesion

Goal: To ensure that the guidewire is optimally positioned (through the target lesion and advanced several cm distally, but not into small distal branches that could lead to perforation).

How: In most CTO PCIs, confirmation of guidewire position is through contralateral contrast injection as described in Section 8.1. Visualization of the CTO vessel with antegrade injections should be avoided in most cases, especially when dissection/reentry strategies are used, in order to minimize formation (or expansion) of an extraplaque hematoma. Intravascular ultrasound can also help confirm intraluminal distal wire position, but could aggravate a perforation if the guidewire has exited the adventitia/vessel structure.

The CTO crossing guidewire should be exchanged for a workhorse guidewire prior to balloon and stent delivery to minimize the risk of distal vessel dissection or perforation from excessive distal guidewire movement. The guidewire tip should be placed into the main vessel and not into a small distal branch.

What can go wrong?

If the guidewire is not placed optimally, a distal vessel perforation can occur during attempts to deliver balloons (and other equipment). Treatment of perforations is discussed in Chapter 26.

9.3.1.2   Step 2. Confirm that the guide catheter is aspirated and flushed

Goal: To ensure that there is no air or debris inside the guide catheter.
How:

1. Back bleed the Y-connector, or
2. Aspirate the guide catheter, and then fill with contrast

What can go wrong?

1. **Distal embolization** (Section 25.2.3)
   Causes:
   1. Air sucked into the guide catheter when using the trapping technique. Use of trapping is very common after crossing in CTO PCI: back-bleeding and clearing the guide catheter is critically important after using the trapping technique.
   2. Air sucked into the guide catheter during balloon withdrawal. This is more likely when withdrawing large diameter balloons but could also occur with withdrawal of other equipment, such as intravascular imaging catheters.
   3. Thrombus or plaque entry into the guide catheter while withdrawing a balloon (may be more likely with bulky, large diameter balloons, or plaque modification balloons, such as the Angiosculpt, Chocolate, Scoreflex, and cutting balloon, Section 30.9.3).
   Prevention:
   1. Guide catheter aspiration followed by flushing, as described above.
   Treatment:
   1. Air embolization is treated with administration of 100% oxygen, possibly aspiration, and administration of epinephrine in case of cardiac arrest. Air embolization usually resolves with supportive measures without needing additional intervention. This is described in detail in Section 25.2.3.3.
   2. Embolization of plaque or thrombus is usually treated with thrombectomy.

9.3.1.3   Step 3. Select balloon type and size

Goal: To choose optimal balloon size and type.
How:
Balloon diameter:

For achieving lesion expansion: For most lesions, the balloon diameter is chosen to match the distal vessel diameter (1:1 ratio). Use of intravascular imaging (Chapter 13) is helpful for determining the vessel size during CTO PCI, as antegrade contrast injections are usually avoided after CTO crossing, especially when using dissection/reentry techniques. Full expansion of an appropriately sized balloon, matched to vessel size with use of intracoronary imaging suggests adequate lesion preparation and ability to safely proceed with stenting. This is particularly important for heavily calcified and restenotic lesions. However, for such lesions, it is even better if adequate modification of calcific plaques can also be visualized on intracoronary imaging following ballooning (i.e., calcium fracture), prior to proceeding with stent deployment.

For crossing: For very tight lesions that are hard to cross, small balloons (≤1.5 mm, Section 30.9.2) are initially used to modify the lesion entry point and allow subsequent delivery of larger balloons to modify the lesion before stent placement.

Balloon length: balloon length should be shorter than the estimated lesion length (to avoid injury to neighboring coronary segments that would then require coverage increasing total stent length).

Balloon compliance: noncompliant balloons are preferred because: (1) their evenly distributed expansion mode decreases the risk of coronary perforation during aggressive predilatation; and (2) they can also be used for postdilatation of the target lesion after stenting. However, they are less deliverable than compliant balloons.

Balloon delivery system: there are two balloon delivery systems, monorail and over-the-wire. Monorail balloons should be used in the vast majority of cases, as they are simpler to use, do not require use of long guidewires, and may be easier to deliver because they have a stiffer shaft.

Balloon type: Some balloons have nitinol wires (such as the Angiosculpt, Chocolate, and Scoreflex) or cutting blades (cutting balloon) aiming to modify the lesion and facilitate expansion (these balloons are called "plaque modification" or "scoring" balloons——Section 30.9.3). Plaque modification balloons are harder to deliver (due to larger profile and less flexibility). The cutting balloon also requires slow inflation and deflation. Another potential indication for the cutting balloon is for treating false lumen hematomas that compress the distal true lumen.[4] Plaque modification balloons can in some cases facilitate expansion of "balloon undilatable" lesions (Section 23.2). The very high-pressure SIS OPN balloon can also be very useful for dilating resistant lesions.

The cutting balloon may also be useful for treating distal false lumen hematomas that compress the distal true lumen.[4]

9.3.1.4   Step 4. Prepare the balloon

9.3.1.5   Step 5. Load the balloon on the guidewire

9.3.1.6   Step 6. Advance the balloon monorail segment through the Y-connector

9.3.1.7   Step 7. Advance the balloon to the tip of the guide catheter

These steps are performed as per standard practice as described in detail in the *Percutaneous Coronary Interventions: a step-by-step approach*.[3]

9.3.1.8 Step 8. Advance the balloon to the target lesion

Goal: To advance the balloon to the target lesion
How:

1. The balloon is advanced towards the lesion using the "independent-hand technique" or the "two-hand technique" (Fig. 9.1).

   Independent hand technique – The right hand holds the Y-connector with the middle and ring finger, and the balloon shaft with the thumb and index finger. The guidewire is placed between the ring and little "pinky" finger of the right hand. This allows the operator to push the balloon with one hand (the right hand) while holding and simultaneously advancing the guide catheter with the left hand to increase support.

   Two-hand technique – The left hand holds the Y-connector and the guidewire and the right hand advances the balloon shaft.
2. The position of the balloon is confirmed using X-ray landmarks (such as clips, previous stents, etc.) and/or contrast injection. **Antegrade contrast injection should be avoided when using dissection/reentry techniques, but contrast injection through the retrograde guide catheter can help clarify equipment position**. Sometimes, temporary proximal occlusion with another balloon or guide catheter extension may be necessary if there is brisk antegrade flow that blocks retrograde contrast filling. Small diameter ($\leq$1.5 mm) and short (6 mm) balloons have one marker in the middle;

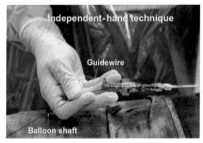 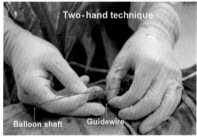

FIGURE 9.1 **The independent hand and the two-hand technique for advancing a balloon.** *Reproduced with permission from Brilakis ES. Manual of Percutaneous Coronary Interventions: a step-by-step approach: Elsevier, 2021.*

hence, they cannot be used to assess the length of the lesion. Larger balloons (≥ 2.0 mm) have two markers, and can thus be used to determine the length of the lesion, ideally along with intracoronary imaging.

X-ray based stent visualization technologies, such as StentBoost, StentViz, and ClearStent, can facilitate exact balloon placement inside previously placed stents.

Challenges:

1. **Balloon unreachable lesions** (balloon shaft is not long enough)[5]
   Causes:
   1. Very distal lesion location, especially in patients with prior coronary bypass grafts and lesions distal to bypass graft anastomoses.
   Prevention:
   1. Use short or shortened (online video: "How to shorten a guide catheter.") guide catheters when treating very distal coronary lesions.
   2. Use balloons with the longest available balloon shaft.
   Treatment:
   1. Use short (or shortened) guide catheters when treating very distal coronary lesions.
   2. Use balloons with the longest available balloon shaft.

2. **Difficult balloon advancement due to proximal vessel disease/tortuosity or balloon uncrossable lesions**
   Causes:
   1. Significant coronary lesions proximal to the target lesion.
   2. Severe proximal vessel or lesion tortuosity.
   3. Severe proximal vessel or lesion calcification.
   4. Severe target lesion diameter stenosis (especially when combined with severe tortuosity and/or calcification).
   5. Use of high-profile and/or noncompliant balloons.
   6. Use of already deployed balloons.

Prevention and treatment (Fig. 9.2)

9.3.1.8.1 Guide catheter

Strong guide catheter support is critical for balloon delivery, especially in complex lesions.

Larger guide catheters (such as 7 or 8 French) provide stronger support than smaller ones (5 or 6 French) and are routinely used in CTO PCI.

Supportive shapes (for example EBU guides provide stronger support than JL4 guides for left coronary interventions, and AL guide catheters provide stronger support than JR4 guide catheters for RCA interventions).

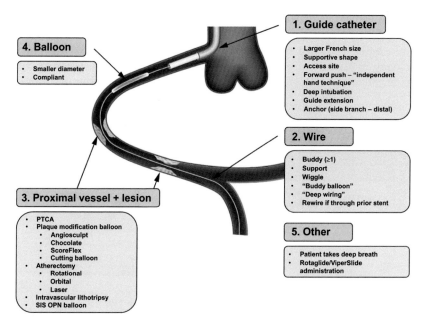

1. Guide catheter
- Larger French size
- Supportive shape
- Access site
- Forward push – "independent hand technique"
- Deep intubation
- Guide extension
- Anchor (side branch – distal)

4. Balloon
- Smaller diameter
- Compliant

2. Wire
- Buddy (≥1)
- Support
- Wiggle
- "Buddy balloon"
- "Deep wiring"
- Rewire if through prior stent

3. Proximal vessel + lesion
- PTCA
- Plaque modification balloon
  - Angiosculpt
  - Chocolate
  - ScoreFlex
  - Cutting balloon
- Atherectomy
  - Rotational
  - Orbital
  - Laser
- Intravascular lithotripsy
- SIS OPN balloon

5. Other
- Patient takes deep breath
- Rotaglide/ViperSlide administration

FIGURE 9.2 **How to achieve balloon delivery to the target lesion**. *Reproduced with permission from Brilakis ES. Manual of Percutaneous Coronary Interventions: a step-by-step approach: Elsevier, 2021.*

Access site: in some patients femoral access provides stronger support than radial access. Long sheaths increase guide catheters support. Left radial access provides stronger support for PCI of the right coronary artery while right radial access provides stronger support for PCI of the left coronary artery.

Forward push of the guide catheter can increase support, but may also in some cases lead to guide prolapse and disengagement. The "**independent hand" technique** (the left hand pushes the guide, while the right hand both fixes the guidewire and advances the balloon, Fig. 9.1) allows a single operator to simultaneously manipulate the guide catheter and advance the stent.

Deep intubation of the guide catheter increases support, but at the same time may cause dissection. Deep intubation is usually done in the RCA or in SVGs and rarely in the left coronary artery.

Guide catheter extensions (Section 30.2) are some of the most commonly used tools for facilitating stent delivery. They not only increase guide catheter support but can also modify the proximal vessel to facilitate stent delivery. They can be advanced distally close to (or through) the target lesion allowing the stent to be delivered through the guide extension without coming in contact with the vessel wall, followed by "unsheathing" the stent by retraction of the guide catheter extension.

The "power guide extension" (also called "tunnel in landslide technique" — TILT[6]) can enhance the support provided by a guide extension by inflating a balloon next to the guide catheter extension to facilitate equipment delivery in severely tortuous or uncrossable lesions (Fig. 9.3). Using an 8-Fr guide catheter, a 5-Fr guide extension is advanced over the wire that crosses

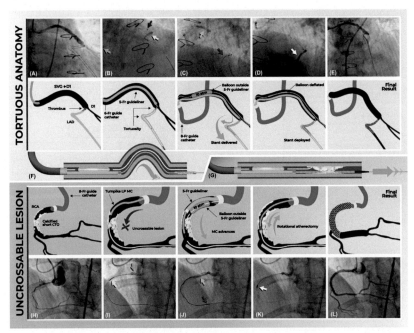

FIGURE 9.3   **Tunnel in landslide technique (TILT)**. Case 1 (*top panels*). **Panel A**: thrombotic occlusion of D1 (grafted by an SVG), with no retrograde flow into the LAD (broken lines). **Panel B**: a balloon (*yellow arrow*) is delivered through a guide catheter extension (GuideLiner, red arrow), but the 6 Fr guide catheter (*green arrow*) fails to provide enough support for stent delivery. **Panel C**: the guide is exchanged for an 8 Fr AL1, the lesion is rewired with two workhorse wires, and a 5 Fr GuideLiner is advanced into D1 to deliver the stent (*pink arrow*), which is facilitated by pinning of the guide extension by a 3.5 mm balloon (*cyan arrow*), advanced over the second guidewire and inflated between the GuideLiner and the graft wall. **Panel D**: stent delivery and implantation are successful (*white arrow*). **Panel E**: final result. **Panel F**: use of TILT in tortuous anatomies.
Case 2 (*bottom panels*). **Panel G**: use of TILT for uncrossable lesions. **Panel H**: short mid-RCA CTO. **Panel I**: after antegrade crossing, the microcatheter (*yellow arrow*) cannot cross the occlusion and the 8 Fr AL1 guide disengages (*green arrow*). **Panel J**: a 5 Fr GuideLiner is advanced to the proximal cap and pinned in place by a 3.0 mm balloon (*cyan arrow*), advanced over a workhorse guidewire and inflated between the GuideLiner and the vessel wall. This provides sufficient support to advance the microcatheter across the lesion to the distal vessel (*pink arrow*). **Panel K**: rotational atherectomy (*white arrow*) is performed. **Panel L**: final result, after implantation of two stents. *Reproduced with permission from Santiago R, Moroni F, Del Rio V, Rodriguez-Escudero J, Azzalini L. The guide extension tunnel in landslide technique (TILT) for equipment delivery in severely tortuous or uncrossable lesions during percutaneous coronary intervention. EuroIntervention 2021;17:e923−e924.*

the lesion. A second wire, advanced to the proximal vessel outside the guide extension, is used to deliver a balloon that is inflated to pin the guide extension against the vessel wall. The device (microcatheter, small balloon) is then advanced over the first guidewire, via the guide catheter extension, across the lesion, thanks to the enhanced support. This technique can also be performed with any guide catheter and guide catheter extension size using the ping-pong guide catheter technique.[6]

Anchor techniques (side branch anchor and distal anchor) can also provide strong support for equipment delivery. The most commonly used is the **side-branch anchoring** technique (Fig. 9.4), in which a balloon is inflated in a proximal side branch of the target vessel.[7–16] Using a long balloon increases support by increasing the "grip" on the vessel. The grip can also be increased by wiping the hydrophilic coating off the balloon before use.

The **distal anchor** technique (Fig. 9.5) is similar to the side-branch anchor technique, except that the balloon is inflated distal to or at the occlusion site within the target artery.[3,17,18] Two guidewires are required for the distal anchor technique, one to deliver the anchor balloon and a second guidewire (which is pinned by the anchor balloon against the vessel wall), for delivering equipment, such as microcatheters, balloons, stents and guide catheter extensions to the lesion.[10,12,19–21] The distal anchor technique requires at least a 6 French guide catheter, but is easier to perform with a 7 or 8 French guide catheter.

The **buddy wire stent anchor** technique (Fig. 9.6) can be used if the proximal vessel needs stenting: a buddy wire can be inserted and a stent deployed over this guidewire, effectively "trapping" the buddy wire, which then provides strong guide catheter support. The jailed wire should be removed prior to deploying more stents in the vessel to avoid wire entrapment.

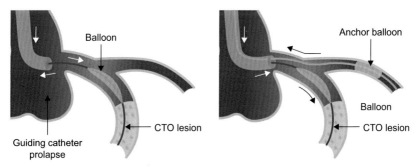

FIGURE 9.4 **Example of side branch anchor to facilitate delivery of a balloon across a chronic total occlusion.**[3] *Reproduced with permission from Brilakis ES.* Manual of coronary chronic total occlusion interventions. a step-by-step approach. *2nd ed. Elsevier; 2017.*

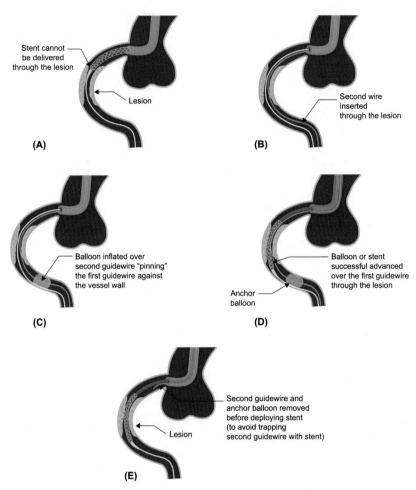

FIGURE 9.5 Illustration of the distal anchor technique. The distal anchor technique can be useful when difficulty is encountered delivering a balloon or stent (or other equipment) through a lesion (**panel A**). A second guidewire is inserted next to the initial guidewire (**panel B**) and a balloon (which is usually easier to deliver than a stent) is delivered distal to the lesion and inflated, "pinning" the initial guidewire against the vessel wall (**panel C**) and enabling stent delivery over the first guidewire (**panel D**). The second guidewire (and balloon) are subsequently withdrawn before stent deployment (**panel E**). *Reproduced with permission from the Brilakis ES. Manual of coronary chronic total occlusion interventions. a step-by-step approach. 2nd ed. Elsevier; 2017.*

Limitations of the anchor techniques:

1. Injury at the site of the anchor balloon inflation, which is usually inconsequential in these small side branches.[17] The risk can be minimized by sizing the balloon 1:1 to the side branch and inflating the anchor balloon at relatively low pressures (4—8 atm).

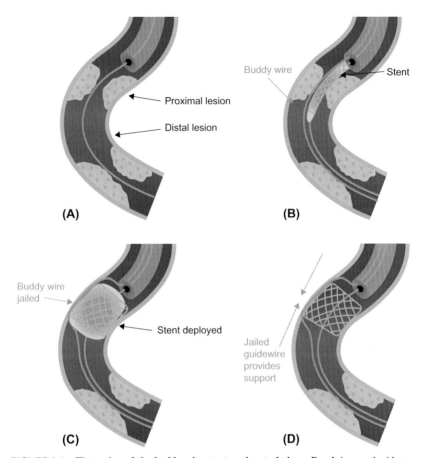

FIGURE 9.6   Illustration of the buddy wire stent anchor technique. **Panel A**: vessel with two coronary lesions, one proximal and one distal. **Panel B**: after inserting a buddy wire a stent is positioned across the proximal lesion. **Panel C**: the stent is deployed "jailing" the buddy wire. **Panel D**: the jailed buddy wire provides extra guide support. *Reproduced with permission from Brilakis ES. Manual of Percutaneous Coronary Interventions: a step-by-step approach: Elsevier, 2021.*

2. Distal dissection can occur with the distal anchor technique, requiring extensive stenting.
3. Larger guide catheters (at least 7 Fr) may be needed for delivering a distal anchor balloon and other equipment.[12,20,21]
4. Use of a side branch anchor may modify the proximal vessel anatomy and hinder antegrade wiring.
5. Rarely, ischemia can occur if large branches (such as diagonals) are used for anchoring. Some patients experience arrhythmias and/or hemodynamic instability with prolonged balloon occlusion in the conus branch of the proximal right coronary artery when used for anchoring. Allowing intermittent deflation of the anchoring balloon may be necessary to relieve ischemia in such cases.

9.3.1.8.2 Wire

Buddy wire(s): One of the simplest ways to facilitate delivery is to use one (or more) buddy wire(s). The buddy wire increases guide catheter support and also straightens the target vessel.

Support wires: Another option is to use support guidewires, such as the Iron Man, Grand Slam, Mailman, or the **Wiggle** wire (Section 30.6.4). Partial withdrawal of the wire while advancing the stent may facilitate advancement.

Buddy balloon: Another option is to **advance a balloon** across the area of resistance to stent advancement, in order to "deflect" the stent during advancement.

If there is resistance to balloon delivery at the location of a previously deployed stent, it is possible that the wire has crossed under a stent strut. **Rewiring through the stent**, ideally after forming a knuckle more proximally (which is easier to do with polymer-jacketed wires), can solve this problem.

Guidewire externalization during retrograde CTO PCI can provide strong support for equipment delivery and also prevents loss of guide catheter position.

Deep wire crossing technique: In the "**deep wiring**" technique a polymer-jacketed guidewire is advanced through the lesion then into the ventricle through arterioluminal communications, which increases wire support.[22] The obvious risk of this technique, is coronary perforation. Deep wiring is used infrequently.

How?

The operator gently advances a polymer-jacketed guidewire through the selected distal coronary artery branches relying on the feelings under fluoroscopic control without contrast, similar to the "surfing technique" for septal collateral crossing during retrograde recanalization of chronic total occlusions.

What can go wrong?

- Unable to advance a guidewire through the selected branch into the ventricular cavity. In this case, the operator must carefully pull the wire back and redirect it to another channel.
- Distal perforation. This most likely happens when the operator continues manipulations with the wire when it is meeting resistance.
- Ventricular extrasystoles or tachyarrhythmias. They usually resolve spontaneously without requiring additional treatment.

9.3.1.8.3 Proximal vessel preparation

If the balloon cannot reach the target lesion due to proximal disease (especially in the setting of tortuosity and calcification), preparation of the proximal vessel (with balloon angioplasty, atherectomy, or intravascular lithotripsy) can facilitate balloon delivery.

9.3.1.8.4 Balloon

Small diameter balloons (such as the 1.0 mm Sapphire Pro in the US and 0.75−1.0 mm balloons outside the US), the Threader (Section 30.9.2), or other 1.20−1.50 balloons are more likely to cross highly resistant lesions. The marker is usually located in the middle of the balloon for ≤ 1.5 mm balloons; hence, longer length is preferred.

Compliant balloons are more flexible, have lower profile and are thus more deliverable compared with noncompliant balloons.

Used balloons are less deliverable and have reduced ability to cross the target lesion. Changing for a **new** balloon may allow reaching and crossing the target lesion.

9.3.1.8.5 Other

Asking the patient to **take a deep breath** could facilitate stent delivery, possibly by straightening the coronary arteries or changing the seating of the guide catheter.

Rotaglide or ViperSlide administration during stent advancement might reduce friction and facilitate delivery

The algorithm for approaching "balloon uncrossable" lesions is discussed in Section 23.1.

What can go wrong?

1. **Wire and guide catheter position loss**
   Causes:
   1. Forceful balloon advancement through areas of resistance.
   Prevention:
   1. Avoid forceful balloon advancement, if possible.
   2. Always monitor the guide catheter and the distal wire position and adjust the effort to advance the balloon accordingly.
   3. Optimal guide catheter selection and coaxial engagement to optimize support.
   4. Use an externalized guidewire for equipment delivery when performing retrograde CTO PCI.
   Treatment:
   1. Re-engage with the guide catheter and rewire the lesion or advance the guidewire.

2. **Loss of distal vessel visualization with antegrade contrast injections**
   Causes:
   1. Balloon advancement through a tight lesion
   2. Dissection

3. Distal embolization
   Prevention:
   1. Avoid use of oversized balloons to reduce the risk of dissection.
   Treatment:
   1 If the reason is balloon advancement through a tight lesion, and if there is no contraindication to injection, such as dissection planes, withdraw the balloon, inject contrast until it reaches distal to the lesion, and advance the balloon; this causes contrast to be entrapped distal to the lesion and allows visualization of the distal vessel. Alternatively, contrast injection in the donor vessel can be performed.
   2 If the reason is dissection, stenting should be performed (Section 25.2.1.)
   3 If the reason is distal embolization, vasodilators should be administered and other measures such as aspiration, may be necessary.

9.3.1.9 Step 9. Inflate balloon

9.3.1.10 Step 10. Deflate the balloon

9.3.1.11 Step 11. Withdraw balloon into the guide catheter

9.3.1.12 Step 12. Remove balloon from the guide catheter

9.3.1.13 Step 13. Check the balloon angioplasty result

These steps are performed as per standard practice, as described in detail in the *Manual of Percutaneous Coronary Interventions: a step-by-step approach.*[3]

9.3.2 Atherectomy

9.3.2.1 Atherectomy: indications and modality selection

Orbital and rotational atherectomy are used infrequently during CTO PCI (in approximately 3%–4% of cases).[23–25] Atherectomy can be used to facilitate crossing of balloon uncrossable lesions or to facilitate expansion of balloon undilatable lesions. Atherectomy can be performed before balloon angioplasty (upfront or primary atherectomy) or if balloon angioplasty fails to achieve lesion expansion (secondary atherectomy).

Atherectomy is usually avoided with extraplaque CTO crossing, but has been successfully used in this setting in refractory cases.[26]

Atherectomy is usually avoided through areas of dissection, although rotational atherectomy has been performed in this setting to treat chronic total occlusions and balloon undilatable lesions.[27,28] Orbital atherectomy should NOT be done within recently placed stents due to risk of entrapment, but has been successfully used in stents that have endothelialized.[29] Orbital atherectomy should also not be performed in areas of dissection. Rotational atherectomy is often preferred to orbital atherectomy in balloon uncrossable lesions due to the forward versus side mechanism of ablation, although orbital atherectomy has also been successfully used in this setting.[30] Orbital atherectomy is often preferred for treating large vessels.

Rotational atherectomy can be performed after retrograde crossing in uncrossable lesions on an externalized Rotawire. The distal coil of the Rotawire must be cut off with scissors after externalization, otherwise the burr cannot be loaded on the wire.

Another option for performing rotational atherectomy in case of retrograde crossing is externalization of a dedicated wire (RG3 or R350) followed by insertion of an antegrade microcatheter into touching position with the retrograde microcatheter. The externalized wire is then replaced by an antegradely inserted Rotawire Drive (rendezvous technique).

9.3.2.2 Preparation for atherectomy

Two key questions need to be answered before proceeding with atherectomy (orbital or rotational):

1. **Need for temporary pacemaker?**

   A temporary pacemaker is not routinely needed for atherectomy of the left anterior descending artery (LAD) or a nondominant circumflex artery.

   For RCA or dominant circumflex: can use temporary pacemaker, or aminophylline (250−300 mg IV over 10 min or 20−40 mg intracoronary bolus[31,32]), or do brief test runs to determine whether the patient develops bradycardia. The advantage of the latter approach (not using a pacemaker or aminophylline) is that atherectomy runs can be tailored (atherectomy stops when the patient develops bradycardia) and there is no risk of right ventricular perforation from the temporary pacemaker lead. To treat acutely symptomatic bradycardia after an atherectomy run without a temporary pacemaker, intravenous atropine boluses can be administered as needed (usually in 0.5 mg increments, up to a maximum dose of 3 mg) (Section 3.7.2). Orbital atherectomy may be less likely to cause bradycardia than rotational atherectomy.

   Bigger burrs and longer runs are more likely to cause bradycardia, as compared with smaller burrs and shorter runs.

2. **Need for hemodynamic support?**

   Atherectomy may lead to complications; hence, prophylactic hemodynamic support should be considered, especially in patients with poor baseline hemodynamics, low ejection fraction, large area of myocardium at risk, severe concomitant valve disease and severe pulmonary hypertension.

9.3.2.3 Rotational atherectomy

Rotational atherectomy (Section 30.9.1) is performed in 11 steps, as follows:

9.3.2.3.1   Step 1: Select burr size

There are multiple burr sizes, ranging from 1.25 to 2.5 mm (Section 30.10.1.1). The 1.5 mm burr is most often used, as it may be less likely to get entrapped as compared with the 1.25 mm burr. Larger burrs (1.75 mm or larger) are infrequently used except for left main lesions, as the goal is to modify the vessel enough to allow balloon expansion rather than full calcium debulking. A burr to artery diameter ratio of 0.5 (<0.6 is recommended by the manufacturer) is usually selected. The minimum guide size required for each burr size is shown in Table 9.1.

The available space within the guide catheter for device advancement may be smaller if the guide catheter is kinked or if it is advanced through highly tortuous vessels (such as highly tortuous subclavian arteries).

9.3.2.3.2   Step 2: engage the target vessel

• Engagement is performed as described in Chapter 5.
• Supportive guide catheters should be used, ensuring coaxial alignment. However, advancement of the Rotablator burr can be challenging through guide catheters with significant bends, such as the Amplatz Left.
• Avoid pressure dampening, as it increases the risk of no-reflow or slow flow.
• Ensure coaxial guide catheter orientation with the target vessel.

9.3.2.3.3   Step 3: wiring

• Rotational atherectomy should only be performed using a dedicated Rotawire Drive guidewire. There are 2 Rotawire Drive guidewire types, the Rotawire Drive Floppy and the Rotawire Drive Extra Support (Section 30.10.1.1.3). The Rotawire Drive floppy is used in most cases with the Rotawire Drive extra support reserved for ostial lesions.
• Although the Rotawire Drive can be used for primary wiring of the target lesion, we recommend wiring the target lesion using a microcatheter and

TABLE 9.1 Compatibility of various size Rotablator burrs with various guide sizes.

Burr size (mm)	Guide size (Fr)
1.25−1.75	≥6
2.0	≥7
2.15−2.25	≥8

workhorse guidewire and then exchanging for the Rotawire Drive over the microcatheter, as the Drive can be difficult to steer and kinks easily.

- The Rotawire Drive guidewire can be torqued using the WireClip torquer (Section 30.10.1.4). It can also prevent the guidewire from spinning during rotational atherectomy if the break is accidentally released by the break defeater.
- If the rotational atherectomy guidewire becomes kinked, it should be replaced for a new one. Atherectomy should not be performed over a kinked guidewire due to increased risk of losing wire position and wire fracture during burr movement.
- The Rotawire extra support should be used for preferential cutting on the outer curvature of the vessel and the Rotawire floppy for nonpreferential cutting.

9.3.2.3.4 Step 4: vasodilator administration

Consider vasodilator administration prior to atherectomy in most patients (unless patient is hypotensive). Nicardipine is most commonly used, but nitroprusside, verapamil, or adenosine can be used as well. A cocktail of verapamil, nitroglycerin, and heparin is often used during atherectomy along with the Rotaglide solution that contains olive oil, egg yolk, phospholipids, sodium deoxycholate, L-histidine, disodium EDTA, sodium hydroxide, and water.

9.3.2.3.5 Step 5: prepare rotational atherectomy device for insertion into the guide catheter

The following checklist (DRAW) should be confirmed prior to insertion of the Rotablator burr into the guide catheter:

D: ensure there is sufficient **D**rip of the flush solution from the tip of the burr.

R: Activate the device outside the body (platforming) and confirm that it rotates at the desired speed (usually 140,000−160,000 **R**pm) (ensure that the burr is not touching a towel or gauze, as this can result in fibers wrapping around the burr).

A: **A**dvance and withdraw the burr using the advancer knob to ensure that the burr is moving smoothly.

W: a gentle pull should be done on the **W**ire to ensure that the brake is active.

Before inserting into the guide catheter the advancer knob is locked approximately 2 cm from the distal end.

9.3.2.3.6 Step 6: deliver burr proximal to lesion

The Rotablator burr can be advanced under fluoroscopy while fixing the Rotawire Drive, or utilizing the Dynaglide mode (which activates the burr at a reduced speed to decrease friction onto the wire) with the wire and wireclip in the break release. This is traditionally performed by two operators, but can also be done by a single operator.[33] Alternatively, if the guide size is big enough, trapping can be used.

The Rotablator burr is advanced $1-2$ cm proximal to the lesion. The advancer knob is then moved back to release any tension in the system. This is important for preventing the burr from jumping forward into the lesion upon initial activation.

9.3.2.3.7 Step 7: perform rotational atherectomy

1. Confirm that the flush solution drip is on.
2. Confirm that the WireClip torquer is on the back end of the wire.
3. Activate the burr proximal to the lesion at the desired rotational speed.
4. Advance the burr using a pecking motion, avoiding decelerations >5000 rpm. Visual, tactile and auditory clues can minimize the risk of decelerations. This reduces the risk of burr entrapment (Section 27.5), and also reduces the size and volume of particles released during atherectomy, and as a result the risk for slow flow/no-reflow.
5. Duration of runs: aim to keep them short, ideally <15 s.
6. Wait for 30 s—1 min between runs.
7. Monitor ECG and pressure during rotablation to detect bradycardia and ST changes due to coronary flow disturbance.
8. Contrast injections can be performed if complications are suspected.
9. Additional rotablation runs are performed until the burr passes easily through the lesion without resistance or decelerations ("polishing" runs). Rotablation should be stopped if a complication occurs, such as slow flow, dissection, or ECG changes, or if the rotablation time exceeds 5 min.

9.3.2.3.8 Step 8: remove burr

The burr can be removed using either the Dynaglide function or the trapping technique (if the guide is big enough to allow insertion of a trapping balloon next to the Rotablator burr).

Option 1: using Dynaglide

1. Dynaglide is activated by pressing the Dynaglide button on the Rotapro Advancer.
2. Confirm that the WireClip torquer is on the back end of the Rotawire.
3. The Rotablator is activated to rotate at $60,000-90,000$ rpm (Dynaglide mode).
4. The brake defeat button is pressed and the burr is retracted while constantly monitoring the distal position of the Rotawire Drive. This is

typically done by two operators, but can also be done by a single operator (by inserting the WireClip torquer in the docking port of the advancer, which keeps the brake defeat engaged).

Option 2: using the trapping technique

1. The burr is withdrawn inside the guide catheter.
2. A balloon is advanced inside the guide catheter distal to the distal end of the burr. Insertion of a trapping balloon is not necessary if a Trapliner is used, but the guide extension decreases the inner lumen diameter, hence compatibility of the Trapliner with the selected burr should be confirmed prior to using it: a 6 Fr guide extension can accommodate a 1.25 mm burr, a 7 Fr guide extension a 1.5 mm burr, and an 8 Fr guide extension a 1.75 mm burr (Figure 30.19).
3. The balloon is inflated "trapping" the Rotawire.
4. The burr is removed (fluoroscopy is only needed during the beginning of burr withdrawal).

9.3.2.3.9   Step 9: check angiographic result

Coronary angiography is performed to identify slow flow/no reflow, dissections, embolization, side branch loss, perforation, or other complications.

9.3.2.3.10   Step 10: replace Rotawire Drive with a standard workhorse guidewire

The Rotawire Drive is exchanged for a standard workhorse guidewire through a microcatheter or over-the-wire balloon, usually using the trapping technique if inserting a short (190−200 cm) guidewire. Alternatively, a workhorse guidewire can be advanced next to the Rotawire Drive in a "parallel wire" fashion if there are no dissections. PCI can be performed over the Rotawire Drive (especially if a complication, such as dissection or perforation, occurs). However, workhorse guidewires are preferred, as they are more torquable, less likely to cause distal vessel injury, and easier to work with compared with the thinner and more likely to kink Rotawire Drive.

9.3.2.3.11   Step 11: Balloon angioplasty

Balloon angioplasty with a 1:1 sized noncompliant balloon should always be performed after atherectomy to ensure that the balloon expands fully. If it does not, repeat atherectomy (often with a larger burr) or other treatments (such as use of plaque modification balloons, intravascular lithotripsy, or the SIS OPN balloon) may be required.

Complications and troubleshooting of rotational atherectomy are discussed in the *Manual of Percutaneous Coronary Interventions: a step-by-step approach*.[3] Rotablator burr entrapment is discussed in Section 27.5.

9.3.2.4   Orbital atherectomy

Orbital atherectomy is performed in 10 steps, as described in detail in the *Manual of Percutaneous Coronary Interventions: a step-by-step approach*[3]:

9.3.2.4.1   Step 1: engage the target vessel

Engagement is performed as described in Chapter 5.

9.3.2.4.2   Step 2: wiring

Orbital atherectomy should only be performed using a dedicated Viperwire (Section 30.10.2.3). The nitinol Viper Advance Flex Tip guidewire (currently used in most orbital atherectomy cases) can be used for primary lesion wiring. The Viper Advance is stiffer and more difficult to steer; hence, we recommend wiring the target lesion using a microcatheter and workhorse guidewire and then exchanging for the Viper Advance over the microcatheter.

9.3.2.4.3   Step 3: vasodilator administration

Consider vasodilator administration prior to atherectomy in most patients (unless the patient is hypotensive). Nicardipine is most commonly used, but nitroprusside, verapamil, or adenosine can be used as well.

9.3.2.4.4   Step 4: test device outside the body

The orbital atherectomy device is connected to the pump and threaded through the back end of the Viper Advance Flex Tip or the Viper Advance guidewire. Before inserting the device through the Y-connector, the following are checked:

- The advancer knob can move easily back and forth.
- ViperSlide is dripping from the tip of the device.
- Test the rotation of the device (brake is pushed down and the on/off button located on top of the crown advancer is pressed and released). During spinning of the device the flow of Viperslide should increase. **Caution**: the crown should be held in the air as it will easily get entangled in towels.
- The crown advancer knob is locked at 1 cm from the fully proximal position by rotating the crown advancer knob to the parallel position.

9.3.2.4.5   Step 5: advance crown proximal to the lesion

The orbital atherectomy crown can be advanced under fluoroscopy while fixing the ViperWire Advance. Alternatively, if the guide size is big enough (Table 9.2) trapping can be used.

Advancement can be done using trapping when 7 French and 8 French guide catheters are used. Trapping can also be done if an 8 French Trapliner is used.

Keeping the knob in the parallel position and the brake up, advance the crown approximately 3−5 mm proximal to the lesion.

TABLE 9.2 Ability to trap the Diamondback crown with the Trapper balloon in various guide catheters.

Guide size	
6 French	No
6 French + Guide extension	No
7 French	Yes
7 French + Guide extension	No
8 French	Yes
8 French + Guide extension	No

9.3.2.4.6   Step 6: perform orbital atherectomy[34]

1. Push brake down. The crown will not spin if the guidewire brake is not locked (except in the GlideAssist mode, which is why the Viperwire should be held (ideally with a torquer) if spinning without the brake on in the GlideAssist mode).

2. Speed: slow speed (80,000 rpm) should be the initial treatment speed in all cases, as it carries lower risk of dissection, perforation or other complications as compared with the high speed (120,000 rpm). Nearly all procedures can be completed on low speed (80,000 rpm). Consider using high speed if the vessel is >3.5 mm in diameter or if the postdilatation balloon does not fully expand. High speed should be avoided in arteries <3.0 mm in diameter.

3. Unlock the crown advancer knob (turn to the perpendicular position) and retract to the fully proximal position to relieve any tension/torque in the driveshaft (which may cause the device to "jump" after activation if not relieved).

4. Push the "on" button on top of the crown advancer knob to start spinning. Wait for 2 s that is required for the crown to orbit at a stable speed.

5. Advance SLOWLY (1 mm/s—even slower if the auditory pitch of the device changes). Do NOT force through lesion. Slowly engage and disengage the lesion to maintain 1:1 motion between the crown and the advancer knob (as viewed via live angiography). The device works both in the antegrade and the retrograde direction. The same 1 mm/s 1:1 motion movement should be maintained with both antegrade and retrograde treatments

6. Duration: limit runs to 10−20 s (device will beep at 25 s). Allow 30−60 s between runs to reduce the risk of spasm and allow for particulate to advance through the microcirculation. At a very minimum, rest periods should be equal to the preceding orbital atherectomy treatment duration.

7. Where to stop: the crown should ideally not be stopped within an obstructive untreated lesion. Instead, the crown should be moved slowly to a distal or proximal position from the target lesion to minimize the risk of: (a) limiting the crown's ability to ramp to speed; and (b) occlusion of the vessel with the crown.[34]
8. Do not allow the crown to get closer than 10−15 mm to the radiopaque distal portion of the ViperWire Advance or ViperWire Advance Flex Tip guidewire to avoid shearing off the guidewire tip.
9. Stop orbital atherectomy when the crown passes easily through the lesion without changes in the sound generated by the device.
10. Total orbital atherectomy time should be <5 min. If >5 min are required, a new orbital atherectomy device should be used.
11. Perform angiography intermittently—definitely before removing the device or if the patient develops chest pain, ECG changes, or hypotension.

9.3.2.4.7 Step 7: remove orbital atherectomy device

The orbital atherectomy crown can be removed either by "walking back" the device under fluoroscopy utilizing the GlideAssist mode, or by using the trapping technique (if the guide is big enough to allow insertion of a trapping balloon next to the orbital atherectomy crown—see Table 9.2).

Option 1: walking device back

1. Lock knob (turn to parallel position).
2. Turn brake up.
3. Remove the device while monitoring the position of the ViperWire Advance guidewire distal tip.

Option 2: GlideAssist

1. Release the brake on the end of the device.
2. Hold the low speed button for 2 s until the light blinks slowly.
3. Press the black button on top of the advancer knob to turn GlideAssist on. The light will flash rapidly.
4. Remove the device while monitoring the position of the ViperWire Advance guidewire distal tip under fluoroscopy.

Option 3: using the trapping technique

1. The crown is withdrawn inside the guide catheter.
2. A balloon is advanced inside the guide catheter but distal to the distal end of the burr.
3. The balloon is inflated "trapping" the ViperWire Advance.
4. The crown is removed (fluoroscopy is only needed during the beginning of crown withdrawal).

9.3.2.4.8   Step 8: perform angiography

Coronary angiography is performed to identify slow flow/no reflow, dissections, embolization, side branch loss, perforation, or other complications.

9.3.2.4.9   Step 9: replace the ViperWire Advance or ViperWire Advance Flex Tip wire with a standard workhorse guidewire

Using the trapping technique, the ViperWire Advance or Viper Wire Advance Flex Tip wire is exchanged for a standard workhorse guidewire using a microcatheter or over-the-wire balloon. PCI can be performed over the ViperWire Advance guidewire (especially if a complication, such as dissection or perforation occurs), but workhorse guidewires are preferred as they are more torquable and less likely to cause distal vessel injury. The ViperWire Advance Flex Tip is easier to use than the ViperWire Advance for equipment delivery.

9.3.2.4.10   Step 10: balloon angioplasty

Balloon angioplasty with a 1:1 sized noncompliant balloon should always be performed after atherectomy to ensure that the balloon expands fully. If it does not, repeat atherectomy or other treatments (such as use of plaque modification balloons) may be required.

9.3.3   Intravascular lithotripsy

Intravascular lithotripsy (Section 30.9.8) is a specialized balloon that can modify heavily calcified lesions and achieve lesion expansion. It has been used with success in both antegrade and retrograde CTO PCI.[35-37] Its main advantage is simplicity: the lithotripsy balloon (sized 1:1 to the target vessel) is inflated at low pressure (up to 4 atm) and a cycle of treatment is administered (up to 10 pulses with each treatment; each pulse lasts 1 s). The balloon should be deflated after each cycle for at least 10 s to allow for coronary perfusion.

The intravascular lithotripsy (IVL) balloon can be challenging to deliver due to its high crossing profile (0.044−0.047"). Use of a guide catheter extension and/or additional lesion preparation with a standard balloon or atherectomy may be needed to allow delivery of the IVL balloon.

9.3.4   Drug-coated balloons

Drug-coated balloons may be useful in small, diffusely diseased vessels when trying to avoid stenting. Drug-coated balloons may also be useful in in-stent restenosis lesions (especially when multiple stent layers are present to minimize the need for additional stent placement) and when performing extraplaque plaque modification (Section 8.3.5.1).[38]

References

1. Zhang M, Matsumura M, Usui E, et al. Intravascular ultrasound-derived calcium score to predict stent expansion in severely calcified lesions. *Circ Cardiovasc Interv* 2021;**14**: e010296.

2. Fujino A, Mintz GS, Matsumura M, et al. A new optical coherence tomography-based calcium scoring system to predict stent underexpansion. *EuroIntervention* 2018;**13**: e2182−9.

3. Brilakis ES. *Manual of coronary chronic total occlusion interventions. A step-by-step approach.* 2nd ed. Elsevier; 2017.

4. Vo MN, Brilakis ES, Grantham JA. Novel use of cutting balloon to treat subintimal hematomas during chronic total occlusion interventions. *Catheter Cardiovasc Interv* 2018;**91**: 53−6.

5. Tajti P, Sandoval Y, Brilakis ES. "Around the world"—how to reach native coronary artery lesions through long and tortuous aortocoronary bypass grafts. *Hellenic J Cardiol* 2018;**59**:354−7.

6. Santiago R, Moroni F, Del Rio V, Rodriguez-Escudero J, Azzalini L. The guide extension tunnel in landslide technique (TILT) for equipment delivery in severely tortuous or uncrossable lesions during percutaneous coronary intervention. *EuroIntervention* 2021;**17**:e923−4.

7. Fujita S, Tamai H, Kyo E, et al. New technique for superior guiding catheter support during advancement of a balloon in coronary angioplasty: the anchor technique. *Catheter Cardiovasc Interv* 2003;**59**:482−8.

8. Hirokami M, Saito S, Muto H. Anchoring technique to improve guiding catheter support in coronary angioplasty of chronic total occlusions. *Catheter Cardiovasc Interv* 2006;**67**:366−71.

9. Kirtane AJ, Stone GW. The Anchor-Tornus technique: a novel approach to "uncrossable" chronic total occlusions. *Catheter Cardiovasc Interv* 2007;**70**:554−7.

10. Matsumi J, Saito S. Progress in the retrograde approach for chronic total coronary artery occlusion: a case with successful angioplasty using CART and reverse-anchoring techniques 3 years after failed PCI via a retrograde approach. *Catheter Cardiovasc Interv* 2008;**71**:810−14.

11. Fang HY, Wu CC, Wu CJ. Successful transradial antegrade coronary intervention of a rare right coronary artery high anterior downward takeoff anomalous chronic total occlusion by double-anchoring technique and retrograde guidance. *Int Heart J* 2009;**50**:531−8.

12. Lee NH, Suh J, Seo HS. Double anchoring balloon technique for recanalization of coronary chronic total occlusion by retrograde approach. *Catheter Cardiovasc Interv* 2009;**73**:791−4.

13. Saito S. Different strategies of retrograde approach in coronary angioplasty for chronic total occlusion. *Catheter Cardiovasc Interv* 2008;**71**:8−19.

14. Surmely JF, Katoh O, Tsuchikane E, Nasu K, Suzuki T. Coronary septal collaterals as an access for the retrograde approach in the percutaneous treatment of coronary chronic total occlusions. *Catheter Cardiovasc Interv* 2007;**69**:826−32.

15. Surmely JF, Tsuchikane E, Katoh O, et al. New concept for CTO recanalization using controlled antegrade and retrograde subintimal tracking: the CART technique. *J Invasive Cardiol* 2006;**18**:334−8.

16. Rathore S, Katoh O, Matsuo H, et al. Retrograde percutaneous recanalization of chronic total occlusion of the coronary arteries: procedural outcomes and predictors of success in contemporary practice. *Circ Cardiovasc Interv* 2009;**2**:124−32.

17. Di Mario C, Ramasami N. Techniques to enhance guide catheter support. *Catheter Cardiovasc Interv* 2008;**72**:505−12.

18. Mahmood A, Banerjee S, Brilakis ES. Applications of the distal anchoring technique in coronary and peripheral interventions. *J Invasive Cardiol* 2011;**23**:291−4.

19. Christ G, Glogar D. Successful recanalization of a chronic occluded left anterior descending coronary artery with a modification of the retrograde proximal true lumen puncture technique: the antegrade microcatheter probing technique. *Catheter Cardiovasc Interv* 2009;**73**:272−5.

20. Mamas MA, Fath-Ordoubadi F, Fraser DG. Distal stent delivery with guideliner catheter: first in man experience. *Catheter Cardiovasc Interv* 2010;**76**:102−11.

21. Fang HY, Fang CY, Hussein H, et al. Can a penetration catheter (Tornus) substitute traditional rotational atherectomy for recanalizing chronic total occlusions? *Int Heart J* 2010;**51**:147−52.

22. Khelimskii D, Badoyan A, Krestyaninov O. The deep-wire crossing technique: a novel method for treating balloon-uncrossable lesions. *J Invasive Cardiol* 2019;**31**:E362−8.

23. Sandoval Y, Brilakis ES. The role of rotational atherectomy in contemporary chronic total occlusion percutaneous coronary intervention. *Catheter Cardiovasc Interv* 2017;**89**:829−31.

24. Azzalini L, Dautov R, Ojeda S, et al. Long-term outcomes of rotational atherectomy for the percutaneous treatment of chronic total occlusions. *Catheter Cardiovasc Interv* 2017;**89**:820−8.

25. Xenogiannis I, Karmpaliotis D, Alaswad K, et al. Usefulness of atherectomy in chronic total occlusion interventions (from the PROGRESS-CTO registry). *Am J Cardiol* 2019;**123**:1422−8.

26. Capretti G, Carlino M, Colombo A, Azzalini L. Rotational atherectomy in the subadventitial space to allow safe and successful chronic total occlusion recanalization: pushing the limit further. *Catheter Cardiovasc Interv* 2018;**91**:47−52.

27. Hussain F, Golian M. Desperate times, desperate measures: rotablating dissections in acute myocardial infarction. *J Invasive Cardiol* 2011;**23**:E226−8.

28. Brinkmann C, Eitan A, Schwencke C, Mathey DG, Schofer J. Rotational atherectomy in CTO lesions: too risky? Outcome of rotational atherectomy in CTO lesions compared to non-CTO lesions. *EuroIntervention* 2018;**14**:e1192−8.

29. Neupane S, Basir M, Tan C, et al. Feasibility and safety of orbital atherectomy for the treatment of in-stent restenosis secondary to stent under-expansion. *Catheter Cardiovasc Interv* 2021;**97**:2−7.

30. Megaly M, Brilakis ES. Primary orbital atherectomy for treating a heavily calcified balloon uncrossable lesion. *Cardiovasc Revasc Med* 2020;**21**:96−9.

31. Murad B. Intracoronary aminophylline for management of bradyarrhythmias during thrombectomy with the AngioJet catheter. *J Invasive Cardiol* 2008;**20**:12A−18AA.

32. Murad B. Intracoronary aminophylline for heart block with AngioJet thrombectomy. *J Invasive Cardiol* 2005;**17**:A4.

33. Lee MS, Wiesner P, Rha SW. Novel technique of advancing the rotational atherectomy device: "Single-Operator" technique. *J Invasive Cardiol* 2016;**28**:183−6.

34. Shlofmitz E, Martinsen BJ, Lee M, et al. Orbital atherectomy for the treatment of severely calcified coronary lesions: evidence, technique, and best practices. *Expert Rev Med Devices* 2017;**14**:867−79.

35. Karacsonyi J, Nikolakopoulos I, Vemmou E, Rangan BV, Brilakis ES, Intracoronary. Lithotripsy: a new solution for undilatable in-stent chronic total occlusions. *JACC Case Rep* 2021;**3**:780−5.

36. Oksnes A, Cosgrove C, Walsh S, et al. Intravascular lithotripsy for calcium modification in chronic total occlusion percutaneous coronary intervention. *J Interv Cardiol* 2021;**2021**:9958035.

37. Yeoh J, Hill J, Spratt JC. Intravascular lithotripsy assisted chronic total occlusion revascularization with reverse controlled antegrade retrograde tracking. *Catheter Cardiovasc Interv* 2019;**93**:1295−7.
38. Ybarra LF, Dandona S, Daneault B, Rinfret S. Drug-coated balloon after subintimal plaque modification in failed coronary chronic total occlusion percutaneous coronary intervention: a novel concept. *Catheter Cardiovasc Interv* 2020;**96**:609−13.

Chapter 10

Stenting

Goal: To deliver and adequately expand a stent, optimally covering the target chronic total occlusion (CTO).

10.1   When to stent?

Stenting with drug-eluting stents (DES) is performed in the vast majority of coronary lesions and especially CTOs because stenting:

1. Prevents vessel recoil and reduces the risk of acute closure, especially in lesions with dissection, rupture, and thrombus.
2. Reduces the risk of restenosis.

   Stenting should **not** be performed in the following scenarios:

1. Poor antegrade flow (unless the poor flow is caused by a dissection). Stenting in the setting of no reflow will worsen it.
2. Inability to expand the target lesion with a balloon.
3. High risk of compromising an important coronary branch that cannot be protected.
4. Distal wire reentry into the distal true lumen of a CTO vessel well beyond the distal cap, as this would result in occlusion of side branches originating proximal to the reentry location. For example stenting should not be performed when performing the STAR procedure (Section 8.3.5.1). Balloon angioplasty should be performed instead, with a repeat PCI attempt after approximately 2 months.
5. Very small target vessel and target lesion (although 2.0 mm DES are currently available [Onyx Frontier]; drug-coated balloons could potentially be used for some of these lesions).

   Stenting should not be performed until after successful wiring of the lesion (as described in Chapter 8) and (in most cases) successful lesion preparation (as described in Chapter 9). Optimal distal wire position should be confirmed prior to stenting.

10.2   How to stent

Several of the stenting steps are similar to the balloon angioplasty steps described in Chapter 9.

10.2.1   Step 1. Confirm that a guidewire is advanced through the target lesion and optimally positioned distally

10.2.2   Step 2. Confirm that the guide catheter is aspirated and flushed

10.2.3   Step 3. Select stent type and size

10.2.3.1   Goal

To choose optimal stent type and size.

10.2.3.2   How

10.2.3.2.1   Stent type

DES versus bare metal stents (BMS): Contemporary DES are currently used in nearly all CTO percutaneous coronary interventions (PCIs), due to better efficacy (less restenosis) and safety (stent thrombosis) as compared with bare metal stents.[1-4]

Stent brand: Thinner strut stents have higher deliverability, but lower visibility. Stent brand selection depends on the target lesion characteristics and local availability and cost.

10.2.3.2.2   Stent size

10.2.3.2.2.1   Stent diameter: Stent diameter can be selected using coronary angiography and/or intravascular imaging (Section 13.3). Intravascular imaging significantly facilitates accurate stent size selection and can also confirm the adequacy of lesion preparation.

In PCI of non-CTO lesions, stent diameter should match the distal reference vessel diameter (as assessed by angiography and/or intravascular imaging). Sizing stents based on the proximal vessel may result in distal vessel dissection or perforation, as the proximal vessel is usually larger than the distal vessel due to normal vessel tapering; it may also lead to side branch occlusion due to carina shift.

In CTO PCI the vessel distal to the CTO often appears small due to low filling pressure and negative remodeling or due to atherosclerosis. Intravascular imaging (Section 13.3) can help evaluate the size and presence of disease in the distal vessel. Stent sizing based on the size of the vessel distal to the CTO carries risk of late-acquired malapposition, due to subsequent lumen enlargement.[5] In one study, 69% of recanalized CTOs had a mean increase in lumen diameter of 0.4 mm over a period of 6 months.[6]

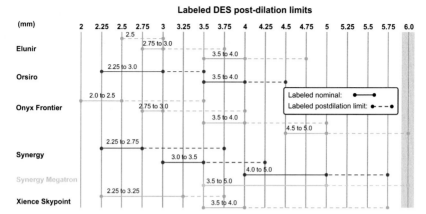

FIGURE 10.1   **Labeled stent postdilatation limits**.

After stenting, the proximal portion of the stent should be postdilated with a larger balloon to match the proximal reference vessel diameter (proximal optimization technique—POT, Section 16.3.1). Aggressive postdilatation should be avoided in heavily calcified vessels due to increased risk of perforation. When the proximal vessel diameter is much larger than the distal vessel diameter (for example in left main lesions or in aneurysmal coronary vessels) it is important to know the limits of expansion of the various stents (Fig. 10.1).

10.2.3.2.2.2 Stent length: The stent should be long enough to cover the entire target lesion, including coronary artery segments proximal or distal to the lesion that were injured with balloon angioplasty. Failure to cover the entire lesion/predilated segment may increase the risk of restenosis. Intravascular imaging can help determine whether there is plaque or negative remodeling in the vessel distal to the CTO. Stenting of the distal vessel should be avoided in case of negative remodeling. Long stents (48 mm in the US and 60 mm in Europe) are currently available and are very useful for CTO PCI, given long lesion length. Stenting areas of dissection caused by CTO PCI can reduce the risk of restenosis and reocclusion and can be facilitated by intravascular imaging (Fig. 10.2).[8]

In CTO PCI extensive stenting of the distal vessel should be avoided in most cases, to minimize the risk of restenosis and stent thrombosis. Balloon angioplasty of the distal vessel (possibly with drug-coated balloons[9−11]) may sometimes be needed, especially if the antegrade flow is impaired.

The desired stent length can be estimated angiographically by using a known length balloon to predilate the lesion. It can also be measured using intravascular imaging with automated pullback (with intravascular ultrasound [IVUS] or optical coherence tomography [OCT]).

Long stents can be challenging to deliver, especially through tortuous and calcified segments. In such lesions, better lesion preparation and increasing

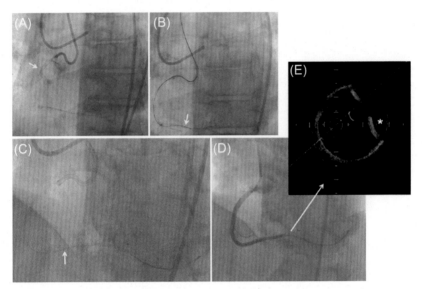

FIGURE 10.2   **Example of distal dissection confirmed with optical coherence tomography**. A right coronary artery CTO (arrow, **panel A**) was crossed in the extraplaque space using antegrade wiring (**panel B**). The Stingray system (arrow, **panel C**) was used enabling successful reentry into the distal true lumen, however subsequent angiography demonstrated suboptimal distal result (**panel D**). Optical coherence tomography demonstrated a distal dissection with compression of the true lumen (asterisk, **panel E**), that was treated with implantation of an additional stent.[7] *Reproduced with permission from Brilakis ES. Manual of coronary chronic total occlusion interventions. A step-by-step approach. 2nd ed. Elsevier; 2017.*

guide catheter support, for example by using a guide catheter extension, can be used to facilitate delivery. Alternatively, delivering >1 shorter stents can facilitate delivery.

10.2.3.2.2.3   Extraplaque versus true lumen stenting: As described in detail in Section 8.2, extensive dissection/reentry crossing strategies (such as the Subintimal Tracking And Reentry—STAR technique) are associated with high restenosis and reocclusion rates and should be only used as a last resort.[12,13] However, outcomes with limited antegrade[14–18] or retrograde[19,20] extraplaque crossing are similar to those of true to true crossing,[21–24] supporting the use of these techniques in contemporary CTO PCI.

10.2.3.3   What can go wrong?

Poor stent size selection may not result in complications until the time of stent delivery and deployment, as follows:

10.2.3.3.1   Inability to deliver stent to the target lesion (see Step 8)

10.2.3.3.2 Perforation (if stent diameter is too big for the target vessel), see Step 13)

10.2.3.3.3 Geographic miss (if stent is too short for the target lesion), see Step 13)

10.2.4   Step 4. Prepare stent balloon

10.2.5   Step 5. Load the stent on the guidewire

10.2.6   Step 6. Advance the stent balloon monorail segment through the Y-connector

10.2.7   Step 7. Advance the stent to the tip of the guide catheter

10.2.8   Step 8. Advance the stent to the target lesion

10.2.8.1   Goal

To advance the stent to target lesion

10.2.8.2   How

1. The stent is advanced towards the target lesion using the "independent-hand" or "two-hand" technique as discussed in Section 9.3.1.8 (Fig. 9.1).
2. The position of the stent is confirmed using X-ray landmarks (such as clips, previous stents, etc.) and/or contrast injection. Antegrade contrast injections should be avoided in CTO PCI, especially when using dissection/reentry techniques, to prevent expansion of the dissection plane. Contralateral contrast injection can help with stent positioning, but sometimes retrograde flow decreases after antegrade ballooning.

Contrast injections should be minimized in patients with chronic kidney disease. Angiographic coregistration with intravascular imaging or physiology systems can be utilized to assist in optimal stent positioning.

10.2.8.3   Challenges

10.2.8.3.1   Failure to reach the lesion with the stent (stent shaft is not long enough)

10.2.8.3.1.1   Causes:
1. Very distal lesion location, especially in patients with prior coronary artery bypass graft surgery and lesions distal to bypass graft anastomoses.

10.2.8.3.1.2   Prevention:
1. Use short (or shortened) guide catheters when treating very distal coronary lesions.

2. Use stents with the longest available shaft (Xience: 145 cm, Promus/Synergy: 144 cm, Onyx Frontier/Elunir: 140 cm)
3. Use guide catheter extensions that may allow safer deep intubation or better aligned guide catheter.

10.2.8.3.1.3 Treatment:
1. Use short (or shortened) guide catheters when treating very distal coronary lesions.
2. Use stents with the longest available shafts.

10.2.8.3.2   Failure to reach the lesion or cross the lesion with the stent

10.2.8.3.2.1 Causes:
1. Proximal vessel: significant disease, tortuosity, and/or calcification.
2. Severe lesion tortuosity.
3. Severe lesion calcification.
4. Severe residual stenosis.
5. Poor lesion preparation.

10.2.8.3.2.2 Prevention:
1. If there is proximal disease, treat the proximal vessel.
2. Good lesion preparation.

10.2.8.3.2.3 Treatment: (Fig. 10.3).
10.2.8.3.2.3.1 Guide catheter Strong guide catheter support is critical for balloon and stent delivery, especially in complex lesions.

Larger guide catheters (such as 7 or 8 French), with **supportive shapes** (for example EBU guides provide much stronger support than JL4 guides for left coronary interventions, and AL1 or 3D Right guide catheters provide stronger support than JR4 guide catheters for RCA interventions), and **femoral access** (as compared with radial) provide stronger support.

For truly complex lesions, many operators increase the likelihood of success by using femoral access with 8 French, supportive shape guide catheters.

Forward push of the guide catheter can increase support, but may also in some cases lead to guide prolapse and disengagement. The "**independent hand" technique** (the left hand pushes the guide catheter, while the right hand both fixes the guidewire and advances the stent) allows a single operator to simultaneously manipulate the guide catheter and advance the stent.

Deep intubation of the guide catheter increases support, but may also cause a dissection. Deep intubation is usually done in the RCA or in SVGs and rarely in the left coronary artery.

Guide catheter extensions (Section 30.2) are among the most commonly used tools for facilitating stent delivery. They not only increase guide

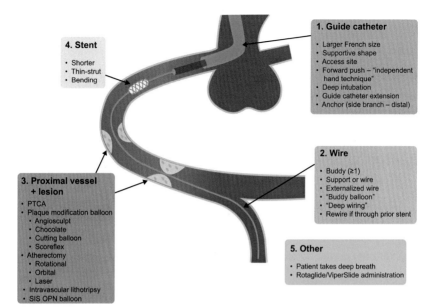

4. Stent
- Shorter
- Thin-strut
- Bending

1. Guide catheter
- Larger French size
- Supportive shape
- Access site
- Forward push – "independent hand technique"
- Deep intubation
- Guide catheter extension
- Anchor (side branch – distal)

2. Wire
- Buddy (≥1)
- Support or wire
- Externalized wire
- "Buddy balloon"
- "Deep wiring"
- Rewire if through prior stent

3. Proximal vessel + lesion
- PTCA
- Plaque modification balloon
  - Angiosculpt
  - Chocolate
  - Cutting balloon
  - Scoreflex
- Atherectomy
  - Rotational
  - Orbital
  - Laser
- Intravascular lithotripsy
- SIS OPN balloon

5. Other
- Patient takes deep breath
- Rotaglide/ViperSlide administration

FIGURE 10.3   **How to facilitate stent delivery to the target lesion.**[25] *Modified with permission from Brilakis ES.* Manual of percutaneous coronary interventions: A step-by-step approach. *Elsevier; 2021.*

catheter support, but may also effectively "modify" the proximal vessel if they are advanced close to, or through, the target lesion (the stent is then advanced through the guide extension without coming in contact with the wall of the target vessel).

Anchor techniques (side branch anchor and distal anchor) can also provide strong support for equipment delivery and are discussed in detail in Section 9.3.1.8.1.

10.2.8.3.2.3.2   Wire One of the simplest ways to facilitate delivery is to use one (or more) **buddy wire(s)**. The buddy wire increases guide support and also straightens the target vessel.

Another option is to use **support** guidewires (such as the IronMan, Grand Slam, or the **Wiggle** wire) (Section 30.6.4).

Use of an **externalized** guidewire provides strong support for equipment delivery and prevents loss of guide catheter and guidewire position.

Partially withdrawal of the wire while advancing the stent may facilitate advancement.

Another option is to **advance a balloon** across the area of resistance to stent advancement, in order to "deflect" the stent during advancement.

In the "**deep wiring**" technique, a polymer-jacketed guidewire is advanced through the lesion then into the ventricle/aorta through arterioluminal communications, which increases wire support. The obvious risk of this technique is coronary perforation.

If there is resistance to stent delivery at the location of a previously deployed stent, it is possible that the wire has crossed under a stent strut. **Rewiring through the stent**, ideally after forming a knuckle more proximally can solve this problem.

10.2.8.3.2.3.3   Proximal vessel + lesion preparation Stent delivery can be challenging, especially through proximal lesions (especially tortuous and calcified) and through severe, calcified lesions. Adequate lesion preparation with balloon angioplasty (with standard and plaque modification balloons) and/or atherectomy or intravascular lithotripsy can facilitate both stent delivery and stent expansion, and minimize the risk of stent loss.

10.2.8.3.2.3.4   Stent **Shorter** and **thinner-strut** stents are easier to deliver. **Bending** the stent may facilitate advancement through tortuosity, but may also increase the risk of stent loss.

10.2.8.3.2.3.5   Other Asking the patient to **take a deep breath** can facilitate stent delivery, possibly by straightening the coronary arteries or changing the seating of the guide catheter.

Rotaglide or ViperSlide administration during stent advancement might reduce friction and facilitate delivery.

10.2.8.4   What can go wrong?

10.2.8.4.1   Wire and guide catheter position loss

10.2.8.4.1.1   Causes:
1. Forceful stent advancement through areas of resistance

10.2.8.4.1.2   Prevention:
1. Avoid forceful stent advancement, if possible.
2. Always monitor the guide catheter and the distal wire position while advancing (or withdrawing stents) and adjust the effort to advance the stent accordingly.

10.2.8.4.1.3   Treatment:
1. Re-engagement with the guide catheter and rewiring of the lesion or advancement of the guidewire.

10.2.8.4.2   Loss of distal vessel visualization

10.2.8.4.2.1   Causes:
1. Balloon advancement through a tight lesion.
2. Dissection.
3. Distal embolization.
4. Spasm.

10.2.8.4.2.2 *Prevention:*
1. Adequate lesion preparation.
2. Nitroglycerin administration.

10.2.8.4.2.3 *Treatment:*
1. If the reason is balloon advancement through a tight lesion, withdraw the balloon, inject contrast until it reaches distal to the lesion, then advance the balloon, which then causes contrast to be entrapped distal to the lesion and allows visualization of the distal vessel.
2. If the reason is dissection, stenting should be performed.
3. If the reason is spasm, intracoronary nitroglycerin should be administered.

10.2.8.4.3   Stent loss
10.2.8.4.3.1   *Causes:*
1. Forceful stent advancement through areas of resistance.
2. Stent deformation during advancement or withdrawal attempts.

10.2.8.4.3.2   *Prevention:*
1. Meticulous lesion preparation.
2. Avoid forceful stent advancement, if possible.
3. Avoid pushing the proximal part of the stent outside the guide catheter if the stent is not advancing through the target coronary artery.
4. Maintain coaxial guide catheter alignment when attempting to withdraw the stent into the guide catheter.

10.2.8.4.3.3   *Treatment:* See Section 27.1.

10.2.9   Step 9. Inflate balloon and deploy stent

10.2.10   Step 10. Deflate stent balloon

10.2.11   Step 11. Withdraw stent balloon into the guide catheter

10.2.12   Step 12. Remove stent balloon from guide catheter

10.2.13   Step 13. Check the stent result

10.2.13.1   *Goal:*

To determine the adequacy of stenting result and the occurrence of any complications.

10.2.13.2 How:

1. Following removal of the stent balloon, the catheter is aspirated and flushed.
2. Optimal guide engagement is confirmed, as the guide may have become disengaged or deeply engaged during balloon removal.
3. Contrast is administered to evaluate the result of the procedure.
4. The "Stent Boost" or "Clear Stent" X-ray modes can help evaluate stent expansion.
5. Intravascular imaging with IVUS or OCT (Section 13.3) or coronary physiology (Section 12.1.2) can be used to assess the procedural result and minimize the use of contrast. Intravascular imaging can help detect areas of underexpansion (CTO Manual online case 18), malapposition, dissection (Fig. 10.2), or incomplete stent coverage of ostial lesions (CTO Manual online case 18), treatment of which may improve outcomes of CTO PCI. OCT can be very useful in limiting the extent of stenting by confirming that the small caliber of the distal coronary vessel is due to chronic hypoperfusion and not due to dissection or atherosclerotic lesions (indicated by the presence of a tri-laminar arterial structure, no evidence of intimal hyperplasia and no atheroma).[5,8]

Two randomized-controlled trials demonstrated improved long-term outcomes with use of IVUS-guided vs. angiography-guided stenting in CTO PCI. The Korean Chronic Total Occlusion InterVention with DES, guided by IVUS (**CTO-IVUS**), trial randomized 402 patients after successful guidewire crossing of the CTO to IVUS versus angiographic guidance. They reported lower 12-month incidence of major adverse cardiac events in the IVUS group (2.6% vs 7.1%; $P = .035$).[26] Similarly, the AIR CTO trial randomized 230 patients after successful guidewire crossing to IVUS- versus angiographic-guided stenting and showed lower 12-month in-stent late lumen loss (0.28 ± 0.48 mm vs 0.46 ± 0.68 mm, $P = .025$) and lower incidence of in-stent restenosis for stents placed in the true lumen (3.9% vs. 13.7%, $P = .021$).[27]

In a contemporary CTO PCI registry intravascular imaging (mainly IVUS) was performed in approximately 1 of 3 cases, for either facilitating crossing and for stent optimization.[28]

Intravascular imaging should be used in nearly all CTO PCIs, as it can significantly improve the outcomes of CTO PCI,[29] but requires education (that has been suboptimal during interventional cardiology fellowship[30]) and continued implementation in day to day practice.

10.2.13.3 What can go wrong?

10.2.13.3.1 Failure to expand (balloon undilatable lesion)

10.2.13.3.1.1 Causes:

1. Severe calcification
2. In-stent restenosis

3. Suboptimal lesion preparation

10.2.13.3.1.2 Prevention:
1. Meticulous lesion preparation, especially for heavily calcified lesions, as described in Chapter 9.

10.2.13.3.1.3 Diagnosis:
1. A waist remains on the balloon during inflation on fluoroscopy.
2. Intravascular imaging.
3. Use of the "Stent Boost" or "Clear Stent" X-ray imaging techniques.

10.2.13.3.1.4 Treatment: See balloon undilatable algorithm (Section 23.2)

10.2.13.3.2 Failure to fully cover the target lesion
10.2.13.3.2.1 Causes:
1. Shorter than needed stent length.
2. Excessive vessel injury during lesion preparation.

10.2.13.3.2.2 Prevention:
1. Optimal stent length selection based on balloon size during predilatation and/or based on intravascular imaging with automated pullback.
2. Avoiding predilatation outside the target coronary lesion.

10.2.13.3.2.3 Diagnosis:
1. Residual stenosis at proximal or distal edge of the stent.

10.2.13.3.2.4 Treatment:
1. Implantation of additional stent(s) to fully cover the target coronary lesion. If the size of the distal vessel is very small, balloon angioplasty (with standard or drug-coated balloons) may suffice.

10.2.13.3.3 Distal edge dissection
10.2.13.3.3.1 Causes:
1. Stent oversizing and/or very high-pressure inflations.
2. Incomplete lesion coverage, especially when the stent edge is implanted over a lipid rich plaque.

10.2.13.3.3.2 Prevention:
1. Avoid stent oversizing and very high-pressure stent balloon inflation.
2. Optimal stent length selection based on balloon size during predilatation and/or based on intervascular imaging with automated pullback.

10.2.13.3.3.3 Diagnosis:
1. Angiographic lucency at the stent edges.
2. If dissection is significant, it may lead to slow flow or no flow.
3. Flap by intravascular ultrasound or by optical coherence tomography.

10.2.13.3.3.4 Treatment:
1. Additional stent implantation (for significant dissections).

10.2.13.3.4 Side branch occlusion

10.2.13.3.4.1 Causes:
1. Jailing of the side branch ostium.
2. Plaque shift and carina shift[31] after stenting.
3. Dissection/reentry strategies.

10.2.13.3.4.2 Prevention:
1. Place guidewire in important side branches before placing a stent.
2. Balloon or stent side branch prior to stenting the main branch.
3. Perform the POT proximal to the carina.

10.2.13.3.4.3 Diagnosis:
1. Slow flow or no flow in side branch.

10.2.13.3.4.4 Treatment:
1. Rewiring followed by balloon angioplasty and/or stenting of the side branch.

10.2.13.3.5 Perforation

The approach to coronary perforations is discussed in Chapter 26.

10.3 Long-term outcomes after stenting for CTO PCI

Restenosis rates after CTO stenting can be relatively high. BMS significantly reduced restenosis compared with balloon angioplasty alone,[32] yet the incidence of restenosis and reocclusion remained very high. In the Total Occlusion Study of Canada 1 (TOSCA-1) trial, the 6-month incidence of restenosis and reocclusion with bare metal stents exceeded 50% and 10%, respectively.[33]

First generation DES significantly reduced restenosis compared with BMS (Table 10.1). The first randomized-controlled trial comparing BMS and DES was the Primary Stenting of Totally Occluded Native Coronary Arteries (**PRISON II**) trial that compared BMS with the sirolimus-eluting stent (SES, Cypher, Cordis). The SES significantly reduced the 6-month incidence of binary angiographic restenosis (from 41% to 11%, $P < .001$), vessel reocclusion (from 13% to 4%, $P < .04$), and the need for new revascularization procedures (from 22% to 8%, $P < .001$) compared with BMS.[35]

TABLE 10.1 Published prospective studies on the clinical and angiographic outcomes with drug-eluting stents in coronary chronic total occlusions.

Author	Year	Stent	n	FU angio time	Prior CABG (%)	Total stent length (mm)	In-stent restenosis (%)	In-segment restenosis (%)	TLR (%)	TVR (%)
ACROSS-TOSCA 4[3,4]	2009	SES	200	6 months	8.5	45.9 (30.2, 62.1)	9.5	12.4	9.8	11.4
PRISON II[35,36]	2006	SES	100	6 months	3	32 ± 15	7	11	4	8
GISSOC II—GISE[37]	2010	SES	78	8 months	6.7	41 ± 18	8.2	9.8	8.1	14.9
CORACTO[38]	2010	SES	48	6 months	NR	45.5 ± 24.8	NR	17.4	NR	10.8
CIBELES[39]	2012	SES	101	9 months	4	47 ± 24	NR	10.5	7.5	11.6
		EES	106	9 months	4.7	50 ± 23	NR	9.1	6.0	7.9
CATOS[40]	2012	SES	80	9 months	NR	44.6 ± 20.2	NR	13.7	NR	13.8
		Endeavor ZES	80	9 months	NR	43.4 ± 21.5	NR	14.1	NR	7.5
PRISON III[41]	2012	SES	60	8 months	5.0	38.4 ± 18.4	2.0	12.0	6.7	8.3
		Endeavor or Resolute ZES	62	8 months	8.1	41.0 ± 19.2	5.5	10.9	4.8	4.8

(Continued)

TABLE 10.1 (Continued)

Author	Year	Stent	n	FU angio time	Prior CABG (%)	Total stent length (mm)	In-stent restenosis (%)	In-segment restenosis (%)	TLR (%)	TVR (%)
ACE-CTO[42]	2015	EES	100	8 months	27	85 ± 34	46	46	37	39
EXPERT-CTO[43]	2015	EES	222	12 months	9.9	52 ± 27	NR	NR	6.3	NR
PRISON IV[44]	2017	Orsiro SES	165	12 months	3.6	52 ± 28	NR	NR	10.5	10.5
		EES	165	12 months	6.7	52 ± 27	NR	NR	4.0	6
CONSISTENT-CTO[21]	2020	Synergy EES	210	12 months	15.7	DART: 96.6 ± 31.6 No DART: 75.4 ± 31.4	NR	14.5%	NR	DART: 10.9 No DART: 3.7
PERSPECTIVE[45]	2020	Resolute Onyx	500	Not routinely performed	29.8	81.5 ± 32.7 mm	NR	NR	1.0%	2.2%

CABG, coronary artery bypass graft surgery; DART, dissection and re-entry techniques; DES, drug-eluting stents; EES, everolimus-eluting stents; FU, follow-up; SES, sirolimus-eluting stents; TLR, target lesion revascularization; TVR, target vessel revascularization.

The benefit persisted at the 5-year followup angiography, although some late catchup in lumen diameter loss was observed in the SES group.[46] Two other studies, the Gruppo Italiano di Studio sullo Stent nelle Occlusioni Coronariche Societá Italiana di Cardiologia Invasiva (**GISSOC II-GISE**)[37] and the **CORACTO** trial[38] showed similar results. Four metaanalyses on DES versus BMS in CTOs were published in 2010−2011,[1−4] all reporting significant reduction in the risk of restenosis, reocclusion and repeat revascularization with DES. DES appeared to be safe in CTOs, although the risk of stent thrombosis was higher with DES in one metaanalysis.[1]

The **CONSISTENT CTO** (Conventional Antegrade versus Sub-Intimal Synergy Stenting in Chronic Total Occlusions) study[21] showed low rates of angiographic restenosis and target vessel revascularization with the Synergy DES.

Second generation DES haven provide incremental benefit compared with the first generation paclitaxel-eluting stent in non-CTO lesions.[47,48] Three randomized clinical trials have compared the first generation SES with the second generation everolimus-eluting stent (EES, Xience V, Abbott Vascular)[39] and zotarolimus-eluting stent (ZES, Medtronic)[40] in CTOs. The Chronic Coronary Occlusion Treated by Everolimus-Eluting Stent (**CIBELES**) compared EES with SES and demonstrated similar restenosis and repeat revascularization rates, with a trend for lower stent thrombosis risk in the EES group (3% vs 0%, $P = .075$). The CAtholic Total Occlusion Study (**CATOS**) trial showed similar angiographic and clinical outcomes with the SES and the Endeavor ZES (Medtronic).[40] In contrast, the **PRISON III** trial, reported higher in-segment late lumen loss at 8-month follow-up angiography with the Endeavor ZES compared with the SES, although rates were similar with the Resolute ZES.[40] In an Italian registry of 802 patients undergoing CTO PCI over 8 years, use of EES was associated with a significantly lower reocclusion rate compared with first generation DES (3.0% vs 10.1%; $P < .001$).[12] However, in the AngiographiC Evaluation of the Everolimus-Eluting Stent in Chronic Total Occlusion (**ACE-CTO**) Study that included very long lesion and stent length and high proportion of patients with prior coronary artery bypass graft surgery, target lesion revascularization rates were significantly higher (37% at 12 months).[42] In the Evaluation of the XIENCE Coronary Stent, Performance, and Technique in Chronic Total Occlusions (**EXPERT CTO**) study of 250 patients (successful CTO recanalization was achieved in 222) from 20 US centers target lesion revascularization at one year was 6.3%.[43] The **PRISON IV** trial randomized 330 patients to either an ultrathin-strut sirolimus-eluting stent (Orsiro SES, Biotronik) with biodegradable polymer or the durable polymer Xience EES and failed to show noninferiority for the primary endpoint of in-segment late lumen loss (0.13 ± 0.63 mm for Orsiro SES vs 0.02 ± 0.47 mm for EES; $P = .08$ for noninferiority).[44] In-stent and in-segment binary restenosis was significantly higher with the Orsiro SES as compared with EES (8.0% vs 2.1%; $P = .028$), with trend for higher risk for target-lesion and target-vessel revascularization (9.2% vs 4.0% [$P = .08$] and 9.2% vs 6.0% [$P = .33$]). The **CONSISTENT CTO** (Conventional Antegrade vs Sub-Intimal Synergy Stenting in Chronic Total Occlusions) study[21] showed low rates of angiographic restenosis and need for target vessel revascularization with the Synergy DES. The need for target vessel revascularization was higher with dissection and reentry techniques.[21]

References

1. Colmenarez HJ, Escaned J, Fernandez C, et al. Efficacy and safety of drug-eluting stents in chronic total coronary occlusion recanalization: a systematic review and *meta*-analysis. *J Am Coll Cardiol* 2010;**55**:1854−66.

2. Saeed B, Kandzari DE, Agostoni P, et al. Use of drug-eluting stents for chronic total occlusions: a systematic review and *meta*-analysis. *Catheter Cardiovasc Interv* 2011;**77**:315−32.

3. Niccoli G, Leo A, Giubilato S, et al. A *meta*-analysis of first-generation drug-eluting vs bare-metal stents for coronary chronic total occlusion: effect of length of follow-up on clinical outcome. *Int J Cardiol* 2011;**150**:351−4.

4. Ma J, Yang W, Singh M, Peng T, Fang N, Wei M. *Meta*-analysis of long-term outcomes of drug-eluting stent implantations for chronic total coronary occlusions. *Heart Lung* 2011;**40**:e32−40.

5. Galassi AR, Tomasello SD, Crea F, et al. Transient impairment of vasomotion function after successful chronic total occlusion recanalization. *J Am Coll Cardiol* 2012;**59**:711−18.

6. Park JJ, Chae IH, Cho YS, et al. The recanalization of chronic total occlusion leads to lumen area increase in distal reference segments in selected patients: an intravascular ultrasound study. *JACC Cardiovasc Interv* 2012;**5**:827−36.

7. Brilakis ES. *Manual of coronary chronic total occlusion interventions. A step-by-step approach.* 2nd ed. Elsevier; 2017.

8. Jaguszewski M, Guagliumi G, Landmesser U. Optical frequency domain imaging for guidance of optimal stenting in the setting of recanalization of chronic total occlusion. *J Invasive Cardiol* 2013;**25**:367−8.

9. Koln PJ, Scheller B, Liew HB, et al. Treatment of chronic total occlusions in native coronary arteries by drug-coated balloons without stenting—a feasibility and safety study. *Int J Cardiol* 2016;**225**:262−7.

10. Cortese B, Buccheri D, Piraino D, Silva-Orrego P, Seregni R. Drug-coated balloon angioplasty: an intriguing alternative for the treatment of coronary chronic total occlusions. *Int J Cardiol* 2015;**187**:238−9.

11. Cortese B, Buccheri D, Piraino D, Silva-Orrego P. Drug-coated balloon angioplasty for coronary chronic total occlusions. An OCT analysis for a "new" intriguing strategy. *Int J Cardiol* 2015;**189**:257−8.

12. Valenti R, Vergara R, Migliorini A, et al. Predictors of reocclusion after successful drug-eluting stent-supported percutaneous coronary intervention of chronic total occlusion. *J Am Coll Cardiol* 2013;**61**:545−50.

13. Kandzari DE, Grantham JA, Lombardi W, Thompson C. Not all subintimal chronic total occlusion revascularization is alike. *J Am Coll Cardiol* 2013;**61**:2570.

14. Mogabgab O, Patel VG, Michael TT, et al. Long-term outcomes with use of the CrossBoss and stingray coronary CTO crossing and re-entry devices. *J Invasive Cardiol* 2013;**25**:579−85.

15. Rinfret S, Ribeiro HB, Nguyen CM, Nombela-Franco L, Urena M, Rodes-Cabau J. Dissection and re-entry techniques and longer-term outcomes following successful percutaneous coronary intervention of chronic total occlusion. *Am J Cardiol* 2014;**114**:1354−60.

16. Amsavelu S, Christakopoulos GE, Karatasakis A, et al. Impact of crossing strategy on intermediate-term outcomes after chronic total occlusion percutaneous coronary intervention. *Can J Cardiol* 2016;**32**(1239):e1−7.

17. Carlino M, Figini F, Ruparelia N, et al. Predictors of restenosis following contemporary subintimal tracking and reentry technique: the importance of final TIMI flow grade. *Catheter Cardiovasc Interv* 2016;**87**:884−92.

18. Azzalini L, Dautov R, Brilakis ES, et al. Procedural and longer-term outcomes of wire- vs device-based antegrade dissection and re-entry techniques for the percutaneous revascularization of coronary chronic total occlusions. *Int J Cardiol* 2017;**231**:78−83.

19. Muramatsu T, Tsuchikane E, Oikawa Y, et al. Incidence and impact on midterm outcome of controlled subintimal tracking in patients with successful recanalisation of chronic total occlusions: J-PROCTOR registry. *EuroIntervention* 2014;**10**:681−8.

20. Saito S, Maehara A, Yakushiji T, et al. Serial intravascular ultrasound findings after treatment of chronic total occlusions using drug-eluting stents. *Am J Cardiol* 2016; **117**:727−34.

21. Walsh SJ, Hanratty CG, McEntegart M, et al. Intravascular healing is not affected by approaches in contemporary CTO PCI: the CONSISTENT CTO study. *JACC Cardiovasc Interv* 2020;**13**:1448−57.

22. Finn MT, Doshi D, Cleman J, et al. Intravascular ultrasound analysis of intraplaque vs subintimal tracking in percutaneous intervention for coronary chronic total occlusions: one year outcomes. *Catheter Cardiovasc Interv* 2019;**93**:1048−56.

23. Song L, Maehara A, Finn MT, et al. Intravascular ultrasound analysis of intraplaque vs subintimal tracking in percutaneous intervention for coronary chronic total occlusions and association with procedural outcomes. *JACC Cardiovasc Interv* 2017;**10**:1011−21.

24. Hasegawa K, Tsuchikane E, Okamura A, et al. Incidence and impact on midterm outcome of intimal vs subintimal tracking with both antegrade and retrograde approaches in patients with successful recanalisation of chronic total occlusions: J-PROCTOR 2 study. *EuroIntervention* 2017;**12**:e1868−73.

25. Brilakis ES. *Manual of percutaneous coronary interventions: a step-by-step approach:*. Elsevier; 2021.

26. Kim BK, Shin DH, Hong MK, et al. Clinical impact of intravascular ultrasound-guided chronic total occlusion intervention with zotarolimus-eluting vs biolimus-eluting stent implantation: randomized study. *Circ Cardiovasc Interv* 2015;**8**:e002592.

27. Tian NL, Gami SK, Ye F, et al. Angiographic and clinical comparisons of intravascular ultrasound-vs angiography-guided drug-eluting stent implantation for patients with chronic total occlusion lesions: two-year results from a randomised AIR-CTO study. *EuroIntervention* 2015;**10**:1409−17.

28. Karacsonyi J, Alaswad K, Jaffer FA, et al. Use of intravascular imaging during chronic total occlusion percutaneous coronary intervention: insights from a contemporary multicenter registry. *J Am Heart Assoc* 2016;**5**.

29. Mintz GS. Back to the future: intravascular imaging to assess and guide CTO PCI procedures. *JACC Cardiovasc Interv* 2020;**13**:1458−9.

30. Flattery E, Rahim HM, Petrossian G, et al. Competency-based assessment of interventional cardiology fellows' abilities in intracoronary physiology and imaging. *Circ Cardiovasc Interv* 2020;**13**:e008760.

31. Kang SJ, Mintz GS, Kim WJ, et al. Changes in left main bifurcation geometry after a single-stent crossover technique: an intravascular ultrasound study using direct imaging of both the left anterior descending and the left circumflex coronary arteries before and after intervention. *Circ Cardiovasc Interv* 2011;**4**:355−61.

32. Agostoni P, Valgimigli M, Biondi-Zoccai GG, et al. Clinical effectiveness of bare-metal stenting compared with balloon angioplasty in total coronary occlusions: insights from a systematic overview of randomized trials in light of the drug-eluting stent era. *Am Heart J* 2006;**151**:682−9.

33. Buller CE, Dzavik V, Carere RG, et al. Primary stenting vs balloon angioplasty in occluded coronary arteries: the Total Occlusion Study of Canada (TOSCA). *Circulation* 1999;**100**:236−42.

34. Kandzari DE, Rao SV, Moses JW, et al. Clinical and angiographic outcomes with sirolimus-eluting stents in total coronary occlusions: the ACROSS/TOSCA-4 (approaches to chronic occlusions with sirolimus-eluting stents/total occlusion study of coronary arteries-4) trial. *JACC Cardiovasc Interv* 2009;**2**:97−106.

35. Suttorp MJ, Laarman GJ, Rahel BM, et al. Primary Stenting of Totally Occluded Native Coronary Arteries II (PRISON II): a randomized comparison of bare metal stent implantation with sirolimus-eluting stent implantation for the treatment of total coronary occlusions. *Circulation* 2006;**114**:921−8.

36. Van den Branden BJ, Rahel BM, Laarman GJ, et al. Five-year clinical outcome after primary stenting of totally occluded native coronary arteries: a randomised comparison of bare metal stent implantation with sirolimus-eluting stent implantation for the treatment of total coronary occlusions (PRISON II study). *EuroIntervention* 2012;**7**:1189−96.

37. Rubartelli P, Petronio AS, Guiducci V, et al. Comparison of sirolimus-eluting and bare metal stent for treatment of patients with total coronary occlusions: results of the GISSOC II-GISE multicentre randomized trial. *Eur Heart J* 2010;**31**:2014−20.

38. Reifart N, Hauptmann KE, Rabe A, Enayat D, Giokoglu K. Short and long term comparison (24 months) of an alternative sirolimus-coated stent with bioabsorbable polymer and a bare metal stent of similar design in chronic coronary occlusions: the CORACTO trial. *EuroIntervention* 2010;**6**:356−60.

39. Moreno R, Garcia E, Teles R, et al. Randomized comparison of sirolimus-eluting and everolimus-eluting coronary stents in the treatment of total coronary occlusions: results from the chronic coronary occlusion treated by everolimus-eluting stent randomized trial. *Circ Cardiovasc Interv* 2013;**6**:21−8.

40. Park HJ, Kim HY, Lee JM, et al. Randomized comparison of the efficacy and safety of zotarolimus-eluting stents vs. sirolimus-eluting stents for percutaneous coronary intervention in chronic total occlusion−CAtholic Total Occlusion Study (CATOS) trial. *Circ J* 2012;**76**:868−75.

41. Van den Branden BJ, Teeuwen K, Koolen JJ, et al. Primary Stenting of Totally Occluded Native Coronary Arteries III (PRISON III): a randomised comparison of sirolimus-eluting stent implantation with zotarolimus-eluting stent implantation for the treatment of total coronary occlusions. *EuroIntervention* 2013;**9**:841−53.

42. Kotsia A, Navara R, Michael TT, et al. The AngiographiC Evaluation of the Everolimus-Eluting Stent in Chronic Total Occlusion (ACE-CTO) Study. *J Invasive Cardiol* 2015;**27**:393−400.

43. Kandzari DE, Kini AS, Karmpaliotis D, et al. Safety and effectiveness of everolimus-eluting stents in chronic total coronary occlusion revascularization: results from the

EXPERT CTO multicenter trial (evaluation of the XIENCE coronary stent, performance, and technique in chronic total occlusions). *JACC Cardiovasc Interv* 2015;**8**:761−9.

44. Teeuwen K, van der Schaaf RJ, Adriaenssens T, et al. Randomized multicenter trial investigating angiographic outcomes of hybrid sirolimus-eluting stents with biodegradable polymer compared with everolimus-eluting stents with durable polymer in chronic total occlusions: the PRISON IV trial. *JACC Cardiovasc Interv* 2017;**10**:133−43.

45. Kandzari DE, Lembo NJ, Carlson HD, et al. Procedural, clinical, and health status outcomes in chronic total coronary occlusion revascularization: results from the PERSPECTIVE study. *Catheter Cardiovasc Interv* 2020;**96**:567−76.

46. Teeuwen K, Van den Branden BJ, Rahel BM, et al. Late catch-up in lumen diameter at five-year angiography in MACE-free patients treated with sirolimus-eluting stents in the Primary Stenting of totally occluded native coronary arteries: a randomised comparison of bare metal stent implantation with sirolimus-eluting stent implantation for the treatment of total coronary occlusions (PRISON II). *EuroIntervention* 2013;**9**:212−19.

47. Kedhi E, Joesoef KS, McFadden E, et al. Second-generation everolimus-eluting and paclitaxel-eluting stents in real-life practice (COMPARE): a randomised trial. *Lancet* 2010;**375**:201−9.

48. Lanka V, Patel VG, Saeed B, et al. Outcomes with first- vs second-generation drug-eluting stents in coronary chronic total occlusions (CTOs): a systematic review and *meta*-analysis. *J Invasive Cardiol* 2014;**26**:304−10.

Chapter 11

Access closure

Achieving hemostasis after chronic total occlusion (CTO) percutaneous coronary intervention (PCI) is important, especially since two arterial access points are used in most cases. The choice of hemostasis technique depends on the type of access (radial vs femoral) and each patient's anatomic characteristics.

For radial access, in most cases a compression band is applied for hemostasis at the end of the procedure. The patent hemostasis technique can minimize the risk of radial artery occlusion. For femoral access use of vascular closure devices can shorten the time to ambulation and potentially reduce the risk of complications, although the latter remains controversial.[1] Preclosing with two Perclose sutures is the most common technique for achieving hemostasis when using an Impella device.

Determining whether to use a vascular closure device versus manual compression, how to use vascular closure devices and how to prevent and treat vascular access complications is discussed in detail in the Manual of Percutaneous Coronary Interventions.[2]

References

1. Robertson L, Andras A, Colgan F, Jackson R. Vascular closure devices for femoral arterial puncture site haemostasis. *Cochrane Database Syst Rev* 2016;**3**:CD009541.
2. Brilakis ES. *Manual of percutaneous coronary interventions: a step-by-step approach:*. Elsevier; 2020.

Chapter 12

Coronary physiology

Coronary physiology has a limited role in chronic total occlusion (CTO) percutaneous coronary intervention (PCI). It cannot be performed in the CTO vessel until after crossing, but it can be used to assess the severity of disease in the donor vessel and for stent optimization (taking into account substantial variation in physiologic assessment after CTO PCI).

Three studies have reported coronary physiology measurements after CTO crossing but before stent implantation. All 3 studies showed that the myocardial territories supplied by a CTO are ischemic, even when extensive collateral circulation is present. In a study of 92 patients who underwent coronary physiology assessment immediately after CTO crossing with a micro-catheter (before balloon angioplasty and stenting), resting ischemia was observed in 78% of patients, and fractional flow reserve (FFR) was <0.80 in all patients.[1] All 50 CTO patients in another similar study had a FFR if < 0.80 regardless of the presence and extent of collateral circulation.[2] All 63 patients assessed in a third similar study had an iFR ≤ 0.89.[3]

12.1   When to do coronary physiology during chronic total occlusion percutaneous coronary intervention?

12.1.1   Before percutaneous coronary intervention

Determine significance of intermediate coronary lesions.

The hemodynamic significance of a lesion in the donor vessel that collateralizes the CTO vessel may change after successful CTO recanalization. With recanalization of the CTO vessel, there is reduction in the area of myocardium supplied by the donor vessel and thus a change in physiologic indices in the donor vessel. In a study by Sachdeva et al., six of nine donor vessels that had baseline ischemia, as assessed by fractional flow reserve measurement (FFR ≤ 0.80) reverted to nonischemic FFR after successful CTO recanalization.[4] The mean increase in donor vessel FFR after CTO recanalization was 0.098 ± 0.04. In another study the mean increase in donor vessel FFR after CTO recanalization was 0.03.[5] Therefore, interpretation of coronary physiology performed for assessing ischemia in a donor vessel supplying collaterals to a CTO territory by FFR should be done with caution, as pressure losses at a

Manual of Chronic Total Occlusion Percutaneous Coronary Interventions.
DOI: https://doi.org/10.1016/B978-0-323-91787-2.00032-0

proximal donor vessel lesion may become hemodynamically nonsignificant once the subtended myocardial territory is reduced by successful CTO PCI.

12.1.2   During percutaneous coronary intervention

1. Post PCI adenosine FFR and non-hyperemic indices are emerging as a tool to determine functionally optimal PCI results,[6] but their utility in CTO PCI remains controversial.
2. The collateral blood supply decreases and physiologic indices increase in the CTO artery post CTO PCI. The change in physiologic indices does not occur immediately but over time. There is a decrease in the microvascular resistance in the myocardium supplied by the CTO vessel and positive remodeling of the CTO artery that contribute to this change.
3. In a study of 34 patients who underwent successful CTO PCI, FFR did not immediately increase, whereas iFR increased from 0.86 ± 0.10 to 0.88 ± 0.10.[7] However, at 4-month followup both FFR and iFR increased. Similar findings were observed in a study of 26 patients in whom FFR increased from 0.82 ± 0.10 after CTO PCI to 0.89 ± 0.07 at 4 months. In a third study, iFR was ≤ 0.89 in 23 of 63 (30%) patients following CTO PCI: stent postdilatation was done in 12 patients and was successful in increasing the iFR to > 0.89 in 10 of them.[3]

Physiologic assessment post CTO PCI should be interpreted with caution as local microvasculature function can take a long period of time to recover after restoration of blood flow in the occluded vessel. In a pilot study of 25 patients, the maximum blood flow in the CTO artery increased by 49% and microvascular resistance decreased by 29% after a mean follow-up of 46 days compared with immediately after the index procedure.[8]

Step-by-step description of how to perform coronary physiologic assessment is presented in the *Manual of Percutaneous Coronary Interventions*.[9]

References

1. Werner GS, Surber R, Ferrari M, Fritzenwanger M, Figulla HR. The functional reserve of collaterals supplying long-term chronic total coronary occlusions in patients without prior myocardial infarction. *Eur Heart J* 2006;**27**:2406–12.
2. Sachdeva R, Agrawal M, Flynn SE, Werner GS, Uretsky BF. The myocardium supplied by a chronic total occlusion is a persistently ischemic zone. *Catheter Cardiovasc Interv* 2014;**83**:9–16.
3. Kayaert P, Coeman M, Drieghe B, et al. iFR uncovers profound but mostly reversible ischemia in CTOs and helps to optimize PCI results. *Catheter Cardiovasc Interv* 2021;**97**:646–55.

4. Sachdeva R, Agrawal M, Flynn SE, Werner GS, Uretsky BF. Reversal of ischemia of donor artery myocardium after recanalization of a chronic total occlusion. *Catheter Cardiovasc Interv* 2013;**82**:E453−8.

5. Ladwiniec A, Cunnington MS, Rossington J, et al. Collateral donor artery physiology and the influence of a chronic total occlusion on fractional flow reserve. *Circ Cardiovasc Interv* 2015;**8**.

6. Hakeem A, Uretsky BF. Role of postintervention fractional flow reserve to improve procedural and clinical outcomes. *Circulation* 2019;**139**:694−706.

7. Mohdnazri SR, Karamasis GV, Al-Janabi F, et al. The impact of coronary chronic total occlusion percutaneous coronary intervention upon donor vessel fractional flow reserve and instantaneous wave-free ratio: implications for physiology-guided PCI in patients with CTO. *Catheter Cardiovasc Interv* 2018;**92**:E139−48.

8. Keulards DCJ, Karamasis GV, Alsanjari O, et al. Recovery of absolute coronary blood flow and microvascular resistance after chronic total occlusion percutaneous coronary intervention: an exploratory study. *J Am Heart Assoc* 2020;**9**:e015669.

9. Brilakis ES. *Manual of percutaneous coronary interventions: a step-by-step approach:*. Elsevier; 2021.

Chapter 13

Intravascular imaging for CTO PCI

Intravascular imaging can: (1) facilitate chronic total occlusion (CTO) crossing; (2) guide the lesion preparation strategy; (3) facilitate stent sizing; and (4) help optimize the CTO PCI result.[1−5] Intravascular imaging should be used in virtually all CTO percutaneous coronary interventions (PCIs).[6]

13.1 Intravascular imaging modality selection

Two types of intravascular imaging are currently clinically available: intravascular ultrasound (IVUS) and optical coherence tomography (OCT). Unlike IVUS, OCT image acquisition requires contrast injection to clear the column of blood from the coronary artery and allow the light beam to reach the vessel wall. Contrast injection is usually not desirable during CTO PCI, especially when performing dissection/reentry (Sections 8.2 and 8.3), as it can lead to propagation of dissections; hence, IVUS is preferred for guiding CTO PCI crossing. IVUS also has higher depth of penetration.[4]

Both IVUS and OCT can be used to assess the result of CTO PCI, but the higher spatial resolution of OCT makes it the preferred modality for detecting dissections and stent strut apposition. Use of either modality requires experience with image acquisition and interpretation.

13.2 Intravascular imaging for CTO crossing

Intravascular imaging (mainly IVUS) can assist with CTO crossing, as follows[4]:

1. Evaluation of proximal cap ambiguity.
2. Evaluation of guidewire position during **antegrade** CTO crossing attempts.
3. Guiding **antegrade** wire reentry into the distal true lumen in case of extraplaque guidewire position.
4. Evaluation of guidewire position during **retrograde** CTO crossing and assisting with true lumen reentry, for example by appropriate balloon size selection for the reverse controlled antegrade and retrograde tracking (reverse CART) technique.
5. Assessment of the risk of side branch occlusion.

Manual of Chronic Total Occlusion Percutaneous Coronary Interventions.
DOI: https://doi.org/10.1016/B978-0-323-91787-2.00010-1
© 2023 Elsevier Inc. All rights reserved.

13.2.1   Resolving proximal cap ambiguity

Proximal cap ambiguity (Section 16.1) is defined as inability to confidently determine the location of the proximal cap and is common in CTOs with a side branch at the proximal cap.[7] Proximal cap ambiguity is present in approximately one-third of CTOs and is associated with lower success rates, higher utilization of the retrograde approach, and lower procedural efficiency.[8–10] Proximal cap ambiguity can sometimes be resolved by better coronary angiography using various projections or by coronary computed tomography angiography. If these approaches fail, three techniques are recommended: IVUS-guided puncture, "move-the-cap" techniques (such as the "scratch and go" and the balloon-assisted subintimal entry [BASE] techniques),[11] and retrograde crossing. The global CTO crossing algorithm supports all three strategies without prioritizing one strategy over another.[7] Instead, the strategy that optimizes safety and increases the likelihood of success should be selected on the basis of CTO anatomy.

IVUS can assist with crossing of CTOs with ambiguous proximal caps if there is a side branch adjacent to the occlusion (Fig. 13.1).[5] IVUS probes with short distance between the imaging sensor and the tip of the catheter (2.5 mm in the short-tip Eagle Eye catheter [Philips] and 8 mm in the AnteOwl WR [Terumo]) are preferred.

The IVUS probe is advanced into the side branch to identify the precise location of the occlusion and the presence of calcium (Fig. 13.2).[5] Any side

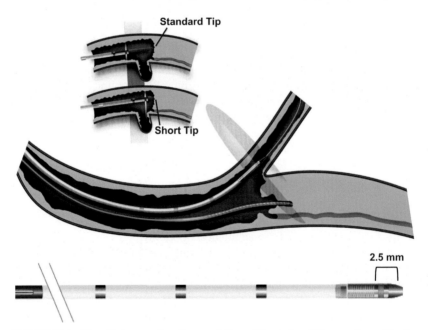

FIGURE 13.1   Use of intravascular ultrasound for guiding entry of the crossing guidewire into the chronic total occlusion. *Reproduced with permission from Philips.*

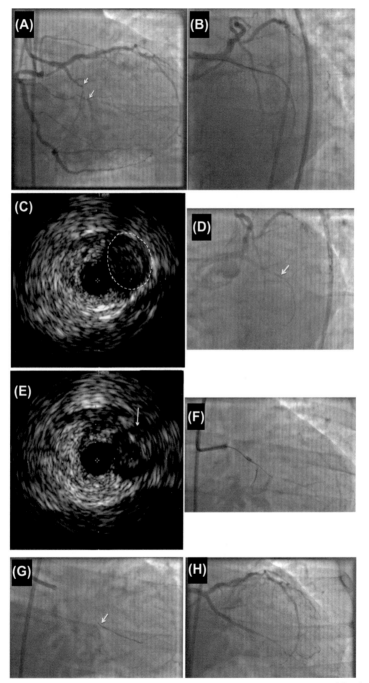

(*Continued*)

branch originating at the site of occlusion can be used, if it is large enough to accept an IVUS probe, unless there is extreme angulation or ostial stenosis (or calcification) preventing the insertion of the IVUS probe.[5] In large vessels, increasing the field of view (from 10 mm to 12 or 14 mm) can help visualize the CTO entry point.

IVUS can be used for crossing the CTO proximal cap using either: (1) continuous real-time imaging, or (2) intermittent serial imaging.

CTO crossing under **real time IVUS guidance** requires positioning the probe at the site with the best view of the stump allowing continuous monitoring of the wire position so that successful entry of the wire, ideally centrally (Fig. 13.3), in the stump can be visualized as it occurs.[5] This technique requires large guide catheters that can simultaneously accommodate the IVUS catheter and a microcatheter: usually 8 French for large microcatheters, such as the Corsair (Asahi Intecc) and Turnpike (Teleflex), or 7 French for smaller microcatheters, such as the FineCross (Terumo), Caravel and Corsair XS (Asahi Intecc) and Turnpike LP (Teleflex).[5] Alternatively, two 6 French guide catheters can be used simultaneously in a ping-pong configuration in the CTO vessel, one for microcatheter insertion and the second for insertion of the IVUS catheter.[12] The bulky IVUS probe may also deflect away guidewires and microcatheters from the stump and preclude simultaneous contrast injections, especially when the side branch is originating at a shallow angle from the occluded vessel. The IVUS probe position may change during crossing attempts, requiring frequent repositioning. Teamwork of two operators, one controlling the IVUS probe position and one managing the wire can facilitate this maneuver.

CTO crossing with intermittent IVUS guidance is performed by puncturing the site where the angiographic roadmap showed the IVUS transducer when the CTO proximal cap was detected, but without the benefit of continuous IVUS image acquisition.[5]

In a study of 22 patients with ostial, stumpless CTOs who underwent PCI under IVUS guidance, procedural success was 77%.[13]

Once the puncture is made where the stump was visualized on IVUS, subsequent IVUS imaging is performed to confirm the intraplaque position of the guidewire.[5] Especially if the side branch is relatively parallel to the

---

◀ FIGURE 13.2 **Example of intravascular ultrasound (IVUS) use to resolve proximal cap ambiguity.** (CTO Manual online case 34). Ostial chronic total occlusion (CTO) of the first obtuse marginal branch (*arrows*, **panel A**). Repeat antegrade crossing attempts were unsuccessful and the guidewire frequently entered the distal circumflex (**panel B**). IVUS demonstrated that the CTO (*yellow circle*, **panel C**) actually originated proximal (*arrow*, **panel D**) to the distal circumflex's apparent origin. During repeat antegrade crossing attempts a Confianza Pro 12 guidewire was utilized and its location within the CTO was confirmed by IVUS (*arrow*, **panel E**) before advancing it through the occlusion (**panel F**). *Reproduced with permission from the* Manual of chronic total occlusion percutaneous coronary interventions. *2nd ed.*

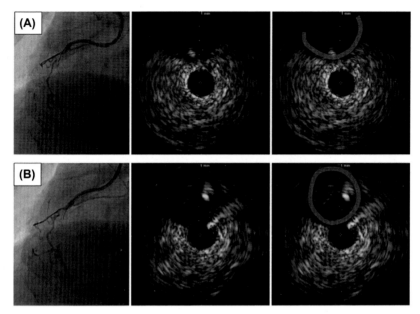

FIGURE 13.3   IVUS-guided puncture of the CTO proximal cap.
Panel A. Proximal RCA CTO. The IVUS probe was inserted into a side branch originating at the proximal cap of the CTO. A guidewire was advanced antegradely, but its position at the level of CTO entry was very close to the EEL (suggesting extraplaque tracking).
Panel B. The guidewire was retrieved and then advanced in a different direction. IVUS showed that the guidewire was far from the EEL, suggesting that the redirected guidewire was tracking intraplaque at the CTO proximal cap. *EEL*, external elastic lamina; *IVUS*, intravascular ultrasound; *RCA*, right coronary artery. *Reproduced with permission from Galassi AR, Sumitsuji S, Boukhris M, et al. Utility of intravascular ultrasound in percutaneous revascularization of chronic total occlusions: an overview. JACC Cardiovasc Interv 2016;9:1979–1991.*

CTO vessel, the wire progression can be monitored successfully for at least the first critical few millimeters.[5] Reinserting the IVUS probe will show whether the highly reflective guidewire is located within the CTO segment and follow it in its first path or, conversely, suggest immediate withdrawal and trying a different entry point.

13.2.2   Evaluation of guidewire position during antegrade crossing attempts

Contralateral injection is the preferred method for determining guidewire position during antegrade CTO crossing attempts. **Microcatheters and other equipment should not be advanced over the guidewire until after confirming that it is located within the vessel architecture.** Sometimes, it may be unclear whether the guidewire is located within the distal true lumen or in a false lumen. In such cases, IVUS can help clarify the wire position (Fig. 13.4). Delivery of the IVUS catheter can be challenging, especially through heavy calcification, sometimes requiring predilatation with a small balloon. The short-tip phased-array IVUS (Eagle-Eye, Philips) is preferred because of the short distance between the tip of the IVUS probe and the sensor, but any IVUS catheter can be used.[5] Advancing an IVUS catheter through the extraplaque space may cause hematoma formation than could hinder reentry.

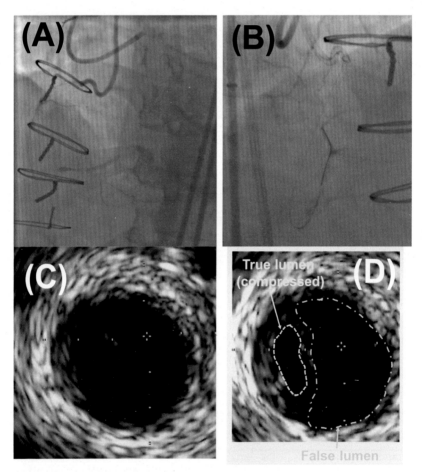

FIGURE 13.4   **Use of IVUS to clarify the guidewire position during antegrade crossing.**
Panel A: right coronary artery CTO.
Panel B: the distal position of the guidewire is unclear (true lumen vs false lumen).
Panels C and D: IVUS shows that the guidewire is located into the distal false lumen.

IVUS can confirm in-stent wire crossing when treating in-stent CTOs, hence preventing stent crushing in case of substrut wire crossing in a segment of the occlusion.

13.2.3 Facilitating antegrade reentry

During antegrade wiring, the guidewire may enter the extraplaque space. One option for reentering into the distal true lumen is to use a reentry system, such as the Stingray balloon, as described in Section 8.2. A second option is to use the parallel wire technique or one of its variations, such as the "see-saw technique" or use of a dual lumen microcatheter. A third option is to attempt directed penetration into the distal true lumen under IVUS guidance.[14,15]

Insertion of the IVUS into the extraplaque space may require predilatation with a small (1.0 to 1.5 mm diameter) balloon. A stiff tip guidewire, such as the Miracle 6 or 12 guidewire (Asahi Intecc) is preferred for inserting the IVUS probe, being careful to avoid inadvertently advancing those wires too far distally. Occasionally, the first stiff tip extraplaque wire enters a side branch beyond the occlusion (e.g., a septal/diagonal in the left anterior descending artery or a right ventricular branch/acute marginal branch in the right coronary artery), still with no access to the distal vessel but allowing safe insertion of the IVUS probe.[5] IVUS can then guide advancement of a second guidewire (such as Gaia 3, Gaia Next 3, Confianza Pro 12 [Asahi Intecc], Infiltrac or Infiltrac Plus [Abbott], or Hornet 14 [Boston Scientific]) (usually advanced through a microcatheter in order to enhance support), until it enters into the distal true lumen. Although this is often called "distal reentry," in most cases it is a form of parallel wiring, as IVUS helps identify a better and more central position for the second guidewire within the occlusion.[14,15] The starting point of the examination is at the proximal cap as identified by IVUS and the second guidewire is advanced in parallel in a central position, finally breaking into the distal true lumen (Fig. 9.5)[5] (Fig. 13.5).

IVUS-guided antegrade reentry is used mainly at expert centers in Japan[14,15] because: (1) it is less predictable than dedicated reentry devices; and (2) insertion of an IVUS catheter in the extraplaque space could cause a hematoma that could hinder subsequent reentry attempts.[5,16] In one case series of twenty patients who had failed both antegrade and retrograde crossing attempts, use of IVUS-guided wiring reentry resulted in 85% success and one case of pericardial tamponade.[17]

13.2.4 Facilitating retrograde wire crossing

IVUS can be used during retrograde crossing attempts for: (1) guiding retrograde guidewire crossing, and (2) facilitating the reverse CART technique.

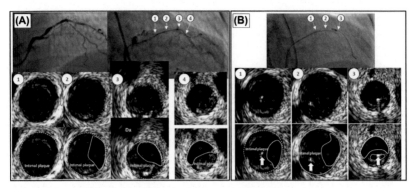

FIGURE 13.5   **IVUS-guided parallel wiring.** Proximal LAD CTO. **Panel A**: a first guidewire was advanced in the LAD and an IVUS probe was inserted to check the guidewire position. IVUS showed that the guidewire was in the extraplaque (subintimal) space in the distal part (2−4), while the guidewire was intraplaque (intimal) proximally (1). **Panel B**: the IVUS probe was left on the first guidewire and a second guidewire was thereafter advanced into the intraplaque space with IVUS guidance, and successfully reached the distal true lumen. The dotted line corresponds to the external elastic lamina, while the continuous line shows the limits of the extraplaque space. *Reproduced with permission from Galassi AR, Sumitsuji S, Boukhris M, et al. Utility of intravascular ultrasound in percutaneous revascularization of chronic total occlusions: an overview.* JACC Cardiovasc Interv *2016;9:1979−1991.*

1. **Retrograde wire crossing from the distal into the proximal true lumen** can be challenging. Antegrade IVUS can confirm the location of retrograde guidewire entry into the proximal true lumen. IVUS use is especially important in ostial LAD or circumflex lesions, in which the retrograde guidewire may cross into the left main through the extraplaque space, potentially compromising the other vessel with catastrophic consequences (Fig. 13.6). Similarly, during PCI of ostial RCA CTOs, the retrograde wire can enter the extraplaque space causing aortic dissection (Fig. 13.7).

2. The reverse CART technique is the most commonly used retrograde dissection/reentry technique (Section 8.3). Reverse CART is performed with balloon inflation over the antegrade guidewire, followed by advancement of the retrograde guidewire into the proximal true lumen. Multiple variations of this technique have been developed.[18] IVUS can facilitate reverse CART by:

   a. Helping determine the position of the antegrade and retrograde guidewires as outlined in Fig. 8.4.33 and Fig. 13.8 and discussed in detail in Section 8.4.4.7.2.2.2.

   b. Allowing optimal balloon sizing for medial disruption. IVUS provides information on true CTO vessel size and plaque composition, thereby reducing the perforation risk.

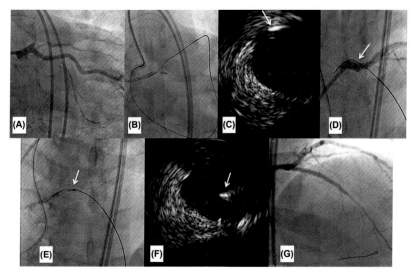

FIGURE 13.6 **IVUS-guided retrograde wire crossing (ostial LAD). Panel A**: ostial LAD CTO with 2 previously failed antegrade attempts. **Panel B**: a retrograde Ultimate 3 Bros (Asahi Intecc) guide-wire was advanced into the left main but could not reenter into the guide catheter. **Panels C and D**: IVUS and angiography confirmed extraplaque (subintimal) position of retrograde guidewire in the left main (*white arrow*). **Panels E and F**: true lumen reentry of Conquest Pro 9 (Asahi Intecc) guidewire, confirmed by IVUS (*white arrow*) with a final successful outcomes **Panel G**: *Reproduced with permission from Galassi AR, Sumitsuji S, Boukhris M, et al. Utility of intravascular ultrasound in percutaneous revascularization of chronic total occlusions: an overview.* JACC Cardiovasc Interv *2016;9:1979–1991.*

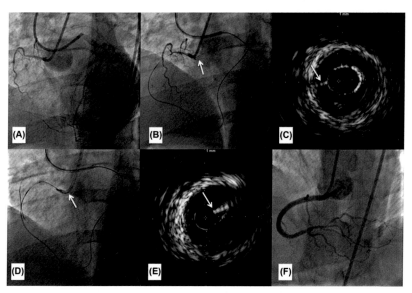

FIGURE 13.7 **IVUS-guided RCA CTO PCI. Panel A**: proximal RCA CTO. **Panel B**: a retrograde Fielder FC guidewire could not be advanced into the antegrade guide catheter. **Panel C**: IVUS-confirmed intraplaque guidewire position (*white arrow*). **Panels D and E**: the retrograde Fielder FC guidewire was placed into the true lumen at the ostial RCA, as confirmed by IVUS (*white arrow*). **Panel E**: fielder FC advancement into the RCA guide catheter. **Panel F**: final result. *Reproduced with permission from Galassi AR, Sumitsuji S, Boukhris M, et al. Utility of intravascular ultrasound in percutaneous revascularization of chronic total occlusions: an overview.* JACC Cardiovasc Interv *2016;9:1979–1991.*

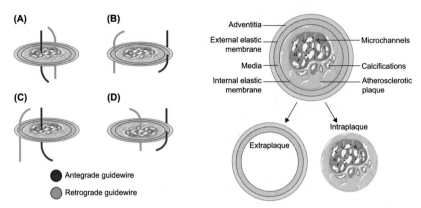

FIGURE 13.8   Possible positions of the antegrade and retrograde guidewire during reverse CART:
Panel A: Antegrade guidewire intraplaque + retrograde guidewire intraplaque.
Panel B: Antegrade guidewire extraplaque + retrograde guidewire extraplaque.
Panel C: Antegrade guidewire intraplaque + retrograde guidewire extraplaque.
Panel D: Antegrade guidewire extraplaque + retrograde guidewire intraplaque (most challenging to overcome).
Reproduced with permission from Galassi AR, Sumitsuji S, Boukhris M, et al. Utility of intravascular ultrasound in percutaneous revascularization of chronic total occlusions: an overview. JACC Cardiovasc Interv *2016;9:1979–1991.*

   c. Facilitating selection of the appropriate position within the CTO vessel to create a connection between the antegrade and the retrograde guidewire (more proximally or more distally) when the initial reverse CART strategy is unsuccessful due to severe calcification (Fig. 13.9). IVUS can help change the location of reentry attempts into a less calcified segment of the target vessel.
   d. After successful reverse CART, it can sometimes be impossible to externalize the retrograde guidewire into the antegrade guide catheter. In this situation, IVUS can show a connection only in the mid-distal portion of the vessel and not in the proximal segment; in such cases using the child-in-mother facilitated technique can make reentry possible in a segment other than the ostium (Fig. 13.10).
   e. After externalization of retrograde guidewire, IVUS-guided stenting helps to avoid contrast injections that could create or worsen antegrade or retrograde subintimal dissection.

In one study that included patients with previously failed antegrade or retrograde true-to-true lumen crossing, IVUS-guided reverse CART resulted in 95.9% and 93.9% technical and procedural success, respectively, with 2% risk of major adverse events.[19]

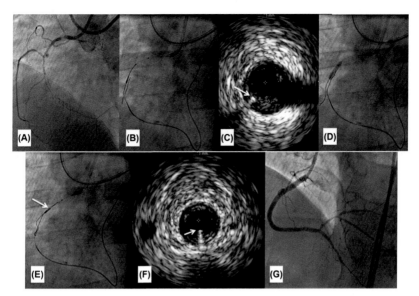

FIGURE 13.9 **IVUS-guided reverse CART. Panel A**: mid RCA CTO. **Panels B and C**: retrograde Ultimate Bros 3 (Asahi Intecc) in the intraplaque space as confirmed by IVUS (*white arrow*). **Panel D**: IVUS-guided reverse CART (balloon sizing). **Panels E and F**: retrograde Fielder FC (Asahi Intecc) guidewire reentry and connection in the same space as the antegrade guidewire, confirmed by IVUS (*white arrow*). **Panel G**: final result. *Reproduced with permission from Galassi AR, Sumitsuji S, Boukhris M, et al. Utility of intravascular ultrasound in percutaneous revascularization of chronic total occlusions: an overview. JACC Cardiovasc Interv 2016;9:1979–1991.*

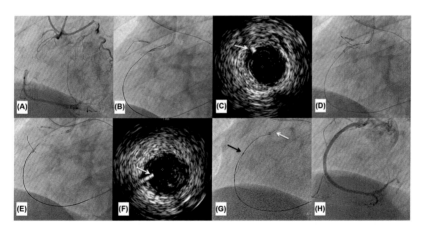

FIGURE 13.10 **IVUS-guided guide extension reverse CART using the child-in-mother facilitated technique. Panel A**: Ostial RCA CTO. **Panel B**: retrograde guidewire is in the false lumen in the RCA ostium. **Panel C**: IVUS confirmed extraplaque position of the retrograde wire (*white arrow*) with extended vessel dissection. **Panel D**: IVUS-guided reverse CART. **Panels E and F**: retrograde guidewire reentry into mid RCA true lumen, with connection between the antegrade and retrograde guidewires, confirmed by IVUS (*white arrow*). **Panel G**: Guideliner (Teleflex) (*black arrow*) assisted reentry of the retrograde guidewire (*white arrow*) into the RCA guide catheter. **Panel H**: final result. *Reproduced with permission from Galassi AR, Sumitsuji S, Boukhris M, et al. Utility of intravascular ultrasound in percutaneous revascularization of chronic total occlusions: an overview. JACC Cardiovasc Interv 2016;9:1979–1991.*

13.2.5   Assessing the risk of side branch occlusion

In CTOs with bifurcations at the proximal or distal cap, there is a risk of side branch occlusion after successful guidewire crossing. Use of IVUS can help detect dissections or severe plaque at the side branch origin that increase the risk of side branch occlusion.

13.3   Intravascular imaging for PCI optimization

Intravascular imaging with IVUS or OCT can be used for optimizing the final result of CTO PCI by:

1. Evaluating the presence and extent of calcium to determine the need of lesion preparation techniques.
2. Assessing the extent of disease and hematoma and selecting adequate landing zones in vessels that are often diffusely diseased.
3. Selecting appropriate stent length and diameter (Section 10.2.3) to minimize the risk of stent undersizing (that can lead to higher rates of restenosis and stent thrombosis) or oversizing (that may lead to dissection or perforation). Intravascular imaging can help determine whether the distal vessel has large plaque burden (and needs to be stented) or is hypoperfused without plaque and will likely increase in size during follow-up without stenting. In long occluded segments, the presence of areas of positive and negative remodeling is frequent and sizing should be tailored accordingly using IVUS to reduce the risk of perforation or malapposition. For this purpose, coregistration that can be performed without contrast injections can be very helpful (Fig. 13.11).

   IVUS can also be useful to identify intramyocardial segments or areas of myocardial bridging in the mid-LAD where stenting is associated with high rates of failure.
4. Confirmation of optimal PCI result [complete lesion coverage with good stent expansion and apposition without significant distal edge dissections (Section 10.2.13; Fig. 10.2)]. Achieving an optimal stenting result can decrease the risk of restenosis and stent thrombosis.[6]
5. Decreasing contrast volume.

   Two randomized controlled clinical trials have demonstrated improved outcomes with use of IVUS during CTO PCI. The Korean Chronic Total Occlusion Intervention with Drug-eluting Stents guided by IVUS (CTO-IVUS) randomized 402 patients after successful guidewire crossing of the CTO to IVUS vs angiographic guidance. After 12 months, the incidence of MACE (major adverse cardiovascular events; composite of death, myocardial infarction and target vessel revascularization) was significantly lower in the IVUS-guided group (2.6% vs 7.1%; $P = .035$).[2]

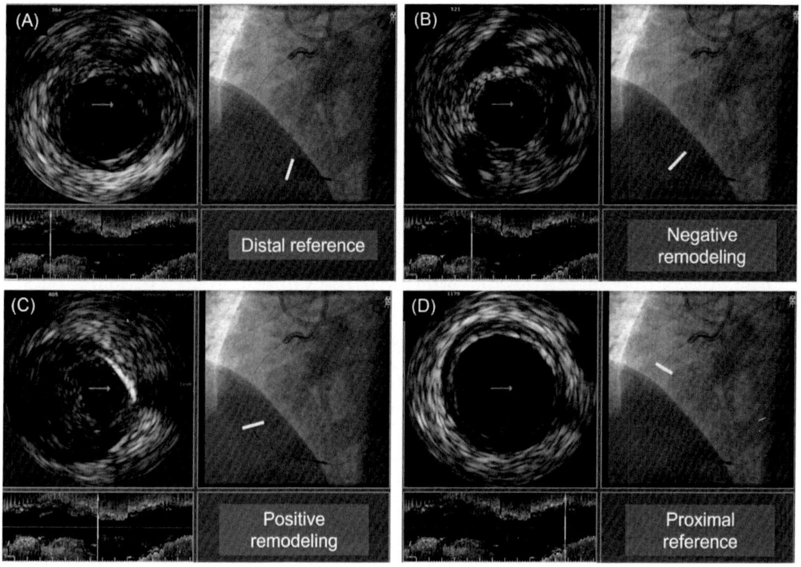

FIGURE 13.11   **Use of IVUS coregistration for balloon and stent sizing in CTO PCI.** Coregistration was performed using the guidewire, without contrast injections. The system indicates the area of the vessel that corresponds to each IVUS cross section. **Panel A**: distal landing zone in a healthy area of the distal RCA. **Panel B**: area of negative remodeling. **Panel C**: positive remodeling and high plaque burden within the occlusion. **Panel D**: proximal landing zone in a healthy area of the vessel in the mid RCA. IVUS is essential to ensure appropriate sizing is vessels with significant changes in size along the segment that needs to be stented. *Courtesy of Dr. Nieves Gonzalo.*

The Comparison of Angiography versus IVUS-guided Stent Implantation for Chronic Total Coronary Occlusion Recanalization (AIR-CTO) trial randomized 230 patients after successful guidewire crossing to IVUS or angiography-guided stenting, and found lower rates of 12-month in-stent late lumen loss (0.28 ± 0.48 mm vs 0.46 ± 0.68 mm, $P = .025$) and restenosis of the "in true lumen" stent (3.9% vs 13.7%, $P = .021$) in the IVUS-guided group. AIR CTO also demonstrated a significantly lower rate of definite/probable stent thrombosis at 2 years (0.9% vs 6.1%, $P = .043$), but no difference in overall MACE at 1-year and 2-year follow-up (18.3% vs 22.6%, $P = .513$ for the first year of follow-up and 21.7% vs 25.2%, $P = .641$ for the second year of follow-up), although the study was not powered for clinical events.[3]

Suboptimal stent optimization is associated with higher risk of subsequent events in CTO-PCI. Among 1396 patients who underwent IVUS-guided intervention in four randomized trials including CTOs and long lesions, stent optimization criteria were minimum stent area >5.5 mm^2 and expansion $>80\%$ of mean reference area. The nonoptimized group had higher MACE compared with the optimized group (4.8% vs 1.9%, log-rank $P = .002$; adjusted hazard ratio 2.95, 95% CI: $1.43-6.06$).[20]

13.4   Frequency of intravascular imaging for CTO PCI

Use of intravascular imaging varies between operators, centers, and countries, due in part to differences in availability and reimbursement. In the Prospective Global Registry for the Study of Chronic Total Occlusion Intervention (PROGRESS-CTO) registry, intravascular imaging was used in 234 of 619 cases: 38% (IVUS in 36%, OCT in 3%, and both in 1.45%). IVUS was used in 206 (39%) cases in the Multicenter Korean CTO Registry and in 554 of 1,166 (47.5%) cases in the Japanese multicenter registry on complications during retrograde approach for chronic total coronary occlusions.[21] IVUS was used in 9.2% of 1582 CTO lesions treated with the retrograde approach between January 2008 and December 2012 at 44 European centers in the EURO CTO registry.[22]

Use of imaging has been steadily increasing[23] and is currently performed in nearly all CTO and complex PCI cases at experienced centers and during live case demonstrations.

References

1. Karacsonyi J, Alaswad K, Jaffer FA, et al. Use of intravascular imaging during chronic total occlusion percutaneous coronary intervention: insights from a contemporary multicenter registry. *J Am Heart Assoc* 2016;**5**.
2. Kim BK, Shin DH, Hong MK, et al. Clinical impact of intravascular ultrasound-guided chronic total occlusion intervention with zotarolimus-eluting vs biolimus-eluting stent implantation: randomized study. *Circ Cardiovasc Interv* 2015;**8**:e002592.
3. Tian NL, Gami SK, Ye F, et al. Angiographic and clinical comparisons of intravascular ultrasound- vs angiography-guided drug-eluting stent implantation for patients with chronic total occlusion lesions: two-year results from a randomised AIR-CTO study. *EuroIntervention* 2015;**10**:1409−17.
4. Xenogiannis I, Tajti P, Karmpaliotis D, et al. Intravascular imaging for chronic total occlusion intervention. *Curr Cardiovasc Imaging Rep* 2018;**11**:31.
5. Galassi AR, Sumitsuji S, Boukhris M, et al. Utility of intravascular ultrasound in percutaneous revascularization of chronic total occlusions: an overview. *JACC Cardiovasc Interv* 2016;**9**:1979−91.
6. Chugh Y, Buttar R, Kwan T, et al. Outcomes of intravascular ultrasound-guided vs angiography-guided percutaneous coronary interventions in chronic total occlusions: a systematic review and *meta*-analysis. *J Invasive Cardiol* 2022;**34**:E310−18.
7. Wu EB, Brilakis ES, Mashayekhi K, et al. Global chronic total occlusion crossing algorithm: JACC state-of-the-art review. *J Am Coll Cardiol* 2021;**78**:840−53.
8. Sapontis J, Christopoulos G, Grantham JA, et al. Procedural failure of chronic total occlusion percutaneous coronary intervention: insights from a multicenter US registry. *Catheter Cardiovasc Interv* 2015;**85**:1115−22.
9. Alessandrino G, Chevalier B, Lefevre T, et al. A Clinical and angiographic scoring system to predict the probability of successful first-attempt percutaneous coronary intervention in patients with total chronic coronary occlusion. *JACC Cardiovasc Interv* 2015;**8**:1540−8.

10. Karatasakis A, Danek BA, Karmpaliotis D, et al. Impact of proximal cap ambiguity on outcomes of chronic total occlusion percutaneous coronary intervention: insights from a Multicenter US Registry. *J Invasive Cardiol* 2016;**28**:391−6.

11. Vo MN, Karmpaliotis D, Brilakis ES. "Move the cap" technique for ambiguous or impenetrable proximal cap of coronary total occlusion. *Catheter Cardiovasc Interv* 2016;**87**:742−8.

12. Nakashima M, Ikari Y, Aoki J, Tanabe K, Tanimoto S, Hara K. Intravascular ultrasound-guided chronic total occlusion wiring technique using 6 Fr catheters via bilateral transradial approach. *Cardiovasc Interv Ther* 2015;**30**:68−71.

13. Ryan N, Gonzalo N, Dingli P, et al. Intravascular ultrasound guidance of percutaneous coronary intervention in ostial chronic total occlusions: a description of the technique and procedural results. *Int J Cardiovasc Imaging* 2017;**33**:807−13.

14. Suzuki S, Okamura A, Iwakura K, et al. Initial outcomes of AnteOwl IVUS-based 3D wiring using the tip detection method for CTO intervention. *JACC Cardiovasc Interv* 2021;**14**:812−14.

15. Okamura A, Iwakura K, Iwamoto M, et al. Tip detection method using the new IVUS facilitates the 3-dimensional wiring technique for CTO intervention. *JACC Cardiovasc Interv* 2020;**13**:74−82.

16. Chou RH, Lai CH, Lu TM. Side-branch and coaxial intravascular ultrasound guided wire re-entry after failed retrograde approach of chronic total occlusion intervention. *Acta Cardiol Sin* 2016;**32**:363−6.

17. Huang WC, Teng HI, Hsueh CH, Lin SJ, Chan WL, Lu TM. Intravascular ultrasound guided wiring re-entry technique for complex chronic total occlusions. *J Interv Cardiol* 2018;**31**:572−9.

18. Matsuno S, Tsuchikane E, Harding SA, et al. Overview and proposed terminology for the reverse controlled antegrade and retrograde tracking (reverse CART) techniques. *EuroIntervention* 2018;**14**:94−101.

19. Dai J, Katoh O, Kyo E, Tsuji T, Watanabe S, Ohya H. Approach for chronic total occlusion with intravascular ultrasound-guided reverse controlled antegrade and retrograde tracking technique: single center experience. *J Interv Cardiol* 2013;**26**:434−43.

20. Kim D, Hong SJ, Kim BK, et al. Outcomes of stent optimisation in intravascular ultrasound-guided interventions for long lesions or chronic total occlusions. *EuroIntervention* 2020;**16**:e480−8.

21. Okamura A, Yamane M, Muto M, et al. Complications during retrograde approach for chronic coronary total occlusion: sub-analysis of Japanese multicenter registry. *Catheter Cardiovasc Interv* 2016;**88**:7−14.

22. Galassi AR, Sianos G, Werner GS, et al. Retrograde recanalization of chronic total occlusions in Europe: procedural, in-hospital, and long-term outcomes from the multicenter ERCTO registry. *J Am Coll Cardiol* 2015;**65**:2388−400.

23. Vemmou E, Khatri J, Doing AH, et al. Impact of intravascular ultrasound utilization for stent optimization on 1-year outcomes after chronic total occlusion percutaneous coronary intervention. *J Invasive Cardiol* 2020;**32**:392−9.

Chapter 14

Hemodynamic support for chronic total occlusion percutaneous coronary intervention

Video 14.1 Hemodynamic support: when and which device.

Video 14.3 Impella step-by-step.

Video 14.4 VA-ECMO cannulation.

14.1 Hemodynamic support: when and what device

Mechanical circulatory support can be used prophylactically before chronic total occlusion (CTO) percutaneous coronary intervention (PCI) (in approximately 4% of cases) or after a complication has occurred (in approximately 1% of cases).[1–3] Fig. 14.1 outlines various parameters that can help determine the need and type of mechanical circulatory support.[4]

14.1.1 Hemodynamics

The most important parameters for determining the need for hemodynamic support are the patient's hemodynamics before, during, or after PCI.

14.1.1.1 Before percutaneous coronary intervention

The most important parameters for determining the need for hemodynamic support are the baseline hemodynamics. In some settings the indication is clear. If a patient develops cardiac arrest during CTO PCI, hemodynamic support with veno-arterial extracorporeal membrane oxygenation (VA-ECMO) is essential to prevent death (VA-ECMO supports both circulation and oxygenation).[5,6] Similarly, hemodynamic support is often needed in patients with cardiogenic shock, although its efficacy in this setting has been

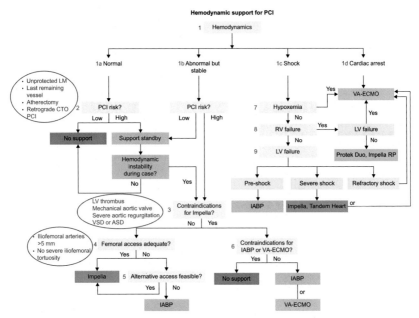

FIGURE 14.1 Determining whether hemodynamic support is needed for PCI and optimal device selection.[4] *ASD*, atrial septal defect; *CTO*, chronic total occlusion; *IABP*, intra-aortic balloon pump; *LM*, left main; *LV*, left ventricle, *PCI*, percutaneous coronary intervention; *RV*, right ventricle; *VA-ECMO*, veno-arterial extracorporeal membrane oxygenation; *VSD*, ventricular septal defect. *Reproduced with permission from Brilakis ES. Manual of percutaneous coronary interventions: a step-by-step approach. Elsevier; 2021.*

controversial.[7,8] In stable patients undergoing elective PCI, determining whether hemodynamic support is needed can be challenging and depends heavily on the anticipated risk of PCI (Section 14.1.2).

While a low left ventricular (LV) ejection fraction might suggest a higher risk of decompensation during complex PCI (e.g., using the retrograde approach or requiring atherectomy), this should not be considered the most relevant (or only) indication for hemodynamic support during CTO PCI. Right heart catheterization can help determine the patient's hemodynamic status and is strongly recommended in patients with low ejection fraction undergoing complex PCI, patients with unclear hemodynamic status, and patients with cardiogenic shock.[9] Low cardiac index (<2.2 l/min/m^2) indicates a low-flow condition. High pulmonary capillary wedge pressure (>18 mmHg) is suggestive of left ventricular failure and low pulmonary artery pulsatility index (PAPi <1.0 in the setting of acute myocardial infarction) is suggestive of right ventricular failure. Patients with severely elevated pulmonary capillary wedge pressure (>25 mmHg) and low mixed venous oxygen saturation ($<50\%$) are at very high risk of periprocedural pulmonary edema and shock.

The Swan Ganz catheter can be left in place during CTO PCI to monitor hemodynamics (pulmonary artery pressure) and determine the need for escalation or de-escalation of support.

14.1.1.2 During and after percutaneous coronary intervention

Hemodynamic support may be needed during or after CTO PCI if the patient develops hemodynamic instability (such as hypotension or acute pulmonary edema). Sometimes the need for hemodynamic support is anticipated, whereas in other cases it may be unexpected (such as in case of acute vessel closure or left main dissection). The outcomes of patients requiring emergent hemodynamic support during the procedure due to a complication have been poor.[1]

14.1.2 Risk of percutaneous coronary intervention

The risk of PCI depends on the location and morphology of the target lesion(s), anticipated PCI techniques, and patient comorbidities.

1. **Lesion location:**
   a. Unprotected left main (especially in patients with occluded or diseased right coronary artery).
   b. Last remaining vessel.
   c. Large area of myocardium at risk.
2. **Lesion morphology:**
   a. Severe calcification (especially if atherectomy is needed) or severe tortuosity (risk of pseudolesion formation with reduction in antegrade flow).
   b. High grade baseline stenosis that carries a high risk of acute vessel occlusion during wiring attempts.
   c. Bifurcation lesion with large side branches with high-grade stenoses that may be challenging to protect.
3. **Procedural plan** (complex, prolonged procedures resulting in significant ischemia):
   a. Atherectomy.
   b. Retrograde CTO PCI.
   c. Large intracoronary thrombus.
4. **Patient comorbidities,** such as:
   a. Ineligibility for coronary artery bypass graft surgery.
   b. Significant valvular heart disease (for example severe aortic stenosis or severe mitral regurgitation) with or without left ventricular dysfunction.
   c. Cardiogenic shock and cardiac arrest.
   d. Chronic kidney disease.

Prophylactic hemodynamic support is in most cases recommended (if feasible) in patients who have *both* abnormal hemodynamics and are planned to undergo PCI of high-risk lesions.[10]

Hemodynamic support is *not used prophylactically* in most patients with (1) abnormal hemodynamics and undergoing low-risk procedure, or (2) normal hemodynamics and undergoing high-risk PCI, but *should be immediately available for use* in case of hemodynamic decompensation. Some operators insert a 4 French femoral arterial sheath prophylactically, so that they can rapidly insert a hemodynamic support device (intra-aortic balloon pump, Impella, or VA-ECMO) if the need arises during the procedure. Prophylactic upfront hemodynamic support may be considered in some of these patients.

14.1.3   Contraindications for Impella

Where available, the most commonly used device for hemodynamic support in CTO and high-risk PCI is the Impella CP device.[11]

Contraindications to Impella use include: (1) mechanical aortic valve; (2) left ventricular thrombus, (3) moderate to severe aortic regurgitation; (4) aortic stenosis with aortic valve area ≤ 0.6 cm^2 (although Impella has been successfully used in such patients, usually after aortic balloon valvuloplasty[12]); (5) severe peripheral arterial disease precluding placement of the Impella system; and (6) atrial septal defect (ASD) or ventricular septal defect (VSD, although the Impella has been successfully used in post myocardial infarction VSD patients[13]).

14.1.4   Femoral access

The Impella CP device requires insertion of a 14 French peel-away sheath, which in turn requires an iliofemoral artery >5 mm in diameter without severe calcification, tortuosity, or obstructive lesions. Intravascular lithotripsy can be used to facilitate insertion of large bore sheaths and catheters through calcified iliofemoral arteries.[14] Most often, hemostasis is achieved by using 2 Perclose devices with the sutures deployed before inserting the Impella sheath. A CTO operator should be familiar with large bore access management.

Sometimes, the Impella sheath is occlusive, potentially causing lower extremity ischemia if the Impella device remains inserted for a prolonged period of time after the procedure. In such cases the ipsilateral superficial femoral artery can be accessed via a 4 Fr or 5 Fr short sheath and a micropuncture kit in antegrade fashion.[15] Contralateral 6 French common femoral arterial access (in a standard retrograde direction) is then obtained and the side arms of the two sheaths are connected using a male-to-male connector. This creates an external femoral-femoral bypass, whereby the blood flows from the contralateral 6 Fr sheath through the side arm into the side arm of the ipsilateral 5 Fr (or larger) antegrade sheath down the ischemic limb providing adequate perfusion.[15] Other combinations can also be used (e.g., radial-femoral or internal ipsilateral/contralateral bypass).[16] The activated clotting time should be kept high (>220 s) and the sheaths flushed every $1-2$ h to minimize the risk of sheath thrombosis.

14.1.5 Alternative access sites

In case of suboptimal femoral access or in case of failure to advance the Impella CP device, alternative access sites can be considered, such as subclavian, axillary and transcaval. Use of alternative access sites requires expertise with large bore access and closure techniques.

14.1.6 Contraindications for veno-arterial extracorporeal membrane oxygenation or intra-aortic balloon pump

1. Severe aortic regurgitation.
2. Aortic dissection.
3. High-risk of bleeding (although hemodynamic support devices can be used without anticoagulation for a short period of time in patients with active bleeding).

14.1.7 Hypoxemia

In patients with cardiogenic shock and hypoxemia, VA-ECMO is used as it can support both oxygenation and circulation. Since VA-ECMO may increase the load to the left ventricle, if prolonged support is needed after PCI, a left-sided Impella or an intra-aortic balloon pump (IABP) might be required to unload the left ventricle and promote recovery.

14.1.8 Right ventricular failure

The risk of right ventricular failure during CTO PCI increases when acute marginal branches become occluded, which is common when performing dissection/reentry in the right coronary artery. Also, preexisting right ventricular failure (e.g., due to chronic pulmonary thromboembolism) might predispose to acute right ventricular failure during CTO PCI. Right ventricular function can be assessed using echocardiography (may not be immediately available in the cardiac catheterization laboratory) or using invasive hemodynamics (right atrial pressure >12 mmHg and Pulmonary Artery Pulsatility Index [PAPi] < 1.0 suggest right ventricular dysfunction).

$$\text{Pulmonary Artery Pulsatility Index (PAPi)} := \frac{\text{sPAP} - \text{dPAP}}{\text{RA}}$$

dPAP, diastolic pulmonary artery pressure; *sPAP*, systolic pulmonary artery pressure; *RA*, right atrial pressure

In patients with cardiogenic shock and isolated right ventricular failure without left ventricular failure, right ventricular support devices may be used, such as the Protek-Duo and Impella RP.

In patients with cardiogenic shock and biventricular failure (that could be due to a complication of CTO PCI, such as coronary perforation[17]), VA-ECMO is usually used, although combined left and right ventricular support devices can also be used in some cases.

14.1.9 Left ventricular failure

In preshock patients (cardiac index >2.0 L/min/m^2), use of IABP may suffice, although continuous hemodynamic monitoring with a Swan Ganz catheter is needed to determine the need for escalating support.[9,18] CTO PCI is rarely, if ever, performed in such patients.

In shock patients with isolated severe LV failure, the Impella CP device is currently the most commonly used device (Tandem Heart can be used at centers experienced in its use and requires a transseptal puncture).

In patients with LV failure and refractory shock (continued hypoperfusion) despite Impella CP or Tandem Heart use, escalation to VA-ECMO or Impella 5.0 or 5.5 may be required. The Impella is often left in place to provide left ventricular unloading (LV venting) during VA-ECMO support as the VA-ECMO increases left ventricular afterload and cardiac mechanical work.

14.2 Hemodynamic support devices

Four devices are currently available in the United States for providing percutaneous left ventricular hemodynamic support: the IABP, the Impella (2.5, CP, 5.0, and 5.5, Abiomed Inc., Danvers, MA), the Tandem Heart (Liva Nova, Pittsburgh, PA), and VA-ECMO.[19] Two devices are available for isolated right ventricular support, the Protek-Duo (LivaNova, London, UK) and the Impella RP (Abiomed). The most commonly used device for CTO PCI is the Impella CP.[1] VA-ECMO may be needed in case of complete circulatory collapse.[5,6] A comparison of the various devices and step-by-step instruction for use of IABP, Impella, VA-ECMO, and Tandem Heart is provided in the *Manual of Percutaneous Coronary Interventions*.[4]

When using an Impella device, the Impella sheath can be used to insert one of the guide catheters required for CTO PCI [Single-access for Hi-risk PCI (SHiP) technique].[20,21] The Impella CP shaft is only 9 French, whereas the Impella CP sheath is 14 French inside diameter (18 French outside diameter). Hence, there is space inside the sheath for inserting up to a 7 French sheath next to the Impella CP shaft by puncturing the Impella CP sheath with a micropuncture needle at 10 o'clock or 2 o'clock position and inserting a long 45 cm 7 French hydrophilic sheath next to the Impella shaft. The SHiP technique reduces the need for additional arterial punctures for inserting the guide catheter. This is of particular importance in CTO PCI because it requires dual access in most cases (the Impella CP sheath is used for one guide catheter and a second arterial access point for the second guide catheter).

A single-access dual-injection technique has been described for performing simple antegrade-only CTO PCI.[22] However, this approach limits the array of techniques that can be performed (such as retrograde approach, anchoring, live intravascular ultrasound guidance for proximal cap puncture, block-and-deliver technique in case of perforation, etc.).

References

1. Danek BA, Basir MB, O'Neill WW, et al. Mechanical circulatory support in chronic total occlusion percutaneous coronary intervention: insights from a Multicenter U.S. Registry. *J Invasive Cardiol* 2018;**30**:81−7.

2. Riley RF, McCabe JM, Kalra S, et al. Impella-assisted chronic total occlusion percutaneous coronary interventions: A multicenter retrospective analysis. *Catheter Cardiovasc Interv* 2018;**92**:1261−7.

3. Neupane S, Basir M, Alqarqaz M, O'Neill W, Alaswad K. High-risk chronic total occlusion percutaneous coronary interventions assisted with tandemheart. *J Invasive Cardiol* 2020;**32**:94−7.

4. Brilakis ES. *Manual of percutaneous coronary interventions: a step-by-step approach.* Elsevier; 2021.

5. Avula V, Karacsonyi J, Hammadah M, Brilakis ES. Venoarterial extracorporeal membrane oxygenation for life-threatening complications of percutaneous coronary and structural heart interventions. *Cardiovasc Revasc Med* 2021.

6. Shaukat A, Hryniewicz-Czeneszew K, Sun B, et al. Outcomes of extracorporeal membrane oxygenation support for complex high-risk elective percutaneous coronary interventions: a single-center experience and review of the literature. *J Invasive Cardiol* 2018;**30**:456−60.

7. Iannaccone M, Albani S, Giannini F, et al. Short term outcomes of Impella in cardiogenic shock: A review and *meta*-analysis of observational studies. *Int J Cardiol* 2021;**324**:44−51.

8. Thiele H, Jobs A, Ouweneel DM, et al. Percutaneous short-term active mechanical support devices in cardiogenic shock: a systematic review and collaborative *meta*-analysis of randomized trials. *Eur Heart J* 2017;**38**:3523−31.

9. Saxena A, Garan AR, Kapur NK, et al. Value of hemodynamic monitoring in patients with cardiogenic shock undergoing mechanical circulatory support. *Circulation* 2020;**141**: 1184−97.

10. Rihal CS, Naidu SS, Givertz MM, et al. SCAI/ACC/HFSA/STS clinical expert consensus statement on the use of percutaneous mechanical circulatory support devices in cardiovascular care: endorsed by the American Heart Assocation, the Cardiological Society of India, and Sociedad Latino Americana de Cardiologia Intervencion; Affirmation of Value by the Canadian Association of Interventional Cardiology-Association Canadienne de Cardiologie d'intervention. *J Am Coll Cardiol* 2015;**2015**(65):e7−e26.

11. O'Neill WW, Kleiman NS, Moses J, et al. A prospective, randomized clinical trial of hemodynamic support with Impella 2.5 vs intra-aortic balloon pump in patients undergoing high-risk percutaneous coronary intervention: the PROTECT II study. *Circulation* 2012; **126**:1717−27.

12. Almalla M, Kersten A, Altiok E, Marx N, Schroder JW. Hemodynamic support with Impella ventricular assist device in patients undergoing TAVI: a single center experience. *Catheter Cardiovasc Interv* 2020;**95**:357−62.

13. Via G, Buson S, Tavazzi G, et al. Early cardiac unloading with ImpellaCP in acute myocardial infarction with ventricular septal defect. *ESC Heart Fail* 2020;**7**:708−13.

14. Di Mario C, Goodwin M, Ristalli F, et al. A prospective registry of intravascular lithotripsy-enabled vascular access for transfemoral transcatheter aortic valve replacement. *JACC Cardiovasc Interv* 2019;**12**:502−4.

15. Kaki A, Blank N, Alraies MC, et al. Access and closure management of large bore femoral arterial access. *J Interv Cardiol* 2018;**31**:969−77.

16. Kaki A, Alraies MC, Kajy M, et al. Large bore occlusive sheath management. *Catheter Cardiovasc Interv* 2019;**93**:678−84.

17. Moroni F, Brilakis ES, Azzalini L. Chronic total occlusion percutaneous coronary intervention: managing perforation complications. *Expert Rev Cardiovasc Ther* 2021;**19**:71−87.

18. Naidu SS, Baran DA, Jentzer JC, et al. *SCAI SHOCK stage classification expert consensus update: a review and incorporation of validation studies: this statement was endorsed by the American College of Cardiology (ACC), American College of Emergency Physicians (ACEP), American Heart Association (AHA), European Society of Cardiology (ESC) Association for Acute Cardiovascular Care (ACVC), International Society for Heart and Lung Transplantation (ISHLT), Society of Critical Care Medicine (SCCM), and Society of Thoracic Surgeons (STS) in December 2021. J Am Coll Cardiol* 2022;**79**:933−946.

19. Brilakis ES. *Manual of coronary chronic total occlusion interventions. A step-by-step approach.* 2nd ed. Elsevier; 2017.

20. Wollmuth J, Korngold E, Croce K, Pinto DS. The single-access for Hi-risk PCI (SHiP) technique. *Catheter Cardiovasc Interv* 2020;**96**:114−16.

21. Hakeem A, Tang D, Patel K, et al. Safety and efficacy of single-access impella for high-risk percutaneous intervention (SHiP). *JACC Cardiovasc Interv* 2022;**15**:347−8.

22. Marmagkiolis K, Caballero JA, Cilingiroglu M, Iliescu C. Single-access dual-injection technique (SADIT) for high-risk PCI with Impella CP. *Catheter Cardiovasc Interv* 2021; **98**:1138−40.

Part B

Complex subgroups

Chapter 15

Ostial chronic total occlusions

Ostial lesions are lesions located within 3−5 mm of the vessel origin.[1,2]

There are two types of ostial lesions: (1) aorto-ostial lesions and (2) branch ostial lesions.[1]

1. Aorto-ostial lesions involve the ostia of the right coronary artery, left main, and aortocoronary bypass grafts (both saphenous vein grafts and arterial grafts, such as internal mammary artery grafts).
2. Branch ostial lesions involve the ostia of branches of the coronary vessels, such as the left anterior descending artery (LAD), ramus, and circumflex (branches of the left main), diagonals (branches of the LAD), obtuse marginals (branches of the circumflex), the posterior descending and posterolateral (branches of the right coronary artery [RCA] in right dominant coronary circulation or the circumflex in left dominant coronary circulation), and ostial lesions of Y-grafts (graft attached to another graft, such as free right internal mammary artery [RIMA] attached to the left internal mammary artery [LIMA]).

15.1 Aorto-ostial chronic total occlusions

Aorto-ostial CTOs originate within 3−5 mm from the vessel ostium.[2] Aorto-ostial CTOs are often complex and require use of the retrograde approach.[3−5]

Flush aorto-ostial CTOs are aorto-ostial occlusions without a stump (right coronary artery, left main, or bypass graft). Such lesions can be challenging to cross, as guide catheters cannot be seated, and antegrade crossing is therefore not feasible, requiring a primary retrograde approach.

Avran et al. developed an algorithm on how to cross aorto-ostial right coronary artery CTOs (Fig. 15.1).

There are two important considerations/challenges in patients with aorto-ostial CTOs:

Manual of Chronic Total Occlusion Percutaneous Coronary Interventions.
DOI: https://doi.org/10.1016/B978-0-323-91787-2.00039-3
© 2023 Elsevier Inc. All rights reserved.

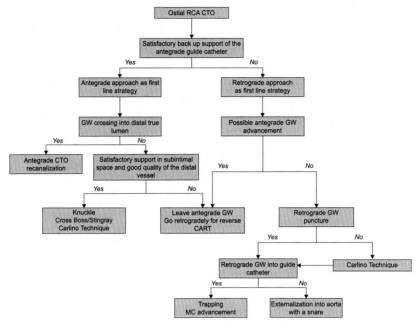

FIGURE 15.1 **Algorithm for the treatment of ostial right coronary artery chronic total occlusions**. *CART*, Controlled antegrade and retrograde tracking; *CTO*, chronic total occlusion; *GW*, guidewire; *MC*, microcatheter; *PCI*, percutaneous coronary intervention; *RCA*, right coronary artery.[6] *Reproduced with permission from Avran A, Boukhris M, Drogoul L, Brilakis ES. An algorithmic approach for the management of ostial right coronary artery chronic total occlusions. Catheter Cardiovasc Interv 2018;92:515−521.*

15.1.1 Determine whether there is a stump that can be used for antegrade crossing of the ostial chronic total occlusion

1. *Angiography in multiple projections, including aortography*

   Some patients are considered to have a flush aorto-ostial CTO when the vessel cannot be engaged. In some of these patients, however, the vessel may be patent, but may have anomalous origin. Various angiographic projections, "cusp angiography" (injection in the coronary cusp), and ascending aortography (typically 20 mL × 3 s injection in the ascending aorta for a total of 60 mL of contrast in the left anterior oblique projection) can help clarify the origin and location of such vessels.

2. *CT angiography*

   Coronary computed tomography angiography (Section 6.4) can provide a definitive answer as to whether the CTO vessel indeed has an aorto-ostial occlusion and provides additional information on the morphology of the occlusion (such as length, calcification, tortuosity), facilitating subsequent CTO recanalization attempts.

15.1.2 Retrograde crossing into the aorta

True flush aorto-ostial CTOs cannot be engaged with a guide catheter (as they do not have a stump), hence retrograde wire crossing into the aorta is necessary (CTO Manual online case 18, 37, 58). Such crossing can be challenging due to frequent severe calcification of the ostial occlusion and occasional tendency of the guidewire to advance into the aortic subintimal space.

How?

Retrograde crossing of a flush aorto-ostial CTO can be facilitated by:

1. *Highly penetrating guidewires*

    Use of stiff tip guidewires supported by the retrograde microcatheter can facilitate penetration into the aorta. Typical guidewires used in this manner include the Gaia and Gaia Next series (usually 2nd and 3rd), the Confianza Pro 12, Astato 20, Hornet 14, Infiltrac and Infiltrac Plus, and the Pilot 200 (CTO Manual online case 18).

2. *The Carlino technique*

    Injection of a small amount of contrast through the tip of the retrograde microcatheter can assist with crossing,[7,8] similar to crossing the antegrade cap, as described in Section 8.3.4.3 (Fig. 15.2).

3. *Transesophageal echocardiography (TEE)-guidance*

    TEE guidance has been successfully used for retrograde crossing of ostial CTOs.[9]

4. *The e-CART technique (electrocautery controlled antegrade and retrograde tracking and dissection)*

    This is a last resort technique, to be used when other techniques fail. It uses the power of electrocautery to penetrate through very challenging occlusions (CTO Manual online case 58).[10,11] E-CART is performed as follows (Fig. 15.3):

    1. The cautery pad is placed on the patient's hip.
    2. The retrograde guidewire is advanced retrograde pointing at the ostium.
    3. The guidewire should be covered by a microcatheter in its entire length, except for the tip.
    4. Typically a Confianza Pro 12 or Astato 20 guidewire is used with the distal 3 mm of the tip amputated (wire amputation is not necessary).
    5. A hemostat is used to clamp the back of the wire to the cautery needle
    6. The cautery power is set at 50 watts on "cut."
    7. The cautery is activated on the guidewire for 1 s under cineangiography to observe the wire "pop through."
    8. Guidance with transesophageal echocardiography may minimize the risk of guidewire perforation.[9]

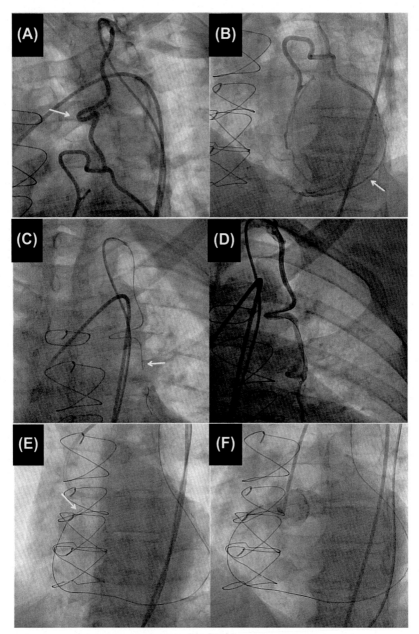

FIGURE 15.2   Left internal mammary artery with significant mid portion tortuosity (arrow, **panel A**), that supplies the distal right coronary artery via epicardial collaterals (arrow, **panel B**) from
(*Continued*)

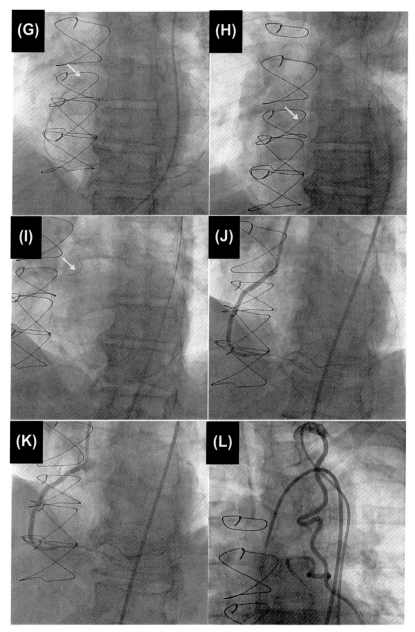

FIGURE 15.2   (Continued).

The e-CART technique should be used with caution, as it can cause a significant perforation or entry in nearby cardiac structures, such as the pulmonary artery. Retrograde electrocautery can often facilitate antegrade wire advancement, providing enough wire overlap to perform CART or reverse CART.

5. *The "Power Flush" technique*

Power Flush consists of a forceful injection of contrast dye directly through the guide catheter positioned against the aortic wall in correspondence to the coronary ostium location (that is identified by advancement of a retrograde guidewire), to gain access to the extraplaque space and proceed with vessel recanalization (Fig. 15.4).[12] In an iteration of this technique named "Nick And Flush," nicking of the aortic wall with a penetrative wire is performed before the Power Flush.

15.1.3   Optimal stent deployment

After crossing aorto-ostial CTOs, stent deployment should be optimized to avoid the following five pitfalls, as discussed in detail in the *Manual of Percutaneous Coronary Interventions*[13]:

1. Missing the ostium (incomplete ostium coverage by the stent).
2. Excessive stent overhang into the aorta.
3. Stent underexpansion.
4. Acute stent recoil.
5. Stent deformation.

◄ the circumflex. After advancement of a Corsair catheter (arrow, **panel C**) over a Sion guidewire (Asahi Intecc) through the LIMA, antegrade flow was preserved (**panel D**). The guidewire and Corsair catheter were advanced to the distal CTO cap (arrow, **panel E**), but retrograde wiring was extremely challenging due to LIMA tortuosity. Only Gaia guidewires (Asahi Intecc) could be advanced to the tip of the Corsair catheter but could not penetrate the occlusion (**panel F**). The Corsair catheter was exchanged for a Finecross catheter and 0.5 cc of contrast was injected advancing slightly more into the occlusion (arrow, **panel G**). Several additional Gaia second guidewires (Asahi Intecc) could not be advanced through the lesion. Intralesion contrast injection was repeated achieving contrast jet entry into the aorta (arrow, **panel H**). An RG3 guidewire (Asahi Intecc) was advanced into the aorta (arrow, **panel I**), snared and externalized (**panel J**), enabling stenting of the right coronary artery CTO with an excellent final result (**panel K**). LIMA flow was not affected during the procedure and no injury was seen in angiography performed after equipment removal (**panel L**).[7] *CTO Manual online case 37. Reproduced with permission from Amsavelu S, Carlino M, Brilakis ES. Carlino to the rescue: Use of intralesion contrast injection for bailout antegrade and retrograde crossing of complex chronic total occlusions. Catheter Cardiovasc Interv 2016;87:1118–1123.*

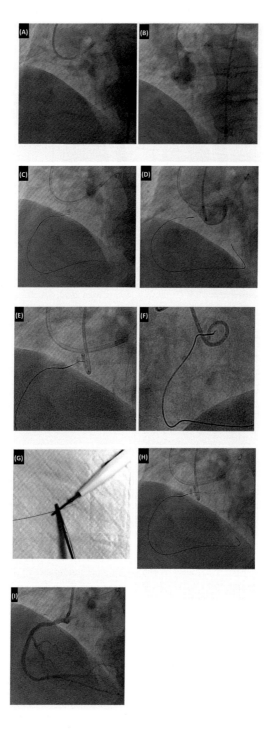

(*Continued*)

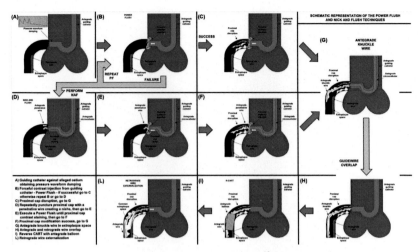

FIGURE 15.4 **Illustration of the Power Flush technique.**[12] *Reproduced with permission from Garbo R, Arioti M, Leoncini M. Power Flush, a novel bail-out technique for stumpless aorto-ostial CTOs. A case-based approach.* Cardiovasc Revasc Med *2022;40S:282−287.*

15.2 Branch ostial chronic total occlusions

Branch ostial lesions are essentially Medina 0.0.1 or 0.1.0. bifurcation lesions and are discussed in Chapter 16.[5] IVUS is often useful for crossing such lesions.[14]

◀ FIGURE 15.3 **Illustration of the e-CART technique.** Initial coronary angiograms showing (**panel A**) the retrograde filling of the distal right coronary artery from the left system and (**panel B**) true ostial occlusion of the right coronary artery. Extraplaque passage of a knuckled guidewire and microcatheter was achieved retrograde to the ostium of the right coronary artery occlusion (**panel C**). A pigtail catheter was placed at the site of the right coronary artery aorto-ostial occlusion (**panel D**). A left anterior oblique 40 projection (**panel E**) and a right anterior oblique 30 projection (**panel F**) were utilized to serve as a target for our retrograde wire to ensure accurate directing of the energized retrograde Confianza Pro 12 (Asahi Intecc). The back end of the guidewire was connected to a unipolar electrosurgery pencil using forceps (**panel G**). Distal crossing tip of the guidewire was energized in cutting mode at 50 W for a 1-s burst, with immediate unimpeded crossing into the lumen of the aorta (**panel H**). Final angiographic result after snaring and externalizing the retrograde wire utilizing standard chronic total occlusion percutaneous coronary intervention techniques in creating a neo-ostium of the right coronary artery (**panel I**).[10] *Reproduced with permission from Nicholson W, Harvey J, Dhawan R. E-CART (electrocautery-assisted re-entry) of an aorto-ostial right coronary artery chronic total occlusion: first-in-man.* JACC Cardiovasc Interv *2016;9:2356−2358.*

References

1. Jokhi P, Curzen N. Percutaneous coronary intervention of ostial lesions. *EuroIntervention* 2009;**5**:511−14.

2. Tajti P, Burke MN, Karmpaliotis D, et al. Prevalence and outcomes of percutaneous coronary interventions for ostial chronic total occlusions: insights from a multicenter chronic total occlusion registry. *Can J Cardiol* 2018;**34**:1264−74.

3. Ojeda S, Luque A, Pan M, et al. Percutaneous coronary intervention in aorto-ostial coronary chronic total occlusion: outcomes and technical considerations in a multicenter registry. *Rev Esp Cardiol (Engl Ed)* 2020;**73**:1011−17.

4. Guelker JE, Bufe A, Blockhaus C, et al. In-hospital outcomes after recanalization of ostial chronic total occlusions. *Cardiovasc Revasc Med* 2020;**21**:661−5.

5. Niizeki T, Tsuchikane E, Konta T, et al. Prevalence and predictors of successful percutaneous coronary intervention in ostial chronic total occlusion. *JACC Cardiovasc Interv* 2022.

6. Avran A, Boukhris M, Drogoul L, Brilakis ES. An algorithmic approach for the management of ostial right coronary artery chronic total occlusions. *Catheter Cardiovasc Interv* 2018;**92**:515−21.

7. Amsavelu S, Carlino M, Brilakis ES. Carlino to the rescue: use of intralesion contrast injection for bailout antegrade and retrograde crossing of complex chronic total occlusions. *Catheter Cardiovasc Interv* 2016;**87**:1118−23.

8. Carlino M, Ruparelia N, Thomas G, et al. Modified contrast microinjection technique to facilitate chronic total occlusion recanalization. *Catheter Cardiovasc Interv* 2016;**87**:1036−41.

9. Corrigan 3rd FE, Karmpaliotis D, Samady H, Lerakis S. Ostial right coronary chronic total occlusion: transesophageal echocardiographic guidance for retrograde aortic re-entry. *Catheter Cardiovasc Interv* 2018;**91**:1070−3.

10. Nicholson W, Harvey J, Dhawan R. E-CART (ElectroCautery-assisted re-entry) of an aorto-ostial right coronary artery chronic total occlusion: first-in-man. *JACC Cardiovasc Interv* 2016;**9**:2356−8.

11. Neupane S, Basir M, Alaswad K. Electrocautery-facilitated crossing (ECFC) of chronic total occlusions. *J Invasive Cardiol* 2020;**32**:55−7.

12. Garbo R, Arioti M, Leoncini M. Power Flush, a novel bail-out technique for stumpless aorto-ostial CTOs. A case-based approach. *Cardiovasc Revasc Med* 2022;**40S**:282−7.

13. Brilakis ES. *Manual of percutaneous coronary interventions: a step-by-step approach.* Elsevier; 2021.

14. Ryan N, Gonzalo N, Dingli P, et al. Intravascular ultrasound guidance of percutaneous coronary intervention in ostial chronic total occlusions: a description of the technique and procedural results. *Int J Cardiovasc Imaging* 2017;**33**:807−13.

Chapter 16

Bifurcations and CTOs

Preserving the patency of side branches during chronic total occlusion (CTO) recanalization can prevent acute complications and improve acute and long-term outcomes.[1–5] The risk of side branch occlusion is higher with use of dissection/reentry, ostial side branch stenosis, proximal main vessel stenosis of $\geq 50\%$ and no protection of the side branch with a wire prior to stenting.[6]

CTOs located at a bifurcation can be challenging to treat because:

1. The location of the proximal cap may be ambiguous;
2. Directing the guidewire at an angulation can be difficult;
3. The guidewire may preferentially enter the patent branch; or
4. The patent branch(es) may become compromised during CTO crossing.

In this chapter we discuss the following CTO percutaneous coronary intervention (PCI) scenarios involving a bifurcation (Fig. 16.1):

1. Proximal cap ambiguity
2. Ostial circumflex and ostial left anterior descending artery (LAD) CTOs
3. Bifurcation at the proximal cap
4. Bifurcation in the CTO body
5. Bifurcation at the distal cap

16.1 Proximal cap ambiguity

CTO *Manual* online cases 2, 5, 16, 30, 32, 33, 34, 45, 47, 49, 51, 55, 56, 69, 80, 88, 93, 104.

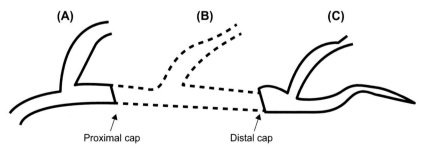

FIGURE 16.1   Potential locations of bifurcations in coronary CTO lesions.

Manual of Chronic Total Occlusion Percutaneous Coronary Interventions.
DOI: https://doi.org/10.1016/B978-0-323-91787-2.00013-7
© 2023 Elsevier Inc. All rights reserved.

Chronic total occlusions and
bifurcations go hand-in-hand

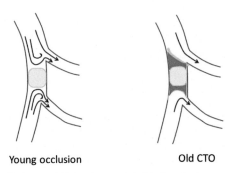

Young occlusion   Old CTO

FIGURE 16.2 **Why CTOs and bifurcations go hand-in-hand.** *Courtesy of Dr. Imre Ungi.*

Proximal cap ambiguity is inability to determine the exact location of the proximal cap of the occlusion, due to presence of obscuring side branches or overlapping branches that cannot be resolved despite multiple angiographic projections. It is encountered in approximately a third of CTO PCIs and is independently associated with technical failure.[7,8] The frequent occurrence of side branches adjacent to CTOs can be explained by hemodynamic mechanisms: turbulent flow and blood stasis generate thrombi proximal and distal to a recent occlusion, until the nearest side branches that stop this prothrombotic cascade (Fig. 16.2).

Fig. 16.3 illustrates several solutions for proximal cap ambiguity, both antegrade and retrograde.

Coronary angiography in multiple projections and coronary computed tomography angiography (CCTA) can sometimes resolve the ambiguity. If not, the following three techniques can be utilized, as highlighted in the global CTO crossing algorithm (Section 8.4):

1. Use of intravascular ultrasound[10]
2. Use of dissection/reentry techniques proximal to the proximal cap (move the cap techniques)[11]
3. A primary retrograde approach[12]

Selecting the optimal approach to proximal cap ambiguity depends on the lesion characteristics and the operator experience and expertise.[13]

16.1.1 Better angiography

CTO *Manual* online, cases 2, 47

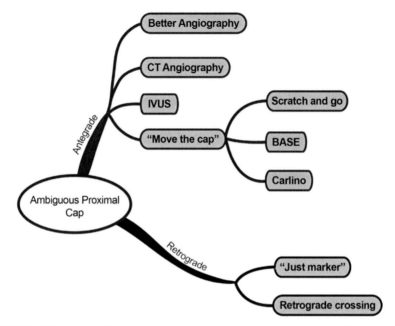

FIGURE 16.3 **Antegrade and retrograde approaches to chronic total occlusions with ambiguous proximal cap.**[9]

High-quality angiography, including dual injections and multiple, possibly steep, angiographic projections may help resolve proximal cap ambiguity. For example:

1. A Vieussens collateral (from the conus branch of the right coronary artery to the left anterior descending artery) may not fill with contrast if the right coronary artery catheter is deeply engaged, or it may have a separate ostium instead of originating from the right coronary artery. Contrast injections with the catheter less deeply engaged in the right coronary artery can allow filling of the Vieussens collateral and help clarify proximal cap ambiguity.
2. The origin of the CTO may overlap with the origin of a side branch. Various angiographic views with different angulation may help separate the branches.

16.1.2 Coronary computed tomography angiography

CCTA can help clarify the course of the occluded vessel (Fig. 16.4), as well as provide information on the presence of calcification and tortuosity.[14–16] CT angiography is particularly useful when the proximal lesion anatomy is unclear (Section 3.3.6). There are ongoing efforts for coregistration of coronary angiography and computed tomography images to facilitate crossing.[17,18]

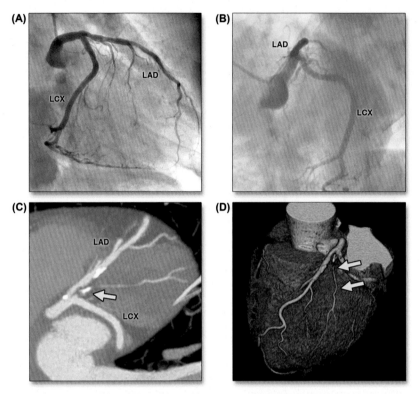

FIGURE 16.4 **Visualization of an ostial occlusion by coronary computed tomography angiography (CCTA). Panels A and B**: invasive angiography in a 53-year-old woman fails to visualize the large intermediate branch that is occluded at its origin. **Panels C and D**: visualization of the occluded intermediate branch by coronary CCTA (arrows). LAD, left anterior descending artery; LCX, left circumflex coronary artery. *Reproduced with permission from Opolski MP, Achenbach S. CT angiography for revascularization of CTO: crossing the borders of diagnosis and treatment. JACC Cardiovasc Imaging 2015;8:846−58.*

16.1.3 Intravascular ultrasound

Intravascular ultrasonography can help clarify the location of the proximal cap, especially when there is a side branch close to the occlusion.[19,20] This is described in detail in Section 13.2.1.

A variation of intravascular ultrasound (IVUS)-guided puncture of the proximal cap is the "Slipstream" technique[21]: an IVUS probe with a sort monorail segment (NaviFocus WR, Terumo) and a dual lumen catheter with small distance between the first and second port (Sasuke, Asahi) are both inserted on the same side branch wire. A second guidewire is advanced through the over-the-wire lumen of the dual lumen microcatheter to penetrate the proximal cap under continuous real-time imaging (Fig. 16.5). Coaxial alignment of the IVUS and

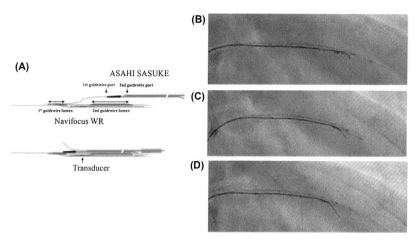

FIGURE 16.5 **Illustration of the "Slipstream" technique (panel A).** The NaviFocusWR intravascular ultrasound (IVUS) catheter is inserted on the side branch guidewire followed by insertion of a Sasuke dual lumen microcatheter over the same guidewire and advancement up to the tip of the IVUS catheter. This enabled successful puncture (**panel B and C**) and wire crossing into the left anterior descending artery (**panel D**). *Reproduced with permission from Kinoshita Y, Fujiwara H, Suzuki T. "Slipstream technique"-New concept of intravascular ultrasound guided wiring technique with double lumen catheter in the treatment of coronary total occlusions.* J Cardiol Cases *2017;16:52−55.*

the microcatheter on a single wire provides strong back up support, facilitating proximal cap puncture.

16.1.4   Move the cap techniques

CTO *Manual* online, cases 2, 16, 32, 45, 59, 61, 82, 83, 100, 104, 126, 134, 146, 153, 166, 168, 175, 181, 193scra.

The "move the cap" techniques use antegrade dissection/reentry to clarify the course of the occluded vessel and achieve crossing.[11] There are three variations of this technique, the "BASE" (balloon-assisted subintimal entry), the "scratch and go" technique and the Carlino technique.[22,23] Each allows the operator to decide on the site of proximal entry into the extraplaque space. The "move the cap" techniques should be avoided if there is an important side branch at the proximal cap.

16.1.4.1   BASE: balloon-assisted subintimal entry (Fig. 16.6)

● **Step 1: Wire the vessel proximal to the CTO**

How?

  ● Use a workhorse guidewire
  ● Check in orthogonal projections that the wire is actually within the intended segment

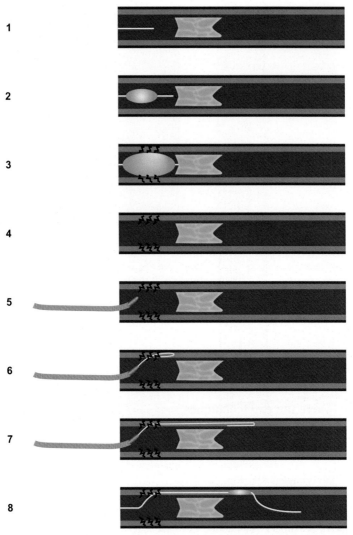

FIGURE 16.6   **Illustration of the balloon-assisted subintimal entry (BASE) technique.**

What can go wrong?

- The wire may advance into a side branch. This should be appreciated using orthogonal projections and corrected before advancing the balloon.

- **Step 2: Advance a balloon proximal to proximal cap**
How?

- Use a slightly oversized compliant balloon (1.1:1 or 1.2:1 balloon:vessel diameter ratio)

- Check in orthogonal projections that the wire and balloon are actually within the intended segment

What can go wrong?
- Balloon may not advance due to severely diseased proximal vessel and/or severe calcification. In such cases predilatation with a smaller balloon or other lesion preparation (for example with atherectomy or laser) may be required to facilitate balloon delivery. Alternatively, a larger or more supportive guide catheter (such as Amplatz 1 for the right coronary artery) or other guide supporting techniques, such as side branch anchor or a guide catheter extension (Section 9.3.1.8) may be needed.

- **Step 3: Balloon inflation**

How?
- At 10−15 atm

What can go wrong?
- Perforation of the proximal vessel, given balloon oversizing, hence high inflation pressures ($\geq$20 atm) should be avoided.

- **Step 4: Contrast injection**

How?
- Through the guide catheter to verify that proximal vessel dissection has occurred.

What can go wrong?
- Propagation of the dissection either downstream (potentially compressing the distal true lumen) or upstream (causing aorto coronary dissection). This can be prevented by gentle injection under fluoroscopic or cineangiographic imaging, use of side-hole guide catheters, or partially disengaging the guide catheter.

- **Step 5: Delivery of microcatheter proximal to proximal cap**

How?
- Over the workhorse guidewire that was used to deliver the angioplasty balloon

What can go wrong?
- Inability to deliver a microcatheter due to tortuosity or calcification (unlikely given prior balloon inflation). If it occurs additional balloon dilatations or increased guide catheter support may be needed.

- **Step 6: Insert a polymer-jacketed guidewire and create a knuckle into the dissection plane**

How?

- Advance a polymer-jacketed guidewire (such as Fielder XT, Fighter, Gladius Mongo, or Pilot 200) through the microcatheter. The wire is advanced by pushing, not turning, to minimize the risk of fracture.

What can go wrong?

- Perforation. The risk of perforation can be minimized by checking the wire course in orthogonal projections.
- Inability to form a knuckle. If this occurs the wire tip may need to be reshaped into an "umbrella handle" that facilitates folding of the wire back on itself.
- Guidewire entrapment in the vessel wall (Fig. 16.7).[24] This is a very infrequent complication. Potential solutions include advancing a second guidewire next to the entrapped wire and perform inflations in order to free the entrapped wire. If the guidewire fractures, IVUS can help ascertain that there is no wire unraveling into the proximal part of the vessel or into the aorta.

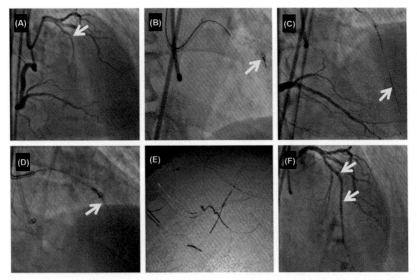

FIGURE 16.7 **Guidewire entrapment during antegrade dissection reentry.** Bilateral coronary angiography demonstrating a chronic total occlusion of the mid left anterior descending artery (arrow, **panel A**). Entrapment of a knuckled Fielder XT guidewire (arrow, **panel B**). The chronic total occlusion was successfully crossed with a Pilot 200 guidewire (arrow, **panel C**) advanced parallel to the entrapped guidewire. After balloon angioplasty was performed around the entrapped guidewire (arrow, **panel D**) it was successfully retrieved (**panel E**) with an excellent final angiographic result (arrow, **panel F**). *Reproduced with permission from the* Manual of chronic total occlusion percutaneous coronary interventions. *2nd ed. Elsevier 2017.*

- **Step 7: CTO crossing** (as described in Section 8.2)

How?
- Once a wire knuckle enters the extraplaque space, it can be advanced through the occluded segment with low risk of causing perforation due to the distensibility of the extraplaque space. Reentry can then be attempted beyond the distal cap of the CTO, ideally proximal to the origin of any large branches.

- **Step 8: Reentry** (as described in Section 8.2)

How?
- In most cases, reentry is achieved as close to the distal cap as possible using the Stingray system (the ReCross dual lumen microcatheter can sometimes also be used). Reentry may be challenging if the distal cap is at the bifurcation of a large branch (e.g., right posterior descending and right posterolateral vessel).

A variation of the BASE technique is the "side-BASE" technique (the balloon is inflated halfway in and halfway out of a side branch at the proximal cap) that aims to prevent guidewire entry into the side branch and also preserve the patency of the side branch (Fig. 16.8).[25]

16.1.4.2 "Scratch and go"
(Figs. 16.9 and 16.10)
  CTO *Manual* online, cases 2, 16, 32, 45, 59, 61, 153

- **Step 1: Wire the vessel proximal to the CTO**

How?
- Use a workhorse guidewire.
- Confirm in orthogonal projections that the wire is within the intended segment proximal to the proximal cap of the CTO.

What can go wrong?
- Wire may advance into a side branch or may curl up on itself. This should be appreciated using orthogonal projections and corrected before advancing the balloon.
- **Step 2: Advance a microcatheter proximal to the proximal cap**

How?
- Use any standard microcatheter, such as the Corsair and Turnpike.
- Check in orthogonal projections that the wire and balloon are actually within the intended segment.

What can go wrong?
- The microcatheter may not advance due to severely diseased proximal vessel and/or severe calcification. In such cases predilatation with a small

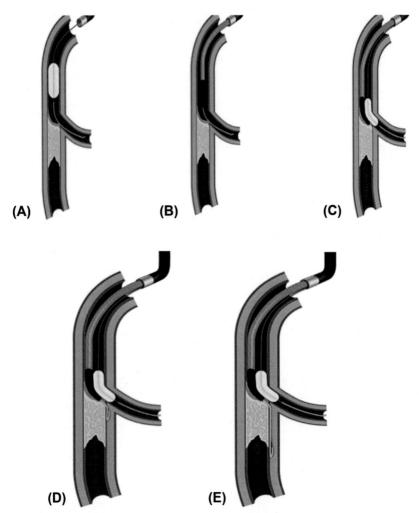

FIGURE 16.8 Side-BASE technique steps. Panel A: create a proximal tear in the vessel by inflating a balloon (sized 1:1 with the proximal vessel). **Panel B**: a microcatheter on a second wire (polymer-coated) is advanced into the proximal vessel. **Panel C**: a different balloon (sized 1:1 with the side branch) is placed partly in the side branch with the proximal segment extending into the proximal vessel and inflated at nominal pressure. This balloon then acts to both deflect and anchor the wire on the microcatheter. **Panel D**: with the side branch balloon inflated the polymer wire is pushed in a knuckle (loop) past the inflated balloon. **Panel E**: the balloon deflects the looped wire passage beyond the proximal cap into the occluded segment, usually within the extraplaque space. *Reproduced with permission from Roy J, Hill J, Spratt JC. The "side-BASE technique": Combined side branch anchor balloon and balloon assisted sub-intimal entry to resolve ambiguous proximal cap chronic total occlusions.* Catheter Cardiovasc Interv *2018;92:E15−E19.*

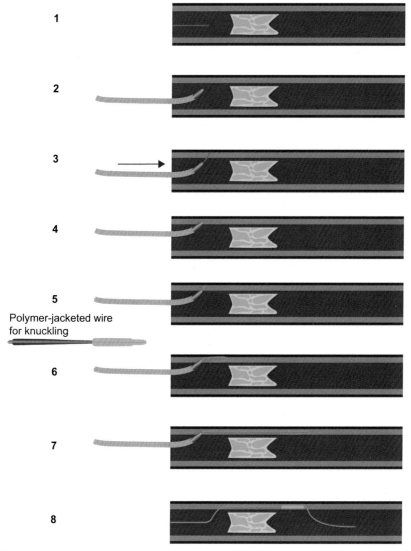

FIGURE 16.9   **Illustration of the "scratch and go" technique.**

balloon may need to be performed first. Alternatively, a larger or more supportive guide catheter (such as Amplatz 1 for the right coronary artery) or other guide supporting techniques (such as side branch anchor, or guide catheter extensions) can be used.

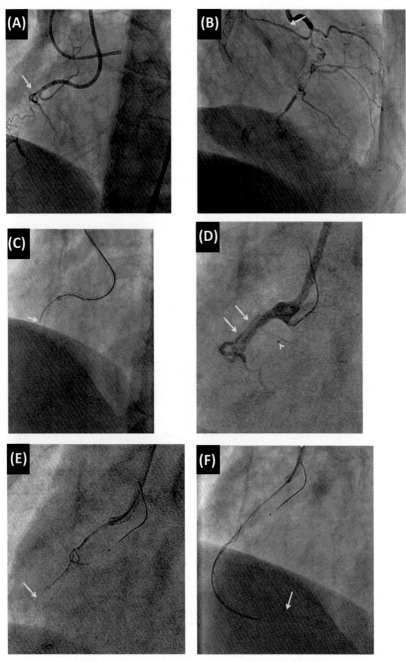

FIGURE 16.10   *(Continued)*

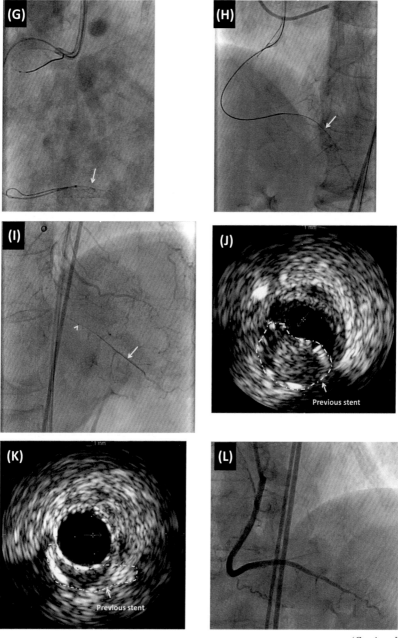

(Continued)

- **Step 3: Insert stiff tip guidewire into the microcatheter and advance it towards the vessel wall**

How?

- Use a stiff tip guidewire, such as the Confianza Pro 12, Gaia 2 or 3, Gaia Next 2 or 3, Infiltrac or Infiltrac Plus, Hornet 14, or Astato 20.
- Shape the wire distal tip at 90° bend and 2−3 mm length.
- Advance the guidewire into the vessel wall proximal to the proximal cap. Advance the wire only 1−2 mm into the wall to minimize the risk of perforation.

What can go wrong?

- The wire may be advanced too far causing perforation. Wire advancement alone is extremely rare to cause a clinically significant perforation, but if the microcatheter follows, clinically significant perforation is possible.

- **Step 4: Advance the microcatheter over the stiff tip guidewire inside the vessel wall**

How?

- Advance the microcatheter over the stiff tip guidewire towards the vessel wall (only a minimal distance, usually 1 mm or less).

What can go wrong?

- Perforation. If the guidewire or microcatheter exit the vessel wall, confirmation that the guidewire has not perforated (by antegrade contrast injection) is needed before advancing the microcatheter.

◀ FIGURE 16.10 **CTO with ambiguous proximal cap and in-stent occluded segment, successfully recanalized using the "scratch and go" technique with crushing of a prior occluded stent**.
Panel A: proximal right coronary artery CTO with ambiguous proximal cap (arrow).
Panel B: lateral view showing the ambiguous proximal cap of the right coronary artery CTO (arrow).
Panel C: "scratch and go" technique: creation of a dissection proximal to the proximal cap with a Confianza Pro 12 guidewire (arrow) through a Corsair microcatheter.
Panel D: proximal vessel dissection (arrows). A side branch anchor balloon has been placed in an acute marginal branch (arrowhead).
Panel E: a knuckled guidewire (arrow) is advanced in the extraplaque space around the proximal cap.
Panel F: the knuckle (arrow) reaches the distal stent.
Panel G: advancement of a CrossBoss catheter and the knuckled wire (arrow) around the previously placed (now occluded) stent.
Panel H: extraplaque guidewire advancement into the right posterior descending artery.
Panel I: successful reentry into the distal true lumen with a Pilot guidewire (arrow) advanced through a Stingray balloon (arrowhead).
Panel J: intravascular ultrasound demonstrating extraplaque crossing around the previously placed stent.
Panel K: intravascular ultrasound after stent implantation showing "crushing" of the previously placed stents.
Panel L: successful recanalization of the right coronary artery CTO.

- **Step 5: Insert a polymer-jacketed guidewire through the microcatheter**

How?
  - Insert a polymer-jacketed guidewire (such as the Fielder XT, Gladius Mongo or Pilot 200) into the microcatheter.
  - Advance the wire (by pushing without turning) to form a knuckle.

What can go wrong?
  - Inability to deliver the microcatheter due to tortuosity or calcification (unlikely given prior balloon inflation). If this occurs additional balloon dilatations or increased guide catheter support may be needed.

- **Step 6: Create a of knuckle into the dissection plane**

- **Step 7: CTO crossing** (as described in Section 8.2)

- **Step 8: Reentry** (as described in Section 8.2)

16.1.4.3   "Carlino" technique for resolving proximal cap ambiguity
(Fig. 16.11)

The Carlino microdissection technique is described in detail in Section 8.2.4.3. The Carlino microdissection technique has multiple other uses, such as to facilitate crossing of "wire uncrossable" lesions and to facilitate forward advancement during antegrade and retrograde extraplaque crossing attempts.

How?
  - The Carlino technique can be used after the microcatheter tip has entered the extraplaque space before, during, or after guidewire advancement attempts, to facilitate dissection and also to confirm that the wire tip is indeed located within the extraplaque space.

- **Step 1: Wire the vessel proximal to the CTO**

- **Step 2: Advance a microcatheter proximal to proximal cap**

- **Step 3: Insert a stiff tip guidewire over the microcatheter and advance towards vessel wall**

- **Step 4: Advance the microcatheter over the stiff tip guidewire inside the vessel wall**
- **Step 5: Inject a small amount of contrast through the microcatheter**

How?
  - Use a small (usually 3 cc Medallion [Merit Medical]) syringe.
  - Inject a small amount ($<0.5-1.0$ mL) of contrast gently under cine-angiographic guidance.[22]

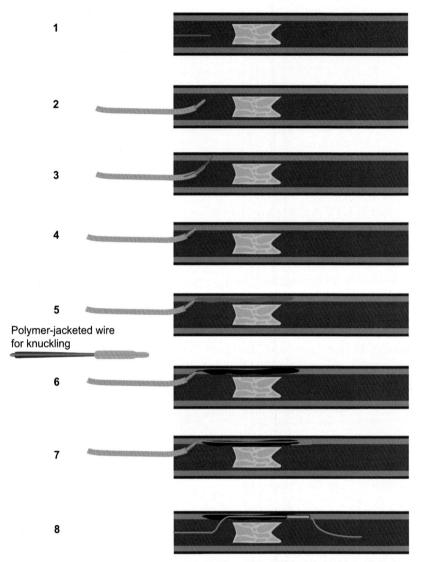

Polymer-jacketed wire
for knuckling

FIGURE 16.11   Illustration of the Carlino technique for resolving proximal cap ambiguity.

What can go wrong?
- Perforation. The risk of perforation can be minimized by injecting a small amount of contrast and meticulous fluoroscopic visualization of the injection.
- Retrograde contrast propagation causing side branch or proximal branch occlusion

- **Step 6: Advance a guidewire into the dissection plane**

  Sometimes, the contrast might dissect into the distal true lumen.

- **Step 7: CTO crossing** (as described in Section 8.2)

- **Step 8: Reentry** (as described in Section 8.2)

  (in case of extraplaque crossing)

16.1.4.4   Retrograde crossing

CTO *Manual* online cases 5, 31, 33, 36, 51, 56, 69, 70, 88, 104

Retrograde crossing can provide an excellent solution to proximal cap ambiguity, as the retrograde guidewire can be advanced either in the proximal true lumen or close to the proximal cap, clarifying the vessel course and "resolving" the proximal cap ambiguity. Moreover, the retrograde guidewire can modify the proximal cap, facilitating antegrade crossing. The Carlino technique can be used through the retrograde microcatheter (CTO *Manual* online, case 62) and retrograde balloons can be used to perform the CART technique.

16.2   Ostial circumflex and ostial left anterior descending artery CTOs

Ostial circumflex and ostial LAD CTOs are examples of branch-ostial CTOs (in contrast to aorto-ostial, such as right coronary artery and left main CTOs). Both ostial circumflex and ostial LAD CTOs can have ambiguous proximal caps, requiring use of the techniques described in Section 16.1 to resolve the ambiguity.

If the retrograde approach is used for such lesions, it may be best to:

1. Use a guide catheter extension to perform reverse CART to minimize the risk of left main dissection (PCI *Manual* online, case 36).
2. Insert a "safety" guidewire into the nonoccluded vessel (circumflex for LAD CTOs and LAD for circumflex CTOs) to maintain access to the vessel in case of dissection or plaque shift during recanalization of the CTO.
3. Use IVUS to confirm that the retrograde wire reentry is distal to the carina to avoid side branch loss after stenting.
4. Direct wiring towards the left main should be avoided whenever possible. If the operator fails to perform antegrade preparation, direct retrograde wiring toward the left main should be performed with extreme caution, and the intraluminal position of the guidewire tip should be confirmed by intravascular ultrasound before further equipment advancement.

Ostial circumflex or LAD CTOs may be challenging to cross due to angulation: use of the Venture catheter (CTO *Manual* online, case 48), or the angulated SuperCross microcatheter (Section 30.6.3) or a dual lumen microcatheter

(Section 30.6.4) can help direct and support the wire to cross the lesion (Fig. 16.12). PCI of circumflex CTOs has lower success rates than PCI of CTOs of the right coronary artery or the left anterior descending artery, likely because of tortuosity and frequent lack of interventional collaterals.[26]

Sometimes, the ostium of an ostial LAD or circumflex CTO can be engaged with a guidewire with a large bend, allowing advancement of a microcatheter and exchange to a penetration or polymer-jacketed guidewire with a small bend.

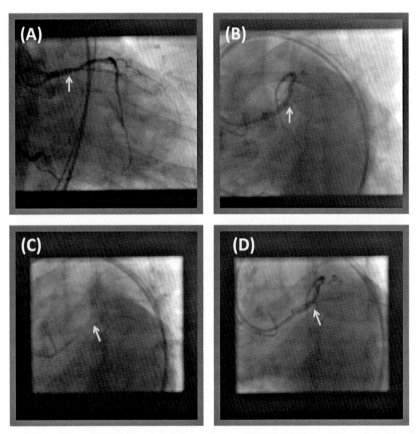

FIGURE 16.12 Bilateral coronary angiography demonstrating an ostial chronic total occlusion of the circumflex (arrows, **panel A and B**). Intravascular ultrasonography was performed to identify the ostium of the circumflex (arrow, **panel C**), followed by insertion of a Venture catheter (arrow, **panel D**), through which a Pilot 200 wire easily crossed the occlusion (arrow, **panel E**). After predilatation, antegrade flow in the circumflex was restored (arrow, **panel F**). Despite multiple balloon predilatations and use of a Guideliner catheter, a 2.5 × 23 mm stent could not be delivered and during attempts to retrieve it into the guide catheter it was dislodged from the balloon into the left main coronary artery. Attempts to retrieve the stent using a 4 mm Gooseneck snare were unsuccessful. The lost stent was crushed by another stent deployed in the left main coronary artery, followed by rewiring and balloon angioplasty of the circumflex with a nice final angiographic result (**panel G**).

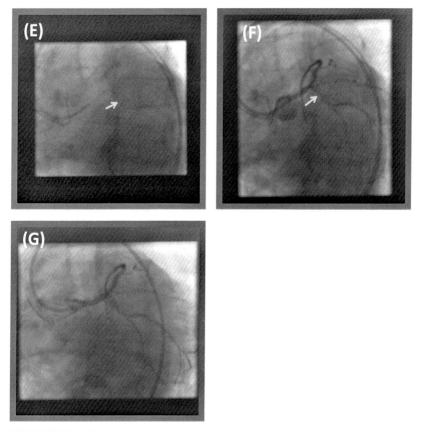

FIGURE 16.12 *(Continued)*

Fig. 16.12 highlights the value of the Venture catheter in highly angulated lesions, especially in ostial circumflex CTOs. The Venture catheter enabled rapid lesion crossing of a very challenging CTO in a tortuous and calcified vessel. In such vessels stent delivery may also be very difficult and may be complicated by stent loss, as in this case. Optimal lesion preparation may minimize the risk of stent loss.

16.3 Bifurcation at the proximal cap

There are two potential challenges associated with bifurcations at the proximal or distal cap (Fig. 16.13):

1. **Inability to wire the occlusion** due to preferential guidewire entry into the patent branch and failure to engage the CTO (Fig. 16.13, panel A) (CTO *Manual* online, case 55).

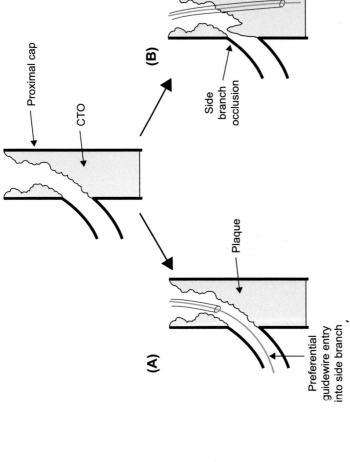

FIGURE 16.13 Challenges associated with CTOs that have a bifurcation at the proximal cap.

2. **Occlusion of the side branch** (Fig. 16.13, panel B). This may be inconsequential for small branches, but occlusion of large branches can lead to periprocedural myocardial infarction[3,27,28] or arrhythmias. Inserting a "safety" guidewire into the patent branch can help maintain access to the vessel in case of dissection or plaque shift during recanalization of the CTO (CTO *Manual* online case 49).

16.3.1 How to cross CTOs with a bifurcation at the proximal cap

(Fig. 16.14)

There are several techniques for successfully crossing CTOs with bifurcation at the proximal cap:

1. Insertion of a guidewire into the side branch that can act as a marker of the side branch origin, facilitating antegrade crossing attempts with another guidewire (Fig. 16.14, panel 1). In addition, the side branch wire can protect the side branch during CTO crossing and stenting.
2. Balloon inflation in the side branch can induce a geometrical shift of the hard plaque in the proximal cap and enable guidewire entry into the CTO (this has been called the "open sesame" technique) (Fig. 16.14, panel 2).[29] A modification of this technique for heavily calcified lesions is to use

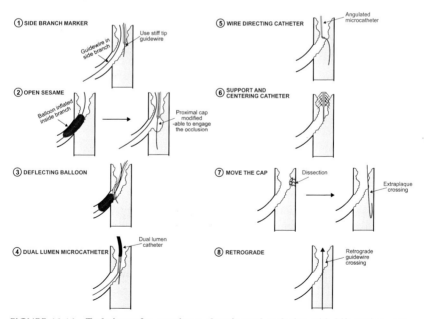

FIGURE 16.14 **Techniques for crossing a chronic total occlusion with bifurcation at the proximal cap**.

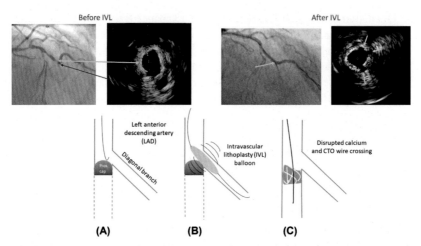

FIGURE 16.15   The mechanism of lithotripsy-assisted proximal cap modification. Calcified proximal cap of a mid-left anterior descending artery (LAD) CTO at the bifurcation of the diagonal (**panel A**). Images show the angiography and intravacular ultrasound before and after intravascular lithotripsy (IVL) in the diagonal branch (yellow and red arrows before IVL). **Panel B**: advancement and inflation of the lithotripsy balloon partially advanced into the diagonal branch with shock impulses delivery at the calcified cap. **Panel C**: fractured and disrupted calcium after intravascular lithotripsy with facilitated wire crossing (yellow arrow after IVL). *Reproduced with permission from Elbasha K, Richardt G, Hemetsberger R, Allali A. Lithoplasty-facilitated proximal cap penetration of a calcified chronic total occlusion coronary lesion. JACC Case Rep 2022;4:44–48.*

a plaque modification balloon (Section 30.9.3) or an intravascular lithotripsy balloon (Section 30.9.8)[30] (Fig. 16.15).

Another modification is to perform POT (proximal optimization technique) by inflating a balloon sized 1:1 to the proximal vessel after placing the distal balloon marker at the carina (Fig. 16.16).[31]

3. "Deflecting balloon" or "blocking balloon": a balloon is inflated at the ostium of the side branch, blocking entry of the guidewire into the side branch, which can then engage the CTO proximal cap (Fig. 16.14, panel 3).
4. Use of a dual lumen microcatheter (Section 30.6.4) to direct the second wire into the CTO proximal cap (Fig. 16.14, panel 4) (CTO *Manual* online, case 55).[32]

   A dual lumen microcatheter helps: (1) enhance the wire penetrating capacity, (2) straighten the proximal vessel bending, (3) change the puncture position by altering the catheter position. When the CTO wire is deeply advanced into the occlusion, the dual lumen microcatheter should be exchanged for a single lumen microcatheter by using the trapping technique to facilitate subsequent wire manipulations. This strategy provides strong guidewire support.

   When the size of side branch is large enough to advance a balloon, strong support of the microcatheter can be achieved as follows: (1) two

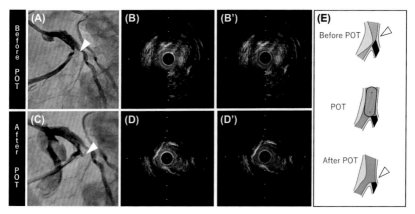

FIGURE 16.16 **Coronary angiography and IVUS findings and scheme of changes in CTO entry before and after use of the proximal optimization technique (POT).** **Panel A**: initial angiography before proximal optimization technique (POT). **Panel B**: intravascular ultrasound (IVUS) before POT. **Panel C**: coronary angiography after POT. **Panel D**: IVUS after POT. **Panel E**: scheme of the changes in CTO entry before and after POT. **Panels A, C, E**: arrowheads indicate the entry of CTO of the side branch. **Panels B′ and D′**: blood flow regions are traced in red on the IVUS images. *Reproduced with permission from Yokoi K, Sonoda S, Yoshioka G, Jojima K, Natsuaki M, Node K. Proximal optimization technique facilitates wire entry into stumpless chronic total occlusion of side branch. JACC Cardiovasc Interv 2021;14:e231−e233.*

floppy wires are advanced into the same side branch; (2) an adequate size rapid exchange balloon is inserted and inflated in the branch to anchor one of the wires; (3) the dual lumen microcatheter is then advanced over the anchored wire, followed by CTO crossing attempts.

5. Use of an angulated microcatheter, such as the Venture and the Supercross microcatheters (Section 30.6.3) (Fig. 16.14, panel 5).[33]
6. Use of a support centering catheter, such as the NovaCross (Section 30.3) to point more directly to the proximal cap (Fig. 16.14, panel 6).
7. Antegrade dissection/reentry techniques (if the side branch is small and its loss is acceptable) ("move the cap,"[11] Section 16.1.4) (Fig. 16.14, panel 7). The side-BASE technique can help preserve the patency of the side branch.[25]
8. Retrograde crossing (Fig. 16.14, panel 8).

Jailing of deformed guidewires in the side branch with a stent should be avoided as it may lead to wire entrapment.[34]

16.4 Bifurcation in the CTO body

Failure to recanalize side branches located within the CTO occlusion segment usually carries fewer adverse consequences than loss of side branches at the proximal or distal cap, since branches originating within the occlusion are filling via collaterals.

Bifurcations located within the CTO occlusion are approached with the same techniques described for bifurcations at the proximal and distal cap.

16.5   Bifurcation at the distal cap

The same challenges associated with bifurcations at the proximal cap also apply to CTOs with bifurcation at the distal cap: (1) difficulty crossing the CTO, as the retrograde wire may preferentially enter into the branch; and (2) risk of side branch occlusion (Fig. 16.17).

If antegrade dissection/reentry is used for crossing a CTO with a bifurcation at the distal cap, reentering immediately proximal or at the bifurcation is ideal for maintaining patency of both branches (CTO *Manual* online case 45). Otherwise, separate reentry into both branches may be needed (CTO Manual online case 80), followed by kissing angioplasty and a two-stent bifurcation stenting technique.[35]

CTOs with a bifurcation at the distal cap can be approached in a primary retrograde or antegrade direction:

16.5.1   Retrograde techniques

CTO *Manual* online, cases 5, 12, 23, 46 (Fig. 16.18)

A primary retrograde approach allows engagement of the CTO from a favorable angle and increases the likelihood of maintaining patency of both branches. However, entering the CTO can be challenging, often requiring advancement of the retrograde microcatheter all the way to the distal cap and use of stiff tip guidewires, such as Gaia, Confianza Pro 12, and Pilot 200.

1. If retrograde crossing is achieved without compromising the origin of the branch vessel, the lesion can be stented while maintaining patency of both branches (Fig. 16.18 panel A). Intravascular ultrasound can confirm that the guidewire entered into the true lumen entry without involving the bifurcation (Fig. 16.19)

2. If retrograde crossing succeeds but causes occlusion of an (important) branch (Fig. 16.18 panel B), the following options exist:

    B1. Accept occlusion of the side branch without further recanalization attempts. This may be the best course of action, if the branch is relatively small or in long procedures requiring large amounts of radiation or contrast (CTO *Manual* online, case 42).

    B2. Attempt antegrade wiring of the side branch, which can be facilitated by use of a dual lumen microcatheter.

    B3. Attempt reentry into the occluded branch, usually using the Stingray system.

    B4. Double retrograde: attempt retrograde crossing of the occluded side branch (CTO *Manual* online, case 64). Penetration with the retrograde wire

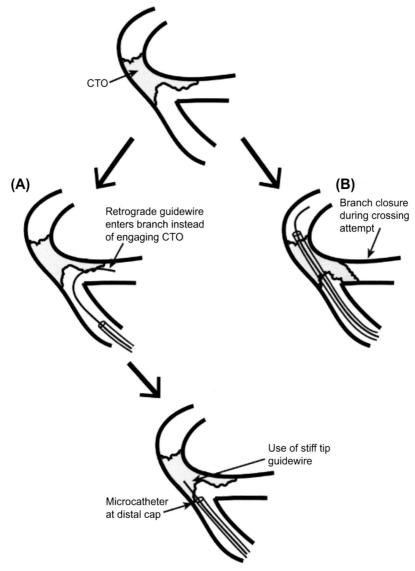

FIGURE 16.17 **Challenges associated with percutaneous coronary intervention of a chronic total occlusion with a bifurcation at the distal cap.**

can be challenging in case of resistant, round shape, distal cap. This anatomy usually diverts the wire into the other branch. A retrograde dual lumen microcatheter can be used to improve the control of the second retrograde wire (Fig. 16.20).

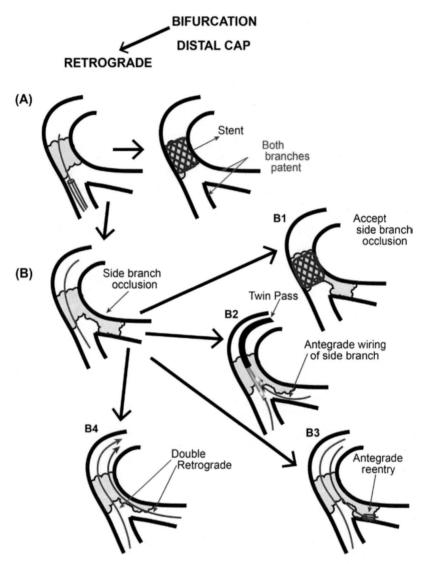

FIGURE 16.18 **Retrograde techniques for crossing a chronic total occlusion with a bifur-cation at the distal cap.**

16.5.2 Antegrade techniques (Fig. 16.21)

CTO *Manual* **online cases** 11, 25, 70, 80.

Antegrade crossing of a CTO with bifurcation at the distal cap carries the risk of compromising one of the branches, especially if dissection/reentry is required. Antegrade crossing can be achieved into the main branch or the distal side branch.

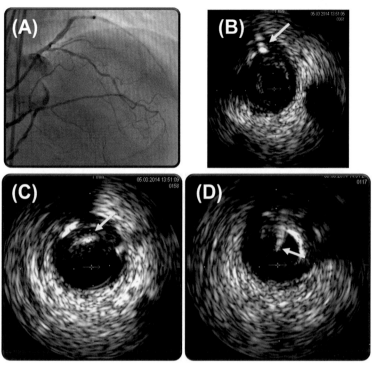

FIGURE 16.19 Use of IVUS placed in the proximal true lumen to determine the position of the retrograde guidewire during retrograde true to true lumen crossing attempts. **Panel A**: left anterior descending artery CTO. **Panel B**: IVUS showing wire position within the vessel wall (arrow). **Panel C**: IVUS showing wire position within the vessel wall (arrow); **Panel D**: IVUS showing proximal true lumen wire position (arrow). *Courtesy of Dr. Andrea Garbo.*

Distal cap puncture with retrograde dual lumen microcatheter

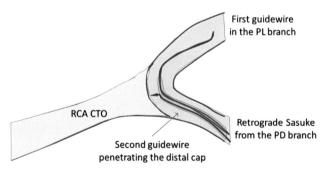

FIGURE 16.20 **Use of a retrograde Sasuke dual lumen microcatheter through septal collaterals for crossing a right coronary artery CTO with distal cap at the crux.** *PD*, posterior descending artery; *PL*, posterolateral; *RCA*, right coronary artery. *Courtesy of Dr. Imre Ungi.*

**BIFURCATION
DISTAL CAP**

ANTEGRADE

(A) WIRE IN DISTAL MAIN VESSEL

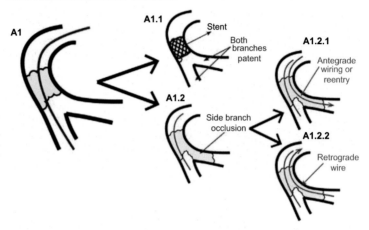

(B)   WIRE IN DISTAL SIDE BRANCH

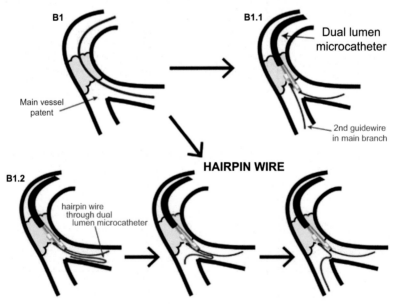

FIGURE 16.21   **Antegrade techniques for crossing a chronic total occlusion with a bifurcation at the distal cap.**

(C)

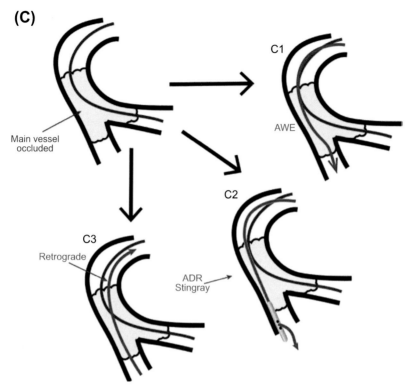

FIGURE 16.21 Continued.

1. **Antegrade crossing into the main branch**.
   A.1.1 In the best-case scenario distal true lumen crossing is achieved without affecting the patency of the side branch, even after stenting.
   A.1.2. If the side branch is compromised after crossing into the distal main branch, the side branch is rewired, either in the antegrade direction (A.1.2.1) or in the retrograde direction (A1.2.2). If the side branch is small or the procedure is challenging, recanalization of the side branch may be deferred or not performed at all (CTO *Manual* online cases 8, 67).

2. **Antegrade crossing into the side branch *maintaining* main branch patency**.
   B.1.1. A second guidewire is advanced into the main branch, ideally using a dual lumen microcatheter in order to maintain access to the side branch. An angulated microcatheter could also be used in such cases.
   B.1.2. Prolonged antegrade wiring attempts with a second guidewire may cause dissection of the other branch ostium and side branch occlusion. In such cases, the reversed guidewire[36,37] (also called "hairpin wire") technique can be useful in directing a guidewire from the side branch into the main

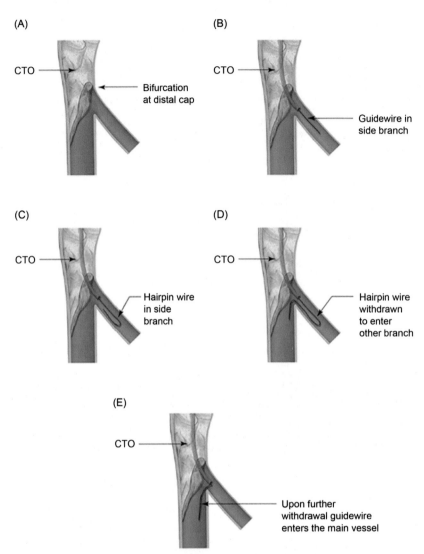

FIGURE 16.22   The reversed guidewire technique for approaching chronic total occlusions (CTOs) after crossing into a side branch at the distal cap. A CTO with a bifurcation at the distal cap (**panel A**) is successfully crossed with a guidewire that enters the side branch (**panel B**). A hairpin guidewire is advanced into the side branch (**panel C**) and withdrawn, entering the main vessel (**panels D and E**).

vessel (CTO *Manual* online, case 71).[38] In this technique, a polymer-jacketed wire is bent approximately 3 cm from the wire tip and a conventional 2–3 mm bent is created at the tip of the hairpin wire in the opposite direction to facilitate side branch entry upon withdrawal (Fig. 8.1.9). The reversed wire is advanced into the side branch (Fig. 16.22, panel C), and

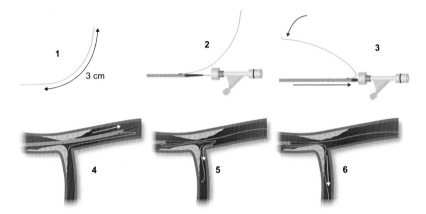

FIGURE 16.23   **The reversed wire technique using a dual lumen microcatheter.**

pulled back (Fig. 16.22, panel D) entering the main branch (Fig. 16.22, panel E). Alternatively, the hairpin wire can be advanced through a dual lumen microcatheter (Fig. 16.23).

What can go wrong?
- Use of the reversed wire technique may cause vessel dissection. Also after the "hairpin" enters the main vessel, further advancement may be challenging due to the bend in the wire.

3. **Antegrade crossing into the side branch** *occluding* **the main branch**.

Crossing into the main vessel is critical in such cases and can be achieved in several ways:

C1. Antegrade wiring, potentially with a dual lumen microcatheter.
C2. Reentry into the main vessel distal true lumen, usually using the Stingray system (CTO *Manual* online cases 21 and 25).
C3. By retrograde crossing through the main vessel (CTO *Manual* online case 7).
In most cases a two-stent bifurcation stenting technique is needed to maintain patency of both branches. These techniques (DK crush, culotte, TAP, etc.) are discussed in detail in the *Manual of Percutaneous Coronary Interventions*.[39]

References

1.   Adachi Y, Kinoshita Y, Murata A, et al. The importance of side branch preservation in the treatment of chronic total occlusions with bifurcation lesions. *Int J Cardiol Heart Vasc* 2021;**36**:100873.
2.   Paizis I, Manginas A, Voudris V, Pavlides G, Spargias K, Cokkinos DV. Percutaneous coronary intervention for chronic total occlusions: the role of side-branch obstruction. *EuroIntervention* 2009;**4**:600−6.

3. Nguyen-Trong PK, Rangan BV, Karatasakis A, et al. Predictors and outcomes of side-branch occlusion in coronary chronic total occlusion interventions. *J Invasive Cardiol* 2016;**28**:168−73.

4. Galassi AR, Boukhris M, Tomasello SD, et al. Incidence, treatment, and in-hospital outcome of bifurcation lesions in patients undergoing percutaneous coronary interventions for chronic total occlusions. *Coron Artery Dis* 2015;**26**:142−9.

5. Nikolakopoulos I, Vemmou E, Karacsonyi J, et al. Percutaneous coronary intervention of chronic total occlusions involving a bifurcation: insights from the PROGRESS-CTO Registry. *Hellenic J Cardiol* 2022.

6. Guo Y, Peng H, Zhao Y, Liu J. Predictors and complications of side branch occlusion after recanalization of chronic total occlusions complicated with bifurcation lesions. *Sci Rep* 2021;**11**:4460.

7. Karatasakis A, Danek BA, Karmpaliotis D, et al. Impact of proximal cap ambiguity on outcomes of chronic total occlusion percutaneous coronary intervention: insights from a multicenter US Registry. *J Invasive Cardiol* 2016;**28**:391−6.

8. Niizeki T, Tsuchikane E, Konta T, et al. Prevalence and predictors of successful percutaneous coronary intervention in ostial chronic total occlusion. *JACC Cardiovasc Interv* 2022.

9. Brilakis ES. *Manual of coronary chronic total occlusion interventions. A step-by-step approach.* 2nd ed. Elsevier; 2017.

10. Ryan N, Gonzalo N, Dingli P, et al. Intravascular ultrasound guidance of percutaneous coronary intervention in ostial chronic total occlusions: a description of the technique and procedural results. *Int J Cardiovasc Imaging* 2017;**33**:807−13.

11. Vo MN, Karmpaliotis D, Brilakis ES. "Move the cap" technique for ambiguous or impenetrable proximal cap of coronary total occlusion. *Catheter Cardiovasc Interv* 2016;**87**:742−8.

12. Brilakis ES, Grantham JA, Rinfret S, et al. A percutaneous treatment algorithm for crossing coronary chronic total occlusions. *JACC Cardiovasc Interv* 2012;**5**:367−79.

13. Wu EB, Brilakis ES, Mashayekhi K, et al. Global chronic total occlusion crossing algorithm: JACC state-of-the-art review. *J Am Coll Cardiol* 2021;**78**:840−53.

14. Opolski MP, Achenbach SCT. Angiography for revascularization of CTO: crossing the borders of diagnosis and treatment. *JACC Cardiovasc Imaging* 2015;**8**:846−58.

15. Dautov R, Abdul Jawad Altisent O, Rinfret S. Stumpless chronic total occlusion with no retrograde option: multidetector computed tomography-guided intervention via bi-radial approach utilizing bioresorbable vascular scaffold. *Catheter Cardiovasc Interv* 2015;**86**:E258−62.

16. Hong SJ, Kim BK, Cho I, et al. Effect of coronary CTA on chronic total occlusion percutaneous coronary intervention: a randomized trial. *JACC Cardiovasc Imaging* 2021;**14**:1993−2004.

17. Opolski MP, Debski A, Borucki BA, et al. First-in-man computed tomography-guided percutaneous revascularization of coronary chronic total occlusion using a wearable computer: proof of concept. *Can J Cardiol* 2016;**32**(829):e11−13.

18. Ghoshhajra BB, Takx RAP, Stone LL, et al. Real-time fusion of coronary CT angiography with x-ray fluoroscopy during chronic total occlusion PCI. *Eur Radiol* 2017;**27**:2464−73.

19. Galassi AR, Sumitsuji S, Boukhris M, et al. Utility of intravascular ultrasound in percutaneous revascularization of chronic total occlusions: an overview. *JACC Cardiovasc Interv* 2016;**9**:1979−91.

20. Karacsonyi J, Alaswad K, Jaffer FA, et al. Use of intravascular imaging during chronic total occlusion percutaneous coronary intervention: insights from a contemporary Multicenter Registry. *J Am Heart Assoc* 2016;5.

21. Kinoshita Y, Fujiwara H, Suzuki T. "Slipstream technique"-New concept of intravascular ultrasound guided wiring technique with double lumen catheter in the treatment of coronary total occlusions. *J Cardiol Cases* 2017;**16**:52−5.

22. Carlino M, Ruparelia N, Thomas G, et al. Modified contrast microinjection technique to facilitate chronic total occlusion recanalization. *Catheter Cardiovasc Interv* 2016;**87**:1036−41.

23. Amsavelu S, Carlino M, Brilakis ES. Carlino to the rescue: use of intralesion contrast injection for bailout antegrade and retrograde crossing of complex chronic total occlusions. *Catheter Cardiovasc Interv* 2016;**87**:1118−23.

24. Danek BA, Karatasakis A, Brilakis ES. Consequences and treatment of guidewire entrapment and fracture during percutaneous coronary intervention. *Cardiovasc Revasc Med* 2016;**17**:129−33.

25. Roy J, Hill J, Spratt JC. The "side-BASE technique": combined side branch anchor balloon and balloon assisted sub-intimal entry to resolve ambiguous proximal cap chronic total occlusions. *Catheter Cardiovasc Interv* 2018;**92**:E15−19.

26. Christopoulos G, Karmpaliotis D, Wyman MR, et al. Percutaneous intervention of circumflex chronic total occlusions is associated with worse procedural outcomes: insights from a Multicentre US Registry. *Can J Cardiol* 2014;**30**:1588−94.

27. Jang WJ, Yang JH, Choi SH, et al. Association of periprocedural myocardial infarction with long-term survival in patients treated with coronary revascularization therapy of chronic total occlusion. *Catheter Cardiovasc Interv* 2016;**87**:1042−9.

28. Lo N, Michael TT, Moin D, et al. Periprocedural myocardial injury in chronic total occlusion percutaneous interventions: a systematic cardiac biomarker evaluation study. *JACC Cardiovasc Interv* 2014;**7**:47−54.

29. Saito S. Open Sesame Technique for chronic total occlusion. *Catheter Cardiovasc Interv* 2010;**75**:690−4.

30. Elbasha K, Richardt G, Hemetsberger R, Allali A. Lithoplasty-facilitated proximal cap penetration of a calcified chronic total occlusion coronary lesion. *JACC Case Rep* 2022;**4**:44−8.

31. Yokoi K, Sonoda S, Yoshioka G, Jojima K, Natsuaki M, Node K. Proximal optimization technique facilitates wire entry into stumpless chronic total occlusion of side branch. *JACC Cardiovasc Interv* 2021;**14**:e231−3.

32. Chiu CA. Recanalization of difficult bifurcation lesions using adjunctive double-lumen microcatheter support: two case reports. *J Invasive Cardiol* 2010;**22**:E99−103.

33. Iturbe JM, Abdel-Karim AR, Raja VN, Rangan BV, Banerjee S, Brilakis ES. Use of the venture wire control catheter for the treatment of coronary artery chronic total occlusions. *Catheter Cardiovasc Interv* 2010;**76**:936−41.

34. Karagoz A, Kilic D, Goktekin O. Devastating consequences of a jailed knuckled wire in CTO PCI of an anomalous RCA. *JACC Case Rep* 2020;**2**:499−502.

35. Tajti P, Doshi D, Karmpaliotis D, Brilakis ES. The "double stingray technique" for recanalizing chronic total occlusions with bifurcation at the distal cap. *Catheter Cardiovasc Interv* 2018;**91**:1079−83.

36. Kawasaki T, Koga H, Serikawa T. New bifurcation guidewire technique: a reversed guidewire technique for extremely angulated bifurcation−a case report. *Catheter Cardiovasc Interv* 2008;**71**:73−6.

37. Suzuki G, Nozaki Y, Sakurai M. A novel guidewire approach for handling acute-angle bifurcations: reversed guidewire technique with adjunctive use of a double-lumen microcatheter. *J Invasive Cardiol* 2013;**25**:48−54.

38. Michael T, Banerjee S, Brilakis ES. Distal open sesame and hairpin wire techniques to facilitate a chronic total occlusion intervention. *J Invasive Cardiol* 2012;**24**:E57—9.

39. Brilakis ES. *Manual of percutaneous coronary interventions: a step-by-step approach.* Elsevier; 2021.

Chapter 17

Left main chronic total occlusions

CTO Manual online case 29, 48, 67, 148, 179, 186.

PCI of left main CTOs is done infrequently (0.45% of all chronic total occlusion (CTO) percutaneous coronary interventions (PCIs) in the PROGRESS-CTO registry).[1] Many of the left main CTO PCIs are done in prior coronary artery bypass graft surgery patients who often have one or more patent graft(s) to the left anterior descending or circumflex artery territories (Fig. 17.1).[1,2]

There are three potential challenges of left main CTO PCI:

1. Treating aorto-ostial occlusions.
2. Treating bifurcations.
3. Risk of hemodynamic compromise, especially in patients without patent grafts to the left main territory.

17.1   Aorto-ostial left main chronic total occlusions

Crossing of aorto-ostial left main CTOs is performed as described in Section 15.1. After crossing, stent deployment should be optimized to ensure that the ostium is covered without excessive stent overhang into the aorta and that the stent is well expanded.

17.2   Treatment of the bifurcation

The left main bifurcates into the circumflex and the left anterior descending artery. Some prior coronary artery bypass graft (CABG) patients may have patent bypass grafts to one or more arteries, hence recanalization is often only needed in non-grafted occluded arteries. In patients without prior CABG or without patent grafts, recanalization of both of the LAD and circumflex is needed and is performed as described in Sections 16.2 and 16.3. Use of the retrograde approach is frequently needed in these cases.[1]

Manual of Chronic Total Occlusion Percutaneous Coronary Interventions.
DOI: https://doi.org/10.1016/B978-0-323-91787-2.00021-6

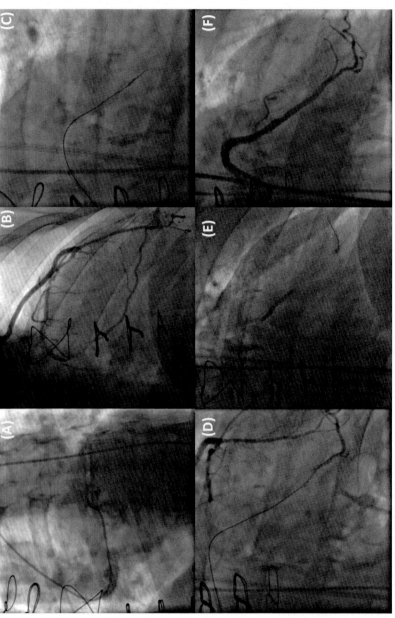

FIGURE 17.1 A 73-year-old patient with prior coronary artery bypass graft surgery and Canadian Cardiovascular Society class III angina on maximal medical therapy was referred for left main chronic total occlusion (CTO) percutaneous coronary intervention. Coronary angiography revealed a patent left internal mammary artery (LIMA) to the mid left anterior descending (LAD) artery and a totally occluded saphenous vein graft to the first obtuse marginal. **Panel A**: ostial left main CTO with blunt stump. **Panel B**: patent LIMA to LAD. The circumflex artery filled by epicardial collaterals from the LAD. **Panel C**: antegrade wiring was attempted, but the guidewire entered the extraplaque space. A Stingray LP balloon was used to reenter into the distal true lumen. **Panel D**: using the "stick and swap" technique and contralateral injection through the LIMA graft, the wire entered into the distal true lumen. **Panel E**: after implantation of three drug-eluting stents, the artery was successfully recanalized. **Panel F**: final result. *Reproduced with permission from Xenogiannis I, Karmpaliotis D, Alaswad K, et al. Left main chronic total occlusion percutaneous coronary intervention: a case series. J Invasive Cardiol 2019;31:E220–E5.*

17.3 Risk of hemodynamic collapse

Patients with unprotected left main CTOs may be at high-risk for hemodynamic decompensation during the procedure. Use of mechanical circulatory support may reduce the risk of hemodynamic compromise, as discussed in Chapter 14.

References

1. Xenogiannis I, Karmpaliotis D, Alaswad K, et al. Left main chronic total occlusion percutaneous coronary intervention: a case series. *J Invasive Cardiol* 2019;**31** E220−E5.
2. Flores-Umanzor E, Martin-Yuste V, Caldentey G, et al. Percutaneous coronary intervention due to chronic total occlusion in the left main coronary artery after bypass grafting: a feasible option in selected cases. *Rev Port Cardiol (Engl Ed)* 2018;**37** 865e1−e4.

Chapter 18

CTO PCI in the prior CABG patient

Most patients with prior coronary artery bypass graft surgery (CABG) have CTOs in their native coronary arteries and/or their bypass grafts.[1]

18.1 Challenges of CTO PCI in prior CABG patients

Chronic total occlusion (CTO) percutaneous coronary intervention (PCI) in prior CABG patients can be challenging due to high-risk patient characteristics (such as chronic kidney disease) and complex anatomy with high prevalence of calcification,[2] high atherosclerotic burden with diffuse disease and distortion of the course of the native vessels due to "tenting" from attachment of grafts, As a result, CTO PCI in prior CABG patients is associated with lower success and higher complication rates.[3-8] Prior CABG is included in three CTO PCI angiographic scores: CL,[9] RECHARGE,[10] and CASTLE[11] (Section 6.3).

Chapter 19 discusses how to approach severely calcified CTOs. Bypass grafts can cause "tenting" of the native coronary arteries they are anastomosed to, which can hinder CTO crossing. Short or shortened guide catheters are recommended for PCI of the distal native vessels through long saphenous vein grafts (SVGs) (including sequential grafts), or through internal mammary artery grafts.[4] Sometimes, triple arterial access is needed for CTO PCI in prior CABG patients, as described in Section 4.1.[12,13]

18.2 Selecting the target vessel in prior CABG patients

In patients presenting with SVG lesions, several operators advocate treating the corresponding native coronary artery instead[4] because of high short- and long-term event rates after SVG PCI.[14,15]

In the 2018 European Society of Cardiology/European Association of CardioThoracic Surgery (ESC/EACTS) guidelines on myocardial revascularization PCI to a native vessel is preferred over PCI of the bypass graft (Class IIa, Level of Evidence: C).[16] However, native coronary artery lesions are often CTOs and can be challenging to recanalize.

In patients with recurrent SVG failure treatment of the native coronary artery CTO is often recommended (Fig. 18.1).[4]

Manual of Chronic Total Occlusion Percutaneous Coronary Interventions.
DOI: https://doi.org/10.1016/B978-0-323-91787-2.00035-6

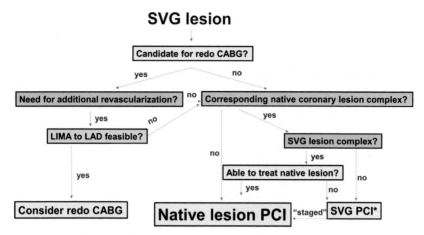

FIGURE 18.1 **Algorithm for selecting the target vessel in prior CABG patients who present with a saphenous vein graft lesion.** *Reproduced with permission from Xenogiannis I, Tajti P, Hall AB, et al. Update on cardiac catheterization in patients with prior coronary artery bypass graft surgery.* JACC Cardiovasc Interv *2019;12:1635−49.*

If a thrombosed SVG cannot be recanalized or if there is high probability of distal embolization and no reflow, PCI of the native coronary artery CTO can sometimes be performed instead.[17] If PCI of the failed SVG is selected, staged PCI of the corresponding native coronary artery should be considered after the initial procedure because SVGs that become occluded due to thrombus have high rates of reocclusion.[18,19] In such cases, stenting the distal SVG anastomosis should be avoided, if possible, as it could hinder subsequent treatment of the native coronary vessel.

In some patients the SVG occludes after recanalization of the native coronary artery CTO. In other patients antegrade flow continues through the SVG and could potentially lead to native stent thrombosis.[20] As a result, some operators advocate routine SVG occlusion after treating the native coronary artery, although this approach remains controversial.[21] Some patients have large SVG aneurysms that **should be occluded at both the proximal and the distal anastomosis** after recanalization of the corresponding native coronary artery to prevent continued blood flow into the aneurysm that can lead to continued aneurysm expansion.

18.3 Retrograde approach through saphenous vein grafts

Saphenous vein grafts (patent and occluded) are very useful for retrograde CTO PCI. Crossing of occluded SVGs should generally only be attempted if the proximal cap has a favorable, tapered morphology to minimize the risk of aortic perforation (Fig. 18.2). Knowing the course of the occluded SVG by reviewing a prior angiogram when the SVG was still patent provides a much better understanding of the course and distal anastomosis of the SVG.

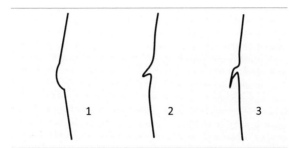

FIGURE 18.2 **Anatomic variants of the proximal cap of occluded saphenous vein grafts (SVGs):** retrograde crossing should not be attempted in morphology #1 due to increased risk of perforation that may require emergent surgery. *Reproduced with permission from Xenogiannis I, Tajti P, Hall AB, et al. Update on cardiac catheterization in patients with prior coronary artery bypass graft surgery.* JACC Cardiovasc Interv *2019;12:1635−49.*

Strong guide catheter support can facilitate retrograde PCI through SVGs. The guide catheter length should ideally be 90 cm long to permit access to the native CTO guide catheter. SVGs to the right coronary artery (RCA) are usually approached with a multipurpose guide catheter. SVGs to the left coronary system are usually approached with an Amplatz Left 1 guide catheter. Often, in order to traverse the SVG occlusion and provide adequate support for the retrograde procedure, a guide catheter extension is used. Two guide catheter extensions can be used simultaneously in a mother−daughter−granddaughter configuration (i.e., a 6 Fr extension through an 8 Fr extension) when multiple extreme bends need to be navigated, for example when performing PCI through angulated saphenous vein grafts (Fig. 18.3) (CTO Manual online Case 87).[22]

A large torqueable microcatheter, such as the Turnpike Spiral and the Corsair is often used to provide increased support and tracking through a SVG occlusion. Usually a stiff, polymer-jacketed guidewire, such as a Pilot 200, is used through the microcatheter. If the SVG has a resistant proximal cap, a penetrating guidewire can be used to puncture the cap and advance for a few millimeters. The microcatheter is then advanced through the cap and the stiff tip guidewire is exchanged for a polymer-jacketed guidewire. Stiff tip, highly penetrating guidewires (such as Confianza Pro 12, Hornet 14, Infiltrac and Infiltrac Plus, and Warrior) should be avoided for crossing occluded SVGs to reduce the risk of perforation. It is not uncommon for the polymer-jacketed guidewire to form a loop inside the SVG CTO, but knuckles should be avoided close to the distal anastomosis. If the wire and microcatheter stall while attempting to cross the SVG occlusion, a guide catheter extension can be advanced over the microcatheter (Fig. 18.4). If the guide extension cannot be advanced, balloon inflation immediately distal to the guide extension (inchworming technique) can be used. As the balloon is deflated, the guide extension is advanced over the balloon. This process can be repeated until the guide catheter extension is in proximity, but not across, the distal anastomosis.

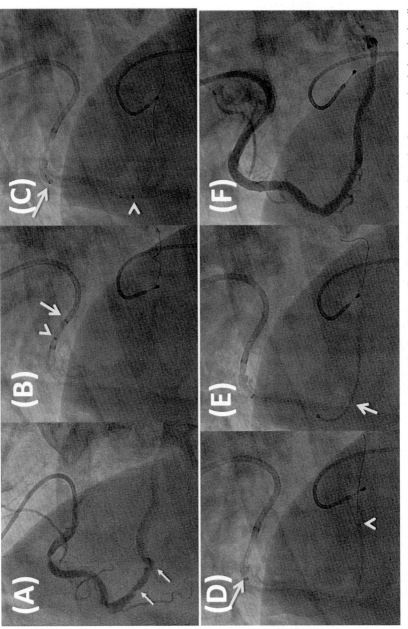

FIGURE 18.3 **Illustration of the "mother–daughter–granddaughter" technique. Panel A:** diagnostic angiography demonstrating lesions in the distal right coronary artery (*arrows*), which was very tortuous. **Panel B:** 6 Fr GuideLiner (*arrowhead*) is advanced inside an 8 Fr GuideLiner (*arrow*) into the guide catheter. **Panel C:** the 6 Fr GuideLiner (arrowhead) and the 8 Fr GuideLiner (*arrow*) are advanced into the right coronary artery. **Panel D:** the 6 Fr GuideLiner (*arrowhead*) is advanced past the distal right coronary artery lesion. **Panel E:** a stent (*arrow*) is delivered through the "mother–daughter–granddaughter" system to the distal right coronary artery. **Panel F:** excellent final result after stent implantation. *Courtesy of Dr. William Nicholson.*

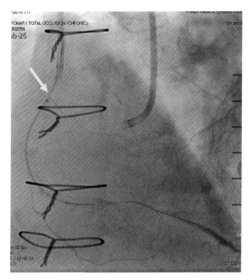

FIGURE 18.4   **Simultaneous use of a guide catheter extension (arrow) and a microcatheter for crossing an occluded SVG.** *Courtesy of Dr. Robert Gallino.*

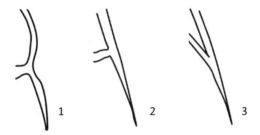

FIGURE 18.5   **Anatomic variants of bypass graft distal anastomoses.** Retrograde guidewire and equipment crossing is more likely to be challenging for morphology #3.

Advancing a guidewire proximal to the SVG distal anastomosis can be challenging to perform if the angulation is unfavorable (Fig. 18.5). Various techniques can be used to facilitate wiring, such as use of angulated (Section 30.6.3) or dual lumen (Section 30.6.4) microcatheters, polymer-jacketed guidewires, the deflection balloon technique, and the reversed guidewire technique, as described in Section 8.2.2.6 challenge c. Sometimes advancing a support guidewire may straighten the distal SVG anastomosis and facilitate retrograde wiring to the distal cap.

18.4   Retrograde approach through internal mammary grafts

Retrograde crossing is usually avoided via in situ internal mammary grafts, as injury of the graft could have catastrophic consequences. If retrograde crossing is done via internal mammary grafts, pressure dampening should be avoided and hemodynamic support should be considered due to increased risk of hemodynamic deterioration.[23] Retrograde crossing should be avoided through highly tortuous internal mammary grafts and through internal mammary grafts that are "stuck" on the chest wall due to adhesions, as straightening of the graft may lead to antegrade flow cessation (due to pseudolesion formation).[24] Use of a guide catheter extension can provide better support (Fig. 18.6) but may also lead to dissection of the internal mammary graft ostium (Fig. 25.5). Short or shortened guide catheters should be used for CTO PCI through internal mammary grafts.

18.5   Saphenous vein graft CTOs

CTO Manual online case 44, 193, 194.

In the 2021 ACC/AHA revascularization guidelines treatment of saphenous vein (SVG) graft CTOs is given a class III (level of evidence C) recommendation, due to high restenosis and repeat revascularization rates.[25] In patients with prior CABG, treatment of a native coronary artery CTO, if feasible, is preferable to treatment of a SVG CTO supplying the same territory.[17,26] The saphenous vein grafts (patent or occluded) can serve as retrograde conduits for recanalization of the native coronary CTO (CTO Manual online case 4).[17,20,27] If native CTO PCI is not possible or if it fails, PCI of the SVG-CTO could be performed (CTO Manual online case 193, 194).[17,28–33]

Crossing occluded SVGs is usually performed with a microcatheter and a stiff, polymer-jacketed guidewire, such as the Pilot 200 or Gladius. Retrograde crossing into the native vessel proximal to the SVG touchdown may be challenging due to acute angulation and may require use of a Venture catheter or the Supercross angulated microcatheter, or the reversed guidewire technique (Section 8.2.2.6, challenge c). Coronary CT angiography may be helpful for understanding the angle of SVG insertion into the target vessel.

18.6   Saphenous vein graft aneurysms

In patients who develop a saphenous vein graft aneurysm, recanalization of the native coronary artery followed by occlusion of the SVG can provide effective treatment of the aneurysm (CTO Manual online: cases 16 and 61).[26] **Both the proximal and the distal ends of the SVG should be occluded to completely stop flow into the aneurysm.**

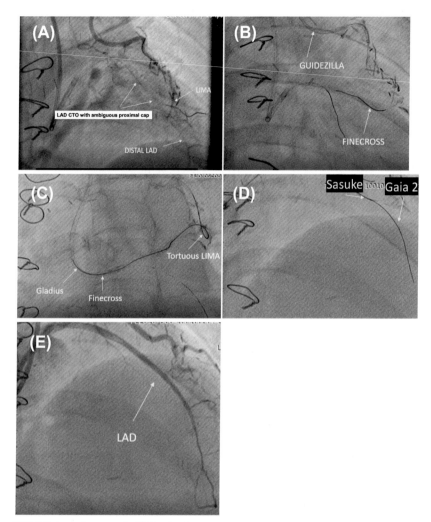

FIGURE 18.6   **Retrograde CTO PCI via a LIMA graft. Panel A**: mid LAD CTO with ambiguous proximal cap. The distal LAD is filling poorly via a LIMA graft. **Panel B**: retrograde LIMA crossing using a guide catheter extension. **Panel C**: successful retrograde crossing of the mid LAD CTO using a Gladius wire through a FineCross microcatheter. **Panel D**: externalization of a RG3 wire, followed by antegrade crossing to the distal LAD using a dual lumen microcatheter. **Panel E**: final result after balloon angioplasty and stenting. *Courtesy of Dr. Sevket Gorgulu.*

18.7   Coronary perforation in prior CABG patients

As discussed in Section 26.5, coronary perforation in prior CABG patients can lead to loculated effusions that can compress various cardiac structures[34] (such as the left atrium[35−37] or the right ventricle[38]). Such loculated effusions can be lethal, as they can be challenging to reach and drain percutaneously. Sometimes, intramyocardial hematomas can develop interfering with cardiac function (causing pseudotamponade or "dry tamponade") and can be challenging to treat.[39] If a hemodynamically significant loculated effusion develops and is not accessible for percutaneous drainage, surgical evacuation should be urgently performed via left thoracotomy, or if feasible via right side minimally invasive thoracoscopic techniques.

Perforations in prior CABG patients should, therefore, be immediately treated (for example with covered stents or coils) to minimize the risk of developing a loculated effusion.[34] Monitoring and serial echocardiography is recommended, as perforations in prior CABG patients may present with delayed tamponade or other hemodynamic or electrical instability.

References

1. Jeroudi OM, Alomar ME, Michael TT, et al. Prevalence and management of coronary chronic total occlusions in a tertiary Veterans Affairs Hospital. *Catheter Cardiovasc Interv* 2014;**84**:637−43.
2. Sakakura K, Nakano M, Otsuka F, et al. Comparison of pathology of chronic total occlusion with and without coronary artery bypass graft. *Eur Heart J* 2014;**35**:1683−93.
3. Farag M, Gue YX, Brilakis ES, Egred M. *Meta*-analysis comparing outcomes of percutaneous coronary intervention of native artery vs bypass graft in patients with prior coronary artery bypass grafting. *Am J Cardiol* 2021;**140**:47−54.
4. Xenogiannis I, Tajti P, Hall AB, et al. Update on cardiac catheterization in patients with prior coronary artery bypass graft surgery. *JACC Cardiovasc Interv* 2019;**12**:1635−49.
5. Tajti P, Karmpaliotis D, Alaswad K, et al. In-hospital outcomes of chronic total occlusion percutaneous coronary interventions in patients with prior coronary artery bypass graft surgery. *Circ Cardiovasc Interv* 2019;**12**:e007338.
6. Megaly M, Abraham B, Pershad A, et al. Outcomes of chronic total occlusion percutaneous coronary intervention in patients with prior bypass surgery. *JACC Cardiovasc Interv* 2020;**13**:900−2.
7. Hernandez-Suarez DF, Azzalini L, Moroni F, et al. Outcomes of chronic total occlusion percutaneous coronary intervention in patients with prior coronary artery bypass graft surgery: insights from the LATAM CTO registry. *Catheter Cardiovasc Interv* 2022;**99**:245−53.
8. Budassi S, Zivelonghi C, Dens J, et al. Impact of prior coronary artery bypass grafting in patients undergoing chronic total occlusion-percutaneous coronary intervention: procedural and clinical outcomes from the REgistry of Crossboss and Hybrid procedures in FrAnce, the NetheRlands, BelGium, and UnitEd Kingdom (RECHARGE). *Catheter Cardiovasc Interv* 2021;**97**:E51−60.
9. Alessandrino G, Chevalier B, Lefevre T, et al. A clinical and angiographic scoring system to predict the probability of successful first-attempt percutaneous coronary intervention in patients with total chronic coronary occlusion. *JACC Cardiovasc Interv* 2015;**8**:1540−8.
10. Maeremans J, Spratt JC, Knaapen P, et al. Towards a contemporary, comprehensive scoring system for determining technical outcomes of hybrid percutaneous chronic total occlusion treatment: the RECHARGE score. *Catheter Cardiovasc Interv* 2018;**91**:192−202.

11. Szijgyarto Z, Rampat R, Werner GS, et al. Derivation and validation of a chronic total coronary occlusion intervention procedural success score from the 20,000-patient EuroCTO Registry: the EuroCTO (CASTLE) score. *JACC Cardiovasc Interv* 2019;**12**:335−42.

12. Tsiafoutis I, Liontou C, Antonakopoulos A, Katsanou K, Koutouzis M, Katsivas A. Triple-access retrograde chronic total occlusion intervention through vein graft and epicardial collaterals. *J Invasive Cardiol* 2020;**32**:E172−3.

13. Michael TT, Banerjee S, Brilakis ES. Role of internal mammary artery bypass grafts in retrograde chronic total occlusion interventions. *J Invasive Cardiol* 2012;**24**:359−62.

14. Brilakis ES, Edson R, Bhatt DL, et al. Drug-eluting stents vs bare-metal stents in saphenous vein grafts: a double-blind, randomised trial. *Lancet* 2018;**391**:1997−2007.

15. Colleran R, Kufner S, Mehilli J, et al. Efficacy over time with drug-eluting stents in saphenous vein graft lesions. *J Am Coll Cardiol* 2018;**71**:1973−82.

16. Neumann FJ, Sousa-Uva M, Ahlsson A, et al. 2018 ESC/EACTS guidelines on myocardial revascularization. *Eur Heart J* 2019;**40**:87−165.

17. Brilakis ES, Banerjee S, Lombardi WL. Retrograde recanalization of native coronary artery chronic occlusions via acutely occluded vein grafts. *Catheter Cardiovasc Interv* 2010;**75**:109−13.

18. Xenogiannis I, Tajti P, Burke MN, Brilakis ES. Staged revascularization in patients with acute coronary syndromes due to saphenous vein graft failure and chronic total occlusion of the native vessel: a novel concept. *Catheter Cardiovasc Interv* 2019;**93**:440−4.

19. Tsiafoutis I, Liontou C, Koutouzis M, Brilakis E. Use of a totally occluded graft as a conduit for retrograde native artery recanalization. *Hellenic J Cardiol* 2021;**62**:490−2.

20. Dautov R, Manh Nguyen C, Altisent O, Gibrat C, Rinfret S. Recanalization of chronic total occlusions in patients with previous coronary bypass surgery and consideration of retrograde access via saphenous vein grafts. *Circ Cardiovasc Interv* 2016;**9**.

21. Wilson SJ, Hanratty CG, Spence MS, et al. Saphenous vein graft sacrifice following native vessel PCI is safe and associated with favourable longer-term outcomes. *Cardiovasc Revasc Med* 2019;**20**:1048−52.

22. Finn MT, Green P, Nicholson W, et al. Mother-daughter-granddaughter double guideliner technique for delivering stents past multiple extreme angulations. *Circ Cardiovasc Interv* 2016;**9**.

23. Tajti P, Karatasakis A, Karmpaliotis D, et al. Retrograde CTO-PCI of native coronary arteries via left internal mammary artery grafts: insights from a multicenter U.S. registry. *J Invasive Cardiol* 2018;**30**:89−96.

24. Lichtenwalter C, Banerjee S, Brilakis ES. Dual guide catheter technique for treating native coronary artery lesions through tortuous internal mammary grafts: separating equipment delivery from target lesion visualization. *J Invasive Cardiol* 2010;**22**:E78−81.

25. Lawton JS, Tamis-Holland JE, Bangalore S, et al. 2021 ACC/AHA/SCAI guideline for coronary artery revascularization: a report of the American College of Cardiology/American Heart Association Joint Committee on Clinical Practice Guidelines. *Circulation* 2022;**145**: e18−e114.

26. Katoh H, Nozue T, Michishita I. A case of giant saphenous vein graft aneurysm successfully treated with catheter intervention. *Catheter Cardiovasc Interv* 2016;**87**:83−9.

27. Nguyen-Trong PK, Alaswad K, Karmpaliotis D, et al. Use of saphenous vein bypass grafts for retrograde recanalization of coronary chronic total occlusions: insights from a multicenter registry. *J Invasive Cardiol* 2016;**28**:218−24.

28. Sachdeva R, Uretsky BF. Retrograde recanalization of a chronic total occlusion of a saphenous vein graft. *Catheter Cardiovasc Interv* 2009;**74**:575−8.

29. Takano M, Yamamoto M, Mizuno K. A retrograde approach for the treatment of chronic total occlusion in a patient with acute coronary syndrome. *Int J Cardiol* 2007;**119**:e22−4.

30. Ho PC, Tsuchikane E. Improvement of regional ischemia after successful percutaneous intervention of bypassed native coronary chronic total occlusion: an application of the CART technique. *J Invasive Cardiol* 2008;**20**:305−8.

31. Brilakis ES, Grantham JA, Thompson CA, et al. The retrograde approach to coronary artery chronic total occlusions: a practical approach. *Catheter Cardiovasc Interv* 2012;**79**:3−19.

32. Garg N, Hakeem A, Gobal F, Uretsky BF. Outcomes of percutaneous coronary intervention of chronic total saphenous vein graft occlusions in the contemporary era. *Catheter Cardiovasc Interv* 2014;**83**:1025−32.

33. Debski A, Tyczynski P, Demkow M, Witkowski A, Werner GS, Agostoni P. How should I treat a chronic total occlusion of a saphenous vein graft? Successful retrograde revascularisation. *EuroIntervention* 2016;**11**:e1325−8.

34. Karatasakis A, Akhtar YN, Brilakis ES. Distal coronary perforation in patients with prior coronary artery bypass graft surgery: the importance of early treatment. *Cardiovasc Revasc Med* 2016;**17**:412−17.

35. Aggarwal C, Varghese J, Uretsky BF. Left atrial inflow and outflow obstruction as a complication of retrograde approach for chronic total occlusion: report of a case and literature review of left atrial hematoma after percutaneous coronary intervention. *Catheter Cardiovasc Interv* 2013;**82**:770−5.

36. Wilson WM, Spratt JC, Lombardi WL. Cardiovascular collapse post chronic total occlusion percutaneous coronary intervention due to a compressive left atrial hematoma managed with percutaneous drainage. *Catheter Cardiovasc Interv* 2015;**86**:407−11.

37. Franks RJ, de Souza A, Di Mario C. Left atrial intramural hematoma after percutaneous coronary intervention. *Catheter Cardiovasc Interv* 2015;**86**:E150−2.

38. Adusumalli S, Morris M, Pershad A. Pseudo-pericardial tamponade from right ventricular hematoma after chronic total occlusion percutaneous coronary intervention of the right coronary artery: successfully managed percutaneously with computerized tomographic guided drainage. *Catheter Cardiovasc Interv* 2016;**88**:86−8.

39. Ezad S, Wardill T, Talwar S. Intramyocardial haematoma complicating chronic total occlusion percutaneous coronary intervention: case series and review of the literature. *Cardiovasc Revasc Med* 2022;**34**:142−7.

Chapter 19

CTO PCI in heavily calcified vessels

CTO Manual Online Cases: 5, 18, 49, 65, 85, 92, 133, 134, 136, 142, 147, 148, 149, 164, 171, 172, 175, 182, 188, 198.

Severe calcification can hinder all stages of chronic total occlusion (CTO) percutaneous coronary intervention (PCI): crossing, equipment delivery, and stent expansion.[1] Calcification is common (especially in older CTOs) and is part of several CTO PCI scores (such as J-CTO,[2] CL,[3] RECHARGE,[4] W-CTO,[5] and CASTLE[6]). Calcification has a larger adverse impact when combined with severe tortuosity. On fluoroscopy, calcification is defined as severe when it can be seen on both sides of the vessel wall, providing an outline of the vessel, on a still frame. Calcification by angiography can be deceiving: for example, calcification may not be located within the occluded lumen, but may be present in the vessel wall where CTO wires and equipment may need to traverse.

Calcification can hinder every step of CTO PCI, as follows.

19.1 Difficulty crossing the proximal or distal CTO cap (wire impenetrable cap)

Heavily calcified proximal (and distal) caps may be challenging or impossible to penetrate with a guidewire. Distal caps may be particularly challenging to cross in patients with prior coronary artery bypass graft surgery, presumably because the distal cap was exposed to systemic arterial pressure when the bypass graft was patent and also because coronary bypass graft surgery may lead to accelerated atherosclerosis and resultant severe calcification in the native circulation.[7] There are several possible strategies for crossing wire impenetrable caps that can be grouped into two major categories: (1) Get through and (2) Go around (Fig. 19.1).

19.1.1 Get through

Similar to the approach to a balloon uncrossable lesion, getting through a wire impenetrable lesion can be achieved via (1) strong support; (2) high penetrating power guidewires or devices; or (3) a combination of (1) and (2). Novel

Manual of Chronic Total Occlusion Percutaneous Coronary Interventions.
DOI: https://doi.org/10.1016/B978-0-323-91787-2.00020-4
© 2023 Elsevier Inc. All rights reserved.

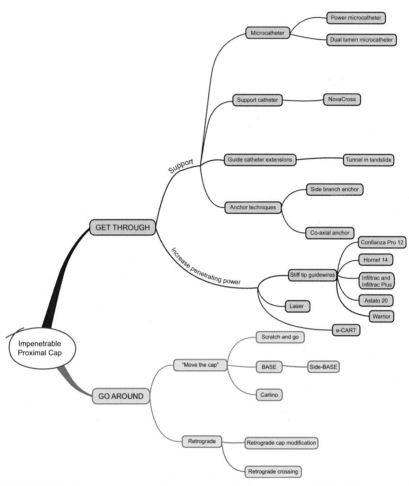

FIGURE 19.1 **Approach to the balloon impenetrable proximal cap.** *BASE*, Balloon-assisted subintimal entry; *e-CART*, ElectroCautery-Assisted ReenTry.

technologies are currently in development and could facilitate crossing of such lesions, such as the Soundbite system,[8] collagenase,[9] or directional lithotripsy.

Strong guidewire support can be achieved via various microcatheters (especially the more supportive Corsair Pro and Turnpike Spiral), support catheters, guide catheter extensions and anchoring techniques, as described in Section 9.3.1.8. A variation of the co-axial anchoring technique is inflating a balloon next to the microcatheter, hence providing support by pressing against the vessel wall ("power" microcatheter, Fig. 19.2). This approach can also be used to power trap a guide catheter extension for even more support (this is called "tunnel in landslide [TILT] technique), but this often requires a second guide catheter system, e.g., ping pong guide.[10]

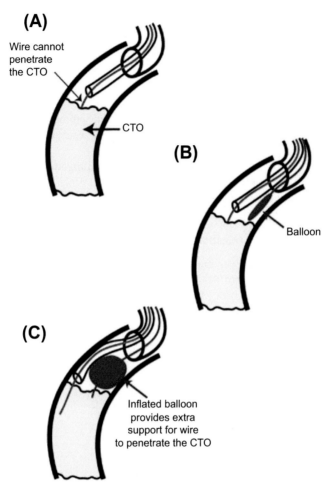

FIGURE 19.2   **Illustration of the power microcatheter technique**. When the support of the microcatheter and guide catheter is not enough for crossing a chronic total occlusion (**panel A**), a balloon (sized 1:1 to the vessel) can be advanced next to the microcatheter (**panel B**) and inflated (**panel C**). Balloon inflation anchors the microcatheter against the vessel wall, providing extra support (power) for guidewire advancement.

High penetration power can be provided via high tip stiffness, tapered-tip guidewires (such as the Confianza Pro 12, Hornet 14, Infiltrac, Infiltrac Plus, Warrior, Stingray and the Astato 20 guidewire [Section 30.6.3.3]). Some operators recommend short laser activation (if the proximal cap anatomy is very clear), however this approach carries increased risk of perforation and is not favored by most CTO operators. Use of the Carlino technique (contrast injection through a microcatheter, Section 8.3.4.3) can help create a dissection plane and allow subsequent wire crossing (CTO Manual online case 37).

The e-CART technique (ElectroCautery-Assisted Re-enTry; Section 15.1.2) has been used for penetrating both distal and proximal calcified caps.[11,12]

19.1.2 Go around

CTO Manual online cases: 12, 13, 46, 49, 57, 60, 62.

Going "around" instead of "through" a wire-impenetrable CTO using dissection/reentry techniques can be a very effective and safe strategy for crossing wire impenetrable CTOs.

As described in the approach to CTOs with ambiguous proximal cap (Section 16.1), extraplaque crossing can be achieved either in the antegrade direction (move the cap techniques) or in the retrograde direction.

19.2 Crossing the occlusion

Wire advancement and penetration can be challenging through calcified lesions. Use of high tip stiffness, tapered tip guidewires (such as Confianza Pro 12, Hornet 14, Warrior, Infiltrac and Infiltrac Plus, and Astato 20; Section 30.6.3.3) can help advance through the occlusion, but should be de-escalated for a softer guidewire as soon as possible to decrease the risk of perforation. Alternative solutions include use of dissection/reentry techniques (both antegrade and retrograde) and use of the more torqueable, intermediate tip stiffness, tapered tip guidewires, (Gaia, Gaia Next and Judo series; Section 30.6.3.2).

19.3 Reentry into the distal true lumen

Severe calcification may hinder guidewire reentry into the distal true lumen after extraplaque guidewire advancement. Selecting a different reentry area (more proximal or distal to the calcification—"bobsled" technique) or using the retrograde approach are potential solutions. Successful puncture of a calcified reentry zone into the true lumen may require a high stiffness, tapered tip guidewire (such as Confianza Pro 12, Hornet 14, Infiltrac, Infiltrac Plus, Stingray, Warrior, and Astato 20; Section 30.6.3.3), often followed by swapping to a stiff polymer-jacketed wire, such as Pilot 200 or Gladius (Section 30.6.2.3), to wire the distal vessel.

19.4 Equipment delivery

Severe calcification may hinder equipment advancement. Careful lesion preparation with high-pressure balloon inflations, plaque modification balloons (Section 30.6.5), very high pressure balloons (Section 30.9.7), rotational/orbital atherectomy[13-15] (Section 30.10) or intravascular lithotripsy[16-20]

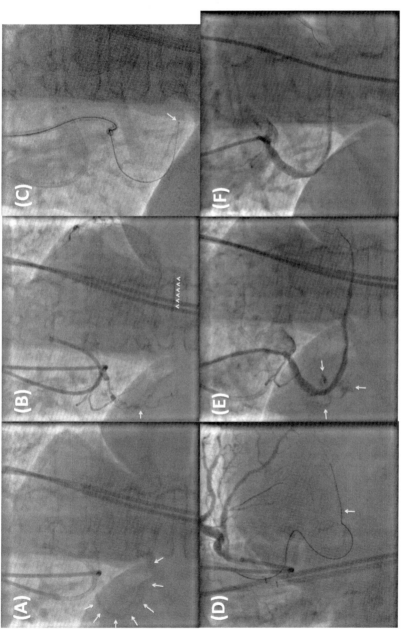

FIGURE 19.3 **Percutaneous coronary intervention of a heavily calcified right coronary artery complicated by perforation** (CTO Manual online case 17). **Panel A**: severely calcified right coronary artery (arrows, cineangiography obtained without contrast injection). **Panel B**: CTO of the mid right coronary artery (arrow) with the right posterior descending artery filling via collaterals (arrowheads). **Panel C**: extraplaque crossing of the CTO with a CrossBoss catheter. **Panel D**: successful reentry into the distal true lumen using a Stingray balloon (arrow). **Panel E**: mid right coronary artery perforation (arrows) after stent implantation. **Panel F**: sealing of the perforation after prolonged balloon inflations.

(Section 30.9.8) can assist with equipment delivery, but additional techniques to increase support (such as use of guide catheter extensions) may be required.

Atherectomy of the extraplaque space has been successfully performed[21–23], but should generally be avoided due to increased risk of perforation. If there is limited space for antegrade guidewire advancement, the tip of the Rotawire can be cut off to allow treatment of the target lesion. Rotational atherectomy has been performed over an externalized RG3 guidewire,[22,23] but this maneuver is off-label and its safety has not yet been established. Intravascular lithotripsy has also been successfully performed in the extraplaque space.[16]

19.5   Stent expansion

Severe calcification may prevent stent expansion. Overly aggressive postdilation of severely calcified lesions should be avoided, as it may lead to perforation (Fig. 19.3). Intravascular imaging is strongly recommended to accurately determine balloon sizing for postdilatation of severely calcified lesions. High pressure postdilatation in the extraplaque space carries increased risk of complications.

Prevention:

1. Ensure that calcified lesions are predilated with noncompliant balloons, sized one to one with the vessel before implanting a stent.
2. If balloons fail to expand, additional pretreatment (using plaque modification balloons, one or more buddy wires, rotational or orbital atherectomy, very high-pressure balloon, or intravascular lithotripsy) should be performed. Stenting should not be done until lesion expansion is confirmed. Intravascular imaging showing fractures in the calcified segments is associated with good stent expansion.

Treatment:

If stent underexpansion is detected after stent placement, several maneuvers can be performed, as outlined in Section 23.2 on balloon undilatable lesions.

References

1. Cosgrove C, Mahadevan K, Spratt JC, McEntegart M. The impact of calcium on chronic total occlusion management. *Interv Cardiol* 2021;**16**:e30.
2. Morino Y, Abe M, Morimoto T, et al. Predicting successful guidewire crossing through chronic total occlusion of native coronary lesions within 30 minutes: the J-CTO (Multicenter CTO Registry in Japan) score as a difficulty grading and time assessment tool. *JACC Cardiovasc Interv* 2011;**4**:213–21.
3. Alessandrino G, Chevalier B, Lefevre T, et al. A clinical and angiographic scoring system to predict the probability of successful first-attempt percutaneous coronary intervention in patients with total chronic coronary occlusion. *JACC Cardiovasc Interv* 2015;**8**:1540–8.

4. Maeremans J, Spratt JC, Knaapen P, et al. Towards a contemporary, comprehensive scoring system for determining technical outcomes of hybrid percutaneous chronic total occlusion treatment: the RECHARGE score. *Catheter Cardiovasc Interv* 2018;**91**:192−202.

5. Khanna R, Pandey CM, Bedi S, Ashfaq F, Goel PK. A weighted angiographic scoring model (W-CTO score) to predict success of antegrade wire crossing in chronic total occlusion: analysis from a single centre. *AsiaIntervention* 2018;**4**:18−25.

6. Szijgyarto Z, Rampat R, Werner GS, et al. Derivation and validation of a chronic total coronary occlusion intervention procedural success score from the 20,000-patient EuroCTO Registry: the EuroCTO (CASTLE) score. *JACC Cardiovasc Interv* 2019;**12**:335−42.

7. Sakakura K, Nakano M, Otsuka F, et al. Comparison of pathology of chronic total occlusion with and without coronary artery bypass graft. *Eur Heart J* 2014;**35**:1683−93.

8. Benko A, Berube S, Buller CE, et al. Novel crossing system for chronic total occlusion recanalization: first-in-man experience with the soundbite crossing system. *J Invasive Cardiol* 2017;**29**:E17−20.

9. Strauss BH, Osherov AB, Radhakrishnan S, et al. Collagenase total occlusion-1 (CTO-1) trial: a phase I, dose-escalation, safety study. *Circulation* 2012;**125**:522−8.

10. Santiago R, Moroni F, Del Rio V, Rodriguez-Escudero J, Azzalini L. The guide extension tunnel in landslide technique (TILT) for equipment delivery in severely tortuous or uncrossable lesions during percutaneous coronary intervention. *EuroIntervention* 2021;17: e923−e924.

11. Nicholson W, Harvey J, Dhawan R. E-CART (ElectroCautery-Assisted Re-enTry) of an aorto-ostial right coronary artery chronic total occlusion: first-in-man. *JACC Cardiovasc Interv* 2016;**9**:2356−8.

12. Neupane S, Basir M, Alaswad K. Electrocautery-Facilitated Crossing (ECFC) of chronic total occlusions. *J Invasive Cardiol* 2020;**32**:55−7.

13. Pagnotta P, Briguori C, Mango R, et al. Rotational atherectomy in resistant chronic total occlusions. *Catheter Cardiovasc Interv* 2010;**76**:366−71.

14. Azzalini L, Dautov R, Ojeda S, et al. Long-term outcomes of rotational atherectomy for the percutaneous treatment of chronic total occlusions. *Catheter Cardiovasc Interv* 2017;**89**:820−8.

15. Xenogiannis I, Karmpaliotis D, Alaswad K, et al. Usefulness of atherectomy in chronic total occlusion interventions (from the PROGRESS-CTO registry). *Am J Cardiol* 2019;**123**:1422−8.

16. Oksnes A, Cosgrove C, Walsh S, et al. Intravascular lithotripsy for calcium modification in chronic total occlusion percutaneous coronary intervention. *J Interv Cardiol* 2021;**2021**:9958035.

17. Karacsonyi J, Nikolakopoulos I, Vemmou E, Rangan BV, Brilakis ES. Intracoronary lithotripsy: a new solution for undilatable in-stent chronic total occlusions. *JACC Case Rep* 2021;**3**:780−5.

18. Garbo R, Di Russo C, Sciahbasi A, Fedele S. The last resort during complex retrograde percutaneous coronary chronic total occlusion intervention: extraplaque intracoronary lithotripsy to externally crush a heavy calcified occluded stent. *Catheter Cardiovasc Interv* 2022;**99**:497−501.

19. Yeoh J, Hill J, Spratt JC. Intravascular lithotripsy assisted chronic total occlusion revascularization with reverse controlled antegrade retrograde tracking. *Catheter Cardiovasc Interv* 2019;**93**:1295−7.

20. Azzalini L, Bellini B, Montorfano M, Carlino M. Intravascular lithotripsy in chronic total occlusion percutaneous coronary intervention. *EuroIntervention* 2019;**15**:e1025−6.

21. Capretti G, Carlino M, Colombo A, Azzalini L. Rotational atherectomy in the subadventitial space to allow safe and successful chronic total occlusion recanalization: pushing the limit further. *Catheter Cardiovasc Interv* 2018;**91**:47−52.

22. Iannaccone M, Colangelo S, Colombo F, Garbo R. Rotational atherectomy with the new RotaPro system over RG3 guidewire in subadventitial retrograde highly calcified CTO PCI. *Catheter Cardiovasc Interv* 2020;**95**:242−4.
23. Kaneko U, Kashima Y, Kanno D, Sugie T, Kobayashi K, Fujita T. Successful rotational atherectomy over RG3 guidewire after failure of various techniques to deliver RotaWire. *Cardiovasc Interv Ther* 2017;**32**:386−91.

Chapter 20

Chronic total occlusion percutaneous coronary intervention in patients presenting with acute coronary syndromes

CTO Manual online cases 11, 102, 103, 137

Approximately 10%−15% of patients presenting with ST-segment eleva-tion acute myocardial infarction have a chronic total occlusion (CTO).[1−3] In patients who present with myocardial infarction, presence of a CTO has been associated with higher risk of out-of-hospital cardiac arrest,[4] higher risk of arrhythmias,[5] and higher early and late mortality[2,3,6−10] (Figure 7.2) likely due to more severe ischemia ("double jeopardy" phenomenon, Figure 7.3).

Some patients may have multiple possible culprit lesions. The following criteria can help determine the culprit lesion(s) (Chapter 7.2.1a):

1. Artery supplies myocardium that corresponds to the area of ST-elevation.
2. Artery supplies myocardium that is hypokinetic by left ventriculography or echocardiography.
3. Thrombus.
4. Contrast staining.
5. Poor or no collaterals. However, some highly stenotic lesions that are chronic may suffer acutely from an abrupt plaque rupture and cause an acute coronary syndrome (ACS); hence, well developed collaterals are not synonymous with a CTO.
6. Ease of wiring. Significant difficulty crossing the occlusion may sometimes (but not always) suggest that the occlusion is chronic.

In most patients presenting with an acute coronary syndrome CTO percu-taneous coronary intervention (PCI) should not be performed during the index procedure. Rarely, ad hoc CTO PCI may be needed, such as in patients who present with an acute coronary syndrome due to failure of a highly

Manual of Chronic Total Occlusion Percutaneous Coronary Interventions.
DOI: https://doi.org/10.1016/B978-0-323-91787-2.00006-X

diseased saphenous vein graft (SVG) that cannot be recanalized.[11] In most such cases, however, PCI of the SVG first, followed by staged PCI of the native coronary artery CTO is preferred, if technically feasible.[12] Based on the results of the CULPRIT-SHOCK trial,[13,14] PCI of the culprit lesion only is recommended in ACS patients presenting with cardiogenic shock. CTO PCI should not be performed ad hoc in such patients.[15]

In ST-segment elevation acute myocardial infarction (STEMI) patients complete revascularization provides better outcomes compared with culprit-only revascularization.[16] CTO PCI is performed as a staged procedure in most such patients. In the EXPLORE trial, PCI of a nonculprit CTO within one week from presentation with STEMI did not improve left ventricular ejection fraction at 4 months.[17] However, EXPLORE had several limitations, including small sample size, low risk population, unblinded design, lack of viability assessment and low CTO PCI success rate (73%).

References

1. Fefer P, Carlino M, Strauss BH. Intraplaque therapies for facilitating percutaneous recanalization of chronic total occlusions. *Can J Cardiol* 2010;**26**(Suppl A):32A−36AA.

2. Claessen BE, Dangas GD, Weisz G, et al. Prognostic impact of a chronic total occlusion in a non-infarct-related artery in patients with ST-segment elevation myocardial infarction: 3-year results from the HORIZONS-AMI trial. *Eur Heart J* 2012;**33**:768−75.

3. Claessen BE, van der Schaaf RJ, Verouden NJ, et al. Evaluation of the effect of a concurrent chronic total occlusion on long-term mortality and left ventricular function in patients after primary percutaneous coronary intervention. *JACC Cardiovasc Interv* 2009;**2**:1128−34.

4. Kosugi S, Shinouchi K, Ueda Y, et al. Clinical and angiographic features of patients with out-of-hospital cardiac arrest and acute myocardial infarction. *J Am Coll Cardiol* 2020; **76**:1934−43.

5. Saad M, Fuernau G, Desch S, et al. Prognostic impact of non-culprit chronic total occlusions in infarct-related cardiogenic shock: results of the randomised IABP-SHOCK II trial. *EuroIntervention* 2018;**14**:e306−13.

6. Hoebers LP, Vis MM, Claessen BE, et al. The impact of multivessel disease with and without a co-existing chronic total occlusion on short- and long-term mortality in ST-elevation myocardial infarction patients with and without cardiogenic shock. *Eur J Heart Fail* 2013;**15**:425−32.

7. Lexis CP, van der Horst IC, Rahel BM, et al. Impact of chronic total occlusions on markers of reperfusion, infarct size, and long-term mortality: a substudy from the TAPAS-trial. *Catheter Cardiovasc Interv* 2011;**77**:484−91.

8. Brilakis ES, Banerjee S, Karmpaliotis D, et al. Procedural outcomes of chronic total occlusion percutaneous coronary intervention: a report from the NCDR (National Cardiovascular Data Registry). *JACC Cardiovasc Interv* 2015;**8**:245−53.

9. O'Connor SA, Garot P, Sanguineti F, et al. *Meta*-analysis of the impact on mortality of noninfarct-related artery coronary chronic total occlusion in patients presenting with ST-segment elevation myocardial infarction. *Am J Cardiol* 2015;**116**:8−14.

10. Shinouchi K, Ueda Y, Kato T, et al. Relation of chronic total occlusion to in-hospital mortality in the patients with sudden cardiac arrest due to acute coronary syndrome. *Am J Cardiol* 2019;**123**:1915−20.

11. Brilakis ES, Banerjee S, Lombardi WL. Retrograde recanalization of native coronary artery chronic occlusions via acutely occluded vein grafts. *Catheter Cardiovasc Interv* 2010; **75**:109–13.

12. Xenogiannis I, Tajti P, Burke MN, Brilakis ES. Staged revascularization in patients with acute coronary syndromes due to saphenous vein graft failure and chronic total occlusion of the native vessel: a novel concept. *Catheter Cardiovasc Interv* 2019;**93**:440–4.

13. Thiele H, Akin I, Sandri M, et al. One-year outcomes after PCI strategies in cardiogenic shock. *N Engl J Med* 2018;**379**:1699–710.

14. Thiele H, Akin I, Sandri M, et al. PCI strategies in patients with acute myocardial infarction and cardiogenic shock. *N Engl J Med* 2017;**377**:2419–32.

15. Braik N, Guedeney P, Behnes M, et al. Impact of chronic total occlusion and revascularization strategy in patients with infarct-related cardiogenic shock: a subanalysis of the culprit-shock trial. *Am Heart J* 2021;**232**:185–93.

16. Elgendy IY, Mahmoud AN, Kumbhani DJ, Bhatt DL, Bavry AA. Complete or culprit-only revascularization for patients with multivessel coronary artery disease undergoing percutaneous coronary intervention: a pairwise and network *meta*-analysis of randomized trials. *JACC Cardiovasc Interv* 2017;**10**:315–24.

17. Henriques JP, Hoebers LP, Ramunddal T, et al. Percutaneous intervention for concurrent chronic total occlusions in patients with STEMI: the EXPLORE trial. *J Am Coll Cardiol* 2016;**68**:1622–32.

Chapter 21

Multiple chronic total occlusions—Prior chronic total occlusion percutaneous coronary intervention failure

21.1 Multiple chronic total occlusions

Revascularization of patients with multiple coronary chronic total occlusions (CTOs) can be challenging to achieve and is associated with increased risk of complications.[1] Coronary artery bypass graft (CABG) is often preferred in such patients, if they are good surgical candidates (Chapter 7).[1] For patients who need revascularization of >1 CTOs, are not good surgical candidates and are referred for percutaneous coronary intervention (PCI), staged CTO PCI is preferred to attempting PCI of multiple CTOs during the same procedure; the latter is associated with longer procedure time, higher contrast and radiation dose and higher risk of complications.[2] In contrast, concomitant treatment of non-CTO lesions at the same time as CTO PCI is not associated with higher risk of complications (but still requires long procedure time and higher contrast and radiation dose).[3,4].

Recanalization of one CTO may provide additional options for crossing the second CTO: for example in patients with left anterior descending artery and right coronary artery (RCA) CTOs, recanalizing one of the two CTOs may allow collateral development that could be used for retrograde crossing of the second CTO (CTO Manual online case 200).

Patients with multiple CTOs may have low ejection fraction and hemodynamic support (Chapter 14) should be considered (CTO Manual online case 148), especially when attempting retrograde crossing through the last remaining vessel.

21.2 Prior chronic total occlusion percutaneous coronary intervention failure

Even when performed at the most experienced centers and by experienced operators, 10%−15% of CTO PCIs fail (Table 7.2).[5] The risk of failure is higher in more complex CTOs (Figure 8.4.9).[6] Prior failure is one of the parameters included in the J-CTO score.[7] In the PROGRESS-CTO registry, a

Manual of Chronic Total Occlusion Percutaneous Coronary Interventions.
DOI: https://doi.org/10.1016/B978-0-323-91787-2.00034-4
479

previously failed CTO PCI attempt was associated with higher angiographic complexity, longer procedural duration, and fluoroscopy time, but not with the success and complication rates of subsequent CTO PCI attempts.[8]

The Global CTO crossing algorithm recommends stopping the CTO PCI procedure if the procedure time is >3 h, if contrast volume is >3 times the estimated glomerular filtration rate, or if the air kerma radiation dose is >5 Gy, unless the procedure is well advanced.[9] The procedure should also be stopped in most cases if a complication occurs or if it becomes evident that advanced crossing strategies, such as retrograde crossing or antegrade dissection and reentry (ADR), are needed for successful crossing, but the operator does not have expertise in those techniques or if there is significant operator or patient fatigue.[9]

The success rates of CTO intervention after previously failed attempts ranges from 64%−88%.[8,10−12]. If CTO crossing attempts fail, a variety of *investment procedures* should be considered if the anticipated benefit exceeds the potential harm.[9] Intentional dilatation of the extraplaque (formerly called "subintimal") space and proximal cap with balloon angioplasty (subintimal plaque modification) may consolidate the success achieved during the initial, failed procedure and improve the likelihood of success with re-attempt.[13,14] In addition, the operator can often use the subintimal tracking and reentry technique (STAR, Chapter 8.2) to establish antegrade flow without stent implantation.[14−17] Once an investment procedure is performed, the operator should not inject contrast since this may result in an hydraulic dissection that could be harmful. These investment procedures may lead to recanalization, as dissections may communicate with the distal vessel. Repeat CTO crossing attempts are usually performed after ≥2 months to allow healing of any created dissections.[18] It may also be best to delay repeat CTO PCI attempts for at least 2 months in patients who receive a large radiation dose (>5 Gy).

Understanding the mechanism of failure is important for optimizing the likelihood of success of subsequent attempts. Coronary CT angiography can help better understand the coronary anatomy and facilitate subsequent crossing efforts.[19,20] If repeat CTO PCI is likely to require advanced CTO crossing techniques, such as ADR and retrograde crossing, and the operator has limited experience in such techniques, referral to a CTO expert center or repeat attempt with a proctor could be considered.[9]

References

1. Kim BS, Yang JH, Jang WJ, et al. Clinical outcomes of multiple chronic total occlusions in coronary arteries according to three therapeutic strategies: bypass surgery, percutaneous intervention and medication. *Int J Cardiol* 2015;**197**:2−7.
2. Tajti P, Alaswad K, Karmpaliotis D, et al. In-hospital outcomes of attempting more than one chronic total coronary occlusion through percutaneous intervention during the same procedure. *Am J Cardiol* 2018;**122**:381−7.

3. Xenogiannis I, Karmpaliotis D, Alaswad K, et al. Impact of concomitant treatment of non-chronic total occlusion lesions at the time of chronic total occlusion intervention. *Int J Cardiol* 2020;**299**:75−80.

4. Riley RF, Sapontis J, Karmpaliotis D, et al. Clinical and health status outcomes among patients treated with single as compared to multivessel angioplasty during chronic total occlusion percutaneous coronary interventions: a report from the OPEN CTO registry. *Coron Artery Dis* 2021;**32**:112−18.

5. Assali M, Buda KG, Megaly M, Hall AB, Burke MN, Brilakis ES. Update on chronic total occlusion percutaneous coronary intervention. *Prog Cardiovasc Dis* 2021;**69**:27−34.

6. Tajti P, Karmpaliotis D, Alaswad K, et al. The hybrid approach to chronic total occlusion percutaneous coronary intervention: update from the PROGRESS CTO registry. *JACC Cardiovasc Interv* 2018;**11**:1325−35.

7. Morino Y, Abe M, Morimoto T, et al. Predicting successful guidewire crossing through chronic total occlusion of native coronary lesions within 30 minutes: the J-CTO (Multicenter CTO Registry in Japan) score as a difficulty grading and time assessment tool. *JACC Cardiovasc Interv* 2011;**4**:213−21.

8. Karacsonyi J, Karatasakis A, Karmpaliotis D, et al. Effect of previous failure on subsequent procedural outcomes of chronic total occlusion percutaneous coronary intervention (from a Contemporary Multicenter Registry). *Am J Cardiol* 2016;**117**:1267−71.

9. Wu EB, Brilakis ES, Mashayekhi K, et al. Global chronic total occlusion crossing algorithm: JACC state-of-the-art review. *J Am Coll Cardiol* 2021;**78**:840−53.

10. Guelker JE, Blockhaus C, Bufe A, et al. In-hospital outcome of re-attempted percutaneous coronary interventions for chronic total occlusion. *Cardiol J* 2021.

11. Tanabe M, Kodama K, Asada K, Kunitomo T. Lesion characteristics and procedural outcomes of re-attempted percutaneous coronary interventions for chronic total occlusion. *Heart Vessel* 2018;**33**:573−82.

12. Fu M, Chang S, Ge L, et al. Reattempt percutaneous coronary intervention of chronic total occlusions after prior failures: a single-center analysis of strategies and outcomes. *J Interv Cardiol* 2021;**2021**:8835104.

13. Wilson WM, Walsh SJ, Yan AT, et al. Hybrid approach improves success of chronic total occlusion angioplasty. *Heart* 2016;**102**:1486−93.

14. Xenogiannis I, Choi JW, Alaswad K, et al. Outcomes of subintimal plaque modification in chronic total occlusion percutaneous coronary intervention. *Catheter Cardiovasc Interv* 2020;**96**:1029−35.

15. Hirai T, Grantham JA, Gosch KL, et al. Impact of subintimal or plaque modification on repeat chronic total occlusion angioplasty following an unsuccessful attempt. *JACC Cardiovasc Interv* 2020;**13**:1010−12.

16. Hirai T, Grantham JA, Sapontis J, et al. Impact of subintimal plaque modification procedures on health status after unsuccessful chronic total occlusion angioplasty. *Catheter Cardiovasc Interv* 2018;**91**:1035−42.

17. Wu NQ, Qian J. Impact of subintimal plaque modification (SPM) technique on failed intervention of coronary artery chronic total occlusion. *Chronic Dis Transl Med* 2021;**7**:135−8.

18. Ybarra LF, Rinfret S, Brilakis ES, et al. Definitions and clinical trial design principles for coronary artery chronic total occlusion therapies: CTO-ARC consensus recommendations. *Circulation* 2021;**143**:479−500.

19. Hong SJ, Kim BK, Cho I, et al. Effect of coronary CTA on chronic total occlusion percutaneous coronary intervention: a randomized trial. *JACC Cardiovasc Imaging* 2021;**14**:1993−2004.

20. Werner GS. Use of coronary computed tomographic angiography to facilitate percutaneous coronary intervention of chronic total occlusions. *Circ Cardiovasc Interv* 2019;**12**:e007387.

Chapter 22

Percutaneous coronary intervention of in-stent chronic total occlusions

Approximately 11%−12% of all chronic total occlusions (CTO) percutaneous coronary intervention (PCIs) are performed in in-stent CTOs.[1,2] Patients with in-stent CTOs often have diabetes mellitus, hypertension, and peripheral vascular disease.[3] When compared with de novo CTOs, in-stent CTOs[4] are less frequently associated with ambiguous caps, calcification, and well-developed collaterals, but are more frequently ostial and have a higher J-CTO score. Such angiographic heterogeneity results in a mixed subset of patients, which may range from simple cases that require only antegrade wiring to very complex ones that can be very challenging to recanalize.[5−10] The extent or not of disease beyond stent borders, presence of a large side branch for left anterior descending artery occlusions and tortuosity for right coronary artery (RCA) occlusions are important factors to consider when assessing the likelihood of success.[3,11] Although similar procedural success can be achieved for in-stent and de novo CTOs,[1,2,12−14] the risk of subsequent restenosis is higher for in-stent CTOs.[15,16]

In-stent CTOs that are located within the prior stent are easier to recanalize and usually only require antegrade crossing. Occlusions extending beyond the proximal and distal ends of the stent have been associated with lower recanalization rates and frequent need for retrograde crossing (Fig. 22.1).[17]

22.1   Approach to in-stent chronic total occlusions

The major differences between approaching in-stent CTOs and de novo CTOs are the following:

1. *Extraplaque (subintimal) crossing techniques are best avoided* for in-stent CTOs, as wire exit behind the stent would lead to "crushing" of the occluded stent after repeat stenting. However, "crushing" of the occluded stent is an acceptable option if intra-stent crossing fails, and has been associated with encouraging short and mid-term outcomes (CTO Manual online cases 32, 54, and 72).[8,18−23]

Manual of Chronic Total Occlusion Percutaneous Coronary Interventions.
DOI: https://doi.org/10.1016/B978-0-323-91787-2.00028-9

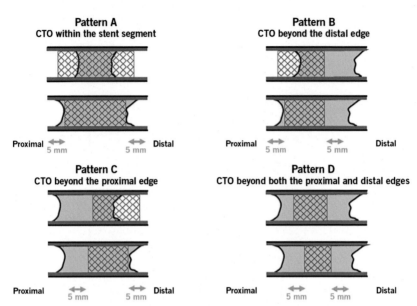

FIGURE 22.1   Occlusion patterns of in-stent chronic total occlusions. According to the occlusion patterns, in-stent chronic total occlusions (CTOs) are divided into the following groups. **Pattern A:** CTO within the stent segment. **Pattern B:** CTO beyond the distal edge. **Pattern C:** CTO beyond the proximal edge. **Pattern D:** CTO beyond both the proximal and distal edges. An edge is defined as an area 5 mm from the stent edge. The stent segment includes the in-stent region as well as the proximal and distal edges. *Reproduced with permission from Sekiguchi M, Muramatsu T, Kishi K, et al. Occlusion patterns, strategies and procedural outcomes of percutaneous coronary intervention for in-stent chronic total occlusion. EuroIntervention 2021;17:e631—e638.*

2. *Use of the CrossBoss catheter* (Section 8.3.4.1) (CTO Manual online cases 8 and 19).[24,25] The CrossBoss catheter is well suited for crossing in-stent CTOs, likely because the stent struts act as a barrier preventing advancement of the CrossBoss catheter behind the occluded stent (Fig. 22.2). In the CrossBoss First trial, crossing of in-stent CTOs was faster with the CrossBoss catheter compared with antegrade wiring.[26] Rarely, the CrossBoss catheter can advance behind stent struts, leading to stent crushing (Fig. 22.3).[27] Occasionally the CrossBoss advancement may stop, requiring redirection with a guidewire (usually a polymer-jacketed guidewire, such as the Pilot 200, Gladius and Gladius Mongo, and Raider).[24]

22.2   Challenges associated with percutaneous coronary intervention of in-stent chronic total occlusions

The following steps can be especially challenging when performing PCI of in-stent CTOs (Fig. 22.4).

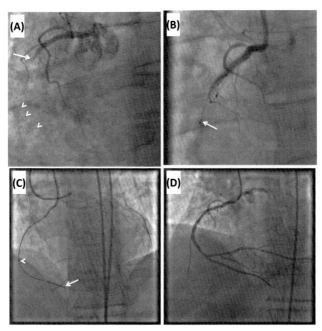

FIGURE 22.2 **Example of crossing an in-stent CTO with the CrossBoss catheter**. Coronary angiography demonstrating a chronic total occlusion of the right coronary artery (arrow, **panel A**) due to in-stent restenosis (arrowheads, **panel A**). The CrossBoss catheter (arrow, **panel B**) was inserted into the lesion and advanced using the "fast spin" technique. The CrossBoss catheter could not be advanced through the stent (arrowhead, **panel C**), but a Confianza Pro 12 wire (arrow, **panel C**) crossed into the distal true lumen, as confirmed by contralateral injection. After stent implantation an excellent final result was achieved (**panel D**). *Reproduced with permission from Papayannis A, Banerjee S, Brilakis ES. Use of the crossboss catheter in coronary chronic total occlusion due to in-stent restenosis. Catheter Cardiovasc Interv 2012;80:E30–E36.*

22.2.1 Penetrating the proximal cap

The proximal cap of in-stent CTOs can be calcified and hard to penetrate. As outlined in Section 19.1 penetrating a tough proximal cap requires use of support techniques and penetrating guidewires, often in combination (Fig. 22.5) (CTO Manual online case 7).

22.2.2 Crossing the occlusion

1. When crossing in-stent occlusions, the prior stent provides a roadmap of the vessel course, potentially facilitating wire advancement.
2. Frequent visualization in orthogonal projections ensures that the guidewire is advancing within the stent struts.
3. Similar to de novo lesions, several guidewires can be used to cross in-stent CTOs, usually starting from soft, tapered polymer-jacketed

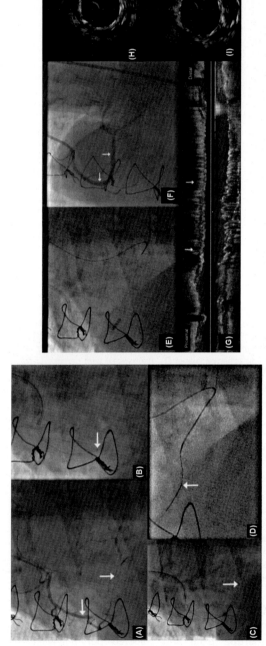

FIGURE 22.3 **Example of CrossBoss catheter exit between stent struts. Panel A:** diagnostic coronary angiography demonstrated occlusive in-stent resteno-sis with ipsilateral and contralateral collaterals (arrows show proximal and distal edges of the previously implanted overlapping stents). **Panel B:** the CrossBoss catheter entered the chronic total occlusion, but stopped advancing after 5–10 mm. **Panel C:** following further rapid rotation, it progressed with relative ease toward the distal vessel. **Panel D:** "rendezvous" in the distal right coronary artery. Favorable alignment and advancement of the retrograde wire into the CrossBoss (the arrow is at the level of the CrossBoss tip). **Panel E:** inflation of a high pressure balloon at the point of stent exit. **Panel F:** final angiographic result. **Panel G:** intravascular ultrasound (IVUS) longitudinal view after implantation of three overlapping Xience stents. **Panel H:** double stent strut layer in the middle right coronary artery. **Panel I:** IVUS image poststent implantation in the distal RCA, also showing the old "crushed" stent. *Reproduced with permission from Ntatsios A, Smith WHT. Exit of CrossBoss between stent struts within chronic total occlusion to subintimal space: completion of case via retrograde approach with rendezvous in coronary. J Cardiology Cases 2014;9:183-186.*

Challenges with PCI of in-stent CTOs.

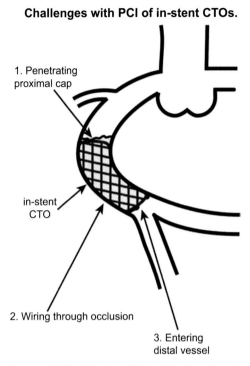

1. Penetrating proximal cap

in-stent CTO

2. Wiring through occlusion

3. Entering distal vessel

FIGURE 22.4 **Challenges of PCI of in-stent CTOs.** *CTO,* Chronic total occlusion; *PCI,* percutaneous coronary intervention.

guidewires and escalating to stiff polymer-jacketed guidewires or stiff tip, non-polymer jacketed guidewires. The Carlino technique (antegrade or retrograde) could help cross the lesion without extensive dissection, especially if the old stent is not undersized. Laser could also be useful.[28]

4. If the guidewire exits the stent struts, there are three options (Fig. 22.6):
   a. Change guidewire (polymer-jacketed wires may be more likely to enter into the extraplaque space).
   b. Change for a CrossBoss catheter (although occasionally the CrossBoss catheter may "track" the course of a previously inserted guidewire).
   c. Continue wiring and reenter distally, then "crush" the stent with a new stent (Fig. 22.3 and CTO Manual online cases 32 and 54).[18–20] Conservative sizing of the new stent, ideally informed by intravascular ultrasound (IVUS) can minimize the risk of perforation.
   d. Parallel wiring.

22.2.3 Entering the distal vessel

When the wire is located in the intraplaque space, guidewires with increasing tip stiffness are used.

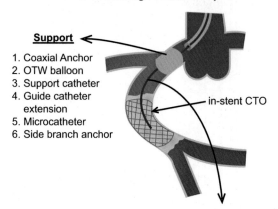

Penetrating Proximal Cap

Support
1. Coaxial Anchor
2. OTW balloon
3. Support catheter
4. Guide catheter extension
5. Microcatheter
6. Side branch anchor

in-stent CTO

Penetrating Guidewire
1. Confianza Pro 12
2. Astato 20, Infiltrac, Infiltrac Plus
3. Gaia Next 2nd, 3rd Judo 1, 3, 6
4. Hornet 14

FIGURE 22.5 Penetrating the proximal cap of in-stent chronic total occlusions.

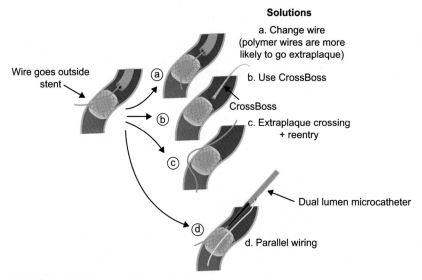

Solutions

Wire goes outside stent

a. Change wire (polymer wires are more likely to go extraplaque)

b. Use CrossBoss

CrossBoss

c. Extraplaque crossing + reentry

Dual lumen microcatheter

d. Parallel wiring

FIGURE 22.6 What to do if the guidewire exits through the stent struts.

When the wire is located in the extraplaque space, the quality of the distal vessel and proximity to a bifurcation can help choose the Stingray system for antegrade reentry or a retrograde approach for reverse CART.

22.2.4 Lesion preparation/repeat stenting

1. Restenosis/reocclusion rates can be very high in in-stent CTOs, especially in patients with multiple overlapping stents.
2. Obtaining information on the previously placed stents can be useful.
3. Intavascular imaging should be used in all in-stent CTOs to ascertain the mechanism of the prior stent failure (fracture/undersizing/underexpansion), determine the presence and extent of calcification and select appropriate stent size.
4. Plaque modification techniques should be used, if needed, to expand the lesion. Intravascular lithotripsy can help treat stent under-expansion due to calcification. Multiple modalities may be needed to maximize the stent minimum lumen area and thereby decrease the risk of repeat in-stent restenosis.
5. Restenting with a DES reduces the risk of restenosis. Of the different DES stent types, everolimus-eluting stents appear to have the best efficacy.[29]
6. Staged repeat angiography should be considered in cases of multiple prior in-stent restenosis/stent occlusion episodes.

References

1. Christopoulos G, Karmpaliotis D, Alaswad K, et al. The efficacy of "hybrid" percutaneous coronary intervention in chronic total occlusions caused by in-stent restenosis: insights from a US multicenter registry. *Catheter Cardiovasc Interv* 2014;**84**:646−51.
2. Azzalini L, Dautov R, Ojeda S, et al. Procedural and long-term outcomes of percutaneous coronary intervention for in-stent chronic total occlusion. *JACC Cardiovasc Interv* 2017;**10**:892−902.
3. Abdel-karim AR, Lombardi WB, Banerjee S, Brilakis ES. Contemporary outcomes of percutaneous intervention in chronic total coronary occlusions due to in-stent restenosis. *Cardiovasc Revasc Med* 2011;**12**:170−6.
4. Shlofmitz E, Iantorno M, Waksman R. Restenosis of drug-eluting stents: a new classification system based on disease mechanism to guide treatment and state-of-the-art review. *Circ Cardiovasc Interv* 2019;**12**:e007023.
5. Abbas AE, Brewington SD, Dixon SR, Boura J, Grines CL, O'Neill WW. Success, safety, and mechanisms of failure of percutaneous coronary intervention for occlusive non-drug-eluting in-stent restenosis vs native artery total occlusion. *Am J Cardiol* 2005;**95**:1462−6.
6. Yang YM, Mehran R, Dangas G, et al. Successful use of the frontrunner catheter in the treatment of in-stent coronary chronic total occlusions. *Catheter Cardiovasc Interv* 2004;**63**:462−8.
7. Ho PC. Treatment of in-stent chronic total occlusions with blunt microdissection. *J Invasive Cardiol* 2005;**17**:E37−9.
8. Lee NH, Cho YH, Seo HS. Successful recanalization of in-stent coronary chronic total occlusion by subintimal tracking. *J Invasive Cardiol* 2008;**20**:E129−32.
9. Werner GS, Moehlis H, Tischer K. Management of total restenotic occlusions. *EuroIntervention* 2009;**5**(Suppl D):D79−83.
10. Lamelas P, Padilla L, Abud M, et al. In-stent chronic total occlusion angioplasty in the LATAM-CTO registry. *Catheter Cardiovasc Interv* 2021;**97**:E34−9.
11. Brilakis ES, Lombardi WB, Banerjee S. Use of the Stingray guidewire and the venture catheter for crossing flush coronary chronic total occlusions due to in-stent restenosis. *Catheter Cardiovasc Interv* 2010;**76**:391−4.
12. de la Torre Hernandez JM, Rumoroso JR, Subinas A, et al. Percutaneous intervention in chronic total coronary occlusions caused by in-stent restenosis. Procedural results and long

term clinical outcomes in the TORO (spanish registry of chronic TOtal occlusion secondary to an occlusive in stent RestenOsis) multicenter registry. *EuroIntervention* 2017;**13**:e219−226.

13. Vemmou E, Quadros AS, Dens JA, et al. In-stent CTO percutaneous coronary intervention: individual patient data pooled analysis of 4 multicenter registries. *JACC Cardiovasc Interv* 2021;**14**:1308−19.

14. Vemmou E, Alaswad K, Karmpaliotis D, et al. Outcomes of percutaneous coronary intervention for in-stent chronic total occlusions: insights from the PROGRESS-CTO Registry. *JACC Cardiovasc Interv* 2020;**13**:1969−71.

15. Rinfret S, Ribeiro HB, Nguyen CM, Nombela-Franco L, Urena M, Rodes-Cabau J. Dissection and re-entry techniques and longer-term outcomes following successful percutaneous coronary intervention of chronic total occlusion. *Am J Cardiol* 2014;**114**:1354−60.

16. Spratt JC, Hung JD. In-stent CTOs: same story with a different conclusion? *EuroIntervention* 2021;**17**:e611−12.

17. Sekiguchi M, Muramatsu T, Kishi K, et al. Occlusion patterns, strategies and procedural outcomes of percutaneous coronary intervention for in-stent chronic total occlusion. *EuroIntervention* 2021;**17**:e631−8.

18. Ohya H, Kyo E, Katoh O. Successful bypass restenting across the struts of an occluded subintimal stent in chronic total occlusion using a retrograde approach. *Catheter Cardiovasc Interv* 2013;**82**:E678−83.

19. Quevedo HC, Irimpen A, Abi Rafeh N. Succesful antegrade subintimal bypass restenting of in-stent chronic total occlusion. *Catheter Cardiovasc Interv* 2015;**86**:E268−71.

20. Roy J, Lucking A, Strange J, Spratt JC. The difference between success and failure: subintimal stenting around an occluded stent for treatment of a chronic total occlusion due to in-stent restenosis. *J Invasive Cardiol* 2016;**28**:E136−8.

21. Tasic M, Sreckovic MJ, Jagic N, Miloradovic V, Nikolic D. Knuckle technique guided by intravascular ultrasound for in-stent restenosis occlusion treatment. *Postepy Kardiol Interwencyjnej* 2015;**11**:58−61.

22. Capretti G, Mitomo S, Giglio M, Carlino M, Colombo A, Azzalini L. Subintimal crush of an occluded stent to recanalize a chronic total occlusion due to in-stent restenosis: insights from a multimodality imaging approach. *JACC Cardiovasc Interv* 2017;**10**:e81−3.

23. Azzalini L, Karatasakis A, Spratt JC, et al. Subadventitial stenting around occluded stents: a bailout technique to recanalize in-stent chronic total occlusions. *Catheter Cardiovasc Interv* 2018;**92**:466−76.

24. Papayannis A, Banerjee S, Brilakis ES. Use of the Crossboss catheter in coronary chronic total occlusion due to in-stent restenosis. *Catheter Cardiovasc Interv* 2012;**80**:E30−6.

25. Wilson WM, Walsh S, Hanratty C, et al. A novel approach to the management of occlusive in-stent restenosis (ISR). *EuroIntervention* 2014;**9**:1285−93.

26. Karacsonyi J, Tajti P, Rangan BV, et al. Randomized comparison of a crossboss first vs standard wire escalation strategy for crossing coronary chronic total occlusions: the CrossBoss First Trial. *JACC Cardiovasc Interv* 2018;**11**:225−33.

27. Ntatsios A, Smith WHT. Exit of CrossBoss between stent struts within chronic total occlusion to subintimal space: completion of case via retrograde approach with rendezvous in coronary. *J Cardiol Cases* 2014;**9**:183−6.

28. Sapontis J, Grantham JA, Marso SP. Excimer laser atherectomy to overcome intraprocedural obstacles in chronic total occlusion percutaneous intervention: case examples. *Catheter Cardiovasc Interv* 2015;**85**:E83−9.

29. Lawton JS, Tamis-Holland JE, Bangalore S, et al. 2021 ACC/AHA/SCAI guideline for coronary artery revascularization: a report of the American College of Cardiology/American Heart Association Joint Committee on Clinical Practice Guidelines. *Circulation* 2022;**145**:e18−e114.

Chapter 23

Balloon uncrossable and balloon undilatable CTOs

CTO PCI manual online cases: 1, 5, 15, 18, 26, 27, 30, 47, 52, 86, 124, 151, 160, 182

PCI manual online cases: 8, 15, 17, 21, 99, 100, 121, 137, 140

23.1 Balloon uncrossable lesions

CTO manual online cases: 1, 5, 15, 18, 27, 30, 31, 47, 49, 52, 53, 57, 73, 124, 126, 155, 157, 168, 178, 179, 181, 200

PCI manual online cases: 17, 64, 99, 137

Balloon uncrossable CTOs are lesions that cannot be crossed with a balloon after successful guidewire crossing.[1] Approximately 6%−9% of CTOs treated with PCI are balloon uncrossable.[2−4] These lesions are also often uncrossable with microcatheters.

Fig. 23.1 outlines a step-by-step algorithm for approaching such lesions.

23.1.1 Small balloon

1. The first step in trying to cross a balloon uncrossable lesion is to use a small balloon (Section 30.9.2).
2. Use single marker, rapid exchange compliant balloons with a low crossing profile (0.75, 0.85, 1.0, 1.20, 1.25, and 1.5 mm in diameter) and long length (15−30 mm). The balloon profile is highest at the marker segment; hence, longer balloons may allow for deeper lesion penetration with the lower profile segment of the balloon.
3. The Sapphire Pro III 1.0 mm (OrbusNeich) and Sapphire 3 0.85 mm (OrbusNeich) the Ryurei 1.0 mm (Terumo, not available in the US), the Zinrai 0.75 mm and Ikazuchi Zero 1.0 mm (Kaneka, not available in the US) and the NIC Nano 0.85 mm balloon (SIS, not available in the US) are the lowest profile balloons currently available and are the first choice for balloon uncrossable lesions.

Manual of Chronic Total Occlusion Percutaneous Coronary Interventions.
DOI: https://doi.org/10.1016/B978-0-323-91787-2.00012-5

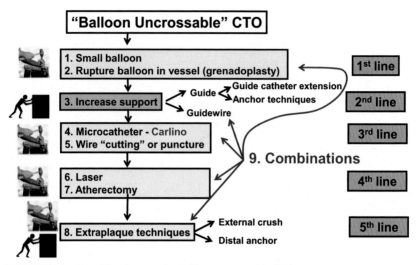

FIGURE 23.1   **Algorithm for crossing balloon uncrossable CTOs.** *Yellow*: plaque modification, *Green*: increase support.

4. Other options are the Threader (1.2 mm, Boston Scientific), and the Takeru (Terumo) balloons that have low crossing profiles and stiff shafts to facilitate advancement though challenging lesions.

5. The Threader (Boston Scientific) has a hydrophilic coating and 0.017″ lesion entry profile. The Threader is available in both rapid exchange and over-the-wire versions. The rapid exchange Threader is preferred to the over-the-wire version for balloon uncrossable lesions, as it has more penetrating capacity (likely due to stiffer shaft). However, the over-the-wire Threader allows guidewire changes and contrast injection.

6. The Blimp scoring balloon (Section 30.9.3.5) may also help cross some of the balloon uncrossable lesions.

7. If the balloon stops advancing, it can be inflated while maintaining forward pressure. This may dilate the entry to the lesion and allow lesion crossing, sometimes even with the same balloon ("balloon wedge"/"leopard-crawl" technique).

8. If the balloon fails to advance, consider balloon rotation. Avoid rotating more than 1−2 times in each direction so as to prevent the balloon from getting "stuck" on the guidewire.

9. If the balloon fails to advance after inflation:
   a. The operator can reshape it using his or her fingers while applying vacuum, or
   b. Try a new small balloon (balloons lose their original profile after inflation), or

c. Try a balloon manufactured by another company, as different crossing profile and tip characteristics may assist in crossing. Rapid exchange balloon catheters allow more pushability into the lesion, or

d. Attempt crossing with a larger 2.5−3.0 mm diameter rapid exchange balloon. Sometimes, inflation with a larger diameter balloon just proximal to the lesion will disrupt the architecture of the lesion entry enough to allow subsequent passage of a small profile balloon or microcatheter.

What can go wrong?

- Guide catheter and guidewire position can be lost during attempts to advance the balloon or microcatheter. Carefully monitor the guide catheter position and stop advancing if the guide catheter starts backing out of the coronary ostium or if the distal wire position is being compromised. Forceful back and forth balloon manipulation can cause ostial dissection, especially if there is a preexisting ostial lesion.

- Guide catheter prolapse through the aortic valve causing hemodynamically significant aortic regurgitation.

- Injury of the distal target vessel can occur (dissection or perforation) due to significant distal guidewire movement ("see-saw" action of wire with forward push and retraction of the balloon), especially when stiff (such as Confianza Pro 12, Asahi) or polymer-jacketed (such as the Pilot 200, Abbott Vascular) guidewires are used.

- Balloon entrapment or fracture can occur within the lesion, although this is rare.

- Sometimes high-pressure inflation of a balloon can damage the hypotube and make the balloon "sticky," hindering balloon removal and potentially leading to loss of guidewire position upon withdrawal. The operator must be aware of this possibility and slowly remove the balloon while checking the position of the wire with fluoroscopy during balloon withdrawal.

23.1.2 Grenadoplasty (intentional balloon rupture; also called balloon assisted microdissection or "BAM")

This is a simple, safe, and often effective technique, that is increasingly being used in the treatment algorithm for balloon uncrossable lesions.[5]

How?

A small (usually 1.0−1.5 mm) balloon is advanced as far as possible into the lesion and inflated at high pressure until it ruptures (Fig. 23.2).[6] When the balloon ruptures, suction should be immediately applied through the inflating device to avoid unnecessary barotrauma to the vessel. The balloon rupture can often sufficiently modify the plaque, resulting in subsequent successful crossing with a new balloon.

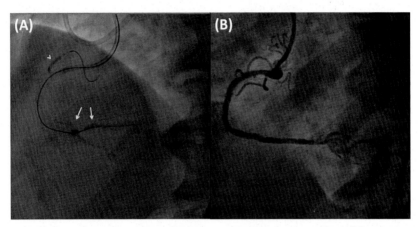

FIGURE 23.2 Illustration of the grenadoplasty technique to cross a distal right coronary artery balloon uncrossable CTO. The lesion could not be crossed despite using an 8 Fr Amplatz guide and an anchor balloon (arrowhead, **panel A**). A 1.2 mm balloon was ruptured with contrast spreading proximally and distally (arrows, **panel A**). A Finecross microcatheter could then be advanced through the lesion followed by wire exchange and a 2.0 mm balloon with an excellent final result (**panel B**). *Courtesy of Dr. Gabriele Gasparini. Reproduced with permission from the* Manual of coronary chronic total occlusion interventions. a step-by-step approach.

What can go wrong?
- Proximal vessel dissection and perforation. This is extremely unlikely when small (1.0−1.5 mm) balloons are used. **Larger balloons (≥2.0 mm) should NOT be used for grenadoplasty**.
- The balloon should be meticulously prepared to empty all air and, hence, minimize the risk of air embolism.
- Watching the indeflator rather than the screen allows more rapid deflation of the balloon immediately upon rupture. This will reduce the chance of pinhole contrast-induced vessel injury from the rupture site of the balloon.
- The operator may encounter difficulty removing the ruptured balloon. In some cases, the ruptured balloon becomes entangled with the guide-wire due to collapse of the hypotube after balloon rupture, requiring removal of both, and as a result loss of guidewire position.[5]
- A guide catheter extension positioned very close to the balloon may protect the proximal vessel during grenadoplasty. In some cases, the contact between the guide extension and the inflated balloon may in itself cause the balloon to rupture ("guide extension facilitated grenadoplasty").

23.1.3 Increase support

Increasing support is critical for crossing a balloon uncrossable lesion and is discussed in detail in Section 9.3.1.1.4. Support can be increased via the guide catheter (for example using a guide catheter extension or anchoring techniques) or via the guidewire.

23.1.3.1   Guide catheter

Better guide catheter support increases the likelihood of successful balloon or microcatheter crossing. Guide support can be increased by using larger guide catheters with supportive shapes, using femoral access, forward push or deep intubation, guide catheter extensions[7] and various anchoring techniques,[8] as described in detail in Section 9.3.1.1. challenge B. Changing guide catheter after wire crossing can be challenging and may lead to wire position loss, which may not be acceptable in some cases (such as chronic total occlusions). Given the complexity of CTO PCI, upfront selection of a supportive guide catheter can circumvent the need to make this exchange.

23.1.3.1.1   Guide catheter extension
How?

A Guideliner (Teleflex), Trapliner (Teleflex), Guidezilla (Boston Scientific), Guidion (IMDS), Telescope (Medtronic), or Boosting (QX Medical) guide catheter extension (Section 30.2) is advanced into the vessel, enhancing guide catheter support and the pushability of balloons/microcatheters. In a randomized trial, use of a 5-in-6 guide catheter extension was more effective and efficient in facilitating the success of transradial PCI for complex coronary lesions, as compared with buddy-wire or balloon-anchoring.[9]

The "power guide extension" (also called "tunnel in landslide technique"[10]) can enhance the support provided by a guide extension by inflating a balloon next to the guide catheter extension to facilitate equipment delivery in severely tortuous or uncrossable lesions.

What can go wrong?
- Guidewire and guide catheter position loss or distal vessel injury during attempts to advance the guide extension.
- Guide catheter extension advancement can cause ostial or mid-target vessel dissection.[7] When a guide catheter extension is advanced distally into a wedged position and a dampened pressure tracing is observed, guide catheter injections should be avoided due to the high risk of hydraulic dissection and damage to the proximal vessel.
- Dislodgement of the guide catheter extension distal marker can also occur.[11]
- Once equipment is advanced through the guide catheter extension, another potential complication is separation (stripping) of the stent from the stent balloon during attempts to advance it through the guide catheter extension proximal "collar." This is more likely to occur when using an undersized guide extension (such as a 6 Fr guide catheter extension inside an 8 Fr guide). This may also occur around the subclavian curve when using radial

access; hence, consider using guide catheter extensions with a long collar in radial access cases. Careful advancement of the stent through the collar under fluoroscopic visualization can help prevent this complication.

-   Prolonged guide extension positioning inside a vessel can lead to ischemia and thrombus formation despite effective anticoagulation due to stagnant blood.

23.1.3.1.2   Anchor strategies

How?

1. **Side branch anchor technique** (Fig. 23.3). A workhorse guidewire is advanced into a side branch (usually a conus or acute marginal branch for the right coronary artery or a diagonal for the left anterior descending artery), followed by a small balloon (usually 1.5 to 2.0 mm in diameter depending on the side branch vessel size). The balloon is inflated usually at 6−8 atm "anchoring" the guide into the vessel and enhancing advancement of balloons or microcatheters. Advancing a second guidewire into the side branch further stabilizes the balloon and improves support. Sometimes, patients may develop chest pain during inflation of the balloon in the side branch.[6,8,12]

2. Buddy wire stent anchor (Section 9.3.1.1). If the proximal vessel requires stenting, a buddy wire can be inserted and a stent deployed over it, effectively "trapping" the buddy wire, which then provides strong guide catheter support.

3. Both antegrade and retrograde anchoring can be performed in challenging retrograde CTO PCI cases.[12]

What can go wrong?

-   Guidewire and guide catheter position loss or distal vessel injury can occur during attempts to advance an anchor balloon.

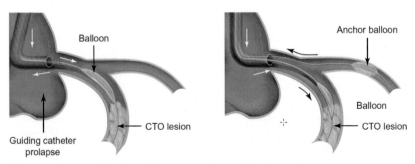

FIGURE 23.3   Illustration of the side branch anchor technique for treating a balloon uncrossable lesion. *Reproduced with permission from the* Manual of coronary chronic total occlusion interventions. a step-by-step approach.

- A side branch anchor can cause injury or dissection of the side branch; however, this is infrequent and usually does not lead to significant adverse consequences.
- Perforation of the side branch may rarely occur. Oversizing of the anchor balloon should be avoided to minimize the risk of side branch perforation and dissection; a workhorse wire should be used to minimize wire-related vessel injuries.
- Potential risks of the "buddy wire stent anchor" technique include: inability to remove the buddy wire or failure to advance equipment through the proximal stent.

23.1.3.2   Guidewire

Guidewire exchange may not be feasible or desirable in difficult to cross lesions, such as chronic total occlusions. However, leaving the original guidewire in place and crossing the lesion with a second guidewire may succeed and can sometimes be performed through a second guide catheter or using a dual lumen microcatheter.[13]

Using supportive guidewires or the deep wiring technique can significantly facilitate crossing of balloon uncrossable lesions, as described in Section 9.3.1.1.1.

Multiple punctures of the proximal cap with a highly penetrating guidewire can often sufficiently modify the lesion to allow subsequent delivery of balloons/microcatheters.

23.1.4   Microcatheter - Carlino

23.1.4.1   Microcatheter advancement

How?

1. The concept behind use of a microcatheter is that advancement of a microcatheter through the lesion can modify the occlusion, enabling subsequent crossing with a balloon.
2. There are several microcatheters that can be utilized as described in Section 30.6.
3. Microcatheters specifically designed for balloon uncrossable lesions (Section 30.6.5):
   a. The Tornus catheter (Asahi Intecc) was designed for advancing through calcified and difficult to penetrate lesions and should be advanced using counterclockwise rotation and withdrawn using clockwise rotation.[14]
   b. The Turnpike Spiral and Turnpike Gold catheters (Teleflex) were also designed with threads to "screw into the lesion" and modify it. In contrast to the Tornus catheter, they are advanced by turning clockwise and withdrawn by turning counterclockwise.

4. Standard microcatheters can also be used:
   a. The Corsair Pro and Corsair Pro XS microcatheters (Asahi Intecc) (Section 30.6.1.1) can be advanced by rotating in either direction (in contrast to the Tornus catheter), although counterclockwise rotation is preferred.
   b. The Turnpike and Turnpike LP catheters (Teleflex) (Sections 30.6.1.1 and 30.6.1.2) can also be rotated in either direction, although clockwise rotation is preferred.
   c. Similarly, the Finecross (Terumo) (Section 30.6.2) can be rotated in either direction, although rotation may be challenging to achieve and there is a risk of tip dislodgement if aggressively torqued.
   d. The Caravel (Asahi Intecc) is a low profile microcatheter, but is not designed for aggressive torqueing which is often done with the Corsair and Turnpike family of microcatheters.
   e. The Corsair Armet (Asahi Intecc) is a full braided metal-tipped peripheral microcatheter with a very low crossing profile that has been used off label in the coronary arteries.
5. If successful advancement of a microcatheter is achieved, a balloon can often subsequently cross the lesion. Alternatively, the guidewire can be exchanged for a more supportive guidewire or an atherectomy wire (PCI Manual online case 64), if the latter is planned as the next lesion preparation step.

What can go wrong?
- Guide catheter and guidewire position may be lost with aggressive advancement of the microcatheters.
- Distal vessel injury can occur from uncontrolled guidewire movement during microcatheter advancement attempts.
- The microcatheter can get damaged if overtorqued, leading to catheter tip entrapment or tip/shaft fracture. If the tip of the microcatheter breaks off it can become entrapped in the lesion. Rotation should not exceed 10 turns before allowing the catheter to "unwind." A guidewire should always be kept within the microcatheter lumen to prevent kinking and possible entrapment.
- Excessive manipulation of the microcatheter can disrupt the device and/ or the guidewire and lock both devices together. This occurs when the tip of the microcatheter is engaged on the lesion and no longer spinning, while the body of the microcatheter is continuously spun (usually in the same direction); this effectively "strangulates" the wire inside the microcatheter leading to potential loss of wire access, as the microcatheter cannot be removed without the wire, requiring withdrawal of both. Checking whether the wire can still move freely back and forth in the microcatheter while doing prolonged attempts at lesion crossing can forewarn the operator when this is starting to occur. A polymer-jacketed guidewire may then be advanced through the track that has been established, allowing the crossing attempts to restart.

23.1.4.2 Carlino

The Carlino technique is described in detail in Section 8.3.4.3.[15,16]

How?

1. A second wire and microcatheter are advanced as close to the proximal cap as possible and the wire is removed.
2. A small amount of contrast (<0.5−1.0 mL) is injected gently through the microcatheter under cineangiography.
3. Contrast injection causes modification of the plaque compliance facilitating subsequent advancement of a balloon or a microcatheter.

What can go wrong?

   1. Perforation, if a large amount of contrast is injected and/or if the catheter is inserted into a small side branch. The risk is low with injection of a small amount of contrast.

23.1.5 Wire cutting or puncture

23.1.5.1 Wire-cutting[17]

Fig. 23.4

How?

1. A second guidewire is advanced through the lesion (which may be challenging to accomplish).
2. A balloon is advanced over the first guidewire, as far as possible into the proximal cap and inflated.
3. The second guidewire is withdrawn while the balloon is inflated, effectively "cutting" the proximal cap and modifying it.
4. After deflation and removal of the original balloon, a new balloon is advanced over the first guidewire, often successfully crossing the modified lesion.
5. A modified version of this technique is the "**see-saw wire-cutting technique.**" In this technique two balloons are advanced over the two guidewires (one balloon over each wire). One of the balloons is first advanced as distally as possible and inflated, pressing the other wire against the proximal cap. The first balloon is then pulled back and the other balloon is advanced distally and inflated, producing a similar cutting effect to modify the cap on the other side. This process is repeated multiple times until one of the balloons crosses the lesion. A retrospective study of 80 patients found this technique to be associated with higher device and procedural success rates and shorter procedure time as compared with use of the Tornus catheter.[18,19]

What can go wrong?

−   Withdrawal of the second guidewire may result in deep intubation of the guide catheter, potentially leading to proximal vessel dissection. This complication can be prevented by carefully watching the guide catheter during withdrawal of the guidewire, and promptly backing the

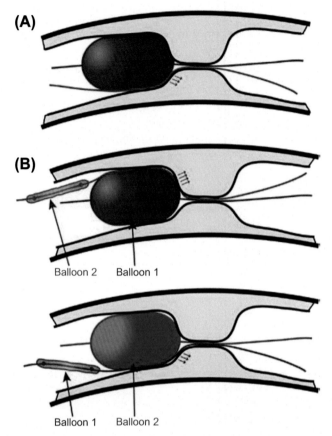

FIGURE 23.4 Illustration of the wire-cutting (**panel A**) and see-saw wire-cutting (**panel B**) techniques. *Reproduced with permission from the* Manual of coronary chronic total occlusion interventions. a step-by-step approach.

guide catheter out of the vessel as needed. If, however, a dissection occurs it could be used for extraplaque wire advancement and extraplaque lesion modification (or extraplaque distal anchor) as described in Sections 23.1.6 and 23.1.7, provided that the operator has experience in these techniques.

23.1.5.2 Wire puncture

How?

1. A stiff guidewire (such as Hornet 14, Infiltrac and Infiltrac Plus, Confianza Pro 12 and Astato 20) is advanced to the lesion through a microcatheter or over-the-wire balloon.

2. The lesion entry is "punctured" or "jabbed" several times with the stiff tip guidewire, which may modify the lesion and facilitate equipment crossing.
3. Advancing a supportive second guidewire through the balloon uncrossable lesion may in itself facilitate lesion crossing with a balloon or microcatheter.
4. Sometimes a Rotawire Drive can be advanced through the "holes" created by the penetrating wire, allowing subsequent lesion modification.

What can go wrong?
− Perforation, although this is unlikely even if the guidewire exits the vessel architecture (unless it is followed by a balloon or microcatheter).

23.1.6 Laser

• Online video Impact of contrast on laser activation

• **CTO Manual online cases** 5, 18, 27, 47, 52, 73

Laser atherectomy (Section 30.10) is frequently used in balloon uncrossable and balloon undilatable lesions, in part because laser atherectomy can be performed over any standard 0.014 inch guidewire in contrast to rotational and orbital atherectomy that require dedicated guidewires.[20,21] It may be best to avoid laser over polymer-jacketed guidewires due to risk of "melting" the polymer.

How?
1. The **0.9 mm excimer laser atherectomy catheter** should be used[21] usually at maximum repetition and fluence levels (repetition rate 80 Hz and fluence of 80 mJ/mm^2).[22] The coronary laser catheter is programmed to minimize the risk of coronary ischemia, with laser activation designed to work with a slow, continuous saline infusion;[23] however, the peripheral "turbo elite" catheter (Philips) is sometimes preferred over the coronary catheter since it does not automatically stop after 10 seconds of treatment. The laser can modify the uncrossable segment and facilitate balloon entry into the lesion.[24] Even if the laser does not cross, multiple laser runs at the proximal cap may disrupt the cap sufficiently for subsequent crossing with a different device.
2. The old laser console requires a **5-minute warming period** as well as calibration prior to use. Anticipating its potential use and setting up the system early can minimize delays and improve the efficiency of the procedure. The new laser console is more compact and has a quicker startup period.
3. **A variation of the above technique** is to perform laser activation while injecting contrast through the guide catheter or via the over-the-wire side port of the laser catheter itself (if an over-the-wire laser is used, which is

more commonly done in peripheral arteries and rarely in the coronary arteries). Laser activation with simultaneous contrast injection can cause profound plaque modification through the acoustico-mechanical effect of the rapidly exploding bubbles. Laser with contrast, however, also increases the risk of vessel injury and/or perforation. As a result, laser with contrast injection is typically reserved for treatment of in-stent balloon undilatable lesions.

What can go wrong?
- Vessel dissection and/or perforation, especially when performing laser in tortuous coronary vessels (PCI Manual online case 55). The risk may be higher when the high laser settings are used (repetition rate 80 Hz and fluence of 80 mJ/mm^2) and when contrast is injected and lower when laser is activated within previously placed stents.
- Laser activation over polymer-jacketed guidewires can result in polymer damage, making the wire "sticky" or "gummy" and as a result hindering removal or advancement of equipment over them.

23.1.7 Atherectomy

How?
1. Rotational (Section 9.3.2.3 and Section 30.9.1) or orbital (Section 9.3.2.4 and Section 30.9.2) atherectomy can greatly facilitate lesion crossing with a balloon. These atherectomy modalities, however, require wire exchange for a dedicated guidewire (Rotawire Drive floppy or Rotawire Drive extra support for rotational atherectomy and ViperWire Advance or ViperWire Advance with Flex Tip for orbital atherectomy), which may not always be feasible through a balloon uncrossable lesion.[25]
2. If no other maneuver is successful in crossing the balloon uncrossable lesion, it is sometimes possible to "bury" a microcatheter as far as possible into the lesion, pull the original wire, and attempt to rewire with a dedicated atherectomy guidewire. If rewiring is successful, then atherectomy can be performed.
3. Rotational atherectomy may be advantageous over orbital atherectomy in balloon uncrossable lesions, as it provides forward, end-on cutting in contrast to orbital atherectomy that provides "sideways" cutting/sanding.
4. Atherectomy will differentially cut calcific tissue but will not cut through elastic tissue, such as the adventitia.
5. Rotational atherectomy has been successfully performed in the extraplaque space with a 1.25 mm burr.[26]

What can go wrong?
- Loss of guidewire position across the lesion (if the guidewire has to be removed and replaced with a dedicated atherectomy guidewire) that may fail to recross the lesion.

- Vessel perforation, which is why a small diameter rotational atherectomy burr (1.25−1.50 mm) should be used for balloon uncrossable lesions. The risk of perforation may be higher with atherectomy in the extraplaque space.
- Burr entrapment upon forceful forward advancement "kokeshi phenomenon." The burr should be advanced in a repetitive and gentle manner avoiding forceful "wedging" into the occlusion with burr deceleration. The risk of entrapment is higher with 1.25 mm burrs. Prevention and management of this complication is discussed in Section 27.5.

23.1.8   Extraplaque (formerly called "subintimal") techniques

CTO Manual online case 1

Extraplaque techniques (Fig. 23.5) can significantly facilitate treatment of "balloon uncrossable" lesions, but the operators should be experienced in the use of these CTO PCI techniques.

23.1.8.1   External cap crush

How?

CTO Manual online case 15

A second guidewire (antegrade or retrograde) is advanced in the extraplaque space around the balloon uncrossable segment of the lesion, as explained in Section 8.2 (Fig. 23.5, panels A, B, and C). A balloon is advanced over the extraplaque guidewire next to the lesion and inflated (usually at 8−10 atm), "crushing" the plaque from the outside (Fig. 23.5, panel D1). This can often sufficiently modify the plaque to allow passage of a balloon over the guidewire that had previously entered the distal true lumen.[27,28] A modification of this technique is to use an intravascular lithotripsy balloon.[29] An alternative option is to stent over the extraplaque guidewire.

23.1.8.2   Extraplaque distal anchor

In a variation of the extraplaque external crush technique entitled "extraplaque distal anchor" a second guidewire is advanced into the false lumen distal to the uncrossable lesion. A balloon is advanced over the extraplaque guidewire distal to the CTO. The balloon is inflated distally, "anchoring" the true lumen guidewire, and enabling antegrade delivery of a balloon over the true lumen guidewire (Fig. 23.5, panel D2).[30]

What can go wrong?

1. The extraplaque techniques require extraplaque wire crossing, which may not always be feasible (for example the guidewire may track side branches).

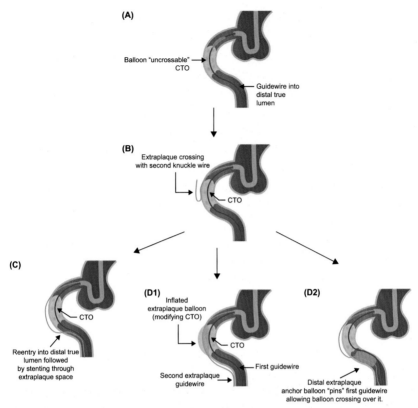

FIGURE 23.5 Illustration of extraplaque crossing techniques for crossing a "balloon uncrossable" CTO. If a balloon cannot cross the lesion after successful guidewire crossing (**panel A**), a second guidewire is advanced into the extraplaque space across the lesion (**panel B**) followed by reentry into the distal true lumen (**panel C**). A balloon can be inflated over the extraplaque guidewire next to the balloon uncrossable lesion to "crush" and modify it (**panel D1**), or distal to the lesion to "anchor" the true lumen guidewire and allow balloon crossing over the true lumen guidewire (**panel D2**). *Modified with permission from the* Manual of coronary chronic total occlusion interventions. A step-by-step approach.

2. Positioning of the extraplaque guidewire within the vessel "architecture" should always be confirmed to prevent perforation, for example if the guidewire enters side branches or if it exits the vessel and enters the pericardium. Confirmation of wire position is usually done using contralateral injection in two orthogonal projections.
3. Small balloons (balloon-to-artery-ratio <1) should be used to minimize the risk of perforation.
4. Extraplaque crossing and balloon inflation in the extraplaque space can create a large hematoma that may compress the distal true lumen.

23.1.9   Simultaneous versus sequential use of strategies

Simultaneous application of lesion modification strategies and techniques that increase guide catheter support can enhance the likelihood of successful balloon crossing. For example the "Anchor-Tornus,"[31] "Proxis-Tornus"[32] and "Anchor-Laser" (Fig. 23.6)[22] techniques have been described for crossing "balloon uncrossable" lesions. In addition to simultaneous application, sequential application of various techniques is used until a final successful outcome is achieved.[30,33] An operator must be creative in using the above strategies in various combinations to achieve the desired goal. This process can be difficult and laborious but also very rewarding.

23.2   Balloon undilatable CTOs

CTO manual online cases: 26, 86, 124, 151, 160, 182, 192

PCI manual online cases: 8, 15, 17, 21, 99, 100, 121, 137, 140

Balloon undilatable CTOs are those CTOs that cannot be expanded despite multiple high-pressure balloon inflations. Balloon undilatable lesions often have severe calcification and are also often balloon uncrossable (Section 23.1). In the Progress-CTO registry, 12% of CTOs were balloon undilatable.[34]

Prevention

"Balloon undilatable" lesions are more frequent in CTOs than non-CTO lesions.[34] It is important to avoid implanting a stent in a "balloon undilatable" lesion. Adequate predilatation with a balloon sized according to the vessel reference diameter is critical before stenting a coronary lesion to ensure proper stent expansion, especially when the lesion is severely calcified. Intravascular imaging can be useful in determining the appropriate predilatation balloon size and plaque characteristics that may benefit from atherectomy or plaque modification.[35,36] If the lesion is resistant, additional predilatation, intravascular lithotripsy and/or atherectomy should be pursued prior to stent deployment.

Treatment

Fig. 23.7 outlines an algorithm for approaching de novo and in-stent "balloon undilatable" lesions.

De novo versus in-stent

A key determinant of treatment options for balloon undilatable lesions is whether the undilatable lesion is de novo or is within a stent (implanted during the same procedure or during a prior procedure). Orbital atherectomy should not be done in recently implanted stents[37] (although it can be done within previously implanted stents[38]) and rotational atherectomy carries increased risk of burr entrapment and stent distortion when performed within recently implanted stents.[39] Laser with simultaneous contrast injection can

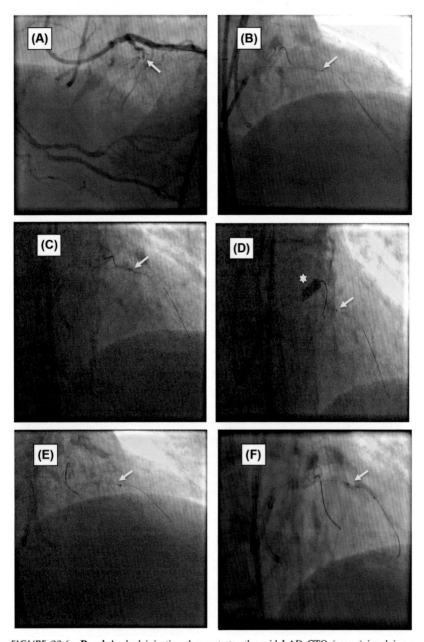

FIGURE 23.6 **Panel A**: dual injection demonstrates the mid LAD CTO (arrow) involving a semiambiguous cap, short length (less than 20 mm), significant calcification and a distal vessel filling mainly via septal collaterals from the right coronary artery. **Panel B**: successful antegrade wiring but a microcatheter (arrow) could not be advanced through the lesion due to severe calcification. **Panels C and D**: use of Threader catheter alone (**Panel C**) and in conjunction with a 4.0 anchor balloon (*yellow star*) in the circumflex stent (**Panel D**). **Panel E**: laser (arrow). **Panel F**: severe balloon underexpansion. **Panel G**: orbital atherectomy (arrow). **Panel H**: adequate balloon expansion. **Panel I**: final angiography after stenting. *Reproduced with permission from Sandoval Y, Lobo AS, Tajti P, Brilakis ES. Laser-assisted orbital or rotational atherectomy: a hybrid treatment strategy for balloon-uncrossable lesions.* Hellenic J Cardiol 2020;61:57−59.

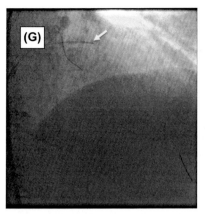

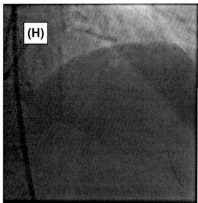

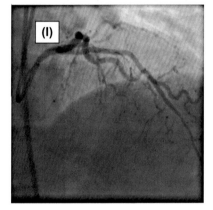

FIGURE 23.6 (*Continued*)

often expand in-stent undilatable lesions (ESLAP = Extra-Stent Laser-Assisted Plaque modification),[20] but it is infrequently used in de novo lesions because it can lead to dissection or perforation.

23.2.1 Likely to respond to balloon?

Heavily calcified lesions may be less likely to respond to balloon dilatation. Intravascular imaging can help determine whether balloon angioplasty is unlikely to be successful. The following four IVUS findings have been associated with stent underexpansion[40]:

1. superficial calcium angle >270 degrees longer than 5 mm
2. 360 degrees of superficial calcium
3. calcified nodule
4. vessel diameter <3.5 mm

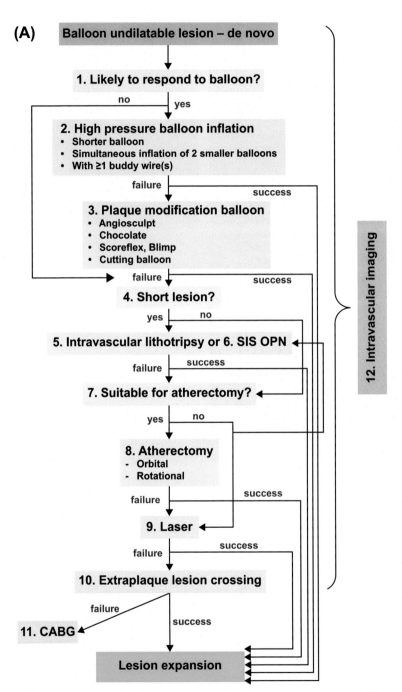

FIGURE 23.7 **Algorithm for dilating a de novo (panel A) or in-stent (panel B) "balloon undilatable" lesion.**

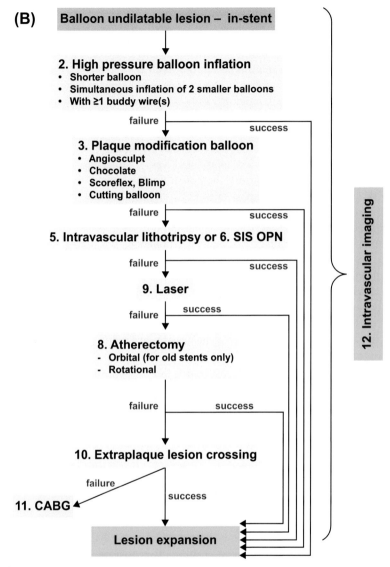

FIGURE 23.7   (Continued)

The following three optical coherence tomography (OCT) findings have been associated with stent underexpansion[41]:

1. maximum calcium angle >180 degrees
2. maximum calcium thickness >0.5 mm; and
3. calcium length >5.0 mm.

23.2.2 High pressure balloon inflation

How?

1. Using noncompliant balloons inflated at high pressures (often up to 28 atm).[42] It is important to be familiar with the rated burst pressure and percentage diameter growth of the various noncompliant balloons.
2. Performing prolonged inflations (30−60 s or more). Alternatively, once at high pressure, quickly reducing and increasing the pressure multiple times may help "break" the resistant lesion.
3. Using slightly undersized balloons (usually by 0.5 mm).
4. Avoiding balloon injury of nonstented vessel segments. All ballooned segments should subsequently be covered with stent(s) to minimize the risk of restenosis (geographic miss).
5. Using shorter balloons might help concentrate the force of the dilatation and achieve the desired balloon expansion inside the lesion.
6. Simultaneous inflation of two undersized balloons positioned side by side within the lesion creates asymmetric pressure, sometimes resulting in lesion expansion.
7. Advancing one[43−45] or more[46] buddy wires through the undilatable lesion, followed by high-pressure balloon inflation: the wires can modify the balloon forces exerted on the vessel wall, leading to plaque modification and expansion (this "focused force" technique is a makeshift version of a cutting balloon or scoring balloon).

What can go wrong?

1. Vessel perforation is best avoided by conservative balloon sizing (in nearly all cases 1:1 or lower balloon-to-artery ratio should be used) and careful balloon preparation that removes all air from the balloon.
2. Balloon rupture leading to vessel dissection and/or perforation. If balloon rupture occurs, immediate angiography should be performed to determine whether a perforation or flow-limiting dissection has occurred.
3. After high-pressure inflation, many balloons become "sticky" (due to partial collapse of the rapid exchange port containing the guidewire due to overinflation of the hypotube) and challenging to remove, sometimes leading to loss of wire position. Fluoroscopy during balloon removal can minimize the risk of losing wire position in such cases.
4. Balloon slippage ("watermelon seeding") is more common with shorter balloons and may injure adjacent coronary segments.

23.2.3 Plaque modification balloons

How?

Plaque modification balloons (Angiosculpt, Chocolate, ScoreFlex, Wolverine cutting balloon; Section 30.9.3) facilitate lesion expansion through application of focused pressure,[47] but are less deliverable and more costly than standard balloons.

What can go wrong?
1. Loss of guidewire and guide catheter position, as the plaque modification balloons may not deliver easily through the balloon undilatable lesion: enhanced guide catheter support techniques (such as anchor techniques and guide catheter extensions) may be needed to deliver the plaque modification balloon through the lesion.
2. Vessel rupture or perforation.
3. Balloon entrapment (especially if the balloon ruptures). Excessive inflation forces or balloon rupture within the lesion may make equipment withdrawal challenging causing fracture of the cutting balloon blade,[48] stent strut avulsion,[49,50] and balloon entrapment.[51−54] The cutting balloon should be inflated slowly, avoiding >14 atm inflation pressure.

23.2.4 Suitable for intravascular lithotripsy?

Short lesions can often be treated with intravascular lithotripsy (the length of the coronary lithotripsy balloon is 12 mm; hence, it is less suitable for long lesions since only 80 pulses can currently be given per balloon). The longer peripheral intravascular lithotripsy catheter can sometimes be used off label in large coronary arteries (PCI Manual online case 63).

23.2.5 Intravascular lithotripsy

The intravascular lithotripsy system (Shockwave Medical, Inc.; Section 30.9.8) uses shockwaves to fracture calcific plaque[55] and has been successfully used for expanding undilatable lesions with circumferential calcification.[56−58] It is particularly well suited for in-stent balloon undilatable lesions due to severe calcification, as it is less likely to disrupt the stent compared with atherectomy. However, intravascular lithotripsy balloons are hard to deliver and are less effective in noncircumferential calcium.

23.2.6 Very high-pressure balloon

The very-high pressure OPN balloon (SIS Medical, Switzerland; Section 30.9.7), is a specialized noncompliant balloon that can deliver very high (>40 atm) pressures.[42]

23.2.7 Suitable for atherectomy?

As described in Chapter 19 the risks of atherectomy (dissection, embolization, perforation, equipment loss or entrapment) increase with increasing tortuosity (relative contraindication) and in the setting of thrombus, dissections, and bypass grafts.

23.2.8 Atherectomy

Rotational (Section 9.3.2.3 and Section 30.9.1) or orbital (Section 9.3.2.4 and Section 30.9.2) atherectomy can be used to expand a "balloon undilatable" lesion. The combination of laser and rotational atherectomy (RASER) has been described for highly resistant lesions[59] and can offer a solution to lesions resistant to either approach applied independently. Similarly, combined rotational atherectomy followed by intravascular lithotripsy (IVL) ("RotaTripsy") has been proposed as an effective solution in this setting.[60−66]

Discovering that a lesion is "balloon undilatable" *after* stenting poses significant challenges, as stent underexpansion can predispose to stent thrombosis and restenosis. Prevention is key and can be achieved by use of intravascular imaging and/or performing careful balloon predilatation with a 1:1 sized balloon and ensuring complete balloon expansion in orthogonal projections before stent implantation. If the balloon fails to fully expand, additional vessel preparation (often with atherectomy, plaque modification or very-high pressure balloon or intravascular lithotripsy) is needed before stent implantation.

Atherectomy ("stentablation")[67−71] can be performed for in-stent undilatable lesions, but is infrequently done due to the risk of stent material or plaque embolization, burr entrapment, and stent damage, necessitating implantation of an additional stent (PCI Manual online case 23).[70] The risk of complications (especially burr entrapment) is higher when performing atherectomy in recently placed stents. To avoid burr entrapment, larger sized burrs (≥ 1.75 mm) and faster rotation should be selectively used for undilatable stents. Moreover, in-stent atherectomy carries a higher risk of restenosis as compared with the use of laser.[67,72] Orbital atherectomy is contraindicated for in-stent lesions, although it has been done successfully within previously implanted stents.[38] In general, atherectomy for in-stent balloon undilatable lesions should be reserved for cases where other strategies, such as high-pressure noncompliant or plaque modification balloon inflations (with and without buddy wires), laser, intravascular lithotripsy, or use of the very-high pressure balloon fail to expand the stent.

23.2.9 Laser

How?
1. The **0.9 mm excimer laser atherectomy catheter** should be used, usually at maximum repetition and fluence levels (repetition rate 80 Hz and fluence of 80 mJ/mm^2).[22,73−77]
2. Laser is activated during saline flushing (or with simultaneous contrast injection if performed within an underexpanded stent).[22,73,76,78]
3. High-pressure balloon inflation is subsequently performed to determine whether lesion modification is sufficient, to allow expansion of the underexpanded lesion.

What can go wrong?

1. Vessel dissection, which is the reason why laser activation with contrast is typically reserved for use within an underexpanded stent.
2. Perforation.

23.2.10   Extraplaque lesion crossing

This may also allow expansion of the lesion, as discussed in the treatment of balloon uncrossable lesions (Sections 23.1.6 and 23.1.7).

23.2.11   CABG

In rare cases when all treatment options fail, coronary artery bypass graft surgery (CABG) may be required and is preferable to leaving an underexpanded stent that has high risk of stent thrombosis and in-stent restenosis.

23.2.12   Intravascular imaging

Intravascular imaging (Chapter 13) is critical for all treatment stages of balloon undilatable lesions. Sometimes coronary angiography may fail to identify balloon undilatable lesions.

Intravascular imaging can detect possible challenges with lesion expansion before treatment, and also ensure that a good result has been achieved after treatment. Some heavily calcified lesions may appear well expanded during balloon inflations but have a small minimum lumen area when evaluated by intravascular imaging. Heavy circumferential calcification favors early use of coronary atherectomy whereas, less calcified lesions are usually initially approached with balloon angioplasty.[34]

References

1. Elrayes MM, Xenogiannis I, Nikolakopoulos I, et al. An algorithmic approach to balloon-uncrossable coronary lesions. *Catheter Cardiovasc Interv* 2021;**97**:E817−25.
2. Karacsonyi J, Karmpaliotis D, Alaswad K, et al. Prevalence, indications and management of balloon uncrossable chronic total occlusions: insights from a contemporary multicenter US registry. *Catheter Cardiovasc Interv* 2017;**90**:12−20.
3. Kovacic JC, Sharma AB, Roy S, et al. GuideLiner mother-and-child guide catheter extension: a simple adjunctive tool in PCI for balloon uncrossable chronic total occlusions. *J Interv Cardiol* 2013;**26**:343−50.
4. Patel SM, Pokala NR, Menon RV, et al. Prevalence and treatment of "balloon-uncrossable" coronary chronic total occlusions. *J Invasive Cardiol* 2015;**27**:78−84.
5. Vo MN, Christopoulos G, Karmpaliotis D, Lombardi WL, Grantham JA, Brilakis ES. Balloon-assisted microdissection "BAM" technique for balloon-uncrossable chronic total occlusions. *J Invasive Cardiol* 2016;**28**:E37−41.
6. Brilakis ES. *Manual of coronary chronic total occlusion interventions. A step-by-step approach.* 2nd ed Elsevier; 2017.

7. Luna M, Papayannis A, Holper EM, Banerjee S, Brilakis ES. Transfemoral use of the GuideLiner catheter in complex coronary and bypass graft interventions. *Catheter Cardiovasc Interv* 2012;**80**:437−46.

8. Di Mario C, Ramasami N. Techniques to enhance guide catheter support. *Catheter Cardiovasc Interv* 2008;**72**:505−12.

9. Zhang Q, Zhang RY, Kirtane AJ, et al. The utility of a 5-in-6 double catheter technique in treating complex coronary lesions via transradial approach: the DOCA-TRI study. *EuroIntervention* 2012;**8**:848−54.

10. Santiago R, Moroni F, Del Rio V, Rodriguez-Escudero J, Azzalini L. The guide extension tunnel in landslide technique (TILT) for equipment delivery in severely tortuous or uncrossable lesions during percutaneous coronary intervention. *EuroIntervention* 2021;**17**:e923−4.

11. Papayannis AC, Michael TT, Brilakis ES. Challenges associated with use of the GuideLiner catheter in percutaneous coronary interventions. *J Invasive Cardiol* 2012;**24**:370−1.

12. Synetos A, Toutouzas K, Latsios G, et al. Proximal anchoring distal trapping technique in a chronic total occlusion unable to cross. *Cardiovasc Revasc Med* 2018;**19**:887−9.

13. Koutouzis M, Avdikos G, Nikitas G, et al. "Ping-pong" technique for treating a balloon uncrossable chronic total occlusion. *Cardiovasc Revasc Med* 2018;**19**:117−19.

14. Fang HY, Lee CH, Fang CY, et al. Application of penetration device (Tornus) for percutaneous coronary intervention in balloon uncrossable chronic total occlusion-procedure outcomes, complications, and predictors of device success. *Catheter Cardiovasc Interv* 2011;**78**:356−62.

15. Carlino M, Ruparelia N, Thomas G, et al. Modified contrast microinjection technique to facilitate chronic total occlusion recanalization. *Catheter Cardiovasc Interv* 2016;**87**:1036−41.

16. Azzalini L, Uretsky B, Brilakis ES, Colombo A, Carlino M. Contrast modulation in chronic total occlusion percutaneous coronary intervention. *Catheter Cardiovasc Interv* 2019;**93**:E24−9.

17. Hu XQ, Tang L, Zhou SH, Fang ZF, Shen XQ. A novel approach to facilitating balloon crossing chronic total occlusions: the "wire-cutting" technique. *J Interv Cardiol* 2012;**25**:297−303.

18. Xue J, Li J, Wang H, et al. Seesaw balloon-wire cutting" technique is superior to Tornus catheter in balloon uncrossable chronic total occlusions. *Int J Cardiol* 2017;**228**:523−7.

19. Li Y, Li J, Sheng L, et al. "Seesaw balloon-wire cutting" technique as a novel approach to "balloon-uncrossable" chronic total occlusions. *J Invasive Cardiol* 2014;**26**:167−70.

20. Karacsonyi J, Armstrong EJ, Truong HTD, et al. Contemporary use of laser during percutaneous coronary interventions: insights from the laser veterans affairs (LAVA) multicenter registry. *J Invasive Cardiol* 2018;**30**:195−201.

21. Karacsonyi J, Alaswad K, Choi JW, et al. Laser for balloon uncrossable and undilatable chronic total occlusion interventions. *Int J Cardiol* 2021;**336**:33−7.

22. Ben-Dor I, Maluenda G, Pichard AD, et al. The use of excimer laser for complex coronary artery lesions. *Cardiovasc Revasc Med* 2011;**12**(69):e1−8.

23. Shen ZJ, Garcia-Garcia HM, Schultz C, van der Ent M, Serruys PW. Crossing of a calcified "balloon uncrossable" coronary chronic total occlusion facilitated by a laser catheter: a case report and review recent four years' experience at the Thoraxcenter. *Int J Cardiol* 2010;**145**:251−4.

24. Niccoli G, Giubilato S, Conte M, et al. Laser for complex coronary lesions: impact of excimer lasers and technical advancements. *Int J Cardiol* 2011;**146**:296−9.

25. Pagnotta P, Briguori C, Mango R, et al. Rotational atherectomy in resistant chronic total occlusions. *Catheter Cardiovasc Interv* 2010;**76**:366−71.
26. Capretti G, Carlino M, Colombo A, Azzalini L. Rotational atherectomy in the subadventitial space to allow safe and successful chronic total occlusion recanalization: pushing the limit further. *Catheter Cardiovasc Interv* 2018;**91**:47−52.
27. Vo MN, Ravandi A, Grantham JA. Subintimal space plaque modification for "balloon-uncrossable" chronic total occlusions. *J Invasive Cardiol* 2014;**26**:E133−6.
28. Christopoulos G, Kotsia AP, Rangan BV, et al. "Subintimal external crush" technique for a "balloon uncrossable" chronic total occlusion. *Cardiovasc Revasc Med* 2017;**18**:63−5.
29. Yeoh J, Hill J, Spratt JC. Intravascular lithotripsy assisted chronic total occlusion revascularization with reverse controlled antegrade retrograde tracking. *Catheter Cardiovasc Interv* 2019;**93**:1295−7.
30. Michael TT, Banerjee S, Brilakis ES. Subintimal distal anchor technique for "balloon-uncrossable" chronic total occlusions. *J Invasive Cardiol* 2013;**25**:552−4.
31. Kirtane AJ, Stone GW. The Anchor-Tornus technique: a novel approach to "uncrossable" chronic total occlusions. *Catheter Cardiovasc Interv* 2007;**70**:554−7.
32. Brilakis ES, Banerjee S. The "Proxis-Tornus" technique for a difficult-to-cross calcified saphenous vein graft lesion. *J Invasive Cardiol* 2008;**20**:E258−61.
33. Sandoval Y, Lobo AS, Tajti P, Brilakis ES. Laser-assisted orbital or rotational atherectomy: a hybrid treatment strategy for balloon-uncrossable lesions. *Hellenic J Cardiol* 2020;**61**:57−9.
34. Tajti P, Karmpaliotis D, Alaswad K, et al. Prevalence, presentation and treatment of 'Balloon Undilatable' chronic total occlusions: insights from a Multicenter US Registry. *Catheter Cardiovasc Interv* 2018;**91**:657−66.
35. Kim BK, Shin DH, Hong MK, et al. Clinical impact of intravascular ultrasound-guided chronic total occlusion intervention with zotarolimus-eluting vs biolimus-eluting stent implantation: randomized study. *Circ Cardiovasc Interv* 2015;**8**:e002592.
36. Tian NL, Gami SK, Ye F, et al. Angiographic and clinical comparisons of intravascular ultrasound- vs angiography-guided drug-eluting stent implantation for patients with chronic total occlusion lesions: two-year results from a randomised AIR-CTO study. *EuroIntervention* 2015;**10**:1409−17.
37. Shaikh K, Kelly S, Gedela M, Kumar V, Stys A, Stys T. Novel use of an orbital atherectomy device for in-stent restenosis: lessons learned. *Case Rep Cardiology* 2016;**2016**:4.
38. Neupane S, Basir M, Tan C, et al. Feasibility and safety of orbital atherectomy for the treatment of in-stent restenosis secondary to stent under-expansion. *Catheter Cardiovasc Interv* 2021;**97**:2−7.
39. Si D, Liu G, Tong Y, He Y. Rotational atherectomy ablation for an unexpandable stent under the guide of IVUS: a case report. *Med (Baltim)* 2018;**97**:e9978.
40. Zhang M, Matsumura M, Usui E, et al. Intravascular ultrasound-derived calcium score to predict stent expansion in severely calcified lesions. *Circ Cardiovasc Interv* 2021;**14**: e010296.
41. Fujino A, Mintz GS, Matsumura M, et al. A new optical coherence tomography-based calcium scoring system to predict stent underexpansion. *EuroIntervention* 2018;**13**:e2182−9.
42. Raja Y, Routledge HC, Doshi SN. A noncompliant, high pressure balloon to manage undilatable coronary lesions. *Catheter Cardiovasc Interv* 2010;**75**:1067−73.
43. Yazdanfar S, Ledley GS, Alfieri A, Strauss C, Kotler MN. Parallel angioplasty dilatation catheter and guide wire: a new technique for the dilatation of calcified coronary arteries. *Cathet Cardiovasc Diagn* 1993;**28**:72−5.

44. Stillabower ME. Longitudinal force focused coronary angioplasty: a technique for resistant lesions. *Cathet Cardiovasc Diagn* 1994;**32**:196−8.
45. Meerkin D. My buddy, my friend: focused force angioplasty using the buddy wire technique in an inadequately expanded stent. *Catheter Cardiovasc Interv* 2005;**65**:513−15.
46. Lindsey JB, Banerjee S, Brilakis ES. Two "buddies" may be better than one: use of two buddy wires to expand an underexpanded left main coronary stent. *J Invasive Cardiol* 2007;**19**:E355−8.
47. Wilson A, Ardehali R, Brinton TJ, Yeung AC, Lee DP. Cutting balloon inflation for drug-eluting stent underexpansion due to unrecognized coronary arterial calcification. *Cardiovasc Revasc Med* 2006;**7**:185−8.
48. Haridas KK, Vijayakumar M, Viveka K, Rajesh T, Mahesh NK. Fracture of cutting balloon microsurgical blade inside coronary artery during angioplasty of tough restenotic lesion: a case report. *Catheter Cardiovasc Interv* 2003;**58**:199−201.
49. Harb TS, Ling FS. Inadvertent stent extraction six months after implantation by an entrapped cutting balloon. *Catheter Cardiovasc Interv* 2001;**53**:415−19.
50. Wang HJ, Kao HL, Liau CS, Lee YT. Coronary stent strut avulsion in aorto-ostial in-stent restenosis: potential complication after cutting balloon angioplasty. *Catheter Cardiovasc Interv* 2002;**56**:215−19.
51. Kawamura A, Asakura Y, Ishikawa S, et al. Extraction of previously deployed stent by an entrapped cutting balloon due to the blade fracture. *Catheter Cardiovasc Interv* 2002;**57**:239−43.
52. Sanchez-Recalde A, Galeote G, Martin-Reyes R, Moreno R. AngioSculpt PTCA balloon entrapment during dilatation of a heavily calcified lesion. *Rev Esp Cardiol* 2008;**61**:1361−3.
53. Giugliano GR, Cox N, Popma J. Cutting balloon entrapment during treatment of in-stent restenosis: an unusual complication and its management. *J Invasive Cardiol* 2005;**17**:168−70.
54. Pappy R, Gautam A, Abu-Fadel MS. AngioSculpt PTCA balloon catheter entrapment and detachment managed with stent jailing. *J Invasive Cardiol* 2010;**22**:E208−10.
55. Brinton TJ, Ali ZA, Hill JM, et al. Feasibility of shockwave coronary intravascular lithotripsy for the treatment of calcified coronary stenoses. *Circulation* 2019;**139**:834−6.
56. Watkins S, Good R, Hill J, Brinton TJ, Oldroyd KG. Intravascular lithotripsy to treat a severely underexpanded coronary stent. *EuroIntervention* 2019;**15**:124−5.
57. Karacsonyi J, Nikolakopoulos I, Vemmou E, Rangan BV, Brilakis ES, Intracoronary. Lithotripsy: a new solution for undilatable in-stent chronic total occlusions. *JACC Case Rep* 2021;**3**:780−5.
58. Oksnes A, Cosgrove C, Walsh S, et al. Intravascular lithotripsy for calcium modification in chronic total occlusion percutaneous coronary intervention. *J Interv Cardiol* 2021;**2021**:9958035.
59. Egred M. RASER angioplasty. *Catheter Cardiovasc Interv* 2012;**79**:1009−12.
60. Gonzalvez-Garcia A, Jimenez-Valero S, Galeote G, Moreno R, Lopez de Sa E, Jurado-Roman A. RotaTripsy": combination of rotational atherectomy and intravascular lithotripsy in heavily calcified coronary lesions: a case series. *Cardiovasc Revasc Med* 2022;**35**:179−84.
61. Gonzalvez-Garcia A, Jimenez-Valero S, Galeote G, Moreno R, Jurado-Roman A. RotaTripsy, not only a bailout strategy for heavily calcified coronary lesions. *Cardiovasc Revasc Med* 2022;**35**:189.
62. Kaur N, Pruthvi CR, Sharma Y, Gupta H. Rotatripsy: synergistic effects of complementary technologies: a case report. *Eur Heart J Case Rep* 2021;**5**.

63. Giacchi G, Contarini M, Ruscica G, Brugaletta S. The "RotaTripsy Plus" approach in a heavily calcified coronary stenosis. *Cardiovasc Revasc Med* 2021;**28S**:203−5.
64. Buono A, Basavarajaiah S, Choudhury A, et al. "RotaTripsy" for severe calcified coronary artery lesions: insights from a real-world multicenter cohort. *Cardiovasc Revasc Med* 2022;**37**:78−81.
65. Aznaouridis K, Bonou M, Masoura C, Kapelios C, Tousoulis D, Barbetseas J. Rotatripsy: a hybrid "drill and disrupt" approach for treating heavily calcified coronary lesions. *J Invasive Cardiol* 2020;**32**:E175.
66. Jurado-Roman A, Gonzalvez A, Galeote G, Jimenez-Valero S, Moreno R. RotaTripsy: combination of rotational atherectomy and intravascular lithotripsy for the treatment of severely calcified lesions. *JACC Cardiovasc Interv* 2019;**12**:e127−9.
67. Ferri LA, Jabbour RJ, Giannini F, et al. Safety and efficacy of rotational atherectomy for the treatment of undilatable underexpanded stents implanted in calcific lesions. *Catheter Cardiovasc Interv* 2017;**90**:E19−24.
68. Vales L, Coppola J, Kwan T. Successful expansion of an underexpanded stent by rotational atherectomy. *Int J Angiol* 2013;**22**:63−8.
69. Hernandez J, Galeote G, Moreno R. Rotational atherectomy: if you do not do it before, you can do it after stenting. *J Invasive Cardiol* 2014;**26**:E122−3.
70. Medina A, de Lezo JS, Melian F, Hernandez E, Pan M, Romero M. Successful stent ablation with rotational atherectomy. *Catheter Cardiovasc Interv* 2003;**60**:501−4.
71. Kobayashi Y, Teirstein P, Linnemeier T, Stone G, Leon M, Moses J. Rotational atherectomy (stentablation) in a lesion with stent underexpansion due to heavily calcified plaque. *Catheter Cardiovasc Interv* 2001;**52**:208−11.
72. Latib A, Takagi K, Chizzola G, et al. Excimer Laser LEsion modification to expand non-dilatable stents: the ELLEMENT registry. *Cardiovasc Revasc Med* 2014;**15**:8−12.
73. Karacsonyi J, Danek BA, Karatasakis A, Ungi I, Banerjee S, Brilakis ES. Laser coronary atherectomy during contrast injection for treating an underexpanded stent. *JACC Cardiovasc Interv* 2016;**9**:e147−8.
74. Fernandez JP, Hobson AR, McKenzie D, et al. Beyond the balloon: excimer coronary laser atherectomy used alone or in combination with rotational atherectomy in the treatment of chronic total occlusions, non-crossable and non-expansible coronary lesions. *EuroIntervention* 2013;**9**:243−50.
75. Badr S, Ben-Dor I, Dvir D, et al. The state of the excimer laser for coronary intervention in the drug-eluting stent era. *Cardiovasc Revasc Med* 2013;**14**:93−8.
76. Sunew J, Chandwaney RH, Stein DW, Meyers S, Davidson CJ. Excimer laser facilitated percutaneous coronary intervention of a nondilatable coronary stent. *Catheter Cardiovasc Interv* 2001;**53**:513−17.
77. Veerasamy M, Gamal AS, Jabbar A, Ahmed JM, Egred M. Excimer laser with and without contrast for the management of under-expanded stents. *J Invasive Cardiol* 2017;**29**:364−9.
78. Egred M. A novel approach for under-expanded stent: excimer laser in contrast medium. *J Invasive Cardiol* 2012;**24**:E161−3.

Chapter 24

Complex patient subgroups

24.1 Prior transcatheter aortic valve replacement (TAVR) patients

Chronic total occlusion (CTO)percutaneous coronary intervention (PCI) in prior TAVR patients can be challenging due to difficulties with coronary engagement. Moreover, prior TAVR patients are often very old and have multiple comorbidities. In the ACTIVATION, trial 235 severe aortic stenosis patients with obstructive coronary artery disease were randomized to PCI versus no-PCI prior to TAVR. PCI prior to TAVR resulted in a higher incidence of bleeding (41.2% vs 21.7%).[1] The majority of bleeding occurred in the first 30 days after TAVR. Therefore, the challenges of coronary engagement after TAVR should be weighed against the increased risk of bleeding if PCI is performed prior to TAVR.

Given frequent difficulty in engaging the coronary arteries in TAVR patients, preprocedural ECG-gated CT angiography can be performed to help understand the 3-dimensional geometric interaction among the valve prosthesis, the aortic root, and coronary ostia to help predict and prepare for potential challenges of coronary reaccess (Fig. 24.1).[2]

However, use of CT angiography in prior TAVR patients has several limitations[2]:

1. CT cannot be performed in urgent or emergent situations, such as in patients with an acute coronary syndrome requiring urgent cardiac catheterization or PCI.
2. CT angiography requires contrast administration that increases the risk of contrast-induced acute kidney injury when giving additional contrast during cardiac catheterization.
3. Motion artifact and image quality may limit the ability to visualize leaflet orientation of the transcatheter valve relative to the coronary ostia, making it difficult to determine whether the commissural post may impede the ability to reaccess the coronary arteries.

Coronary engagement should be performed with extreme care, aiming for coaxial positioning and avoiding deep coronary artery intubation.

24.1.1 Evolut-PRO CoreValve

CTO PCI Manual case 113

Manual of Chronic Total Occlusion Percutaneous Coronary Interventions.
DOI: https://doi.org/10.1016/B978-0-323-91787-2.00026-5

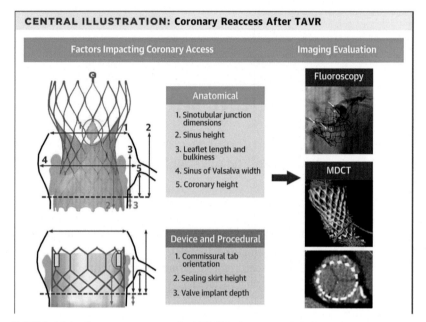

FIGURE 24.1 **Coronary reaccess after TAVR**. Summary of factors impacting coronary access and imaging evaluation after TAVR. *MDCT*, multidetector computed tomography; *TAVR*, transcatheter aortic valve replacement.[2] *Reproduced with permission from Yudi MB, Sharma SK, Tang GHL, Kini A. Coronary angiography and percutaneous coronary intervention after transcatheter aortic valve replacement. J Am Coll Cardiol 2018;71:1360−1378.*

Engagement depending on depth of valve implantation

If the Evolut-PRO CoreValve is positioned optimally (skirt below coronary ostia), it is feasible to engage the coronary artery in a coaxial manner, assuming the native aortic valve leaflets will not interfere with the path to the coronary ostium (Fig. 24.2, panel A).

If the valve is deployed high (Fig. 24.2 panel B) coronary obstruction would not occur due to the narrow waist of the valve and sufficient sinus of Valsalva width.[2] However, selective coronary angiography would be difficult in this scenario and would have to occur from a diamond above the ostium, given that the supra-annular valve and its covered segment (e.g., sealing skirt) would be above the level of the ostium. Careful wiring of the coronary artery and advancement of a guide extension catheter over a balloon (with or without further advancement of the guide catheter) may allow safer PCI in cases of longer distance or uneven level of the lowest patent "diamond" and the coronary ostium. A straighter catheter with a short tip, such as a Judkins right (JR) 4, could be used in this scenario, even for left main artery engagement.[2]

Engagement depending on the position of the transcatheter valve commissures in relation to those of the native aortic valve

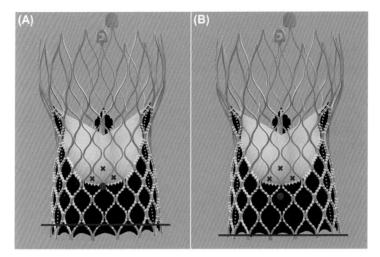

FIGURE 24.2   **Self-expanding valve and coronary access depending on level of implantation across the annulus.** Red dot represents the location of the coronary ostium in relation to the valve frame, and the red line represents the annular plane. The red x's depict the closest diamonds that can be used to access the coronary ostium. An optimally positioned Evolut-R (Medtronic, Galway, Ireland) (**panel A**) would make coronary access potentially easier than one with a higher implant (**panel B**).[2] *Reproduced with permission from Yudi MB, Sharma SK, Tang GHL, Kini A. Coronary angiography and percutaneous coronary intervention after transcatheter aortic valve replacement. J Am Coll Cardiol 2018;71:1360–1378.*

The circumferential sealing skirt of the Evolut-PRO CoreValve is 13 mm in height (14 mm in the 34-mm Evolut-R), however it rises up to 26 mm at the commissural insertion point (Fig. 24.3). If a commissure ends up being positioned directly in front of the coronary ostium coaxial, engagement of the coronary ostia would be challenging, if not impossible.[2] Engagement also depends on the width of the sinus of Valsalva that determines the space between the valve frame and the coronary ostia; the wider the sinus the more room there is to manipulate a catheter toward the coronary ostia. A narrow sinus would require a very acute angle for the catheter to be pointing toward the ostia for a nonselective coronary angiogram. If selective engagement is required, a coronary wire would have to be manipulated into the coronary artery, and the guide, or a guide catheter extension, would then have to be railed into the ostium. This represents the most difficult scenario: a valve commissure overlying a low coronary ostium in a patient with a narrow sinus of Valsalva. This description does not account for the native aortic leaflet height and severity of calcification facing the left and right sinuses. A tall and bulky leaflet may extend beyond the 13- or 14-mm sealing skirt of the repositionable Evolut-PRO self-expanding valve and would likely further add to the challenge of coronary reaccess based on this scenario.[2]

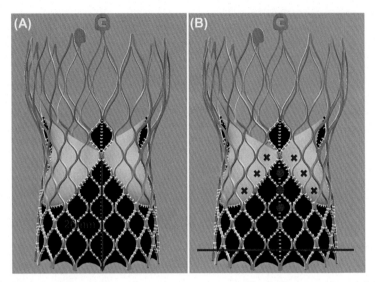

FIGURE 24.3 **Self-expanding valve and coronary access.** If ostia lines up with commissural post. Red line represents the annular plane. The three red dots depict coronary ostia heights of approximately 10, 14, and 18 mm above the annular plane, respectively. The red x's depict the closest diamonds that can be used to access the coronaries. The commissural post of an Evolut-R (Medtronic, Galway, Ireland) is 26 mm in height (**panel A**). Depending on the height of coronary ostia, a different catheter and approach is necessary for coronary reaccess, when the ostium faces the side of the commissural post (**panel B**).[2] *Reproduced with permission from Yudi MB, Sharma SK, Tang GHL, Kini A. Coronary angiography and percutaneous coronary intervention after transcatheter aortic valve replacement. J Am Coll Cardiol 2018;71:1360−1378.*

The ALIGN-ACCESS study of 206 patients suggests that intentional implantation of supra-annular transcatheter heart valves (THVs) to achieve favorable commissural alignment significantly improves the likelihood of successful coronary engagement. Nonetheless, even in commissural aligned supra-annular THVs, engagement of the right coronary artery (RCA) and left main was nonselective in 26% of patients, while cannulation of at least one coronary artery was deemed unfeasible in 11% of misaligned supra-annular THVs.[3]

Catheter selection

Left coronary artery: smaller catheters, such as a JL3.5 or JL3, are frequently used to engage the left main. On the contrary, engagement of the RCA can usually be managed with a JR4 catheter. There are reports of XB guide catheter kinking and entrapment through the valve diamonds (Fig. 24.4).[4] The following measures may minimize the risk of catheter entrapment:

1. Crossing the stent frame perpendicularly through a diamond at the same level of the coronary ostium.

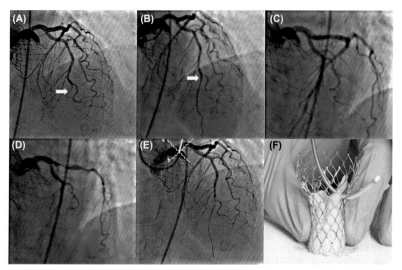

FIGURE 24.4 **Guide catheter entrapment**. **Panel A**: left coronary angiography showing total occlusion of the distal left anterior descending artery (arrow). **Panel B**: left coronary angiography post−percutaneous coronary intervention and drug-eluting stent deployment (arrow). **Panels C and D**: dissection of the left main and left anterior descending coronary arteries. **Panel E**: (a) interrupted line showing guide catheter engagement into the left coronary ostium at an acute angle with the vertical axis of the stent frame; (b) interrupted line showing crossing of the stent frame at a perpendicular angle. **Panel F**: ex vivo simulation using an extra back-up 3.5 guide catheter, showing catheter entrapment within the stent frame when crossed at an acute angle.[4] *Reproduced with permission from Harhash A, Ansari J, Mandel L, Kipperman R. STEMI after TAVR: procedural challenge and catastrophic outcome. JACC Cardiovasc Interv 2016;9:1412−1413.*

2. Using catheters with favorable geometry (e.g., Judkins Left (JL) for the left main).
3. Using a balloon and/or a guidewire to back the catheter out of the coronary ostium.

Right coronary artery: JR4 or Ikari right catheters are usually used. A JR4.5, JR5, or Amplatz right 2 catheter may be preferable if the sinus width is large, creating a larger distance from the valve frame to the ostium.[2]

24.1.2 SAPIEN 3

CTO PCI Manual case 197

Coronary engagement is often easier in patients who have a SAPIEN valve compared with a self-expanding valve. In the ALIGN-ACCESS study, coronary cannulation was feasible in all patients with a SAPIEN valve (albeit 5% nonselective), and significantly more likely to be successful than in both commisurally aligned and, even more so, misaligned supra-annular THVs.[3]

Engagement depending on depth of valve implantation (Fig. 24.5).

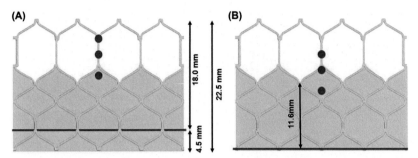

FIGURE 24.5   Balloon-expandable valve and coronary ostia based on depth of implant.
Red dots represent the different locations of the coronary ostium in relation to the valve frame
of a 29 mm Sapien 3 valve (Edwards Lifesciences, Irvine, CA), and the red line represents the
annular plane. An optimally positioned Sapien 3 valve (Edwards Lifesciences, Irvine, CA)
(**panel A**) would make coronary access potentially easier than one with a higher implant
(**panel B**), where the coronary ostium will be located below the seal skirt. Tall native leaflet or
bulky calcium at the leaflet tip may further increase difficulty of coronary access in a high valve
implant.[2] *Reproduced with permission from Yudi MB, Sharma SK, Tang GHL, Kini A.
Coronary angiography and percutaneous coronary intervention after transcatheter aortic valve
replacement. J Am Coll Cardiol 2018;71:1360–1378.*

24.2   Heart failure patients

CTO PCI can be performed in patients with heart failure with high success
rates,[5,6] but may carry increased risk of complications, including higher periprocedural mortality.[6] Several retrospective studies have demonstrated that successful CTO revascularization can improve left ventricular systolic function,[7–16]
provided that the CTO-supplied myocardium is viable[12,13] and the vessel
remains patent during follow-up.[10,11] However, two randomized-controlled trials
(EXPLORE[17] and REVASC[18]) did not demonstrate improvement in ejection
fraction post CTO PCI, but the baseline ejection fraction was normal in both
studies.

Patients with heart failure who require CTO PCI may benefit from preprocedural optimization, intraprocedural monitoring, and use of hemodynamic support.

24.2.1   Preprocedural optimization and planning

Optimizing the patient's hemodynamic status before CTO PCI (for example
with aggressive diuresis) could reduce the risk of periprocedural hemodynamic deterioration.

Working together with the advanced heart failure team can be invaluable
for coordinating care, selecting the need and type of hemodynamic support,
and determining bailout options (such as ventricular assist devices) if a complication or hemodynamic decompensation occurs during the procedure.

24.2.2 Intraprocedural monitoring

Performing right heart catheterization before and during CTO PCI can help manage the patient's hemodynamic status (for example by helping determine if the patient needs intravenous fluid administration or diuresis) and allow early detection of adverse hemodynamic changes.

24.2.3 Hemodynamic support

Use of hemodynamic support may reduce the risk of hemodynamic collapse and improve the safety of the procedure in high-risk patients. As described in Chapter 14, the decision on when to use hemodynamic support and device selection is based on coronary anatomy, hemodynamic status, and arterial access:

1. *Coronary anatomy.* Hemodynamic support is often recommended for challenging anatomy, especially with large areas of myocardium at risk, such as PCI through the last remaining vessel, PCI of the left main in a left dominant system (CTO Manual online case 67) or in patients with occluded right coronary arteries, use of the retrograde approach (especially when retrograde crossing of a left internal mammary artery is performed), and use of atherectomy.
2. *Hemodynamic status.* Factors favoring use of hemodynamic support include low ejection fraction, increased pulmonary capillary wedge pressure, severe mitral regurgitation, and severe pulmonary hypertension.

Use of hemodynamic support during CTO PCI is associated with several unique challenges:

1. *Arterial access.* All hemodynamic support devices require large (13−18 Fr) arterial access (except for intra-aortic balloon pump that requires 8 Fr access). Such access is commonly obtained in the femoral artery. In patients with severe peripheral arterial disease peripheral intervention can be performed to facilitate insertion of a hemodynamic support device, subclavian access can be obtained (percutaneously or surgically), or transcaval access can be used. CTO PCI can then be performed using femoral and radial access, biradial access, or by doing a double stick of the other femoral artery. When using an Impella device, the Impella sheath can be used to insert one guide catheter (Single-access for Hi-risk PCI [SHiP] technique).[19,20]
2. *Overcrowding in the aorta.* When the Impella device is used for hemodynamic support, its location in the aortic root can interfere with guide engagement of the coronary arteries. Usually the Impella device is inserted first, followed by the right coronary artery guide catheter, leaving insertion of the left coronary artery guide catheter for the end. If the Impella is affecting the guide catheter position, it may be better to reposition the guide catheter before CTO PCI is started.

References

1. Patterson T, Clayton T, Dodd M, et al. ACTIVATION (PercutAneous Coronary inTervention prIor to transcatheter aortic VAlve implantaTION): a randomized clinical trial. *JACC Cardiovasc Interv* 2021;**14**:1965−74.

2. Yudi MB, Sharma SK, Tang GHL, Kini A. Coronary angiography and percutaneous coronary intervention after transcatheter aortic valve replacement. *J Am Coll Cardiol* 2018;**71**:1360−78.

3. Tarantini G, Nai Fovino L, Scotti A, et al. Coronary access after transcatheter aortic valve replacement with commissural alignment: the ALIGN-ACCESS study. *Circ Cardiovasc Interv* 2022;**15**:e011045.

4. Harhash A, Ansari J, Mandel L, Kipperman R. STEMI after TAVR: procedural challenge and catastrophic outcome. *JACC Cardiovasc Interv* 2016;**9**:1412−13.

5. Galassi AR, Boukhris M, Toma A, et al. Percutaneous coronary intervention of chronic total occlusions in patients with low left ventricular ejection fraction. *JACC Cardiovasc Interv* 2017;**10**:2158−70.

6. Simsek B, Kostantinis S, Karacsonyi J, et al. Outcomes of chronic total occlusion percutaneous coronary intervention in patients with reduced left ventricular ejection fraction. *Catheter Cardiovasc Interv* 2022;**99**:1059−64.

7. Melchior JP, Doriot PA, Chatelain P, et al. Improvement of left ventricular contraction and relaxation synchronism after recanalization of chronic total coronary occlusion by angioplasty. *J Am Coll Cardiol* 1987;**9**:763−8.

8. Danchin N, Angioi M, Cador R, et al. Effect of late percutaneous angioplastic recanalization of total coronary artery occlusion on left ventricular remodeling, ejection fraction, and regional wall motion. *Am J Cardiol* 1996;**78**:729−35.

9. Van Belle E, Blouard P, McFadden EP, Lablanche JM, Bauters C, Bertrand ME. Effects of stenting of recent or chronic coronary occlusions on late vessel patency and left ventricular function. *Am J Cardiol* 1997;**80**:1150−4.

10. Sirnes PA, Myreng Y, Molstad P, Bonarjee V, Golf S. Improvement in left ventricular ejection fraction and wall motion after successful recanalization of chronic coronary occlusions. *Eur Heart J* 1998;**19**:273−81.

11. Piscione F, Galasso G, De Luca G, et al. Late reopening of an occluded infarct related artery improves left ventricular function and long term clinical outcome. *Heart* 2005;**91**:646−51.

12. Baks T, van Geuns RJ, Duncker DJ, et al. Prediction of left ventricular function after drug-eluting stent implantation for chronic total coronary occlusions. *J Am Coll Cardiol* 2006;**47**:721−5.

13. Kirschbaum SW, Baks T, van den Ent M, et al. Evaluation of left ventricular function three years after percutaneous recanalization of chronic total coronary occlusions. *Am J Cardiol* 2008;**101**:179−85.

14. Cheng AS, Selvanayagam JB, Jerosch-Herold M, et al. Percutaneous treatment of chronic total coronary occlusions improves regional hyperemic myocardial blood flow and contractility: insights from quantitative cardiovascular magnetic resonance imaging. *JACC Cardiovasc Interv* 2008;**1**:44−53.

15. Werner GS, Surber R, Kuethe F, et al. Collaterals and the recovery of left ventricular function after recanalization of a chronic total coronary occlusion. *Am Heart J* 2005;**149**:129−37.

16. Cardona M, Martin V, Prat-Gonzalez S, et al. Benefits of chronic total coronary occlusion percutaneous intervention in patients with heart failure and reduced ejection fraction: insights from a cardiovascular magnetic resonance study. *J Cardiovasc Magn Reson* 2016;**18**:78.

17. Henriques JP, Hoebers LP, Ramunddal T, et al. Percutaneous intervention for concurrent chronic total occlusions in patients with STEMI: the EXPLORE trial. *J Am Coll Cardiol* 2016;**68**:1622−32.

18. Mashayekhi K, Nuhrenberg TG, Toma A, et al. A randomized trial to assess regional left ventricular function after stent implantation in chronic total occlusion: the REVASC trial. *JACC Cardiovasc Interv* 2018;**11**:1982−91.

19. Wollmuth J, Korngold E, Croce K, Pinto DS. The Single-access for Hi-risk PCI (SHiP) technique. *Catheter Cardiovasc Interv* 2020;**96**:114−16.

20. Hakeem A, Tang D, Patel K, et al. Safety and efficacy of single-access impella for high-risk percutaneous intervention (SHiP). *JACC Cardiovasc Interv* 2022;**15**:347−8.

Part C

Complications

Chapter 25

Acute vessel closure

CTO PCI Manual online cases 22, 38, 50, 66, 69, 92, 97, 98, 112, 125, 165, 171

Acute vessel closure is defined as (partial or complete) decrease in antegrade coronary flow during or immediately after a percutaneous coronary intervention. At first glance, acute vessel closure may seem as a nonissue for CTO PCI, since the CTO vessel is occluded at baseline. However, CTO PCI can result in occlusion of a non-CTO vessel (Fig. 25.1) resulting in a large area of severe ischemia, hemodynamic compromise, cardiac arrest, and death. The risk is highest when using the retrograde approach, and when the donor vessel is the last remaining vessel, which is more frequent in prior CABG patients.

Similarly, development of an acute coronary syndrome in patients with a preexisting CTO is more likely to lead to cardiac arrest ("double jeopardy" phenomenon, Figure 7.3)[1] (CTO Manual online case 165).

Fig. 25.2 presents a step-by-step algorithm for approaching acute vessel closure.[2] Step 1 is to maintain guidewire position (if a guidewire was already inserted in the target vessel). The second and third steps are performed simultaneously. Step 2 is to determine the cause of dissection and treat accordingly. Step 3 is to carefully monitor the patient's hemodynamics and provide hemodynamic support if needed.

25.1 Maintain guidewire position

Having a guidewire within an acutely occluded vessel significantly facilitates management. **A workhorse safety guidewire should, therefore, be prophylactically inserted into the CTO donor vessel(s) to facilitate prompt treatment in case of acute vessel closure.**

Acute vessel closure can cause significant patient and operator stress, potentially leading to inadvertent loss of position of a previously delivered guidewire (such as the safety wire). **Avoid losing guidewire position at all cost**, as it can have catastrophic consequences, especially if the acute vessel closure was caused by a dissection. Guidewire position loss can be the result of guide manipulations; hence, pay continuous attention to the guide position.

Manual of Chronic Total Occlusion Percutaneous Coronary Interventions.
DOI: https://doi.org/10.1016/B978-0-323-91787-2.00003-4

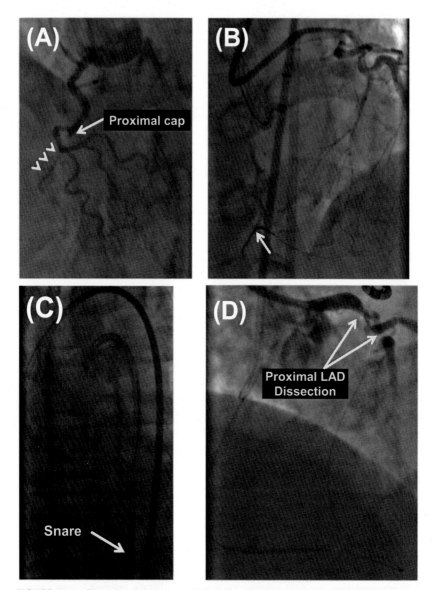

FIGURE 25.1 **Example of donor vessel dissection during retrograde CTO PCI.** PCI of a right coronary artery (RCA) CTO (**panel A**). After a failed antegrade crossing attempt, retrograde crossing was performed (**panel B**) and the retrograde guidewire was externalized (**panel C**). During RCA stenting over the externalized guidewire (**panel D**), the patient developed severe chest pain and hypotension due to proximal left anterior descending artery (LAD) dissection (**panel D**). The LAD was immediately stented (**panel E**) with restoration of antegrade flow and stabilization of the patient (**panels F and G**). After removal of the entrapped retrograde guidewire and stenting of the right coronary artery an excellent final angiographic result was achieved (**panel H**) (CTO Manual online case 22).

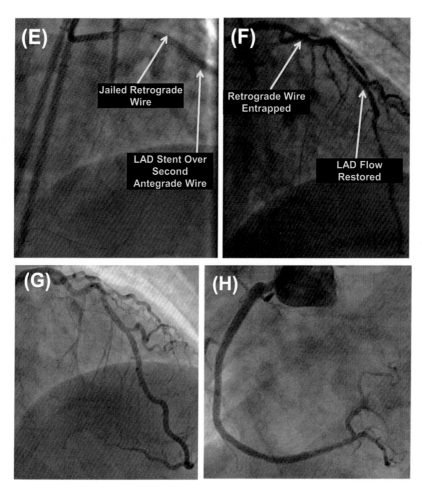

FIGURE 25.1   (Continued).

In summary, remain calm and maintain true lumen wire position (if the wire was placed in the true lumen prior to the complication) to successfully manage acute vessel closure.

25.2   Determine the cause of acute vessel closure and treat accordingly

Treatment of acute vessel closure depends on the cause (Table 25.1). The key differentiation is between dissection (that requires stenting, CTO Manual online cases 112, 125, 165) and distal embolization (that may worsen from stenting and requires physical removal of thrombus or debris and/or vasodilator administration; CTO Manual online case 19).

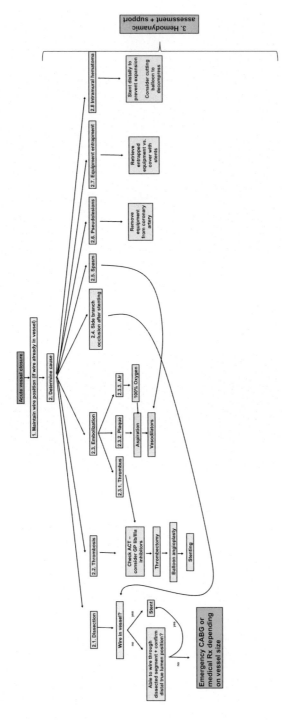

FIGURE 25.2 Approach to acute vessel closure. *Reproduced with permission from: Brilakis ES. Manual of percutaneous coronary interventions: a step-by-step approach. Elsevier; 2021.*

TABLE 25.1 Causes of acute vessel closure.

Mechanism	Risk factors	Diagnosis	Treatment
1. Dissection	• SCAD • Iatrogenic from guide or guide catheter extension, guidewire, lesion preparation, stent • Severe tortuosity	Angiography OCT IVUS	• Stenting • STAR or Stingray-based reentry as bailout (Section 8.3.5)
2. Thrombosis	• Suboptimal anticoagulation and antiplatelet Rx • Suboptimal stent implantation	Angiography IVUS OCT	• Optimize anticoagulation/ antiplatelet Rx • Thrombectomy • Balloon angioplasty • Stenting
3.1. Embolization—thrombus	• Suboptimal catheter preparation • Suboptimal balloon/stent/ indeflator preparation • Thrombus formation in sheath	Angiography IVUS OCT	• Optimize anticoagulation/ antiplatelet Rx • Thrombectomy • Balloon angioplasty • Stenting
3.2. Embolization—plaque	• Lipid-rich lesions • Atherectomy	Angiography IVUS OCT	• Aspiration • Vasodilators • Balloon angioplasty (if epicardial embolization)
3.3. Embolization—air	• Inadequate catheter preparation	Angiography ECG changes	• 100% oxygen • Aspiration • Vasodilators • Intracoronary epinephrine
4. Side branch occlusion after stenting	• Ostial disease • Dissection • Large plaque burden	Angiography ECG changes	• Rewire branch • Balloon angioplasty, including kissing balloon inflation • Stenting
5. Spasm	• Catheter manipulation	Angiography + response to vasodilators	• Vasodilators
6. Pseudolesions	• Severe tortuosity	Angiography	• Remove equipment from tortuous vessel

(Continued)

TABLE 25.1 (Continued)

Mechanism	Risk factors	Diagnosis	Treatment
7. Equipment entrapment	• Calcification • Tortuosity • Poor lesion preparation • Rotational atherectomy • Equipment advancement through stents	Angiography	• Equipment retrieval (Chapter 27) • Stenting over entrapped equipment
8. Intramural hematoma	• SCAD • Large lipid rich plaque • Use of cutting balloons	IVUS OCT	• Stent distally first, then proximally • Cutting balloon • Medical therapy

25.2.1 Dissection

In coronary dissections there is separation of the various layers of the coronary arterial wall. Angiographically, a dissection appears as a linear or spiral filling defect within the vessel lumen, although in severe cases complete vessel occlusion may occur (Chapter 6). During CTO PCI, dissection can occur in either or both (CTO PCI Manual online case 165) the donor vessel (often with catastrophic consequences; CTO PCI Manual online cases 22, 97) and the CTO vessel (CTO PCI Manual online case 125) (Fig. 25.3).

The consequences of dissection depend on:

1. the severity of coronary flow obstruction (none, partial, or complete cessation of coronary flow), and
2. the location of the dissection:
    a. Donor vessel dissection can cause severe ischemia.
    b. More proximal dissections, such as in the left main (CTO Manual online case 165), have more severe consequences as they cause a larger area of ischemia, potentially leading to arrhythmias and hemodynamic compromise.
    c. Dissection of small branches, is unlikely to cause major hemodynamic changes.

25.2.1.1   Causes (* = particularly important for CTO PCI)

Donor vessel dissection:

1. Guide-induced coronary dissections.*
2. Equipment advancement through a significant donor vessel lesion during retrograde crossing attempts.*

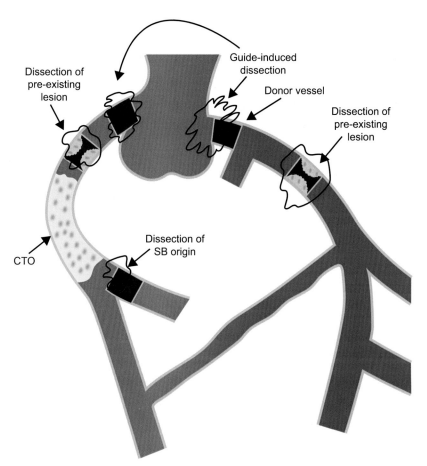

FIGURE 25.3 **Potential locations of dissection during CTO PCI.** Dissection can occur in various locations in both the donor vessel and the CTO vessel. *SB*, side branch.

CTO vessel dissection:

1. Use of dissection/reentry crossing techniques.
2. Disruption of side branches during CTO crossing.
3. Subintimal stent implantation. Occasionally, false lumen distal position of the guidewire may not be appreciated and stents may be inadvertently deployed within the extraplaque space, obstructing the outflow of the vessel (Fig. 25.4).[3] After this occurs, the patient may remain asymptomatic[4,5] or may develop ST-segment elevation due to side branch occlusion.[3]

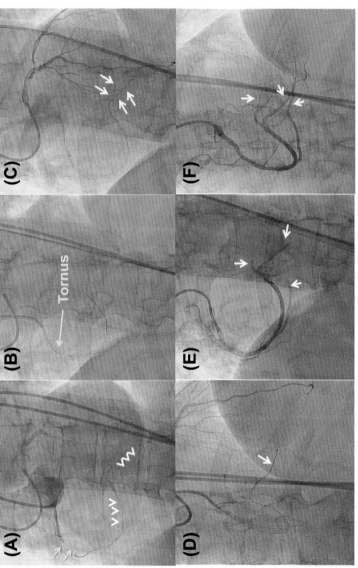

FIGURE 25.4 **Example of distal vessel dissection due to false lumen stenting.** Coronary angiography demonstrating a chronic total occlusion of the mid right coronary artery (arrows, **panel A**). The distal right coronary artery and the right posterior descending artery (arrowheads, **panel A**) were filling via collaterals from the left anterior descending artery. The right coronary artery was crossed antegradely with a Pilot 200 wire, however no other catheter, such as the Tornus catheter (arrow, **panel B**) could cross the occlusion. The right coronary artery occlusion was crossed with a second Pilot 200 wire (arrows, **panel C**). Although contralateral injection suggested intraluminal distal wire position (arrow, **panel D**), after stenting antegrade flow in the acute marginal branch, right posterior descending artery and right posterolateral branch (arrows, **panel E**) ceased. After rewiring and balloon angioplasty of the acute marginal branch, right posterior descending artery and right posterolateral branch (arrows, **panel F**), antegrade coronary flow was restored in all three vessels. *Reproduced with permission from Patel VG, Banerjee S, Brilakis ES. Treatment of inadvertent subintimal stenting during intervention of a coronary chronic total occlusion. Interv Cardiol 2013;5:165—169.*

Either CTO or donor vessel dissection:

1. Preexisting disease of the vessel ostium.*
2. Contrast injection despite dampened pressure waveform.*
3. Noncoaxial guide position.*
4. Excessive guide catheter movement that is more likely to occur with guidewire externalization during the retrograde approach.
5. Wire crossing attempts through severely stenotic and tortuous lesions.
6. Lesion preparation (balloon angioplasty, atherectomy, laser, especially when using oversized balloons).
7. Balloon rupture.
8. Heavily calcified and tortuous lesions.
9. Stenting.

25.2.1.2 Prevention

Donor vessel dissection:

1. Insert a workhorse safety guidewire into the donor vessel, both to stabilize the guide catheter, but also to allow rapid treatment of the donor vessel in case of acute donor vessel occlusion.
2. Do NOT use side hole guide catheters in the donor vessel, especially in the left main, as they can mask pressure dampening and flow compromise, potentially leading to severe ischemia.
3. Avoid use of the retrograde approach through diffusely diseased donor vessels. Consider intracoronary imaging to investigate the anatomy of the donor vessel in advance. If the donor artery requires PCI, treat the donor vessel prior to advancing equipment for retrograde crossing.
4. Keep the retrograde guidewire encased by a microcatheter or over-the-wire (OTW) balloon during all manipulations.[6]
5. Convert the retrograde system to a fully antegrade system either by using the kissing microcatheter technique, or by using the tip-in technique (advance an antegrade microcatheter over the retrograde wire into the distal vessel and then exchange the retrograde wire for an antegrade workhorse guidewire).
6. When performing dual injection, remove the donor vessel catheter as soon as it is not needed after CTO crossing.
7. Have a high threshold for use of the left internal mammary artery (LIMA) graft for retrograde CTO PCI because LIMA dissection can occur (Fig. 25.5; CTO Manual online case 50).

CTO vessel dissection:

1. Disconnect the manifold to prevent inadvertent contrast injection of the CTO vessel when using dissection/reentry crossing techniques.*

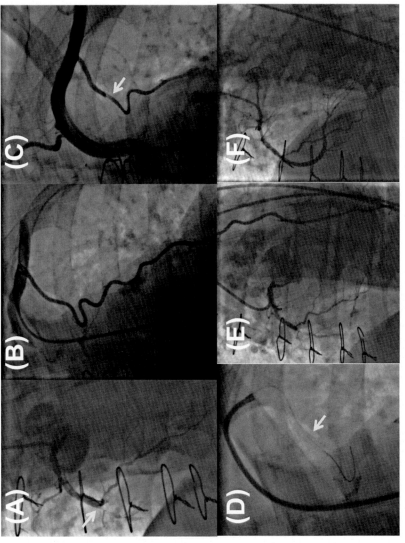

FIGURE 25.5 **Example of LIMA graft dissection**. PCI of a right coronary artery (RCA) CTO (**panel A**) was attempted with dual injection via a catheter inserted in the left internal mammary artery graft (LIMA, **panel B**). Engagement of the LIMA graft was challenging due to an acute take-off angle. A Guideliner catheter was used to engage the LIMA causing ostial LIMA dissection with decreased antegrade flow (**panel C**). LIMA flow was restored after stenting (**panel D**), enabling continuation of the procedure (**panel E**), with a final successful outcome (**panel F**).

2. Advance a guidewire into important side branches after CTO crossing to reduce the risk of acute closure after stenting.*

3. Confirm true lumen wire position prior to stent implantation.* The first step in preventing extraplaque (subintimal) stent deployment is to have a high threshold of suspicion. For example, apparent intraluminal position distal to the CTO when using a wire knuckle or very aggressive guidewires ($\geq$12 gr tapered tip wires, for example) should always raise the concern that the wire might actually be in the extraplaque space close to the lumen, and should be followed by an adequate check before proceeding to balloon dilatation and stenting.

To confirm that the wire has entered the distal true lumen before balloon dilatation and stenting the following methods can be used:

1. **Contralateral injection**. This is the most commonly used method and is crucial for nearly all CTO procedures, even when most collaterals are ipsilateral, because ipsilateral collaterals may become compromised during crossing attempts.[7,8] After CTO crossing it is recommended to confirm intraluminal position of the guidewire in two orthogonal angiographic projections

2. **Contrast injection through a microcatheter**. Routine use of this maneuver is discouraged, since antegrade contrast injection through a microcatheter always entails the risk of extraplaque (subintimal) space "staining" and dissection propagation if the wire is not in the distal true lumen, which can then hinder subsequent reentry attempts. In selected cases, controlled microcatheter tip injections can be performed with care to verify intraluminal location, always checking for the "back bleeding sign" (blood coming out of the microcatheter after waiting for at least 30 s after withdrawing the guidewire).

3. **Intravascular imaging**. Intravascular ultrasound (particularly with a short-tip IVUS probe that is less likely to extend the suspected dissection) can be of great help in establishing an extraplaque wire course (Fig. 25.6).[9] Optical coherence tomography (OCT) has recently been reported as an alternative,[10] but is hampered by the limited penetration of OCT imaging, the distance of the OCT lens from the catheter tip, and the need to perform contrast or dextran[11] injections that may propagate an extraplaque (subintimal) dissection.

4. **Observing the wire movement into distal branches**. This is suggestive of true lumen position,[12] but may also be misleading as the wire (especially polymer-jacketed, knuckled, or stiff tip wires) can also advance extraplaque into side branches. Exchanging for a workhorse guidewire can facilitate the process, as the latter is less likely to advance into side branches and luminal movement is easier to discern.

5. Transduce pressure through the microcatheter: an arterial pressure waveform suggests distal true lumen crossing.

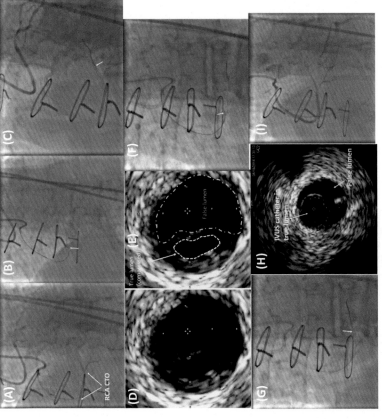

FIGURE 25.6 **Use of intravascular ultrasound to achieve and confirm successful reentry into the distal true lumen** (CTO Manual online case 35). **Panel A:** right coronary artery chronic total occlusion. **Panel B:** an antegrade knuckled guidewire is advanced past the distal cap. **Panel C:** using the Stingray balloon a guide-wire is advanced into the distal right coronary artery (arrow), however it is not 100% clear that it is located within the distal true lumen. **Panels D and E:** intravascular ultrasound demonstrates that the guidewire is into the false lumen, not the distal true lumen. **Panel F:** reentry is attempted again using the Stingray system. **Panel G:** a second guidewire is advanced distally and appears to be in a different location than the first guidewire. **Panel H:** intravascular ultrasound confirms that the second guidewire is within the distal true lumen, whereas the first guidewire remains within the false lumen. **Panel I:** final result after stenting.

Either CTO or donor vessel dissection:

1. Pay close attention to guide catheter position (especially during equipment manipulations) and to the pressure waveform (avoid dampening and promptly correct it if it occurs).
2. Guidewire externalization creates a wire loop with outstanding support for equipment delivery. However, this support comes at the expense of communicating to the contralateral guide catheter any traction during PCI, potentially causing deep guide engagement and vessel injury. During externalization, meticulous attention should be paid to the contralateral catheter at all times.*
3. Achieve and maintain coaxial catheter position.*
4. Do not inject contrast if the pressure waveform is dampened.*
5. Avoid aggressive wiring strategies through nonocclusive lesions; do not use polymer-jacketed or stiff-tip guidewires as workhorses.
6. Avoid oversized balloons and very-high pressure balloon inflations.
7. Prepare calcified lesions prior to stenting.

25.2.1.3 Treatment (* = of special importance for CTO PCI)

1. **Avoid contrast injections**; if they are absolutely needed they should be performed nonselectively with the catheter disengaged from the coronary ostium to avoid propagation of the dissection.*
2. **If wire position is maintained into the distal true lumen: STENT!** (CTO PCI Manual online cases 69, 97, 125)
   a. Deploy a stent promptly, especially in proximal dissections, such as left main dissections, to minimize the duration of ischemia and the risk of cardiac arrest.[13,14]
   b. In cases of dissection of the donor vessel during retrograde CTO PCI, the operator faces the problem of hardware in the donor vessel. Treatment options include:
      i. Withdrawing the retrograde wire, leaving the retrograde microcatheter beyond the acute occlusion, and exchanging for a regular PCI wire for stenting, avoiding the advancement of a new wire through a dissected segment.
      ii. Using the externalized retrograde wire as a platform for stenting the acutely occluded donor vessel (provided it is anatomically feasible, for example in a short, limited dissection)
      iii. Performing PCI with a second wire, jailing the externalized retrograde wire with the stent (CTO PCI Manual online case 22).
      Options ii and iii are preferred if CTO PCI is virtually completed (pending only CTO stenting, for example), while option i is probably the best choice if the CTO has not yet been crossed. Option ii requires great caution during withdrawal of the jailed retrograde wire through the collateral channels (it may require protection with an antegrade microcatheter).

An overall assessment of the patient's safety is mandatory at the time of deciding the best possible choice (for example, contralateral femoral artery access may be required for intra-aortic balloon pump insertion, making option i the preferred option).

c. If a stent cannot be delivered due to significant tortuosity, calcification, or other reasons, balloon angioplasty or using a guide catheter extension may be needed first (CTO PCI Manual online case 171).

d. Sometimes prolonged balloon angioplasty alone may restore antegrade flow but stenting is preferable for preventing vessel reocclusion.*

3. **If there is no wire into the distal true lumen: wire into the distal true lumen then stent the dissected segment**:

a. Workhorse guidewires are preferred as polymer-jacketed guidewires are more likely to enter into the dissection planes (CTO Manual online cases 22, 50, 165).

b. In case of proximal dissection, change the position of the guide catheter to minimize the likelihood of the guidewire entering the dissection plane.

c. Confirm distal true lumen guidewire position with IVUS prior to stenting. Alternatively a microcatheter can be advanced distally and contrast injected, but if the wire is extraplaque (subintimal), contrast injection will enlarge the extraplaque space and make reentry into the distal true lumen more challenging. A safer strategy is to advance the microcatheter distally and then attempt to advance a workhorse guidewire. If the guidewire is in the true lumen it will reach the distal vessel without resistance and will often enter side branches. If the guidewire is in the extraplaque space, significanct resistance will often be encountered when attempting to advance it.*

d. In patients who are tolerating the dissection well, CTO PCI techniques (such as use of the Stingray balloon[15,16] or retrograde crossing[17]) can be used to achieve wire crossing into the distal true lumen (CTO PCI Manual online cases 38, 92, 98).

4. **If there is no wire into the distal true lumen and wiring attempts fail: refer for emergent CABG**. Other options include: (1) STAR with deferred stenting (Section 8.3.5.1); or (2) medical therapy if the occluded vessel is small (such as a diagonal branch) and the patient is tolerating the event well.* STAR should be avoided in the LAD due to risk of multiple side branch occlusion; CABG is preferred in such cases.

5. **If acute vessel closure is due to extraplaque stent deployment, re-entering the distal true lumen can be achieved using several strategies**[18]

a. Stingray system or the ReCross dual lumen microcatheter.[19]

b. Retrograde crossing of the target vessel.[8]

c. Wire-based techniques, such as the STAR (subintimal tracking and reentry) technique[20] the LAST (limited antegrade subintimal tracking) technique, and the mini-STAR[21] technique, in which the area of extraplaque dissection is limited by reentering into the true lumen as close as possible to the distal cap without propagating the dissection into the distal part of the vessel, as described in Section 8.3. However, wire-based reentry is discouraged due to unpredictability and potential for enlarging the dissection; use of the Stingray system or the ReCross microcatheter or the retrograde approach are preferred instead.

    d. IVUS guided reentry (Fig. 25.7).

These same techniques can also be employed to reenter into the true lumen in cases of acute vessel closure due to dissection during non-CTO PCI (Figs. 25.8 and 25.9).[15]

6. **Small, nonflow limiting dissections sometimes do not require treatment**. Compared with IVUS, OCT (Chapter 13) is more sensitive in identifying dissections.[22] Proximal edge dissections after stenting often do not require treatment, as the stent prevents forward propagation of the dissection.[22]

25.2.1.4 Aortocoronary dissection

CTO Manual online cases 10, 112, 118

Aortocoronary dissection is a rare complication that can occur with any PCI, but is more common with CTO PCI (especially retrograde procedures) (the incidence was 0.8% to 1.8% in two series[23,24]) and most commonly occurs in the right coronary artery[25] (Fig. 25.10).[2] The dissection may be limited to the coronary sinus, but may extend to the proximal ascending aorta or even beyond the ascending aorta.[26]

Causes[24]:

1. Preexisting disease of the vessel ostium.
2. Contrast injection despite dampened pressure waveform.
3. Noncoaxial guide catheter position.
4. Deep coronary engagement and utilization of aggressive guide catheters, such as 8 French Amplatz catheters.
5. Excessive guide catheter movement that is more likely to occur with guidewire externalization during the retrograde approach.
6. Predilatation of the coronary ostium.
7. Balloon rupture.
8. Retrograde wire advancement into the extraplaque and subaortic space during retrograde crossing attempts.
9. Antegrade attempts to recanalize aorto-ostial CTOs.

Prevention:

1. Use anchor techniques (Section 9.3.1.1) as an alternative to aggressive guide catheter intubation to enhance guide catheter support.

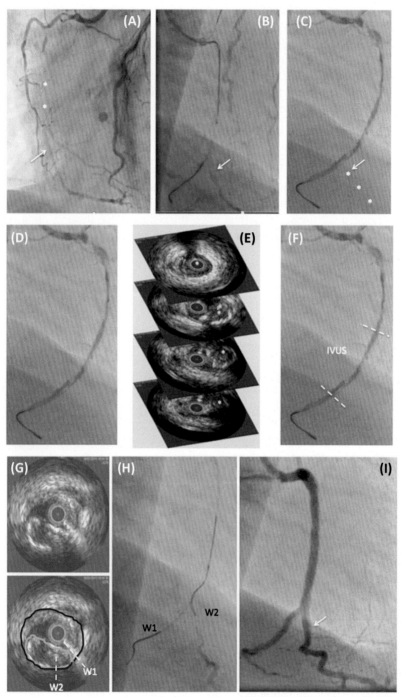

(Continued)

2. When using guide extensions, avoid delivery into the coronary artery directly over the guidewire. Delivery over a balloon, for example using the "inchworming" technique (Section 30.2.3) is less likely to injure the vessel. Use caution when injecting through a guide catheter extension, as there is increased risk of hydraulic vessel dissection that can propagate backwards causing aortocoronary dissection.
3. Use of guide catheters with side holes in occluded right coronary arteries may decrease the risk of vessel injury, but may also provide a false sense of security, as the pressure waveform may appear normal, but ischemia and aortocoronary dissection can still occur.
4. Avoid power injectors when treating the proximal segment of a CTO; manual injections and contralateral injections from the donor vessel are preferred.

Treatment:

1. *STOP INJECTING* contrast into the coronary artery (as injections can expand the dissection).
2. *Stent the ostium* of the dissected coronary artery with a stent that can expand to a diameter that will seal the dissection. The stent should protrude 1−2 mm into the aorta to fully cover the ostium of the dissected vessel. Avoid use of too short stents that may result in stent embolization, since a portion of the stent will not be secured in the vessel.
3. Use IVUS to guide stent placement and ensure complete ostial coverage.[27]
4. If contrast injection is considered to be absolutely essential for checking the status of the distal vessel, it should ideally be performed through a dual lumen microcatheter or an aspiration thrombectomy catheter (that allows better vessel filling, but is bulkier and harder to deliver) advanced into the distal vessel.[28]
5. If the aortocoronary dissection is large, perform serial noninvasive imaging (with computed tomography or transesophageal echocardiography) to ensure that the dissection has stabilized and/or resolved (Fig. 25.11).[24] This is of

◀ FIGURE 25.7   **Use of intravascular ultrasonography (IVUS) to facilitate side branch occlusion assessment and treatment**. Antegrade PCI was planned in a first attempt to recanalize a long CTO located in the mid segment of the RCA (asterisks, **panel A**). Adequate progress was made using the parallel wire technique with a polymer-jacketed guidewire and a blunt tip coil wire down to the posterolateral branch (**Panel B**). However, antegrade injections after predilatation with a 1.5 mm balloon revealed occlusion of the posterior descending artery (asterisks, **panel C**). Intravascular ultrasound of the mid RCA and crux (**panels E and F**) revealed subadventitial course of the wire located in the posterolateral artery with compression of the vessel structures at the level of the RCA crux (stars in IVUS shown in **panel E**). IVUS-guided reentry to the PDA with a new wire (W2) was performed (IVUS shadows of both guidewires are shown in **panel G**). This allowed successful advancement of W2 into the PDA (**panel H**) with a good result using a provisional stenting technique. *Courtesy of Dr. Javier Escaned.*

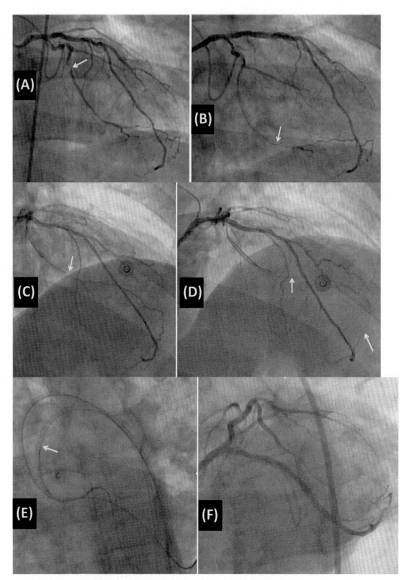

FIGURE 25.8   **Use of the retrograde approach to treat acute vessel closure** (CTO Manual online case 38). **Panel A**: severe lesion of the second obtuse marginal branch (arrow). **Panel B**: extraplaque guidewire crossing (arrow). **Panel C**: acute vessel closure (arrow). **Panel D**: successful retrograde crossing into the second obtuse marginal branch via an epicardial collateral from the distal left anterior descending artery (arrows). **Panel E**: after reverse CART was performed the retrograde guidewire and the Corsair catheter was inserted into a second guide catheter (arrow), using the ping-pong guide catheter technique. **Panel F**: successful recanalization of the second obtuse marginal branch after stenting with loss of a side branch proximal to the original stenosis.

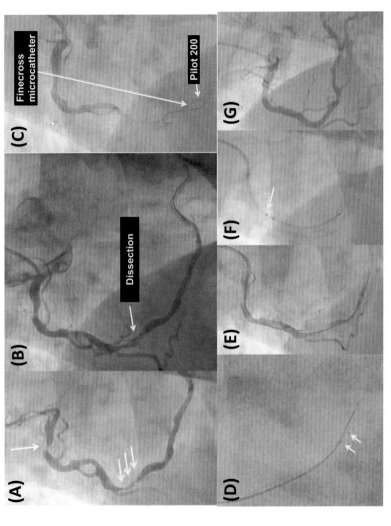

FIGURE 25.9 **Use of antegrade dissection/reentry to treat acute vessel closure.** Coronary angiography demonstrating a tortuous right coronary artery with a proximal (arrow, **panel A**) and mid (multiple arrows, **panel A**) lesions. Mid right coronary artery dissection after balloon predilatation (arrow, **panel B**). Guidewire position and antegrade flow were lost after an unsuccessful attempt for stent delivery. After failure to advance a guidewire through the dissected segment, a knuckle was formed with a Pilot 200 guidewire (Abbott Vascular) (arrow, **panel C**) and advanced around the dissected segment. Using a Stingray balloon (Boston Scientific, arrows, **panel D**) and guidewire, distal true lumen reentry was achieved (**panel D**). Using a Guideliner catheter (Teleflex) (arrow, **panel F**) two stents were successfully delivered with an excellent final angiographic result (**panel G**). *Reproduced with permission from Martinez-Rumayor AA, Banerjee S, Brilakis ES. Knuckle wire and stingray balloon for recrossing a coronary dissection after loss of guidewire position. JACC Cardiovasc Interv 2012;5:e31−e32.[15]*

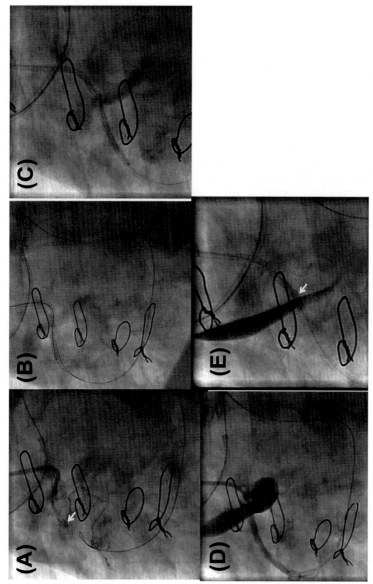

FIGURE 25.10 **Illustration of an aortocoronary dissection during a retrograde CTO intervention.** A retrograde CTO intervention was performed to recanalize a proximal right coronary artery CTO (arrow, **panel A**), using the reverse controlled antegrade and retrograde tracking (CART) technique (**panel B**). Staining of the aortocoronary junction was observed with test injections during stent placement (**panel C**), that expanded with cine angiography (**panel D**). Stenting of the right coronary artery ostium was performed (arrow, **panel E**) without further antegrade contrast injections. The patient had an uneventful recovery. This case illustrates the importance of stopping antegrade contrast injections and stenting the vessel ostium if aortocoronary dissection occurs, in order to seal the dissection flap at the entry point of the dissection. *Courtesy of Dr. Parag Doshi. Reproduced with permission from the Brilakis ES. Manual of coronary chronic total occlusion interventions. A step-by-step approach. 2nd ed. Elsevier; 2017.*

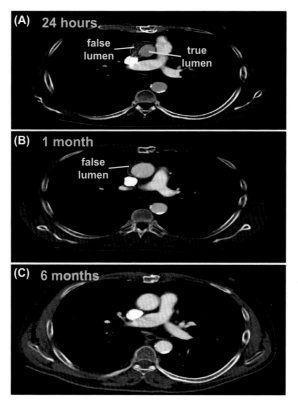

FIGURE 25.11 **Computed tomography (CT) followup of an aortocoronary dissection.**
Panel A: twenty-four hours after the procedure, CT angiogram examination showing dislocation
of intimal calcification with an eccentric double lumen. **Panel B**: one-month CT control examination demonstrating almost total resolution of the dissected thrombosed lumen. **Panel C**: six-
month CT control examination demonstrating total resolution of the dissection. *Reproduced with
permission from Boukhris M, Tomasello SD, Marza F, Azzarelli S, Galassi AR. Iatrogenic aortic
dissection complicating percutaneous coronary intervention for chronic total occlusion.* Can J
Cardiol *2015;31:320−327.*

particular importance if the dissection involves the ascending aorta. The
blood pressure should be carefully controlled after the procedure.
6. Emergency surgery is rarely needed, except in patients who develop aortic regurgitation, tamponade due to rupture into the pericardium, or
   extension of the dissection (>40 mm from the coronary ostia has been
   proposed as a cutoff[29]) (Fig. 25.12).[23−25,30] The surgical outcomes of
   patients with iatrogenic aortic dissections are better than the outcomes of
   patients with spontaneous aortic dissections.[31]

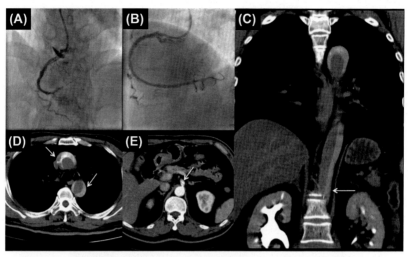

FIGURE 25.12 **Example of aortocoronary dissection extending into the descending aorta after CTO intervention**. Angiography demonstrating proximal long segment dissection of the right coronary artery, extending to sinus of Valsalva (**panels A and B**). After stenting with a 3.5 × 24 mm bare metal stent, the final angiogram revealed limited dissection to the sinus of Valsalva (**panel B**). Computed tomography imaging demonstrated a type A aortic dissection extending from the ascending aorta to the suprarenal abdominal level (**panels C and D**) with involvement of the aortic arch and celiac trunk (**panel E**). *Reproduced with permission from Liao MT, Liu SC, Lee JK, Chiang FT, Wu CK. Aortocoronary dissection with extension to the suprarenal abdominal aorta: a rare complication after percutaneous coronary intervention. JACC Cardiovasc Interv 2012;5:1292–1293.*

25.2.2 Thrombosis

25.2.2.1 Causes

1. Inadequate anticoagulant and antiplatelet treatment.
2. Suboptimal lesion treatment (such as stent underexpansion).
3. Hypercoagulable state, including heparin-induced thrombocytopenia.
4. Venous access failure resulting in inadequate intravenous administration of the anticoagulation agent.

25.2.2.2 Prevention

1. Maintain therapeutic ACT (>300 s for antegrade and >350 s for retrograde CTO PCI). The ACT should be checked every 20–30 min, which is more easily accomplished by inserting a small venous sheath and delegating this task to a nurse, with a clear message that any drop in ACT below the prespecified safety level should be communicated and corrected with additional boluses of unfractionated heparin.
2. Flush regularly all guide and diagnostic catheters to prevent in-catheter thrombosis. Back-bleed the guide catheter after performing balloon trapping to minimize the risk of air embolization.

3. Avoid bivalirudin.
4. Meticulous lesion preparation to achieve excellent stent expansion.
5. Use intravascular imaging to confirm excellent CTO PCI result.

25.2.2.3   Treatment

1. Ensure optimal anticoagulation and antiplatelet therapy is administered.
2. Thrombectomy may be needed in cases of large thrombus burden.
3. Occasionally, stenting may trap thrombus and restore antegrade flow.

25.2.3   Embolization

CTO Manual online case 19

Embolization, if severe, can cause profound hemodynamic compromise and require prompt treatment. Embolization may be due to thrombus, plaque/debris, and air.

25.2.3.1   Thrombus embolization

25.2.3.1.1   Causes

Causes are the same as described in Section 25.2.2.1.
Embolization can also occur if thrombus develops in the arterial sheath and is then picked up by catheters advanced through the sheath.

25.2.3.1.2   Prevention

Prevention is as described in Section 25.2.2.2.
Routine aspiration, followed by flushing (using a different syringe) of the arterial sheath prior to inserting a catheter is important, as is periodic aspiration and flushing of the guide catheters.

25.2.3.1.3   Treatment

Treatment is as described in Section 25.2.2.3. Distal thrombectomy may be needed for removal of the embolized thrombus.

25.2.3.2   Plaque embolization

25.2.3.2.1   Causes

1. Atherectomy (that is performed in about 3% of CTO PCI cases).
2. Lipid-rich plaque.[32]
3. Treatment of saphenous vein graft lesions (Chapter 18).
4. Inadequate cleaning of the catheter after advancement through the aorta, resulting in aortic or iliac plaque embolization.
5. Aspiration thrombectomy or initial balloon predilation of a culprit lesion in STEMI.

25.2.3.2.2 Prevention

1. Avoidance of atherectomy; if atherectomy is needed consider using orbital atherectomy or lower speed rotational atherectomy (140,000−150,000 rpm) that may be associated with lower risk of distal embolization and no reflow.
2. Use of embolic protection devices for saphenous vein graft lesions.
3. Thoroughly clear catheter after advancement through the aorta before performing contrast injection.
4. Vasodilators (to facilitate passage of debris through the microcirculation).

25.2.3.2.3 Treatment

1. Stenting should be avoided in cases of plaque embolization, as it can result in release of more plaque into the coronary circulation and worsening of the no reflow phenomenon. Covered stents may prevent distal embolization but carry high risk of restenosis and stent thrombosis.
2. Aspiration thrombectomy can be performed to remove thrombus and remaining debris. (CTO manual online case 19).
3. If aspiration fails, laser could be used.
4. Vasodilators can be used to facilitate passage of any remaining debris through the microcirculation. Vasodilators should ideally be administered through a microcatheter or aspiration catheter to ensure that the maximum effect is achieved in the embolized vessel.

25.2.3.3 Air embolization

Air embolization can be a dramatic event. Large air embolization can potentially lead to cardiac arrest[33] or death. CTO PCI carries increased risk of air embolization given often prolonged procedure duration and repeated use of the trapping technique.

25.2.3.3.1 Causes

1. Suboptimal manifold (or automated injector device) preparation.
2. Poor connections of the manifold.
3. Pressure dampening due to ostial disease or because the tip of the catheter is against the aortic wall.
4. Use of large devices that may entrain air into the guide catheter.
5. *Use of the trapping technique without back bleeding the catheter afterwards.*
6. Rupture of a poorly prepared balloon that contains air.

25.2.3.3.2 Prevention

1. Meticulous preparation of the manifold (or automated injector device).
2. Meticulous clearing of the manifold after advancement through the aorta.

3. *Always back bleed the hemostatic connector after using the trapping technique or after multiple equipment insertions.*
4. Meticulous preparation of balloons to remove all air.
5. Meticulous preparation of other large devices inserted through the guide catheter.

25.2.3.3.3 Treatment

1. Administer 100% oxygen using a nonrebreather mask (helps with resorption of the air).
2. Advance a coronary workhorse wire to the distal vessel and into different side branches. The tip of the wire can rupture any remaining bubbles in the distal circulation.
3. Aspirate through the guide catheter,[34] through a guide catheter extension, or through an aspiration thrombectomy catheter[35,36] (if a large amount of air has embolized).
4. Intracoronary epinephrine if the patient develops cardiac arrest.[33]
5. Hemodynamic deterioration can be very rapid, requiring rapid intervention such as mechanical circulatory support.
6. Large systemic air embolization could be treated by hyperbaric chamber, whereas placing the patient in Trendelenburg position is controversial as it may increase cerebral edema.[37,38]

25.2.4 Side branch occlusion after stenting

Side branch occlusion (Chapter 16) can occur during CTO PCI, especially when dissection/reentry strategies are used, and is associated with higher frequency of post PCI myocardial infarction.[18,25]

25.2.4.1 Causes

1. Use of dissection/reentry strategies in vessels with side branches at the proximal or distal CTO cap.
2. Stenting a diseased segment of the main vessel may cause plaque shift and acute side branch occlusion, especially if there is preexisting disease in the side branch.

25.2.4.2 Prevention

1. When treating CTOs that involve a bifurcation (for example, those involving the right coronary artery crux), a careful analysis of collateral support should be performed before the procedure, to determine whether both branches have independent collateral support.
2. Avoid use of (antegrade or retrograde) dissection/reentry strategies when a bifurcation is present at the proximal or distal CTO cap.
3. Whenever possible, protect the side branch with a second guidewire.

25.2.4.3 Treatment

1. Antegrade wiring of the occluded branch (which may be challenging if dissection/reentry strategies were used for crossing). This may require use of a guidewire with higher tip load, such as the Gaia, Gaia Next, or Judo family of wires.
2. Retrograde recanalization of the occluded branch (if collaterals to that branch exist).
3. Use of IVUS may facilitate the identification of the cause of side branch occlusion (for example, the presence of an extraplaque track at the level of the side branch ostium) and rewiring (Fig. 25.7).
4. The "side power-knuckle" technique (Fig. 25.13) that is performed as follows: (1) the tip of a microcatheter is advanced to the side branch origin, as identified on IVUS; (2) a balloon sized 1:1 with the main vessel is inflated in the main vessel across the side branch ostium; and (3) a nontapered stiff polymer-jacketed wire is advanced into the side branch.

If a dissection/reentry strategy is used, minimize the extent of the extraplaque dissection by reentering into the true lumen at the most proximal location possible, usually using the Stingray system, as described in Section 8.2.[40] Moreover, using the CrossBoss catheter may minimize the extent of dissection and facilitate reentry attempts. The presence of a coronary bifurcation at the distal CTO cap may favor use of a primary retrograde approach that often has lower risk of side branch occlusion as compared with antegrade crossing attempts.

Side Power Knuckle technique (SPK)

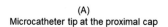

(A)
Microcatheter tip at the proximal cap

(B)
Inflation of 1:1 sized balloon in the main vessel and advancing a stiff polymer-jacked knuckled wire into the occlusion

FIGURE 25.13 **Illustration of the side-power-knuckle technique.**[39] *Reproduced with permission from Megaly M, Basir MB, Brilakis E, Alaswad K. Side power knuckle and antegrade-antegrade dissection re-entry: techniques to overcome difficulties in chronic occlusion revascularization. JACC Cardiovasc Interv 2022;15:e13—e15.*

25.2.5   Spasm

Spasm is infrequently the cause of acute complete flow cessation, but can contribute in multiple other settings of acute vessel occlusion, such as in the setting of dissection and distal embolization.

25.2.5.1   Causes

1. Deep guide catheter engagement.
2. Aggressive guide catheter manipulations.
3. Coronary artery instrumentation (wiring, ballooning, stenting).

25.2.5.2   Prevention

1. Avoid deep engagement and aggressive catheter manipulations.
2. Administer nitroglycerine after coronary engagement, before performing coronary angiography.

25.2.5.3   Treatment

1. Intracoronary nitroglycerine (usual dose is 100−300 mcg; may need to repeat administration).
2. Intracoronary calcium channel blockers.

25.2.6   Pseudolesions

25.2.6.1   Causes

1. Catheter and wire insertion through highly tortuous vessels, such as LIMA grafts.

25.2.6.2   Prevention

1. Avoid instrumentation of highly tortuous vessels.
2. Use soft body guidewires and microcatheters when treating highly tortuous vessels.

25.2.6.3   Treatment

1. Remove equipment from the tortuous vessel. Prior to removal, ensure that no dissection has occurred (for example by using intravascular ultrasound).
2. Withdrawing the guidewire until only the soft distal portion remains across the lesion can restore the vessel anatomy, allowing the differentiation of pseudolesions while maintaining distal wire position.
3. Exchange stiff guidewires for softer body guidewires using a microcatheter.

25.2.7   Equipment entrapment

Equipment loss and entrapment is discussed in detail in Chapter 27.

25.2.8   Intramural hematoma

Post-PCI intramural hematomas can cause extensive compression of the true lumen leading to acute vessel closure.[41]

25.2.8.1   Causes

1. Spontaneous coronary artery dissection.
2. Complication of balloon angioplasty or stenting.

25.2.8.2   Prevention

1. Proper balloon and stent sizing.
2. Avoid very high stent deployment pressures.
3. Avoid protrusion of the postdilatation balloon from the proximal and distal stent edge.

25.2.8.3   Treatment

1. Stent distally first (to prevent distal propagation of the hematoma) and then more proximally.
2. Alternatively, a cutting balloon can be used to decompress the hematoma (Fig. 25.14).[42,43]
3. Small hematomas may be treated conservatively, especially if the inciting tear has already been sealed / stented.

25.3   Hemodynamic support

CTO Manual online case 165

Acute closure of a major coronary vessel can rapidly lead to arrhythmias and hemodynamic compromise, potentially necessitating the use of hemodynamic support, as described in Chapter 14.

In case of acute vessel closure, anticipate hemodynamic collapse. Notify medical staff to prepare a hemodynamic support device, such as intra-aortic balloon pump, Impella CP, or VA-ECMO, ensure femoral artery access (in radial procedures), prepare drugs, etc., while you concentrate in re-opening the vessel. Also contact cardiac surgery, in case percutaneous treatment fails and emergency surgery is required. Having a second operator can greatly facilitate management of acute vessel closure.

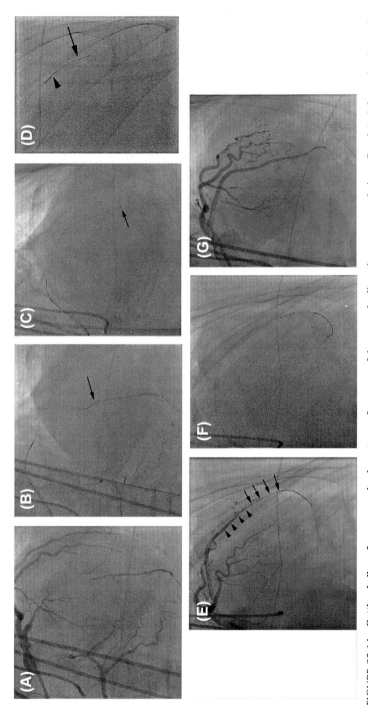

FIGURE 25.14 Cutting balloon for compressive hematoma after successful antegrade dissection reentry technique. Panel A: left anterior descending (LAD) chronic total occlusion (CTO). **Panel B:** Pilot 200 guidewire knuckle in the extraplaque space (arrow). **Panel C:** CrossBoss catheter advanced in the extraplaque space (arrow). **Panel D:** Stingray balloon (arrow head) assisted Pilot 200 (arrow) reentry into the true lumen. **Panel E:** compressive hematoma (arrows) after stenting of LAD (arrow heads). **Panel F:** 2.25 × 10 mm cutting balloon to a compressed segment of the LAD. **Panel G:** resolution of compressive hematoma after cutting balloon inflation at two affected segments of LAD. *Reproduced with permission from Vo MN, Brilakis ES, Grantham JA. Novel use of cutting balloon to treat subintimal hematomas during chronic total occlusion interventions. Catheter Cardiovasc Interv 2018;91:53–56.*

References

1. Kosugi S, Shinouchi K, Ueda Y, et al. Clinical and angiographic features of patients with out-of-hospital cardiac arrest and acute myocardial infarction. *J Am Coll Cardiol* 2020;**76**:1934−43.

2. Brilakis ES. *Manual of percutaneous coronary interventions: a step-by-step approach.* Elsevier; 2021.

3. Patel VG, Banerjee S, Brilakis ES. Treatment of inadvertent subintimal stenting during intervention of a coronary chronic total occlusion. *Interv Cardiol* 2013;**5**:165−9.

4. Omurlu K, Ozeke O. Side-by-side false and true lumen stenting for recanalization of the chronically occluded right coronary artery. *Heart Vessel* 2008;**23**:282−5.

5. Krivonyak GS, Warren SG. Compression of a subintimal or false lumen stent by stenting in the true lumen. *J Invasive Cardiol* 2001;**13**:698−701.

6. Ge JB, Zhang F, Ge L, Qian JY, Wang H. Wire trapping technique combined with retrograde approach for recanalization of chronic total occlusion. *Chin Med J (Engl)* 2008;**121**:1753−6.

7. Singh M, Bell MR, Berger PB, Holmes Jr. DR. Utility of bilateral coronary injections during complex coronary angioplasty. *J Invasive Cardiol* 1999;**11**:70−4.

8. Brilakis ES, Grantham JA, Rinfret S, et al. A percutaneous treatment algorithm for crossing coronary chronic total occlusions. *JACC Cardiovasc Interv* 2012;**5**:367−79.

9. Banerjee S, Master R, Brilakis ES. Intravascular ultrasound-guided true lumen re-entry for successful recanalization of chronic total occlusions. *J Invasive Cardiol* 2010;**22**:608−10.

10. Schultz C, van der Ent M, Serruys PW, Regar E. Optical coherence tomography to guide treatment of chronic occlusions? *J Am Coll Cardiol Interv* 2009;**2**:366−7.

11. Frick K, Michael TT, Alomar M, et al. Low molecular weight dextran provides similar optical coherence tomography coronary imaging compared to radiographic contrast media. *Catheter Cardiovasc Interv* 2014;**84**:727−31.

12. Hussain F. Distal side branch entry technique to accomplish recanalization of a complex and heavily calcified chronic total occlusion. *J Invasive Cardiol* 2007;**19**:E340−2.

13. Abdel-Karim AR, Gadiparthi C, Banerjee S, Brilakis ES. Catastrophic left main coronary artery occlusion following diagnostic coronary angiography: salvage by emergency left main coronary artery stenting. *Acute Card Care* 2011;**13**:170−3.

14. Koza Y, Tas H, Sarac I. Successful management of an iatrogenic left main coronary artery occlusion during coronary angiography: a case report and brief review. *Cardiovasc Revasc Med* 2019;**20**:432−5.

15. Martinez-Rumayor AA, Banerjee S, Brilakis ES. Knuckle wire and stingray balloon for recrossing a coronary dissection after loss of guidewire position. *JACC Cardiovasc Interv* 2012;**5**:e31−2.

16. Shaukat A, Mooney M, Burke MN, Brilakis ES. Use of chronic total occlusion percutaneous coronary intervention techniques for treating acute vessel closure. *Catheter Cardiovasc Interv* 2018;**92**:1297−300.

17. Kotsia A, Banerjee S, Brilakis ES. Acute vessel closure salvaged by use of the retrograde approach. *Interv Cardiol* 2014;**6**:145−7.

18. Michael TT, Papayannis AC, Banerjee S, Brilakis ES. Subintimal dissection/reentry strategies in coronary chronic total occlusion interventions. *Circ Cardiovasc Interv* 2012;**5**:729−38.

19. Wosik J, Shorrock D, Christopoulos G, et al. Systematic review of the bridgepoint system for crossing coronary and peripheral chronic total occlusions. *J Invasive Cardiol* 2015;**27**:269−76.

20. Colombo A, Mikhail GW, Michev I, et al. Treating chronic total occlusions using subintimal tracking and reentry: the STAR technique. *Catheter Cardiovasc Interv* 2005;**64**:407−11.

21. Galassi AR, Tomasello SD, Costanzo L, et al. Mini-STAR as bail-out strategy for percutaneous coronary intervention of chronic total occlusion. *Catheter Cardiovasc Interv* 2012;**79**:30−40.

22. Prati F, Romagnoli E, Burzotta F, et al. Clinical Impact of OCT findings during PCI: the CLI-OPCI II study. *JACC Cardiovas Imag* 2015;**8**:1297−305.

23. Shorrock D, Michael TT, Patel V, et al. Frequency and outcomes of aortocoronary dissection during percutaneous coronary intervention of chronic total occlusions: a case series and systematic review of the literature. *Catheter Cardiovasc Interv* 2014;**84**:670−5.

24. Boukhris M, Tomasello SD, Marza F, Azzarelli S, Galassi AR. Iatrogenic aortic dissection complicating percutaneous coronary intervention for chronic total occlusion. *Can J Cardiol* 2015;**31**:320−7.

25. Carstensen S, Ward MR. Iatrogenic aortocoronary dissection: the case for immediate aortoostial stenting. *Heart Lung Circ* 2008;**17**:325−9.

26. Gomez-Moreno S, Sabate M, Jimenez-Quevedo P, et al. Iatrogenic dissection of the ascending aorta following heart catheterisation: incidence, management and outcome. *EuroIntervention* 2006;**2**:197−202.

27. Abdou SM, Wu CJ. Treatment of aortocoronary dissection complicating anomalous origin right coronary artery and chronic total intervention with intravascular ultrasound guided stenting. *Catheter Cardiovasc Interv* 2011;**78**:914−19.

28. Al Salti Al Krad H, Kaminsky B, Brilakis ES. Use of a thrombectomy catheter for contrast injection: a novel technique for preventing extension of an aortocoronary dissection during the retrograde approach to a chronic total occlusion. *J Invasive Cardiol* 2014;**26**:E54−5.

29. Dunning DW, Kahn JK, Hawkins ET, O'Neill WW. Iatrogenic coronary artery dissections extending into and involving the aortic root. *Catheter Cardiovasc Interv* 2000;**51**:387−93.

30. Liao MT, Liu SC, Lee JK, Chiang FT, Wu CK. Aortocoronary dissection with extension to the suprarenal abdominal aorta: a rare complication after percutaneous coronary intervention. *JACC Cardiovasc Interv* 2012;**5**:1292−3.

31. Rylski B, Hoffmann I, Beyersdorf F, et al. Iatrogenic acute aortic dissection type A: insight from the German Registry for Acute Aortic Dissection Type A (GERAADA). *Eur J Cardiothorac Surg* 2013;**44**:353−9.

32. Papayannis AC, Abdel-Karim AR, Mahmood A, et al. Association of coronary lipid core plaque with intrastent thrombus formation: a near-infrared spectroscopy and optical coherence tomography study. *Catheter Cardiovasc Interv* 2013;**81**:488−93.

33. Prasad A, Banerjee S, Brilakis ES. Images in cardiovascular medicine. Hemodynamic consequences of massive coronary air embolism. *Circulation* 2007;**115**:e51−3.

34. Sinha SK, Madaan A, Thakur R, Pandey U, Bhagat K, Punia S. Massive coronary air embolism treated successfully by simple aspiration by guiding catheter. *Cardiol Res* 2015;**6**:236−8.

35. Patterson MS, Kiemeneij F. Coronary air embolism treated with aspiration catheter. *Heart* 2005;**91**:e36.

36. Yew KL, Razali F. Massive coronary air embolism successfully treated with intracoronary catheter aspiration and intracoronary adenosine. *Int J Cardiol* 2015;**188**:56−7.

37. Shaikh N, Ummunisa F. Acute management of vascular air embolism. *J Emerg Trauma Shock* 2009;**2**:180−5.

38. McCarthy CJ, Behravesh S, Naidu SG, Oklu R. Air embolism: practical tips for prevention and treatment. *J Clin Med* 2016;5.

39. Megaly M, Basir MB, Brilakis E, Alaswad K. Side power knuckle and antegrade-antegrade dissection re-entry: techniques to overcome difficulties in chronic occlusion revascularization. *JACC Cardiovasc Interv* 2022;**15**:e13−15.

40. Lombardi WL. Retrograde PCI: what will they think of next? *J Invasive Cardiol* 2009;**21**:543.

41. Maehara A, Mintz GS, Bui AB, et al. Incidence, morphology, angiographic findings, and outcomes of intramural hematomas after percutaneous coronary interventions: an intravascular ultrasound study. *Circulation* 2002;**105**:2037−42.

42. Vo MN, Brilakis ES, Grantham JA. Novel use of cutting balloon to treat subintimal hematomas during chronic total occlusion interventions. *Catheter Cardiovasc Interv* 2018;**91**:53−6.

43. Alsanjari O, Myat A, Cockburn J, Karamasis GV, Hildick-Smith D, Kalogeropoulos AS. A case of an obstructive intramural haematoma during percutaneous coronary intervention successfully treated with intima microfenestrations utilising a cutting balloon inflation technique. *Case Rep Cardiol* 2018;**2018**:4875041.

Chapter 26

Perforation

CTO PCI Manual online cases: 3, 13, 17, 26, 39, 40, 41, 42, 63, 89, 90, 112, 118, 119, 125, 131, 134, 139, 140, 146, 167, 169, 171, 172, 173, 174, 176, 177, 188, 189, 193, 194, 200

PCI Manual online cases: 6, 16, 20, 45, 55, 124, 133

Coronary perforation is one of the most feared complications of chronic total occlusion (CTO) percutaneous coronary intervention (PCI).[1,2] Coronary perforation is more common in CTO than in non-CTO PCI for many reasons[3]:

1. During CTO PCI equipment is advanced through occluded (and often long) coronary segments that are poorly or not visualized. That is why dual injection (Chapter 6) is critical to minimize the risk of perforation during CTO PCI.
2. CTO PCI commonly uses polymer-jacketed (Section 30.6.2) and high tip stiffness guidewires (Section 30.6.3).
3. CTOs often have proximal cap ambiguity.
4. During retrograde CTO PCI equipment is advanced through often small size and tortuous collaterals that may rupture. However, 61% of perforations during retrograde CTO PCI in the OPEN CTO registry were actually antegrade.[4]
5. Use of dissection and reentry techniques may weaken the vessel wall, increasing susceptibility to perforation with subsequent interventional procedures.
6. CTOs often have high complexity: calcification, tortuosity, bifurcations, etc.

Although coronary perforations are common in CTO PCI,[4] many perforations (especially wire perforations) do not have serious consequences. Although coronary perforations requiring an interventional treatment occur in approximately 5% to 8% of CTO PCI procedures[5−8] the risk of tamponade is low (approximately 0.3%[9]) although it is higher with retrograde CTO PCI (1.3%).[10,11] In contrast to PCI of non-CTO vessels, occlusion of a perforated target vessel in CTO PCI usually does not cause myocardial ischemia, which allows testing sequential strategies, preparing hardware, etc.

Because pericardial effusions from coronary perforations may accumulate rapidly, they often cause hypotension, sometimes necessitating emergency pericardiocentesis (and rarely cardiac surgery). Sometimes perforation may not lead to classic tamponade, but instead create a loculated effusion (especially in prior coronary artery bypass graft surgery (CABG) patients, Section 26.6)[12−16] causing compression of cardiac chambers, intramyocardial hematoma,[17] or intracavitary bleeding.[18]

Manual of Chronic Total Occlusion Percutaneous Coronary Interventions.
DOI: https://doi.org/10.1016/B978-0-323-91787-2.00019-8

563

26.1 Perforation classification

Based on CTO-ARC, each perforation is given two descriptors: the first descriptor denotes the location of the perforation, and the second descriptor denotes its severity.[19]

There are three main perforation locations with important implications regarding management: (1) large vessel perforation, (2) distal vessel perforation, and (3) collateral vessel perforation, in either a septal or an epicardial collateral (Figs. 26.1 and 26.2).[20–22] Most coronary perforations are large vessel perforations, followed by distal vessel perforations.[23]

The severity of coronary perforations has traditionally been graded using the Ellis classification (Fig. 26.1)[24]:

- Class I: a crater extending outside the lumen only, in the absence of linear staining angiographically suggestive of dissection.
- Class II: Pericardial or myocardial blush without a $\geq$ 1 mm exit hole.
- Class III: Frank streaming of contrast through a $\geq$ 1 mm exit hole.
- Class III-cavity spilling: Perforation into an anatomic cavity chamber, such as the coronary sinus, the right ventricle, the left ventricle, etc. Some of these perforations, such as perforations into the left ventricle (**PCI Manual online case 16**), do not require treatment.

There is two scores for predicting the risk of coronary perforation during CTO PCI, the OPEN-CLEAN perforation score (Section 6.3.2) and the PROGRESS-

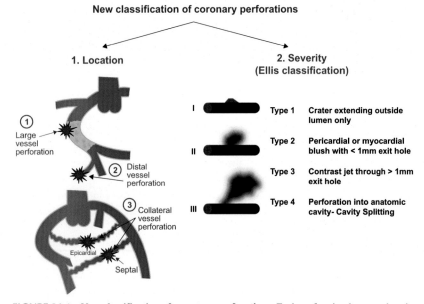

FIGURE 26.1 **New classification of coronary perforations**. Each perforation has two descriptors, the first one denoting the location of the perforation and the second one denoting the severity of the perforation.

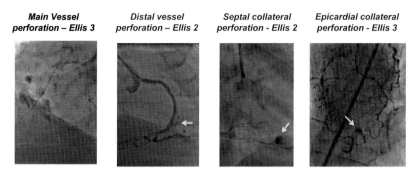

FIGURE 26.2. **Examples of coronary perforations using the new perforation classification.** *Reproduced with permission from Ybarra LF, Rinfret S, Brilakis ES, Karmpaliotis D, Azzalini L, Grantham JA, et al. Definitions and clinical trial design principles for coronary artery chronic total occlusion therapies: CTO-ARC consensus recommendations. Circulation. 2021;143:479—500.*

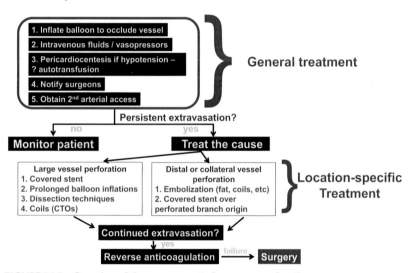

FIGURE 26.3 **Overview of the management of coronary perforations.**

CTO perforation score.[25] The OPEN-CLEAN score is computed from five parameters: prior coronary artery bypass graft (CABG), occlusion length, ejection fraction (EF), age and severe calcification. The score range is 0—7 and higher score is associated with higher risk of perforation. Patients with OPEN-CLEAN score >4 have >8% risk of perforation and should ideally be treated at centers experienced in CTO PCI.[26] The PROGRESS-CTO perforation score is computed from five parameters: age, moderate/severe calcification, blunt/no stump, use of antegrade dissection and re-entry, and use of the retrograde approach.

26.2   General treatment of perforations

(Fig. 26.3)[27]

The following five general measures apply to the treatment of all perforation types (treatments specific to each perforation location are described in the following sections):

1. **Balloon inflation** proximal to or at the site of the perforation to stop the bleeding.
   - This is the first step in the management of all perforations and should be performed **immediately** after recognition of the perforation to prevent accelerated accumulation of blood in the pericardial space and cardiac tamponade.
   - The balloon should be same size as the vessel and inflated at low pressure (usually $<8-10$ atm) to ensure occlusion of antegrade flow, without over stretching the vessel and expanding the perforation site. Longer balloons may increase the likelihood of sealing.
   - Perform coronary angiography to confirm cessation of bleeding through the perforation site. Sometimes the balloon is undersized and needs to be changed for a larger balloon. During retrograde crossing, it is possible to have bleeding through retrograde flow, hence bilateral angiography is needed to confirm hemostasis.
   - Balloon inflation is often prolonged, sometimes lasting $>30-60$ min. Prolonged balloon inflations can cause ischemia, especially when the perforation is in the CTO donor vessel, but are usually well tolerated in the CTO vessel (Fig. 26.4).
   - In some cases, prolonged balloon inflation may achieve sealing of the perforation (particularly if the perforation is less severe such as Ellis class I or II) (**CTO Manual online case 17, 40**).
2. Administration of **intravenous fluids and vasopressors** (and atropine if the patient develops bradycardia due to a vagal reaction).
3. **Pericardiocentesis**
   Video 26.5. How to do pericardiocentesis.
   - Pericardiocentesis should be promptly performed if the patient develops tamponade. However, prompt diagnosis of the perforation along with balloon inflation significantly reduces the volume of blood that enters the pericardium, decreasing the likelihood of developing tamponade.
   - A pericardiocentesis tray should be readily available to expedite treatment.
   - Appropriate timing for performing pericardiocentesis: hemodynamic instability requires immediate pericardiocentesis, yet smaller size pericardial effusions may be best managed conservatively, as the elevated pericardial pressure due to entrance of blood into the pericardial space may externally compress the perforation site and reduce the risk of further bleeding.
   - Carefully observe the arterial waveform to detect **pulsus paradoxus** that suggests impaired cardiac filling and is the precursor of cardiac arrest.

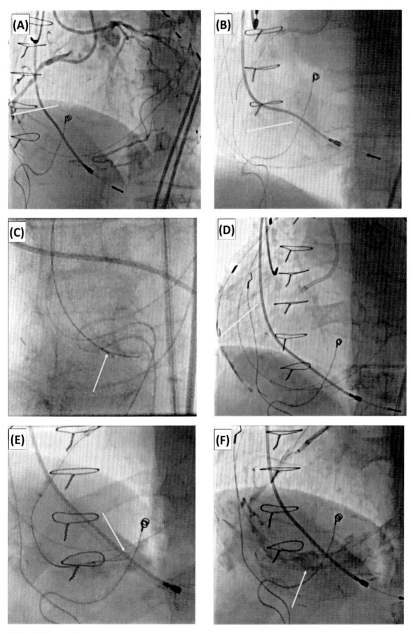

FIGURE 26.4   Continued.

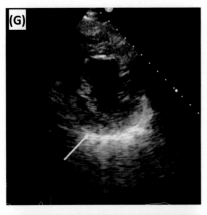

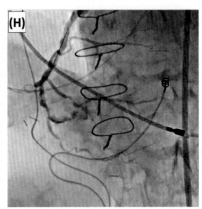

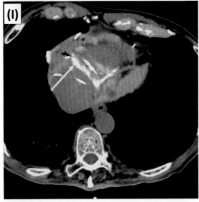

FIGURE 26.4   **Prolonged balloon inflation for treating a perforation.**

Panel A: percutaneous coronary intervention (PCI) of a RCA CTO (arrow) was attempted in a 75-year-old patient with medically refractory stable angina and prior coronary artery bypass graft surgery.

Panel B: antegrade wire escalation was attempted, but a Pilot 200 and multiple other guidewires inserted through a Corsair Pro XS microcatheter failed to advance.

Panel C: a Gladius Mongo wire was knuckled and advanced distally, followed by the "double blind stick and swap" technique with a Stingray balloon and an Astato 20, a Gladius Mongo, and a Pilot 200 wire without success.

Panel D: retrograde crossing strategy through the occluded SVG graft with a Corsair XS and Turnpike Spiral and multiple guidewires failed. The epicardial collateral from the LAD was too small and tortuous for retrograde crossing.

Panel E: a knuckled Gladius Mongo and Fielder XTA guidewire reached the distal RCA.

Panel F: perforation of the distal RCA: a 3.0 × 15 mm balloon was inflated for several minutes and further CTO crossing attempts were aborted.

Panel G: echocardiography did not show a pericardial effusion.

Panel H: final angiography.

Panel I: postprocedural chest computed tomography revealed a small inferior hemopericardium and extravasated contrast and air bubbles around the calcified occluded distal right coronary artery. The patient was managed conservatively and had an uneventful recovery.

- Pericardiocentesis can sometimes be performed using X-ray guidance due to contrast exit into the pericardial space. Echocardiography remains important for assessing the size of pericardial effusion, guiding the pericardiocentesis (if time allows) and evaluating the result of pericardiocentesis. Availability of a portable echo machine in the cath lab could help expedite treatment.
- An echocardiographic contrast agent, administered intravenously or intracoronary, can be useful for detecting ongoing bleeding into the pericardial space.[28]
- Perforations in patients with prior coronary artery bypass graft surgery might lead to pleural instead of pericardial effusions.[29]
- Autotransfusion of the aspirated blood (aspirate through the pericardial drain usually with 60 cc syringe, then return the blood through a venous sheath) may reduce the need for red blood cell transfusion, especially in cases with significant and difficult to stop pericardial bleeding.

4. **Cardiac surgery notification + ask for help**
   - Although emergency surgery is rarely required for treatment of coronary perforations, early notification of the surgeons may facilitate subsequent treatment, if pericardial bleeding continues despite percutaneous management attempts.[28] In a recent US survey of 1,094 coronary perforations reported among 335 hospitas, the incidence of emergency surgery was 7.9% for Type III/IIICS perforations versus 2.2% for Type I/II perforations ($P < .001$).[30]
   - Additional staff (nurses, technicians, physicians) could help optimally manage the case: for example one physician may perform pericardiocentesis or obtain 2nd arterial access while another physician is delivering a balloon to achieve hemostasis.

5. **Obtain 2nd arterial access**: A 2nd arterial access is part of most CTO PCIs, but if the procedure is performed through a single guide catheter, consider obtaining a second arterial access once a perforation occurs (especially large vessel perforation), as additional bulky equipment is likely to be needed for sealing the perforation. The second access can be used to introduce a second guide catheter ("ping pong technique") with specific hardware to treat the perforation if needed, while maintaining hemostasis at the site of perforation with an inflated balloon through the first guide catheter. Anchoring of the guidewire from the second catheter with the occluding balloon may also facilitate delivery of covered stents. Hemostasis should be confirmed using contrast injection.

6. **Reversal of anticoagulation**.
   - Reversal of anticoagulation should be considered if there is continued bleeding into the pericardium despite general and lesion-specific treatments. It is usually performed for distal vessel perforations with slow bleeding into the pericardium. Reversal of anticoagulation should NOT be done in most large vessel perforations, as it is unlikely to

achieve hemostasis and can lead to vessel thrombosis if equipment (guidewires, balloons, etc.) remains within the coronary arteries.

- Do NOT reverse anticoagulation with protamine until after removal of interventional equipment and complete evacuation of blood from the pericardial space. Reversing the effect of heparin carries the risk of guide and/or target vessel thrombosis, as well as thrombosis of the pericardial blood making it inaccessible to percutaneous drainage.
- Protamine dose for heparin reversal is 1 mg per 100 units of heparin (maximum dose 50 mg), administered at a rate not to exceed 5 mg per minute.[31] Protamine administration may cause anaphylactic reactions in patients treated with NPH insulin in the past or with a history of fish allergy.[32]
- If glycoprotein IIb/IIIa inhibitors or cangrelor (Section 3.5.2) were used they should be immediately discontinued if a perforation occurs. However, administration of these medication should be avoided in CTO PCI, as they can convert an otherwise insignificant wire perforations to a potentially life-threating pericardial effusion.

26.3 Large vessel perforation

CTO Manual online cases: 1, 3, 39, 40, 63, 89, 90, 118, 134, 140, 146, 167, 169, 171, 173, 174, 189, 193
PCI Manual online cases: 45, 55

26.3.1 Causes

1. Wire exit from the vessel during lesion crossing attempts, followed by inadvertent advancement of equipment (such as balloons or microcatheters) into the pericardial space. Whereas wire perforation alone seldom causes blood extravasation and pericardial effusion (because it creates a very small, self-sealing hole), catheter/balloon advancement over the wire enlarges the hole, increasing the risk of bleeding. Accordingly, confirmation of distal wire position (usually done with contralateral injection) is critical before advancing other equipment over the guidewire (**CTO Manual online case 118**). Occasionally, the contrast extravasation may not occur until after a stent is placed over the perforated area.
2. Implantation of oversized stents or high-pressure balloon inflations (especially in heavily calcified vessels with extraplaque crossing or saphenous vein grafts) (**CTO Manual online case 17, 40**).
3. Balloon rupture (when balloon rupture occurs, an angiogram should be obtained immediately after removal of the ruptured balloon to determine whether vessel perforation has occurred) (**CTO Manual online case 39**).
4. High-pressure balloon inflation in heavily calcified vessels.
5. Atherectomy.

26.3.2   Prevention

1. Avoid use of oversized stents and balloons (intravascular imaging can help guide balloon and stent size selection as outlined in Chapter 13).
2. Avoid very high pressure balloon inflations, especially in heavily calcified vessels.
3. Always confirm guidewire position within the vessel "architecture" (intraplaque or extraplaque space) before advancing other equipment.

26.3.3   Treatment

- Inflate a balloon proximal to or at the site of the perforation to stop the bleeding, as described under "general treatment" above.
- Sometimes, it can be challenging to localize the perforation site. Use of long occlusion balloons can facilitate achieving hemostasis.

26.3.3.1   Covered Stents

- For severe large vessel perforations, if extravasation persists despite prolonged balloon inflations, or if prolonged balloon inflations are not feasible due to ischemia, **cover the perforation site with a covered stent**.
- **Two covered stents are available in the US:** the PK Papyrus (Section 30.18.1.2) and the Graftmaster Rx (Section 30.18.1.1)[19,20] (Table 26.1). The PK Papyrus is compatible with 5 French guide catheters (and 6 French guide extensions) for 2.5−4.0 mm diameters or 6 French guides (7 or 8 French guide extensions) for 4.5 and 5.0 mm diameters.[33] The Graftmaster is compatible with 6 French guide catheters (or 8 French guide extensions) for 2.8−4.0 mm diameters or 7 French guide catheters for 4.5 and 4.8 mm diameters. The 5 French guide catheter compatible Begraft covered stent (Section 30.18.1.3) is available outside the US).[34,35]

TABLE 26.1 Comparison of the Graftmaster Rx and the PK Papyrus covered stents.

	Graftmaster Rx	PK Papyrus
Design	Two stents (sandwich)	Single stent
Material	ePTFE	Polyurethane
Guide needed	6 French (7 French for 4.5 and 4.8 mm stents)	5 French (6 French for 4.5 and 5.0 mm stents)
Available diameters (mm)	2.8, 3.5, 4.0, 4.5, 4.8	2.5, 3.0, 3.5, 4.0, 4.5, 5.0
Available lengths (mm)	16, 19, 26	15, 20, 26

● The cath lab staff should know where the covered stents are stored, to minimize any delays in treatment.

COVERED STENT DELIVERY TECHNIQUES: Depending on the size of the guide catheter and the type and size of the selected covered stent, delivery of the covered stent could be achieved using (a) a single guide catheter (**"block and deliver"** technique[36]) (Fig. 26.5) or (b) two guide catheters (**dual guide catheter, also called "ping pong" guide catheter, or "dueling" guide catheter technique**) (Fig. 26.6). The goal of both techniques is to minimize bleeding into the pericardium while preparing for covered stent delivery and deployment. If the balloon used for hemostasis and the covered stent can fit through a single (usually 8 French for the Graftmaster covered stents and 7 French for PK Papyrus) guide catheter, then the single guide catheter technique is used, otherwise two guide catheters are required.

26.3.3.1.1 Single guide catheter technique

for delivering a covered stent (Fig. 26.5).[27]

26.3.3.1.2 Dual guide catheter technique (Fig. 26.6).[37]

● **Covered stent delivery**: covered stents (especially the Graftmaster) are bulkier than standard DES and require excellent guide catheter support and possibly other maneuvers such as use of a guide extension (**CTO Manual online case 131, 174**) or distal anchor balloon for delivery. When delivery is challenging there is a risk of the covered stent coming off the delivery balloon. Guide catheter extensions can facilitate covered stent delivery, but compatibility should be checked before using them with a particular covered stent. After deployment, a covered stent should be postdilated aggressively to achieve good expansion and reduce the risk of stent thrombosis. If there are residual dissections beside the covered stents they should be sealed by additional stenting, as a residual dissection can be a reentry point for bleeding. Some operators advocate implantation of DES inside covered stents to reduce the risk of restenosis, but at present there are no data to support this approach.

● **Covered stents for treating perforations at a bifurcation**: in perforations involving a bifurcation, delivery of a covered stent into one of the branches will occlude the other branch. Occlusion of small side branches is often well tolerated by the patient. When large branches are occluded, such as the circumflex during sealing of perforations involving the left main or ostial LAD, a Papyrus drug-eluting stent can be deployed across the side branch origin after a wire is placed into the side branch; a stiff guidewire can then be used to puncture into the side branch (using the jailed guidewire for guidance) restoring antegrade flow, but this is a complex procedure requiring

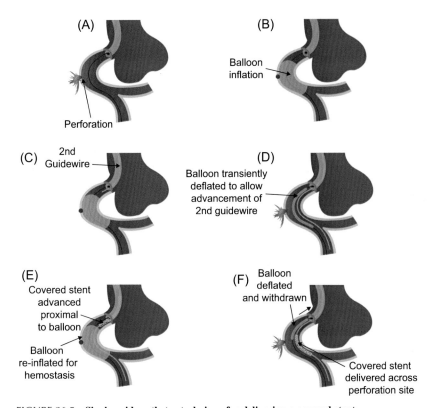

FIGURE 26.5   Single guide catheter technique for delivering a covered stent.
Panel A: a large vessel perforation has occurred.
Panel B: a balloon is inflated at the perforation site stopping pericardial bleeding.
Panel C: a second guidewire is advanced to the tip of the guide catheter.
Panel D: the balloon used for hemostasis is transiently deflated to allow advancement of the second guidewire.
Panel E: the balloon that is over the first guidewire is re-inflated and a covered stent is advanced to the tip of the guide catheter. The covered stent is advanced towards the inflated balloon that acts as a distal anchor, facilitating delivery of the covered stent to the perforation site.
Panel F: the balloon that is over the first guidewire is deflated and the covered stent is delivered across the perforation site. If the covered stent cannot be delivered to the perforation site, the balloon is re-inflated to prevent pericardial bleeding until other techniques to facilitate delivery are employed. The inflated blocking balloon may provide strong support, facilitating delivery of the covered stent.
Panel G: the first guidewire and the hemostasis balloon are removed.
Panel H: the covered stent is deployed and postdilated (the Graftmaster requires high-pressure postdilatation to achieve hemostasis).
Panel I: the perforation is sealed. Sometimes implantation of another stent proximal or distal may be necessary to seal any residual dissection that can serve as reentry point for bleeding.
Reproduced with permission from the *Manual of percutaneous coronary interventions.*

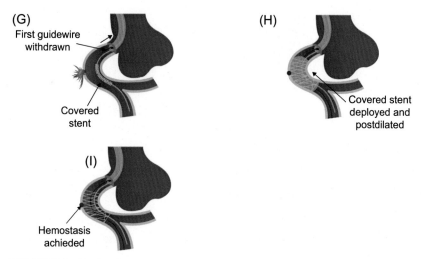

(G) First guidewire withdrawn

Covered stent

(H) Covered stent deployed and postdilated

(I) Hemostasis achieded

FIGURE 26.5   Continued.

significant experience and expertise.[38,39] A stent often needs to be placed in the jailed branch after fenestration of the covered stent.

- Retrograde crossing through saphenous vein grafts provides additional options for treating coronary perforations with retrograde delivery of balloons and covered stents, thus the ping-pong guide catheter technique is not necessary.[29]
- Occasionally, the PK Papyrus can develop a pinhole perforation, since it is covered with a highly elastic polyurethane membrane.[40]

26.3.3.2   Prolonged balloon inflations

If a covered stent cannot be delivered to the perforation site, prolonged balloon inflation could lead to hemostasis (**CTO Manual online case 17** and **40**), especially in Ellis 2 perforations of in-stent lesions. Heparin should not be reversed, until after hemostasis is achieved and equipment is removed from the coronary artery. If pericardial bleeding continues despite prolonged balloon inflations, emergency cardiac surgery may be required.

If the patient develops ischemia (which is particularly important for perforations in the donor vessel during CTO PCI) a microcatheter can be advanced distal to the occlusion balloon through which the patient's own blood is injected to relieve ischemia.[41] A perfusion balloon (Ringer, Teleflex) is currently undergoing clinical trial evaluation.

26.3.3.3   Dissection techniques for treating a large vessel perforation

An alternative treatment for large vessel perforations is to create an extraplaque dissection plane (proximal or distal to the perforation) that could seal the

Dual guide catheter technique
for covered stent delivery

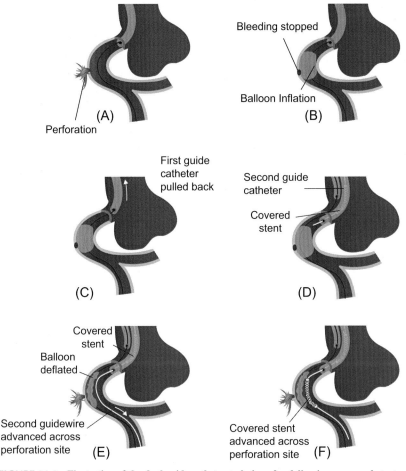

FIGURE 26.6   **Illustration of the dual guide catheter technique for delivering a covered stent.**
Panel A: a large vessel perforation has occurred.
Panel B: a balloon is inflated at the perforation site stopping pericardial bleeding
Panel C: the guide catheter is pulled back into the aorta.
Panel D: a second guide catheter (ideally 7 French or larger for better support) is advanced to the ostium of the perforated vessel. A second guidewire is inserted into the 2nd guide catheter along with a covered stent.
Panel E: the balloon that is over the first guidewire is deflated to allow advancement of the second guidewire across the perforation site.
Panel F: the covered stent is delivered to the perforation site. If the covered stent cannot be delivered to the perforation site, the blocking balloon is re-inflated to prevent pericardial bleeding until other techniques to facilitate covered stent delivery are employed.
Panel G: the first guidewire and the balloon achieving hemostasis are removed.
Panel H: the covered stent is deployed and postdilated (the Graftmaster requires high-pressure postdilation to achieve hemostasis).
Panel I: the perforation is sealed.
Reproduced with permission from the *Manual of percutaneous coronary interventions*.

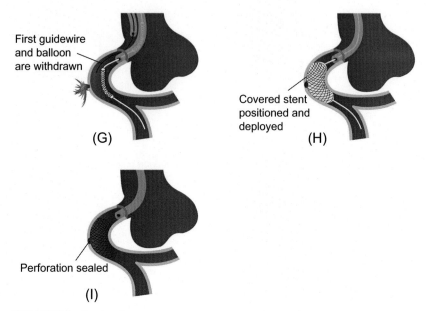

First guidewire
and balloon
are withdrawn

(G)

Covered stent
positioned and
deployed

(H)

Perforation sealed

(I)

FIGURE 26.6   Continued.

perforation (**CTO Manual online case 118, 146**) (Fig. 26.7).[42-44] This approach requires significant expertise in CTO PCI techniques.

26.3.3.4   Coils for treating a large vessel perforation

Large vessel perforations during attempts to cross a CTO vessel may not be treatable with a covered stent is the guidewire is not placed into the distal true lumen. Coils can be used to seal such perforations without causing ischemia (as the perforated vessel was occluded at baseline, **CTO Manual online case 176**, Fig. 26.8).

26.4   Distal vessel perforation

CTO Manual online cases: 26, 41, 42, 43, 112, 125, 131, 134, 139, 140, 174
PCI Manual online cases: 6, 20, 124

Distal vessel perforations can sometimes be difficult to diagnose, especially when collimation is used to minimize radiation exposure, limiting visualization of the guidewire tip. In distal vessel perforations blood flow into the pericardium can be slow, hence tamponade may not occur until several hours after the procedure has ended.[45] Coronary angiography should be performed at the end of every CTO PCI with careful review to detect any perforation before removing the guidewires. Patients with distal vessel perforation should be monitored closely and should not receive a glycoprotein IIb/IIIa inhibitor.

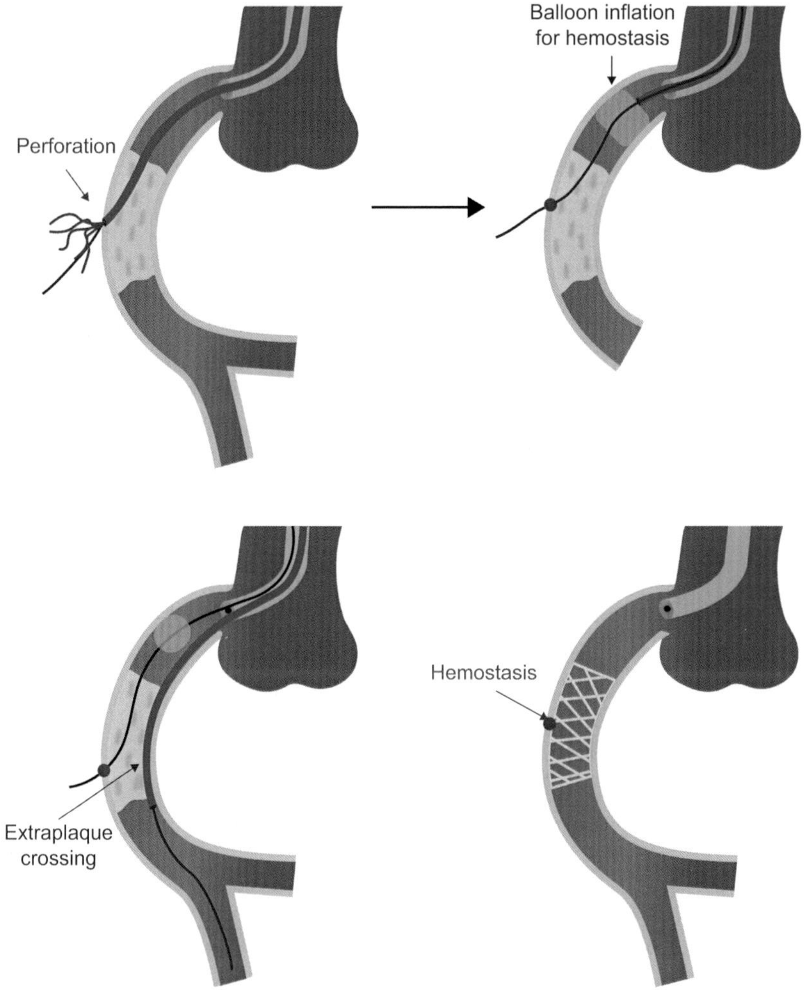

FIGURE 26.7   **Use of dissection techniques for sealing a large vessel perforation.**
Panel A: a large vessel perforation has occurred.
Panel B: a balloon is inflated at the perforation site stopping pericardial bleeding.
Panel C: the CTO is crossed using (antegrade or retrograde) using extraplaque (subintimal) techniques.
Panel D: sealing of the perforation after extraplaque (subintimal) stenting.

26.4.1   Causes

1. Inadvertent advancement of a guidewire, balloon and/or microcatheter into a distal small branch. Stiff tip, tapered, and polymer-jacketed guide-wires are more likely to cause such perforations. Even though they are safer, knuckled guidewires can still exit the vessel and cause a perfora-tion when advanced to a side-branch or distal small vessel.

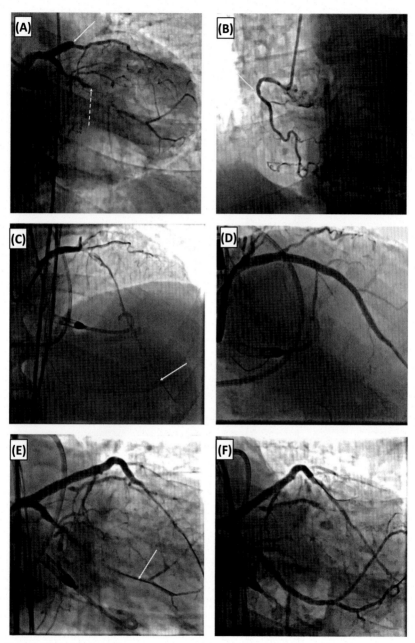

FIGURE 26.8   Continued.

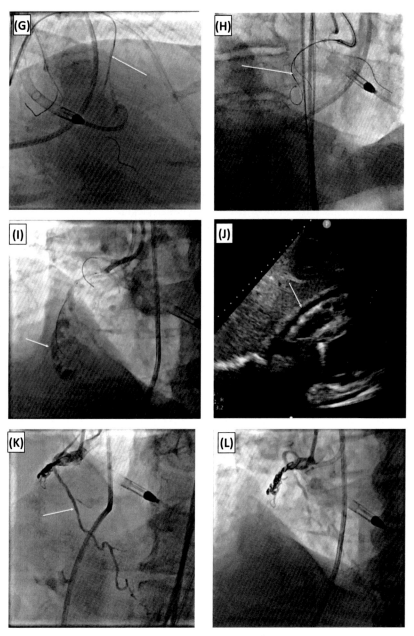

FIGURE 26.8   Continued.

2. Uncontrolled advancement of a hydrophilic-coated or polymer-jacketed wire to a distal small branch during over-the-wire device exchanges may cause distal vessel perforation. Use of balloon trapping (first choice) or wire extensions (second choice) are preferred whereas Nanto's maneuver (saline injection through the microcatheter—"hydraulic exchange," Section 3.7.2) should be avoided.

26.4.2   Prevention

Distal wire perforation can be prevented by:

1. Paying meticulous attention to distal guidewire position during attempts to deliver equipment or when multiple guidewires are being used simultaneously, especially when stiff tip or polymer-jacketed wires are used, as these wires are more likely to perforate if they are advanced into a small branch compared with workhorse guidewires. If collimation is used, the location of the guidewire tip should be periodically checked to ensure that distal migration has not occurred. For LAD wires it is safer to place the tip in a septal instead of a diagonal branch.
2. Using the trapping technique to minimize wire movement during equipment exchanges.

◄ FIGURE 26.8   **Coil embolization for treating a large vessel perforation of a CTO vessel. (CTO Manual online** case **176).**

Panel A: a 67-year-old man presented with non-ST-elevation myocardial infarction and was referred for complex high-risk percutaneous coronary intervention after being turned down for coronary artery bypass graft surgery. Angiography showed 3 CTOs with an ejection fraction of 16% on cardiac MRI. The solid arrow shows the LAD CTO and the dashed arrow the circumflex (LCX) CTO.

Panel B: right coronary artery (RCA) chronic total occlusion (CTO) (arrow).

Panel C: a Sion Blue wire was advanced through a Corsair microcatheter (Asahi) across the LAD CTO. An Impella CP device (Abiomed) was used for hemodynamic support.

Panel D: LAD after intravascular ultrasound (IVUS) guided deployment of a 3.0 × 38 mm and a 4.0 × 40 mm drug eluting stent (DES).

Panel E: the LCX CTO was successfully crossed with a Gladius Mongo wire via a Corsair microcatheter.

Panel F: final result after LCX stenting with two DES (2.5 × 30 mm and 3.0 × 22 mm).

Panel G: retrograde attempt to reach the PDA failed, as there was no connection between the LAD and the distal RCA.

Panel H: antegrade wire escalation, followed by antegrade dissection and reentry with multiple "knuckles."

Panel I: perforation of the proximal RCA.

Panel J: echocardiography showed a small pericardial effusion.

Panel K: two 2.5 mm × 4 cm Axium coils (Medtronic) were delivered through a Caravel microcatheter but failed to seal the perforation. The perforation was successfully sealed with a 4.0 mm × 15 cm Ruby coil (Penumbra). Flow in the ipsilateral collateral remained intact, filling the distal RCA territory.

Panel L: successful occlusion of the RCA perforation. The patient was discharged on the 2nd day post procedure.

3. Exchanging a stiff tip or polymer-jacketed guidewire for a workhorse guidewire immediately after confirmation of successful crossing.
4. Exchange of devices should always be done in non-collimated mode to visualize distal wire tip movement and prevent perforations.

26.4.3   Treatment (Figs. 26.9)

STEP 1: Inflate a balloon proximal to perforation

As in all coronary perforations the first management step is to inflate a balloon proximal to the perforation site to stop bleeding into the pericardium (Section 26.2. and Figs. 26.9, panel A and B). Pericardiocentesis may be required if the patient develops hypotension. Notifying cardiac surgery could expedite management in case percutaneous treatment fails.

STEP 2: Assess for continued pericardial bleeding

If balloon inflation seals the perforation, observation and heparin reversal (after removal of equipment from the coronary artery) may be all that is needed. Sometimes, suction applied through a microcatheter may collapse the vessel leading to hemostasis.[47] However, in most cases definitive treatment with embolization or a covered stent is preferred to minimize the risk of late re-opening and late tamponade.

STEP 3: Decide about embolization or covered stent implantation

Embolization is the most common treatment for distal vessel perforation and can usually be achieved using fat or coils (Fig. 26.9, panel C1-C7). There are reports of embolization using thrombin, thrombus,[48] microparticles, portion of a guidewire[49] or of a balloon,[50–52] or other material, such as sutures[53,54] and gelfoam. Embolization can in most cases be achieved through a single guide catheter using the "block and deliver" technique.[36,55] The "block and deliver" technique consists in simultaneous delivery of a balloon and a microcatheter through a single $\geq$ 6-French guide catheter, obviating the need for a second guide catheter. An important benefit of this technique is the ability to assess sealing of the perforation before and after embolization by tip injections from the microcatheter, without the need to deflate the proximal occluding balloon.[55] The minimum size required for various microcatheters to allow use of the "block and deliver" technique is shown in Table 26.2.

Embolization may not be feasible in some cases, for example when the perforated branch is too small or too angulated to allow wiring and delivery of a microcatheter. In such cases an alternative treatment strategy is implantation of a **covered stent** across the ostium of the perforated branch.

In rare cases in which neither embolization nor covered stent delivery are feasible **prolonged balloon inflations** may lead to hemostasis, otherwise cardiac surgery may be required. Cardiac surgery was required in only 3% of

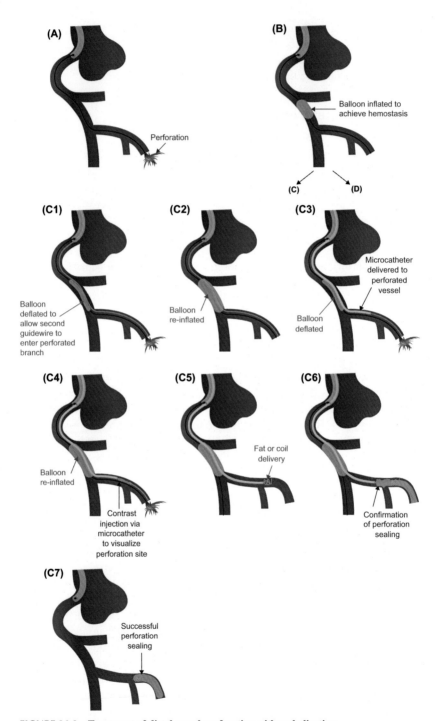

FIGURE 26.9 **Treatment of distal vessel perforation with embolization.**

(*Continued*)

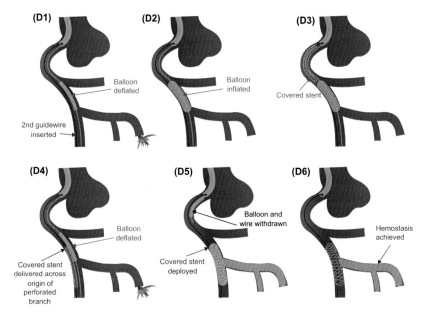

FIGURE 26.9   Continued.

◀ **Panel A**: distal vessel perforation with active bleeding into the pericardium.
Panel B: a balloon (blocking balloon) is inflated proximal to the perforation site to stop pericardial bleeding.
Panel C1: the blocking balloon is temporarily deflated to allow advancement of a second guidewire into the perforated branch.
Panel C2: the blocking balloon is re-inflated to stop pericardial bleeding.
Panel C3: the blocking balloon is transiently deflated to allow delivery of a microcatheter into the perforated vessel. The blocking balloon could also be an over-the-wire balloon through which thrombin (mixed with contrast to make it visible under X-ray) is injected while preventing backflow. The blocking balloon is then deflated to confirm that the perforation is sealed.
Panel C4: the blocking balloon is re-inflated. Injection of contrast is performed through the microcatheter to clarify the location of the perforation.
Panel C5: fat or a coil (or other material) is delivered through the microcatheter proximal to the perforation site. If the coil position is considered satisfactory the coil is released.
Panel C6: contrast is injected through the microcatheter to determine whether the perforation has been sealed (sometimes sealing is delayed for a few minutes after coil delivery).
Panel C7: successful sealing of the perforation.
Panel D1: the blocking balloon is temporarily deflated to allow advancement of a second guidewire into the main vessel.
Panel D2: the blocking balloon is re-inflated to stop pericardial bleeding.
Panel D3: a covered stent is advanced over the second guidewire proximal to the blocking balloon (which acts as distal anchor).
Panel D4: the blocking balloon is deflated and the covered stent is advanced across the ostium of the perforated vessel.
Panel D5: the covered stent is deployed. High pressure postdilation with a non-compliant balloon should be performed in nearly all cases.
Panel D6: successful sealing of the perforated branch. This needs to be confirmed with contralateral injection to rule out retrograde filling of the perforated branch.[46]
Reproduced with permission from the *Manual of percutaneous coronary interventions*.

TABLE 26.2 Microcatheters and their compatibility with the "Block and Deliver" technique using guiding catheters of different diameters.

Manufacturer	Microcatheter	Length (cm)	Coil diameter compatibility	Block and deliver compatibility (Fr)
Acrostak	M-CATH	135	≤ 0.014"	8
Asahi	CORSAIR PRO	135/150	≤ 0.014"	≥7
	CARAVEL	135/150	≤ 0.014"	≥6
	STRIDESMOOTH	125/150	≤ 0.018"	≥7
Boston Scientific	MAMBA/MAMBA FLEX	135/150	≤ 0.014"	8
Cordis	RAPID TRANSIT	70/150/170	≤ 0.018"	≥7
IMDS	NHANCER PRO X	135/155	≤ 0.014"	≥6
Orbus Neich	TELEPORT	135/150	≤ 0.014"	≥6
	TELEPORT CONTROL	135/150	≤ 0.014"	≥7
Stryker Neurovascular	EXCELSIOR SL10	150	≤ 0.014"	≥6
	EXCELSIOR XT17	150	≤ 0.014"	≥6
	EXCELSIOR 1018	150	≤ 0.018"	≥6
Terumo	PROGREAT	110/130/150	≤ 0.018"	8
	FINECROSS	130/150	≤ 0.014"	≥6
Teleflex	TURNPIKE	135/150	≤ 0.014"	8
	TURNPIKE LP	135/150	≤ 0.014"	7
	TURNPIKE SPIRAL	135/150	≤ 0.014"	8

Source: Courtesy of Dr. Gabriele Gasparini.

1762 coronary perforations reported by the British Cardiovascular Society between 2006 and 2013.[56]

26.4.3.1   Embolization of distal vessel perforation

(Fig. 26.9. panels C1−C7).

Step 1 (panel C1): The blocking balloon is temporarily deflated to allow advancement of a second guidewire into the perforated branch.

Step 2 (Panel C2): The blocking balloon is re-inflated to stop pericardial bleeding.

Step 3 (Panel C3): The blocking balloon is transiently deflated to allow delivery of a microcatheter into the perforated vessel.

Fat can be delivered through any microcatheter, but for coils the microcatheter size is important. Most commercially available coils, such as the Azur (Terumo), Interlock (Boston Scientific) and Micronester (Cook), are compatible with 0.018 inch microcatheters, such as the Progreat (Terumo) or the Renegade (Boston Scientific), and **cannot** be delivered through standard 0.014 inch microcatheters with the exception of the Finecross. There are neurovascular coils compatible with 0.014 inch microcatheters, such as the Axium coils (Medtronic) (Fig. 26.10) that can be delivered through any 0.014 inch microcatheter (Section 30.17.2).

Step 4 (Panel C4): The blocking balloon is re-inflated. Injection of contrast is performed through the microcatheter to clarify the location of the perforation.

FIGURE 26.10   **Example of Axium coils stocked in the cardiac catheterization laboratory for the treatment of coronary perforations.**

TABLE 26.3 Advantages and disadvantages of fat versus coil embolization for treating distal coronary perforations.

	Fat	Coil
Visibility	0/ +	+
Controlled delivery	0	+
Catheter needed for delivery	Any microcatheter	Bigger microcatheter
Availability	Universal	Often limited
Cost	0	High

If the microcatheter is too proximal it can be repositioned so that the extent of the proximal vessel occluded with the coil is minimized.

Step 5. Embolization

Embolization is most commonly done using fat or coils. Fat is preferred in most cases (except for very large perforations) because of universal availability, low cost and biologic compatibility, however delivery is not as controlled as when a coil is used (Table 26.3).

26.4.3.1.1   Fat embolization

(Fig. 26.11)

The starting point for coil or fat delivery is advancing a microcatheter just proximal to the perforation site.

Fat embolization Step 1. Fat can be harvested by advancing a hemostat in the femoral arteriotomy site (Fig. 26.11 panels A–C). Larger pieces can be cut into smaller ones using a scalpel.

Fat embolization Step 2. The harvested fat can be dipped into contrast for 1–2 min to absorb contrast and become visible under X-ray.

Fat embolization Step 3. The harvested fat is loaded into the microcatheter (Fig. 26.11 panels D–F) that has been advanced just proximal to the perforation. Loading the fat into the microcatheter can be challenging because fat has low density and floats on water. Turning the microcatheter hub upside down can facilitate this step (Fig. 26.12 panel E).

Fat embolization Step 4. The fat is injected through the microcatheter by flushing the microcatheter with saline or contrast through a 2 cc syringe (Fig. 26.11 panel G).

The process may need to be repeated, as several fat pieces may be required to seal the distal vessel perforation.

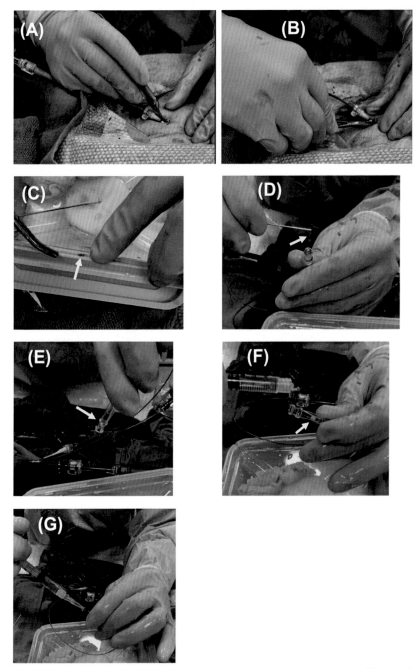

(Continued)

26.4.3.1.2   Coil embolization

(Fig. 26.12)

Online video **"How to deliver and deploy an Axium detachable coil"**

Since coiling is very infrequent in the cardiac catheterization laboratory, achieving familiarity with how to deliver and deploy a coil before a complication occurs can significantly facilitate management. Alternatively, obtaining help from an interventional radiologist (radiologists have extensive experience with coiling and embolization) can be very helpful.

1. Coil essentials: as described in Section 30.17.2 the most important characteristics of coils are:

   a. microcatheter compatibility (0.018 vs 0.014 inch), and

   b. mechanism of release (pushable vs detachable).

For coronary perforation management it is ideal to have coils that are 0.014 inch compatible (and hence can be delivered and deployed through standard microcatheters) and detachable (allowing accurate delivery to the desired location).

Table 26.4 describes various 0.014 microcatheter compatible coils currently available in the US. Having only one or two types of coils is usually sufficient. In our laboratory we currently use the smaller sizes of the Axium coils (Fig. 26.10).

Similar to fat embolization, the starting point for coil delivery is advancing a microcatheter just proximal to the perforation site. Sometimes, delivering the coil in a small bifurcation proximal to the perforation may help avoid advancement of the coil outside the artery through the perforation.

Coil embolization Step 1. The coil is inserted into the microcatheter (Fig. 26.12, panels A-D). The coil size depends on the size of the perforated vessel: for small distal coronary vessels the 2 mmx2 cm Axium coil is usually used.

Coil embolization Step 2. The coil is advanced into the microcatheter (Fig. 26.12, panel E), then the delivery sheath is removed (Fig. 26.12, panel F).

◀ FIGURE 26.11   **Illustration of fat embolization.**

Panel A: the skin entry point of the femoral sheath is enlarged with a scalpel.

Panel B: a hemostat is inserted through the skin nick, it is opened and closed to catch subcutaneous fat and is then removed along with a piece of fat.

Panel C: piece of fat harvested through the femoral arteriotomy site.

Panel D: the harvested fat is placed at the hub of the microcatheter that will be used for delivery to the perforation site.

Panel E: the microcatheter is turned upside down and gently tapped to allow for the fat to advance (fat floats in water).

Panel F: the fat has advanced more distally in the microcatheter.

Panel G: a 2 cc syringe filled with normal saline or contrast is connected to the microcatheter hub, followed by injection of the fat into the perforated coronary segment.

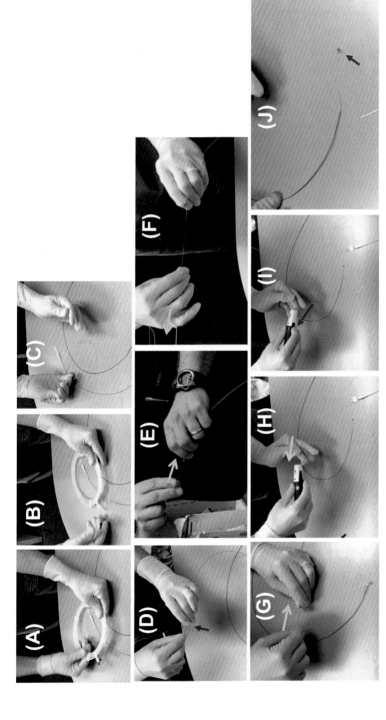

FIGURE 26.12 How to perform coil embolization with an Axium coil.

Panel A: axium detachable coil within the dispenser track after removal from the sterile package.

Panels B + C: remove the Axium detachable coil from the dispenser track.

Panel D: insert the distal end of the introducer sheath into the hub of the microcatheter until the sheath is firmly seated.

Panel E: push the Axium detachable coil into the microcatheter.

Panel F: stop approximately 15 cm from the distal end of the implant pusher (i.e. the proximal guidewire to which the coil is attached). Pull back and remove the introducer sheath.

Panel G: advance the Axium detachable coil until it reaches the desired position within the coronary vessel.

Panel H: insert the back end of the coil pusher (wire that is attached to the coil distally) into the instant detacher (*device that releases the coil).

Panel I: to release the coil, retract the thumb-slide of the instant detacher back until it stops and clicks, and slowly allow the thumb-slide to return to its original position.

Panel J: the coil has been released.

TABLE 26.4 Commercially available neurovascular coils in the US (compatible with 0.014 inch microcatheters). All neurovascular coils are detachable.

Coil name	Manufacturer	Description	Detachment system
Axium	Medtronic	Bare platinum coil with or without PGLA or Nylon microfilaments enlaced through the coil	Axium I.D. (Instant Detacher) (mechanical)
Hydrocoil (HES) MicroPlex (MCS)	Microvention	**HES**: Bare platinum coil combined with an expanding hydrogel polymer **MCS**: Bare platinum coil with various shapes and softness profiles	V-Grip (thermo-mechanical)
Orbit Cerecyte	Codman	**Orbit**: Bare platinum coil with various shapes and softness profiles **Cerecyte**: Bare platinum coil with PGA member within coil core	EnPower (thermomechanical)
Target	Stryker	Bare platinum coil with various shapes and softness profiles	InZone (electrolytic)

Coil embolization Step 3. The coil is advanced out of the tip of the microcatheter (Fig. 26.12, panel G). This will result in deployment for pushable coils.

Coil embolization Step 4. The detachable coil is moved until optimal positioning and configuration is achieved (Fig. 26.12, panel G).

Coil embolization Step 5. The detachable coil is connected to the deployment device and released (Fig. 26.12, panel H−J).

Step 6 (Panel C6): Contrast is injected through the microcatheter to determine whether the perforation has been sealed (sometimes sealing is delayed after coil delivery).

Sealing of the perforation can sometimes take several minutes, for various reasons:

- the coil or harvested fat is not large enough.
- it can take time for thrombus to form, especially in anticoagulated patients.

If bleeding continues through the perforation, additional fat pieces or coils are delivered. Sometimes both fat and coil embolization may be needed if either modality alone fails to seal the perforation (**PCI Manual online case 124**).[57]

Step 7 (Panel C7): Angiography is performed through the guide to confirm complete sealing of the perforation.

26.4.3.2 Covered Stent for treating distal vessel perforation

When the perforated vessel is too small or too tortuous, advancing a guidewire into it may not be feasible (Fig. 26.9). Such cases could be treated with coiling of a more proximal larger branch, but if the perforated vessel is originating from a large vessel, occlusion of that vessel can be undesirable. An alternative solution is implantation of a covered stent over the origin of the perforated branch (**PCI Manual online case 20, CTO Manual online cases 42, 174**).[58]

Step 1 (Panel D1). The blocking balloon is temporarily deflated to allow advancement of a second guidewire into the main vessel.
Step 2 (Panel D2): The blocking balloon is re-inflated to stop pericardial bleeding.
The goal of the blocking balloon is to stop bleeding through the perforation site into the pericardium, reducing the risk of tamponade. Prompt use of a blocking balloon may limit the size of a pericardial effusion, potentially obviating the need for pericardiocentesis.

Step 3 (Panel D3): A covered stent is advanced over the second guidewire proximal to the blocking balloon (which acts as distal anchor).
The second guidewire allows for delivery of a covered stent while maintaining hemostasis by the inflated (blocking) balloon. The blocking balloon provides extra support of the second guidewire (distal anchor), facilitating delivery of the covered stent.

Step 4 (Panel D4): The blocking balloon is deflated and the covered stent is advanced across the ostium of the perforated vessel.
The blocking balloon and its guidewire are usually not removed until after the covered stent has reached the perforation site.

Panel D5: The covered stent is deployed. High pressure postdilatation should be performed in nearly all cases.
High pressure postdilatation is important for covered stents (especially the Graftmaster stent) given they are hard to expand (the Graftmaster consists of two bare metal stents with a PTFE sandwiched in-between).

Panel D6: Successful sealing of the perforated branch. If the perforated vessel was filling via contralateral collaterals prior to the perforation, contralateral injection should be performed to rule out retrograde bleeding.

26.5 Collateral vessel perforation

CTO Manual online case 13

Perforation of an epicardial collateral branch is a serious complication of retrograde CTO PCI, as it can rapidly lead to tamponade and may be particularly difficult to control.[59,60] In contrast, perforation of septal collaterals is unlikely to have adverse consequences,[61,62] although septal hematomas,[63,64] right ventricular wall hematomas,[65] and even dry tamponade[48] have been reported following septal wire perforation.

Collateral vessel perforation may occur before, during and after collateral vessel instrumentation with guidewires or devices. A meticulous technique is recommended to ensure its prevention, detection and management.

26.5.1 Septal collateral perforation

Septal rupture/hematoma has been reported to occur in up to 6.9% of cases in a single series of patients treated with a retrograde approach.[66] Septal hematomas appear as an echo free space in the interventricular septum on transthoracic echocardiography, can cause arrhythmias and chest pain, and usually resolve spontaneously.[67] Careful attention should be paid to the collateral branch course, as a collateral that appears to be septal, may in reality be epicardial. Perforation into the coronary sinus[68] or a cardiac vein (Fig. 26.13) during attempts to cross a septal collateral usually do not require treatment. Perforation into a cardiac chamber usually does not cause complications, however balloon dilatation or advancement of additional equipment should be avoided.

26.5.2.1.1 Causes

1. Aggressive guidewire maneuvers for septal crossing.
2. Advancing a microcatheter over a guidewire that was advanced to an extraluminal location.
3. Forceful pushing of a retrograde microcatheter over an already crossed wire against resistance. A dissecting loop can be formed and migrate towards the apex, lacerating the septum.
4. Selection of a very thin or tortuous septal channel.
5. Dilatation of the septal channel.

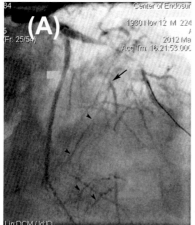
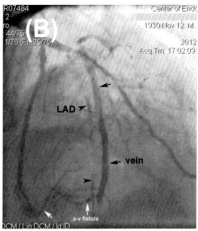
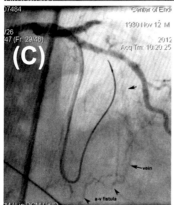
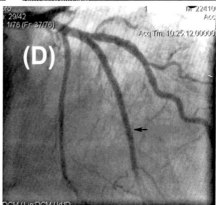

FIGURE 26.13 Arterio-venous fistula creation during attempts to cross a septal collateral.
Panel A: retrograde wiring attempts through ipsilateral distal septal collaterals (arrow).
Panel B: perforation of a septal collateral into a cardiac vein, creating an arterio-venous fistula.
Panel C: successful crossing of a more proximal septal collateral.
Panel D: occlusion of the arterio-venous fistula at the end of the procedure.
Courtesy of Dr. Avtandil Banunashvili.

26.5.1.2   Prevention

1. Selection of the largest and least tortuous interventional septal channel.
2. Caution with tip injections of contrast in collateral channels if a wedged position of the microcatheter is suspected. The "back bleeding" sign observed at the hub of the microcatheter may help in preventing injections that may cause barotrauma and rupture. In the absence of back bleeding, withdraw the microcatheter tip to a slightly more proximal location before injecting contrast.

3. Avoid advancement of the microcatheter, until the guidewire position (within the collateral vessel or the distal true lumen) has been ascertained.
4. Withdrawal of guidewires and microcatheters from collaterals after completion of CTO recanalization should be performed after collateral perforation has been ruled out. Bilateral injection while maintaining the retrograde wire position through the collateral vessel facilitates treatment of the septal perforation, if needed.
5. Pay attention on the microcatheter shaft while advancing it, and use a rotation-compatible microcatheter (such as the Corsair Pro over the Finecross) to minimize longitudinal pushing force.

26.5.1.3 Treatment

1. No specific treatment is required in most cases, unless the patient develops hemodynamic decompensation, as most septal perforations are self-limiting.[62]
2. Advancing a microcatheter can often control bleeding.
3. Negative pressure can be applied from a wedged microcatheter to collapse the collateral channel and seal the rupture.
4. If the patient becomes hemodynamically unstable ("dry" tamponade or acute LV outflow tract obstruction), coiling of the ruptured septal will not stabilize the patient, as the culprit hematoma requires time for reabsorption. Stopping further leakage with coiling is, therefore, not an immediate solution. Instead, attempts to drain the hematoma into the right or left ventricular cavity should be done, advancing a curved-tip stiff tip wire through a microcatheter into the septal collateral to create a puncture.

26.5.2 Epicardial collateral perforation

Epicardial collateral perforation is riskier than septal collateral perforation, as it can rapidly lead to tamponade. Hence, wiring epicardial collaterals should only be performed by operators experienced in the retrograde approach if no other collateral option are available (septal collaterals and SVGs are preferred over epicardial collaterals). Epicardial collateral wiring is NOT safer in patients with prior coronary bypass graft surgery or other surgery requiring opening of the pericardial sac, as bleeding may cause a loculated hematoma that can compress various cardiac chambers and cause hypotension, shock or death[12,14,16]; such hematomas may require drainage under computed tomography guidance[14,16] or surgical evacuation.[69] Contrast guidance is essential for epicardial collateral crossing: "surfing" epicardial collaterals should be avoided.

26.5.2.1   Causes

1. Aggressive guidewire and microcatheter advancement, especially through tortuous epicardial collaterals.
2. Advancement of microcatheters or other equipment after guidewire perforation has occurred.
3. Tip injection of contrast using a microcatheter placed in the epicardial collateral.

26.5.2.2   Prevention

1. Epicardial collaterals should be wired using a "contrast-guided" technique with selective angiography using the microcatheter (tip injections). "Surfing" should **not** be performed in epicardial collaterals.
2. Use soft guidewires (the Suoh 03 is most commonly used) and advance them gently.
3. Use small and soft microcatheters through epicardial collaterals, such as the Caravel and Turnpike LP or Corsair Pro XS microcatheters.
4. In contrast to septal collaterals, epicardial collaterals should **never** be dilated. However, microcatheters can (and should) be used in epicardial collaterals paying careful attention to avoid catheter advancement in front of the guidewire.
5. Retrograde wiring via ipsilateral collaterals can cause significant strain and injury of the collateral.[70] In such cases, the "tip-in" technique can be used to advance an antegrade guidewire across the CTO, avoiding wire externalization and reducing the risk of collateral perforation.
6. Ensure that the portion of the retrograde guidewire located in the collateral is covered by a microcatheter throughout the procedure.
7. Before removal of the guidewire from a collateral, angiography should be performed to rule out perforation, as a perforation is much easier to treat if guidewire access to the collateral is maintained.

26.5.2.3   Treatment

Treatment varies for atrioventricular (AV) groove and non-AV groove perforations.

26.5.2.3.1   Non atrioventricular-groove collateral perforations

1. General perforation treatment measures should be employed, as described in Section 26.2.
2. Epicardial channel perforations carry high risk of tamponade and should be promptly treated.
3. Prolonged inflation of a small balloon as well as negative pressure applied from a wedged microcatheter can sometimes lead to collapse of the collateral channel and sealing of the perforation.

4. Advancing a microcatheter through the perforated epicardial channels can sometimes lead to hemostasis (**CTO Manual online case 13**).

5. If bleeding continues, the perforation may need to be embolized/coiled (Fig. 26.14). Embolization or coiling should ideally be performed on both sides of the perforation, as blood flow can continue retrogradely in spite of occluding the antegrade limb of the collateral.[59]

Step 1. Collateral wiring (panel A of Fig. 26.14)

The first step when epicardial collateral perforation is diagnosed is to maintain guidewire access (if the guidewire is still within the collateral), otherwise the collateral is wired with a guidewire from both sides of the perforation.

What can go wrong? Collateral wiring may fail in case of extreme tortuosity or angulation. In such cases a covered stent can often be deployed sealing the origin on both ends of the perforated collateral.

Step 2. Microcatheter delivery (panel B of Fig. 26.14)

A microcatheter is advanced from both ends of the collateral adjacent to the perforation.

Step 3. Embolization (panel C and D of Fig. 26.14)

Embolization can be achieved with thrombin or with coils.

Thrombin embolization[71]: Thrombin is mixed with a small amount of contrast to allow visualization of delivery within the perforated segment and injected slowly to avoid spilling over into the main vessel. A small volume (0.2 to 0.3 mL) of thrombin is injected slowly through each microcatheter.

What can go wrong? Extreme care should be taken when injecting thrombin to seal a collateral vessel perforation, as thrombin could cause a large myocardial infarction if it leaks into the main coronary vessel. Moreover, if the perforation occurs before recanalization of the CTO, occlusion of the collateral with thrombin might lead to ischemia or infarction of the myocardial territory supplied by the collateral, unless additional collaterals exist.

Coil embolization: Coils are delivered and deployed from both sides of the perforation, as described in detail in Section 26.4.

Step 4. Confirmation of perforation sealing (panel E of Fig. 26.14)

6. Embolization from both sides of the perforation may not be feasible in cases in which the CTO cannot and has not been recanalized. Although in most such cases selective embolization of the collateral channel proximal to the perforation is sufficient to cause hemostasis (after proximal coil implantation, retrograde intra-channel pressure from an occluded vessel is low), bilateral angiography is mandatory to rule out retrograde bleeding, which might require cardiac surgery for treatment.

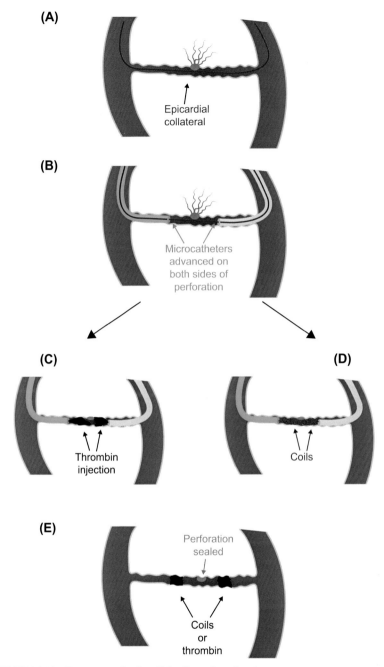

FIGURE 26.14   **Treatment of epicardial collateral perforations.**
Panel A: epicardial collateral perforation. Guidewire access within the collateral is preserved.
Panel B: a microcatheter is advanced on each side of the perforated collateral.
Panel C: thrombin is injected from the two microcatheters
Or
Panel D: coils are delivered from the two microcatheters.
Panel E: the epicardial collateral perforation is sealed.
Reproduced with permission from Brilakis ES. Manual of coronary chronic total occlusion inter-
ventions. A step-by-step approach. 2nd ed. Elsevier; 2017.

26.5.2.3.2 Atrioventricular-groove collateral perforations

1. General perforation treatment measures and occlusion of the perforated collateral should be performed, as described for non-AV groove collateral perforations.
2. AV groove collateral perforation may result in left atrial wall hematoma, acute mitral regurgitation due to mitral annulus deformity, and pulmonary vein obstruction. Intubation and CT-guided drainage of the hematoma or emergency cardiac surgery may be required.

26.6 Perforation in patients with prior coronary artery bypass graft surgery carries very high risk

CTO Manual online cases 42 and 43

Although in the past prior coronary bypass graft surgery was considered protective from tamponade in patients in whom perforation occurs, we currently know that loculated effusions can develop in these patients and can compress various cardiac structures[12] (such as the left atrium[13-15] or the right ventricle[16]). Such loculated effusions can be lethal, as they can be challenging to reach and drain percutaneously. Sometimes intramyocardial hematomas can develop and can also be challenging to treat.[72]

Therefore, perforations in prior CABG patients should be immediately treated (for example with covered stents or coils) to minimize the risk of developing a loculated effusion.[12]

References

1. Patel SM, Menon RV, Burke MN, Jaffer FA, Yeh RW, Vo M, et al. Current perspectives and practices on chronic total occlusion percutaneous coronary interventions. *J Invasive Cardiol* 2018;**30**:43−50.
2. Azzalini L, Poletti E, Ayoub M, Ojeda S, Zivelonghi C, La Manna A, et al. Coronary artery perforation during chronic total occlusion percutaneous coronary intervention: epidemiology, mechanisms, management, and outcomes. *EuroIntervention* 2019;**15**:e804−11.
3. Moroni F, Brilakis ES, Azzalini L. Chronic total occlusion percutaneous coronary intervention: managing perforation complications. *Expert Rev Cardiovasc Ther* 2021;**19**:71−87.
4. Rathore S, Matsuo H, Terashima M, Kinoshita Y, Kimura M, Tsuchikane E, et al. Procedural and in-hospital outcomes after percutaneous coronary intervention for chronic total occlusions of coronary arteries 2002 to 2008: impact of novel guidewire techniques. *JACC Cardiovasc Interv* 2009;**2**:489−97.
5. Hirai T, Nicholson WJ, Sapontis J, Salisbury AC, Marso SP, Lombardi W, et al. A detailed analysis of perforations during chronic total occlusion angioplasty. *JACC Cardiovasc Interv* 2019;**12**:1902−12.
6. Kandzari DE, Grantham JA, Karmpaliotis D, Lombardi W, Moses JW, Nicholson W, et al. Safety and efficacy of dedicated guidewire and microcatheter technology for chronic total coronary occlusion revascularization: principal results of the Asahi Intecc Chronic Total Occlusion Study. *Coron Artery Dis* 2018;**29**:618−23.

7. Kandzari DE, Lembo NJ, Carlson HD, Kalynych A, Spertus JA, Gibson CM, et al. Procedural, clinical, and health status outcomes in chronic total coronary occlusion revascularization: results from the PERSPECTIVE study. *Catheter Cardiovasc Interv* 2020;**96**:567−76.

8. Kandzari DE, Alaswad K, Jaffer FA, Brilakis E, Croce K, Kearney K, et al. Safety and efficacy of dedicated guidewire, microcatheter, and guide catheter extension technologies for chronic total coronary occlusion revascularization: primary results of the Teleflex Chronic Total Occlusion Study. *Catheter Cardiovasc Interv* 2021;**99**:263−70.

9. Patel VG, Brayton KM, Tamayo A, Mogabgab O, Michael TT, Lo N, et al. Angiographic success and procedural complications in patients undergoing percutaneous coronary chronic total occlusion interventions: a weighted *meta*-analysis of 18,061 patients from 65 studies. *JACC Cardiovasc Interv* 2013;**6**:128−36.

10. Karmpaliotis D, Karatasakis A, Alaswad K, Jaffer FA, Yeh RW, Wyman RM, et al. Outcomes with the use of the retrograde approach for coronary chronic total occlusion interventions in a contemporary multicenter US Registry. *Circ Cardiovasc Interv* 2016;9.

11. Danek BA, Karatasakis A, Karmpaliotis D, Alaswad K, Yeh RW, Jaffer FA, et al. Development and validation of a scoring system for predicting periprocedural complications during percutaneous coronary interventions of chronic total occlusions: the prospective global registry for the study of chronic total occlusion intervention (PROGRESS CTO) complications score. *J Am Heart Assoc* 2016;5.

12. Karatasakis A, Akhtar YN, Brilakis ES. Distal coronary perforation in patients with prior coronary artery bypass graft surgery: the importance of early treatment. *Cardiovasc Revasc Med* 2016;**17**:412−17.

13. Aggarwal C, Varghese J, Uretsky BF. Left atrial inflow and outflow obstruction as a complication of retrograde approach for chronic total occlusion: report of a case and literature review of left atrial hematoma after percutaneous coronary intervention. *Catheter Cardiovasc Interv* 2013;**82**:770−5.

14. Wilson WM, Spratt JC, Lombardi WL. Cardiovascular collapse post chronic total occlusion percutaneous coronary intervention due to a compressive left atrial hematoma managed with percutaneous drainage. *Catheter Cardiovasc Interv* 2015;**86**:407−11.

15. Franks RJ, de Souza A, Di Mario C. Left atrial intramural hematoma after percutaneous coronary intervention. *Catheter Cardiovasc Interv* 2015;**86**:E150−2.

16. Adusumalli S, Morris M, Pershad A. Pseudo-pericardial tamponade from right ventricular hematoma after chronic total occlusion percutaneous coronary intervention of the right coronary artery: successfully managed percutaneously with computerized tomographic guided drainage. *Catheter Cardiovasc Interv* 2016;**88**:86−8.

17. Kawana M, Lee AM, Liang DH, Yeung AC. Acute right ventricular failure after successful opening of chronic total occlusion in right coronary artery caused by a large intramural hematoma. *Circ Cardiovasc Interv* 2017;10.

18. Mertens A, Dalal P, Ashbrook M, Hanson I. Coil embolization of coronary-cameral fistula complicating revascularization of chronic total occlusion. *Case Rep Cardiol* 2018;**2018**:4.

19. Ybarra LF, Rinfret S, Brilakis ES, Karmpaliotis D, Azzalini L, Grantham JA, et al. Definitions and clinical trial design principles for coronary artery chronic total occlusion therapies: CTO-ARC consensus recommendations. *Circulation.* 2021;**143**:479−500.

20. Brilakis ES, Karmpaliotis D, Patel V, Banerjee S. Complications of chronic total occlusion angioplasty. *Interv Cardiol Clin* 2012;**1**:373−89.

21. Xenogiannis I, Brilakis ES. Advances in the treatment of coronary perforations. *Catheter Cardiovasc Interv* 2019;**93**:921—2.

22. Tajti P, Xenogiannis I, Chavez I, Gossl M, Mooney M, Poulose A, et al. Expecting the unexpected: preventing and managing the consequences of coronary perforations. *Expert Rev Cardiovasc Ther* 2018;**16**:805—14.

23. Shaukat A, Tajti P, Sandoval Y, Stanberry L, Garberich R, Burke MN, et al. Incidence, predictors, management and outcomes of coronary perforations. *Catheter Cardiovasc Interv* 2019;**93**:48—56.

24. Ellis SG, Ajluni S, Arnold AZ, Popma JJ, Bittl JA, Eigler NL, et al. Increased coronary perforation in the new device era. incidence, classification, management, and outcome. *Circulation.* 1994;**90**:2725—30.

25. Hirai T, Grantham JA, Sapontis J, Nicholson WJ, Lombardi W, Karmpaliotis D, et al. Development and validation of a prediction model for angiographic perforation during chronic total occlusion percutaneous coronary intervention: OPEN-CLEAN perforation score. *Catheter Cardiovasc Interv* 2022;**99**:280—5.

26. Kostantinis S, Simsek B, Karacsonyi J, Alaswad K, Jaffer FA, Khatri JJ, et al. Development and validation of a scoring system for predicting clinical coronary artery perforation during percutaneous coronary intervention of chronic total occlusions: the PROGRESS-CTO perforation score. *EuroIntervention* 2022 Oct 24; EIJ-D-22-00593. Available from: https://doi.org/10.4244/EIJ-D-22-00593. Epub ahead of print. https://www.orosapparel.com/pages/orostechnology

27. Brilakis ES. *Manual of coronary chronic total occlusion interventions. A step-by-step Approach.* 2nd ed. Elsevier; 2017.

28. Bagur R, Bernier M, Kandzari DE, Karmpaliotis D, Lembo NJ, Rinfret S. A novel application of contrast echocardiography to exclude active coronary perforation bleeding in patients with pericardial effusion. *Catheter Cardiovasc Interv* 2013;**82**:221—9.

29. Karacsonyi J, Koike H, Fukui M, Kostantinis S, Simsek B, Brilakis ES. Retrograde treatment of a right coronary artery perforation. *JACC Cardiovasc Interv* 2022;**15**:670—2.

30. Kandzari D, Waksman R. CRT-100.99 clinical experience of the PK papyrus covered stent in patients with coronary artery perforations. *JACC: Cardiovas Interv* 2022;**15**:S26.

31. Kataruka A, Daniels D, Maynard C, Kearney KE, Mahmoud AM, Doll JA, et al. Protamine utilization and clinical outcomes for coronary artery perforation in chronic total occlusion procedures. *J Am Coll Cardiology* 2020;**75**:1293.

32. Stewart WJ, McSweeney SM, Kellett MA, Faxon DP, Ryan TJ. Increased risk of severe protamine reactions in NPH insulin-dependent diabetics undergoing cardiac catheterization. *Circulation.* 1984;**70**:788—92.

33. Kandzari DE, Birkemeyer RPK. Papyrus covered stent: device description and early experience for the treatment of coronary artery perforations. *Catheter Cardiovasc Interv* 2019;**94**:564—8.

34. Briguori C, Nishida T, Anzuini A, Di Mario C, Grube E, Colombo A. Emergency polytetrafluoroethylene-covered stent implantation to treat coronary ruptures. *Circulation.* 2000;**102**:3028—31.

35. Romaguera R, Waksman R. Covered stents for coronary perforations: is there enough evidence? *Catheter Cardiovasc Interv* 2011;**78**:246—53.

36. Tarar MN, Christakopoulos GE, Brilakis ES. Successful management of a distal vessel perforation through a single 8-French guide catheter: combining balloon inflation for bleeding control with coil embolization. *Catheter Cardiovasc Interv* 2015;**86**:412—16.

37. Ben-Gal Y, Weisz G, Collins MB, Genereux P, Dangas GD, Teirstein PS, et al. Dual catheter technique for the treatment of severe coronary artery perforations. *Catheter Cardiovasc Interv* 2010;**75**:708—12.

38. Adusumalli S, Gaikwad N, Raffel C, Dautov R. Treatment of rotablation-induced ostial left circumflex perforation by papyrus covered stent and its fenestration to recover the left anterior descending artery during CHIP procedure. *Catheter Cardiovasc Interv* 2019;**93**:E331−6.
39. Davies RE, Cheney AE, McCabe JM, Alaswad K, Lombardi WL. A novel hybrid approach to the treatment of a left main coronary artery aneurysm. *JACC Case Rep* 2020;**2**:1675−8.
40. Warisawa T, Kawase Y, Itakura R, Akashi YJ, Matsuo H. Pinhole perforation of polyurethane membrane-covered stent detected by optical coherence tomography. *JACC Cardiovasc Interv* 2022;**15**:1089−91.
41. Ishihara S, Tabata S, Inoue T. A novel method to bail out coronary perforation: microcatheter distal perfusion technique. *Catheter Cardiovasc Interv* 2015;**86**:417−21.
42. Xenogiannis I, Tajti P, Nicholas Burke M, Brilakis ES. An alternative treatment strategy for large vessel coronary perforations. *Catheter Cardiovasc Interv* 2019;**93**:635−8.
43. Kartas A, Karagiannidis E, Sofidis G, Stalikas N, Barmpas A, Sianos G. Retrograde access to seal a large coronary vessel balloon perforation without covered stent implantation. *JACC Case Rep* 2021;**3**:542−5.
44. Doshi D, Hatem R, Masoumi A, Lembo NJ, Karmpaliotis D. The self-made covered stent technique for treating coronary artery perforations. *J Cardio Case Rep* 2020;**3**. Available from: https://doi.org/10.15761/JCCR.1000141.
45. Stathopoulos IA, Kossidas K, Garratt KN. Delayed perforation after percutaneous coronary intervention: rare and potentially lethal. *Catheter Cardiovasc Interv* 2014;**83**:E45−50.
46. Brilakis ES. *Manual of percutaneous coronary interventions: a step-by-step approach.* Elsevier; 2021.
47. Yasuoka Y, Sasaki T. Successful collapse vessel treatment with a syringe for thrombus-aspiration after the guidewire-induced coronary artery perforation. *Cardiovasc Revasc Med* 2010;**11**(263):e1−3.
48. Matsumi J, Adachi K, Saito S. A unique complication of the retrograde approach in angioplasty for chronic total occlusion of the coronary artery. *Catheter Cardiovasc Interv* 2008;**72**:371−8.
49. Hartono B, Widito S, Munawar M. Sealing of a dual feeding coronary artery perforation with homemade spring guidewire. *Cardiovasc Interv Ther* 2015;**30**:347−50.
50. Sobieszek G, Zieba B. Balloon fragment technique used to close distal coronary vessel perforation. *J Invasive Cardiol* 2020;**32**:E370−2.
51. Shah SV, Shah VK, Thakur MV, Sembhi KL. Half-cut balloon, a novel, ultra-rapid, and cheap technique to seal a coronary perforation. *JACC Cardiovasc Interv* 2020;**13**:e177−8.
52. Agac MT, Vatan MB, Tatli E. A novel closed-loop balloon-stent technique for vessel closure. *JACC Cardiovasc Interv* 2022;**15**:e27−9.
53. Li Y, Wang G, Sheng L, Xue J, Sun D, Gong Y. Silk suture embolization for sealing distal coronary artery perforation: report of two cases. *Rev Cardiovasc Med* 2015;**16**:165−9.
54. Sheng L, Gong YT, Sun DH, Li Y. Successful occluding by absorbable sutures for epicardial collateral branch perforation. *J Geriatr Cardiol* 2018;**15**:653−6.
55. Garbo R, Oreglia JA, Gasparini GL. The balloon-microcatheter technique for treatment of coronary artery perforations. *Catheter Cardiovasc Interv* 2017;**89**:E75−83.
56. Kinnaird T, Kwok CS, Kontopantelis E, Ossei-Gerning N, Ludman P, deBelder M, et al. British Cardiovascular Intervention S and the National Institute for Cardiovascular Outcomes R. Incidence, determinants, and outcomes of coronary perforation during percutaneous coronary intervention in the United Kingdom Between 2006 and 2013: an analysis of 527 121 cases from the British Cardiovascular Intervention Society Database. *Circ Cardiovasc Interv* 2016;**9**.

57. Guddeti RR, Kostantinis ST, Karacsonyi J, Brilakis ES. Distal coronary perforation sealing with combined coil and fat embolization. *Cardiovasc Revasc Med* 2022;**40S**:222−4.

58. Sandoval Y, Lobo AS, Brilakis ES. Covered stent implantation through a single 8-french guide catheter for the management of a distal coronary perforation. *Catheter Cardiovasc Interv* 2017;**90**:584−8.

59. Boukhris M, Tomasello SD, Azzarelli S, Elhadj ZI, Marza F, Galassi AR. Coronary perforation with tamponade successfully managed by retrograde and antegrade coil embolization. *J Saudi Heart Assoc* 2015;**27**:216−21.

60. Ngo C, Christopoulos G, Brilakis ES. Conservative management of an epicardial collateral perforation during retrograde chronic total occlusion percutaneous coronary intervention. *J Invasive Cardiol* 2016;**28**:E11−12.

61. Lee NH, Seo HS, Choi JH, Suh J, Cho YH. Recanalization strategy of retrograde angioplasty in patients with coronary chronic total occlusion -analysis of 24 cases, focusing on technical aspects and complications. *Int J Cardiol* 2010;**144**:219−29.

62. Araki M, Murai T, Kanaji Y, Matsuda J, Usui E, Niida T, et al. Interventricular septal hematoma after retrograde intervention for a chronic total occlusion of a right coronary artery: echocardiographic and magnetic resonance imaging-diagnosis and follow-up. *Case Rep Med* 2016;**2016**:8514068.

63. Lin TH, Wu DK, Su HM, Chu CS, Voon WC, Lai WT, et al. Septum hematoma: a complication of retrograde wiring in chronic total occlusion. *Int J Cardiol* 2006;**113**:e64−6.

64. Abdel-Karim AR, Vo M, Main ML, Grantham JA. Interventricular septal hematoma and coronary-ventricular fistula: a complication of retrograde chronic total occlusion intervention. *Case Rep Cardiol* 2016;**2016**:8750603.

65. Ghobrial MSA, Egred M. Right ventricular wall hematoma following angioplasty to right coronary artery occlusion. *J Invasive Cardiol* 2019;**31**:E66.

66. Sianos G, Barlis P, Di Mario C, Papafaklis MI, Buttner J, Galassi AR, et al. European experience with the retrograde approach for the recanalisation of coronary artery chronic total occlusions. A report on behalf of the euroCTO club. *EuroIntervention* 2008;**4**:84−92.

67. Fairley SL, Donnelly PM, Hanratty CG, Walsh SJ. Images in cardiovascular medicine. Interventricular septal hematoma and ventricular septal defect after retrograde intervention for a chronic total occlusion of a left anterior descending coronary artery. *Circulation* 2010;**122**:e518−21.

68. Sachdeva R, Hughes B, Uretsky BF. Retrograde approach to a totally occluded right coronary artery via a septal perforator artery: the tale of a long and winding wire. *J Invasive Cardiol* 2010;**22**:E65−6.

69. Marmagkiolis K, Brilakis ES, Hakeem A, Cilingiroglu M, Bilodeau L. Saphenous vein graft perforation during percutaneous coronary intervention: a case series. *J Invasive Cardiol* 2013;**25**:157−61.

70. Mashayekhi K, Behnes M, Valuckiene Z, Bryniarski L, Akin I, Neuser H, et al. Comparison of the ipsi-lateral vs contra-lateral retrograde approach of percutaneous coronary interventions in chronic total occlusions. *Catheter Cardiovasc Interv* 2017;**89**:649−55.

71. Kotsia AP, Brilakis ES, Karmpaliotis D. Thrombin injection for sealing epicardial collateral perforation during chronic total occlusion percutaneous coronary interventions. *J Invasive Cardiol* 2014;**26**:E124−6.

72. Ezad S, Wardill T, Talwar S. Intramyocardial haematoma complicating chronic total occlusion percutaneous coronary intervention: case series and review of the literature. *Cardiovasc Revasc Med* 2022;**34**:142−7.

Chapter 27

Equipment loss and entrapment

Various types of equipment, such as stents, guidewires, and catheters can be lost or entrapped[1] during CTO PCI either within or outside the coronary artery.[2] Equipment loss or entrapment and the ensuing retrieval attempts can lead to occlusion or perforation of the coronary vessel or to device embolization, for example to the brain causing a stroke. Device entrapment is a more serious complication than device loss, and may require emergency surgery for removal.[2] In this chapter we discuss how to prevent, diagnose and treat such complications. Entrapment of the Rotablator burr is discussed in Section 27.5.

27.1 Stent loss or entrapment

CTO Manual online cases 74, 122, 128, 167, 171
PCI Manual online cases 10, 93, 108, 134
Video 27.1 Stent loss

27.1.1 Causes

1. Coronary tortuosity and calcification.[3,4]
2. Poor vessel preparation prior to attempting stent delivery. In the ROTAXUS trial, stent loss occurred in 2% of lesions in the no atherectomy vs. 0.5% of lesions in the rotational atherectomy arm of the study.[5] Poor vessel preparation may result in stent deformation during attempts to deliver the stent, followed by stent loss when attempting to retrieve the stent inside the guide catheter (Figs. 27.1 and Figs. 27.2).
3. Stent delivery through another stent.[6]
4. Direct stenting.
5. Use of small (such as 5 French) guide catheters.
6. Forceful withdrawal of the stent inside the guide catheter (or inside a guide catheter extension) when resistance is felt.[3]
7. Use of guide catheter extensions. Stent loss can occur both during advancement of the stent if the stent catches the proximal collar, or during stent withdrawal, especially when the stent is deformed.
8. Attempting to deliver a stent via a collateral during retrograde CTO PCI (that can predispose to both stent loss[7] and wire entrapment[8]).

Manual of Chronic Total Occlusion Percutaneous Coronary Interventions.
DOI: https://doi.org/10.1016/B978-0-323-91787-2.00023-X
© 2023 Elsevier Inc. All rights reserved.

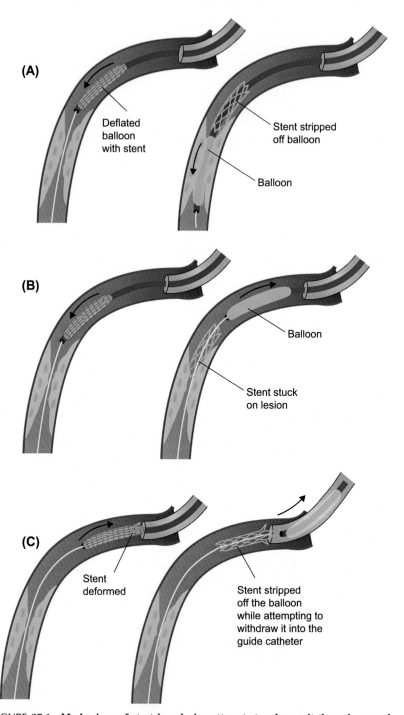

FIGURE 27.1 Mechanisms of stent loss during attempts to advance it through a poorly prepared lesion. Panel A: the stent is stripped off the balloon at the proximal end of the target lesion. **Panel B**: the stent becomes entrapped in the lesion, resulting in separation from the balloon during withdrawal. **Panel C**: the stent becomes deformed during delivery attempts and is stripped off the balloon during attempts to withdraw it into the guide catheter.

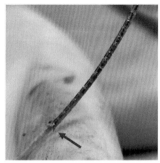

FIGURE 27.2 **Images of a deformed stent.** Arrows point to the site of deformation.

27.1.2 Prevention

1. Avoid direct stenting, especially in tortuous and calcified vessels. Given high level of complexity, direct stenting should *not* be performed in CTO PCI with rare exceptions.
2. Meticulous vessel preparation, often using atherectomy in calcified lesions, and intracoronary imaging to confirm adequate lesion expansion before attempting stent delivery.
3. Stent from distal to proximal. Sometimes, however, the need for distal stenting does not arise until after a proximal stent is deployed, for example in cases of distal edge dissection.[2]
4. Use a guide catheter extension for delivering stents in tortuous and calcified vessels or across recently deployed stents.
5. Avoid forceful stent advancement. If there is significant difficulty advancing a stent, stop delivery attempts and perform additional lesion preparation and/or use a guide catheter extension.
6. When using a guide catheter extension, always place the external push rod under a towel at the side of the Y-connector to reduce the risk of the guidewires "wrapping around" the guide catheter extension delivery rod.
7. Do not apply force if resistance is felt while advancing a stent through a guide catheter extension. Instead, remove the guide catheter extension and reinsert it, paying particular attention to avoiding wrapping of the guidewire around the guide extension delivery rod. Inserting the stent

into the guide catheter extension before inserting into the guide catheter and advancing the guide extension/stent assembly together into the guide catheter reduces the risk of stent dislodgement or loss.

8. Avoid small (such as 5 French) guide catheters when treating complex lesions, as smaller guide catheters have less room to allow for withdrawal of deformed stents. CTO PCI should generally be performed using 7 or 8 French guide catheters to facilitate equipment delivery and combined use of various equipment.

27.1.3 Treatment

27.1.3.1 Location of the lost stent

The location where the stent is lost determines the subsequent steps (Fig. 27.3). If a stent is lost in a coronary artery it should be either retrieved or deployed/crushed. If a stent is lost in a noncritical location in the peripheral circulation (such as the lower extremity or pelvic vessels), it can often be left in place without attempting retrieval (**CTO Manual online case** 171).[4]

Occasionally, the lost stents may be difficult to visualize, especially thin strut stents in obese patients with calcified or previously stented coronary

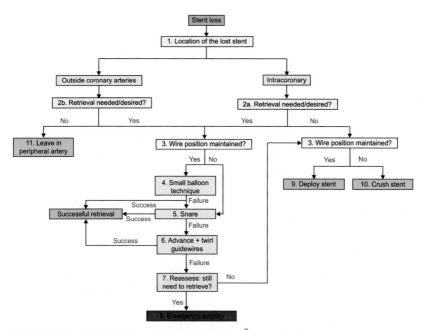

FIGURE 27.3 **Algorithm for approaching stent loss**.[9]
Reproduced with permission from the Brilakis ES. Manual of percutaneous coronary interventions: a step-by-step approach. Elsevier; 2021.

arteries. In such cases, intravascular ultrasonography (IVUS) may facilitate localization of the stent (**CTO Manual online case** 74).[2]

27.1.3.2 Retrieval needed/desired

Lost stents do *NOT* always need to be retrieved. If the stent is lost in a noncritical location of the coronary artery, stent deployment (if wire position is maintained through the stent) or stent crushing (if wire position through the stent has been lost) can be a faster and safer approach (**CTO Manual online case** 167). Similarly, if the stent is lost in a noncritical location of the peripheral circulation (such as the lower extremities), the risks of retrieval attempts often outweigh any potential benefit.[4] Conversely, if the lost stent is located within a critical coronary artery location (such as left main or major bifurcation) or peripheral artery location (such as cerebral or renal arteries), retrieval is required.

27.1.3.3 Wire position maintained

Maintaining wire position through the lost stent can greatly facilitate retrieval attempts (by using the small balloon technique), and also allows for easier deployment if selected. If wire position is lost it is often impossible to rewire the lost stent, limiting retrieval options to snares or the "guidewire twirling" technique. Similarly, loss of wire position prohibits deployment, making stent crushing the only option if retrieval is not possible or desired.

27.1.3.4 Small balloon technique

The small-balloon technique[4,10] can be used when a stent is dislodged from its delivery balloon, but guidewire access is maintained through the lost stent (Fig. 27.4). A small (1.0−1.5 mm) balloon is advanced through the stent, inflated distal to the stent, and withdrawn often bringing together the lost stent (**CTO Manual online case** 128).

Sometimes, it may be difficult to advance the balloon through the stent, possibly pushing the stent more distally in the vessel. If the balloon is partially advanced through the stent it can be inflated in the proximal-mid part of the lost stent, followed by withdrawal that sometimes retrieves the lost stent.

If the stent is not significantly deformed and the guide catheter is large, the balloon and the stent can be retrieved inside the guide catheter and removed. If the stent is severely deformed and cannot be withdrawn inside the guide catheter, the inflated balloon, lost stent, and the guide catheter can sometimes be removed "en bloc" from the body (**PCI Manual online case** 10).

27.1.3.5 Snares

Snaring is an effective technique for retrieving lost or entrapped equipment. Available snares are discussed in Section 30.18.5.1.

Stent retrieval: small balloon technique

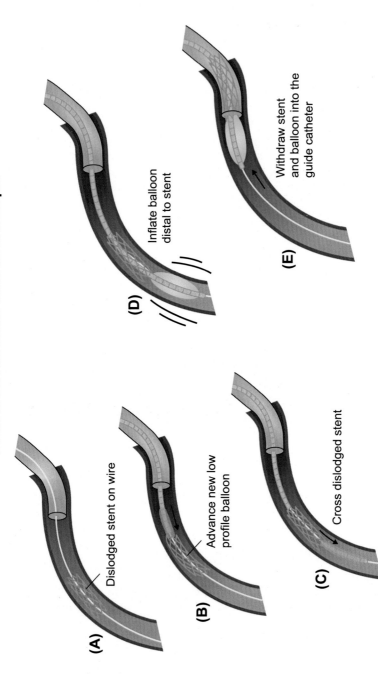

FIGURE 27.4 **Illustration of the small balloon technique.** for retrieving a lost stent if wire position is maintained through the stent. **Panel A:** a stent is dislodged from the stent delivery balloon, but remains over the guidewire. **Panel B:** a new low profile balloon (1.0-1.5 mm in diameter) is advanced through the lost stent, under fluoroscopy to ensure that the stent is not displaced further distally by the balloon. **Panel C:** the balloon is advanced through the lost stent. **Panel D:** the balloon is inflated distal to the lost stent. **Panel E:** the inflated balloon is withdrawn along with the lost stent under fluoroscopy into the guide catheter.
Reproduced with permission from the Brilakis ES. Manual of percutaneous coronary interventions: a step-by-step approach. Elsevier; 2021.

27.1.3.5.1 Stents lost in the coronary circulation

Once a loop snare is positioned around the stent to be snared, the snare wire is held still and the snare catheter advanced forward until the lasso loop secures the stent to be snared, followed by withdrawal of the assembly into the guide catheter (Fig. 27.5). However, attempts to remove a deformed stent may lead to vessel injury, such as dissection and perforation (**CTO Manual online case** 167).

For retrieval of stents lost in the coronary circulation the Micro-Elite snare (Teleflex) (which is 0.014 inch in diameter, has loop size of 2−7 mm, is 180 cm in length and does not require a delivery catheter) or the 2 mm En Snare (Merit Medical) are commonly utilized.

27.1.3.5.2 Stents lost in the peripheral circulation

For retrieving stents lost in the peripheral circulation an 18−30 mm Ensnare through a JR4 or multipurpose diagnostic or guide catheter is often the snare of choice, because of the 3 loop design that facilitates object retrieval. However, it should be used with caution as the snare wires may still cause vessel injury.[2]

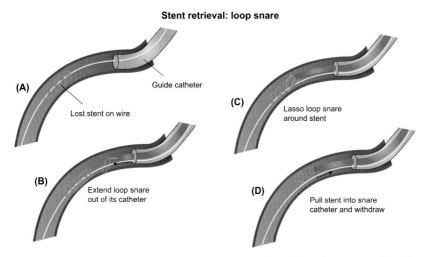

Stent retrieval: loop snare

(A) Guide catheter

Lost stent on wire

(B) Extend loop snare out of its catheter

(C) Lasso loop snare around stent

(D) Pull stent into snare catheter and withdraw

FIGURE 27.5 **Retrieval of a lost stent in the coronary circulation using a snare. Panel A:** a stent is dislodged from its delivery balloon but remains on the 0.014 inch guidewire. **Panel B:** a loop snare is advanced over the guidewire into the target vessel. **Panel C:** the snare is positioned around the lost stent. **Panel D:** the snare is tightened capturing the lost stent and is withdrawn into the guide catheter.

Reproduced with permission from the Brilakis ES. Manual of percutaneous coronary interventions: a step-by-step approach. Elsevier; 2021.

27.1.3.6 Guidewire "twirling"

In this technique, two or more 0.014 inch guidewires are advanced through the lost stent, and rotated several times in order to entangle their distal ends (Fig. 27.6). The guidewires are then withdrawn, often bringing the lost stent with them.

27.1.3.7 Reassess need for retrieval

If attempts to retrieve a stent lost in the coronary circulation fail, the need for retrieval may need to be reassessed, considering the risks and benefits associated with further retrieval attempts, including surgery versus deploying or crushing the stent. In most cases, deploying or crushing the stent is the fastest and safest approach.

27.1.3.8 Emergency surgery

Surgical retrieval of lost stents can be very challenging and carries high risk of complications. It should, therefore, only be done under extenuating circumstances.

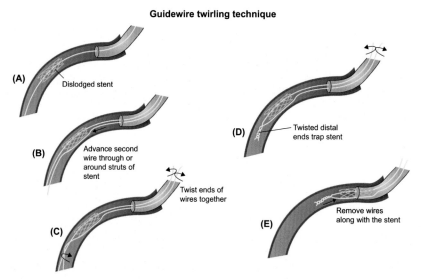

FIGURE 27.6 **Illustration of the guidewire "twirling" technique**. **Panel A**: a stent is dislodged from its delivery balloon but remains over the coronary guidewire. **Panel B**: a second guidewire is advanced through or around the struts of the lost stent. **Panel C**: both guidewires are rotated rapidly until their ends become intertwined (**panel D**). **Panel E**: the guidewires are withdrawn into the guide catheter along with the lost stent.
Reproduced with permission from the Brilakis ES. Manual of percutaneous coronary interventions: a step-by-step approach. Elsevier; 2021.

27.1.3.9   Stent deployment

If a stent is lost inside the coronary circulation and wire position is maintained, deployment may be the safest and fastest treatment strategy (**PCI Manual online cases** 108, 134). Crossing the lost stent with a balloon can sometimes be challenging: in such cases a small balloon is usually advanced through the lost stent, followed by increasingly larger balloons until stent deployment is optimized. Sometimes, attempts to advance a balloon through a lost stent may lead to more distal displacement of the stent. If a balloon cannot advance through the lost stent (usually due to stent deformation), crushing of the stent could be performed as described in Section 27.1.3.10.

If the stent is lost in the proximal segment of a coronary artery, an alternative approach is to advance a guide catheter or guide catheter extension over the lost stent, followed by (partial or complete) stent deployment inside the guide catheter or guide catheter extension, followed by removal of all equipment as a unit (Fig. 27.7).

27.1.3.10   Stent crushing

If wire position is not maintained within the lost stent and retrieval is not feasible or desired, crushing the stent with another stent can be the strategy of choice, unless the stent is located in a critical location, such as the left main (Fig. 27.8). A coronary guidewire is advanced around the lost stent, a balloon is used to crush the stent against the coronary artery wall, and another stent is placed, "excluding" the lost stent from the coronary circulation. Crushing the lost stent with a balloon before placing an additional stent is preferred, as balloons are more deliverable than stents and less likely to cause further distal displacement of the lost stent.

Crushing the stent without placing another stent should be avoided as the crushed stent may subsequently migrate more distally (**CTO Manual online case** 167). Avoid inadvertently passing through one or more stent struts with the second wire, especially if the lost stent has been partially deployed. This should be suspected if balloon delivery is challenging around the lost stent and rewiring should be performed, ideally using a knuckled polymer-jacketed guidewire. If balloon delivery remains challenging after rewiring, consider rewiring again or using a Wiggle guidewire.

Meticulous attention should be given to completely appose the stent struts to the vessel wall to avoid limitation of blood flow through the coronary artery, ideally using intracoronary imaging to confirm complete exclusion of the "crushed" stent from the coronary circulation.[3,11] Both stent crushing and stent deployment carry a risk of restenosis, which is, however, much lower than the potential risk of aggressive stent retrieval attempts. Intentional stent crushing is often performed during PCI of in-stent CTOs with encouraging results.[12,13] In some cases of very distal embolization of the lost stent, observation without additional attempts to retrieve it may be preferred (**CTO Manual online case** 171).

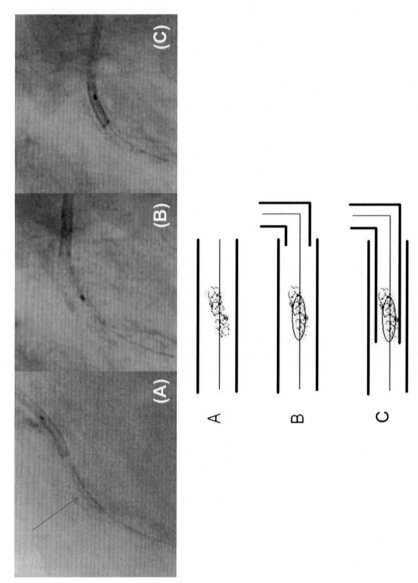

FIGURE 27.7 Stent deployment inside a guide or guide catheter extension. Panel A: an undeployed stent is lost the proximal right coronary artery. **Panel B:** a low profile balloon is inserted inside the undeployed stent. **Panel C:** after advancement of the guide catheter or a guide catheter extension, the stent is deployed inside the catheter and then removed from the coronary artery.
Courtesy of Dr. Konstantinos Marmagkiolis.

Stent crushing

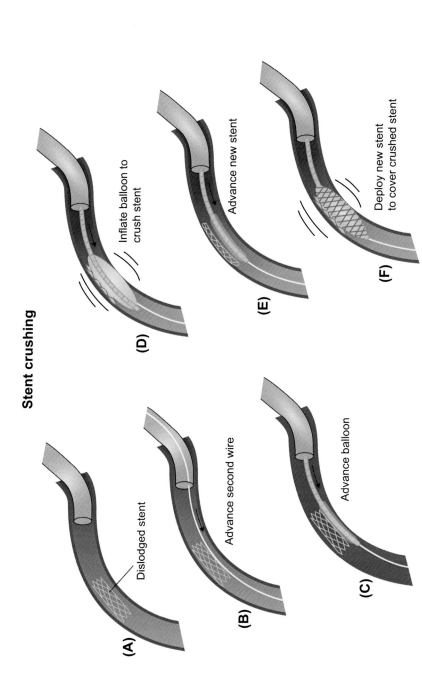

FIGURE 27.8 **Crushing of a dislodged stent. Panel A**: a stent has been dislodged from its delivery balloon and remains inside the coronary artery. **Panel B**: a guidewire is advanced alongside the lost stent. **Panel C**: a balloon is advanced over the guidewire next to the lost stent. **Panel D**: the balloon is inflated "crushing" the lost stent against the vessel wall. **Panel E**: a new stent is advanced alongside the crushed stent. **Panel F**: the new stent is deployed covering the lost stent. *Reproduced with permission from the Brilakis ES. Manual of percutaneous coronary interventions: a step-by-step approach. Elsevier; 2021.*

27.1.3.11 Leave in peripheral artery

If the lost stent is removed from the coronary artery into the iliac or femoral vessels but cannot subsequently be removed through the sheath or the vessel wall, it could be crushed against the iliac or femoral artery wall using a peripheral stent. Distal stent embolization into a small arterial branch (such as a femoral artery branch) may be left untreated, since the distally embolized stents appear to have a benign clinical course. None of the 12 patients with distal stent embolization in three series[4,10,14] had clinical sequela; hence, the risk of extraction usually outweighs the risk of local complications.[3,4]

27.2   Guidewire entrapment and fracture

CTO Manual online case 24
PCI Manual online cases 32, 33, 88, 90

Guidewires are the most commonly entrapped devices during CTO PCI.[1] Guidewire entrapment and fracture is an infrequent and potentially preventable complication, but can have devastating consequences, especially if the guidewire unravels.[2]

27.2.1   Causes

Guidewire entrapment

1. Jailing of guidewire during bifurcation stenting.
2. Guidewire deformation, for example during aggressive attempts for guidewire knuckling (**CTO Manual online case** 24) or during collateral crossing attempts.[8]
3. Withdrawal of a guidewire with a loop at its tip through a proximal stent (**PCI Manual online case** 33).[2]
4. Use of a buddy wire together with a filter-based embolic protection device. Inadvertent stenting over the buddy wire can result in entrapment of the filter, that can be very challenging or impossible to remove.[2]
5. Severe lesion calcification.
6. Aortic prosthetic valve: there is a report of a retrograde guidewire becoming entrapped in an aortic bioprosthesis.[1]

Guidewire fracture

1. Aggressive pulling of an entrapped guidewire.
2. Guidewire over-rotation.
3. Atherectomy over a kinked guidewire.
4. Atherectomy next to a buddy wire.

27.2.2 Prevention

Guidewire entrapment

1. Some advocate to avoid jailing the side branch guidewire during bifurcation stenting; however, the risk of side branch occlusion exceeds the risk of guidewire entrapment in most cases.
2. Avoid aggressive guidewire manipulation, especially in small branches.
3. Straighten the guidewire tip before removing through a stent (**PCI Manual online case** 33)

Guidewire fracture

1. Do NOT pull hard if a guidewire is difficult to retrieve!
2. Do not perform atherectomy over a kinked guidewire. Replace the guidewire with a new one before proceeding with atherectomy.

27.2.3 Treatment

Guidewire entrapment

27.2.3.1 Do NOT pull hard!

Pulling hard may lead to damage of the entrapped guidewire and cause guidewire fracture or unraveling that can be very challenging to treat and may require cardiac surgery (Fig. 27.9). Aggressive pulling of the guidewire may also damage the proximal vessel (cutting-like effect), especially in tortuous vessels.

27.2.3.2 Balloon or microcatheter

Advance a microcatheter or balloon over the entrapped guidewire, as far as possible, then pull the guidewire gently. The balloon or microcatheter allow more focused application of the withdrawal force, facilitating retrieval. A guide catheter extension could also be advanced as distally as possible over the entrapped guidewire, followed by inflation of a balloon inside it "trapping" the wire, followed by withdrawal of the guide extension/guidewire assembly. If the entrapped guidewire is inside a side branch, a wire should be placed inside the main vessel, as proximal stent strut deformation may occur.

27.2.3.3 Inflate balloon

If step 2 (Fig. 27.9) fails, inflate a balloon advanced as far distally as possible over the entrapped guidewire and attempt to withdraw again. In most cases this will allow release of the entrapped guidewire.

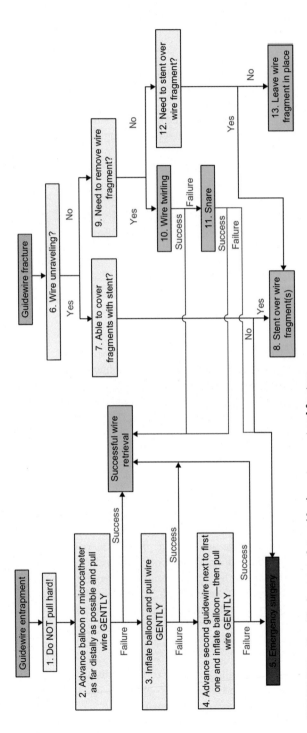

FIGURE 27.9 **Algorithmic approach to guidewire entrapment and fracture.**
Reproduced with permission from the Brilakis ES. Manual of percutaneous coronary interventions: a step-by-step approach. Elsevier; 2021.

27.2.3.4 Second guidewire + balloon

If step 3 (Fig. 27.9) fails, advance a second guidewire next to the entrapped guidewire and inflate a balloon in order to help "free" the entrapped guidewire.[15]

27.2.3.5 Emergency surgery

If all retrieval attempts fail, surgery may be required. A "last resort" option would be to pull "hard" on the guidewire (having a balloon or microcatheter in place) hoping that the guidewire will fracture without unraveling.[16]

Guidewire fracture

27.2.3.6 Wire unraveling

The key consideration in managing a fractured guidewire is whether guidewire unraveling has occurred. Guidewire unraveling is much more challenging to treat than a "clean" fracture, as the guidewire distal spring coil may unravel creating a metal "bird's nest" that could predispose to thrombosis.[2] In cases of guidewire fracture, it is important to perform intravascular ultrasonography to confirm that no wire coil unraveling has occurred.[8]

27.2.3.7 Able to cover wire fragments with a stent?

If guidewire unraveling has occurred, subsequent treatment depends on the location of the unraveled guidewire fragments. If they are extending into the aorta, or if they are located in a critical coronary location (such as the left main or other major bifurcation), emergency surgery is needed to remove those fragments and prevent future thromboembolic complications (both coronary and systemic). Attempts to retrieve the unraveled guidewire fragment should generally be avoided (because they can cause even more unraveling), although successful percutaneous wire fragment removal has been described.[8] A snare can be used to break the unraveled guidewire as close as possible to the coronary ostium (or even within the coronary artery) to allow subsequent stenting over the wire fragment.

27.2.3.8 Stenting over wire fragments

If the unraveled fragments can be "trapped" behind a stent, stenting is performed. Intravascular imaging is important to confirm complete coverage of the wire fragments.

27.2.3.9 Need to remove guidewire fragments?

If there is no guidewire unraveling after guidewire fracture (confirmed by intravascular ultrasonography), the key question is whether the guidewire fragment needs to be removed. Small wire fragments located in small distal branches or in collateral vessels may be best left untreated. If, however, there are long wire fragments, especially if located within a large coronary artery, retrieval is preferred.

27.2.3.10 "Wire twirling" technique

In this technique one or more coronary guidewires are advanced next to the guidewire fragment and twisted several times (Fig. 27.10). The wires may intermingle, allowing subsequent retrieval.

27.2.3.11 Snaring

Use of small coronary snares (as described in Section 27.1.3.5) may allow retrieval of the wire fragments. If it fails emergency surgery may be required for removal.[15]

27.2.3.12 Need to stent over wire fragment?

Similar to Section 27.2.5.2 if the wire fragment is in a distal location in a small branch, it is usually left in place. If, however, it is located more centrally, stenting

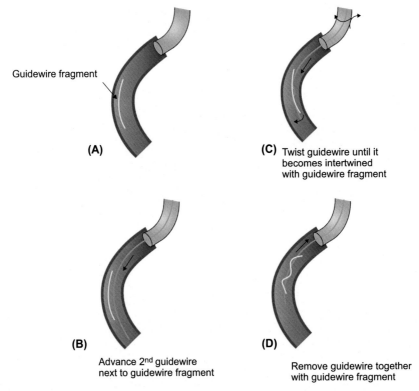

FIGURE 27.10 Illustration of the "wire twirling" "technique" for removing a guidewire fragment. Panel A: a guidewire fragment is located within a coronary artery. **Panel B**: a guidewire is advanced next to the guidewire fragment. **Panel C**: the guidewire is rotated until it becomes intertwined with the guidewire fragment. **Panel D**: the guidewire is removed together with the guidewire fragment.

Reproduced with permission from the Brilakis ES. Manual of percutaneous coronary interventions: a step-by-step approach. Elsevier; 2021.

can "trap" the guidewire against the arterial wall, preventing future migration and reducing the likelihood of intracoronary thrombus formation.[8]

27.2.3.13   Leave wire fragment in situ

This is done for small wire fragments located distally, as described in Section 27.2.5.7.[15] Consider prolonged dual antiplatelet therapy for such cases.

27.3   Balloon entrapment or shaft fracture

CTO Manual online case 84, 131, 169, 171
PCI Manual online case 8

Balloons can get entrapped and fractured. Subsequent treatment depends on the type and location of the entrapped and/or fractured balloon.

27.3.1   Causes

Balloon entrapment

1. Balloon rupture (resulting in the balloon getting "stuck" in the lesion).
2. Balloon failure to deflate (discussed in Sections 27.3.3.11−27.3.3.14).
3. Balloon interaction with a previously placed stent (which is more likely when plaque modification balloons are used). Occasionally previously placed stents have been extracted while retrieving cutting balloons.[17−19]

Balloon shaft fracture

1. Kinking of the (balloon or stent) catheter shaft.

27.3.2   Prevention

Balloon entrapment

1. Consider primary atherectomy for very heavily calcified coronary lesions.
2. Avoid very high-pressure balloon inflation in highly calcified lesions, as it may lead to balloon rupture and entrapment.

Balloon fracture

1. If the shaft of a catheter kinks, discard the catheter and use a new one.

27.3.3   Treatment

Balloon entrapment

27.3.3.1   Failure to deflate

If the reason for entrapment is failure to deflate, specific steps (Sections 27.3.3.11−27.3.3.14) are followed (Fig. 27.11).

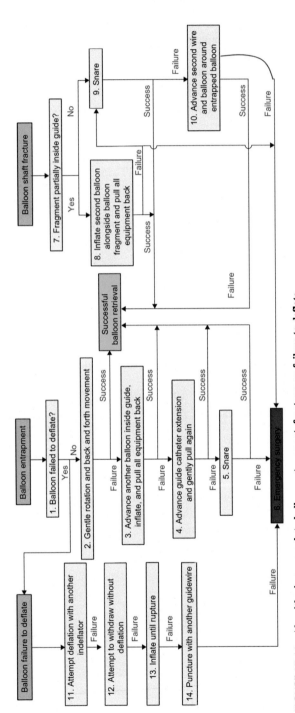

FIGURE 27.11 Algorithmic approach to balloon entrapment, fracture, or failure to deflate.
Reproduced with permission from the Brilakis ES. Manual of percutaneous coronary interventions: a step-by-step approach. Elsevier; 2021.

27.3.3.2 Gentle rotation and back and forth movement

Sometimes balloon rotation may help rewrap the balloon and facilitate retrieval. Similarly, gentle back and forth movements may help free the entrapped balloon.

27.3.3.3 Inflate second balloon inside guide catheter and pull

Another balloon is inserted inside the guide catheter as close to the guide tip as possible, but without exiting the guide catheter. The balloon is inflated at high pressure ($\geq$20 atm) and the guide catheter along with both balloons (the entrapped one and the second one that is inflated) is retracted as a unit.

27.3.3.4 Guide catheter extension

Cut the balloon shaft and advance a guide catheter extension as close as possible to the entrapped balloon, followed by pulling the entrapped balloon. Using a Trapliner may be advantageous, as it allows simultaneous trapping of the entrapped balloon inside the guide catheter.

27.3.3.5 Snare

Advance a snare as close to the balloon as possible, tighten it, and withdraw.

27.3.3.6 Surgery

Emergency surgery may be needed if all efforts to retrieve the balloon fail.[20]

Balloon fracture

27.3.3.7 Balloon fragment inside guide?

If the balloon shaft breaks, the key question is where the fracture occurred and whether a portion of the retained balloon fragment is inside the guide catheter. One way to tell is by comparing the retrieved balloon catheter fragment with a new balloon catheter. If the retained balloon fragment is inside the guide catheter, it can potentially be trapped with another balloon as described in step 8 (Fig. 27.11), followed by removal of both the guide catheter and the balloon fragment. If not, then the balloon fragment can be snared.

27.3.3.8 Inflate second balloon inside guide catheter and retract

The second balloon is inflated at high pressure ($\geq$20 atm), trapping the balloon fragment. The entire assembly (guide catheter, balloon fragment and second inflated balloon) are then withdrawn under fluoroscopy while maintaining the balloon inflation.[21]

27.3.3.9 Snares

Snaring of the retained balloon catheter fragment is performed, followed by withdrawal.

27.3.3.10 Wire and balloon around the entrapped balloon

A guidewire is advanced around the entrapped balloon (usually through a second guide catheter), followed by balloon angioplasty around the entrapped balloon fragment aiming to free the fragment from the coronary artery wall. The wire "twirling" technique can also be used (Fig. 27.12).

Balloon failure to deflate

27.3.3.11 Attempt deflation with another indeflator

Attempt deflation with another indeflator (ideally filled with saline) or with a Luer-lock syringe (also filled with saline).

27.3.3.12 Attempt to withdraw without deflation

If the size of the balloon is smaller than the reference diameter of the vessel, an attempt could be made to withdraw it into the guide catheter without deflation.

27.3.3.13 Inflate at high pressure to rupture the balloon

If the balloon diameter is smaller than the vessel diameter, it can be inflated at high pressure until it ruptures.

27.3.3.14 Puncture the inflated balloon with a guidewire

An over-the-wire balloon is advanced over a second guidewire just proximal to the failing to deflate balloon and inflated at low pressure (2−4 atm). The wire is then removed and a stiff tip guidewire (such as Astato 20, Confianza Pro 12, Infiltrac or Infiltrac Plus, or Hornet 14) or alternatively the back end of a guidewire is advanced through the OTW balloon aiming to puncture the "failing to deflate" balloon. Use of an inflated over-the-wire (OTW) balloon minimizes the risk of vessel injury by centering the wire and helping direct it towards the failing to deflate balloon.

Alternatively, the balloon shaft can be cut and a guide catheter extension advanced and gently pushed against the failing to deflate balloon. A stiff guidewire or the back end of a guidewire can then be advanced through the guide catheter extension to puncture the balloon, minimizing the risk of proximal vessel injury.

If all attempts to remove the failing to deflate balloon fail, emergency surgery may be required.

Balloon entrapment may also compromise coronary flow and result in hemodynamic instability, potentially requiring hemodynamic support.

27.4 Microcatheter entrapment or fracture

CTO Manual online case 87

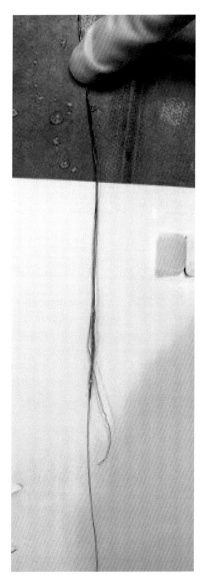

FIGURE 27.12 **Successful retrieval of a fractured balloon shaft using the wire "twirling" technique.**
Courtesy of Dr. Konstantinos Marmagkiolis.

27.4.1 Causes

Microcatheter entrapment

1. Aggressive microcatheter advancement through complex (tortuous and calcified) lesions.

2. Advancement of two microcatheters, or a retrograde microcatheter and an antegrade balloon over the externalized guidewire with both ends of the microcatheters meeting and "interlocking."
3. Microcatheter manipulations without having a guidewire inside the microcatheter lumen.

Microcatheter fracture

1. Aggressive manipulation, especially of softer tip microcatheters (such as the Caravel) through complex lesions.[22] The Mamba microcatheters are the least likely to have tip fracture.
2. Over-rotating a microcatheter: this may compromise the structural integrity of the microcatheter and hinder retrieval (and advancement) attempts. Do not perform more than 10 microcatheter rotations if its tip does not move.

27.4.2   Prevention

Microcatheter entrapment

1. Avoid aggressive manipulation of microcatheters through highly complex and calcified lesions. If there is resistance in moving the guidewire through the microcatheter (microcatheter fatigue), the microcatheter should be replaced for a new one.
2. When performing retrograde CTO PCI over an externalized guidewire, the tip of the antegrade equipment should not come in contact with the tip of the retrograde microcatheter.
3. Avoid microcatheter over-rotation.
4. Always keep a guidewire within the microcatheter lumen during micro-catheter manipulations.

Microcatheter fracture

1. Avoid aggressive manipulation of microcatheters (especially soft-tip, such as the Caravel) through highly complex and calcified lesions or across stent struts.[22] Microcatheters with stronger tip construction (such as the Mamba or Teleport) may have lower risk of tip fracture.[23]

27.4.3   Treatment

Retrieving the entrapped and/or fractured microcatheter is important for pre-venting acute and chronic complications. Retrieval is performed using similar techniques, as those described for retrieving a fractured balloon catheter, such as snares, use of guide catheter extensions and "trapping" the microcatheter inside the guide, and wiring through the area of entrapment with another guidewire and performing balloon angioplasty or sometimes laser[24] or rotational atherectomy. If retrieval attempts fail, emergency surgery may be required.

In cases of microcatheter tip fracture, retrieval can be very challenging or impossible. In such cases, stenting over the tip fragment may be the fastest and safest solution (Fig. 27.13). Sometimes, the microcatheter tip becomes "fused" to the guidewire, hence guidewire removal also removes the microcatheter tip.

A similar approach can be used for fracture of distal tip of guide catheter extensions (**CTO Manual online case** 171).

27.5 Rotablator burr entrapment

Causes

1. Forceful burr advancement with significant decelerations.[25]
2. Use of small burrs (especially the 1.25 mm burr—this is often called the "kokeshi" effect).
3. Rotablation within previously deployed stents.
4. Rotablation in very tortuous vessels.

Prevention

1. Gentle burr advancement with short, nonforceful contact with the lesion ("pecking" the lesion).
2. Use 1.5 mm or larger burrs (1.25 mm burrs are more likely to advance through the lesion and then fail to come back through the lesion).
3. Do not stop the burr distal to the lesion.
4. Do not start or stop rotablation within the target lesion.

Treatment (Fig. 27.14)

1. Straighten out the guide, ensure coaxial alignment, and try again to remove the entrapped burr.
2. If rotation is still possible, using the Dynaglide setting may facilitate burr withdrawal.
3. A second wire can be advanced next to the entrapped burr (intraplaque or extraplaque), followed by balloon dilatation in an effort to dig the burr out.[26]
4. The Rotablator shaft is cut, and a guide extension is advanced over it.[27–29]
5. A snare can also be advanced over the Rotablator shaft.[27]
6. The Rotawire can be withdrawn (gently) possibly helping bring back the burr (the 0.014 inch tip of the Rotawire is larger than the 0.009 lumen of the burr).
7. If hemodynamics permit, vasodilators can be administered.
8. If all attempts fail, emergency cardiac surgery may be required.[30]

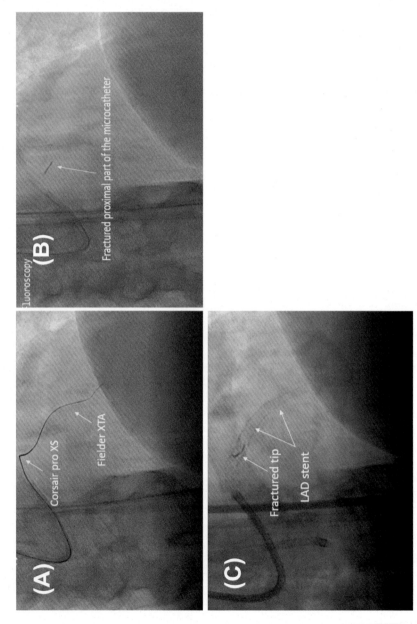

FIGURE 27.13 Stenting over a fractured microcatheter tip. Panel A: excessive pushing and rotation of a Corsair Pro XS microcatheter. **Panel B**: the microcatheter became entrapped within the calcified CTO and the microcatheter tip fractured during retrieval. **Panel C**: the microcatheter tip was stented against the vessel wall after successful crossing and ballooning of the CTO.
Courtesy of Dr. Sevket Gorgulu.

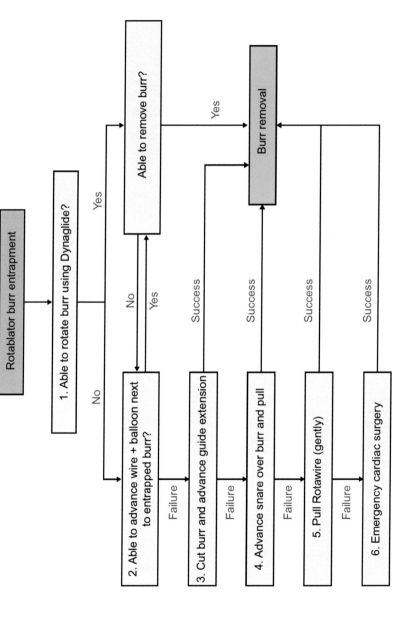

FIGURE 27.14 **Approach to entrapped Rotablator burr.**
Reproduced with permission from the Brilakis ES. Manual of percutaneous coronary interventions: a step-by-step approach. Elsevier; 2021.

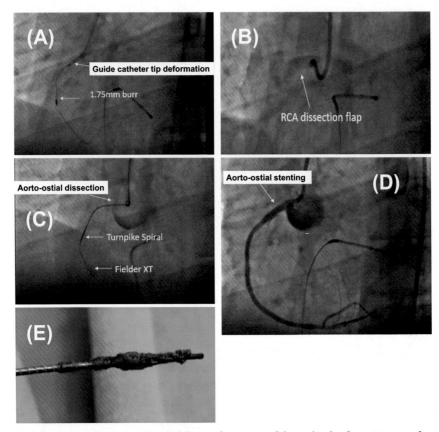

FIGURE 27.15 **Extensive vessel injury after successful retrieval of an entrapped Rotablator burr. Panel A**: entrapment of a 1.75 mm Rotablator burr in the mid RCA. Forceful withdrawal caused deformation of the guide catheter tip. **Panel B**: proximal RCA occlusion due to dissection causing acute myocardial infarction and hemodynamic compromise. **Panel C**: aorto-ostial dissection. Successful reentry with a Turnpike Spiral and a Fielder XT guidewire. **Panel D**: successful recanalization of the RCA after implantation of multiple stents. **Panel E**: image of the Turnpike Spiral microcatheter.
Courtesy of Dr. Sevket Gorgulu.

Extensive stenting may be required due to vessel injury after successful retrieval of the entrapped burr (Fig. 27.15).

References

1. Gasparini GL, Sanz-Sanchez J, Regazzoli D, et al. Device entrapment during percutaneous coronary intervention of chronic total occlusions: incidence and management strategies. *EuroIntervention* 2021;**17**:212−19.

2. Iturbe JM, Abdel-Karim AR, Papayannis A, et al. Frequency, treatment, and consequences of device loss and entrapment in contemporary percutaneous coronary interventions. *J Invasive Cardiol* 2012;**24**:215–21.
3. Brilakis ES, Garrat KN. Device loss during percutaneous coronary intervention: incidence, complications, and retrieval methods. In: Ellis SG, Holmes DRJ, editors. *Strategic approaches in coronary intervention.* Lippincott, Williams and Wilkins; 2005. p. 325–31.
4. Brilakis ES, Best PJ, Elesber AA, et al. Incidence, retrieval methods, and outcomes of stent loss during percutaneous coronary intervention: a large single-center experience. *Catheter Cardiovasc Interv* 2005;**66**:333–40.
5. Abdel-Wahab M, Richardt G, Joachim Buttner H, et al. High-speed rotational atherectomy before paclitaxel-eluting stent implantation in complex calcified coronary lesions: the randomized ROTAXUS (Rotational Atherectomy Prior to Taxus Stent Treatment for Complex Native Coronary Artery Disease) trial. *JACC Cardiovasc Interv* 2013;**6**:10–19.
6. Kozman H, Wiseman AH, Cook JR. Long-term outcome following coronary stent embolization or misdeployment. *Am J Cardiol* 2001;**88**:630–4.
7. Utsunomiya M, Kobayashi T, Nakamura S. Case of dislodged stent lost in septal channel during stent delivery in complex chronic total occlusion of right coronary artery. *J Invasive Cardiol* 2009;**21**:E229–33.
8. Sianos G, Papafaklis MI. Septal wire entrapment during recanalisation of a chronic total occlusion with the retrograde approach. *Hellenic J Cardiol* 2011;**52**:79–83.
9. Brilakis ES. *Manual of percutaneous coronary interventions: a step-by-step approach.* Elsevier; 2021.
10. Eggebrecht H, Haude M, von Birgelen C, et al. Nonsurgical retrieval of embolized coronary stents. *Catheter Cardiovasc Interv* 2000;**51**:432–40.
11. Candilio L, Mitomo S, Carlino M, Colombo A, Azzalini L. Stent loss during chronic total occlusion percutaneous coronary intervention: optical coherence tomography-guided stent 'crushing and trapping'. *Cardiovasc Revasc Med* 2017;**18**:531–4.
12. Capretti G, Mitomo S, Giglio M, Carlino M, Colombo A, Azzalini L. Subintimal crush of an occluded stent to recanalize a chronic total occlusion due to in-stent restenosis: insights from a multimodality imaging approach. *JACC Cardiovasc Interv* 2017;**10**:e81–3.
13. Azzalini L, Karatasakis A, Spratt JC, et al. Subadventitial stenting around occluded stents: a bailout technique to recanalize in-stent chronic total occlusions. *Catheter Cardiovasc Interv* 2018;**92**:466–76.
14. Alfonso F, Martinez D, Hernandez R, et al. Stent embolization during intracoronary stenting. *Am J Cardiol* 1996;**78**:833–5.
15. Danek BA, Karatasakis A, Brilakis ES. Consequences and treatment of guidewire entrapment and fracture during percutaneous coronary intervention. *Cardiovasc Revasc Med* 2016;**17**:129–33.
16. Karacsonyi J, Martinez-Parachini JR, Danek BA, et al. Management of guidewire entrapment with laser atherectomy. *J Invasive Cardiol* 2017;**29**:E61–2.
17. Pappy R, Gautam A, Abu-Fadel MS. AngioSculpt PTCA balloon catheter entrapment and detachment managed with stent jailing. *J Invasive Cardiol* 2010;**22**:E208–10.
18. Sanchez-Recalde A, Galeote G, Martin-Reyes R, Moreno R. AngioSculpt PTCA balloon entrapment during dilatation of a heavily calcified lesion. *Rev Esp Cardiol* 2008;**61**:1361–3.
19. Giugliano GR, Cox N, Popma J. Cutting balloon entrapment during treatment of in-stent restenosis: an unusual complication and its management. *J Invasive Cardiol* 2005;**17**:168–70.

20. Lorusso R, De Cicco G, Ettori F, Curello S, Gelsomino S, Fucci C. Emergency surgery after saphenous vein graft perforation complicated by catheter balloon entrapment and hemorrhagic shock. *Ann Thorac Surg* 2008;**86**:1002—4.

21. Karacsonyi J, Sasi V, Ungi I, Brilakis ES. Management of a balloon shaft fracture during subintimal retrograde chronic total occlusion percutaneous coronary intervention due to in-stent restenosis. *J Invasive Cardiol* 2018;**30**:E64—6.

22. Megaly M, Sedhom R, Pershad A, et al. Complications and failure modes of coronary microcatheters. *EuroIntervention* 2021;**17**:e436—8.

23. Vemmou E, Nikolakopoulos I, Xenogiannis I, et al. Recent advances in microcatheter technology for the treatment of chronic total occlusions. *Expert Rev Med Devices* 2019;**16**:267—73.

24. Tajti P, Ayoub M, Loffelhardt N, Mashayekhi K. Management of microcatheter fracture in complex percutaneous coronary intervention with laser atherectomy. *Cardiovasc Revasc Med* 2021;**28S**:208—11.

25. Fouquet O, Gomez M, Abi-Khalil W, Binuani P. Open heart surgery for a stuck rotablator. *Eur J Cardiothorac Surg* 2016;**49**:e149.

26. Tanaka Y, Saito S. Successful retrieval of a firmly stuck rotablator burr by using a modified STAR technique. *Catheter Cardiovasc Interv* 2016;**87**:749—56.

27. Chiang CH, Liu SC. Successful retrieval of an entrapped rotablator burr by using a guideliner guiding catheter and a snare. *Acta Cardiol Sin* 2017;**33**:96—8.

28. Imamura S, Nishida K, Kawai K, Hamashige N, Kitaoka H. A rare case of Rotablator((R)) driveshaft fracture and successful percutaneous retrieval of a trapped burr using a balloon and GuideLiner((R)). *Cardiovasc Interv Ther* 2017;**32**:294—8.

29. Cunnington M, Egred M. GuideLiner, a child-in-a-mother catheter for successful retrieval of an entrapped rotablator burr. *Catheter Cardiovasc Interv* 2012;**79**:271—3.

30. Sulimov DS, Abdel-Wahab M, Toelg R, Kassner G, Geist V, Richardt G. Stuck rotablator: the nightmare of rotational atherectomy. *EuroIntervention* 2013;**9**:251—8.

Chapter 28

Other complications: hypotension, radiation skin injury, contrast-induced acute kidney injury

28.1 Hypotension

CTO Manual online cases 69, 85, 92, 129, 146
PCI Manual online cases 23, 27, 40, 45, 72, 105, 127, 140
Video 28.1 Hypotension

Continuous monitoring of the pressure and electrocardiographic tracing is critical for enhancing the safety of CTO PCI (Section 2.2). Awareness of the differential diagnostic algorithm if hypotension occurs during angiography or PCI, can facilitate rapid decision making and initiation of corrective action.

28.1.1 Causes

The appearance of low blood pressure on hemodynamic monitoring does not necessarily mean that the patient's systemic blood pressure is low, as "hypotension" could be due to technical issues (Fig. 28.1). Therefore, when hypotension occurs, it should be immediately assessed to determine whether the patient is truly hypotensive or not. If the patient's systemic blood pressure is indeed low, the differential diagnosis should be immediately considered to allow prompt diagnosis and treatment.

28.1.1.1 False hypotension (systemic pressure is normal, but appears low on the arterial pressure tracing)
1. Hemostatic valve of the Y-connector is open.
2. Connection between catheter and pressure transducer is open or has air in the line.
3. Pressure dampening. This is common in CTO PCI, especially in patients with ostial coronary lesions and diffusely diseased vessels and with use of large guide catheters, such as 8 French. Pressure dampening can be masked when

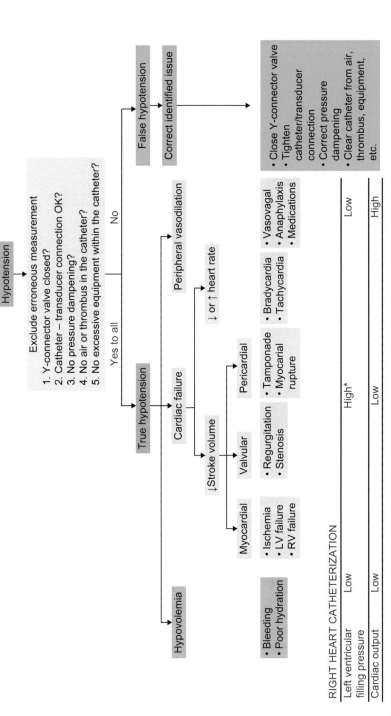

FIGURE 28.1 Differential diagnosis of hypotension. *Left ventricular filling pressure can be low in cases of right ventricular failure, for example, in cases of loculated effusions in the right atrium or right ventricle.[1] Reproduced with permission from the Brilakis ES. Manual of percutaneous coronary interventions: a step-by-step approach. Elsevier; 2021.*

side hole guide catheters are used, hence the latter should never be used to engage the left main coronary artery (with the exception of ostial left main CTOs), as they can mask ischemia and lead to hemodynamic collapse.

4. Catheter obstruction by air, thrombus, or contrast (contrast can cause pressure dampening, especially in 4 and 5 French catheters). The catheter should be aspirated. If aspiration fails, the catheter should be removed **without performing any injection**.
5. Bulky equipment within the catheter.

28.1.1.2 True hypotension

There are three major causes of true hypotension: hypovolemia, cardiac failure, and peripheral vasodilation (Fig. 28.1). If the cause of hypotension is not immediately apparent, right heart catheterization can be very useful for determining the cause of hypotension.

28.1.1.2.1 Hypovolemia

Hypovolemia is most commonly caused by bleeding. Inspection of the access sites may reveal a hematoma. Fluoroscopy of the bladder may demonstrate the "indented bladder" sign that suggests retroperitoneal hematoma.

28.1.1.2.2 Cardiac failure

Cardiac output is the product of (stroke volume) x (heart rate). Decreased stroke volume may be due to: (1) left or right ventricular dysfunction, which is most commonly caused by ischemia; (2) valvular abnormalities, such as acute valvular regurgitation; or (3) pericardial tamponade.

Amplatz catheters (and deeply curved EBU catheters) may push the aortic cusp open and cause acute aortic regurgitation (Fig. 28.2); simple guide catheter repositioning can immediately correct the hypotension.

Both tachy- and brady-arrhythmias can also reduce the cardiac output.

28.1.1.2.3 Peripheral vasodilation

Peripheral vasodilation may be due to a vasovagal reaction, a systemic anaphylactic reaction, or medication administration, such as nitroglycerin (especially if the patient was recently exposed to phosphodiesterase inhibitors) or verapamil (intracoronary or intra-arterial for preventing radial spasm).

28.1.2 Prevention

Preventing *false hypotension* can be accomplished using meticulous technique:

1. Careful assessment and verification of all connections.
2. Meticulous care in clearing the catheter after insertion.
3. Avoiding deep catheter engagement causing dampening.

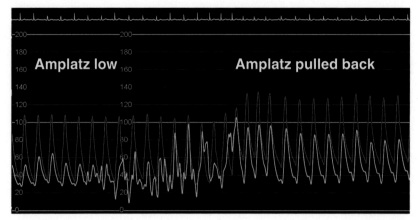

FIGURE 28.2 Blood pressure is lower in the left half of the screen due to deep advancement of an Amplatz catheter on the aortic valve. Blood pressure increased after withdrawal of the Amplatz catheter (right half of the screen). The green pressure tracing has dampening. *Reproduced with permission from the Brilakis ES. Manual of percutaneous coronary interventions: a step-by-step approach. Elsevier; 2021.*

Preventing true hypotension can be achieved via careful planning of the procedure and attention to each step, as follows:

1. Careful arterial access technique and use of radial access, as well as avoiding excessive anticoagulation can decrease the risk of bleeding.
2. Adequate preprocedural hydration.
3. Considering prophylactic hemodynamic support in high-risk patients undergoing CTO PCI, as described in Chapter 14.
4. Preventing prolonged coronary artery occlusion will minimize ischemia.
5. Avoiding perforations and the resultant tamponade.
6. Adequate premedication for patients with contrast allergy will decrease the risk of anaphylactic shock (Section 3.3).
7. Early diagnosis and correction of tachy- and brady-arrhythmias.
8. Adequate sedation and hydration may minimize the risk of vasovagal reaction.
9. Avoiding excessive doses of nitroglycerin and other vasodilators.

28.1.3 Treatment

Immediate and complete treatment of the underlying cause is best for correcting hypotension, but may not always be feasible, for example, in patients who develop acute vessel closure that cannot be recanalized. Therefore, concomitant implementation of measures that can increase the systemic blood pressure and maintain systemic perfusion is often needed.

28.1.3.1 Maintain systemic perfusion

While treating the underlying cause of hypotension, administration of vaso-pressors and inotropes and initiation of mechanical circulatory support (Chapter 14) may be necessary in some patients. When performing CTO PCI in high-risk heart failure patients, a central venous sheath can facilitate administration of vasopressors or inotropes during CTO PCI. If the patient develops cardiac arrest, cardiopulmonary resuscitation is performed (ideally using an automated system, such as the Lucas system), followed by veno-arterial extracorporeal membrane oxygenation (VA-ECMO) initiation if the patient does not promptly recover.

28.1.3.2 Treat underlying cause

Hypovolemia: normal saline administration and blood transfusion in case of bleeding.

Myocardial dysfunction: this is usually caused by ischemia and can be improved with coronary revascularization and in some cases with mechanical circulatory support.

Valvular disease: although acute mitral or aortic regurgitation may benefit from vasodilator administration, they are often poorly tolerated requiring urgent or emergent surgery. Acute mechanical circulatory support may also improve hemodynamics: an intra-aortic balloon pump can be useful in the setting of acute mitral regurgitation and the TandemHeart for acute aortic insufficiency resulting in cardiogenic shock.

Pericardial disease: tamponade is treated with pericardiocentesis. If the patient has myocardial rupture (usually in the setting of acute myocardial infarction) emergency surgery is needed.

Arrhythmias: arrhythmias occurring in the cardiac catheterization labora-tory are usually due to ischemia; hence, treatment of ischemia will improve or correct them. Arrhythmias often occur after reperfusion (post-reperfusion arrhythmias). Bradycardia (for example due to atrioventricular block) is often caused by medication administration, such as adenosine (Section 3.2.4), how-ever the duration of action of adenosine is very short.

Peripheral vasodilation: epinephrine for anaphylactic shock and fluid administration.

28.2 Radiation skin injury

Video 28.2 Preventing radiation-related complications

Excessive radiation dose can result in deterministic complications (e.g., radiation skin injury), but can also lead to stochastic complications (e.g., can-cer or birth defects).[2] Technological developments can help reduce the operator's (and sometimes also the patient's) radiation dose.[3]

28.2.1   Causes

Deterministic radiation effects, such as skin injury and cataracts, correlate directly with the *air kerma (AK) dose* to a particular skin area.[4]

The following AK dose thresholds are important to remember[5]:

- *< 5 Gray:* below this threshold skin injury is unlikely to occur.
- *5−10 Gray:* skin injury is possible.
- *10−15 Gray:* skin injury is likely, requiring physicist assessment of the case.
- *> 15 Gray:* this is considered a sentinel event by the Joint Commission and requires reporting to the regulatory authorities in the United States.

28.2.2   Prevention

Preventing radiation skin injury can be achieved by limiting the overall radiation dose and by rotating the image receptor (using different imaging views during the procedure) to distribute the radiation dose to various skin areas (Fig. 28.3).

Before the procedure	During the procedure

1. Planning
    1. Develop a plan prior to the procedure based on prior studies and all clinical information
 	
 2. Equipment
    1. New X-Ray machines
    2. Zero Gravity (operator only)
    3. Robotic PCI (operator only)
    4. EGGNEST (operator only)
    5. RAMPART IC (operator only)
	3. Do not use radiation
	    1. No pedal "lag time"
	    2. Use balloon and wire markers when inserting them into the guide catheter
	    3. Use the trapping technique for equipment exchanges
	    4. Use "Fluoro Store" for documenting guidewire manipulations and balloon or stent inflations
	    5. Use intravascular imaging
	 4. Minimize radiation dose
	    1. Fluoroscopy rate: 3.75-7.5 fps
	    2. Low magnification
	    3. Collimation, especially when increasing the field of view
	    4. Avoid extreme angles of the image receptor
	    5. Position patient close to image receptor
	    6. Re-position image receptor often
	    7. Real time dose rate monitoring – make adjustments if >20 mGy/min
	    8. Real time patient radiation dose monitoring
	    9. Real time personnel dose monitoring
	 5. Shielding and distance (operator only)
	    1. Take a step back
	    2. Use manifold tubing extensions
	    3. Place fixed radiation shields close to the patient and not too high
	    4. Use disposable shields

FIGURE 28.3   How to prevent radiation injury during cardiac catheterization. *Modified with permission from the Brilakis ES. Manual of percutaneous coronary interventions: a step-by-step approach. Elsevier; 2021.*

Reducing patient (and operator) radiation dose can be achieved both before and during the procedure:

Before the procedure

28.2.2.1 Planning

Careful planning of the procedure can prevent unnecessary steps, facilitate procedural success and minimize contrast and radiation dose. Careful procedural planning is especially important for complex procedures, such as CTO PCI.[6] For example, carefully studying the angiogram and developing a strategic plan[7] can facilitate CTO crossing. Prior knowledge of the type and location of aortocoronary bypass grafts can expedite bypass graft engagement. Catheterization in very obese patients should be performed with newer X-ray systems.

28.2.2.2 Equipment

28.2.2.2.1 X-ray machines

Newer X-ray machines achieve satisfactory image quality with lower radiation dose.[8,9] Using low radiation settings (low fluoroscopy and cineangiography frames per second and dose rate and lower copper filter thickness) can reduce radiation dose for both the patient and the operator.[10-12]

28.2.2.2.2 Zero gravity (reduces operator dose only)

The Zero Gravity ceiling-suspended lead (Biotronik) not only provides radiation protection to the operator, but also obviates the need for wearing lead and the associated orthopedic injuries.

28.2.2.2.3 Robotic PCI (reduces operator dose only)

Robotic PCI (CorPath, Corindus) allows near elimination of operator radiation dose,[13] but is currently available only at a few centers.

28.2.2.2.4 EggNest (reduces operator dose only)

The EggNest-XR System (Egg Medical) includes various shields that reduce scatter radiation.

28.2.2.2.5 Rampart IC (reduces operator dose only)

The Rampart IC system (Rampart IC) includes several shields that reduce scatter radiation.

During the procedure

28.2.2.3 Do not use radiation unless absolutely necessary

28.2.2.3.1 No pedal "lag time"

The "heavy foot" syndrome is defined as using X-ray when it is not needed, for example when the operator is not looking at the screen! A "lag time" in releasing the pedal is common, especially in early stages of training, and should be a major focus for learning and improvement.

28.2.2.3.2 Use balloon and wire markers

When advancing equipment through the guide catheter, use the balloon, wire, and stent shaft markers to determine if the device is close to the tip of the guide catheter, at which time fluoroscopy is needed. Knowing the length of the guide catheter is critical, since using the more proximal marker in 90 cm long guide catheters may result in the device exiting the guide catheter before the marker reaches the Y-connector.

28.2.2.3.3 Use trapping technique for equipment exchanges

The trapping technique (Section 8.2.2.9.1) is of particular importance for CTO PCI as it allows secure equipment exchanges while minimizing the use of X-ray.

28.2.2.3.4 Fluoro store function

Cine-angiography exposes the patient to $\approx$ 10x higher dose compared with fluoroscopy and is not reflected in the fluoroscopy time. The "fluoro store" or "fluoro save" function, is available in most modern X-ray equipment and should be used instead of cine to document wire manipulations and balloon and stent inflations.

28.2.2.3.5 Use of intravascular imaging

Intravascular imaging can limit the need for both contrast and radiation administered during PCI.

28.2.2.4 Minimize radiation dose

28.2.2.4.1 Fluoroscopy frame rate: 6–7.5 fps

Fluoroscopy at 6 to 7.5 frames per second (or even 3.75 frames per second in thin patients or for right heart catheterization) provides in nearly all cases adequate image quality, while reducing radiation dose by half compared with 15 fps fluoroscopy and should be routinely used.[14]

28.2.2.4.2 Low magnification

Lower magnification requires less radiation exposure. Similar to using 7.5 fps, using lower magnification requires a "learning curve" to adjust to the change in image size, although lower magnification images still appear quite large in contemporary large screens.

28.2.2.4.3 Collimation, especially when increasing field of view

Collimation reduces the size of the skin area exposed to radiation and reduces the overall dose area product dose received by the patient, even though the total air kerma is not changed. Collimation allows a smaller skin exposure in any one projection lessening potential skin injury from overlapping exposures when imaging angles are changed.[15]

A caveat of collimation is that some equipment (for example the tips of guidewires or guide catheters) may not be included in the field of view, requiring intermittent monitoring to ensure that no significant changes have occurred (for example excessive distal migration of a guidewire that can lead to distal vessel perforation or deep engagement of the guide catheter that can lead to aortocoronary dissection).[15]

28.2.2.4.4 Avoid extreme angles

When performing PCI, the working angle of the image receptor should be minimized. Steep angles, such as greater than 30 degrees from anteroposterior (AP), are associated with significantly higher radiation exposure due to penetration through more tissue, hence less steep angles are preferred.[15] This increased dose is often not recognized by the operator and reflects the automatic dose increase by the equipment to maintain image quality. *The AP (anteroposterior) projection may not be the optimal view, because the spine is included in the field*, hence slightly angulated views are preferable.[16]

The right anterior oblique projection can result in less operator radiation exposure, but can be challenging for mid right coronary artery wiring, although it is excellent for working in the mid left anterior descending artery.

28.2.2.4.5 Position patient close to image receptor

The table should be placed as high as possible and the image receptor as close to the patient, as possible (Fig. 28.4).[2]

28.2.2.4.6 Reposition image receptor often

Using multiple angles during fluoroscopy and cine-angiography is critical during long procedures to minimize radiation exposure of the same entry point through the patient's skin.[15]

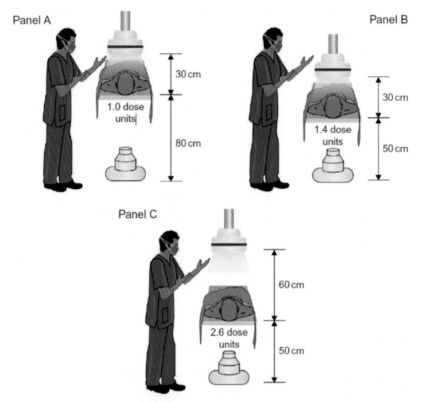

Panel A

30 cm

1.0 dose units|

80 cm

Panel B

30 cm

1.4 dose units

50 cm

Panel C

60 cm

2.6 dose units

50 cm

FIGURE 28.4 Example of optimal table positioning to minimize patient (and operator) radiation exposure. Panel A: the physician performs the procedure with the patient table elevated and the image receptor close to the patient (total distance from the X-ray tube to the detector = 110 cm). **Panel B:** the physician employs a lower table setting but maintains the image receptor close to the patient's chest (total distance from the X-ray tube to the detector = 80 cm). Because of the closer proximity to the X-ray tube, the dose rate to the patient at the beam entrance port will be about 40% higher. **Panel C:** the physician employs a low table height but has elevated the image receptor (total distance from the X-ray tube to the detector = 110 cm). The skin dose to the patient is 260% that of the patient on panel A. (The image generated by the configuration on panel C is 40%−50% larger owing to geometric magnification caused by the elevated image receptor). If the procedure on panel A required a 3 Gy skin dose, the same procedure employing the panel B configuration would require 4.2 Gray, whereas the one performed employing the panel C configuration would require 7.8 Gray.[2] *Reproduced with permission from Writing Committee M, Hirshfeld JW, Jr, Ferrari VA, et al. 2018 ACC/HRS/ NASCI/SCAI/SCCT expert consensus document on optimal use of ionizing radiation in cardiovascular imaging-best practices for safety and effectiveness, part 1: radiation physics and radiation biology: a report of the American College of Cardiology Task Force on Expert Consensus Decision Pathways Developed in Collaboration with mended hearts. Catheter Cardiovasc Interv 2018;92:203−221.*

28.2.2.4.7 Real time dose rate monitoring

The radiation dose rate (displayed by most contemporary X-ray systems) can provide real time feedback on the radiation dose being administered, prompting changes if the dose rate is high (> 20 mGy/min).

28.2.2.4.8 Real time patient dose monitoring

The operator should continually monitor the *cumulative air kerma dose*: consider stopping the procedure if >5 Gray air kerma dose is administered and consider staging the procedure for a later time. Each catheterization laboratory should have a protocol for alerting the operator on radiation dose (e.g., announcing the dose used every 1 Gray and/or every 30 min).

28.2.2.4.9 Real time personnel dose rate monitoring

Real time operator radiation dose monitors (such as Dose Aware, Philips) can reduce operator dose by approximately 30% by providing real-time radiation exposure feedback that should prompt the operator to implement techniques that can reduce radiation dose.[17]

28.2.2.5 Shielding and distance

These measures only reduce operator dose, which is very important given daily exposure of operators to ionizing radiation over many years.

28.2.2.5.1 Distance—Take a step back

Operator and staff must maximize their distance from the X-ray tube, as radiation dose decreases exponentially with distance (inverse square law).[18] All appendages—operators' and patients'—should be out of the imaging field!

28.2.2.5.2 Use manifold tubing extensions

Manifold tubing extensions allow the operators to inject contrast while maintaining a longer distance from the X-ray tube.

28.2.2.5.3 Place fixed radiation shields close to the patient and not too high

The radiation shields should be placed close to the patient at the point of maximal scatter (where the X-rays enter the patient's back) to minimize radiation scatter.

28.2.2.5.4 Disposable shields

Additional disposable radiation absorbing pads can be placed in the sterile field to reduce scatter radiation, such as the RadPad shields (Worldwide Innovations & Technologies) (Section 30.18) that should be placed over the patient's abdomen.[19] Disposable shields should not be placed in the field of view, as this will increase the output of the X-ray tube, due to less radiation received by the detector.

28.2.3   Treatment

All patients who receive >5 Gray air kerma dose should be followed up within a month and their skin examination documented in the medical record. During physical examination, the back of the patients should be inspected to detect any radiation injury (Fig. 28.5); if such an injury is diagnosed, patients should be referred to specialists (dermatologist, plastic surgeons) for further evaluation and treatment. If the patient is not aware of this risk and develops erythema and local discomfort, he/she may see a dermatologist who may biopsy the lesion, potentially leading to a nonhealing ulcer.

For >10 Gray air kerma dose, a qualified physicist should promptly calculate peak skin dose and the patient's skin should be examined in 2–4 weeks.

The Joint Commission identifies peak skin doses >15 Gray as a sentinel event; hospital risk management and regulatory agencies need to be contacted within 24 h.

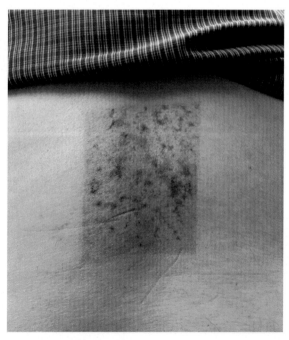

FIGURE 28.5   **Example of radiation skin injury**. *Courtesy of Dr. Brian Jefferson.*

28.3   Contrast-induced acute kidney injury (CI-AKI)

PCI Manual online case 31, 36
Video 28.3 How to prevent contrast-induced acute kidney injury

28.3.1   Causes

There are several risk factors for CI-AKI (formerly called contrast-induced nephropathy [CIN])[20]:

1. Chronic kidney disease. In patients with advanced chronic kidney disease (CKD), defined as an estimated glomerular filtration rate (eGFR) <30 mL/min/1.73 m^2, the incidence of CI-AKI can be as high as 27%.[20] CKD patients often have complex occlusions with heavy calcification. Compared with patient without CKD, CTO PCI in CKD patients has lower success and higher complication rates.[21–23]
2. Contrast volume. A contrast-volume-to-creatinine clearance (CV/CrCl) ratio >2 has been identified as an independent predictor of CI-AKI in patients with an eGFR <30 mL/min/1.73 m^2.[24]
3. Hypotension in various settings: cardiogenic shock, acute heart failure, acute coronary syndrome.
4. Advanced age (>75 years).
5. Diabetes.
6. Anemia.
7. Low ejection fraction.
8. Volume depletion.

There are several scores that can be used to identify patients at high-risk for AKI, such as the Mehran risk score,[25] the Blue Cross Blue Shield of Michigan Cardiovascular Collaborative (BMC2) model,[26] or the National Cardiovascular Data Registry Cath-PCI registry AKI prediction Model.[27]

28.3.2   Prevention

Several strategies can be used to minimize the risk of CI-AKI before, during, and after CTO PCI.

Before procedure

1. Hydration. The single most important preprocedural (and postprocedural) measure to reduce the occurrence of CI-AKI is hydration. Hydration should be done by intravenous administration of normal saline, because no other solutions (e.g., bicarbonate, half-normal saline) have been shown to be advantageous.[20] Administer pre- and post-procedural hydration depending on left ventricular function, the presence of severe valvular disease, and/or hemodynamic status. Hydration strategies tailored on patients' volume status

appear to offer an advantage over guideline-supported fixed-rate hydration for CI-AKI prevention after percutaneous coronary intervention.[28] The ESC/EACTS revascularization guidelines recommend 1 mL/kg/h starting 12 h before and continued for 24 h after the procedure, except for patients with left ventricular ejection fraction $\leq 35\%$ or NYHA >2 for whom 0.5 mL/kg/h is recommended.[29] The POSEIDON trial used 5 mL/kg/h if LVEDP was <13 mmHg, 3 mL/kg/h if LVEDP was $13-18$ mmHg, and 1.5 mL/kg/h if LVEDP was >18 mmHg; 3 mL/kg/hour were used prior to the procedure and the sliding scale regimen was started during the procedure and continued during the procedure and for 4 h postprocedure.[30]

2. High dose statins.
3. Discontinue nephrotoxic medication, such as NSAIDS, if possible.

During procedure

1. Limiting contrast volume. In advanced CKD patients, administering contrast volume less than the creatinine clearance is ideal to minimize the risk of CI-AKI. There are several contrast saving options during PCI (Table 28.1).[20]

TABLE 28.1 Contrast sparing strategies for PCI.

1. Use 5 French catheters with no side holes for coronary angiogram.
2. Display previous coronary angiograms on cath lab monitors (if available) to avoid acquiring new diagnostic images.
3. Use biplane or rotational angiography.
4. Limit the volume of contrast per injection (ideally, $2-3$ mL/injection).
5. Use diluted contrast media.
6. Test injections should be avoided:
   a. Enter coronary artery ostium with guidewire to confirm guide catheter engagement.
   b. If unable to wire side branches, use intravascular ultrasound in live-view mode.
7. Use stent enhancement techniques, for example, StentBoost (Philips), and ClearStent (Siemens).
8. Use increased acquisition rates (15 or 25 frames/s) to improve image quality during diagnostic angiogram and to evaluate the final result.
9. Allow for elimination of contrast from the guide catheter by back bleeding or aspirating before inserting and advancing new equipment.
10. Use additional guidewires to create a roadmap of the target vessel and its side branches, or use dedicated software, for example, Dynamic 3D Roadmap (Philips).
11. Extensive use of intravascular ultrasound, dextran-based optical coherence tomography, and coronary physiology testing.
12. If zero-contrast PCI is performed, a transthoracic echocardiogram should be done before and after the procedure to exclude development of a pericardial effusion that could be due to perforation.

Source: Reproduced with permission from Almendarez M, Gurm HS, Mariani J, Jr, et al. Procedural strategies to reduce the incidence of contrast-induced acute kidney injury during percutaneous coronary intervention. JACC Cardiovasc Interv 2019;12:1877−1888.

2. If needed, hemodynamic support may reduce the risk of CI-AKI by reducing the incidence and duration of hemodynamic instability.
3. Use of iso-osmolar contrast media.
4. When using contrast injection systems (such as the ACIST device) avoid using the system to flush with saline proximal to the Y-connector, other than prior to starting the procedure, as this may add additional contrast to fill the tubing.
5. Backbleeding contrast out of all guides and catheters prior to equipment insertion or removal from the body to prevent contrast in the catheter from being pushed into the body.

After procedure

1. Hydration is the key post procedure action to reduce the incidence of AKI.
2. Serum creatinine should be monitored for at least 48 h post procedure to detect CI-AKI.

28.3.3 Treatment

Once CI-AKI is established, there is no specific treatment, hence the goal is prevention.[20] In most cases renal function returns to baseline, however in some cases CI-AKI may require dialysis.

References

1. Brilakis ES. *Manual of percutaneous coronary interventions: a step-by-step approach.* Elsevier; 2021.
2. Writing Committee M, Hirshfeld Jr. JW, Ferrari VA, et al. 2018 ACC/HRS/NASCI/SCAI/ SCCT expert consensus document on optimal use of ionizing radiation in cardiovascular imaging-best practices for safety and effectiveness, part 1: radiation physics and radiation biology: a report of the American College of Cardiology Task Force on Expert Consensus Decision Pathways Developed in Collaboration with mended hearts. *Catheter Cardiovasc Interv* 2018;**92**:203−21.
3. Brilakis ES. Innovations in radiation safety during cardiovascular catheterization. *Circulation* 2018;**137**:1317−19.
4. Chambers CE. Radiation dose in percutaneous coronary intervention OUCH did that hurt? *JACC Cardiovasc Interv* 2011;**4**:344−6.
5. Brilakis ES, Patel VG. What you can't see can hurt you! *J Invasive Cardiol* 2012;**24**:421.
6. Brilakis ES, Mashayekhi K, Tsuchikane E, et al. Guiding principles for chronic total occlusion percutaneous coronary intervention. *Circulation* 2019;**140**:420−33.
7. Wu EB, Brilakis ES, Mashayekhi K, et al. Global chronic total occlusion crossing algorithm: JACC state-of-the-art review. *J Am Coll Cardiol* 2021;**78**:840−53.
8. Christopoulos G, Christakopoulos GE, Rangan BV, et al. Comparison of radiation dose between different fluoroscopy systems in the modern catheterization laboratory: Results from bench testing using an anthropomorphic phantom. *Catheter Cardiovasc Interv* 2015;**86**:927−32.

9. McNeice AH, Brooks M, Hanratty CG, Stevenson M, Spratt JC, Walsh SJ. A retrospective study of radiation dose measurements comparing different cath lab x-ray systems in a sample population of patients undergoing percutaneous coronary intervention for chronic total occlusions. *Catheter Cardiovasc Interv* 2018;**92**:E254−61.

10. Werner GS, Yaginuma K, Koch M, et al. Reducing fluoroscopic and cineangiographic contribution to radiation exposure for chronic total coronary occlusion interventions. *Cardiovasc Revasc Med* 2022;**36**:58−64.

11. Werner GS, Avran A, Mashayekhi K, et al. Radiation exposure for percutaneous interventions of chronic total coronary occlusions in a multicenter registry: the influence of operator variability and technical set-up. *J Invasive Cardiol* 2021;**33**:E146−54.

12. Werner GS, Yaginuma K, Koch M, et al. Modulated radiation protocol achieves marked reduction of radiation exposure for chronic total coronary occlusion intervention. *Catheter Cardiovasc Interv* 2021;**97**:1196−206.

13. Madder RD, VanOosterhout S, Mulder A, et al. Impact of robotics and a suspended lead suit on physician radiation exposure during percutaneous coronary intervention. *Cardiovasc Revasc Med* 2017;**18**:190−6.

14. Abdelaal E, Plourde G, MacHaalany J, et al. Effectiveness of low rate fluoroscopy at reducing operator and patient radiation dose during transradial coronary angiography and interventions. *JACC Cardiovasc Interv* 2014;**7**:567−74.

15. Brilakis ES. *Manual of coronary chronic total occlusion interventions. a step-by-step approach.* 2nd ed. Elsevier; 2017.

16. Agarwal S, Parashar A, Bajaj NS, et al. Relationship of beam angulation and radiation exposure in the cardiac catheterization laboratory. *JACC Cardiovasc Interv* 2014;**7**:558−66.

17. Christopoulos G, Papayannis AC, Alomar M, et al. Effect of a real-time radiation monitoring device on operator radiation exposure during cardiac catheterization: the radiation reduction during cardiac catheterization using real-time monitoring study. *Circ Cardiovasc Interv* 2014;**7**:744−50.

18. Christopoulos G, Makke L, Christakopoulos G, et al. Optimizing radiation safety in the cardiac catheterization laboratory: a practical approach. *Catheter Cardiovasc Interv* 2016;**87**:291−301.

19. Shorrock D, Christopoulos G, Wosik J, et al. Impact of a disposable sterile radiation shield on operator radiation exposure during percutaneous coronary intervention of chronic total occlusions. *J Invasive Cardiol* 2015;**27**:313−16.

20. Almendarez M, Gurm HS, Mariani Jr. J, et al. Procedural strategies to reduce the incidence of contrast-induced acute kidney injury during percutaneous coronary intervention. *JACC Cardiovasc Interv* 2019;**12**:1877−88.

21. Azzalini L, Ojeda S, Demir OM, et al. Recanalization of chronic total occlusions in patients with vs without chronic kidney disease: the impact of contrast-induced acute kidney injury. *Can J Cardiol* 2018;**34**:1275−82.

22. Tajti P, Karatasakis A, Danek BA, et al. In-hospital outcomes of chronic total occlusion percutaneous coronary intervention in patients with chronic kidney disease. *J Invasive Cardiol* 2018;**30**:E113−21.

23. Moroni F, Spangaro A, Carlino M, Baber U, Brilakis ES, Azzalini L. Impact of renal function on the immediate and long-term outcomes of percutaneous recanalization of coronary chronic total occlusions: a systematic review and *meta*-analysis. *Int J Cardiol* 2020;**317**:200−6.

24. Gurm HS, Dixon SR, Smith DE, et al. Renal function-based contrast dosing to define safe limits of radiographic contrast media in patients undergoing percutaneous coronary interventions. *J Am Coll Cardiol* 2011;**58**:907−14.

25. Mehran R, Aymong ED, Nikolsky E, et al. A simple risk score for prediction of contrast-induced nephropathy after percutaneous coronary intervention: development and initial validation. *J Am Coll Cardiol* 2004;**44**:1393−9.

26. Gurm HS, Seth M, Kooiman J, Share D. A novel tool for reliable and accurate prediction of renal complications in patients undergoing percutaneous coronary intervention. *J Am Coll Cardiol* 2013;**61**:2242−8.

27. Tsai TT, Patel UD, Chang TI, et al. Contemporary incidence, predictors, and outcomes of acute kidney injury in patients undergoing percutaneous coronary interventions: insights from the NCDR Cath-PCI registry. *JACC Cardiovasc Interv* 2014;**7**:1−9.

28. Moroni F, Baldetti L, Kabali C, et al. Tailored vs standard hydration to prevent acute kidney injury after percutaneous coronary intervention: network *meta*-analysis. *J Am Heart Assoc* 2021;**10**:e021342.

29. Neumann FJ, Sousa-Uva M, Ahlsson A, et al. 2018 ESC/EACTS guidelines on myocardial revascularization. *Eur Heart J* 2019;**40**:87−165.

30. Brar SS, Aharonian V, Mansukhani P, et al. Haemodynamic-guided fluid administration for the prevention of contrast-induced acute kidney injury: the POSEIDON randomised controlled trial. *Lancet* 2014;**383**:1814−23.

Chapter 29

Vascular access complications

Chronic total occlusion (CTO) percutaneous coronary intervention (PCI) can have similar types of vascular access complications as non-CTO PCI, as described in detail in the *Manual of Percutaneous Coronary Interventions*.[1] The risk of access site complications is likely higher with CTO PCI than non-CTO PCI because of dual arterial access and use of large size catheters and sheaths (7 or 8 French). The risk of vascular access complications can be reduced by use of meticulous access technique and radial access, as discussed in Chapter 4.

Reference

1. Brilakis ES. *Manual of percutaneous coronary interventions: a step-by-step approach.* Elsevier; 2021.

Manual of Chronic Total Occlusion Percutaneous Coronary Interventions.
DOI: https://doi.org/10.1016/B978-0-323-91787-2.00029-0
© 2023 Elsevier Inc. All rights reserved.

Part D

Equipment

Chapter 30

Equipment

Introduction

Dedicated equipment is necessary for performing chronic total occlusion (CTO) percutaneous coronary intervention (PCI). Although many operators would like to have everything available, the reality is that equipment costs and space limitations require prioritization. Here are some criteria to use when deciding on the "must have" equipment for CTO PCI:

1. At least one item that fulfills each of the requisite steps of CTO PCI (e.g., septal crossing, wire externalization, snaring, etc.) should be available, and
2. The operator should be familiar with the equipment, understand its strengths and limitations and be willing to use it when needed (otherwise it will expire on the shelf). In some cases, such as covered stents and coils, equipment expiration is to some extent expected given the low frequency of complications requiring their use.

Table 30.1 classifies PCI equipment into 22 categories.[1-5]

30.1 Sheaths

As described in Chapter 4, both femoral and radial access can be used for CTO PCI in various configurations (femoral-femoral, femoral-radial, radial-radial). Large guide catheters and sheaths are preferred for CTO PCI through either access point.

30.1.1 Femoral access

30.1.1.1 Sheath diameter

Seven or 8 French sheaths are preferred for femoral access for CTO PCI.

Larger sheaths are needed for hemodynamic support devices: 7–9 French for intra-aortic balloon pump (although intra-aortic balloon pumps [IABPs] can also be inserted without a sheath), 13 French for Impella 2.5, 14 French for Impella CP, and 15–17 French for the arterial cannula of VA-ECMO.

TABLE 30.1 Equipment needed for PCI.

Category no.	Equipment	
1.	Sheaths	1. Femoral access 2. Radial access
2.	Catheters	1. Diagnostic 2. Guide
3.	Guide catheter extensions	1. Guideliner V3 2. Trapliner 3. Guidezilla II 4. Telescope 5. Guidion 6. LiquID 7. GuidePlus II 8. Boosting catheter 9. Heartrail 5 in 6 system
4.	Support catheters	1. NovaCross
5.	Y connectors	1. Y-connectors with hemostatic valve
6.	Microcatheters	1. Large 2. Small 3. Angulated 4. Dual lumen 5. Plaque modification
7.	Guidewires	**0.014 inch** 1. Workhorse 2. Polymer-jacketed 3. Stiff-tip without polymer jacket 4. Support 5. Atherectomy 6. Externalization 7. Pressure wire **0.035 inch** 1. Soft-tip 2. Polymer-jacketed
8.	Embolic protection devices	1. Filterwire 2. Spider FX
9.	Balloons	1. Standard balloons 2. Small balloons 3. Plaque modification balloons 4. Trapping balloons 5. Ostial FLASH 6. Drug-coated balloons (DCB) 7. Very-high pressure balloon 8. Coronary lithotripsy balloon

(*Continued*)

TABLE 30.1 (Continued)

Category no.	Equipment	
10.	Atherectomy	1. Rotational 2. Orbital
11.	Laser	
12.	Thrombectomy	1. Aspiration thrombectomy 2. Penumbra
13.	Ostial lesion equipment	1. Ostial pro 2. Ostial FLASH
14.	Stents	1. Bare metal stents 2. Drug-eluting stents 3. Covered stents (discussed in complication management equipment)
15.	Arterial closure	1. Angioseal 2. Perclose 3. Other
16.	CTO PCI dissection/ re-entry equipment	1. CrossBoss catheter 2. Stingray balloon and wire 3. ReCross
17.	Intravascular imaging	1. IVUS 2. OCT
18.	Complication management	1. Covered stents 2. Coils 3. Pericardiocentesis kit 4. Snares (gooseneck, 3-loop)
19.	Radiation protection	1. Radiation scatter shields 2. Zero Gravity system
20.	Hemodynamic support	1. Intra-aortic balloon pump 2. Impella CP 3. Tandem Heart 4. VA-ECMO 5. External compression devices
21.	Contrast management	1. Dyevert system
22.	Brachytherapy	1. Novoste beta radiation system

30.1.1.2   Sheath length

The usual femoral sheath length is 10 cm. In many cases, longer sheaths (such as 25- or 45-cm long) are preferred, for example in obese patients,

patients with significant iliac or aortic tortuosity, or when strong guide catheter support is needed. Longer sheaths provide better guide catheter support and torque response compared with shorter ones.

The tip of the 10-cm sheath usually reaches the origin of the internal iliac artery, the 25-cm sheath usually reaches the distal aortic bifurcation, and the tip of the 45-cm long sheath usually reaches the level of the diaphragm (Figs. 30.1 and 30.2). Although there is increased risk of thrombus formation within longer sheaths, this is rarely an issue, especially during retrograde CTO PCI, given the high goal ACT (> 350 s) for this procedure.

30.1.2 Radial access

30.1.2.1 Sheath diameter

Seven French sheaths are preferred for radial access unless the radial artery is too small to accommodate them. Thin wall sheaths, such as the Glidesheath Slender (Terumo) and the Prelude Ideal (Merit Medical), reduce the likelihood of spasm but may be more prone to kinking.

A sheathless guide system (Eaucath, Asahi Intecc) allows CTO PCI with 7.5 French guides through an arterial puncture equivalent to that created by a 5 French sheath. An alternative approach is to use standard 7 or 8 French guide catheters delivered through a short 7 or 8 French sheath or inserted using the Railway system (Cordis), that allows the sheathless insertion of

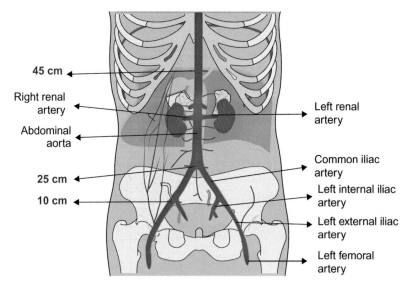

FIGURE 30.1 **The location of the sheath tip depends on the sheath length.** *Reproduced with permission from Brilakis ES. Manual of percutaneous coronary interventions: a step-by-step approach. Elsevier; 2021.*

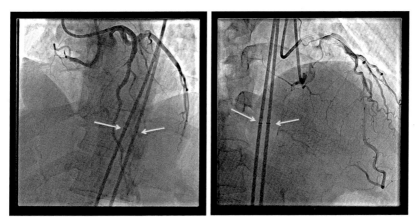

FIGURE 30.2   **Location of the distal tip of two 45-long femoral sheaths.**

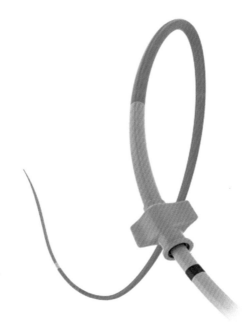

FIGURE 30.3   **The Railway system for sheathless insertion of guide catheters.** © *2022 Cordis. All Rights Reserved.*

any guide. The Railway is a dilator that is inserted through the guide catheter facilitating entry into the radial artery (Fig. 30.3).

Similarly, a 125 cm long diagnostic catheter can be used as a dilator to insert a 90 cm or 100 cm guide catheter that is two French sizes larger in diameter (Fig. 30.4).

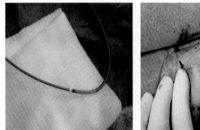

FIGURE 30.4 Use of a 5 French, 125 cm long, Cook Select catheter to insert a 7 French 100 cm long EBU 3.5 guide. *Courtesy of Dr. Jaikirshan Khatri. Reproduced with permission from Brilakis ES.* Manual of percutaneous coronary interventions: a step-by-step approach. *Elsevier; 2021.*

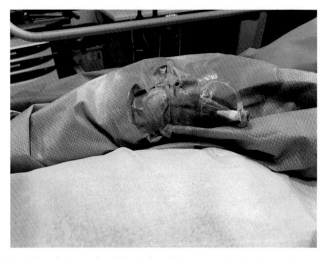

FIGURE 30.5 Use of a long sheath for left radial access. The sheath partially protrudes outside the skin, facilitating catheter insertion and exchanges. *Courtesy of Dr. Stéphane Rinfret. Reproduced with permission from Brilakis ES.* Manual of percutaneous coronary interventions: a step-by-step approach. *Elsevier; 2021.*

Alternatively, the Balloon Assisted Tracking (BAT) technique can be used for sheathless insertion of guide catheters.[6]

30.1.2.2 Sheath length

The usual radial sheath length is 10 cm; however, longer hydrophilic sheaths (such as 16 or 25 cm long) can be used, especially when using left radial access (the sheath may be left partially outside the radial artery to facilitate equipment insertion) (Fig. 30.5). Left distal radial access is ergonomically more comfortable for the patient and operator. Longer R2P (radial to peripheral) sheaths (75 and 85 cm long) can be very useful in patients with severe subclavian tortuosity, but are only currently available in 6 French diameter.

30.2   Catheters

30.2.1   Guide catheter outer diameter

Choosing the right guide catheter size is a critical step in complex PCI. The "Complex PCI Solutions" app (only available for phones) can be used to choose guide size based on planned equipment utilization.

Eight French guide catheters are most commonly used for CTO PCI in the US, whereas 7 Fr guides are most commonly used in Europe and Canada. Compared with smaller caliber guide catheters, 8 Fr guide catheters provide enhanced support and improve vessel visualization.

Six or seven French guide catheters in the donor vessel often suffice for retrograde crossing. Six French guide catheters often provide adequate support for delivering retrograde gear and may reduce the risk of donor artery dissection. Therefore, many operators use 6 Fr catheters from either the femoral or radial approach for retrograde access. Some operators consistently insert the right coronary artery guide catheter through the right femoral artery and the left coronary artery guide catheter through the left femoral artery when using bifemoral access to maintain consistency and minimize the chance of accidental injections made through the wrong guide catheter.

30.2.2   Guide catheter length

Most guide catheters are 100 cm long; however, *shorter* guide catheters (usually 90 cm long, which are commercially available from multiple manufacturers) may be needed for retrograde CTO PCI or when attempting to deliver equipment to very distal target lesions (for example, when delivering equipment to native lesions through bypass grafts).[7] Most internal mammary (IM) guide catheters that are designed for PCI in left internal mammary artery (LIMA) grafts are 90 cm long for this reason. Shorter (80 cm long) guides may not reach the coronary ostia in some patients and are not commonly used. If short guide catheters are not available, any guide catheter can be shortened, as described in Section 30.2.3.

Longer guide catheters (125 cm long) are also available and may be needed to reach the coronary arteries in very tall patients or patients with very tortuous aortas.

Occasionally a vessel may not be able to be engaged despite using various guide catheters, requiring primary retrograde crossing (**CTO Manual online case** 18).

30.2.3 Shortening the guide catheter

With the availability of long externalization guidewires, such as the RG3 and R350, and short (90 cm) guide catheters, guide catheter shortening is rarely performed. Guide catheter shortening may, however, still be needed for some retrograde cases via bypass grafts or apical collaterals, as in such cases retrograde microcatheters might not be long enough to reach the antegrade guide catheter.

If premanufactured short guide catheters are not available, a 100 cm long guide can be shortened using the following technique (Fig. 30.6)[8] (**Video: How to shorten a guide catheter**):

1. The guide catheter is inserted into the body to engage the target coronary artery and the length of the guide that is outside the femoral sheath is marked.
2. The guide is removed from the body and the marked segment is cut using sterile scissors and removed (Fig. 30.6, panels A-C).
3. A sheath (one Fr size smaller than the guide catheter, i.e. 6 French sheath for a 7 French guide catheter, 7 French sheath for a 8 French guide catheter, etc.) is cut to create a 3−4 cm connecting segment for the two guide catheter pieces (Fig. 30.6 panels D and E). Both ends of this connecting segment are flared with a dilator (of equal size to the guide) to facilitate insertion (Fig. 30.6, panel F).
4. This connecting sheath segment is used to re-connect the proximal and distal guide catheter pieces minus the portion that was removed to shorten the guide catheter (Fig. 30.6, panels G and H—final result in panel I). Placing a Tegaderm (3 M) over the connection site may help prevent accidental disconnection.

A limitation of shortened guide catheters is that they have poor torque transmission during vessel engagement and guide manipulations (especially during long procedures).

30.2.4 Side holes

Guide catheters with side holes (Fig. 30.7) can be used for ostial right coronary artery or ostial bypass graft lesions because they can prevent pressure dampening, may allow antegrade flow into the vessel, and may decrease the risk of hydraulic dissection during antegrade contrast injection. Side-hole guide catheters may, however, provide a false sense of security, as hydraulic dissections can still occur upon injection. Dampening of the pressure is desired when antegrade dissection-reentry techniques are used in order to minimize antegrade flow and extraplaque hematoma expansion. A strategy of thoughtful, active guide manipulation is often chosen over the use of side hole guide catheters, depending on the preference and comfort of the operator.

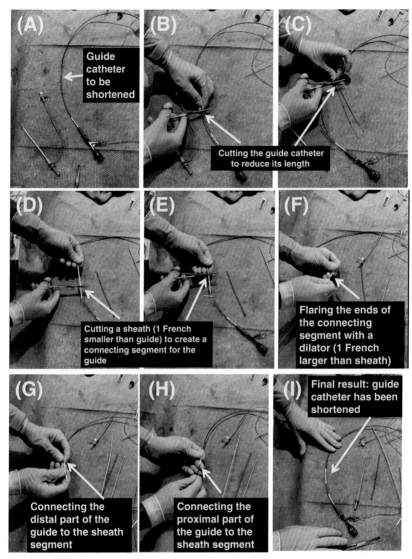

FIGURE 30.6   Overview of the guide shortening technique.

In contrast, engagement of an unprotected left main coronary artery with side-hole guide catheters should be avoided (with the exception of ostial left main CTOs), as suboptimal guide catheter position may not be recognized, leading to decreased antegrade left main flow, global ischemia, and hemodynamic collapse.

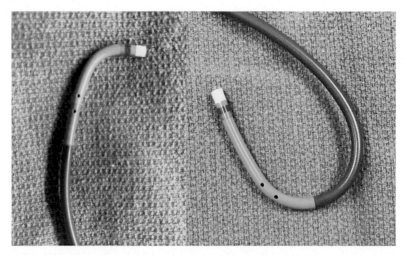

FIGURE 30.7 **Example of guide catheters with side holes.** *Courtesy of Dr. William Nicholson. Reproduced with permission from Brilakis ES.* Manual of coronary chronic total occlusion interventions. a step-by-step approach. *2nd ed. Elsevier; 2017.*

Another disadvantage of side-hole guide catheters is that they lead to higher contrast use and image quality degradation due to escape of contrast into the aorta through the side holes during contrast injection.[3] If no side-hole guides are available, an 18−23 gauge needle or a scalpel can be used to create side holes in any guide catheter, followed by flushing with saline before use. Side holes made by hand may prevent advancement of a guide catheter extension within the modified guide catheter and can also weaken it and lead to kinking.

30.2.5 Guide catheter shapes

Strong guide catheter support is essential for CTO PCI. The "must have" guides are those with supportive shapes, such as the XB and EBU for the left coronary artery and AL or 3D Right for the right coronary artery.

30.3 Guide catheter extensions

30.3.1 Guide catheter extension types

Six guide catheter extensions are currently available in the US (Table 30.2).

- Guideliner V3 catheter (Teleflex, Fig. 30.8)
- Trapliner (Teleflex, Fig. 30.9) is a rapid exchange guide catheter extension with a guidewire trapping balloon.

TABLE 30.2 Overview of guide catheter extensions.

Name	Sizes	Internal diameter	Total length	Distal cylinder length
Guideliner V3	5 Fr 5.5 Fr 6 Fr 7 Fr 8 Fr	0.046" (1.17 mm) 0.051" (1.30 mm) 0.056" (1.42 mm) 0.062" (1.57 mm) 0.071" (1.80 mm)	150 cm	25 cm XL: 40 cm
Trapliner	6 Fr 7 Fr 8 Fr	0.056" (1.42 mm) 0.062" (1.57 mm) 0.071" (1.80 mm)	150 cm	13 cm
Guidezilla II	6 Fr 7 Fr 8 Fr	0.057" (1.45 mm) 0.063" (1.60 mm) 0.072" (1.83 mm)	145 cm	25 cm XL: 40 cm
Guidion	5 Fr 6 Fr 7 Fr 8 Fr	0.041" (1.04 mm) 0.056" (1.42 mm) 0.062" (1.57 mm) 0.071" (1.80 mm)	150 cm	25 cm
Telescope	6 Fr 7 Fr	0.056" (1.42 mm) 0.062" (1.57 mm)	150 cm	25 cm
Heartrail II	5 in 6 system	0.059" (1.50 mm)	120 cm	No distal cylinder—Heartrail is a 120 cm long catheter inserted through a standard guide catheter
liquID	6 Fr 7 Fr	0.061" (1.54 mm) 0.071" (1.80 mm)	150 cm	15 cm
Guideplus II	ST EL	0.052" (1.33 mm) 0.056" (1.43 mm)	150 cm	25 cm

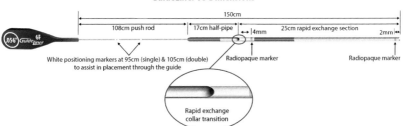

GuideLiner V3 Dimensions

FIGURE 30.8   **Illustration of the Guideliner V3 guide catheter extension.** *Image courtesy of Teleflex Incorporated.* © *2022 Teleflex Incorporated. All rights reserved.*

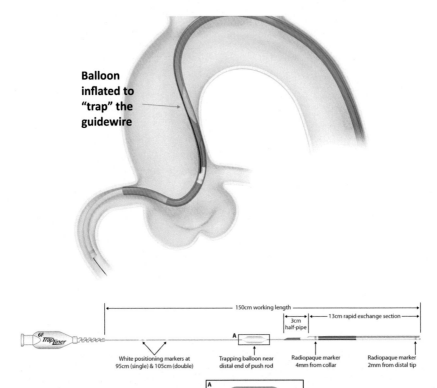

FIGURE 30.9 Illustration of the Trapliner guide catheter extension. *Image courtesy of Teleflex Incorporated.* © *2022 Teleflex Incorporated. All rights reserved.*

- Guidezilla II (Boston Scientific, Fig. 30.10)
- Telescope (Medtronic, Fig. 30.11).
- liquID (Seigla Medical).

The Heartrail 5 in 6 system (Terumo, Fig. 30.13)[9] the Guidion (Interventional Medical Device Solutions, Fig. 30.12), the GuidePlus II (Nipro), and the Boosting catheter (QXMedical) are not available in the US but are available in other countries.

All guide catheter extensions (with the exception of the Heartrail 5 in 6 system) consist of a push rod and a distal cylinder (25 cm long in the Guideliner V3, the Guidezilla II, the Guidion and the Telescope, and 13 cm in the Trapliner) that is advanced into the coronary artery. They are manufactured in various sizes (Table 30.2) to fit multiple guide catheters, resulting in an inner diameter that is approximately 1 Fr smaller than the nominal size of the guide catheter. In addition to the pushing

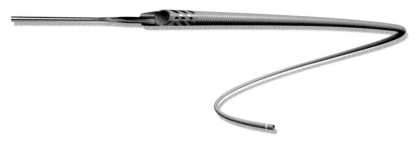

FIGURE 30.10 **Illustration of the Guidezilla II guide catheter extension**. *Image provided courtesy of Boston Scientific © 2022 Boston Scientific Corporation or its affiliates. All rights reserved.*

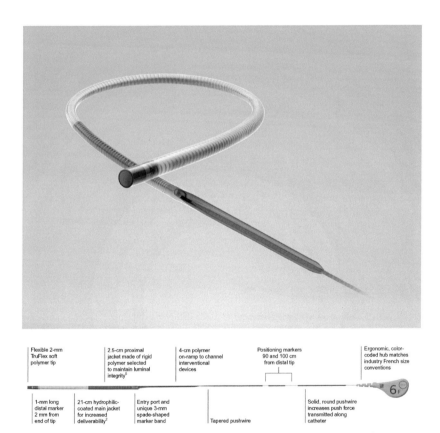

Flexible 2-mm TruFlex soft polymer tip	2.5-cm proximal jacket made of rigid polymer selected to maintain luminal integrity[5]	4-cm polymer on-ramp to channel interventional devices	Positioning markers 90 and 100 cm from distal tip	Ergonomic, color-coded hub matches industry French size conventions
1-mm long distal marker 2 mm from end of tip	21-cm hydrophilic-coated main jacket for increased deliverability[2]	Entry port and unique 3-mm spade-shaped marker band	Tapered pushwire	Solid, round pushwire increases push force transmitted along catheter

FIGURE 30.11 **Illustration of the Telescope guide catheter extension**. *Used with permission by Medtronic, Inc.*

Guidion

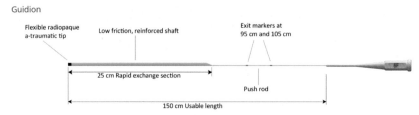

FIGURE 30.12 **Illustration of the Guidion guide catheter extension**. *Reproduced with permission from IMDS.*

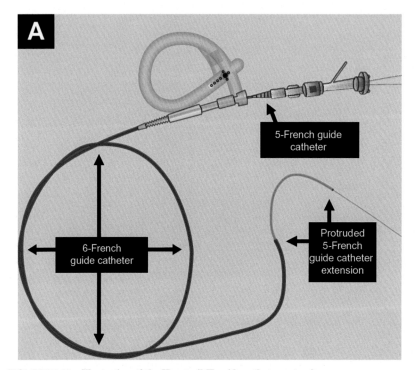

FIGURE 30.13 **Illustration of the Heartrail II guide catheter extension**.

rod and cylinder, the Trapliner also has a balloon proximal to the proximal collar that allows trapping equipment (Fig. 30.9).

30.3.2 Guide catheter extension uses

1. Facilitating equipment delivery (balloons, stents, etc.), is the major indication for using a guide catheter extension.[10] Although coaxial alignment of the guide catheter is ideal, the guide catheter extension may be

particularly effective in facilitating vessel engagement and equipment delivery when guide coaxial alignment is not possible, for example in anomalous coronary arteries (Fig. 30.14)[11] or in internal mammary artery grafts.[12]

2. Facilitate coronary artery engagement in challenging clinical scenarios, for example in patients with dilated ascending aorta or post-TAVR.[13]

3. Perform thrombectomy, by advancing the guide catheter extension into the target vessel and aspirating through the guide catheter.[14]

4. Facilitate retrieval of entrapped equipment, such as Rotablator burrs (see Section 27.5).[15]

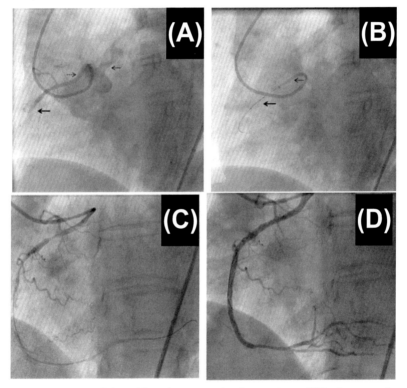

FIGURE 30.14   Use of the guide catheter extension for treating a CTO of an anomalous right coronary artery. Panel A: chronic total occlusion (arrow) of an anomalous right coronary artery arising from the left sinus of Valsalva. **Panel B**: a Guideliner catheter was placed over a Fielder guidewire with the support of an uninflated balloon kept in the proximal right coronary artery for better support. **Panel C**: the CTO was crossed with a Confianza Pro 9 wire with the support of the Guideliner. **Panel D**: successful recanalization of the right coronary artery with TIMI 3 flow. *Reproduced with permission from Senguttuvan NB, Sharma SK, Kini A. Percutaneous intervention of chronic total occlusion of anomalous right coronary artery origi-nating from left sinus—use of mother and child technique using guideliner.* Indian Heart J *2015;67 Suppl. 3:S41−S42.*

5. Facilitate the reverse controlled antegrade and retrograde tracking (reverse CART) technique ("guide catheter extension reverse CART," Fig. 8.3.32) during retrograde CTO interventions. A guide catheter extension is advanced through the antegrade guide catheter to reduce the distance that the retrograde guidewire needs to traverse.[16] Guide catheter extensions can also be used from the retrograde side to increase support for retrograde equipment delivery.
6. Create a "homemade" snare (KAM-snare), that consists of a wire loop being trapped by an inflated balloon at the distal portion of a guide catheter extension.[17]
7. Reduce hematoma formation during antegrade dissection and reentry by pressure dampening and reducing antegrade flow.

30.3.3 Guide catheter extensions tips and tricks

1. To minimize the risk of the guidewire wrapping around the guide catheter extension push rod after insertion,[18] the external push rod should be placed under a towel at the side of the Y-connector (Fig. 30.15).
2. Advancing the guide catheter extension may be easier to achieve by inflating a balloon half-way inside the guide catheter extension distal tip

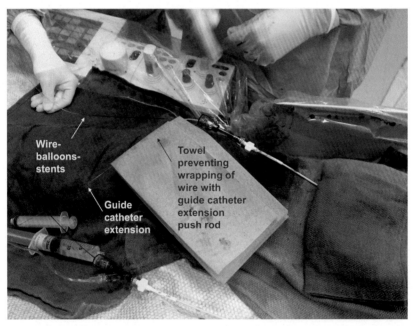

FIGURE 30.15 **Guide catheter extension manipulation to minimize the risk for guidewire "wrap around" the guide catheter extension push rod.** *Reproduced with permission from Brilakis ES. Manual of coronary chronic total occlusion interventions. a step-by-step approach. 2nd ed. Elsevier; 2017.*

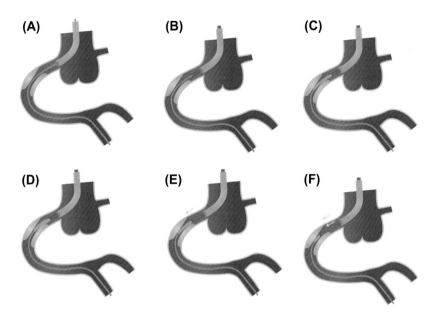

FIGURE 30.16 Delivery of a guide catheter extension to the target coronary segment using the "inchworming technique." Panel A: guide catheter in position. **Panel B**: guide catheter extension is advanced to the tip of the guide catheter. **Panel C**: a small balloon (usually 2.0 mm) is advanced halfway in and halfway out of the guide catheter extension. **Panel D**: the balloon is inflated, usually at low pressure (6−8 atm). **Panel E**: the balloon is deflated. **Panel F**: while the balloon is being deflated the guide catheter extension is advanced forward. *Reproduced with permission from Brilakis ES.* Manual of percutaneous coronary interventions: a step-by-step approach. *Elsevier; 2021.*

and the vessel (Fig. 30.16). The guide catheter extension is then advanced upon balloon deflation ("inchworming" technique). Advancement over a balloon catheter or microcatheter is preferred to advancement over a 0.014″″ coronary wire to minimize the risk of catching a plaque edge and causing a dissection. (**CTO PCI Manual online case** 44).

Online video: How to deliver a guide catheter extension using the "inchworming" technique

3. Additional strategies for delivering guide catheter extensions include:
   a. Guide extension balloon assisted tracking (inflation of a balloon protruding from the guide catheter extension at nominal pressure followed by advancement of both the guide catheter extension and the balloon over the coronary to the desired coronary location[19]); and
   b. Distal anchor technique (balloon inflated distally in the target vessel, providing support for advancing the guide extension).[20]
4. The "inchworming" technique is generally preferred over the above methods for delivering a guide catheter extension, as it centers the guide catheter extension, and is hence less likely to cause vessel trauma.

Attempts to advance guidewires through a guide that contains a guide catheter extension smaller than the size of the guide (for example a 6 French extension within an 8 French guide), should be avoided, or only performed under direct fluoroscopy, as the wire is likely to advance between the cylinder of the guide catheter extension and the guide catheter wall (Fig. 30.17).

Even when the guide extension is of similar size to the guide (for example a 7 French guide catheter extension within a 7 French guide) there is a risk of wire damage when trying to advance it into the guide extension cylinder. Using a microcatheter to enter the guide extension cylinder with the wire withdrawn into the microcatheter can reduce this risk.

5. The proximal collar of the guide catheter extension should not be advanced outside the guide catheter.
6. Trapping of equipment using a guide catheter extension is difficult: (1) the trapping balloon needs to be placed proximal to the proximal collar; and (2) the equipment needs to be retracted to this location for successful trapping to occur. When the need for trapping is anticipated, use of the Trapliner is preferred.
7. Guide catheter extensions are very flexible and can advance even through highly tortuous lesions.[21]

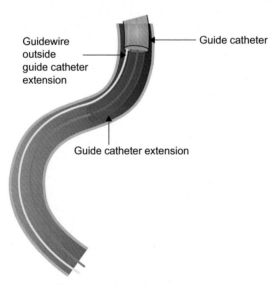

FIGURE 30.17 **Guidewire inserted within the guide catheter but outside the guide catheter extension**. *Reproduced with permission from Brilakis ES.* Manual of percutaneous coronary interventions: a step-by-step approach. *Elsevier; 2021.*

8. Two guide catheter extensions can be used simultaneously in a "mother-daughter-granddaughter" configuration (i.e. a 6 Fr extension through an 8 Fr extension) when multiple extreme bends need to be navigated (Fig. 18.3).[22]

9. Although distal to proximal stenting is preferred, "proximal to distal" stenting is a viable option and can be facilitated by inserting the guide catheter extension through the proximal stent.[23]

10. An alternative strategy to using a guide catheter extension is deep intubation of the guide catheter.[24] Usually 5 or 6 French guide catheters are used for deep intubation.

11. Compatibility of the Guideliner with various equipment is shown in Fig. 30.18.

30.3.4 Guide catheter extension related complications

1. Deformation of guidewires, stents, or other equipment can occur during *advancement* through the guide catheter extension collar (Fig. 30.19, panel A).[25] In some cases, it may be necessary to advance the stent or other equipment into the guide catheter extension *outside* the body and then introduce everything as a single unit into the guide catheter. If any resistance is encountered during equipment advancement when using a guide catheter extension, fluoroscopic guidance should be used to visualize the equipment as it enters to proximal port of the guide catheter extension.

	Caravel	Corsair	Corsair Pro	Minnie	Super Cross Angled Tip	Super Cross Straight Tip	Turnpike	Turnpike Gold	Turnpike LP	Turnpike Spiral	TwinPass	TwinPass Torque	Venture RX	Venture OTW
	ASAHI	ASAHI	ASAHI	Teleflex	Teleflex	Teleflex	Teleflex	Teleflex	Teleflex	Teleflex	Teleflex	Teleflex	Teleflex	Teleflex
Max OD -->	2.6Fr (0.85mm, 0.034")	2.8Fr (0.93mm, 0.037")	2.8Fr (0.93mm, 0.037")	3.1Fr (1.02mm, 0.040")	3.2Fr (1.07mm, 0.042")	2.5Fr (0.84mm, 0.033")	2.9Fr (0.97mm, 0.038")	2.9Fr (0.97mm, 0.038")	2.9Fr (0.97mm, 0.038")	2.9Fr (0.97mm, 0.038")	3.4Fr (1.14mm, 0.045")	3.5Fr (1.17mm, 0.046")	4.1Fr (1.37mm, 0.054")	4 Fr (1.32mm, 0.052")
5F Guideliner Effective ID =0.044" (1.12 mm)	✓	✓	✓	✓	✓	✓	✓	✓	✓	✓	No	No	No	No
5.5F Guideliner Effective ID = 0.051" (1.30 mm)	✓	✓	✓	✓	✓	✓	✓	✓	✓	✓	No	No	No	No
6F Guideliner Effective ID = 0.056" (1.42 mm)	✓	✓	✓	✓	✓	✓	✓	✓	✓	✓	✓	✓	✓	✓
7F Guideliner Effective ID = 0.062" (1.57 mm)	✓	✓	✓	✓	✓	✓	✓	✓	✓	✓	✓	✓	✓	✓
8F Guideliner Effective ID = 0.071" (1.80 mm)	✓	✓	✓	✓	✓	✓	✓	✓	✓	✓	✓	✓	✓	✓

	Rotablator Burr 1.25mm	Rotablator Burr 1.5mm	Rotablator Burr 1.75mm	Rotablator Burr 2.0mm	CSI OA
5F Guideliner	No	No	No	No	No
5.5F Guideliner	✓	No	No	No	✓
6F Guideliner	✓	No	No	No	✓
7F Guideliner	✓	✓	No	No	✓
8F Guideliner	✓	✓	✓	No	✓

Stent Sizes	
5F	3.0 mm
5.5F	3.5 mm
6F	4.0 mm
7F	
8F	

FIGURE 30.18 **Guideliner compatibility with other equipment**. *Courtesy of Dr. Bilal Murad. Reproduced with permission from Brilakis ES. Manual of percutaneous coronary interventions: a step-by-step approach. Elsevier; 2021.*

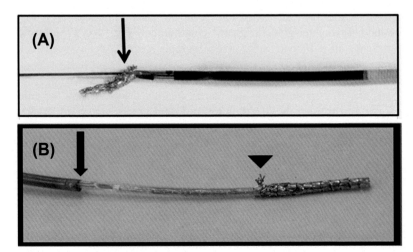

FIGURE 30.19   Complications of stent delivery through guide catheter extensions. Panel A: stent deformation while attempting to deliver it through a Guideliner catheter. **Panel B:** stent deformation while attempting to retrieve an undeployed stent into the distal tip of a Guidezilla catheter. The tip of the Guidezilla prolapsed on itself (arrow) during attempted stent retrieval, resulting in catching the proximal edge of the stent and causing deformation (arrowhead). *Panel A. Reproduced with permission from Papayannis AC, Michael TT, Brilakis ES. Challenges associated with use of the GuideLiner catheter in percutaneous coronary interventions. J Invasive Cardiol 2012;24:370_1. Panel B. Courtesy of Dr. William Nicholson. Reproduced with permission from Brilakis ES. Manual of coronary chronic total occlusion interventions. a step-by-step approach. 2nd ed. Elsevier; 2017.*

2. Deformation of stents or other equipment can also occur while *withdrawing* the equipment back into the distal tip of the catheter after a failed attempt to advance to the target lesion (Fig. 30.19, panel B). Hence, any withdrawn stent after a failed delivery attempt should be closely inspected for damage.

3. Coronary or aortocoronary dissection. Guide extensions should not be advanced through tortuous or diseased anatomy without use of a balloon rail or the inchworming technique to minimize the risk of vessel dissection. Deep advancement of the guide catheter extension may cause coronary dissection,[10] especially if contrast is injected while the pressure is dampened. Every effort should be undertaken to minimize pressure dampening. If dampening occurs, it is essential to verify that adequate antegrade flow is preserved and that no vessel injury has occurred before proceeding with the intervention.[10] Injection through a guide catheter extension with dampened pressure waveform may cause dissection that can propagate either antegrade or retrograde (Fig. 30.20). In general, injections though deeply engaged guide catheter extensions should be avoided due to risk of dissection.

   In some cases, use of a guide catheter extension may lead to acute vessel closure that can be challenging or impossible to correct.[26]

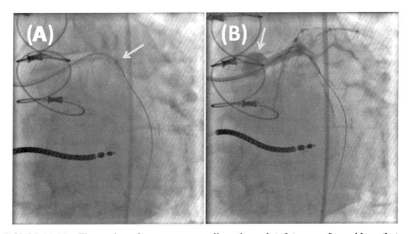

FIGURE 30.20   **Illustration of aortocoronary dissection related to use of a guide catheter extension.** Retrograde dissection caused by contrast injection through a guide catheter extension with dampened waveform. **Panel A:** guide catheter extension with its tip (arrow) deep-seated into the circumflex artery. **Panel B:** retrograde dissection into the aortic root (arrow) from contrast injection through the deep-seated guide catheter extension. *Reproduced with permission from Brilakis ES. Manual of coronary chronic total occlusion interventions. a step-by-step approach. 2nd ed. Elsevier; 2017.*

4. Because the guide catheter extension decreases the original guide inner size by approximately 1 French (for example the inner diameter of a 5 French Launcher guide catheter is 0.058 inch and that of a 6 French Guideliner is 0.056 inch), special attention to pressure dampening and to activated clotting time (ACT) is needed to decrease the risk of thrombus formation.
5. Guide catheter extension distal advancement: hold the guide catheter extension push rod during forceful contrast injection to minimize the risk of "ejecting" the guide catheter extension into the vessel![27]
6. Guide catheter extension fracture.[28]
7. Longitudinal stent deformation upon forceful advancement of a guide catheter extension through a previously deployed stent.[29,30]

30.4   Support catheters

The NovaCross catheter can deeply intubate the vessel (similar to a guide catheter extension), while at the same time providing additional support contact with the vessel wall, by using self-expanding nitinol wires. The NovaCross catheter (Nitiloop) has a 10 mm long flexible Nitinol element which upon axial compression deforms by curving outwards several helical struts (Fig. 30.21), increasing support.[31]

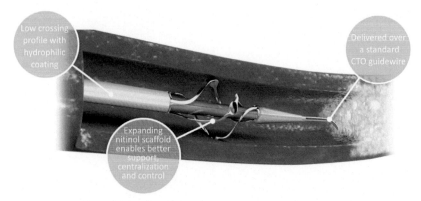

FIGURE 30.21   **The NovaCross catheter.** *Reproduced with permission by Nitiloop Ltd.*

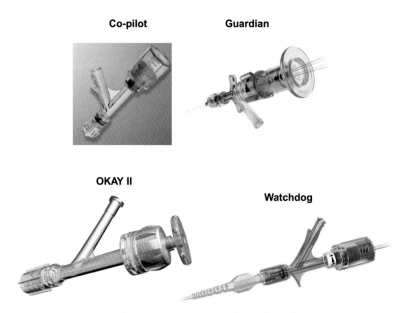

FIGURE 30.22   **Types of Y-connectors with hemostatic valves available in the United States.** *Image of Co-pilot courtesy of Abbott Vascular. ©2022 Abbott. All rights reserved. Image of Watchdog provided courtesy of Boston Scientific © 2022 Boston Scientific Corporation or its affiliates. All rights reserved. Image of OKAY II: Copyright 2022 InfraRedx, Inc. All rights reserved. Image of Guardian courtesy of Teleflex Incorporated. © 2022 Teleflex Incorporated. All rights reserved.*

30.5 Y-connectors with hemostatic valves

Using a Y-connector with a hemostatic valve (Fig. 30.22), such as the Co-Pilot (Abbott Vascular), Guardian (Teleflex), OKAY II (Nipro), Watchdog (Boston Scientific), and PhD (Merit Medical) can help minimize blood loss from back bleeding (which is particularly important for larger guide catheters, such as 8 Fr). Y-connectors with a hemostatic valve are also easier to use compared with standard rotating hemostatic valves.

30.6   Microcatheters

As their name implies, microcatheters (μικρός = small + καθετήρας = catheter) are small (usually 1.8 to 3.2 French in diameter) catheters.

Although over-the-wire balloons can be used to support guidewire advancement instead of a microcatheter, microcatheters are preferred to over-the-wire balloons because:

1. They allow better understanding of the distal tip position (a marker is placed at the microcatheter tip, whereas in small balloons (≤1.5 mm diameter) the marker is located in the middle of the balloon) (Fig. 30.23).[32]
2. Are more flexible and track better than over-the-wire balloons.
3. Have less tendency to kink than over-the-wire balloons (kinking of balloon shaft prohibits future wire exchanges and often necessitates balloon catheter and wire removal and replacement with new gear, losing the crossing progress achieved). Over-the-wire balloons, however, do provide better support than many microcatheters and are significantly cheaper.

Several microcatheters are commercially available and can be classified into five categories (Fig. 30.24 and Table 30.3)[33]:

1. **Large outer diameter single lumen microcatheters** (such as the Corsair [Asahi Intecc][34], Turnpike and Turnpike Spiral [Teleflex], Teleport Control [OrbusNeich], Mamba [Boston Scientific], and M-CATH [Acrostak]).

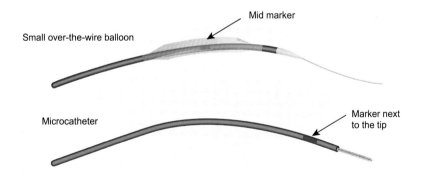

FIGURE 30.23   Comparison of over-the-wire balloons and microcatheters for facilitating wiring. *Reproduced with permission from Brilakis ES. Manual of coronary chronic total occlusion interventions. a step-by-step approach. 2nd ed. Elsevier; 2017.*

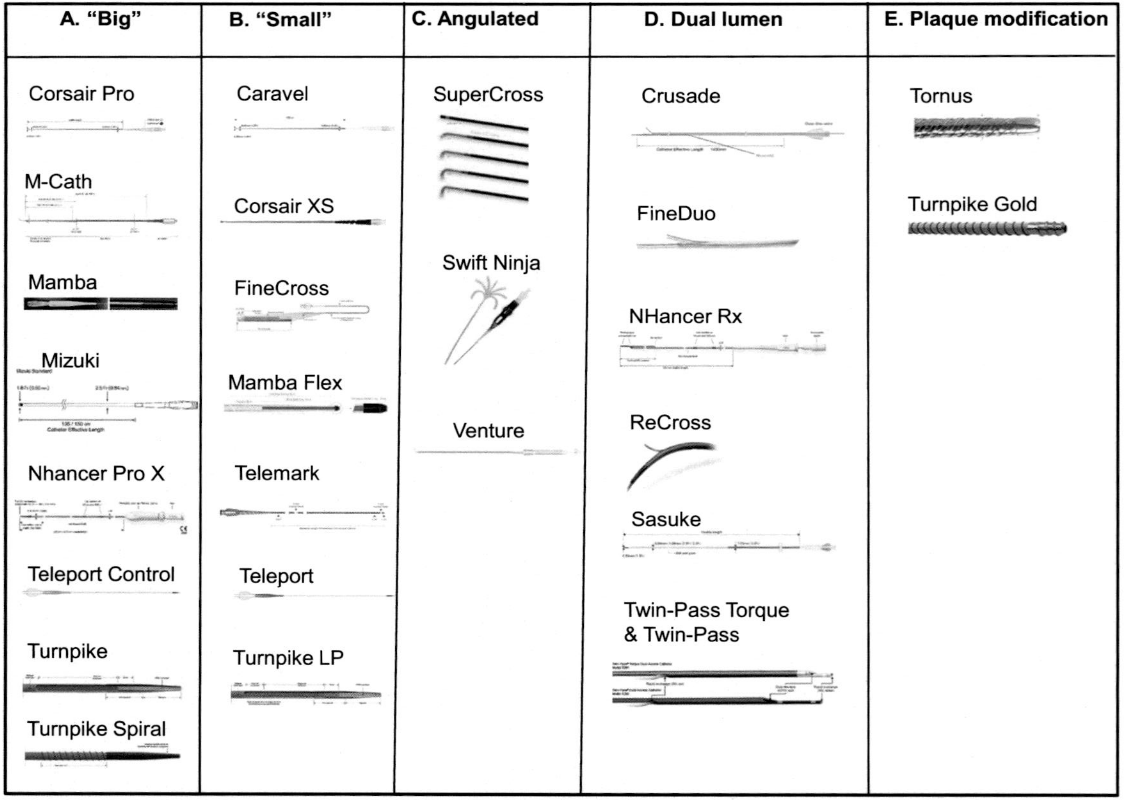

A. "Big"	B. "Small"	C. Angulated	D. Dual lumen	E. Plaque modification
Corsair Pro	Caravel	SuperCross	Crusade	Tornus
M-Cath	Corsair XS	Swift Ninja	FineDuo	Turnpike Gold
Mamba	FineCross	Venture	NHancer Rx	
Mizuki	Mamba Flex		ReCross	
Nhancer Pro X	Telemark		Sasuke	
Teleport Control	Teleport		Twin-Pass Torque & Twin-Pass	
Turnpike	Turnpike LP			
Turnpike Spiral				

FIGURE 30.24 Classification of microcatheters used during PCI and CTO PCI into five categories.

TABLE 30.3 Characteristics of the currently available microcatheters.

A. Large outer diameter, single lumen, microcatheters

	Corsair Pro	Turnpike and Turnpike Spiral	Teleport control	Mamba	M-cath	NHancer ProX NX3	Mizuki	Mizuki FX
Catheter Length (cm)	135/150 cm	135/150 cm (only 135 cm for the Spiral)	135/150 cm	135 cm	135 cm	135/155 cm	135/150 cm	135/150 cm
Prox. O.D.	2.8 F	2.9 F	2.7 F	2.9 F	3.3 F	3.2 F	3.2 F	2.5 F
Distal O.D.	2.6 F	2.6 F	2.1 F	2.4 F	2.25 F	2.0 F	1.8 F	1.8 F
Tip Entry O.D.	1.3 F	1.6 F	0.0190" (1.45 F)	1.4 F	1.6 F	1.5 F	1.7 F	1.7 F
Inner Lumen I.D.	0.0177"	0.0175"	0.0175"	0.023"	0.016"	0.020" with distal tip taper to 0.017"	0.022" with distal tip taper to 0.018"	0.022" with distal tip taper to 0.017"
Tip Length	5 mm	11 mm (10 mm for low profile)	6 mm	3 mm	2.5 mm	6 mm	1 mm	1 mm

(*Continued*)

TABLE 30.3 (Continued)

A. Large outer diameter, single lumen, microcatheters

	Corsair Pro	Turnpike and Turnpike Spiral	Teleport control	Mamba	M-cath	NHancer ProX NX3	Mizuki	Mizuki FX
Tip Material	Tungsten	Mixture of tungsten and polyurethane	Mixture of Tungsten and Pebax		Pebax	Tungsten polymer blend	Pebax	Pebax
Tip Construction	5 mm tungsten, 0.8 mm platinum coil	5 layer construction including dual-layer coils and a braid	Radiopaque. Dura tapered	Radiopaque, integrated tip with Tapered Coil		Integrated three layer hybrid tip design	Platinum ring marker 1.0 mm	Platinum ring marker 1.0 mm
Exterior Coating	Hydrophilic (60 cm from distal tip)	Hydrophilic coating (distal 60 cm including tip)	Hydrophilic (60 cm of distal catheter including tip)	HydroPass Coating	Hydrophilic (43 cm from distal tip)	Hydrophilic (60 cm for 155 cm length and 25 cm for 135 cm length)	Hydrophilic (70 cm from distal tip)	Hydrophilic (70 cm from distal tip)

B. Small outer diameter microcatheters.

	Turnpike LP	Corsair Pro XS	Teleport	Mamba Flex	Caravel	FineCross	Telemark
Catheter Length (cm)	135/150 cm	135/150 cm	135/150 cm	135/150 cm	135/150 cm	135/150 cm	135/150 cm

(Continued)

TABLE 30.3 (Continued)

B. Small outer diameter microcatheters.

	Turnpike LP	Corsair Pro XS	Teleport	Mamba Flex	Caravel	FineCross	Telemark
Prox. O.D.	2.9 F	2.9 F	2.6 F	2.9 F	2.6 F	2.6 F	2.6 F
Distal O.D.	2.2 F	2.1 F	2.0 F	2.4 F	1.9 F	1.8 F	1.9 F
Tip Entry O.D.	1.6 F	1.3 F	0.0190" (1.45 F)	1.4 F	1.4 F	1.8 F	1.4 F
Inner Lumen I.D.	0.0175"	0.015"	0.017"	0.023"	0.017" (distal) 0.022" (proximal)	0.021"	

C. Dual lumen microcatheters.

	TwinPass Torque	TwinPass	Sasuke	NHancer Rx	ReCross
Catheter Length (cm)	135 cm	135 cm	145 cm	135 cm	140 cm
Prox. O.D.	3.5/3.5 F	3.4/2.7 F	3.2 F	2.6 F	3.4 F
Distal O.D.	2.1 F	2.0 F	2.5/3.3 F	2.3 F	2.3/3.3 F
Tip Entry O.D.	2.1 F	2.0 F	1.5 F	1.5 F	1.5 F

(Continued)

TABLE 30.3 (Continued)

C. Dual lumen microcatheters.

	TwinPass Torque	TwinPass	Sasuke	NHancer Rx	ReCross
Inner Lumen I.D.	0.015″ (RX) 0.0155″ (OTW Distal) 0.0165″ (OTW Proximal)	0.016″ (RX) 0.0165″ (OTW)	0.016″ (tip) 0.017″ (shaft)	0.019″ tip and shaft lumen	0.019″ tip and shaft lumen
Tip Length	7 mm	20 mm	4 mm	5 mm	5 mm
Tip Material	37D Pebax	47D Pebax	Tungsten	Tungsten	Tungsten
Tip Construction	37D Pebax/ 40D Pebax, two eccentric Pt/Ir marker bands—one on each lumen; OTW lumen features 10° guidewire kick-out angle	47D Pebax with two Pt/Ir marker bands	5 mm tungsten, 0.8 mm platinum coil	Tungsten polymer both ports	Tungsten polymer all three ports
Exterior Coating	Hydrophilic Coating (distal 25 cm)	Hydrophilic Coating (distal 18 cm)	Hydrophilic Coating (38 mm length)	Hydrophilic (NDurance)	
Distance of OTW lumen port from tip	7 mm	20 mm	6.5 mm		8 and 12 mm

2. **Small outer diameter single lumen microcatheters,** such as the Finecross (Terumo), Caravel and Corsair XS (Asahi Intecc), Mamba Flex (Boston Scientific) and Turnpike LP (Teleflex).
3. **Angulated microcatheters,** such as the SuperCross (Teleflex), Venture (Teleflex) and Swift-Ninja (Merit Medical).
4. **Dual lumen microcatheters,** such as the TwinPass and TwinPass Torque (Teleflex), Sasuke (Asahi Intecc), Crusade (Kaneka), FineDuo (Terumo), NHancer Rx (IMDS), and ReCross (IMDS).
5. **Plaque modification microcatheters,** such as the Tornus (Asahi Intecc), and Turnpike Gold (Teleflex).

The major role of each microcatheter category[35] is shown in Fig. 30.25. Moreover, microcatheters:

1. Provide better support and increase the wire tip stiffness, enhancing its penetration capacity. Microcatheters should be routinely used for guidewire support during CTO interventions,[36] but they can also be useful in other scenarios, such as wiring through tortuosity, angulation, or areas of severe calcification. They can also be used during bifurcation stenting, after rewiring a side branch through stent struts.[37]

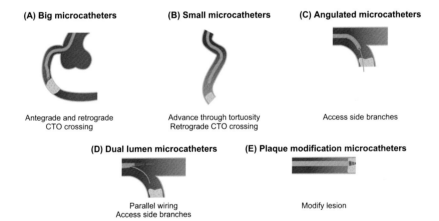

FIGURE 30.25   **Main uses of coronary microcatheters in chronic total occlusion percutaneous coronary intervention. Panel A**: Large single lumen microcatheters are used for antegrade and retrograde CTO crossing. **Panel B**: small single lumen microcatheters are used to cross tortuous vessels and collaterals during retrograde CTO crossing. **Panel C**: angulated microcatheters are used for accessing side branches. **Panel D**: dual lumen microcatheters are used for parallel wiring or for accessing a side branch. **Panel E**. plaque modification microcatheters are used to modify plaque, usually in balloon uncrossable lesions. *Reproduced with permission from Brilakis ES. Manual of percutaneous coronary interventions: a step-by-step approach. Elsevier; 2021.*

2. Allow reshaping of the guidewire tip.
3. Facilitate guidewire exchanges.
4. Protect the proximal part of the vessel (and/or the collateral vessel in case of retrograde crossing) from guidewire-induced injuries.
5. Allow injection of medications or contrast, either in the distal true lumen, or within the intraplaque or extraplaque space during chronic total occlusion interventions (Carlino technique).[38]
6. Allow delivery of coils, fat, or thrombin in case of perforation.[39] However, 0.014 inch microcatheters (all coronary microcatheters are 0.014 inch) except the FineCross cannot be used to deliver 0.018 inch coils (Section 30.18.2). Larger microcatheter, such as the Progreat (2.4 Fr, Terumo), Renegade (2.5 Fr, Boston Scientific) and Transit (2.5 Fr, Cordis) can be used for delivery of 0.018 inch coils. This is why it is important for cardiac catheterization laboratories to have 0.014 inch microcatheter compatible coils available for treating coronary perforations (such as the Axium coils, Medtronic). Alternatively, larger microcatheters should be available.

30.6.1 Large single lumen microcatheters

Large microcatheters are usually used for guidewire support during CTO PCI but can serve any other purpose as described above. Large microcatheters can be spun to facilitate advancement, although overspinning should be avoided to prevent microcatheter structural damage (Section 27.4).

30.6.1.1 Corsair Pro

The Corsair Pro microcatheter (Asahi Intecc, Fig. 30.26) was developed as a septal channel dilator to facilitate retrograde CTO PCI.[40] The Corsair "Shinka" shaft is constructed with eight thin wires wound with two larger

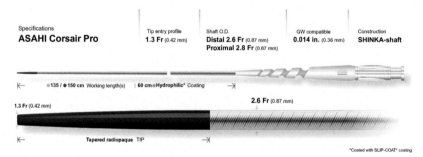

FIGURE 30.26 **Illustration of the Corsair Pro microcatheter**. *Reproduced with permission from Asahi Intecc.*

wires, which facilitates torque transmission. The inner lumen is lined with a polymer that enables contrast injection and facilitates wire advancement. The distal 60 cm of the catheter are coated with a hydrophilic polymer to enhance crossability. The tip is tapered and soft and is loaded with tungsten powder to enhance visibility.

Two Corsair lengths are currently available (135 cm long with light blue proximal hub and 150 cm long with dark blue proximal hub). The Corsair catheter can be advanced by rotating in either direction, although it is braided to have better torque transmission with counterclockwise rotation.

30.6.1.2 Turnpike and Turnpike Spiral

The Turnpike (Teleflex) has a dual layer bidirectional coil (Fig. 30.27) that facilitates torque transmission, improves flexibility, and prevents kinking. It also has a soft, tapered tip facilitating collateral branch crossing.

The Turnpike catheter is produced in four versions (Fig. 30.28): Turnpike, Turnpike LP, Turnpike Spiral and Turnpike Gold. Turnpike is the standard catheter with a 1.6 French outside diameter at the distal tip and 2.6 French outside distal shaft diameter. The Turnpike Spiral has a distal nylon coil and is preferred by many operators for antegrade CTO crossing. The Turnpike and Turnpike LP catheters can be rotated in either direction. In contrast, the Turnpike Spiral and Gold should be rotated clockwise to advance and counter-clockwise for withdrawal (opposite direction compared with the Tornus catheter).

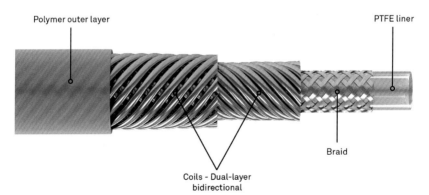

Polymer outer layer

PTFE liner

Braid

Coils - Dual-layer bidirectional

FIGURE 30.27 **Construction of the Turnpike microcatheter.** *Image courtesy of Teleflex Incorporated. © 2022 Teleflex Incorporated. All rights reserved.*

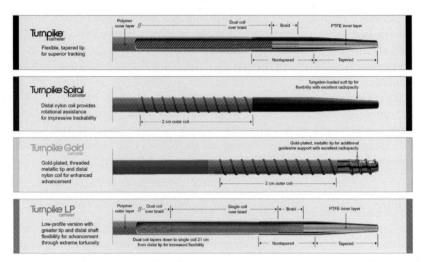

FIGURE 30.28 **Illustration of the Turnpike microcatheters.** *Image courtesy of Teleflex Incorporated. © 2022 Teleflex Incorporated. All rights reserved.*

30.6.1.3 Teleport Control

The Teleport and Teleport Control microcatheters (OrbusNeich, Fig. 30.29) have a tungsten radiopaque short tip, an inner stainless-steel body of hybrid braiding and coil (Hybracoil) construction, an ultra-thin outer jacket from nylon transitioning into Pebax to facilitate greater distal flexibility, and a lubricious hydrophilic coating in the distal 60 cm. The hybrid braiding consists of 2 round and 14 flat wires outside a flat coil. The Teleport Control has a higher distal crossing profile compared with the Teleport. Both the Teleport and Teleport Control microcatheters are available in 135 and 150 cm length.

30.6.1.4 Mamba

The Mamba and Mamba Flex microcatheters (Boston Scientific, Fig. 30.30) have a flexible, tapered coil formed by 11 wires that are tightly wound on the proximal end to provide stiffness, torque, and pushability, and then taper to allow for a lower profile and flexibility at the distal end. They both have a durable, lubricious hydrophilic coating on the distal 60 cm. The coil support extends to 0.5 mm from the tip, improving the tip support and reducing the risk of tip separation, at the cost of lower tip flexibility.

The Mamba has a higher distal crossing profile (0.032″, 0.81 mm) and is designed to provide strong guidewire support for antegrade crossing. It is only available in 135 cm length. The Mamba Flex has a lower distal crossing profile (0.028″, 0.71 mm) and can be used for both retrograde and antegrade

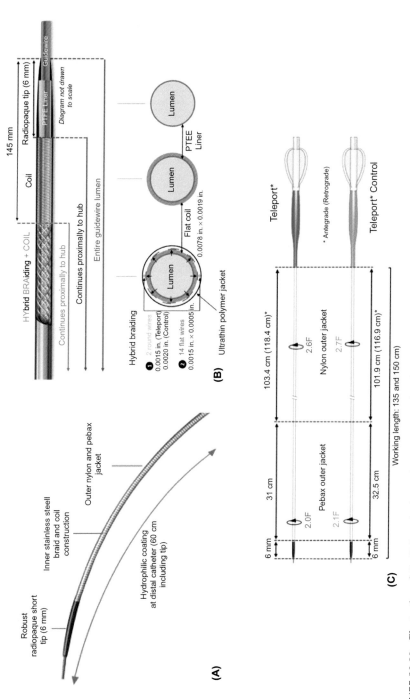

FIGURE 30.29 Illustration of the Teleport microcatheters construction. Panel A: main characteristics of the Teleport microcatheter. Panel B: construction of the Teleport microcatheter in the shaft and the tip. Panel C: comparison of the Teleport and the Teleport Control microcatheters. *Image used with permission of OrbusNeich Medical Company Limited.*

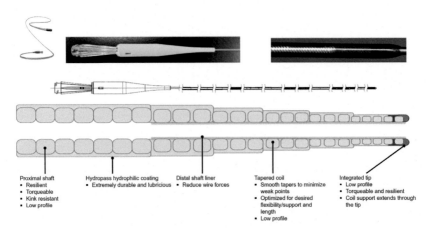

Proximal shaft
- Resilient
- Torqueable
- Kink resistant
- Low profile

Hydropass hydrophilic coating
- Extremely durable and lubricious

Distal shaft liner
- Reduce wire forces

Tapered coil
- Smooth tapers to minimize weak points
- Optimized for desired flexibility/support and length
- Low profile

Integrated tip
- Low profile
- Torqueable and resilient
- Coil support extends through the tip

FIGURE 30.30 **Illustration of the Mamba and Mamba Flex microcatheters.** *Image provided courtesy of Boston Scientific © 2022 Boston Scientific Corporation or its affiliates. All rights reserved.*

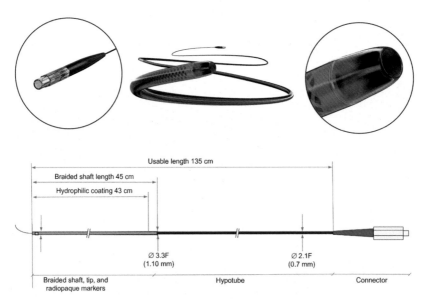

Usable length 135 cm

Braided shaft length 45 cm

Hydrophilic coating 43 cm

Ø 3.3F (1.10 mm)

Ø 2.1F (0.7 mm)

Braided shaft, tip, and radiopaque markers

Hypotube

Connector

FIGURE 30.31 **Illustration of the M-CATH.** *Image used with permission of Acrostak.*

crossing, especially through areas of tortuosity. The Mamba Flex is available in both 135 and 150 cm lengths.

30.6.1.5 M-CATH

The M-CATH(Acrostak, Fig. 30.31) is another microcatheter with low crossing profile and is currently only available in 135 cm length. Its tip is resistant to deformation; hence, it is particularly suitable for heavily calcified occlusions. The M-CATH Flexy is available in 135 and 150 cm lengths.

30.6.1.6 NHancer Pro X

The NHancer Pro X catheter (IMDS, Fig. 30.32) is available in 135 and 155 cm lengths, and has a soft, tapered tip and tip to hub variable braid.

30.6.1.7 Mizuki and Mizuki FX

The Mizuki catheter (Kaneka, Fig. 30.33) is available in two versions, one with standard tip stiffness (Mizuki) and one with a more flexible tip (Mizuki FX). It has hydrophilic coating and fluoro-resin inner surface for lubricity.

30.6.2 Small outer diameter single lumen microcatheters

Small outer diameter microcatheters are easier to deliver through small, tortuous vessels, such as collaterals used for retrograde CTO interventions.

Small outer diameter microcatheters can be subdivided into those with coils in their wall (smaller versions of the large outer diameter microcatheters Turnpike, Corsair Pro, Teleport Control, and Mamba) that can be spun for advancement and those without coils (Caravel, Finecross, Telemark), which have smaller outer diameter but are not designed to be spun.

30.6.2.1 Turnpike LP

The Turnpike LP (low profile) (Fig. 30.28) is similar to the Turnpike catheter but has a lower profile.

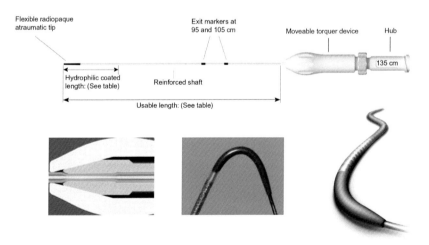

FIGURE 30.32 **Illustration of the NHancer Pro X microcatheter**. *Reproduced with permission from IMDS.*

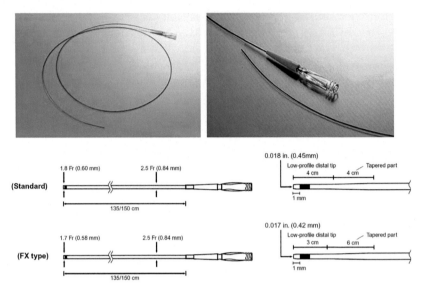

FIGURE 30.33 **Illustration of the Mizuki and Mizuki FX microcatheters.** *Reproduced with permission from Kaneka.*

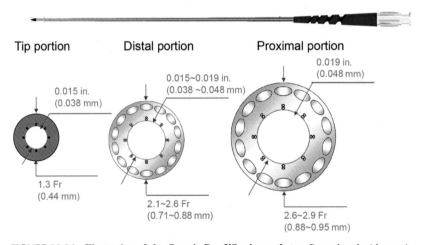

FIGURE 30.34 **Illustration of the Corsair Pro XS microcatheter.** *Reproduced with permission from Asahi Intecc.*

30.6.2.2 Corsair Pro XS

The Corsair Pro XS (extra small) (Fig. 30.34) is similar to the Corsair Pro catheter but has a lower profile.

30.6.2.3   Teleport

The Teleport catheter is similar to the Teleport Control (Fig. 30.29) but has a lower distal crossing profile.

30.6.2.4   Mamba Flex

The Mamba Flex is similar to the Mamba (Fig. 30.30) but has a lower distal crossing profile.

30.6.2.5   Caravel

The Caravel microcatheter (Fig. 30.35) was developed to advance through small and tortuous collaterals. It has a very low distal tip profile (1.4 Fr), and low distal shaft profile (1.9 Fr) with a hydrophilic coating. It also has a braided shaft. It was designed to advance with forward push but could also be gently rotated to cross challenging collaterals.

The Caravel was not designed to withstand aggressive rotation and advancement. Such an approach can strain the distal tip connection to the shaft of the microcatheter and result in tip fracture and separation (Fig. 30.36)[32] (**CTO PCI Manual case** 87).

FIGURE 30.35   **Illustration of the Caravel microcatheter**. *Reproduced with permission from Asahi Intecc.*

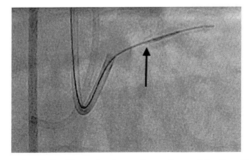

FIGURE 30.36   **The tip of the Caravel microcatheter (arrow) broke off from the remainder of the body of the catheter while attempting torqueing through a calcified stenotic segment of the left anterior descending artery.** *Courtesy of Dr. William Nicholson. Reproduced with permission from Brilakis ES. Manual of coronary chronic total occlusion interventions. a step-by-step approach. 2nd ed. Elsevier; 2017.*

30.6.2.6 Finecross

The Finecross (Terumo) microcatheter is very flexible and has a low crossing profile (1.8 Fr distal tip). It has a stainless-steel braid (to enhance torquability) and a distal marker located 0.7 mm from the tip (Fig. 30.37). Although the Finecross is mainly advanced using forward push, many operators are using a combination of push and rotation to facilitate advancement. Moreover, the Finecross can be used to deliver 0.018 inch coils (such as Tornado coils, Cook Medical) to treat perforations.

30.6.2.7 Telemark

The Telemark (Surmodics, distributed by Medtronic, Fig. 30.38) microcatheter is very flexible and has a low crossing profile (1.4 Fr distal tip outer diameter and 1.9 Fr distal shaft outer diameter). It has markers located 1 and 5 mm from the distal tip.

30.6.3   Angulated microcatheters

30.6.3.1 Venture

CTO PCI Manual online cases 48, 96, 97, 127.
PCI Manual online case 57.

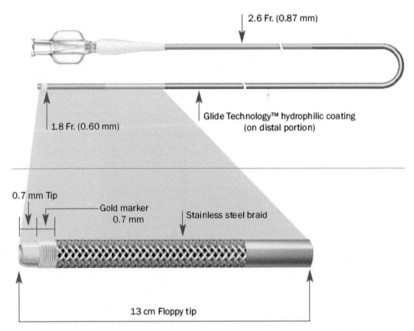

FIGURE 30.37 **Illustration of the Finecross microcatheter.** *Reproduced with permission from Terumo Medical Corporation.*

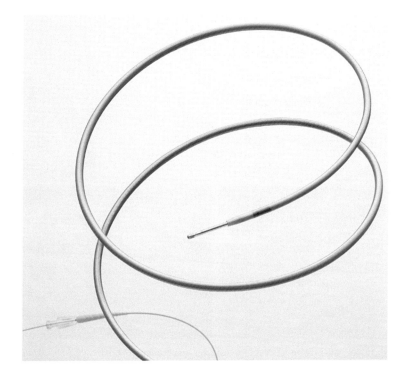

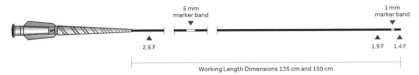

FIGURE 30.38   **Illustration of the Telemark microcatheter.** *Reproduced with permission from Surmodics/Medtronic.*

The Venture catheter (Teleflex, Fig. 30.39) has an 8 mm radiopaque torqueable distal tip that has a bend radius of 2.5 mm.[41−46] The tip can be deflected up to 90 degrees by clockwise rotation of a thumb wheel on the external handle. With rotation of the entire catheter, steering in all planes is possible. It is compatible with 6 Fr guide catheters and with 0.014″ guidewires. Both rapid exchange and over-the-wire versions are available, but the over-the-wire Venture catheter is preferred for CTO PCI, as it allows for wire exchanges.

Venture tips and tricks:

1. The Venture catheter has a deflectable tip, which can be utilized to assist with accessing difficult side branch vessels. As shown in Fig. 30.40, the catheter design allows the operator to rotate the tip

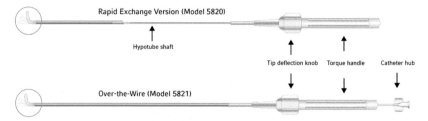

FIGURE 30.39 **Illustration of the Venture microcatheter.** *Image courtesy of Teleflex Incorporated.* © *2022 Teleflex Incorporated. All rights reserved.*

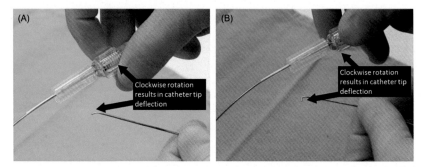

FIGURE 30.40 **Illustration of Venture catheter manipulation.** *Courtesy of Dr. William Nicholson. Reproduced with permission from Brilakis ES. Manual of coronary chronic total occlusion interventions. a step-by-step approach. 2nd ed. Elsevier; 2017.*

deflector twist knob to transmit increasing tip deflection to the distal tip of the microcatheter.

2. The Venture catheter is delivered to the target vessel in a straight configuration over a workhorse guidewire (Fig. 30.41, panel A & B). Once it reaches the target coronary segment the workhorse guidewire is withdrawn inside the Venture catheter (Fig. 30.41, panel C) and the tip deflector twist knob is rotated clockwise to deflect the catheter tip. The deflected catheter is rotated and withdrawn until it points to the proximal cap (Fig. 30.41, panel D), followed by guidewire advancement into the lesion (Fig. 30.41, panel E). The Venture catheter can then be straightened and removed, leaving the guidewire in place (Fig. 30.41, panel F).

3. A classic example for use of the Venture catheter is for crossing ostial circumflex CTOs (Fig. 30.42).[43]

4. The Venture catheter can also prevent the guidewire from prolapsing into a side branch.[46]

5. During retrograde CTO PCI, the Venture catheter can be used to enable wiring of collateral branches with difficult, acutely angulated takeoffs.

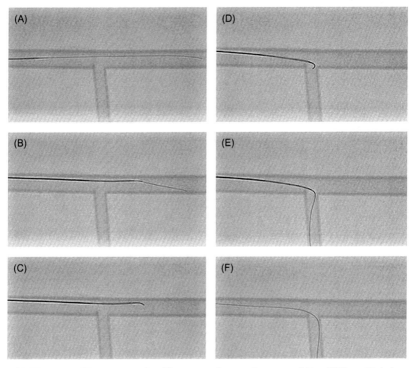

FIGURE 30.41   **How to use the Venture catheter**. *Courtesy of Dr. William Nicholson. Reproduced with permission from Brilakis ES. Manual of coronary chronic total occlusion interventions. a step-by-step approach. 2nd ed. Elsevier; 2017.*

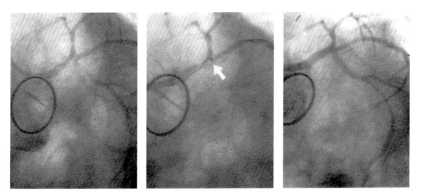

FIGURE 30.42   **Example of Venture catheter use to cross an ostial circumflex CTO**. *Reproduced with permission from McNulty E, Cohen J, Chou T, Shunk K. A "grapple hook" technique using a deflectable tip catheter to facilitate complex proximal circumflex interventions.* Catheter Cardiovasc Interv *2006;67:46−48.*

6. Removal of the Venture catheter using a "trapping balloon technique" requires 8 Fr guide catheters, given its larger profile as compared with an over-the-wire balloon or other microcatheters.[46] For the same reason an 8 Fr guide catheter is needed to perform the parallel-wire technique, when one of the wires is inserted through the Venture catheter.

7. The Venture catheter is stiff, which can be both an advantage and disadvantage, as it can provide extra support, but can also predispose to vessel injury. Since the bend radius is 2.5 mm, special care must be exercised when deflecting the tip in <2.5 mm diameter arteries.

8. The Venture catheter bend should be released, and the tip straightened during advancement or removal to prevent vessel damage.

30.6.3.2 SuperCross

CTO PCI Manual online cases 22, 83, 91, 110, 111, 134, 135.
PCI Manual online cases 41, 62.

The SuperCross (Fig. 30.43) (Teleflex) is a microcatheter that is manufactured with either a straight tip or with various distal tip angulation (45, 90, and 120 degrees) and can facilitate guidewire advancement through areas of tortuosity. The SuperCross microcatheters can be helpful in scenarios similar to those benefiting from use of the Venture catheter. The 120 degree bend can be very helpful in retrograde CTO cases through reverse angled collaterals, in which the retrograde wire is biased to advance towards the

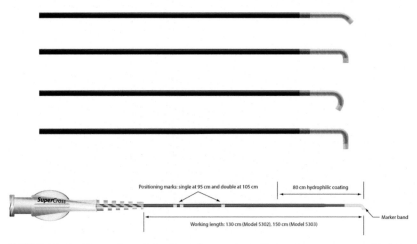

FIGURE 30.43 Illustration of the SuperCross microcatheter. *Image courtesy of Teleflex Incorporated. © 2022 Teleflex Incorporated. All rights reserved.*

distal vessel. Similarly, the SuperCross 120 degree bend microcatheter can be very useful when trying to treat a lesion upstream of a saphenous vein graft distal anastomosis.

30.6.3.3   Swift Ninja

The Swift Ninja (Fig. 30.44) (Merit Medical) is a straight tip catheter that articulates up to 180 degrees in opposing directions.

30.6.4   Dual lumen microcatheters

The Twin Pass (Teleflex, Fig. 30.45), Twin Pass Torque (Teleflex, Fig. 30.45), Sasuke (Asahi Intecc, Fig. 30.46), NHancer Rx (IMDS, Fig. 30.47) and ReCross (IMDS, Fig. 30.48) dual lumen microcatheters are currently available in the US.

The Twin Pass torque has a braided shaft that facilitates torqueing and positioning of the over-the-wire port, which has a 10 degrees exit angle and is located 1.5 mm distal to the proximal marker.

FIGURE 30.44   **Illustration of the Swift Ninja microcatheter.** © *Merit Medical, Reprinted by Permission.*

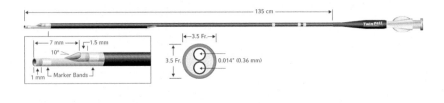

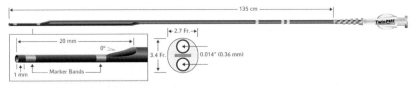

FIGURE 30.45   Illustration of the TwinPass and TwinPass Torque dual lumen microcatheters. *Image courtesy of Teleflex Incorporated.* © *2022 Teleflex Incorporated. All rights reserved.*

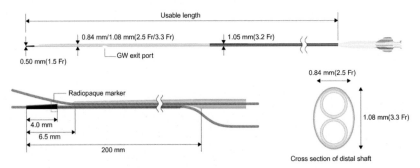

FIGURE 30.46   Illustration of the Sasuke dual lumen microcatheter. *Reproduced with permission from Asahi Intecc.*

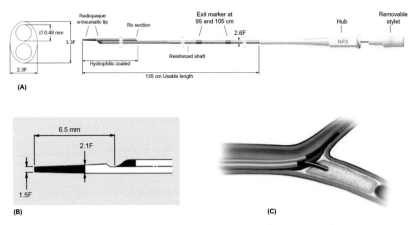

FIGURE 30.47   Illustration of the NHancer Rx dual lumen microcatheter. Panel A: NHancer Rx microcatheter design. **Panel B**: NHancer Rx microcatheter distal shaft design. **Panel C**: illustration of the NHancer Rx use in a CTO located at a bifurcation. *Reproduced with permission from IMDS.*

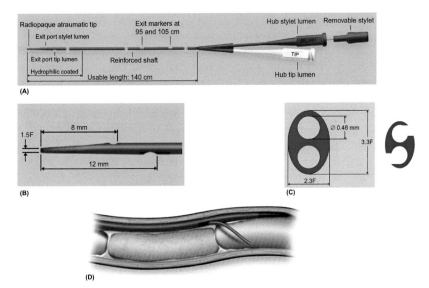

FIGURE 30.48   **Illustration of the ReCross dual lumen microcatheter. Panel A:** ReCross microcatheter design. **Panel B:** ReCross microcatheter distal shaft design. **Panel C:** ReCross microcatheter
transverse view of the shaft and the exit port positioning. **Panel D:** illustration of use of the ReCross
microcatheter for reentry into the distal true lumen. *Reproduced with permission from IMDS.*

The Sasuke microcatheter (Asahi Intecc, Fig. 30.46) is a dual lumen
microcatheter with the same tapered soft tip used in the Corsair and Caravel
microcatheters and an oval design with a double stainless-steel core in the
proximal shaft to ensure kink resistance. The over-the-wire exit port is
located 6.5 mm from the tip and is highly visible, facilitating guidewire
manipulations. It is 145 cm long with a hydrophilic coating in the most distal
38 cm.

The NHancer Rx (IMDS, Roden, The Netherlands, Fig. 30.47) is a dual
lumen microcatheter with an over-the-wire lumen and a rapid exchange
lumen. The NHancer Rx has a removable stylet in the over-the-wire lumen
that increases pushability. The distal shaft design is oval and provides the
smallest dual lumen crossing profile (2.3 F) currently available. This very
small crossing profile makes catheter trapping using a conventional balloon
possible in a 6 French guide catheter. The distal tip and over-the-wire lumen
exit port are designed with soft tungsten filled material allowing visibility of
the exit port on X-ray. Shaft braiding reinforcement improves the torque and
lumen integrity performance.

The ReCross microcatheter (IMDS, Fig. 30.48) is the only dual lumen
microcatheter with two over-the-wire lumens instead of a combination of a
rapid exchange lumen and an over-the-wire lumen, as is the case with the

other dual lumen microcatheters. As a result, guidewire exchange is feasible through both lumens. Another unique feature of the ReCross is that the over-the-wire lumen that leads to the tip-exit has a second exit port located 12 mm proximal from the tip. This exit port is oriented 180 degrees opposite from the exit port of the second OTW-lumen that is located 8 mm from the tip. As the shape of the ReCross is oval, it will tends to turn with its broader side towards the coronary lumen, similar to a Stingray balloon, hence one of the opposite proximal exit ports may allow puncture for distal reentry.

Two other dual lumen microcatheters are available outside the US: the Crusade (Kaneka, Fig. 30.49), and the FineDuo (Terumo, Fig. 30.50).

Except for the ReCross, all dual lumen microcatheters consist of a rapid exchange delivery system in the distal segment and an over-the-wire lumen that runs the length of the catheter. A radiopaque marker band identifies the distal tip of each lumen; the distal band corresponds to the exit point of the rapid exchange segment and the proximal band marks the exit point of the over-the-wire lumen.

Dual lumen microcatheters have multiple uses in CTO and non-CTO PCI (Fig. 30.51):

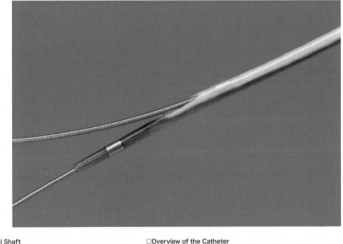

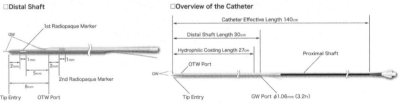

FIGURE 30.49 **Illustration of the Crusade dual lumen microcatheter.** *Reproduced with permission from Kaneka.*

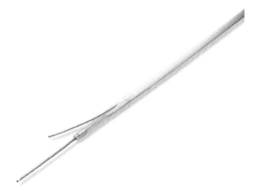

FIGURE 30.50   **Illustration of the FineDuo dual lumen microcatheter.** *Reproduced with permission from Terumo Medical Corporation.*

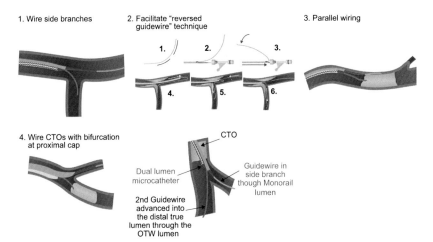

FIGURE 30.51   **The various uses of the dual lumen microcatheters.** *Reproduced with permission from Brilakis ES. Manual of percutaneous coronary interventions: a step-by-step approach. Elsevier; 2021.*

CTO PCI:

1. Parallel wiring.
2. Wiring CTOs with a side branch adjacent to the proximal cap.
3. Wiring the distal vessel if guidewire crossing is achieved into a side branch distal to the lesion (CTO or nonocclusive): instead of pulling back and redirecting the guidewire (risking inability to recross the occlusion), the dual lumen microcatheter enables wiring the distal main vessel without losing access to the side branch.

4. Wire septal branches during retrograde crossing attempts in CTO PCI.
5. Antegrade wiring of the distal true lumen if the externalized retrograde guidewire crosses a collateral in close proximity to the distal cap, precluding safe antegrade dilatation of the CTO over the retrograde wire.

Non-CTO PCI:

1. Inserting a second guidewire after successfully advancing another guidewire to the desired coronary location (for example, after successfully crossing the target lesion with a standard 0.014 inch guidewire, a dual lumen microcatheter can be used to insert a Rotafloppy Drive wire across the target lesion).
2. Wiring the side branch of a bifurcation, including wiring through jailed side branches during bifurcation stenting.
3. Facilitate the reversed guidewire (otherwise called "hairpin" guidewire) technique (Section 8.2.2.6 part 3).[47-49]

Dual lumen microcatheters tips and tricks:

1. Dual lumen microcatheters may, sometimes, be challenging to deliver. Some of them come with a stiffening mandrel (NHancer Rx and ReCross) that, when loaded into the OTW lumen can provide support and pushability during catheter insertion.
2. Controlling the direction of the side port can be challenging and may require removing the microcatheter and re-inserting it.
3. Once wiring is achieved, maintaining distal wire positions while withdrawing the dual lumen microcatheter can be difficult and is best achieved with the assistance of a trapping balloon.

30.6.5   Plaque modification microcatheters

The Tornus and Tornus 88Flex microcatheters (Fig. 30.52) consist of 8 stainless steel wires stranded in a coil.[50] The Tornus is available in 2 sizes (2.1 and 2.6 Fr), with the latter providing more guidewire support. It has a platinum marker located 1 mm from the tip. Unlike other microcatheters, the Tornus does not have an inner polymer and, as a result, injection of contrast cannot be done through it, but it provides strong support. The Tornus is advanced by counterclockwise rotation and withdrawn by clockwise rotation. To avoid catheter kinking and unraveling of the stranded steel wires, no more than 20 rotations should be done in any direction.

Similar to the Tornus, the Turnpike Spiral and Turnpike Gold microcatheters (Fig. 30.28) can be used to modify a resistant lesion. In contrast to Tornus, the Turnpike Spiral and Gold are advanced using clockwise rotation.

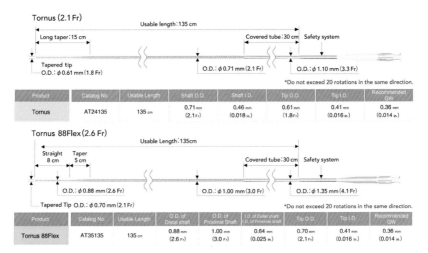

FIGURE 30.52 Illustration of the Tornus (2.1 French) and Tornus 88 Flex (2.6 French) microcatheters. *Reproduced with permission from Asahi Intecc.*

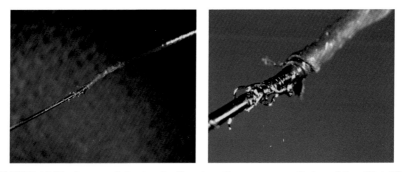

FIGURE 30.53 Images of the tip of a Corsair catheter permanently bound to a Pilot 200 guidewire with destruction of the wire's polymer jacket and entanglement of the wire coil by the tip of the Corsair catheter. *Courtesy of Dr. William Nicholson. Reproduced with permission from Brilakis ES. Manual of coronary chronic total occlusion interventions. a step-by-step approach. 2nd ed. Elsevier; 2017.*

30.6.6 Microcatheter-related complications

1. Microcatheter over-rotation could cause catheter deformation and entrapment, fracture proximal to the catheter tip, or result in the wire binding to the microcatheter (microcatheter "fatigue") (Fig. 30.53)[32] (Section 27.4).
2. Contrast can be injected through the microcatheter (except for the Tornus) for distal vessel visualization, but the microcatheter should subsequently be flushed with normal saline to minimize guidewire "stickiness." Rarely the wire may get "stuck" requiring removal of both the microcatheter and the guidewire.

3. If difficulty is encountered while attempting to advance a microcatheter or manipulate a guidewire after prolonged use, the cause may be "microcatheter fatigue": in such cases the microcatheter should be exchanged for a new microcatheter.
4. During retrograde CTO PCI, after wire externalization and during antegrade gear delivery, the tip of antegrade equipment (such as balloons and stents) should not come in contact with the tip of the retrograde microcatheter catheter over the same guidewire to avoid "interlocking" and equipment entrapment.

30.7 Guidewires

Proper selection and use of 0.014 inch guidewires is critical for CTO PCI.

An 0.035 or 0.038 inch guidewire is also essential for obtaining arterial access and advancing the guide catheters to the coronary or bypass graft ostia.

The 0.014 inch coronary guidewires can be classified into seven categories:

1. Workhorse
2. Polymer-jacketed
3. Stiff-tip
4. Support
5. Atherectomy
6. Externalization
7. Pressure wires

Workhorse, polymer-jacketed, stiff-tip guidewires, and externalization guidewires are essential for CTO PCI.

30.7.1 Workhorse guidewires

Workhorse guidewires are used for most non-CTO PCIs. They are also used during CTO PCI after successful crossing of the occlusion to minimize the risk of distal vessel injury or perforation during equipment delivery. The key characteristics of workhorse guidewires are:

1. Soft tip (usually 1 gram or less tip load), and
2. No polymer jacket (in contrast to polymer-jacketed guidewires, discussed in Section 30.7.2).

Other characteristics of workhorse guidewires include:

1. *Hydrophilic coating.* Several workhorse guidewires have a hydrophilic coating (which is different than a full polymer jacket). As anticipated,

hydrophilic coating increases deliverability but decreases tactile feedback
and increases the risk of vessel injury.

2. *Material*: most wires are made of stainless steel. Nitinol wires have better
   tip shape retention but can be harder to shape.
3. *Construction of the tip*: most guidewires have a wire coil around a central
   core. The central coil might extend to the tip of the wire (core to tip) or a
   shaping ribbon might be added to allow easier tip shaping, which is commonly used in nitinol wires. Some wires have multiple cores (ACT-ONE
   technology), such as the Sion (Fig. 30.54), Sion blue, and Suoh 03
   (Fig. 30.55) (Asahi Intecc), whereas other wires have dual coils, such as
   the Samurai and Samurai RC (Fig. 30.56, Boston Scientific) that improve
   handling.
4. Although often not considered a "workhorse" wire, the Suoh 03 (Fig. 30.55)
   is the softest tip guidewire currently available (0.3 gr tip load). The Suoh 03
   has a 19 cm long coil and a hydrophilic coating in its distal 52 cm and is the
   wire of choice for crossing epicardial collaterals.

30.7.2 Polymer-jacketed guidewires

The distal section of the polymer-jacketed guidewires is fully covered by a
very slippery polymer.

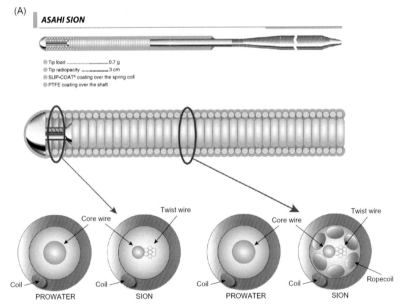

FIGURE 30.54 **Construction of the Sion guidewire**. *Reproduced with permission from Asahi Intecc.*

FIGURE 30.55 **Illustration of the Suoh 03 guidewire.** *Reproduced with permission from Asahi Intecc.*

There are three major groups of polymer-jacketed guidewires (Fig. 30.57 and Table 30.4).

Soft, nontapered polymer jacketed guidewires are used during both CTO and non-CTO PCI, whereas soft tapered and stiff-tip polymer jacketed guidewires are usually used in CTO PCI:

- Stiff-tip polymer jacketed wires are often used during both antegrade and retrograde wiring and for reentry into the distal true lumen through the Stingray balloon or the ReCross microcatheter (stick and swap technique, Section 8.2)
- Polymer-jacketed wires are used for knuckling for extraplaque CTO crossing.

Polymer-jacketed wires may carry increased risk of complications, such as distal vessel perforation and/or extraplaque tracking causing dissection. Polymer-jacketed wires should, therefore, only be used when necessary and should be exchanged for a workhorse wire before delivering balloons and stents.

30.7.2.1   Soft, nontapered polymer-jacketed guidewires

The Sion black and Fielder FC (Fig. 30.58) are the most commonly used soft tip, nontapered polymer-jacketed guidewires. Sion black has the ACT ONE technology and is more torqueable but also more costly.

Soft, nontapered, polymer-jacketed wires are very useful for crossing tortuous coronary segments. Sion black is frequently used for crossing both septal and epicardial collaterals.

SAMURAI Guidewire

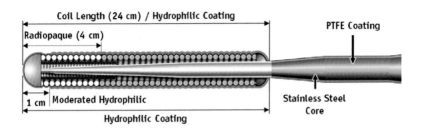

Tip Diameter (inch)	Tip Load (gf)	Tip Radiopacity (cm)	Tip Style	Coil / Covering	Coil Length (cm)	Coating	Core Material
.014	0.5	4	Inner Coil Technology	Spring Coil	20	Moderated Hydrophilic	Stainless Steel

SAMURAI RC Guidewire

Versatile wire with Inner Coil Technology for frontline cases, visible channels, or collateral crossing.

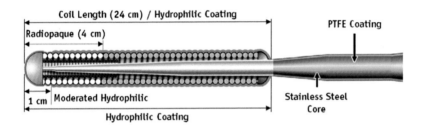

Tip Diameter (inch)	Tip Load (gf)	Tip Radiopacity (cm)	Tip Style	Coil / Covering	Coil Length (cm)	Coating	Core Material
0.014	1.2	4	Inner Coil Technology	Spring Coil	24	Hydrophilic	Stainless Steel

Inner Coil Technology (ICT) - A stainless steel inner coil affixed directly to the distal portion of the stainless steel core enhances the **shape retention** and **durability** of the distal tip, reduces **whipping**, and provides **exceptional torqueability**.

Inner Coil Technology

FIGURE 30.56 **Illustration of the Samurai and Samurai RC guidewires.** *Image provided courtesy of Boston Scientific. © 2022 Boston Scientific Corporation or its affiliates. All rights reserved.*

Polymer-jacketed guidewires

Soft non-tapered	Soft tapered	Stiff-tip
Choice PT	**Bandit**	**Gladius**
Fielder FC	**Fielder XT**	**Gladius Mongo**
Pilot 50	**Fielder XT-A**	**Pilot 200**
Sion black	**Fielder XT-R**	**PT 2**
Whisper	**Fighter**	**Raider**

FIGURE 30.57 **Classification of the polymer-jacketed guidewires.**

TABLE 30.4 Overview of polymer-jacketed guidewires.

Tip style	Commercial name	Tip stiffness (grams)	Manufacturer	Properties
Soft, non-tapered	Choice PT	1.5	Boston Scientific	Non-CTO PCI: Wiring through tortuosity. CTO PCI: collateral vessel crossing.
	Choice PT Floppy	2.1	Boston Scientific	
	Fielder FC	0.8	Asahi Intecc	
	Sion Black	0.8	Asahi Intecc	
	Pilot 50	1.5	Abbott Vascular	
	Whisper LS, Ms, ES	0.8, 1.0, 1.2	Abbott Vascular	
Soft, tapered	Bandit	0.8	Teleflex	Front-line wire for antegrade CTO crossing. Can also be used for knuckle wire formation and for retrograde crossing. Fielder XT-R is designed for retrograde collateral crossing.
	Fielder XT Fielder XT-A Fielder XT-R	0.8 1.0 0.6	Asahi Intecc	
	Fighter	1.2	Boston Scientific	
Stiff-tip	Gladius	3	Asahi Intecc	Non-CTO PCI: difficult to penetrate lesions. CTO PCI: Antegrade crossing, especially when the course of the occluded vessel is unclear. Also useful for knuckle wire formation (Gladius Mongo is specifically designed for knuckling) and for reentry into true lumen during antegrade dissection and reentry.
	Gladius Mongo	3 (designed for knuckling)	Asahi Intecc	
	Pilot 150 | 200	2.7 | 4.1	Abbott Vascular	
	PT2 Moderate Support	2.9	Boston Scientific	
	Raider	4.0	Teleflex	

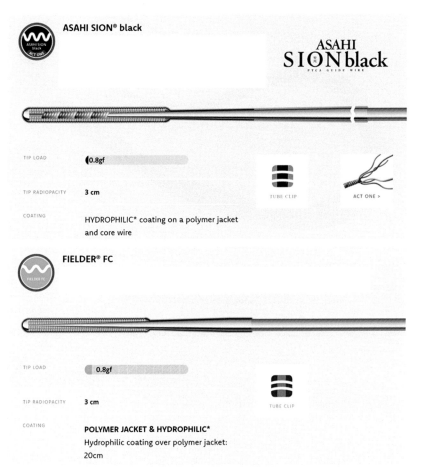

FIGURE 30.58   Illustration of the Sion black and Fielder FC wires. *Reproduced with permission from Asahi Intecc.*

30.7.2.2   Soft, tapered polymer-jacketed guidewires

The Bandit (Fig. 30.59), Fielder XT, Fielder XT-A, Fielder XT-R (Fig. 30.60) and the Fighter (Fig. 30.61) are soft, tapered polymer-jacketed guidewires. These wires are usually the first choice for antegrade wiring and are also frequently used for knuckling.

30.7.2.3   Stiff, polymer-jacketed guidewires

The Gladius (Fig. 30.62), Gladius Mongo (Fig. 30.63), Pilot 200 (Fig. 30.64) and the Raider (Fig. 30.65) are stiff-tip, nontapered polymer-jacketed

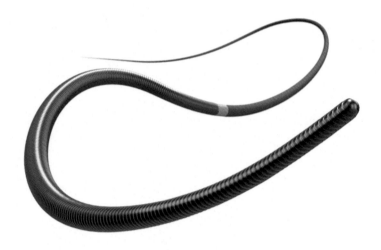

Bandit™ Guidewire

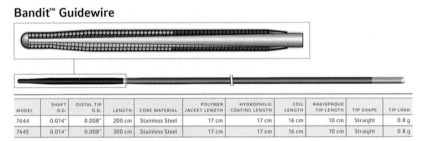

MODEL	SHAFT O.D.	DISTAL TIP O.D.	LENGTH	CORE MATERIAL	POLYMER JACKET LENGTH	HYDROPHILIC COATING LENGTH	COIL LENGTH	RADIOPAQUE TIP LENGTH	TIP SHAPE	TIP LOAD
7444	0.014"	0.008"	200 cm	Stainless Steel	17 cm	17 cm	16 cm	10 cm	Straight	0.8 g
7445	0.014"	0.008"	300 cm	Stainless Steel	17 cm	17 cm	16 cm	10 cm	Straight	0.8 g

FIGURE 30.59 **Illustration of the Bandit wire.** *Image courtesy of Teleflex Incorporated.* © *2022 Teleflex Incorporated. All rights reserved.*

guidewires. These wires are used for both antegrade and retrograde crossing as well as during antegrade dissection/reentry (swap wires for Stingray-based reentry). They can also be used for knuckling: the Pilot 200, Gladius and Raider guidewires make large knuckles, whereas the Gladius Mongo is designed to create a small knuckle. The Gladius Mongo ES (extra support wire) has a stiffer body, shorter spring coil length (3.0 instead of 8.5 cm for the Gladius), and shorter coating length (10 cm vs 41 cm for the Gladius).

30.7.3 Stiff-tip guidewires without a polymer jacket

Stiff-tip guidewires without a polymer jacket are classified based on: (1) whether they are tapered at the tip (usually to 0.008 or 0.009″); and (2) tip stiffness: intermediate (<9 g) or high (≥9 g). There are four major groups of stiff-tip guidewires (Fig. 30.66 and Table 30.5).

Fielder XT

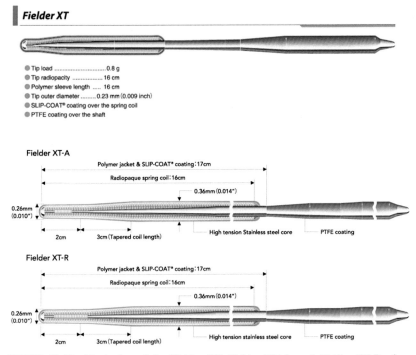

- Tip load ....0.8 g
- Tip radiopacity .... 16 cm
- Polymer sleeve length ..... 16 cm
- Tip outer diameter ....0.23 mm (0.009 inch)
- SLIP-COAT® coating over the spring coil
- PTFE coating over the shaft

Fielder XT-A

Polymer jacket & SLIP-COAT® coating : 17 cm
Radiopaque spring coil : 16 cm
0.36mm (0.014")
0.26mm (0.010")
2cm   3cm (Tapered coil length)   High tension Stainless steel core   PTFE coating

Fielder XT-R

Polymer jacket & SLIP-COAT® coating : 17 cm
Radiopaque spring coil : 16 cm
0.36mm (0.014")
0.26mm (0.010")
2cm   3cm (Tapered coil length)   High tension stainless steel core   PTFE coating

FIGURE 30.60   **Illustration of the Fielder XT, Fielder XT-A, and Fielder XT-R wires.**
Reproduced with permission from Asahi Intecc.

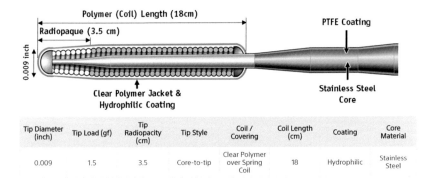

Polymer (Coil) Length (18cm)
Radiopaque (3.5 cm)
PTFE Coating
0.009 inch
Clear Polymer Jacket &
Hydrophilic Coating
Stainless Steel
Core

Tip Diameter (inch)	Tip Load (gf)	Tip Radiopacity (cm)	Tip Style	Coil / Covering	Coil Length (cm)	Coating	Core Material
0.009	1.5	3.5	Core-to-tip	Clear Polymer over Spring Coil	18	Hydrophilic	Stainless Steel

FIGURE 30.61   **Illustration of the Fighter wire.** *Image provided courtesy of Boston Scientific. ©2022 Boston Scientific Corporation or its affiliates. All rights reserved.*

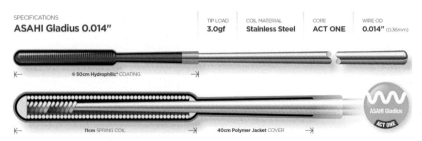

FIGURE 30.62 **Illustration of the Gladius wire**. *Reproduced with permission from Asahi Intecc.*

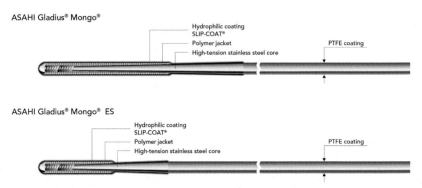

FIGURE 30.63 **Illustration of the Gladius Mongo wire**. *Reproduced with permission from Asahi Intecc.*

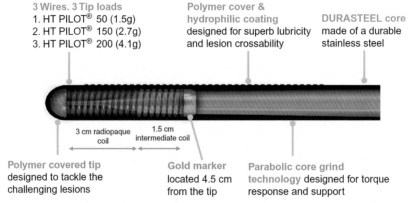

FIGURE 30.64 **Illustration of the Pilot family of wires**. *Courtesy of Abbott Vascular.* © *2022 Abbott. All Rights Reserved.*

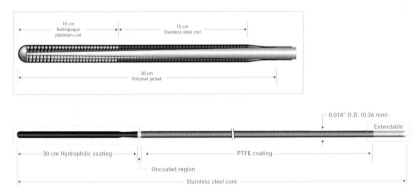

FIGURE 30.65 **Illustration of the Raider wire**. *Image courtesy of Teleflex Incorporated.* © *2022 Teleflex Incorporated. All rights reserved.*

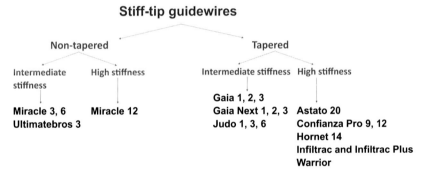

FIGURE 30.66 **Classification of stiff-tip guidewires without a polymer jacket**.

Stiff-tip guidewires without a polymer jacket are designed for crossing resistant proximal and distal caps during CTO PCI.

30.7.3.1 Non-tapered stiff guidewires

The Miraclebros (Fig. 30.67) family of wires has nontapered tip and no hydrophilic coating and is infrequently used at present for antegrade wiring. Because it provides strong support for equipment delivery, the Miraclebros 12 wire is often used to facilitate delivery of the Stingray balloon to the reentry zone.

The UltimateBros 3 (Fig. 30.68) wire has nontapered tip and good torque response and is sometimes used for antegrade wiring.

TABLE 30.5 Overview of stiff-tip guidewires without a polymer jacket.

Tip style	Commercial name	Tip stiffness (grams)	Manufacturer	Properties
Non-tapered, intermediate stiffness	Miracle 3	3	Asahi Intecc	Antegrade wiring. Facilitate delivery of Stingray balloon to reentry zone
	Miracle 6	6	Asahi Intecc	
	Ultimatebros	3	Asahi Intecc	
Non-tapered, high stiffness	Miracle 12	12	Asahi Intecc	Antegrade wiring. Facilitate delivery of Stingray balloon to reentry zone
Tapered, intermediate stiffness	Gaia 1	1.7	Asahi Intecc	Antegrade and retrograde wiring
	Gaia 2	3.5	Asahi Intecc	
	Gaia 3	4.5	Asahi Intecc	
	Gaia Next 1	2	Asahi Intecc	
	Gaia Next 2	4	Asahi Intecc	
	Gaia Next 3	6	Asahi Intecc	
	Judo 1	1	Boston Scientific	
	Judo 3	3	Boston Scientific	
	Judo 6	6	Boston Scientific	
Tapered, High stiffness	Astato 20	20	Asahi Intecc	Penetration of highly resistant caps
	Confianza Pro	9	Asahi Intecc	
	Confianza Pro 12	12	Asahi Intecc	
	Hornet 10	10	Boston Scientific	
	Hornet 14	14	Boston Scientific	
	Infiltrac	11	Abbott Vascular	
	Infiltrac Plus	14	Abbott Vascular	
	Stingray	12	Boston Scientific	
	Warrior	14	Teleflex	

30.7.3.2   Tapered, intermediate tip stiffness guidewires

Tapered-tip, intermediate tip stiffness guidewires include the Gaia family of wires (Fig. 30.69), Gaia Next family of wires (Fig. 30.70) and the Judo

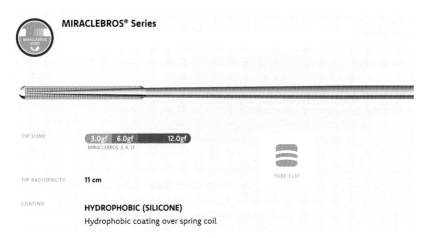

MIRACLEBROS® Series

TIP LOAD	3.0gf 6.0gf 12.0gf MIRACLEBROS 3, 6, 12
TIP RADIOPACITY	11 cm
COATING	**HYDROPHOBIC (SILICONE)** Hydrophobic coating over spring coil

TUBE CLIP

FIGURE 30.67   **Illustration of the Miraclebros family of wires**. *Reproduced with permission from Asahi Intecc.*

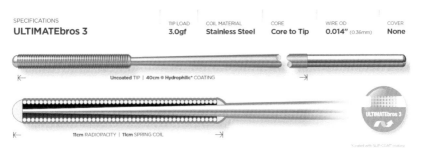

SPECIFICATIONS	TIP LOAD	COIL MATERIAL	CORE	WIRE OD	COVER
ULTIMATEbros 3	**3.0gf**	**Stainless Steel**	**Core to Tip**	**0.014"** (0.36mm)	**None**

Uncoated TIP | 40cm ● Hydrophilic* COATING

11cm RADIOPACITY | 11cm SPRING COIL

ULTIMATEbros 3

*Coated with SLIP-COAT coating

FIGURE 30.68   **Illustration of the UltimateBros 3 wire**. *Reproduced with permission from Asahi Intecc.*

family of wires (Fig. 30.71) and are commonly used for both antegrade and retrograde crossing. They are often used after failure of polymer-jacketed tapered tip guidewires to cross the CTO, if the course of the vessel is well understood.

The Gaia guidewires (Fig. 30.69) are guidewires with ACT ONE technology designed to enhance torquability and maneuverability, but require slower, more meticulous and precise manipulation compared with other CTO guidewires.[51−53] They have a micro-cone tapered tip for smooth lesion entry. The tip core is round instead of flat and has a composite core consisting of stainless-steel tube rope coil with traditional spring-coil on the exterior (composite core: dual coil construction). The distal coil consists of six wires

FIGURE 30.69 **Illustration of the Gaia family of wires.** *Reproduced with permission from Asahi Intecc.*

instead of one and has ten times more torque force, enabling excellent 1:1 torque transmission to avoid torque "whip." The distal 1 mm tip is preshaped to a 45-degree angle. The modification in core thickness allows the wire tip to deflect away from hard tissue such as calcium and extraplaque layers for better intraplaque tracking. The Gaia wires are available in three types (Gaia First, Gaia Second, and Gaia Third) with increasing tip stiffness and tapered tip diameter (the Gaia First tapers to 0.010″, the Gaia Second to 0.011″, and the Gaia Third to 0.012″).

The Gaia Next family of wires (Fig. 30.70) have a micro-cone tip for directional control that is more tapered than the Gaia wires. They also

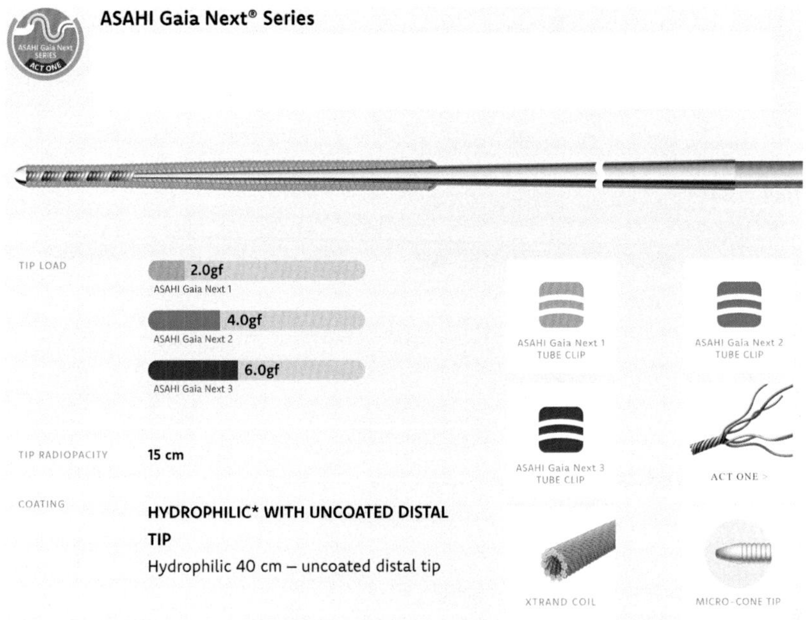

ASAHI Gaia Next® Series

TIP LOAD

2.0gf
ASAHI Gaia Next 1

4.0gf
ASAHI Gaia Next 2

6.0gf
ASAHI Gaia Next 3

ASAHI Gaia Next 1
TUBE CLIP

ASAHI Gaia Next 2
TUBE CLIP

TIP RADIOPACITY   **15 cm**

ASAHI Gaia Next 3
TUBE CLIP

ACT ONE

COATING

HYDROPHILIC* WITH UNCOATED DISTAL TIP

Hydrophilic 40 cm – uncoated distal tip

XTRAND COIL

MICRO-CONE TIP

FIGURE 30.70   **Illustration of the Gaia Next family of wires**. *Reproduced with permission from Asahi Intecc.*

incorporate the XTRAND coil technology, in which the main outer coil of the guidewire is made of wires woven together to form a rope, which prevents potential fracture. The XTRAND coil surrounds the guidewire core and the ACT ONE inner coil of the Gaia family and provides additional torqueing control and penetration power. Similar with the Gaia wires the Gaia Next wires come preshaped with a 1 mm tip and they have 40 cm hydrophilic coating distally that prevents entrapment within the plaque.

The Judo family of wires (Fig. 30.71) taper to 0.008 inch and have hydrophilic coating in the distal 15 cm.

30.7.3.3   Tapered, high tip stiffness guidewires

These are highly penetrating guidewires for crossing highly resistant proximal and distal caps. Such wires are the Confianza Pro and Confianza Pro 12 (Fig. 30.72), Hornet and Hornet 14 (Fig. 30.73), Infiltrac and Infiltrac Plus, Warrior (Fig. 30.74), and Astato 20 (Fig. 30.75). The Stingray guidewire is a stiff guidewire with a 20 cm distal radiopaque segment and a 0.009″ tapered tip with a 0.0035″ distal prong designed for facilitating reentry into the true lumen through the Stingray balloon, as discussed in Section 30.16.2. The closer the microcatheter to the guidewire tip, the higher the penetrating power of the guidewire.

JUDO 1 Guidewire

Soft intraluminal crossing wire for antegrade microchannels.

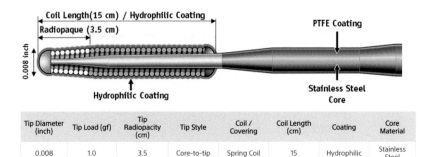

Tip Diameter (inch)	Tip Load (gf)	Tip Radiopacity (cm)	Tip Style	Coil / Covering	Coil Length (cm)	Coating	Core Material
0.008	1.0	3.5	Core-to-tip	Spring Coil	15	Hydrophilic	Stainless Steel

JUDO 3 Guidewire

Immediate intraluminal crossing wire for fibro-calcific lesions.

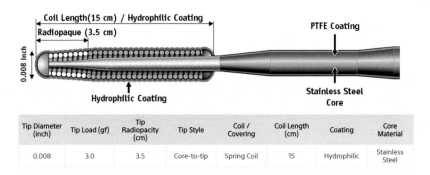

Tip Diameter (inch)	Tip Load (gf)	Tip Radiopacity (cm)	Tip Style	Coil / Covering	Coil Length (cm)	Coating	Core Material
0.008	3.0	3.5	Core-to-tip	Spring Coil	15	Hydrophilic	Stainless Steel

JUDO 6 Guidewire

Extra penetration with excellent steerability in tight lesions.

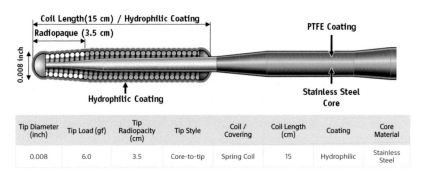

Tip Diameter (inch)	Tip Load (gf)	Tip Radiopacity (cm)	Tip Style	Coil / Covering	Coil Length (cm)	Coating	Core Material
0.008	6.0	3.5	Core-to-tip	Spring Coil	15	Hydrophilic	Stainless Steel

FIGURE 30.71 **Illustration of the Judo family of wires.** *Image provided courtesy of Boston Scientific.*

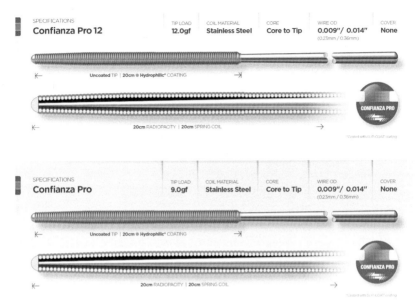

FIGURE 30.72   **Illustration of the Confianza Pro family of wires**. *Reproduced with permission from Asahi Intecc.*

30.7.4   Support guidewires

Support guidewires have strong, supportive bodies and soft tips. They are designed to facilitate equipment delivery (or even guide catheter exchanges) (Table 30.6). Support wires include the BHW, Extra S'port, All Star, and Iron Man (Abbott Vascular); Sion Blue extra support and Grand Slam (Asahi Intecc); and Mailman (Boston Scientific).

The Wiggle wire (Fig. 30.76; Abbott Vascular) is a unique support wire with three deflections in the distal portion: the deflections start 6 cm from the tip and go over the next 6 cm giving three waves, each having an amplitude of 3 mm. These deflections reduce side-wall bias and center the delivery portion of the wire. The Wiggle wire can be especially useful for equipment delivery through previously placed stents and also for delivery through tortuosity and calcification. The Wiggle wire should be not used for primary wiring: instead another guidewire should be used first, followed by exchange for the Wiggle wire over a microcatheter or over-the-wire balloon.

If a Wiggle guidewire is not available, bends can be placed manually into the distal portion of a workhorse guidewire to mimic the Wiggle wire construction.

HORNET 10 Guidewire

Penetration wire with excellent control.

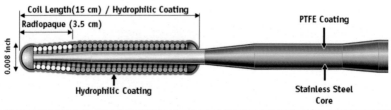

Tip Diameter (inch)	Tip Load (gf)	Penetration Force (gf/mm2)	Tip Radiopacity (cm)	Tip Style	Coil / Covering	Coil Length (cm)	Coating	Core Material
0.008	10	308	3.5	Core-to-tip	Spring Coil	15	Hydrophilic	Stainless Steel

HORNET 14 Guidewire

Penetration wire with excellent control and more penetration force.

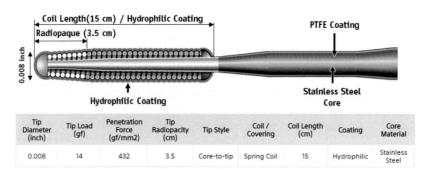

Tip Diameter (inch)	Tip Load (gf)	Penetration Force (gf/mm2)	Tip Radiopacity (cm)	Tip Style	Coil / Covering	Coil Length (cm)	Coating	Core Material
0.008	14	432	3.5	Core-to-tip	Spring Coil	15	Hydrophilic	Stainless Steel

FIGURE 30.73 **Illustration of the Hornet family of wires.** *Image provided courtesy of Boston Scientific.* © *2022 Boston Scientific Corporation or its affiliates. All rights reserved.*

30.7.5 Atherectomy guidewires

Atherectomy requires dedicated guidewires, i.e. the Rotawire Drive (floppy and extra support, Section 30.10.1.3) for rotablation and the ViperWire (Advance and Advance with Flex Tip, Section 30.10.2.3) for orbital atherectomy.

30.7.6 Externalization guidewires

The RG3 (330-cm long, Fig. 30.77, Asahi Intecc) and R350 (350-cm long, Fig. 30.78, Teleflex) wires are used for externalization in retrograde CTO PCI.

Warrior™ Guidewire

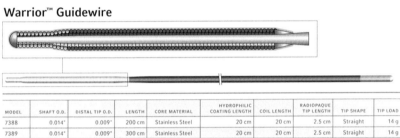

MODEL	SHAFT O.D.	DISTAL TIP O.D.	LENGTH	CORE MATERIAL	HYDROPHILIC COATING LENGTH	COIL LENGTH	RADIOPAQUE TIP LENGTH	TIP SHAPE	TIP LOAD
7388	0.014"	0.009"	200 cm	Stainless Steel	20 cm	20 cm	2.5 cm	Straight	14 g
7389	0.014"	0.009"	300 cm	Stainless Steel	20 cm	20 cm	2.5 cm	Straight	14 g

FIGURE 30.74   Illustration of the Warrior wire. *Image courtesy of Teleflex Incorporated.* © *2022 Teleflex Incorporated. All rights reserved.*

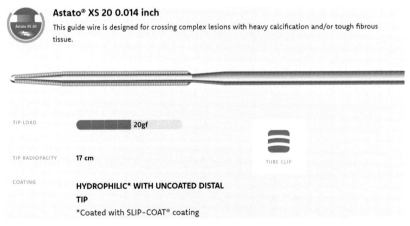

Astato® XS 20 0.014 inch

This guide wire is designed for crossing complex lesions with heavy calcification and/or tough fibrous tissue.

| TIP LOAD | 20gf |
| TIP RADIOPACITY | **17 cm** |

TUBE CLIP

| COATING | **HYDROPHILIC* WITH UNCOATED DISTAL TIP**
*Coated with SLIP–COAT® coating |

FIGURE 30.75   Illustration of the Astato 20 wire. *Reproduced with permission from Asahi Intecc.*

TABLE 30.6 Description of support coronary guidewires.

Wire category	Tip style	Commercial name	Tip stiffness	Manufacturer	Properties
Extra support guidewires	Soft, nontapered	BHW (nitinol) Extra S'port All Star Iron Man	0.8 gr 0.9 gr 0.8 gr 1.0 g	Abbott Vascular	190 cm and 300 cm long
		Grand Slam Sion blue extra support	0.7 g 0.5 g	Asahi Intecc	180 and 300 cm long
		Mailman	0.8 g	Boston Scientific	182 and 300 cm long
Wiggle wire (3 deflection starting 6 cm from the tip)	Soft, nontapered	Wiggle	1.0 g	Abbott Vascular	190 cm and 300 cm long

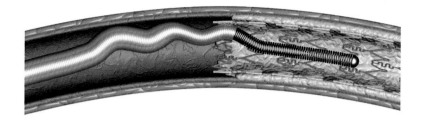

FIGURE 30.76   **Illustration of the Wiggle wire**. *Courtesy of Abbott Vascular. ©2022 Abbott. All Rights Reserved.*

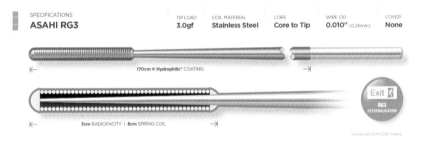

FIGURE 30.77   **Illustration of the RG3 wire**. *Reproduced with permission from Asahi Intecc.*

30.7.7   Pressure wires

Pressure wires, such as the OmniWire (Philips), PressureWire X (Abbott Vascular), Comet (Boston Scientific), Optowire (Opsens) and the Navvus rapid exchange FFR microcatheter can be used for physiologic assessment as discussed in Chapter 12.

30.7.8   0.035- or 0.038-inch guidewires

Using 0.035- or 0.038-inch guidewires is necessary for obtaining arterial access and for advancing catheters to the coronary ostium. There are different 0.035-inch guidewires with different tip stiffness and shape, body stiffness, and coatings (with vs without polymer jacket).

Standard 0.035-inch or 0.038-inch guidewires have a J-tip and are used in most procedures.

Soft tip 0.035-inch guidewires, such as the Bentson (Cook Medical), the Wholey wire (Medtronic), and the Hi-Torque Versacore wire (Abbott) are

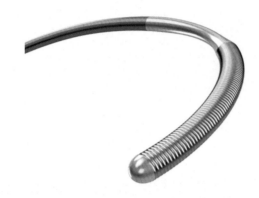

R350™ Guidewire

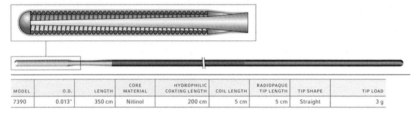

MODEL	O.D.	LENGTH	CORE MATERIAL	HYDROPHILIC COATING LENGTH	COIL LENGTH	RADIOPAQUE TIP LENGTH	TIP SHAPE	TIP LOAD
7390	0.013″	350 cm	Nitinol	200 cm	5 cm	5 cm	Straight	3 g

FIGURE 30.78 **Illustration of the R350 wire**. *Image courtesy of Teleflex Incorporated.* © *2022 Teleflex Incorporated. All rights reserved.*

useful for crossing diseased and tortuous peripheral vessels. The steerable STORQ wire (Cordis) can be very useful when more precise tip control is required.

Polymer-jacketed 0.035-inch guidewires also facilitate crossing diseased and tortuous vessels but may be more likely to create a subintimal dissection compared with non-polymer-jacketed wires. The Glidewire (Terumo) is the most commonly used. The Glidewire Advantage has a polymer-jacketed tip with a stiff nonjacketed body. The Glidewire baby-J has a 1.5 mm distal J-tip radius (vs 3 mm for standard J-tip wires) and is specifically designed for use with radial access.

Stiff body 0.035-inch guidewires, such as the Amplatz Superstiff (Boston Scientific), Supracore (Abbott Vascular), and Lunderquist (Cook Medical) facilitate sheath insertion and equipment advancement through areas of calcification and tortuosity. They usually have soft tips and are inserted over a diagnostic catheter.

30.8   Embolic protection devices

Embolic protection devices are used very rarely in CTO PCI. They are discussed in the *Manual of Percutaneous Coronary Interventions*.[5]

30.9   Balloons

Balloons are the main method for plaque modification during PCI, but they serve several other roles, such as expanding the stent, delivering drugs to the vessel wall, trapping equipment into the guide catheter, and performing intravascular lithotripsy. There are eight key balloon categories (Fig. 30.79):

1. Standard balloons.
2. Small balloons.
3. Plaque modification balloons.
4. Trapping balloons.
5. Ostial FLASH balloon (described in Section 30.13.2).
6. Drug-coated balloons (DCBs).
7. Very-high pressure balloon.
8. Intracoronary lithotripsy balloon.

30.9.1   Standard balloons

Standard balloons are classified into semicompliant, and noncompliant. Semicompliant balloons are more flexible and easier to deliver, whereas, noncompliant balloons are harder to deliver but do not expand much upon high pressure inflation and are typically used for stent postdilatation and for treating balloon undilatable lesions (Section 23.1).

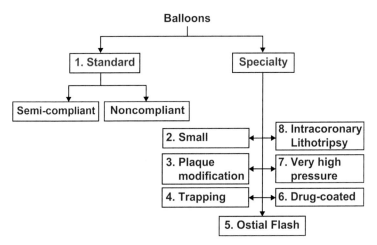

FIGURE 30.79   Balloon classification. *Reproduced with permission from Brilakis ES.* Manual of percutaneous coronary interventions: a step-by-step approach. *Elsevier; 2021.*

30.9.2   Small balloons

Small balloons are important for "challenging to cross lesions", such as balloon uncrossable lesions (Section 23.1) and jailed side branches (Chapter 16).

The Sapphire II PRO 1.0 mm balloon (Fig. 30.80, OrbusNeich) is the lowest profile balloon currently available in the US and is currently the first choice for balloon uncrossable lesions in the US.

Other 1.0 mm balloons that are not currently available in the US are the Ikazuchi Zero (Fig. 30.81) and Zinrai (Kaneka, Zinrai is also available in 0.75 mm diameter) and the Ryurei 1.0 mm (Terumo).

The Nano 0.85 mm balloon (SIS) is the lowest profile balloon currently available in Europe.

The Threader microdilatation catheter (Boston Scientific, Fig. 30.82) combines a 1.20x12 mm balloon at its tip with a kink resistant shaft that enhances deliverability. The Threader is available in both rapid exchange

FIGURE 30.80   **Illustration of the Sapphire II PRO 1.0 mm balloon.** *Image used with permission of OrbusNeich Medical Company Limited.*

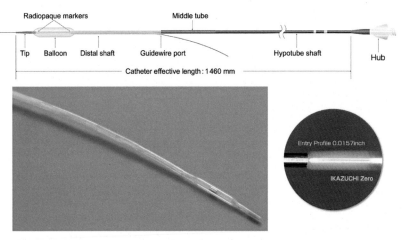

FIGURE 30.81   **Illustration of the Ikazuchi Zero balloon.** *Reproduced with permission from Kaneka.*

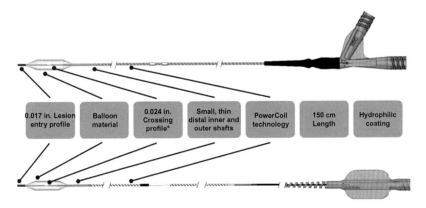

*Crossing profile is defined as the maximum diameter found between the proximal end of the balloon and the distal tip of the catheter.

Threader PowerCoil Technology

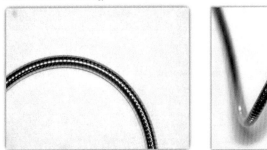

FIGURE 30.82 **The Threader microdilatation catheter.** *Image courtesy Boston Scientific ©* *2022 Boston Scientific Corporation or its affiliates. All rights reserved.*

and over-the-wire systems, although the rapid exchange system provides more support and is preferred by most operators.

30.9.3 Plaque modification balloons

These balloons are designed to "score" or modify the lesion to facilitate expansion and prevent "watermelon seeding." Four such balloons are currently available in the US: Angiosculpt (Philips), Chocolate XD (Teleflex), Scoreflex (OrbusNeich), and the Wolverine cutting balloon (Boston Scientific); another such balloon is only available in Europe: Blimp (IMDS).

30.9.3.1 Angiosculpt

The AngioSculpt scoring balloon catheter (Philips) (Fig. 30.83) is composed of a semicompliant balloon encircled by three nitinol spiral struts to score the target lesion upon balloon inflation. Compared with the cutting balloon,

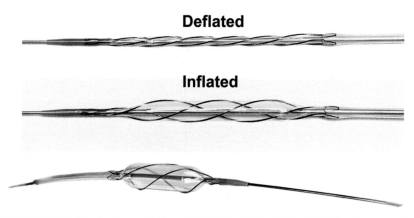

FIGURE 30.83   **The Angiosculpt balloon.** *© 2022 Koninklijke Philips N.V. All rights reserved.*

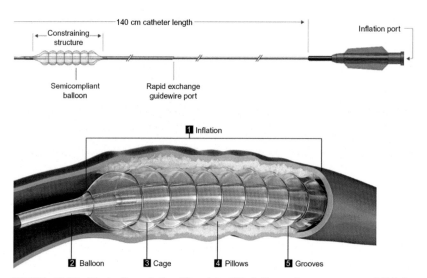

FIGURE 30.84   **Illustration of the Chocolate XD balloon.** *Image courtesy of Teleflex Incorporated. © 2022 Teleflex Incorporated. All rights reserved.*

the AngioSculpt balloon has a lower crossing profile and produces more scoring marks per millimeter of plaque.

30.9.3.2   Chocolate XD

In contrast to the Angiosculpt, the Chocolate XD balloon (Teleflex, Fig. 30.84) is a nitinol-caged balloon. As the balloon inflates, the cage causes the balloon to form a series of segmented pillows and grooves along the entire lesion. The pillows apply force to create small dissections that facilitate effective dilatation. The grooves relieve the stress and stop dissections from propagating.

30.9.3.3   Scoreflex NC

The Scoreflex NC (OrbusNeich, Fig. 30.85) is a scoring balloon that uses a 0.011″ nitinol integral wire, as well as the conventional guidewire to score lesions. It has excellent deliverability due to a low crossing profile and dual coatings for minimal friction; hydrophilic and hydrophobic coatings. The balloon is noncompliant, and has a rated burst pressure (RBP) of 20 atm.

30.9.3.4   Cutting balloon

The Wolverine cutting balloon (Boston Scientific, Fig. 30.86) contains atherotomes that protrude 0.005″ from the surface of the balloon and modify the target lesion by creating three or four endovascular radial incisions through the fibrocalcific tissue, thus allowing further expansion with conventional balloons.[54] In addition to helping expand challenging lesions the cutting balloon has been used for releasing intramural hematomas that may develop post stenting, after reentry,[55] or in the setting of spontaneous coronary dissection. Cutting balloon inflation should be done very slowly (1 atm every 5 s). The cutting balloon has a high crossing profile.

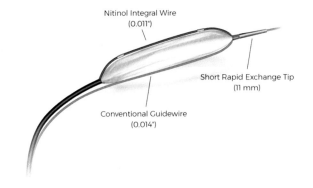

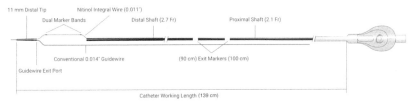

FIGURE 30.85   Illustration of the Scoreflex NC balloon. *Image used with permission of OrbusNeich Medical Company Limited.*

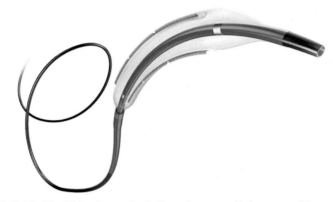

FIGURE 30.86 **The Wolverine cutting balloon**. *Image provided courtesy of Boston Scientific* © *2022 Boston Scientific Corporation or its affiliates. All rights reserved.*

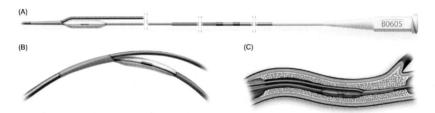

FIGURE 30.87 **The Blimp balloon**. *Reproduced with permission from IMDS.*

30.9.3.5 Blimp balloon

The Blimp scoring balloon catheter (IMDS, Fig. 30.87) has a short distal monorail segment with a 0.6 mm diameter, 5 mm long balloon adjacent to the coronary guidewire. The device has a high burst pressure (nominal pressure 25 atm and RBP 30 atm) that increases its ability to open a stenosis. The Blimp balloon scores the lesions by exerting force on the plaque through the guidewire, which is back-loaded from the tip of the device and exits just at the distal side of the balloon through the Rx port and is positioned along the balloon.

30.9.4 Trapping balloons

Trapping (Section 8.2.2.9.1) can be performed with standard balloons, however dedicated trapping balloons (that do not have a wire lumen and are, therefore, resistant to kinking and deformation and have a lower profile) are also available. Two trapping balloons are currently available in the US, the Trapper (Boston Scientific, Fig. 30.88), and the TrapIT (IMDS, Fig. 30.89).

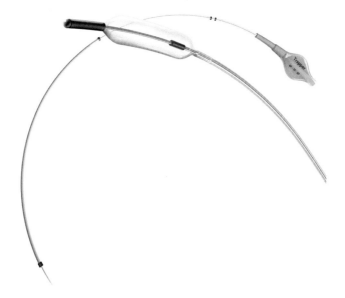

FIGURE 30.88 **The Trapper balloon.** *Image provided courtesy of Boston Scientific © 2022 Boston Scientific Corporation or its affiliates. All rights reserved.*

The Kusabi (Kaneka) trapping balloon is available in other countries. Trapping can also be performed with the Trapliner guide catheter extension (Section 30.3.1).

30.9.5 Ostial FLASH

The Ostial FLASH balloon is designed to flare aorto-ostial stents and is discussed in Section 30.13.2.

30.9.6 Drug-coated balloons

Drug-coated balloons (DCBs) have a drug (paclitaxel and sirolimus are most commonly used) that is released to the vessel wall upon balloon inflation, decreasing the risk of restenosis. They are used most often for treating in-stent restenosis but have also been used in bifurcation lesions and small vessels.[56] Coronary DCBs are not currently available in the US, although peripheral DCBs have been used off label in some patients with large coronary vessels.

30.9.7 Very high-pressure balloon

The OPN NC balloon (SIS Medical, Switzerland), is a specialized noncompliant balloon that can deliver very high (> 40 atm) pressures (Section 23.2.10).[57]

(A)

(B)

FIGURE 30.89   **The TrapIT balloon**. *Reproduced with permission from IMDS.*

30.9.8   Intracoronary lithotripsy balloon

The intracoronary lithotripsy system (Shockwave Medical, Inc., Fig. 30.90) uses ultrasonic shockwaves to overcome extreme lesion resistance. The intracoronary lithotripsy balloon can be very useful for treating balloon undilatable lesions (Section 23.2.9). However, it is bulkier than compliant and noncompliant balloons, hence it may need to used in conjunction with other modalities, such as rotational atherectomy.

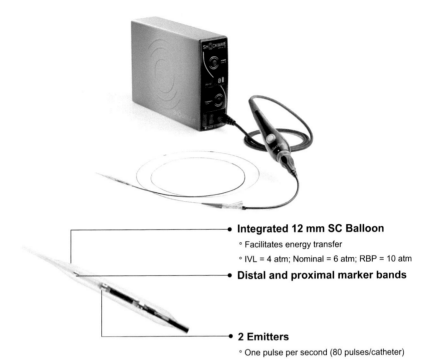

Integrated 12 mm SC Balloon
  ◦ Facilitates energy transfer
  ◦ IVL = 4 atm; Nominal = 6 atm; RBP = 10 atm
Distal and proximal marker bands

2 Emitters
  ◦ One pulse per second (80 pulses/catheter)

FIGURE 30.90   **The intracoronary lithotripsy system.** *IVL*, intravascular lithotripsy; *RBP*, rated burst pressure; *SC*, semi compliant. *Reproduced with permission from Shockwave Medical Inc.*

30.10   Atherectomy

30.10.1   Rotational atherectomy

Rotational atherectomy (Section 9.3.2.3) is performed using the Rotapro system that has six components:

1. Burr.
2. Advancer (which can be purchased preconnected to the burr).
3. Rotawire (there are 2 types: Rotawire Drive Floppy and Rotawire Drive Extra Support).
4. WireClip Torquer.
5. Console.
6. Rotaglide.

30.10.1.1   Burr

The Rotablator burr (Fig. 30.91) has an elliptical shape and is nickel coated. Its distal edge is coated with 2,000–3,000 microscopic diamond crystals

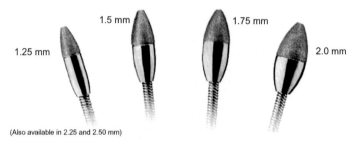

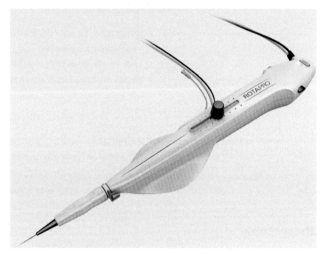

between 20–30 μm in size that protrude 5 μm from the surface. The proximal edge has no coating, which explains the potential risk of device entrapment (atherectomy can only happen with the leading edge of the burr during advancement but not during withdrawal of the burr). There are 8 burr sizes (1.25, 1.50, 1.75, 2.00, 2.15, 2.25, 2.38, 2.5 mm) although the 1.25 and 1.50 mm burrs are the most commonly used. The burr to artery ratio should be <0.7.

30.10.1.2 Advancer

The Rotablator advancer (Fig. 30.92) is used by the operator to advance and withdraw the burr during rotablation.

ROTAWIRE™ Drive Floppy

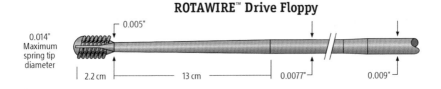

0.014"
Maximum
spring tip
diameter

0.005"

2.2 cm

13 cm

0.0077"

0.009"

ROTAWIRE™ Drive Extra Support

0.014"
Maximum
spring tip
diameter

0.005"

2.8 cm

5 cm

0.009"

FIGURE 30.93   **Illustration of the Rotawire Drive Floppy and Rotawire Drive Extra Support wires.** *Image provided courtesy of Boston Scientific* © *2022 Boston Scientific Corporation or its affiliates. All rights reserved.*

The driveshaft of the Rotablator burr is attached in the front of the advancer (burrs can be purchased preconnected with the advancer) and the Rotawire exits from the back of the advancer. There is a knob at the top of the advancer that is used for advancing and withdrawing the burr. Pushing the button on the knob of the Rotapro advancer can activate and stop rotation. There is a brake defeat button at the back to allow for the wire brake to be released and a docking port for the WireClip torquer.

The advancer has three connection ports:

1. Connection to the compressed gas cylinder port that rotates the drive shaft of the burr.
2. Rotaglide or saline infusion port.
3. Fiber-optic cable port to measure the rotating speed of the burr.

30.10.1.3   Rotawire Drive

Rotational atherectomy should only be performed using a dedicated Rotawire Drive. There are two versions of the Rotawire Drive: the Rotawire Drive Floppy and the Rotawire Drive Extra Support (Fig. 30.93 and Table 30.7).

Both guidewires are 330 cm in length with a 0.009″ shaft and 0.014″ tip that does not allow the Rotablator burr to go past the wire. The Rotawire Drive Floppy has a longer taper than the Rotawire Drive Extra Support. The Rotawire Drive Floppy is used in the vast majority of cases. The Rotawire Drive Extra Support may be helpful in ostial lesions to better align the guide catheter with the vessel.

TABLE 30.7 Characteristics of the Rotawire Drive Floppy and the Rotawire Drive extra support.

Model/ description	Length	Tip length	Flexibility	Spring tip diameter	Maximum diameter
ROTAWire Drive Extra Support Guide Wire	330 cm	2.8 cm	Stiff	0.009″	0.014″
ROTAWire Drive Floppy Guide Wire	330 cm	2.2 cm	Flexible	0.009″	0.014″

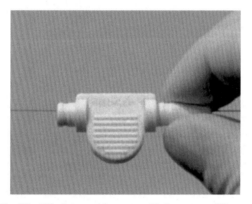

FIGURE 30.94 **The WireClip torquer.** *Image provided courtesy of Boston Scientific © 2022 Boston Scientific Corporation or its affiliates. All rights reserved.*

30.10.1.4 WireClip torquer

The Rotawire can be torqued using the WireClip torquer (Fig. 30.94).

It can also prevent the guidewire from spinning during rotational atherectomy when performing rotational atherectomy while the brake is defeated.

30.10.1.5 Console

The Rotapro console (Fig. 30.95) has several indicators, the most important of which are:

1. Tachometer (speed of rotation in rpm).
2. Dynaglide indicator.
3. Deceleration indicator.

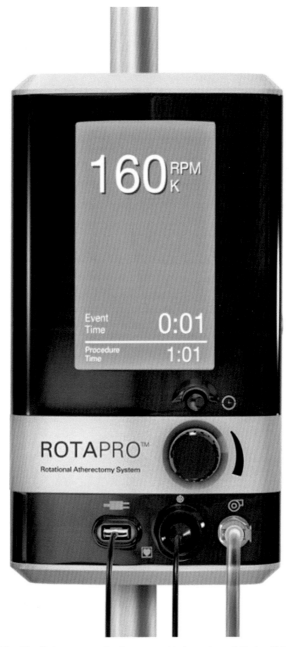

FIGURE 30.95   **The Rotapro console**. *Image provided courtesy of Boston Scientific* © *2022 Boston Scientific Corporation or its affiliates. All rights reserved.*

30.10.1.6   Rotaglide

The Rotaglide is a solution made of olive oil, egg yolk, phospholipids, sodium deoxycholate, L-histidine, disodium EDTA, and sodium hydroxide and is continuously infused through the drive shaft to lubricate the system and reduce friction and heat generation.

It is contraindicated in patients who are allergic to eggs, in whom normal saline can be infused, often with verapamil and nitroglycerine added to the infused saline solution. Normal saline is used instead of Rotaglide in countries where Rotaglide is not available.

30.10.2   Orbital atherectomy

Orbital atherectomy is performed as described in Section 9.3.2.4.

The orbital atherectomy system has four components (Fig. 30.96):

30.10.2.1   Diamondback 360 Coronary Orbital Atherectomy Device

The orbital atherectomy device (OAD) is a hand-held, over-the-wire device that includes a sheath-covered drive shaft and a diamond-coated crown. The diamond coating on the crown provides an abrasive surface that reduces coronary plaque within the coronary arteries. The GlideAssist feature facilitates advancing and retracting the OAD crown over the guidewire.

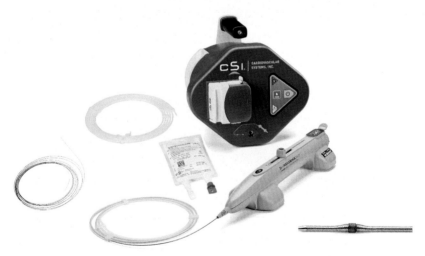

FIGURE 30.96   **Illustration of the orbital atherectomy system components.** © *2022 Cardiovascular Systems, Inc. Image is used with permission from Cardiovascular Systems, Inc. CSI,* Diamondback 360®, ViperWire *Advance and ViperSlide are trademarks of Cardiovascular Systems, Inc.*

The OAD is electrically powered and consists of the following:

- Crown.
- Crown advancer knob.
- Drive shaft.
- Sheath (covering the drive shaft up to the crown).
- Electrical power cord.
- Tubing for connecting with the orbital atherectomy system pump.
- On/off button at the top of the advancer knob.
- Two speed control buttons (80,000 or 120,000 rpm).
- The speed control buttons can also be used to activate the GlideAssist function (rotation of the crown at 5,000 rpm).
- Crown advancement measurement indicators, and
- Manual guidewire brake.

30.10.2.2 Pump

The OAS pump (Fig. 30.97) provides the saline pumping mechanism and power to the OAD. It is reusable, and portable and attaches to a standard five-wheel rolling intravenous pole or table mount pole. The OAS pump

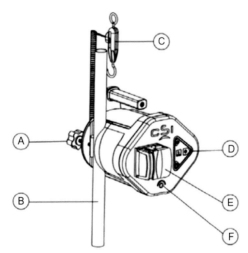

A. IV pole screw clamp
B. IV pole or table mount (not included)
C. Low saline level sensor and connector cord
D. Control panel
E. OAS pump door
F. OAD connection

FIGURE 30.97 **Illustration of the orbital atherectomy pump.** ©*2022 Cardiovascular Systems, Inc. Image is used with permission from Cardiovascular Systems, Inc. CSI, Diamondback 360®, ViperWire Advance and ViperSlide are trademarks of Cardiovascular Systems, Inc.*

includes a built-in, audible 25 s spin time notification, OAS power and priming buttons, and status indicators.

30.10.2.3 ViperWire guidewire

The OAD is designed to track and spin only over the CSI ViperWire coronary guidewire. There are two ViperWires:

1. **ViperWire Advance,** which is a moderately stiff, stainless-steel wire, with a silicone coating, epoxy proximal tip bond, and a radiopaque distal spring tip. Its diameter is 0.012″ in the shaft and 0.014″ at the tip.
2. **ViperWire Advance with Flex Tip,** which is a nitinol guidewire with 1 gram tip load. Its body is more flexible, reducing wire bias, and has excellent torque response.

30.10.2.4 ViperSlide

The ViperSlide contains 10% soybean oil, 1.2% egg yolk phospholipid, 2.25% glycerin, sodium hydroxide and water for injection. Use 20 mL ViperSlide per liter of saline.

30.11 Laser

Online video: "Impact of contrast on laser activation"
CTO Manual online cases: 5, 18, 27, 47, 52, 73, 86, 155
PCI Manual online cases: 23, 35, 43, 55, 64, 67, 88, 99, 117

Excimer laser coronary atherectomy (ELCA, Philips, Fig. 30.98) can be used to treat balloon uncrossable (Section 23.1) or undilatable (Section 23.2)

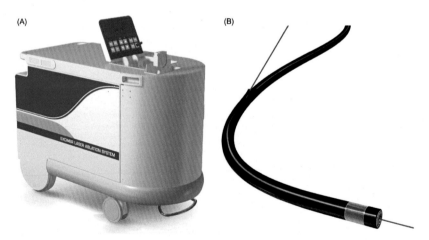

(A)　　　　　　　　　　　(B)

FIGURE 30.98 The laser console **(panel A)** and catheter **(panel B)**. © *2022 Koninklijke Philips N.V. All rights reserved.*

lesions (including in-stent undilatable lesions) and occasionally intracoronary thrombus.

ELCA uses ultraviolet energy (wavelength: 308 nm) delivered by a xenon-chlorine pulsed laser catheter with pulse frequency of $25-80$ Hz and fluence of $30-80$ mJ/mm^2. ELCA uses energy for disruption and disintegration of the molecular bonds within the atherosclerotic plaque in a highly controlled manner through ablation rather than burning. Since laser catheters can be advanced over any 0.014″ guidewire, laser is highly valuable for the treatment of balloon uncrossable (and also balloon undilatable) lesions.[58] In some instances, the laser catheter can facilitate lesion crossing by modifying the proximal cap without actually crossing the lesion.

The laser catheters currently available for coronary use range in diameter from 0.9 mm to 2.0 mm and are available as rapid exchange (0.9, 1.4, 1.7, and 2.0 mm) or over-the-wire (0.9 mm). The catheter used in coronary lesions is usually the 0.9 mm X-80 catheter that is compatible with 6 French guide catheters. Since the coronary laser catheters allow lasing for only 10 s before the laser is switched off for 5 s, many operators prefer to use the 0.9-mm Turbo-Elite catheter off-label that has no lasing duration limits and allows repetition up to 80 Hz. Laser is usually performed with saline infusion, although laser has been used with simultaneous contrast injection to expand underexpanded stents (Section 23.1.7).[59]

30.12   Thrombectomy devices

Thrombectomy is rarely performed during CTO PCI (CTO Manual online case 19)—it is discussed in the *Manual of Percutaneous Coronary Interventions*.[5]

30.13   Aorto-ostial lesion equipment

Treatment of aorto-ostial CTOs is described in Chapter 15.

There are two devices specifically designed for facilitating stenting of aorto-ostial lesions, the Ostial Pro (designed to facilitate accurate stent placement) and the Ostial FLASH (designed to achieve flaring of the portion of the stent hanging out in the aorta).

30.13.1   Ostial pro

The Ostial Pro (Merit Medical, Fig. 30.99) is a nitinol device with distal, self-expanding legs that are advanced just distal to the tip of the guide

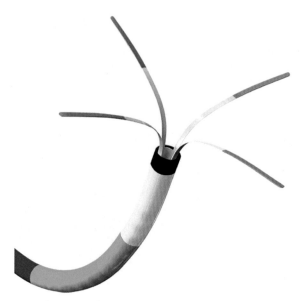

FIGURE 30.99   **The Ostial Pro system.** © *Merit Medical, Reprinted by Permission.*

catheter to prevent entry of the guide catheter into the target vessel, mark the plane of the aortic wall, and align the tip of the guide catheter with the aorto-ostial plane.[60,61]

30.13.2   Ostial FLASH

The Ostial FLASH balloon flares the proximal stent struts against the aortic wall and facilitates expansion and re-engagement with a catheter (Fig. 30.100). Ostial FLASH is a dual balloon angioplasty catheter with a larger proximal low pressure balloon and a higher pressure distal balloon.[62,63] There are three markers that can be visualized under fluoroscopy: (1) a proximal marker that marks the proximal end of the anchoring balloon and should be located in the aorta, (2) a mid-marker that marks the proximal end of the angioplasty balloon and should be placed at the vessel ostium, and (3) a distal marker that marks the distal end of the angioplasty balloon. The distal balloon length is 8 mm and is available in diameters from 3.0 to 4.5 mm. The proximal anchoring balloon can expand up to 14 mm. The distal balloon is inflated with an inflating device whereas the proximal balloon is inflated with a 1 cc syringe.

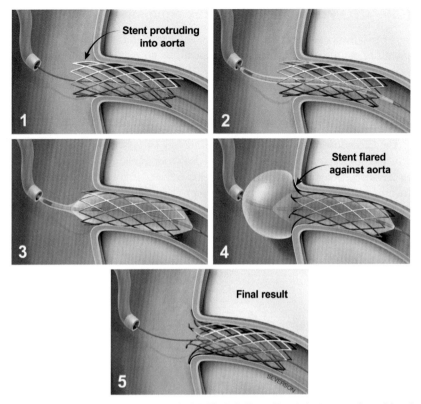

FIGURE 30.100   **Illustration of the Ostial Flash balloon. Panel 1**: the stent is positioned approximately 1−2 mm from the vessel ostium. **Panel 2**: the Ostial Flash balloon is positioned with the proximal marker in the aorta outside the guide catheter, middle marker at the ostium, and distal marker inside the stent. **Panel 3**: the distal balloon is inflated inside the stent with an inflating device. **Panel 4**: the proximal balloon is inflated with a 1 cc syringe to flare the stent struts against the aorta. **Panel 5**: final result. *Reproduced with permission from Nguyen-Trong PJ, Martinez Parachini JR, Resendes E, et al. Procedural outcomes with use of the flash ostial system in aortocoronary ostial lesions.* Catheter Cardiovasc Interv *2016;88:1067−74.*

30.14   Stents

Latest generation drug-eluting stents should be used in CTO PCI as described in Chapter 10.

30.15   Vascular closure devices

Use of vascular closure devices can be used to achieve hemostasis at the end of the procedure. Step-by-step description of vascular closure is provided the Manual of PCI.[5]

30.16   CTO PCI dissection/re-entry equipment

The CrossBoss catheter and Stingray balloon and guidewire (Boston Scientific) can be used for antegrade dissection and reentry during CTO PCI as described in Section 8.2.

30.16.1   CrossBoss

CTO PCI Manual online cases 8, 9, 16, 17, 19, 25, 32, 40, 50, 180.

The CrossBoss catheter (Fig. 30.101, Section 8.2.4.1) is a stiff, metallic, over-the-wire catheter with a 1 mm blunt, rounded, hydrophilic-coated distal tip that can advance through the occlusion when the catheter is rotated rapidly using a proximal torque device ("fast spin" technique). If the catheter enters the extraplaque space, it creates a limited dissection plane making reentry into the distal true lumen easier. The risk of perforation is low provided that the CrossBoss catheter is not advanced into side branches. If the CTO is crossed through the extraplaque space, the Stingray LP balloon and guidewire can be used to assist with reentry into the distal true lumen, as described below.[64–66]

30.16.2   Stingray

CTO PCI Manual online cases 9, 16, 17, 21, 25, 27, 32, 34, 35, 43, 45, 48, 72, 79, 80, 82, 83, 92, 93, 97, 98, 100, 105, 120, 129, 131, 135, 142, 144, 146, 148, 149, 152, 154, 155, 159, 161, 164, 165, 166, 168, 173, 181, 184, 187, 189, 191, 198
PCI Manual online cases 40, 105, 120, 128, 135

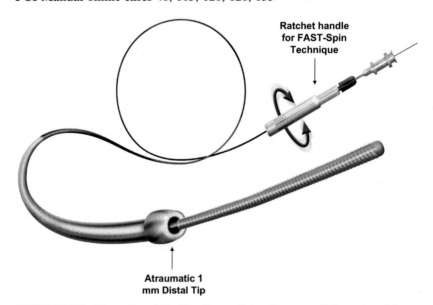

Ratchet handle
for FAST-Spin
Technique

Atraumatic 1
mm Distal Tip

FIGURE 30.101   **Illustration of the CrossBoss catheter**. *Image provided courtesy of Boston Scientific © 2022 Boston Scientific Corporation or its affiliates. All rights reserved.*

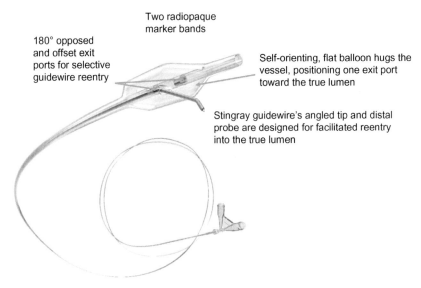

180° opposed
and offset exit
ports for selective
guidewire reentry

Two radiopaque
marker bands

Self-orienting, flat balloon hugs the
vessel, positioning one exit port
toward the true lumen

Stingray guidewire's angled tip and distal
probe are designed for facilitated reentry
into the true lumen

FIGURE 30.102 **Illustration of the Stingray LP catheter and the Stingray guidewire**.
Image provided courtesy of Boston Scientific © 2022 Boston Scientific Corporation or its affili-
ates. All rights reserved.

The Stingray LP (low-profile) balloon is a 2.5 mm wide when inflated
and 10 mm in length and has a flat shape with two side exit ports. Upon
low-pressure (4 atm) inflation it orients one exit port automatically towards
the true lumen, especially when the space created by dissection is small
(Fig. 30.102).[67] The Stingray guidewire is a stiff guidewire with a 20 cm distal radiopaque segment and a 0.009″ tapered tip with a 0.0035″ distal prong.
The Stingray guidewire can be directed towards one of the two side ports of
the Stingray LP balloon under fluoroscopic guidance or blindly (double blind
stick and swap technique) to reenter into the distal true lumen.[64–66]

30.17   Intravascular imaging

Intravascular ultrasound (IVUS) and optical coherence tomography (OCT)
are the currently available intravascular imaging modalities to guide PCI.
Chapter 13 provides detailed information on how to use intravascular imaging, including a comparison of IVUS and OCT (Section 13.2).

30.17.1   Intravascular ultrasound (Section 13.4)

There are three types of IVUS systems:

1. Solid-state, phased array system (Eagle Eye Platinum, Philips). This
   catheter is also available with a short tip (2.5 mm tip-to imaging distance),
   which may be allow more distal advancement during CTO crossing.

2. Rotational systems, such as the Revolution and Refinity (45 MHz, Philips), OptiCross HD (60 MHz, Boston Scientific), and the Kodama HD IVUS (40−60 MHz, ACIST).
3. Dual imaging systems:
   a. DualPro (InfraRedx), that combines IVUS (35−65 MHz) with NIRS (near-infrared spectroscopy), providing simultaneous IVUS and NIRS assessment. Large lipid core plaques as identified by NIRS have been associated with increased risk of distal embolization and no-reflow (Section 25.2.3.2).
   b. Novasight hybrid system that provides simultaneous IVUS and OCT imaging (Conavi Medical).

The solid state IVUS system does not require preparation and is hence faster to use but is more difficult to deliver compared with the rotational systems, has lower resolution, and does not allow length measurements, unless coregistration is performed (Section 13.3).

30.17.2 Optical coherence tomography (Section 13.3)

OCT provides high-resolution images that greatly facilitate interpretation, but requires contrast injection during acquisition to clear the blood from the coronary artery. There are two OCT imaging systems in the US (Dragonfly Optis Imaging catheter, Abbott Vascular and Gentuity HF-OCT Imaging System, Nipro) and one combined IVUS-OCT system (Novasight, Conavi Medical). OCT is infrequently used for CTO crossing due to the requirement for contrast injection but can be useful for stenting optimization.

30.18 Complication management

Covered stents, coils, Amplatzer vascular plugs, pericardiocentesis tray, and snares are needed to treat perforations and equipment loss or entrapment. This equipment should be available in all cardiac catheterization laboratories.

30.18.1 Covered stents

CTO PCI Manual online cases 26, 39, 40, 42, 63, 89, 90, 131, 134, 167, 169, 171, 174, 177, 189, 193
PCI Manual online cases 45, 55, 80, 110, 133.

As of 2023, two coronary covered stents are commercially available in the US: the Graftmaster Rx (Abbott Vascular) and PK Papyrus (Biotronik). They are both approved through a humanitarian device exemption[68] for use in large vessel perforations. Another covered stent, the BeGraft coronary stent (Bentley InnoMed GmbH, Hechingen, Germany), is available in Europe.

FIGURE 30.103 **Illustration of the Graftmaster Rx covered stent**. *Courtesy of Abbott Vascular. ©2019 Abbott. All Rights Reserved.*

30.18.1.1 Graftmaster

1. The Graftmaster Rx (Fig. 30.103) consists of two stainless steel stents with a middle layer of ePTFE.
2. It is bulky and difficult to deliver; hence excellent guide catheter support is important.
3. It is available in diameters of 2.8 to 4.8 mm and lengths between 16 and 26 mm.
4. It requires a 6 French guide catheter for the 2.8 mm to 4.0 mm stents and a 7 French guide catheter for the 4.5 and 4.8 mm stents.
5. The Graftmaster Rx may be difficult to advance through previously deployed stents, necessitating techniques such as distal anchoring and use of delivery catheters, such as an 8 French guide catheter extension.
6. Minimum inflation pressure is 15 atm, but even higher pressures for up to 60 s (and use of IVUS) are preferred to ensure adequate stent expansion.
7. After expansion the stent may shorten up to 1.6 mm on each side (for a total of 3.2 mm at nominal pressure, which is 15 atm). Hence, an adequate overlap of stents is essential to cover long segments of perforation.
8. Use of a *dual catheter ("ping-pong guide") technique* (Section 26.3) is often required to minimize bleeding into the pericardium while preparing for covered stent delivery and deployment,[69] although delivery of both a balloon and a covered stent is feasible through 8 French guide catheters.
9. Postdilatation of the shoulders of the stent may be necessary to fully oppose the stent to the vessel wall if extravasation persists behind the stent despite covering the perforation.

30.18.1.2 PK Papyrus

1. The PK Papyrus (Fig. 30.104) covered coronary stent system is a balloon-expandable covered stent mounted on a rapid-exchange delivery catheter.[70] The stent is manufactured by covering the PRO-Kinetic Energy ultrathin strut, amorphous silicon carbide-coated cobalt chromium stent with an electrospun polymeric matrix composed of siloxane-based polyurethane. More specifically, the electropsun cover consists of individual nonwoven fibers approximately 2μm in thickness. When the stent is fully expanded, the cover is approximately 90 μm in thickness.[70]
2. The device size matrix ranges from 2.5 mm to 5.0 mm stent diameters and lengths of 15, 20, and 26 mm. The PK Papyrus stent can be

FIGURE 30.104 **Illustration of the PK Papyrus covered stent.** *Image used with permission from BIOTRONIK, Inc.*

postdilated to a maximum stent expansion diameter of 3.50 mm for the 2.5 and 3.0 mm stents; 4.65 mm for the 3.5 and 4.0 mm stents; and 5.63 mm for the 4.5 and 5.0 mm stents. PK Papyrus is compatible with 5 French guide catheters for 2.5−4.0 mm diameters or 6 French for 4.5 and 5.0 mm diameters.[70]

3. PK Papyrus is easier to deliver compared with the Graftmaster but can occasionally be dislodged from the balloon during attempts to deliver it to the perforation site.

30.18.1.3   BeGraft

The BeGraft coronary stent (Bentley InnoMed GmbH) is a cobalt chrome (L-605), open-cell platform covered with a single-layer of a PTFE membrane (thickness of $89 \pm 25\mu m$), which is clamped at the proximal and distal stent ends.[71] The single layer design results in a crossing profile between 1.1−1.4 mm with a guide catheter compatibility of 5 French for all sizes.

30.18.2   Coils

Online video: "How to deliver and deploy and Axium coil"

Coils should be available for use in case of distal (Section 26.4) or collateral vessel perforation. They can also be used to stop a large vessel perforation (Section 26.3, CTO Manual online case 176) by occluding the vessel. Coils are permanent embolic agents that can be deployed either through 0.014″ compatible microcatheters [neurovascular coils (Table 26.3)], such as Axium, Medtronic, (Fig. 30.105 or Smart coil, Penumbra) or through the Finecross or larger 0.018″ microcatheters (standard coils, such as Interlock, Boston Scientific, Fig. 30.106).[72] Coils are usually made of stainless steel or platinum alloys and some of them have polymers or synthetic wool or Dacron fibers attached along

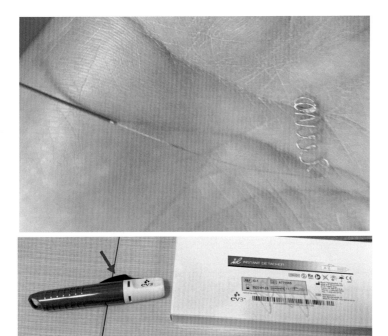

FIGURE 30.105   Illustration of the Axium detachable coil system (Medtronic), that is compatible with 0.014″ microcatheters.

the length of the wire to increase thrombogenicity. Once advanced into the target vessel, the coils assume a preformed shape, sealing the perforation. Particular attention needs to be made when coiling branches to prevent the coil from prolapsing into the main vessel.

The coil is released pulling the lever (arrow) on the delivery device.

Coils tips and tricks[73]:

1. Since coils are used very infrequently in the cardiac catheterization laboratory, it is important for each operator to be familiar with the principles underlying their use and with 1−2 specific coil types, so that coils can be delivered rapidly in case of perforation.
2. There are two broad categories of coils according to the *mechanism of release*: pushable and detachable. *Pushable* coils are inserted into a microcatheter and pushed with a coil pusher or the front end of a guidewire until they exit into the vessel, hence deployment can be unpredictable and is irreversible. *Detachable* coils are released using a dedicated release device once their position into the target vessel is confirmed; conversely if their position is not satisfactory, they can be retrieved. Detachable coils are preferred for treating perforations, as they allow optimal and predictable positioning.

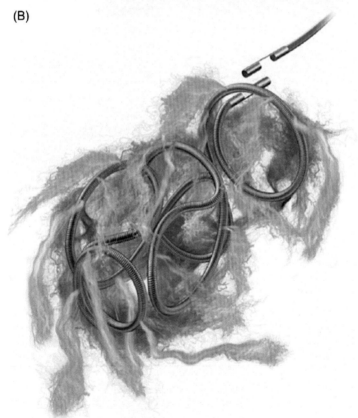

FIGURE 30.106   **Example of a detachable coil that can be used for embolization in case of distal coronary perforation (Interlock, Boston Scientific). Panel A** demonstrates deployment of the coil whereas **panel B** illustrates the coil configuration after delivery. *Image provided courtesy of Boston Scientific © 2022 Boston Scientific Corporation or its affiliates. All rights reserved.*

3. There are also two broad categories of coils according to the *size of the delivery microcatheter. Coils compatible with 0.014″ microcatheters (such as Axium, Medtronic) are preferred*, as they can be delivered through any standard microcatheter used for CTO PCI (such as the Corsair, Caravel, Turnpike) without requiring change to a larger microcatheter. Coils compatible with 0.018″ microcatheters (such as the Interlock [Boston Scientific], Azur [Terumo], and Micronester [Cook]), cannot be delivered through the standard microcatheters used during CTO PCI with the exception of the Finecross, and require change to a larger microcatheter, such as the Progreat (Terumo), Renegade (Boston Scientific), or Transit (Cordis).

4. Simultaneous balloon inflation (to stop bleeding into the pericardium) and coil delivery (through a microcatheter) can be achieved through a single guide catheter ("block and deliver" technique; Section 26.4).[74,75]

30.18.3   Amplatzer vascular plugs

Another method for occluding a vessel (either emergently in case of perforation or electively, for example a saphenous vein graft after recanalization of the native coronary artery) is the Amplatzer vascular plugs. The Amplatzer vascular plugs are disks made of a mesh of braided nitinol that are attached to a PTFE-coated delivery wire with a stainless-steel micro screw, which allows the operator to release the plug into the final position by rotating the cable in a counterclockwise fashion using a supplied torque device. The plug can be retrieved and readjusted as needed before final release.[76] There are four types of Amplatzer vascular plugs (Table 30.8). The Amplatzer Vascular Plug II is used in most cases when saphenous vein grafts need to be occluded.

Size selection: the diameter of the Amplatzer Vascular Plug is selected to be 30%−50% larger than the vessel diameter at the occlusion site. Moreover, the operator should ensure that the occlusion site has sufficient length to accommodate the deployed device length without obstructing other vessels or anatomical structures.

30.18.4   Pericardiocentesis tray

Pericardiocentesis can be performed using a standard 18 gauge needle, a J-tip 0.035 inch guidewire and a standard pigtail catheter, however having all equipment assembled in a premanufactured pericardiocentesis kit can facilitate and speed up the procedure. Use of a micropuncture needle kit and echocardiographic guidance (if time allows) can increase the safety of the procedure.

30.18.5   Snares

Snares are often needed to retrieve entrapped or lost equipment. They are also used for capturing the retrograde guidewire during retrograde CTO PCI.

TABLE 30.8 Overview of the Amplatzer vascular plugs.

	AVP I	AVP II	AVP III	AVP IV
	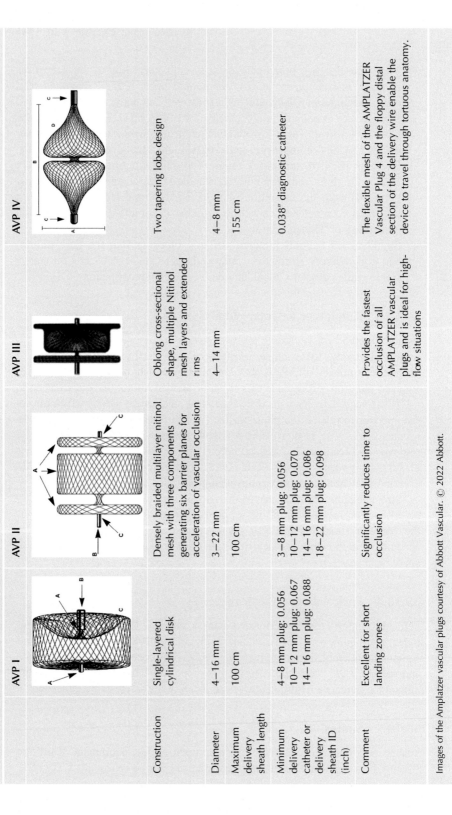			
Construction	Single-layered cylindrical disk	Densely braided multilayer nitinol mesh with three components generating six barrier planes for acceleration of vascular occlusion	Oblong cross-sectional shape, multiple Nitinol mesh layers and extended rims	Two tapering lobe design
Diameter	4–16 mm	3–22 mm	4–14 mm	4–8 mm
Maximum delivery sheath length	100 cm	100 cm		155 cm
Minimum delivery catheter or delivery sheath ID (inch)	4–8 mm plug: 0.056 10–12 mm plug: 0.067 14–16 mm plug: 0.088	3–8 mm plug: 0.056 10–12 mm plug: 0.070 14–16 mm plug: 0.086 18–22 mm plug: 0.098		0.038″ diagnostic catheter
Comment	Excellent for short landing zones	Significantly reduces time to occlusion	Provides the fastest occlusion of all AMPLATZER vascular plugs and is ideal for high-flow situations	The flexible mesh of the AMPLATZER Vascular Plug 4 and the floppy distal section of the delivery wire enable the device to travel through tortuous anatomy.

Images of the Amplatzer vascular plugs courtesy of Abbott Vascular. © 2022 Abbott.

30.18.5.1   Snare types

There is a wide variety of commercially available snares for retrieving equipment from the coronary and peripheral circulation or snaring the retrograde guidewire, such as the Amplatz Goose Neck snare (Medtronic), the Microsnare Elite (Teleflex) and the En Snare (Merit Medical)[77] (Fig. 30.107).

These snares consist of a wire loop typically made of nitinol, that is advanced through a microcatheter (or alternatively through a diagnostic or guide catheter), positioned around the lost device, and then pulled back, trapping the device against the catheter. The catheter/loop assembly is subsequently removed from the body, along with the lost device. Three-loop (tulip) snares are preferred, as they have three overlapping loops (instead of one in the Goose Neck snare); hence, increasing the likelihood of retrieving the lost device.

If a commercially manufactured snare is not available, a loop snare can be created in the catheterization laboratory using an exchange-length coronary guidewire and a multipurpose catheter through the distal tip of which the guidewire tip is reinserted (Fig. 30.108).

Another way to create a "homemade" snare ("KAM-snare") is by using a wire, a balloon and a guide catheter extension: the snare consists of a wire loop being trapped by an inflated balloon (at $8-12$ atm) at the distal site of a guide catheter extension (Fig. 30.109).[7]

Online video: How to make a homemade snare

30.18.5.2   Snaring technique (Fig. 30.110)

The snare is withdrawn and collapsed into the introducer tool (Fig. 30.110, panel A, B, and C). It is then introduced into the antegrade guide catheter (Fig. 30.110, panel D and E) and advanced until it exits from the distal tip of

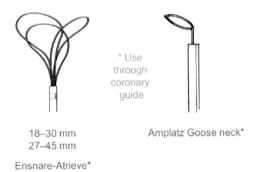

* Use through coronary guide

18–30 mm
27–45 mm

Amplatz Goose neck*

Ensnare-Atrieve*

FIGURE 30.107   **Illustration of three-loop and single-loop snares**. *Reproduced with permission from Brilakis ES. Manual of coronary chronic total occlusion interventions. a step-by-step approach. 2nd ed. Elsevier; 2017.*

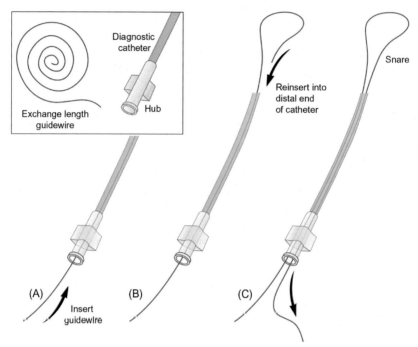

FIGURE 30.108 **How to create a loop snare using a catheter and a long 0.014″ coronary guidewire.** *Reproduced with permission from Brilakis ES.* Manual of percutaneous coronary interventions: a step-by-step approach. *Elsevier; 2021.*

the guide catheter (Fig. 30.110, panel F). The snare is then positioned close to the lost device and is withdrawn into the diagnostic or guide catheter until the lost device is captured between the snare and the catheter tip. The lost device is then withdrawn.

30.18.5.3 Snaring tips and tricks

1. The size of the snare depends on the size of the vascular space that contains the lost or entrapped device. For example, small (2−4 mm) snares are used within the coronary arteries and large snares (27−45 mm or 18−30 mm) are used in the aorta.
2. Each snare comes with a delivery sheath, which is usually discarded. For snaring the retrograde guidewire the snare is inserted through the antegrade guide catheter. For snaring a lost device, the snare is advanced through a diagnostic or guide catheter and withdrawn, trapping the lost device between the loops of the snare and the tip of the catheter.
3. Do not discard the snare collapsing tool, as it is necessary for re-introducing the snare into the guide catheter, if needed.

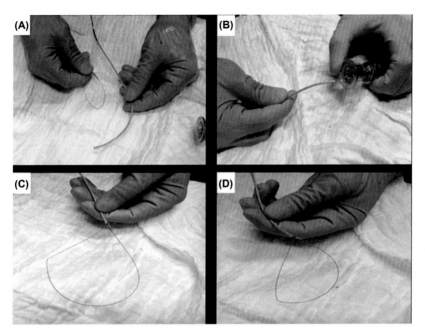

FIGURE 30.109   **How to make a "KAM-snare"**. The "KAM-snare" consists of a guidewire being inserted within the monorail lumen of a conventional angioplasty balloon. The distal end of the guidewire is then shaped as a loop (**panel A**). The looped guidewire with the railed balloon are introduced into a guide catheter extension at the proximal entry site. Hence, the balloon entraps the distal returning end of the guidewire loop within the guide catheter extension (**panel B**). The "KAM-snare" is inserted into the proximal site of the Y-connector (**panel B**). By either pulling or pushing the exterior proximal end of the guidewire the diameter of the "KAM-snare" either increases or decreases (**panels C and D**). *Courtesy of Dr. Kambis Mashayekhi.*

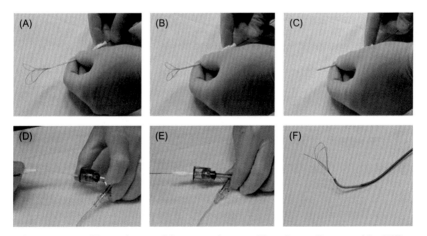

FIGURE 30.110   **How to insert a 3-loop snare into a guide catheter**. *Courtesy of Dr. William Nicholson. Reproduced with permission from Brilakis ES. Manual of coronary chronic total occlusion interventions. a step-by-step approach. 2nd ed. Elsevier; 2017.*

4. Ideally the retrograde microcatheter should be advanced into the aorta before attempting to snare the retrograde guidewire to minimize the risk of the guidewire being retracted back into the CTO. Furthermore, this precaution will protect the proximal coronary artery from traction-induced trauma while snaring the retrograde wire, creating tension.
5. It may be easier to snare the retrograde guidewire if it is positioned in the brachiocephalic artery.
6. The floppy radiopaque part of the externalization guidewire (RG3 or R350) is the safest to snare, followed by careful sweeping into the antegrade guide.
7. Alternatively, if a short retrograde guide catheter has been used, some operators prefer to snare 300-cm long standard guidewires (such as Pilot 200). If a standard guidewire needs to be snared, then the ideal snaring location is *just proximal* to the radiopaque part to avoid fracture or unraveling of the distal segment of the guidewire.
8. Short 180 cm wires should never be snared. If done inadvertently, the wire should be pulled out from the antegrade guide catheter, without attempting to retrieve it from the donor artery, as the snared segment of the wire may become kinked, precluding removal via the retrograde guide catheter without also removing the support catheter.

30.19   Radiation protection

Every effort should be undertaken to minimize patient and operator dose, as described in detail in Section 28.2. In addition to fixed shields, there are several radiation protection pads that decrease scatter radiation from the patient, such as the RadPad (Worldwide Innovations & Technologies, Inc), and the Zero Gravity ceiling suspended radiation protection system (Biotronik) (Fig. 30.111).

30.20   Hemodynamic support devices

CTO PCI Manual online cases 11, 14, 20, 31, 46, 126, 136, 176 (Impella), 29, 67, 138, 146, 165, 171, 176 (VA-ECMO)
PCI Manual online cases 45 (IABP), 31, 36, 41, 59, 63, 73, 91, 92, 105, 110, 120, 126, 140 (Impella), 27, 40, 51, 72 (VA-ECMO)

Hemodynamic support can be used prophylactically, after occurrence of a complication during PCI, or in patients with cardiogenic shock. Given the potential for severe, life-threatening complications during CTO and complex PCI, the availability of hemodynamic support devices is vital as part of comprehensive high-risk and complex PCI programs.[78] Hemodynamic support is discussed in Chapter 14 and in the *Manual of Percutaneous Coronary Interventions.*[5]

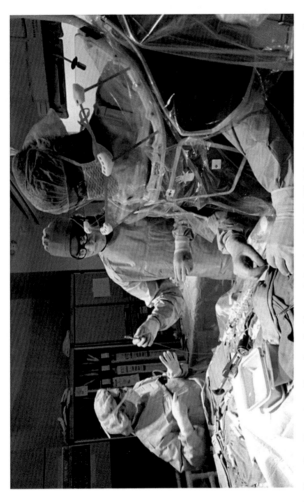

FIGURE 30.111 **The Zero Gravity system.** *Reproduced with permission from Brilakis ES. Manual of percutaneous coronary interventions: a step-by-step approach. Elsevier; 2021.*

30.21   Contrast management

Prevention of acute kidney injury after cardiac catheterization is discussed in Section 28.3. Pre-procedural hydration and reducing contrast volume are the hallmarks of prevention. One way to reduce contrast volume is the Dyevert Plus (Osprey Medical, Minneapolis, MN) that prevents excess contrast during manual injection. The fraction of contrast that would not contribute to coronary opacification, but rather would reflux into the aortic root, is diverted into a reservoir chamber. The system also provides real time measurement of the contrast volume administered to the patient. Initial studies suggest 20%−40% reduction in contrast volume with use of the device.[79,80]

30.22   Brachytherapy

Coronary brachytherapy is useful for treatment of recurrent in-stent restenosis.[81] Beta radiation with a strontium-90/yttrium source is the only currently available system, but is only available in a limited number of centers. Patients who undergo brachytherapy should receive indefinite dual antiplatelet therapy to minimize the risk of stent thrombosis.

30.23   The "CTO−CHIP cart"

Having a dedicated "CTO-CHIP cart" (Fig. 30.112) with all commonly used CTO and PCI equipment (including equipment for managing complications, such as covered stents and coils) can facilitate and expedite performance of these procedures. The CTO-CHIP cart should be readily available during the procedure, re-stocked regularly, and known to the operators and the catheterization laboratory staff.

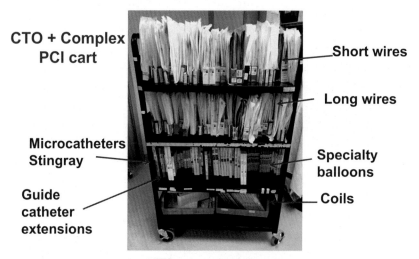

FIGURE 30.112   **Example of a "CTO and complex PCI cart".**

References

1. Brilakis E.S. The essential equipment for CTO interventions. Cardiology today's intervention 2013; May−June 2013.

2. Joyal D, Thompson CA, Grantham JA, Buller CEH, Rinfret S. The retrograde technique for recanalization of chronic total occlusions: a step-by-step approach. *JACC Cardiovasc Interv* 2012;**5**:1−11.

3. Brilakis ES, Grantham JA, Thompson CA, et al. The retrograde approach to coronary artery chronic total occlusions: a practical approach. *Catheter Cardiovasc Interv* 2012;**79**:3−19.

4. Brilakis ES, Grantham JA, Rinfret S, et al. A percutaneous treatment algorithm for crossing coronary chronic total occlusions. *JACC Cardiovasc Interv* 2012;**5**:367−79.

5. Brilakis ES. *Manual of percutaneous coronary interventions: a step-by-step approach.* Elsevier; 2021.

6. Agelaki M, Koutouzis M. Balloon-assisted tracking for challenging transradial percutaneous coronary intervention. *Anatol J Cardiol* 2017;**17**:E1.

7. Tajti P, Karatasakis A, Karmpaliotis D, et al. Retrograde CTO-PCI of native coronary arteries via left internal mammary artery grafts: insights from a Multicenter U.S. Registry. *J Invasive Cardiol* 2018;**30**:89−96.

8. Wu EB, Chan WW, Yu CM. Retrograde chronic total occlusion intervention: tips and tricks. *Catheter Cardiovasc Interv* 2008;**72**:806−14.

9. Takahashi S, Saito S, Tanaka S, et al. New method to increase a backup support of a 6 French guiding coronary catheter. *Catheter Cardiovasc Interv* 2004;**63**:452−6.

10. Luna M, Papayannis A, Holper EM, Banerjee S, Brilakis ES. Transfemoral use of the GuideLiner catheter in complex coronary and bypass graft interventions. *Catheter Cardiovasc Interv* 2012;**80**:437−46.

11. Senguttuvan NB, Sharma SK, Kini A. Percutaneous intervention of chronic total occlusion of anomalous right coronary artery originating from left sinus—use of mother and child technique using guideliner. *Indian Heart J* 2015;**67**(Suppl 3):S41−2.

12. Vishnevsky A, Savage MP, Fischman DL. GuideLiner as guide catheter extension for the unreachable mammary bypass graft. *Catheter Cardiovasc Interv* 2018;**92**:1138−40.

13. Williams R, Shouls G, Firoozi S. Mother-and-child telescopic guide-catheter extension to identify severe left main stem disease in a patient with a severely dilated aortic root. *J Invasive Cardiol* 2018;**30**:E71−2.

14. Stys AT, Stys TP, Rajpurohit N, Khan MA. A novel application of GuideLiner catheter for thrombectomy in acute myocardial infarction: a case series. *J Invasive Cardiol* 2013;**25**:620−4.

15. Sakakura K, Taniguchi Y, Tsukui T, Yamamoto K, Momomura SI, Fujita H. Successful removal of an entrapped rotational atherectomy burr using a soft guide extension catheter. *JACC Cardiovasc Interv* 2017;**10**:e227−9.

16. Xenogiannis I, Karmpaliotis D, Alaswad K, et al. Comparison between traditional and guide-catheter extension reverse controlled antegrade dissection and retrograde tracking: insights from the PROGRESS-CTO registry. *J Invasive Cardiol* 2019;**31**:27−34.

17. Yokoi K, Sumitsuji S, Kaneda H, et al. A novel homemade snare, safe, economical and size-adjustable. *EuroIntervention* 2015;**10**:1307−10.

18. Hashimoto S, Takahashi A, Yamada T, et al. Spontaneous rotation of the monorail-type guide extension support catheter during advancement of a curved guiding catheter: the potential hazard of twisting with the coronary guidewire. *Cardiovasc Interv Ther* 2018;**33**:379−83.

19. Elbarouni B, Moussa M, Kass M, Toleva O, Vo M, Ravandi A. GuideLiner Balloon Assisted Tracking (GBAT): a new addition to the interventional toolbox. *Case Rep Cardiol* 2016;**2016**:6715630.

20. Andreou C, Karalis I, Maniotis C, Jukema JW, Koutouzis M. Guide extension catheter stepwise advancement facilitated by repeated distal balloon anchoring. *Cardiovasc Revasc Med* 2017;**18**:66−9.

21. Repanas TI, Christopoulos G, Brilakis ES. "Candy Cane" guide catheter extension for stent delivery. *J Invasive Cardiol* 2015;**27**:E169−70.

22. Finn MT, Green P, Nicholson W, et al. Mother-daughter-granddaughter double guideliner technique for delivering stents past multiple extreme angulations. *Circ Cardiovasc Interv* 2016;9.

23. Mamas MA, Fath-Ordoubadi F, Fraser DG. Distal stent delivery with guideliner catheter: first in man experience. *Catheter Cardiovasc Interv* 2010;**76**:102−11.

24. Aznaouridis K, Bonou M, Masoura K, Vaina S, Vlachopoulos C, Tousoulis D. Successful stent delivery through a slaloming coronary path. *J Invasive Cardiol* 2019;**31**:E43.

25. Papayannis AC, Michael TT, Brilakis ES. Challenges associated with use of the GuideLiner catheter in percutaneous coronary interventions. *J Invasive Cardiol* 2012;**24**:370−1.

26. Duong T, Christopoulos G, Luna M, et al. Frequency, indications, and outcomes of guide catheter extension use in percutaneous coronary intervention. *J Invasive Cardiol* 2015;**27**:E211−15.

27. Chang YC, Fang HY, Chen TH, Wu CJ. Left main coronary artery bidirectional dissection caused by ejection of guideliner catheter from the guiding catheter. *Catheter Cardiovasc Interv* 2013;**82**:E215−20.

28. Chen Y, Shah AA, Shlofmitz E, et al. Adverse events associated with the use of guide extension catheters during percutaneous coronary intervention: reports from the Manufacturer and User Facility Device Experience (MAUDE) database. *Cardiovasc Revasc Med* 2019;**20**:409−12.

29. Alkhalil M, Smyth A, Walsh SJ, et al. Did the use of the Guideliner V2(TM) guide catheter extension increase complications? A review of the incidence of complications related to the use of the V2 catheter, the influence of right brachiocephalic arterial anatomy and the redesign of the V3(TM) guideliner and clinical outcomes. *Open Heart* 2016;**3**:e000331.

30. Waterbury TM, Sorajja P, Bell MR, et al. Experience and complications associated with use of guide extension catheters in percutaneous coronary intervention. *Catheter Cardiovasc Interv* 2016;**88**:1057−65.

31. Walsh S, Dudek D, Bryniarski L, et al. Efficacy and safety of novel novacross microcatheter for chronic total occlusions: first-in-human study. *J Invasive Cardiol* 2016;**28**:88−91.

32. Brilakis ES. *Manual of coronary chronic total occlusion interventions. A step-by-step approach.* 2nd ed. Elsevier; 2017.

33. Vemmou E, Nikolakopoulos I, Xenogiannis I, et al. Recent advances in microcatheter technology for the treatment of chronic total occlusions. *Expert Rev Med Devices* 2019;**16**:267−73.

34. Kandzari DE, Grantham JA, Karmpaliotis D, et al. Safety and efficacy of dedicated guidewire and microcatheter technology for chronic total coronary occlusion revascularization: principal results of the Asahi Intecc Chronic Total Occlusion Study. *Coron Artery Dis* 2018;**29**:618−23.

35. Mohandes M, Rojas S, Guarinos J, et al. Efficacy and safety of Tornus catheter in percutaneous coronary intervention of hard or balloon-uncrossable chronic total occlusion. *ARYA Atheroscler* 2016;**12**:206−11.

36. Brilakis ES, Mashayekhi K, Tsuchikane E, et al. Guiding principles for chronic total occlusion percutaneous coronary intervention. *Circulation* 2019;**140**:420−33.

37. Fujimoto Y, Iwata Y, Yamamoto M, Kobayashi Y. Usefulness of Corsair microcatheter to cross stent struts in bifurcation lesions. *Cardiovasc Interv Ther* 2014;**29**:47−51.

38. Amsavelu S, Carlino M, Brilakis ES. Carlino to the rescue: use of intralesion contrast injection for bailout antegrade and retrograde crossing of complex chronic total occlusions. *Catheter Cardiovasc Interv* 2016;**87**:1118−23.

39. Fischell TA, Moualla SK, Mannem SR. Intracoronary thrombin injection using a microcatheter to treat guidewire-induced coronary artery perforation. *Cardiovasc Revasc Med* 2011;**12**:329−33.

40. Tsuchikane E, Katoh O, Kimura M, Nasu K, Kinoshita Y, Suzuki T. The first clinical experience with a novel catheter for collateral channel tracking in retrograde approach for chronic coronary total occlusions. *JACC Cardiovasc Interv* 2010;**3**:165−71.

41. McClure SJ, Wahr DW, Webb JG. Venture wire control catheter. *Catheter Cardiovasc Interv* 2005;**66**:346−50.

42. Naidu SS, Wong SC. Novel intracoronary steerable support catheter for complex coronary intervention. *J Invasive Cardiol* 2006;**18**:80−1.

43. McNulty E, Cohen J, Chou T, Shunk KA. "Grapple hook" technique using a deflectable tip catheter to facilitate complex proximal circumflex interventions. *Catheter Cardiovasc Interv* 2006;**67**:46−8.

44. Aranzulla TC, Colombo A, Sangiorgi GM. Successful endovascular renal artery aneurysm exclusion using the Venture catheter and covered stent implantation: a case report and review of the literature. *J Invasive Cardiol* 2007;**19**:E246−53.

45. Aranzulla TC, Sangiorgi GM, Bartorelli A, et al. Use of the Venture wire control catheter to access complex coronary lesions: how to turn procedural failure into success. *EuroIntervention* 2008;**4**:277−84.

46. Iturbe JM, Abdel-Karim AR, Raja VN, Rangan BV, Banerjee S, Brilakis ES. Use of the venture wire control catheter for the treatment of coronary artery chronic total occlusions. *Catheter Cardiovasc Interv* 2010;**76**:936−41.

47. Ide S, Sumitsuji S, Kaneda H, Kassaian SE, Ostovan MA, Nanto S. A case of successful percutaneous coronary intervention for chronic total occlusion using the reversed guidewire technique. *Cardiovasc Interv Ther* 2013;**28**:282−6.

48. Kawasaki T, Koga H, Serikawa T. New bifurcation guidewire technique: a reversed guidewire technique for extremely angulated bifurcation−a case report. *Catheter Cardiovasc Interv* 2008;**71**:73−6.

49. Suzuki G, Nozaki Y, Sakurai M. A novel guidewire approach for handling acute-angle bifurcations: reversed guidewire technique with adjunctive use of a double-lumen microcatheter. *J Invasive Cardiol* 2013;**25**:48−54.

50. Fang HY, Lee CH, Fang CY, et al. Application of penetration device (Tornus) for percutaneous coronary intervention in balloon uncrossable chronic total occlusion-procedure outcomes, complications, and predictors of device success. *Catheter Cardiovasc Interv* 2011;**78**:356−62.

51. Tomasello SD, Giudice P, Attisano T, Boukhris M, Galassi AR. The innovation of composite core dual coil coronary guide-wire technology: a didactic coronary chronic total occlusion revascularization case report. *J Saudi Heart Assoc* 2014;**26**:222−5.

52. Galassi AR, Ganyukov V, Tomasello SD, Haes B, Leonid B. Successful antegrade revascularization by the innovation of composite core dual coil in a three-vessel total occlusive

disease for cardiac arrest patient using extracorporeal membrane oxygenation. *European Heart Journal* 2014;**35**:2009.

53. Khalili H, Vo MN, Brilakis ES. Initial experience with the gaia composite core guidewires in coronary chronic total occlusion crossing. *J Invasive Cardiol* 2016;**28**:E22−5.

54. Barbato E, Shlofmitz E, Milkas A, Shlofmitz R, Azzalini L, Colombo A. State of the art: evolving concepts in the treatment of heavily calcified and undilatable coronary stenoses—from debulking to plaque modification, a 40-year-long journey. *EuroIntervention* 2017;**13**:696−705.

55. Vo MN, Brilakis ES, Grantham JA. Novel use of cutting balloon to treat subintimal hematomas during chronic total occlusion interventions. *Catheter Cardiovasc Interv* 2018;**91**:53−6.

56. Megaly M, Rofael M, Saad M, et al. Outcomes with drug-coated balloons in small-vessel coronary artery disease. *Catheter Cardiovasc Interv* 2019;**93**:E277−86.

57. Raja Y, Routledge HC, Doshi SN. A noncompliant, high pressure balloon to manage undilatable coronary lesions. *Catheter Cardiovasc Interv* 2010;**75**:1067−73.

58. Karacsonyi J., Karatasakis A., Danek B.A., Banerjee S., Brilakis E.S. *Laser applications in the coronaries. Textbook of atherectomy.* HMP Communications; 2016.

59. Karacsonyi J, Danek BA, Karatasakis A, Ungi I, Banerjee S, Brilakis ES. Laser coronary atherectomy during contrast injection for treating an underexpanded stent. *JACC Cardiovasc Interv* 2016;**9**:e147−8.

60. Fischell TA, Malhotra S, Khan S. A new ostial stent positioning system (Ostial Pro) for the accurate placement of stents to treat aorto-ostial lesions. *Catheter Cardiovasc Interv* 2008;**71**:353−7.

61. Fischell TA, Saltiel FS, Foster MT, Wong SC, Dishman DA, Moses J. Initial clinical experience using an ostial stent positioning system (Ostial Pro) for the accurate placement of stents in the treatment of coronary aorto-ostial lesions. *J Invasive Cardiol* 2009;**21**:53−9.

62. Nguyen-Trong PJ, Martinez Parachini JR, Resendes E, et al. Procedural outcomes with use of the flash ostial system in aorto-coronary ostial lesions. *Catheter Cardiovasc Interv* 2016;**88**:1067−74.

63. Desai R, Kumar G. Flash ostial balloon in right internal mammary artery percutaneous coronary intervention: a novel approach. *Cureus* 2017;**9**:e1537.

64. Werner GS. The BridgePoint devices to facilitate recanalization of chronic total coronary occlusions through controlled subintimal reentry. *Expert Rev Med Devices* 2011;**8**:23−9.

65. Brilakis ES, Lombardi WB, Banerjee S. Use of the Stingray guidewire and the venture catheter for crossing flush coronary chronic total occlusions due to in-stent restenosis. *Catheter Cardiovasc Interv* 2010;**76**:391−4.

66. Brilakis ES, Badhey N, Banerjee S. "Bilateral knuckle" technique and Stingray re-entry system for retrograde chronic total occlusion intervention. *J Invasive Cardiol* 2011;**23**: E37−9.

67. Michael TT, Papayannis AC, Banerjee S, Brilakis ES. Subintimal dissection/reentry strategies in coronary chronic total occlusion interventions. *Circ Cardiovasc Interv* 2012;**5**:729−38.

68. Romaguera R, Waksman R. Covered stents for coronary perforations: is there enough evidence? *Catheter Cardiovasc Interv* 2011;**78**:246−53.

69. Ben-Gal Y, Weisz G, Collins MB, et al. Dual catheter technique for the treatment of severe coronary artery perforations. *Catheter Cardiovasc Interv* 2010;**75**:708−12.

70. Kandzari DE, Birkemeyer R. PK Papyrus covered stent: device description and early experience for the treatment of coronary artery perforations. *Catheter Cardiovasc Interv* 2019;**94**:564−8.

71. Kufner S, Schacher N, Ferenc M, et al. Outcome after new generation single-layer polyte-trafluoroethylene-covered stent implantation for the treatment of coronary artery perforation. *Catheter Cardiovasc Interv* 2019;**93**:912−20.
72. Pershad A, Yarkoni A, Biglari D. Management of distal coronary perforations. *J Invasive Cardiol* 2008;**20**:E187−91.
73. Brilakis ES, Karmpaliotis D, Patel V, Banerjee S. Complications of chronic total occlusion angioplasty. *Interventional Cardiology Clinics* 2012;**1**:373−89.
74. Tarar MN, Christakopoulos GE, Brilakis ES. Successful management of a distal vessel perforation through a single 8-French guide catheter: combining balloon inflation for bleeding control with coil embolization. *Catheter Cardiovasc Interv* 2015;**86**:412−16.
75. Garbo R, Oreglia JA, Gasparini GL. The balloon-microcatheter technique for treatment of coronary artery perforations. *Catheter Cardiovasc Interv* 2017;**89**:E75−83.
76. Lopera JE. The amplatzer vascular plug: review of evolution and current applications. *Semin Intervent Radiol* 2015;**32**:356−69.
77. Malik SA, Brilakis ES, Pompili V, Chatzizisis YS. Lost and found: coronary stent retrieval and review of literature. *Catheter Cardiovasc Interv* 2018;**92**:50−3.
78. Kirtane AJ, Doshi D, Leon MB, et al. Treatment of higher-risk patients with an indication for revascularization: evolution within the field of contemporary percutaneous coronary intervention. *Circulation* 2016;**134**:422−31.
79. Gurm HS, Mavromatis K, Bertolet B, et al. Minimizing radiographic contrast administration during coronary angiography using a novel contrast reduction system: a multicenter observational study of the DyeVert plus contrast reduction system. *Catheter Cardiovasc Interv* 2019;**93**:1228−35.
80. Tajti P, Xenogiannis I, Hall A, et al. Use of the dyevert system in chronic total occlusion percutaneous coronary intervention. *J Invasive Cardiol* 2019;**31**:253−9.
81. Negi SI, Torguson R, Gai J, et al. Intracoronary brachytherapy for recurrent drug-eluting stent failure. *JACC Cardiovasc Interv* 2016;**9**:1259−65.

Part E

How to develop a CTO PCI program

Chapter 31

How to build a successful chronic total occlusion program

A strong commitment and multidisciplinary collaboration are critical for developing a successful chronic total occlusion (CTO) percutaneous coronary intervention (PCI) program. This is best achieved using a multifaceted approach, addressing four areas: (1) CTO operator; (2) catheterization laboratory staff, equipment and policies; (3) administration; and (4) referring physicians and patients.

31.1 Operator

31.1.1 Is CTO PCI for you?

Start with why
Simon Sinek

Should you embark on the trip of learning CTO PCI? This is a challenging question with no easy answer. It requires significant introspection and thought. Here are some factors that may be useful in this decision making:

1. *Commitment*

    Commitment is key for going through the learning curve of CTO PCI and for continuing to develop skills to improve the safety, efficiency and success over time. The CTO operator is committed to and passionate about helping each patient by achieving excellent results, even in very challenging cases. Commitment is especially important when facing failures, complications, and challenging work environments, that are expected in CTO PCI.

2. *Procedural skills and clinical judgment*

    Procedural skills and clinical judgment can and will be developed and refined while learning CTO PCI, but operators should be already performing complex PCI and have robust technical skills. For example, operators should not be attempting retrograde crossing via epicardial collaterals without being experienced in nonepicardial retrograde cases and in doing pericardiocentesis. Similarly, operators should not attempt retrograde crossing through the last remaining vessel or through an internal

Manual of Chronic Total Occlusion Percutaneous Coronary Interventions.
DOI: https://doi.org/10.1016/B978-0-323-91787-2.00031-9

mammary artery graft, unless they are experienced with and have access to left ventricular support devices.

3. *Career stage*

Career stage is less important than operator plasticity. It is important to develop skillsets, which means that the operators need to be vulnerable and open about admitting what they do not know. Operators should be open to learn new concepts and techniques. They also need to be able to deal with complications, failures and professional challenges that result from these events. Whether the operators are early, mid or late career, CTO PCI can be learned, as long as they have dedication and openness to build a skillset and program.

Several programs have embraced "double scrubbing" in which two operators perform CTO PCI together. Working together allows more advanced and established CTO PCI operators to teach and proctor colleagues.

4. *PCI volume*

CTO PCI is best suited to high-volume operators, as procedural volume correlates with skills and outcomes.[1,2] However, operators need to have availability and time to add more procedures that often take longer time to perform and time to learn. Operators who are already busy may need to give up some cases to make room for CTO PCI and the education necessary to become a successful operator. Increasing PCI volume may not be the optimal incentive for doing CTO PCI and being successful.

5. *Approach to failure and complications*

Even the best operators in the world have failures and complications. Failure can be highly frustrating and demoralizing, especially given the significant effort that goes into planning and executing each case. Being able to accept failure and apply the lessons that failure provides, is a critical step for the CTO operator. New operators will need a coach or mentor to review cases, critique attempts, and help them improve. Reattempt cases are often successful, however, this is about the patient; hence, early stage operators should be willing to refer some cases to more experienced operators to ensure that the patients get the care they need.

6. *Improve overall PCI skills*

CTO operators develop several skills that translate into all aspects of non-CTO PCI. CTO PCI can significantly enhance the operator's armamentarium for treating complex non-CTO lesions.

7. *Time availability*

Time is needed to attend courses, read, and get proctored. Also, early in the learning curve CTO PCI cases can be long, often lasting 2−4 h per case. Most successful CTO PCI operators have given up something to become successful.

8. *State-of-the-art techniques*

To achieve the best possible outcomes, CTO PCI operators should continually improve the overall PCI skillset, beyond techniques specific to CTO PCI, such as use of ultrasound and transradial access, intracoronary imaging, coronary physiology, mechanical circulatory support including large bore access and closure techniques, and radiation management, among others.

There are wrong reasons for wanting to do CTO PCI:

1. *Boosting the ego*

   Being a competent CTO operator can improve self-esteem, but helping the patient should be the main driving force, especially since failures and complications are certain to arise.

2. *Income generation*

   Given procedure complexity, and time and effort required, income generation is not the reason for doing CTO PCI, since successful procedures can be lengthy, and unsuccessful procedures are billed at the diagnostic catheterization level in the US. However, acquiring a new skillset can be valuable in today's job market.

31.1.2 Learning CTO PCI: what is the goal?

There are four distinct stages for learning CTO PCI (Fig. 31.1).

Learning starts with mastering *antegrade* techniques; first antegrade wiring (Section 8.1, stage 1) then antegrade dissection and reentry (Section 8.2, stage 2).

Retrograde techniques (Section 8.3) are initially learned by using septal collaterals and bypass grafts, which are safer and easier to cross (stage 3),

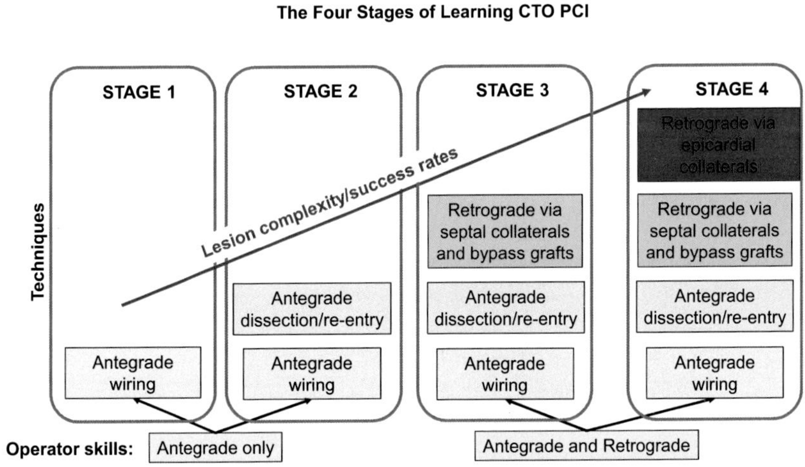

The Four Stages of Learning CTO PCI

FIGURE 31.1 **The four stages of learning CTO PCI.**[3] *Reproduced with permission from Azzalini L, Brilakis ES. Ipsilateral vs. contralateral vs. no collateral (antegrade only) chronic total occlusion percutaneous coronary interventions: What is the right choice for your practice? Catheter Cardiovasc Interv 2017;89:656—57.*

followed by use of the more challenging (and risky) epicardial (and ipsilateral) collaterals (stage 4).

Many operators may initially or permanently choose to remain "antegrade only" operators,[4] given the rapid increase in complexity and risk associated with the use of retrograde techniques. As long as they understand the strengths and limitations of each approach, whether an operator is antegrade-only or antegrade and retrograde is a matter of personal choice. With continued practice, some operators who initially chose to only do antegrade techniques may elect to add the retrograde approach to their skillset.

31.1.3   Learning CTO PCI: fellowship or "on the job" training?

Learning CTO PCI can be achieved either through a fellowship program (1-year non-ACGME accredited program) or through "on the job" training. Most operators currently train for CTO PCI while practicing[5] (Table 31.1).

The advantages of formal fellowship training include the concentrated experience and exposure to large case volumes and highly complex cases, prolonged direct working relationships with advanced CTO operators (that facilitates learning how to explain the risks and benefits of the procedure to the patients, how to plan and perform the procedure and how to manage CTO PCI complications), and the opportunity to get heavily involved in CTO PCI research. Disadvantages include immature catheterization and angioplasty skills (most fellowships are done after the conclusion of the formal interventional training), and limited availability of CTO/CHIP (complex higher risk and indicated procedures[8]) fellowships. Whether through formal training or "on the job" training, development of the CTO PCI skillset will only occur if the operator is dedicated to learning new techniques. The mentors, teachers, coaches should have the skillsets that the trainee trying to

TABLE 31.1 Comparison of CTO PCI training through a formal training program[6,7] or through "on-the-job" training.

	Fellowship program	On the job
Availability	Limited	Wide
Mastering of basic PCI skills	+ +	+ +
Concentrated experience	+ + +	+
Exposure to highly complex cases	+ + +	+
Development of mentoring relationships with advanced CTO operators	+ + +	+ +
Research opportunities	+ + +	+

learn and enough "hands on" volume should be available to continue skillset development.

Both pathways can provide excellent training, however both require great sacrifice and dedication to be successful.

31.1.4   Learning CTO PCI: books, internet, meetings, proctorships

The following tools can assist an interventionalist to evolve into a successful CTO operator:

1. Reading CTO-related literature (all interventional journals; *Catheterization and Cardiovascular Interventions, Journal of Invasive Cardiology, Eurointervention* and *JACC Cardiovascular Interventions* all provide detailed articles on the technical and clinical aspects of CTO PCI).
2. Participating in online CTO-related education: this book provides links to several recorded CTO PCI and PCI cases on YouTube (http://www.cto-manual.org and http://www.pcimanual.org). Also, http://www.tctmd.com and http://www.incathlab.com are outstanding websites providing basic to advanced CTO PCI education. In some cases, success may hinge on a nuance of a technique that an operator may have never performed before but became aware of it through a course, the internet, or the literature.
3. Observing CTO interventions at experienced CTO PCI centers either in-person or through live/prerecorded case transmissions.
4. Attending CTO PCI courses and meetings (Table 31.2).
5. Getting proctored by experienced CTO interventionalists: "on-the-job" training is invaluable for learning CTO PCI techniques.
6. Purposeful practicing: as with any procedure, the more skillsets learned and the more purposeful the trainee is in learning techniques and intra-procedural algorithms, the better CTO operator he/she will become!
7. Working with another interventionalist during CTO PCI (either scrubbed or in the control room), if feasible, allows for real-time feedback and adaptation of the procedural plan.

31.1.5   Learning CTO PCI: what to put particular emphasis on

1. Meticulous procedural planning: understanding the CTO anatomy and the possible crossing strategies facilitates efficient and confident conversion within the CTO crossing algorithms (Section 8.4).
2. Carefully selecting patients who are likely to benefit from CTO PCI, as outlined in Chapter 7.
3. Deferring or referring highly complex CTOs early in the learning curve. Start by treating lower complexity CTOs (J-CTO score 0−1) so as to maximize success to both boost the staff morale and to secure support from administration.

TABLE 31.2 CTO PCI-related meetings.

	US	International
January		IMC-LIVE, Jeddah, Saudi Arabia and Monastir, Tunisia
February	CTO Plus, New York City	
March	CRT CTO Academy, Washington DC	China Interventional Therapeutics (CIT), Beijing, China
		Szeged CTO, Szeged, Hungary
April	ACC	TCTAP, Seoul, South Korea
May	SCAI	EuroPCR, Paris, France
		OPTIMAL CTO Turin CHIP & CTO, Turin, Italy
June	C3 Global Summit, Orlando, FL	CTO Club, Nagoya, Japan
		Multi-Level CTO (ML-CTO), Nice, France Joint Interventional meeting (JIM), Milan, Italy
July	Cardiovascular Innovations, Austin, TX	
	Seattle Complications meeting, Seattle, WA	
August		Encore Seoul, Seoul, Korea
September	TCT	EuroCTO Swiss CTO Summit
October		Complex Cardiovascular Therapeutics (CCT), Kobe, Japan
		Occlusions Made Good (OMG) in Cordoba or Barcelona, Spain
November		TOBI, Milan, Italy
December		

4. Focusing and practicing the basics of CTO PCI, as summarized in the "Global guiding principles of CTO PCI."[9]
5. Persistence: committing time and energy is required for CTO PCI. Per Dr. Bill Lombardi, one of the fathers of CTO interventions in North America, "you either do CTO PCI, or you don't—there is no such

thing as trying." In other words some CTO interventions can be challenging and demanding, but the key to success is persistence. With increasing experience, the procedures become faster and success rates increase.[10]

6. Being creative and adaptable: every CTO is unique and may require a different, tailored, treatment approach. Being creative requires knowledge of the current state of the field (Global CTO algorithm,[11] etc.) to prevent "reinventing the wheel."

7. Critically reviewing all cases: always ask "what could I have done differently"?

8. Learning from failures: unlike non-CTO interventions, CTO PCI failure is not uncommon, especially early in the learning curve. Failed procedures should not be a source of discouragement, but should rather stimulate constructive evaluation and learning.[12] Discussing failed cases with other operators can be fruitful, as can be re-attempting these cases with a proctor or referring them to more experienced centers. Knowing when to fail is also important: it is better to fail without complication than "trying too hard" and having a (sometimes) catastrophic complication. Be open to feedback.

9. Publishing challenging or unique CTO PCI cases, or the overall outcomes of the CTO PCI program.

10. Keeping track of procedural outcomes, for example by creating a local CTO PCI database or by joining a CTO registry, such the Prospective Global Registry for the Study of Chronic Total Occlusion percutaneous coronary interventions (PROGRESS CTO, clinicaltrials.gov Identifier: NCT02061436, http://www.progresscto.org), the Euro CTO Club registry (https://www.eurocto.eu/), the LATAM registry (https://latamctoregistry.com/), the Japan CTO Club (https://cct.gr.jp/ctoclub/) and the Asia Pacific CTO Club (http://www.apcto.club).

11. Participation in new studies on CTO interventions.

31.2 Team

The importance of building a CTO PCI team, procuring the necessary equipment, and implementing appropriate policies cannot be overemphasized. The team consists of the cath lab staff, the recovery room staff, partners, cardiothoracic surgeons, intensivists and advanced heart failure specialists, and administration.

The following actions can help:

1. Staff education, including:
   a. Lectures to the cath lab staff on the indications, complexity, and complications of CTO PCI.
   b. Educating the non-cath lab healthcare staff about the process and outcomes of CTO PCI.

   c. Identifying specific cath lab personnel "champions" who are:
     i. Interested in developing further expertise in CTO PCI.
     ii. Interested in routinely being involved in CTO cases (which helps in building experience and achieving excellent outcomes).

2. Obtaining the necessary infrastructure and equipment (Chapter 30).
   a. At least two cath lab rooms (so that emergencies can go to the second room if CTO PCI is performed in the first room).
   b. Cardiac computed tomography.
   c. On-site cardiac surgery.

3. Establishing CTO-specific protocols for:
   a. Radiation (as described in detail in Chapter 29):
     i. Utilizing 3.75 to 7.5 frame-per-second fluoroscopy.
     ii. Continuously monitoring radiation dose.
     iii. Stopping the procedure if crossing has not been achieved after 5 Gray air kerma dose.
     iv. Following up patients who receive >5 Gray air kerma dose to detect any skin injury.
   b. Anticoagulation:
     i. Repeating activating clotting time (ACT) every 30 min.
     ii. Goal ACT >300 s for antegrade cases.
     iii. Goal ACT >350 s for retrograde cases.

4. Establishing "CTO days," which allows an early start, and uninterrupted and concentrated focus on CTO PCI procedures: the operators should know that they have no other commitments for several hours; hence, prolonged treatment attempts are feasible, if necessary. Moreover, dedicated "CTO days" can improve staff acceptance of starting a CTO program, facilitate visits by proctors or clinical specialists and also allow referring cardiologists to visit.

5. Performing challenging cases as a team: having two interventionalists in the procedure may improve the likelihood of success.

31.3   Administration

The following steps can assist with securing support from administration in order to build a CTO PCI program:

1. Highlighting the need for such a program, by showing the number of patients who could benefit from CTO PCI and the associated halo-effect.

2. Demonstrating to administration that CTO PCI is both feasible and is not a "money-losing" proposition. Economic analyses from the Piedmont Heart Institute demonstrate similar contribution margins for CTO and non-CTO PCI.[5]

3. Highlighting the institutional benefits of a CTO PCI program:
   a. Developing regional and/or national reputation for doing complex interventions.

b. Increasing internal procedural volume (not only of CTOs, but also indirectly as CTO patients often have multivessel nonocclusive disease that must be treated separately).

c. Increasing outside referrals.

31.4   Referring physicians and patients: increasing awareness of your CTO program

Once the CTO program has been developed and good outcomes are achieved, increasing the awareness of the CTO PCI program can increase referrals of patients who may benefit from these procedures. The emphasis should be on patients' benefit and outcomes and less on obtaining referrals. Focus and culture of doing what is best for a patient in the long run will always improve referrals to a program. Don't be afraid to refer a case to a "competitor" or another program, as it shows others that the focus is on the patient and not on the need to get referrals, fame, or financial benefit to an institution.

Premature promotion may hurt the program. Projecting oneself as a CTO PCI expert and tackling very complex occlusions before reaching maturity can be counterproductive, as it can be frustrating for referring MDs to see repeat failures (and possibly complications) of their referred cases. It is best to work quietly "behind the scene" building the CTO PCI skills before aggressively promoting the program. At the same time, it is important to share successes with colleagues, effectively opening their eyes on what can be achieved with CTO PCI.

Increasing awareness of the CTO PCI program can be achieved by:

1. Educating referring physicians (both cardiologists and general practitioners). It is imperative for referring physicians to understand the rationale and potential clinical benefits of CTO PCI.
2. Presenting CTO PCI cases at case conferences, grand rounds, and roundtables. These presentations can illustrate that many of the previously considered "undoable" procedures are actually feasible and can provide significant benefit to the patients.
3. Educating the patients. Many patients with severe anginal (or angina equivalent) symptoms are very motivated to find treatment options themselves. Online posting of patient brochures focusing on CTO intervention, as well as video testimonials (with appropriate patient consent) can be powerful educational tools.

31.5   Why CTO PCI will make you a better interventionalist? (this section is reproduced with permission from the CTO Corner of Cardiology Today Interventions)

The goal of CTO interventions (and any procedure or intervention) is to benefit the patient. In addition to benefiting the patient, performing CTO

PCI can make you a better interventionalist. The following sections show you how.

31.5.1   The importance of growth

Performing CTO interventions offers a refreshing new perspective on PCI: there is a definite (and pretty substantial, especially in the beginning) chance of failure and there is a need to master multiple skills and techniques in order to succeed. CTO PCI trains interventionalists to treat patients with complex coronary anatomy safely and effectively. As a result, CTO PCI becomes a powerful motivator for learning and applying new (and often challenging) techniques.

31.5.2   Better angiogram evaluation skills

Angiographic evaluation can often be brief. Given the superb deliverability of current equipment, we often focus mainly on determining the diameter and the length of the stent we are going to implant. When we are approaching a CTO, however, things may not be quite as simple. We look at the vessel proximal to the lesion to determine if there is tortuosity and calcification that may hinder the advancement of equipment; whether there are lesions that need to be treated before attempting to treat the CTO; and whether there are proximal side branches that could be used to perform the "side-branch" anchor technique to facilitate equipment delivery. We use multiple projections to determine where the CTO actually starts (proximal cap) and whether the entry point to the CTO is tapered or blunt. We evaluate the length of the occlusion, as the longer it is the more likely that prolonged crossing attempts and the use of advanced crossing methods (retrograde and antegrade dissection/reentry) will be required. We examine the vessel distal to the occlusion to determine if reentry would be easy in case of extraplaque guidewire crossing and whether distal side branches are at risk for occlusion. Finally, we examine the presence, size, and tortuosity of collaterals to determine if the retrograde approach is feasible and safe. Given the frequent complexity of CTO PCI we are forced to look at the angiogram in multiple projections using dual injections in order to better understand the anatomy and devise several alternative approaches in case the initially chosen one fails. CTO PCI teaches us the value of careful and detailed planning and the value of understanding in depth each patient's unique coronary anatomy. It is not uncommon to spend 15−30 min looking at the angiogram. This is time well spent as it can make the difference between success and failure later in the case. Learning the "CTO way" of angiogram evaluation "spills over" when evaluating other coronary lesions and enhances the likelihood of successful, efficient, and safe treatment.

31.5.3 Familiarity with complex lesions and techniques

The days after a "CTO day" are so much fun—because everything feels so easy! When operators learn how to deal with the complexities of CTO PCI, treating nonocclusive lesions becomes pretty straightforward! The CTO interventionalists know how to achieve excellent guide support by using 8 French guides (and are not afraid to use 8 French guides when needed), or by using guide catheter extensions and anchor techniques. They often use 45-cm long sheaths to minimize the impact of iliac tortuosity on guide catheter support. They understand how to optimally use microcatheters (and use them for non-CTO complex cases) and which guidewire works best for each task. The extraplaque space stops being a forbidden zone—it becomes an ally, especially for treating long calcified lesions.

31.5.4 Better understanding of equipment

Not all guidewires are created equal! This is especially true when it comes to CTO guidewires. Understanding the differences between the stiff tip, highly penetrating guidewires (such as the Hornet 14, Warrior, Infiltrac and Infiltrac Plus, and the Confianza family of guidewires) and polymer-jacketed guidewires (such as the Pilot 200, Gladius, and Gladius Mongo, Fielder XT, XT-A, and XT-R and Sion black) is important for their optimal use. Such wires are often invaluable for crossing high-grade subtotal lesions, or even 100% occlusions in the setting of acute myocardial infarction. The basic principles of CTO interventions (never advance balloons and microcatheter before you confirm that the guidewire is in the distal true lumen or at least in the vessel "architecture") apply to all interventions, including acute myocardial infarction interventions. In several primary PCI cases, dual injections have been used to confirm the guidewire position during attempts to cross the acute occlusion.

31.5.5 Increased PCI volume

PCI (and coronary artery bypass graft surgery) volumes have been declining. CTO PCI is an area with strong potential to grow.

31.5.6 Improved workflow

By its very nature, CTO PCI requires continuous and prompt adjustment of the procedural strategy. If one technique is not successful within a short period of time, an adjustment is performed (rather than continuing with the same failing technique). The adjustment could be small (such as changing the shape of the guidewire tip) or major (such as changing from an antegrade to a retrograde approach or vice versa). Early change is at the heart of the "hybrid algorithm" and maximizes the likelihood of success by minimizing the time (and radiation) used in failing approaches. However, the basic principle of trying an alternative strategy if the initially selected one is not working applies to any interventional cardiology procedure. Constantly thinking about the steps ahead allows preparation of the necessary equipment and smooth and timely transitions. Workflow is improved and outcomes are improved.

31.5.7 Participation in a community

The CTO PCI community is very strong, perhaps because everyone realizes that they need the help and advice of everyone else in order to be successful. As a result, it is not uncommon to call colleagues for advice in the middle of a challenging case and there is constant interaction and learning, both at meetings and in the social media. There is always more to learn and there is always room for improvement.

31.5.8 Complication management

Planning for CTO PCI means also planning to treat complications. Although perforation is of special concern with CTO PCI, it can occur with any PCI. Hence, the planning and emphasis on preventing and treating complications required for CTO PCI directly helps improve the safety of non-CTO PCI. It forces operators to familiarize themselves with the use of covered stents and coils, thrombin, or microcatheters. There are cases of acute vessel closure during non-CTO PCI that were salvaged using CTO PCI techniques, both antegrade dissection/reentry[13] and the retrograde approach.[14,15] There are also equipment entrapment cases that are approached by using antegrade dissection/reentry techniques.[16] These patients would have, otherwise, required emergent cardiac surgery. They would have been grateful if they realized that emergent surgery was avoided because of operator experience with CTO PCI techniques!

31.5.9 Radiation management

Minimizing radiation exposure is critical for the success of CTO interventions. If too much radiation is used early in the case, crossing attempts will have to stop and the procedure will fail. Expert CTO operators are obsessive/compulsive about minimizing patient radiation exposure. They routinely use 6–7.5 frames per second fluoroscopy and continually monitor the amount of radiation administered to the patient. They also know that the AK (air kerma) is the number to watch, and they know how to make sense of the numbers (after 5 Gray air kerma radiation dose, the risk of radiation skin injury increases; if crossing has not been achieved within 5 to 7 Gray, the case must stop). These skills become second nature and directly affect non-CTO interventions. Many operators currently use 7.5 frames per second fluoroscopy in every case, not just CTO PCI. They also minimize use of cine angiography and consistently use the "fluoro store" function to document balloon and stent inflations.

31.5.10   Humility

Last, but not least, CTO PCI provides a constant lesson on the importance of humility and respect. Respect for the lesion and respect for the patient. CTO PCI teaches us that failure is always a distinct possibility and that we never know it all. Acknowledging our limitations and failures in front of the patients and their families helps us be humble and also to be grateful for what we can accomplish.

In the end, who benefits most from the improved procedural skills that CTO PCI operators develop? The patients themselves, who receive better, safer, and more efficient treatment. Perhaps one day the question: "Do you treat coronary CTOs?" will be increasingly asked by patients in need of a coronary intervention. That day may be closer than we think.

In summary, development of a successful CTO PCI program requires a concerted and coordinated effort that combines operator and staff development and education of administration and referring physicians. There remains great need for CTO PCI in the US and worldwide, and developing a high-level program can offer an excellent therapeutic option to an under-treated patient population.

References

1. Brilakis ES, Banerjee S, Karmpaliotis D, et al. Procedural outcomes of chronic total occlusion percutaneous coronary intervention: a report from the NCDR (National Cardiovascular Data Registry). *JACC Cardiovasc Interv* 2015;**8**:245−53.
2. Galassi AR, Sianos G, Werner GS, et al. Retrograde recanalization of chronic total occlusions in europe: procedural, in-hospital, and long-term outcomes from the multicenter ERCTO registry. *J Am Coll Cardiol* 2015;**65**:2388−400.
3. Azzalini L, Brilakis ES. Ipsilateral vs. contralateral vs. no collateral (antegrade only) chronic total occlusion percutaneous coronary interventions: What is the right choice for your practice? *Catheter Cardiovasc Interv* 2017;**89**:656−7.
4. Rinfret S, Joyal D, Spratt JC, Buller CE. Chronic total occlusion percutaneous coronary intervention case selection and techniques for the antegrade-only operator. *Catheter Cardiovasc Interv* 2015;**85**:408−15.
5. Karmpaliotis D, Lembo N, Kalynych A, et al. Development of a high-volume, multiple-operator program for percutaneous chronic total coronary occlusion revascularization: procedural, clinical, and cost-utilization outcomes. *Catheter Cardiovasc Interv* 2013;**82**:1−8.
6. Kalra A, Bhatt DL, Kleiman NS. A 24-month interventional cardiology fellowship: learning motor skills through blocked repetition. *JACC Cardiovasc Interv* 2017;**10**:210−11.
7. Kalra A, Bhatt DL, Pinto DS, et al. Accreditation and funding for a 24-month advanced interventional cardiology fellowship program: a call-to-action for optimal training of the next generation of interventionalists. *Catheter Cardiovasc Interv* 2016;**88**:1010−15.

8.   Kirtane AJ, Doshi D, Leon MB, et al. Treatment of higher-risk patients with an indication for revascularization: evolution within the field of contemporary percutaneous coronary intervention. *Circulation* 2016;**134**:422−31.

9.   Brilakis ES, Mashayekhi K, Tsuchikane E, et al. Guiding principles for chronic total occlusion percutaneous coronary intervention. *Circulation* 2019;**140**:420−33.

10. Brilakis E.S. The why and how of CTO interventions. Cardiology today's intervention January/February 2012.

11. Wu EB, Brilakis ES, Mashayekhi K, et al. Global chronic total occlusion crossing algorithm: JACC state-of-the-art review. *J Am Coll Cardiol* 2021;**78**:840−53.

12. Brilakis ES. How Stoic principles can help when performing complex interventions. *EuroIntervention* 2021;**17**:e364−6.

13. Martinez-Rumayor AA, Banerjee S, Brilakis ES. Knuckle wire and stingray balloon for recrossing a coronary dissection after loss of guidewire position. *JACC Cardiovasc Interv* 2012;**5**:e31−2.

14. Azemi T, Fram DB, Hirst JA. Bailout antegrade coronary reentry with the stingray balloon and guidewire in the setting of an acute myocardial infarction and cardiogenic shock. *Catheter Cardiovasc Interv* 2013;**82**:E211−4.

15. Patel VG, Zankar A, Brilakis E. Use of the retrograde approach for primary percutaneous coronary intervention of an inferior ST-segment elevation myocardial infarction. *J Invasive Cardiol* 2013;**25**:483−4.

16. Tanaka Y, Saito S. Successful retrieval of a firmly stuck rotablator burr by using a modified STAR technique. *Catheter Cardiovasc Interv* 2016;**87**:749−56.

Part F

Appendices

Appendix 1

Name	Type	Manufacturer	Pages
All Star	Guidewire (support)	Abbott Vascular	
Amplatz goose neck snare	Snare	Medtronic	
Amplatz superstiff	0.035 in. wire (stiff body)	Boston Scientific	
Amplatzer vascular plug	Vascular plugs	Abbott Vascular	
Angiosculpt	Plaque modification balloon	Philips	
AngioSeal	Vascular closure device	Terumo	
Astato XS 20	Guidewire (tapered tip, high tip stiffness)	Asahi Intecc	
Atrieve	Snare (3 loop)	Angiotech	
Axium	0.014 in. coils	Medtronic	
Azur	0.018 in. coils	Terumo	
Bandit	Guidewire (polymer-jacketed with soft, tapered tip)	Teleflex	
Bentson	0.035 guidewire (soft tip)	Cook Medical	
BHW	Guidewire (support)	Abbott Vascular	
Blimp	Plaque modification balloon	Interventional Medical Device Solutions	
Boosting catheter	Guide catheter extension	QXMedical	
Caravel	Microcatheter (low profile)	Asahi Intecc	
Chocolate	Plaque modification balloon	Teleflex	
Choice PT floppy	Guidewire (polymer-jacketed with soft, nontapered tip)	Boston Scientific	
Confianza Pro 12	Guidewire (tapered tip, high tip stiffness)	Asahi Intecc	
Co-Pilot	Y-connector with hemostatic valve	Abbott Vascular	
Corsair Pro	Microcatheter	Asahi Intecc	
Corsair Pro XS	Microcatheter (lower profile than Corsair Pro)	Asahi Intecc	
CrossBoss	Blunt tip microcatheter for antegrade dissection and reentry	Boston Scientific	
Cross-it 100XT	Guidewire	Abbott Vascular	
Crosswire NT	Guidewire	Terumo	
Crusade	Dual lumen microcatheter	Kaneka Medix Corporation	
Diamondback 360	Orbital artherectomy system	CSI	
DualPro	Combined intravascular ultrasound and near-infrared spectroscopy catheter	InfraRedx	

Name	Description	Manufacturer
Dyevert Plus	Contrast saving system	Osprey Medical
Eagle Eye	Intravascular ultrasound catheter (solid state)	Philips
Eagle Eye Short Tip	Intravascular ultrasound catheter with short tip	Philips
Eaucath	Sheathless guide catheter	Asahi Intecc
ELCA	Excimer laser coronary artherectomy	Philips
EN Snare	Snare (3 loop)	Merit Medical
Extra S'port	Guidewire (support)	Abbott Vascular
Fielder FC	Guidewire (polymer-jacketed with soft, nontapered tip)	Asahi Intecc
Fielder XT	Guidewire (polymer-jacketed with soft, tapered tip)	Asahi Intecc
Fielder XT-A	Guidewire (polymer-jacketed, with soft, tapered tip, for antegrade crossing)	Asahi Intecc
Fielder XT-R	Guidewire (polymer-jacketed, with soft, tapered tip, for retrograde collateral crossing)	Asahi Intecc
Fighter	Guidewire (polymer-jacketed with soft, tapered tip)	Boston Scientific
Finecross	Microcatheter	Terumo
FineDuo	Dual lumen microcatheter	Terumo
Gaia (1st, 2nd, 3rd)	Guidewire (tapered tip, intermediate tip stiffness)	Asahi Intecc
Gaia Next (1st, 2nd, 3rd)	Guidewire (tapered tip, intermediate tip stiffness)	Asahi Intecc
Gladius	Guidewire (polymer-jacketed, non-tapered, high tip stiffness)	Asahi Intecc
Gladius Mongo	Guidewire (polymer-jacketed, non-tapered, high tip stiffness, designed to knuckle at the tip)	Asahi Intecc
Glidesheath Slender	Thin wall sheath for radial access	Terumo
Glidewire	0.035 in. wire (polymer-jacketed)	Terumo
Glidewire Advantage	0.035 in. wire (polymer-jacketed with stiff body)	Terumo
Glidewire baby-J	0.035 in. wire (polymer-jacketed with 1.5 mm distal J-tip radius)	Terumo
Graftmaster Rx	Covered stent (rapid-exchange)	Abbott Vascular
Grand Slam	Guidewire (support)	Asahi Intecc
Guardian	Y-connector with hemostatic valve	Teleflex
Guideliner V3	Guide catheter extension	Teleflex
GuidePlus II	Guide catheter extension	Nipro

(Continued)

Name	Type	Manufacturer	Pages
Guidezilla II	Guide catheter extension	Boston Scientific	
Guidion	Guide catheter extension	Interventional Medical Device Solutions	
Heartrail 5 in 6	Guide catheter extension	Terumo	
Hornet 10, 14	Guidewire (tapered tip, high tip stiffness)	Boston Scientific	
Ikazuchi Zero	1.0 mm balloon	Kaneka Medix Corporation	
Impella (2.5, CP, 5.0, 5.5)	Left ventricular assist device	Abiomed Inc.	
Infiltrac, Infiltrac Plus	Guidewire (tapered tip, high-tip stiffness)	Abbott Vascular	
Interlock	0.018 in. coils	Boston Scientific	
IronMan	Guidewire (support)	Abbott Vascular	
Judo (1, 3, 6)	Guidewire (tapered tip, soft to intermediate tip stiffness)	Boston Scientific	
Kodama	Intravascular ultrasound catheter (rotational)	ACIST medical	
Kusabi	Trapping balloon	Kaneka Medix Corporation	
LiquID	Guide catheter extension	Seigla Medical	
Lunderquist	0.035 in. wire (stiff body)	Cook Medical	
M-Cath	Microcatheter	Acrostak	
Mailman	Guidewire (support)	Boston Scientific	
Mamba	Microcatheter	Boston Scientific	
Mamba Flex	Microcatheter	Boston Scientific	
Micronester	0.018 in. coils	Cook Medical	
Microsnare Elite	Snare	Teleflex	
MiracleBros 3, 4.5, 6, 12	Guidewire (nontapered, intermediate to high tip stiffness)	Asahi Intecc	
Mizuki	Microcatheter	Kaneka Medix Corporation	
Mizuki FX	Microcatheter (more flexible than Mizuki)	Kaneka Medix Corporation	
Nano 0.85	0.85 mm balloon	SIS Medical	
Navvus	Rapid exchange pressure transducing microcatheter	ACIST Medical	
Nhancer ProX	Microcatheter	Interventional Medical Device Solutions	
Nhancer Rx	Dual lumen microcatheter	Interventional Medical Device Solutions	
NovaCross	Support catheter	Nitiloop	

Device	Description	Manufacturer
Novasight	Combined intravascular ultrasound and optical coherence tomography catheter	Conavi Medical
OKAY II	Y-connector with hemostatic valve	Nipro
OmniWire	Pressure wire	Philips
OPN NC	Very high pressure balloon	SIS Medical
OptiCross	Intravascular ultrasound catheter (rotational)	Boston Scientific
Optowire	Pressure wire	Opsens
Ostial Pro	Stent positioning device for ostial lesions	Merit Medical
Ostial Flash	Specialized balloon for treatment of ostial lesions	Cardinal Health
Perclose	Vascular closure device	Abbott Vascular
PhD	Y-connector with hemostatic valve	Merit Medical
Pilot 50, 150, 200	Guidewire (polymer-jacketed, non-tapered with increasing tip stiffness)	Abbott Vascular
PK Papyrus	Covered stent	Biotronik
Prelude Ideal	Thin wall sheath for radial access	Merit Medical
PressureWire X	Pressure wire	Abbott Vascular
Progreat	Microcatheter (0.018 in. – usually used for coil delivery)	Terumo
Progress 40, 80, 120, 140T, 200T	Guidewire (stiff, tapered tip)	Abbott Vascular
ProVia 3, 6, 9	Guidewire (moderate to high tip stiffness, non-tapered tip)	Medtronic
PT Graphix Intermediate	Guidewire (polymer-jacketed)	Boston Scientific
PT2 Moderate Support	Guidewire (polymer-jacketed, non-tapered, high tip stiffness)	Boston Scientific
Quick Cross	Microcatheter	Spectranetics
R350	Guidewire (350 cm long for externalization)	Teleflex
R2P	Radial to Peripheral sheaths (long sheaths for radial access)	Terumo
RadPad	Disposable radiation shield	Worldwide Innovations & Technologies
Raider	Guidewire (polymer-jacketed, non-tapered, high tip stiffness)	Teleflex
Railway	Dilator for sheathless insertion of guide catheters	Cordis
ReCross	Microcatheter (dual lumen with two over-the-wire lumens)	Interventional Medical Device Solutions
Refinity	Intravascular ultrasound catheter (rotational)	Philips

(Continued)

Name	Type	Manufacturer	Pages
Renegade	Microcatheter (0.018 in. – usually used for coil delivery)	Boston Scientific	
RG3	Guidewire (330 cm long for externalization)	Asahi Intecc	
Rotablator	Rotational atherectomy system	Boston Scientific	
Rotaglide	Lubricant solution for rotational atherectomy	Boston Scientific	
RotaWire Drive Floppy and Extra support	Guidewire (for rotational atherectomy)	Boston Scientific	
Ryurei 1.0	1.0 mm balloon	Terumo	
Runthrough	Guidewire (workhorse)	Terumo	
Samurai	Guidewire (workhorse)	Boston Scientific	
Samurai RC	Guidewire (workhorse – also for retrograde collateral crossing)	Boston Scientific	
Sapphire II Pro	1.0 mm balloon	OrbusNeich	
Sasuke	Microcatheter (dual lumen)	Asahi Intecc	
Scoreflex	Plaque modification balloon	OrbusNeich	
Shockwave	Intravascular lithotripsy balloon	Shockwave Medical	
Shuttle	Sheath	Cook Medical	
Sion	Guidewire (composite core, soft tip)	Asahi Intecc	
Sion black	Guidewire (polymer-jacketed, nontapered, soft t p)	Asahi Intecc	
Sion blue	Guidewire (workhorse)	Asahi Intecc	
Sion blue extra support	Guidewire (support)	Asahi Intecc	
Stingray LP balloon	Balloon for reentry into true lumen	Boston Scientific	
Stingray wire	Guidewire (tapered tip, high-tip stiffness)	Boston Scientific	
Storq	0.035 guidewire	Cordis	
Suoh 03	Guidewire (soft 0.3 g tip, excellent for epicardial collateral crossing)	Asahi Intecc	
Supercross	Microcatheter (with angled tip)	Teleflex	
Supracore	0.035 in. wire (stiff body)	Abbott Vascular	
Swift-Ninja	Microcatheter (with angled tip)	Merit Medical	
Tandem Heart	Left ventricular assist device	Cardiac Assist Inc.	
Tegaderm	Sterile cover	3M	
Telemark	Microcatheter	Medtronic/Surmodics	

Name	Description	Manufacturer
Teleport	Microcatheter	OrbusNeich
Teleport Control	Microcatheter	OrbusNeich
Telescope	Guide catheter extension	Medtronic
Threader	Microdilation catheter	Boston Scientific
Tornus	Microcatheter (for "balloon uncrossable" lesions)	Asahi Intecc
Tracker Excel 14	Microcatheter	Boston Scientific
Transit	0.021 in. microcatheter (usually used for coil delivery)	Cordis
Trap it	Trapping balloon	Interventional Medical Device Solutions
Trapliner	Guide catheter extension that incorporates a trapping balloon	Teleflex
Trapper	Trapping balloon	Boston Scientific
Turbo Elite	Laser catheter (without lasing duration limit)	Philips
Turnpike	Microcatheter	Teleflex
Turnpike Gold	Microcatheter (gold plated threaded tip, for antegrade approach)	Teleflex
Turnpike LP	Microcatheter (lower profile of the distal tip and shaft)	Teleflex
Turnpike Spiral	Microcatheter (nylon coil distally, for antegrade crossing)	Teleflex
Twin-Pass	Dual lumen microcatheter	Teleflex
Twin-Pass Torque	Dual lumen microcatheter	Teleflex
Ultimate 3	Guidewire (nontapered, intermediate tip stiffness)	Asahi Intecc
Venture	Microcatheter (with deflectable distal tip)	Teleflex
Versacore	0.035 in. Guidewire (soft tip)	Abbott Vascular
ViperSlide	Lubricant solution for orbital atherectomy	CSI
ViperWire Advance	Guidewire (for orbital atherectomy)	CSI
Warrior	Guidewire (tapered tip, high tip stiffness)	Teleflex
Watchdog	Y-connector with hemostatic valve	Boston Scientific
Whisper LS, MS, ES	Guidewire (polymer-jacketed with soft, nontapered tip)	Abbott Vascular
Wholey	0.035 in wire (soft tip)	Medtronic
Wiggle wire	Guidewire (with curved distal shaft for facilitating equipment delivery)	Asahi Intecc
Wolverine	Cutting balloon	Boston Scientific
Zero Gravity	Radiation protection system	Biotronik
Zinrai	0.75 mm balloon	Kaneka Medix Corporation

Appendix 2

Acronym	Full name	Description	Page (main explanation)	Pages (mentioned)
CART	Controlled Antegrade and Retrograde Tracking	Technique for reentry into the distal true lumen after extraplaque CTO crossing during the retrograde approach: a balloon is inflated over the retrograde guidewire creating a space into which an antegrade guidewire is advanced.		
Anchor-Tornus technique	Anchor-Tornus technique	Combined use of the Tornus microcatheter and side-branch anchoring to cross a balloon uncrossable CTO.		
Antegrade Balloon Puncture	Antegrade Balloon Puncture	Variation of the reverse CART technique: the antegrade balloon remains inflated during retrograde crossing attempts and is "punctured" by the retrograde wire, which is then advanced while the antegrade balloon is retracted under fluoroscopy (the latter technique is also called "transit balloon technique").		
Antegrade Microcatheter Probing	Antegrade Microcatheter Probing	After retrograde guidewire crossing the retrograde microcatheter is advanced into the antegrade guide catheter, followed by the removal of the retrograde guidewire and intubation of the microcatheter with an antegrade wire.		
"Back Bleeding Sign"	Back Bleeding Sign	Blood coming out of the microcatheter after aspirating for at least 30 seconds after withdrawing the guidewire; it suggests (but does not prove) distal true lumen position.		
BAM	Balloon Assisted Microdissection or "Grenadoplasty"	Technique for crossing balloon uncrossable lesions: a small ($\leq$1.5 mm in diameter) balloon is advanced into the lesion as far as possible and inflated to high pressure until it ruptures. Balloon rupture can modify the proximal cap and facilitate crossing with another balloon.		
BASE	Balloon Assisted Subintimal Entry	BASE is one of the "move the cap" techniques used in cases of proximal cap ambiguity: a slightly oversized balloon is inflated proximal to the CTO, in order to create a dissection, through which a knuckled guidewire is advanced extraplaque around the proximal cap.		
BAT	Balloon Assisted Tracking technique	Technique for advancing catheters through tortuous radial arteries: a balloon is advanced halfway in and halfway out the tip of the guide catheter and inflated. The guide catheter/inflated balloon assembly is then advanced through the area of tortuosity. The BAT technique can also be used for sheathless insertion of guide catheters.		

"Block and Deliver"	Block and Deliver	Technique for managing coronary perforations. A balloon is advanced proximal to or at the site of perforation and inflated to prevent continued bleeding into the pericardium. A covered stent or a microcatheter (for coil delivery) is advanced proximal to this "blocking" balloon. The blocking balloon is transiently deflated, followed by advancement of the covered stent or microcatheter for sealing the perforation. A large guide catheter is needed for delivering covered stents, whereas smaller guide catheters suffice for delivering coils.
"Bobsled"	Bobsled	Bobsled refers to changing the location of reentry attempts during antegrade dissection and reentry, using the Stingray balloon. Reentry is usually attempted at a healthier, straighter, and larger vessel segment.
"Bridge" or "Rendezvous"	Bridge or Rendezvous	Technique for inserting an antegrade wire through a CTO after successful retrograde crossing. After retrograde guidewire crossing, the retrograde microcatheter is inserted into the antegrade guide catheter and aligned with an antegrade microcatheter, followed by insertion of an antegrade guidewire into the retrograde microcatheter.
"Buddy Wire Stent Anchor"	Buddy Wire Stent Anchor	Technique for increasing guide catheter support that can be used if the proximal vessel needs stenting: a buddy wire can be inserted and a stent deployed over this guidewire, effectively "trapping" the buddy wire, which then provides strong guide catheter support.
"Confluent Balloon"	Confluent Balloon	Variation of the CART technique in which both the antegrade and the retrograde balloons are inflated simultaneously in a kissing fashion to cause the extraplaque space to become confluent, allowing wire passage through the CTO.
Carlino technique	Carlino technique	Extraplaque contrast injection in which a small amount of contrast (<1.0 mL) is injected into the extraplaque space during cine-angiography. The Carlino technique is often used to resolve proximal cap or vessel course ambiguity and for treating balloon uncrossable lesions.
Contrast-Guided STAR	Contrast-Guided Subintimal Tracking and Reentry	Variation of the STAR technique in which contrast is injected through a microcatheter inserted into the proximal cap or within the extraplaque space to create/visualize a dissection plane and facilitate subsequent guidewire advancement.

(Continued)

Acronym	Full name	Description	Page (main explanation)	Pages (mentioned)
"Deflecting Balloon" or "Blocking Balloon" technique	Deflecting Balloon or Blocking Balloon technique	Technique for facilitating crossing when there is a side branch near the proximal or distal cap. A balloon is inflated at the ostium of the side branch, "blocking" entry of the guidewire into it and facilitating advancement of the guidewire through the target lesion.		
Directed Reverse CART	Directed Reverse Controlled Antegrade and Retrograde Tracking	Variation of reverse CART technique: a small (2.0–2.5 mm) antegrade balloon is used to facilitate crossing of the retrograde guidewire into the antegrade true lumen. Using a small balloon (instead of a larger one as is done in the conventional reverse CART technique) minimizes the size of dissection and vessel injury.		
"Double Blind Stick and Swap"	Double Blind Stick and Swap	Technique for reentering into the distal true lumen using the Stingray balloon after extraplaque guidewire crossing. It is similar to the "stick and swap" technique, in which a stiff guidewire (such as the Stingray guidewire) is used to create an exit channel towards the distal true lumen, followed by exchange for a polymer-jacketed guidewire for completing the reentry. In stick and swap, contralateral contrast injection is used to determine the location of the distal true lumen relative to the Stingray balloon. In the "double blind stick and swap" technique there is no contrast injection; instead a puncture is performed using a stiff tip wire on both sides of the Stingray balloon, followed by advancement of a polymer-jacketed guidewire on both sides of the Stingray balloon until reentry is achieved.		
DRAFT	Deflate, Retract and Advance into the Fenestration Technique	Variation of the reverse CART technique that requires two operators: the antegrade balloon is withdrawn by one operator while the other operator advances the retrograde guidewire through the space created by the balloon being retracted until it enters into the antegrade guide catheter.		
e-CART	Electrocautery Controlled Antegrade and Retrograde Tracking and Dissection	Variation of the reverse CART technique used when the retrograde guidewire cannot penetrate the proximal cap and an antegrade wire cannot be advanced (usually in flush aorto-ostial lesions). A retrograde stiff guidewire (usually a Confianza Pro 12 or Astato 20) is advanced as far as possible into the occlusion over a microcatheter, followed by cautery activation in order to burn through the impenetrable tissue into the aorta.		

"Fast spin" CrossBoss technique	Fast spin CrossBoss technique	Technique used for advancing a CrossBoss microcatheter: the catheter is rotated rapidly using the proximal torque device until it advances through the occlusion.
"Finish with the Boss"	Finish with the Boss	Technique for minimizing the extent of extraplaque dissection performed using a knuckled guidewire. Extraplaque advancement of the knuckled guidewire is stopped proximal to the distal cap; the knuckled wire is exchanged for a CrossBoss catheter to complete the last part of extraplaque crossing. The CrossBoss catheter has smaller profile than a knuckled guidewire decreasing the likelihood of extraplaque hematoma formation that can hinder reentry into the distal true lumen.
Guide extension-assisted reverse CART	Guide extension-assisted reverse Controlled Antegrade and Retrograde Tracking and dissection	Variation of the reverse CART technique: a guide extension catheter is advanced on the proximal guidewire to form a proximal target for the retrograde guidewire to enter.
"Hairpin" technique; also called "Reversed Guidewire" technique	Hairpin technique; also called Reversed Guidewire technique	Technique for wiring highly angulated vessels. A polymer jacketed guidewire is bent approximately 3 cm from the wire tip and the knuckle is advanced through the introducer into the coronary artery. Upon withdrawal the guidewire tip enters into the angulated side branch.
IVUS-guided CART	Intravascular Ultrasound-Guided Controlled Antegrade and Retrograde Tracking	Variation of the reverse CART technique: intravascular ultrasound is used to determine the location of the antegrade and retrograde guidewire and to allow precise sizing of the antegrade balloon. Intravascular ultrasound allows safe use of larger balloons, which in turn increases the likelihood of successful retrograde wire crossing. IVUS can also help determine whether significant recoil occurs after antegrade balloon inflation.
J-CTO score	Japan Chronic Total Occlusion Score	Five-point score for prediction of the likelihood of successful guidewire crossing within the first 30 minutes of crossing attempts. It was developed from the Multicenter Chronic Total Occlusion Registry in Japan. The five variables are: blunt stump, CTO calcification, CTO tortuosity, occlusion length $\geq$ 20 mm, and prior failed attempt.
Jet exchange, also called hydraulic exchange, or Nanto technique	Jet exchange, also called hydraulic exchange, or Nanto technique	Technique for removing a microcatheter while maintaining guidewire position. It is performed by connecting an inflating device over the back end of the microcatheter, inflating it at high pressure, and removing the microcatheter, while the antegrade flow keeps the guidewire in position. However, the trapping technique is more reliable and is preferred for over-the-wire system exchanges.

(Continued)

Acronym	Full name	Description	Page (main explanation)	Pages (mentioned)
"Just marker" technique	Just marker technique	Variation of the retrograde technique. The retrograde wire is advanced to the distal cap and acts as a marker of the distal true lumen position, facilitating antegrade crossing attempts.		
"Kissing Wire" technique	Kissing Wire technique	Variation of the retrograde technique that involves manipulation of both antegrade and retrograde wires within the occluded segment until crossing is achieved.		
"Knuckle Boss" technique	Knuckle Boss technique	Technique for preventing entry of the CrossBoss catheter into side branches. The CrossBoss catheter is withdrawn proximal to the origin of the side branch, and a knuckled wire is advanced past the side branch (the larger size of the knuckle often prevents it from entering the side branch).		
LAST	Limited Antegrade Subintimal Tracking	Wire-based technique for reentering into the distal true lumen after extraplaque crossing. Usually a stiff tip guidewire with a 70–90 degree angle is manipulated until it enters into the distal true lumen. Because it is unpredictable, the LAST technique is currently used infrequently. The Stingray system is preferred for achieving distal true lumen reentry.		
Mini-STAR	Mini Subintimal Tracking and Reentry	Variation of the STAR technique in which a polymer-jacketed guidewire (such as the Fielder FC or XT) is used to reenter into the distal true lumen immediately after the occlusion rather than further down the distal vessel. Similar with the LAST technique, mini-STAR is currently used infrequently, with the Stingray system being the preferred strategy for distal true lumen reentry.		
"Mother – Daughter – Granddaughter" technique	Mother – Daughter – Granddaughter technique	Simultaneous use of two guide catheter extensions, (i.e., a 6 Fr extension through an 8 Fr extension) when multiple extreme bends need to be navigated.		
"Move The Cap" techniques	Move The Cap techniques	These techniques use antegrade dissection and reentry to clarify the course of the occluded vessel (proximal cap ambiguity) and facilitate crossing. The following techniques are included in this category: "balloon-assisted subintimal entry" (BASE), "scratch and go" and the Carlino technique.		
"Open Sesame" technique	Open Sesame technique	Technique for facilitating crossing when there is a side branch at the proximal cap. Balloon inflation is performed in the side branch inducing a geometrical shift of the proximal cap plaque, which in turn enables guidewire entry into the CTO.		

Parallel Wire technique	Parallel Wire technique	During antegrade wiring if the initial guidewire enters the extraplaque space, it is left in place and a new guidewire is inserted next to the original guidewire to facilitate crossing into the distal true lumen.
"Ping-Pong" guide	Ping-Pong guide	Two guide catheters are simultaneously engaging the same target vessel. One guide is pulled back to enable engagement with the other guide catheter and vice versa.
PROGRESS CTO score	Prospective Global Registry for the Study of Chronic Total Occlusion Intervention score	Scoring system that uses four variables (proximal cap ambiguity, moderate/severe tortuosity, circumflex artery CTO, and absence of "interventional" collaterals) to create a 4-point score that helps predict technical success.
Proxis-Tornus technique	Proxis-Tornus technique	Combination of Tornus microcatheter through a Proxis device (increased support) to cross balloon uncrossable CTOs. Since the Proxis device is no longer commercially available a guide catheter extension can be used instead of the Proxis catheter.
RASER technique	Rotablation and Laser technique	Technique for treating balloon uncrossable and balloon undilatable lesions in which laser is used first to facilitate advancement of a rotational atherectomy wire (either directly or using a microcatheter for exchange), followed by rotational atherectomy.
Reverse CART	Reverse Controlled Antegrade and Retrograde Tracking	Reverse CART is the opposite of the CART technique: a balloon is inflated over the antegrade guidewire creating a space into which a retrograde guidewire is advanced. At present reverse CART is the most commonly used technique for retrograde reentry.
"Reverse Wire Trapping" technique	Reverse Wire Trapping technique	Technique for delivering an antegrade guidewire through a CTO after successful retrograde guidewire crossing. The retrograde guidewire is snared, followed by withdrawal of the retrograde guidewire by pulling the antegrade snare through the CTO into the distal true lumen.
"Scratch and Go" technique	Scratch and Go technique	This is one of the "move the cap" techniques. A stiff guidewire is advanced towards the vessel wall into the extraplaque space proximal to the CTO ("scratching" the wall). A microcatheter follows the wire into the extraplaque space. The stiff guidewire is exchanged for a polymer-jacketed guidewire that is advanced to form a knuckle that then crosses the occlusion.
"See-Saw Wire Cutting" technique	See-Saw Wire Cutting technique	Modified version of the wire cutting technique for treating balloon uncrossable lesions. Two balloons are advanced over two guidewires to the proximal cap of the uncrossable lesion. The first balloon is inserted as far as possible into the lesion and inflated effectively "cutting" the proximal cap with the second guidewire. The process is repeated with the second balloon modifying the cap on the other side until the lesion is successfully crossed with a balloon.

(Continued)

Acronym	Full name	Description	Page (main explanation)	Pages (mentioned)
"See-Saw" technique	See-Saw technique	This is a variation of the parallel wire technique: during antegrade wire escalation if the guidewire enters the extraplaque space it is left in place and a second guidewire is advanced over a second microcatheter next to the first guidewire to cross the occlusion. Two microcatheters are used in the see-saw technique vs only one in the parallel wire technique.		
Side-BASE		A variation of the BASE technique in which the balloon is inflated halfway in and halfway out of a side branch at the proximal cap. Side-BASE aims to prevent guidewire entry into the side branch and also preserve the patency of the side branch		
"Septal Surfing"	Septal Surfing	Technique for retrograde guidew re crossing through septal collaterals. It is different than the contrast-guided technique, in which contrast injection is used to determine the course of the septal collaterals and help guide the guidewire; in the "surfing" technique, the guidewire is advanced blindly back and forth through the collateral until it crosses into the distal true lumen.		
"Side Branch Anchor" technique	Side Branch Anchor technique	Technique for increasing guide catheter support. A workhorse guidewire is advanced into a side branch proximal to the target lesion, followed by inflation of a small balloon into the side branch. The size of the balloon is selected to match the size of the collateral vessel. The side branch balloon is inflated to 6–8 atm "anchoring" the guide catheter into the vessel.		
STAR	Subintimal Tracking And Reentry	STAR is the original antegrade dissection/reentry technique that was described by Antonio Colombo. A knuckled polymer-jacketed guidewire is advanced through the extraplaque space until it spontaneously reenters into the distal true lumen, usual y at a bifurcation.		
Stent Reverse CART	Stent Reverse Controlled Antegrade and Retrograde Tracking and Dissection	This is a variation of the reverse CART technique: a stent is deployed from the proximal true lumen into the extraplaque space to facilitate retrograde wiring into the stent. It is used infrequently as a last resort because it is irreversible and may sometimes hinder reentry, for example when the retrograde guidewire crosses through the stent struts.		

Stick and Drive	Stick and Drive	This is the classic technique for reentering into the distal true lumen using the Stingray balloon. The Stingray guidewire is advanced without rotation through the side port of the Stingray balloon so as to puncture back into the true lumen. After confirmation of distal true lumen position with contralateral injection the Stingray guidewire is rotated 180 degrees and advanced further down into the vessel. The stick and swap technique is preferred in most cases, especially when the distal vessel is of small caliber, tortuous, of diffusely diseased, as the stiff tip Stingray guidewire may cause injury of the distal vessel.
Stick and Swap	Stick and Swap	Technique for reentry into the true lumen using the Stingray balloon: an initial puncture is performed using a stiff tip guidewire to create a connection with the distal true lumen. The Stingray wire is removed and a Pilot 200 (or similar polymer-jacketed, stiff tip) guidewire is advanced through the same side port into the "tunnel" created by the stiff tip wire to enter the distal true lumen.
STRAW	Subintimal TRAnscatheter Withdrawal technique	Aspiration of hematoma that develops during antegrade dissection/reentry crossing to facilitate reentry. STRAW can be performed either through the Stingray balloon itself, or ideally through another microcatheter or over-the-wire balloon advanced proximal to the Stingray balloon
Subintimal distal anchoring technique	Subintimal distal anchoring technique	Technique for treating "balloon uncrossable" lesions after successful guidewire crossing. A second guidewire is advanced extraplaque through the occlusion and a balloon is advanced over the extraplaque guidewire to "anchor" the true lumen guidewire, over which a balloon can then be advanced through the occlusion.
"Tip In" technique	Tip In technique	After retrograde guidewire crossing into the antegrade guide catheter an antegrade microcatheter is advanced over the retrograde guidewire in the distal part of the guide catheter and then through the occlusion, followed by insertion of an antegrade guidewire and antegrade delivery of balloons and stents. This technique results in less strain on the collaterals, however, unlike guidewire externalization, loss of guidewire position may occur.
"Wire Cutting" technique	Wire Cutting technique	Technique for treating balloon uncrossable or undilatable lesions. A second guidewire is advanced through the occlusion and a balloon is inflated over the original guidewire while pulling the second guidewire, "scoring" and modifying the lesion.

Index

Note: Page numbers followed by "*f*" and "*t*" refer to figures and tables, respectively.